Edited by
Subhash Chandra Singh, Haibo Zeng,
Chunlei Guo, and Weiping Cai

Nanomaterials

Related Titles

Byrappa, K., Adschiri, T.

Hydrothermal Technology for Nanomaterials Processing

2013
ISBN: 978-0-470-17795-2

Kannatey-Asibu, E.

Principles of Laser Materials Processing

2009
ISBN: 978-0-470-17798-3

Fuchs, H. (ed.)

Nanotechnology
Volume 6: Nanoprobes

2009
ISBN: 978-3-527-31733-2

Vollath, D.

Nanomaterials
An Introduction to Synthesis, Properties
and Applications

2008
ISBN: 978-3-527-31531-4

Leng, Y.

Materials Characterization
Introduction to Microscopic and
Spectroscopic Methods

2008
ISBN: 978-0-470-82298-2

Misawa, H., Juodkazis, S. (eds.)

3D Laser Microfabrication
Principles and Applications

2006
ISBN: 978-3-527-31055-5

Cremers, D. A., Radziemski, L. J.

Handbook of Laser-Induced Breakdown Spectroscopy

2006
ISBN: 978-0-470-09299-6

Rao, C. N. R., Müller, A., Cheetham, A. K. (eds.)

The Chemistry of Nanomaterials
Synthesis, Properties and Applications

2004
ISBN: 978-3-527-30686-2

Edited by Subhash Chandra Singh, Haibo Zeng, Chunlei Guo,
and Weiping Cai

Nanomaterials

Processing and Characterization with Lasers

**WILEY-
VCH**

WILEY-VCH Verlag GmbH & Co. KGaA

The Editors

Dr. Subhash Chandra Singh
Dublin City University
School of Physical Sciences
9 Dublin-Glasnevin
Ireland

Prof. Haibo Zeng
Nanjing University of Aeronautics
and Astronautics
State Key Laboratory of Mechanics and
Control of Mechanical Structures
Key Laboratory for Intelligent Nano
Materials and Devices of the Ministry
of Education
College of Material Science and Technology
Yudao Street 29, Nanjing 210016
People's Republic of China

Prof. Chunlei Guo
University of Rochester
The Institute of Optics
275, Hutchison Road
Rochester, NY 14627-0186
USA

Prof. Weiping Cai
Institute of Solid State Physics
Chinese Academy of Sciences
ShuShanHu Road 350
Hefei, Anhui 230031
People's Republic of China

Library of Congress Card No.: applied for

**British Library Cataloguing-in-Publication
Data**
A catalogue record for this book is available
from the British Library.

**Bibliographic information published by the
Deutsche Nationalbibliothek**
The Deutsche Nationalbibliothek
lists this publication in the Deutsche
Nationalbibliografie; detailed bibliographic
data are available on the Internet at
<http://dnb.d-nb.de>.

© 2012 Wiley-VCH Verlag & Co. KGaA,
Boschstr. 12, 69469 Weinheim, Germany

Composition   Laserwords Private Ltd.,
Chennai, India
Printing and Binding   Markono Print Media
Pte Ltd, Singapore
Cover Design   Formgeber, Eppelheim

Print ISBN: 978-3-527-32715-7
ePDF ISBN: 978-3-527-64685-2
ePub ISBN: 978-3-527-64684-5
mobi ISBN: 978-3-527-64683-8
oBook ISBN: 978-3-527-64682-1

Printed in Singapore
Printed on acid-free paper

I would like to dedicate this book to my parents, Mrs. Savitri Devi and Mr. Ravindra Nath Singh, my wife Mrs. Jyoti Singh and my new born daughter Subhra.

Contents

Preface

Lasers and nanomaterials are both highly emergent and hot topics of research recent days. Lasers have shown their potential applications not only in the processing of nanoscaled materials but also in their characterizations since the 1960s. Cutting, drilling, alloying, welding, defect creation inside the bulk, and so on, are some conventional applications of lasers in bulk material processing and are the subject of several books, while laser-based processing methods for nanoscaled materials are less dealt with by authors of books and editors.

Availability of a wide range of lasers with power option from milliwatt to petawatt, wavelength selectivity from soft X-ray to microwave, pulse widths from millisecond to attosecond, and repetition rates from a few hertz to mega hertz and continuous research and development on lasers have fueled research and development in the area of laser processing of nanostructured materials and their characterizations. Lasers have the utility not only in the processing of nanostructures but they can also modify the size, shape, phase, morphology, and hence the properties of the nanostructured materials. All the methods of laser processing of nanostructures are almost simple, quick, one-step and green, and produce materials having surfaces free from chemical contamination. Such materials are highly important for biological and medical applications, where purity of the materials is of highest impact.

There are a number of laser-based nanomaterial processing methods that can produce 0D, 1D, 2D, and 3D nanostructures in the gaseous as well as in liquid phases, and can produce nano-/microstructures at the selective sites of bulk solid materials. Pulsed laser deposition, laser vaporization controlled condensation (LVCC), laser pyrolysis, laser chemical vapor deposition, photolithography, laser-induced direct surface writing for nano-/microfabrication, two-photon polymerization, laser-induced forward transfer (LIFT), laser ablation in liquids (LALs), laser-induced melting and fragmentation for resizing and reshaping of particles, laser-induced photodissociation of liquid precursors, and so on are some laser-based approaches of generation of nanoscaled materials.

Characterization of nanomaterials remains incomplete without the use of lasers. Laser excitation provides information about structural, compositional, electrical, optical, thermal, and lasing properties of nanomaterials. Raman, photoluminescence (PL), laser-induced breakdown spectroscopy (LIBS), laser ablation inductively coupled plasma mass spectroscopy (LA-ICPMS), light dynamic scattering, laser-photoacoustic spectroscopy (PAS), fluorescence correlation spectroscopy (FCS), ultrafast laser spectroscopy, laser-induced thermal pulses for space charge measurements, laser scanning microscopy, coherent diffractive imaging, and so on are some laser-based characterization techniques of nanoscaled materials.

The intention behind this book is that it serves as a platform for the state-of-art laser-based nanomaterials processing and characterization techniques. This book will be an effective medium to help retain scientists and researchers in the field of laser material processing and characterization. Optical and lasing characterization of nanomaterials using PL and Raman investigation for structural and size determination will be highly helpful in the development and industrialization of photonic devices and inexpensive lasing materials.

The researchers who are using lasers for other purposes might be promoted to do research in the filed of laser-based nanomaterial processing and characterizations, while beginners who have just entered the field will be guided effectively. Moreover, UG and PG students might be stimulated to start their research career in this field.

The contents of the book are arranged in the following manner.

The *first three chapters* are devoted to the basic introduction of lasers, nanomaterials, and interaction of lasers with atoms molecules and clusters. Chapter 1 sheds light on the history of laser development in short, basic construction, and principles of lasing, different types of active media for lasing, representative active media and lasers from each category and their operations, characteristics of laser light, and modification in the basic laser structure such as mode locking, pulse shaping, and so on.

Chapter 2 starts with the origin and historical development and introduction of nanomaterials, flows through the band theory of solids, quantum confinement, defects and imperfections of nanomaterials, metal, semiconductor, and insulator nanoparticles (NPs), various synthesis methods and techniques of nanomaterials characterizations, self- and induced assembly as well as aggregation and agglomeration, application of lasers in the synthesis, modification, and characterization of nanomaterials and ends with summary and future prospects.

Chapter 3 forms a bridge between lasers and materials. It deals with the interaction of lasers with atoms, molecules, and clusters. This chapter starts with introduction, flows through laser–atom interaction, laser–molecule interaction, high-pressure atomic physics, strongly coupled plasmas, laser cluster production and interaction, aerosol monitoring, and ends with the conclusion and outlook.

Chapters 4–13 are partitioned into two parts. Part I has been classified into three chapters based on the technique and ablation environment. For example, Chapter 4 is a group of subchapters devoted to the gas-phase laser-based processing techniques, Chapter 5 has sub-chapters related to laser-based nano-/microfabrication, and Chapter 6 has a *collection of* subchapters associated with liquid-phase laser-based nanomaterial processing techniques. Chapters 7–13 are grouped into Part II. Chapter 7 describes Raman spectroscopy, while Chapter 8 is devoted to dynamic light scattering (DLS). PL and fluorescence-based characterization techniques are given under Chapter 9, and Chapter 10 describes PAS for material characterization. Chapter 11 discusses ultrafast spectroscopy of nanomaterials, while Chapter 12 describes nonlinear optical spectroscopy of nanomaterials. Laser-based thermal pulse technique for the polarization and space charge characterization is the subject of Chapter 13.

Chapter 4 is collection of four subchapters (subchapters 4.1–4.4) related to gas-phase laser-based materials processing. Subchapter 4.1 describes synthesis and analysis of nanostructured thin films prepared by laser ablation of metals in vacuum, while subchapter 4.2 deals with the fabrication of nanostructures with pulsed laser ablation in a furnace, also known as high-temperature pulsed laser deposition (HTPLD). Ablation of metals under high-pressure ambience causes synthesis of NPs, which assist in the fabrication of one-dimensional nanostructure, is termed as high-pressure pulsed laser deposition (HPPLD), or nanoparticle-assisted pulsed laser deposition (NAPLD), and is the subject of subchapter 4.3. Removal of material from the target surface using laser vaporization and its transport to the substrate through controlled condensation using a temperature gradient between target and substrate is known as *laser vaporization controlled condensation*, and is the subject of subchapter 4.4.

Chapter 5 is subdivided in the four subchapters. Subchapter 5.1 deals with femtosecond laser nanostructuring and nanopatterning on metals. It starts with the introduction and flows through the basic principle of surface nanostructuring by femtosecond laser, periodic structuring by femtosecond laser, nanostructure-textured microstructures, single and array nanoholes, and ends with the application of femtosecond laser-induced surface structures and the summary. Subchapters 5.2–5.4 are devoted to the **LIFT** approach of transfer of film from a transparent substrate to the other substrates. Subchapter 5.2 is a review of the characteristics and features of various films transferred by LIFT with light shed on the principles and methods. Subchapter 5.3 describes basic fundamentals and processes involved in LIFT and discusses effects of various laser and target film parameters on the morphology of fabricated patterns. Application of LIFT in the fabrication of devices is the subject of subchapter 5.4. It begins with the introduction, discusses the various LIFT techniques such as traditional and modified LIFT, and ends with the conclusion and future aspects.

LALs for particle generation are cheaper, simpler, and quite recent than the gas-phase ablation, and is highly emerging in recent years. Chapter 6 presents liquid-assisted laser ablation for generation of particles and comprises four subchapters (subchapters 6.1–6.4). Subchapter 6.1 deals with the fundamentals of LALs, basic differences between LALs and gas phase, advantages of liquid-phase ablation over the gas phase, laser irradiation of liquid suspended particles for resizing and reshaping, and so on. Reactive laser ablation, described in subchapter 6.2, produces NPs that have elemental contributions from targets as well as from the liquid medium, is useful for the synthesis of oxide, hydroxide, carbide, and nitride NPs. Subchapter 6.3 presents liquid-phase laser ablation of fourth group elements (C, Si, Ge) for the synthesis of their elemental and compound nanomaterials.

Raman spectroscopy, Chapter 7, is an important nondestructive laser-based characterization technique that provides information about the structure, size, and shape, and electronic-, and defect-related properties of nanomaterials. Chapter 7 consists of two subchapters. Chapter 7.1 describes the fundamentals of Raman spectroscopy and some case studies and applications of Raman spectroscopy in the characterization of devices. Size and shape of nanomaterials affect the characteristics of Raman spectra, and is the subject of Chapter 7.2.

When a beam of monochromatic light passes through the colloidal solution of NPs, it gets scattered after the interaction with the moving particles under Brownian motion, and is termed as *dynamic light scattering*. The characteristic of the scattered light depends on the size of the particle and the wavelength of the incident beam. Chapter 8 describes size determination of particles using DLS.

PL/fluorescence is another important and nondestructive technique for the characterization of nanomaterials, which provides information about size, morphology, bandgap, defect, and crystallinity of nanomaterials. It tests the properties of nanomaterials for their possible applications in the fabrication of light-emitting diodes, solar cells, lasers, and as a fluorescence marker. Some of the PL-/fluorescence-based characterization techniques are presented in Chapter 9. This chapter encompasses three subchapters (subchapters 9.1–9.3). Chapter 9.1 presents a basic understanding of PL spectroscopy initiated by the introduction and experimental arrangements, flows through the applications of general PL spectroscopy on nanomaterial ensembles, and application of PL spectroscopy on single nanomaterials, and ends with the conclusion. FCS, presented in subchapter 9.2, is a new laser spectroscopic technique for single molecule detection that has recently been applied, *in vitro* and *in vivo*, to study the dynamical behavior of NPs in solution and the NP–cell interactions inside the biological environment. Subchapter 9.3 is focused on the time-resolved spectroscopy of nanomaterials, which is able to diagnose femtosecond and picosecond time-scaled dynamical processes.

PAS, described in Chapter 10, is a nondestructive and flexible spectroscopic tool, which offers an easy way to obtain the optical absorption spectra of any kind of samples. In the present age of nanotechnology, PAS has great importance in characterization of nanomaterials, since nanomaterials scatter light significantly and have large electron–phonon coupling.

Ultrafast spectroscopy of nanomaterials, described in Chapter 11, determines the lifetimes of fast dynamical processes such as the transition time of electrons, carrier dynamics, times for phonon–phonon, electron–phonon, and electron–electron interactions, and so on. It can determine the rate of reaction, electronic processes involved during the synthesis, and functionalization of nanoparticles.

Nonlinear spectroscopy of nanostructured materials, described in Chapter 12, and its novel applications in optoelectronics, optical switchers and limiters, as well as in optical computers, optical memory, and nonlinear spectroscopy, has attracted much attention in recent days. High- and low-order nonlinearity in refractive indices, susceptibility, and conversion efficiency in higher harmonic generation through laser-produced plasma on the surface of nanostructured materials are the subject of this chapter. Influence of laser ablation parameters on the optical and nonlinear optical characteristics of colloidal solutions of semiconductor NPs, high-order harmonic generation in silver NP-contained plasma, and studies of low- and high-order nonlinear optical properties of $BaTiO_3$ and $SrTiO_3$ NPs are the main topics of Chapter 12.

Chapter 13 presents applications of laser-generated thermal pulses in the polarization and space charge profiling of some polymer films and nanomaterials. This chapter starts with the basic overview and history of thermal techniques for polarization and space charge depth profiling, passes through the theoretical foundation, data analysis, and experimental techniques, and finishes with some applications on polymers and nanomaterials.

Subash Chandra Singh

List of Contributors

Weiping Cai
Key Laboratory of Materials
Physics
Institute of Solid State Physics
Chinese Academy of Sciences
350 Shushanghu Road
Hefei
Anhui 230031
China

Bingqiang Cao
University of Jinan
School of Materials Science and
Engineering
106 Jiwei road
Jinan 250022
Shandong
China

Filippo Causa
University of Naples Federico II
Interdisciplinary Research Centre
on Biomaterials
P.le Tecchio 80
80125 Naples
Italy

Avinash Chandra Pandey
University of Allahabad
Nanotechnology Application
Centre
Allahabad
India

Guoxin Chen
National University of Singapore
Department of Electrical and
Computer Engineering
4 Engineering Drive 3, 117576
Singapore

Tow Chong Chong
Singapore University of
Technology and Design
287 Ghim Moh Road #04-00
279623
Singapore

Yashashchandra Dwivedi
Universidade de São Paulo
Departamento de Física e Ciência
dos Materiais
Instituto de Física de São Carlos
Caixa Postal 369, 13560-970
São Carlos - SP
Brasil

M. Samy El-Shall
Virginia Commonwealth
University
Department of Chemistry
College of Humanities and
Sciences
1001 West Main Street
Richmond
VA 23284-2006
USA

Rashid Ashirovich Ganeev
Academy of Sciences of
Uzbekistan
Institute of Electronics
33 Dormon Yoli Street
Akademgorodok
Tashkent 100125
Uzbekistan

Haibo Gong
University of Jinan
School of Materials Science and
Engineering
106 Jiwei road
Jinan 250022
Shandong
China

Ram Gopal
University of Allahabad
Department of Physics
Laser spectroscopy &
Nanomaterials lab.
Allahabad 211002
India

Chunlei Guo
University of Rochester
The Institute of Optics
275 Hutchison Road
Rochester
NY 14627-0186
USA

Ruiqian Guo
Fudan University
Laboratory of Advanced Materials
2205 Songhu Road
Shanghai 200438
China

Alan M. Heins
University of Rochester
The Institute of Optics
275 Hutchison Road
Rochester
NY 14627-0186
USA

Jung-Il Hong
Daegu Gyeongbuk Institute of
Science and Technology (DGIST)
Daegu 711-873
Korea

and

School of Materials Science and
Engineering
Georgia Institute of Technology
Atlanta 30332
USA

Minghui Hong
National University of Singapore
Department of Electrical and
Computer Engineering
4 Engineering Drive 3 117576
Singapore

Aditya Kumar Singh
Department of Ceramic
engineering
IT, BHU
Varanasi
India

Haruhisa Kato
National Institute of Advanced
Industrial Science and
Technology (AIST)
Tsukuba Central 5
Higashi 1-1-1
Tsukuba
Japan

Kaushal Kumar
Italian Institute of Technology
Centre for Advanced
Biomolecules for Healthcare
(CRIB)
80125 Naples
Italy

and

Department of Applied Physics
Indian School of Mines
Dhanbad 826004
India

Thomas Lippert
Paul Scherrer Institute
Materials Group, General Energy
Research Department
OFLB U110 CH-5232
Villigen-PSI
Switzerland

Patrick J. McNally
Dublin City University
Nanomaterials Processing
Laboratory
The Rince Institute
School of Electronic Engineering
Dublin 9
Ireland

Axel Mellinger
Central Michigan University
Department of Physics
222 Dow Science Complex
Mount Pleasant
MI 48859
USA

Matthias Nagel
EMPA
Laboratory for Functional
Polymers
Swiss Federal Laboratories for
Materials Science and Technology
Überlandstrasse 129
CH-8600 Dübendorf
Switzerland

Paolo Antonio Netti
Italian Institute of Technology
Centre for Advanced
Biomolecules for Healthcare
(CRIB)
80125 Naples
Italy

Tatsuo Okada
Kyushu University
School of Information Science
and Electrical Engineering
744 Motooka, Nishi-ku
Fukuoka 819-0395
Japan

Hironobu Sakata
Tokai University
Department of Optical and
Imaging Science & Technology
4-1-1 Kitakaname
Hiratsuka
Kanagawa 259-1292
Japan

Vahit Sametoglu
University of Alberta
Department of Electrical and
Computer Engineering
Edmonton
Alberta, T6G 2V4
Canada

Luigi Sanguigno
Italian Institute of Technology
Centre for Advanced
Biomolecules for Healthcare
(CRIB)
80125 Naples
Italy

Vasant G. Sathe
University Campus
UGC-DAE Consortium for
Scientific Research
Khandwa Road
Indore-452017
Madhya Pradesh
India

Rajeev Singh
University of Allahabad
Department of Electronics &
Communication
Allahabad 211002
Uttar Pradesh
India

Subhash Chandra Singh
National Centre for Plasma
Science and Technology & School
of Physical Sciences
Dublin City University
Dublin-9
Ireland

Ying Yin Tsui
University of Alberta
Department of Electrical and
Computer Engineering
Edmonton
Alberta T6G 2V4
Canada

Anatoliy Vorobyev
University of Rochester
The Institute of Optics
275 Hutchinson Road
Rochester
NY 14627
USA

Moriaki Wakaki
Tokai University
Department of Optical and
Imaging Science & Technology
4-1-1 Kitakaname
Hiratsuka
Kanagawa 259-1292
Japan

Qing Wang
University of Alberta
Department of Electrical and
Computer Engineering
Edmonton
Alberta T6G 2V4
Canada

Rusen Yang
University of Minnesota-Twin
cities
Department of Mechanical
Engineering
111 Church Street SE
Minneapolis
MN 55455
USA

Shikuan Yang
Key Laboratory of Materials
Physics
Institute of Solid State Physics
Chinese Academy of Sciences
Hefei
Anhui 230031
China

Haibo Zeng
Nanjing University of
Aeronautics and Astronautics
State Key Laboratory of
Mechanics and Control of
Mechanical Structures
Key Laboratory for Intelligent
Nano Materials and Devices of
the Ministry of Education
College of Material Science and
Technology
Nanjing 210016
China

1
Lasers: Fundamentals, Types, and Operations

Subhash Chandra Singh, Haibo Zeng, Chunlei Guo, and Weiping Cai

The acronym LASER, constructed from *Light Amplification by Stimulated Emission of Radiation*, has become so common and popular in every day life that it is now referred to as *laser*. Fundamental theories of lasers, their historical development from milliwatts to petawatts in terms of power, operation principles, beam characteristics, and applications of laser have been the subject of several books [1–5]. Introduction of lasers, types of laser systems and their operating principles, methods of generating extreme ultraviolet/vacuum ultraviolet (EUV/VUV) laser lights, properties of laser radiation, and modification in basic structure of lasers are the main sections of this chapter.

1.1
Introduction of Lasers

1.1.1
Historical Development

The first theoretical foundation of LASER and MASER was given by Einstein in 1917 using Plank's law of radiation that was based on probability coefficients (Einstein coefficients) for absorption and spontaneous and stimulated emission of electromagnetic radiation. *Theodore Maiman* was the first to demonstrate the earliest practical laser in 1960 after the reports by several scientists, including the first theoretical description of *R.W. Ladenburg* on stimulated emission and negative absorption in 1928 and its experimental demonstration by *W.C. Lamb* and *R.C. Rutherford* in 1947 and the proposal of *Alfred Kastler* on optical pumping in 1950 and its demonstration by *Brossel, Kastler*, and *Winter* two years later. *Maiman's* first laser was based on optical pumping of synthetic ruby crystal using a flash lamp that generated pulsed red laser radiation at 694 nm. Iranian scientists *Javan* and *Bennett* made the first gas laser using a mixture of He and Ne gases in the ratio of 1 : 10 in the 1960. *R. N. Hall* demonstrated the first diode laser made of gallium arsenide (GaAs) in 1962, which emitted radiation at 850 nm, and later in the same year *Nick Holonyak* developed the first semiconductor visible-light-emitting laser.

Nanomaterials: Processing and Characterization with Lasers, First Edition.
Edited by Subhash Chandra Singh, Haibo Zeng, Chunlei Guo, and Weiping Cai.

1.1.2
Basic Construction and Principle of Lasing

Basically, every laser system essentially has an active/gain medium, placed between a pair of optically parallel and highly reflecting mirrors with one of them partially transmitting, and an energy source to pump active medium. The gain media may be solid, liquid, or gas and have the property to amplify the amplitude of the light wave passing through it by stimulated emission, while pumping may be electrical or optical. The gain medium used to place between pair of mirrors in such a way that light oscillating between mirrors passes every time through the gain medium and after attaining considerable amplification emits through the transmitting mirror.

Let us consider an active medium of atoms having only two energy levels: excited level E_2 and ground level E_1. If atoms in the ground state, E_1, are excited to the upper state, E_2, by means of any pumping mechanism (optical, electrical discharge, passing current, or electron bombardment), then just after few nanoseconds of their excitation, atoms return to the ground state emitting photons of energy $h\nu = E_2 - E_1$. According to *Einstein's* 1917 theory, emission process may occur in two different ways, either it may induced by photon or it may occur spontaneously. The former case is termed as *stimulated emission*, while the latter is known as *spontaneous emission*. Photons emitted by stimulated emission have the same frequency, phase, and state of polarization as the stimulating photon; therefore they add to the wave of stimulating photon on a constructive basis, thereby increasing its amplitude to make lasing. At thermal equilibrium, the probability of stimulated emission is much lower than that of spontaneous emission $(1 : 10^{33})$, therefore most of the conventional light sources are incoherent, and only lasing is possible in the conditions other than the thermal equilibrium.

1.1.3
Einstein Relations and Gain Coefficient

Consider an assembly of N_1 and N_2 atoms per unit volume with energies E_1 and $E_2 (E_2 > E_1)$ is irradiated with photons of density $\rho_\nu = N\, h\nu$, where $[N]$ is the number of photons of frequency ν per unit volume. Then the stimulated absorption and stimulated emission rates may be written as $N_1 \rho_\nu B_{12}$ and $N_2 \rho_\nu B_{21}$ respectively, where B_{12} and B_{21} are constants for up and downward transitions, respectively, between a given pair of energy levels. Rate of spontaneous transition depends on the average lifetime, τ_{21}, of atoms in the excited state and is given by $N_2 A_{21}$, where A_{21} is a constant. Constants B_{12}, B_{21}, and A_{21} are known as *Einstein coefficients*. Employing the condition of thermal equilibrium in the ensemble, Boltzmann statistics of atomic distribution, and Planck's law of blackbody radiation, it is easy to find out $B_{12} = B_{21}$, $A_{21} = B_{21}(8\pi h\nu^3/c^3)$, known as *Einstein relations*, and ratio, $R = \exp(h\nu/kT) - 1$, of spontaneous and stimulated emissions rates. For example, if we have to generate light of 632.8 nm ($\nu = 4.74 \times 10^{14}$ Hz) wavelength at room temperature from the system of He−Ne, the ratio of spontaneous and stimulated emission will be almost 5×10^{26}, which shows that for getting strong lasing one has to think apart from the thermal equilibrium. For shorter wavelength, laser, ratio of spontaneous to stimulated emission is larger, ensuring that it is more difficult to produce UV light using the principle of stimulated emission compared to the IR. Producing intense laser beam or amplification of light through stimulated emission requires higher rate of stimulated emission than spontaneous emission and self-absorption, which is only possible for $N_2 > N_1$ (as $B_{12} = B_{21}$) even though $E_2 > E_1$ (opposite to the Boltzmann statistics). It means that one will have to create the condition of *population inversion* by going beyond the thermal equilibrium to increase the process of stimulated emission for getting intense laser light.

If a collimated beam of monochromatic light having initial intensity I_0 passes through the mentioned active medium, after traveling length x, intensity of the beam is given by $I(x) = I_0 e^{-\alpha x}$, where α is the absorption coefficient of the medium, which is proportional to the difference of N_1 and N_2. In the case of thermal equilibrium $N_1 \gg N_2$ the irradiance of the beam will decrease with the length of propagation through the medium. However, in the case of population inversion, $(N_2 > N_1) - \alpha$, will be positive and the irradiance of the beam will increase exponentially as $I(x) = I_0 e^{kx}$, where k is the gain coefficient of the medium and may be given by $k = (nN_d h\nu_{21} B_{21})/c$, where N_d is $N_2 - N_1$, c is speed of light, and n is refractive index of the medium.

1.1.4
Multilevel Systems for Attaining Condition of Population Inversion

Considering the case of two energy level system under optical pumping, we have already discussed that $B_{12} = B_{21}$, which means that even with very strong pumping, population distribution in upper and lower levels can only be made equal. Therefore, optical as well as any other pumping method needs either three or four level systems to attain population inversion. A three level system (Figure 1.1a) irradiated by intense light of frequency ν_{02} causes pumping of large number of atoms from lowest energy level E_0 to the upper energy level E_2. Nonradiative decay of atoms from E_2 to E_1 establishes population inversion between E_1 and E_0 (i.e., $N_1 > N_0$), which is practically possible if and only if atoms stay for longer time in the state E_1 (metastable state, i.e., have a long lifetime) and the transition from E_2 to E_1 is rapid. If these conditions are satisfied, population inversion will be achieved between E_0 and E_1, which makes amplification of photons of energy $E_1 - E_0$ by stimulated emission. Larger width of the E_2 energy level could make possible absorption of a wider range of wavelengths to make pumping more effective, which causes increase in the rate of stimulated emission. The three level system needs very high pumping power because lower level involved in the lasing is the ground state of atom; therefore more than half of the total number of atoms have to be pumped to the state E_1 before achieving population inversion and in each of the cycle, energy used to do this is wasted. The pumping power can be greatly reduced if the lower level involved in the lasing is not ground state, which requires at least a four level system (Figure 1.1b). Pumping transfers atoms from ground state to E_3, from where they decay rapidly into the metastable state E_2 to make N_2 larger than

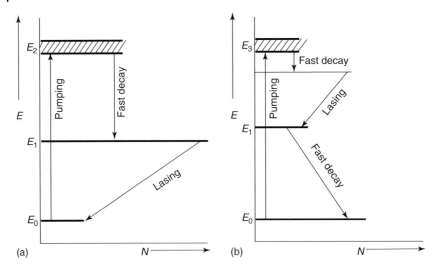

Figure 1.1 Energy level diagram for (a) three- and (b) four level laser systems.

N_1 to achieve the condition of population inversion between E_2 and E_1 at moderate pumping.

1.1.5
Threshold Gain Coefficient for Lasing

Laser beam undergoes multiple oscillations (through active medium) between pair of mirrors to achieve considerable gain before it leaves the cavity through partially reflecting mirror. Laser oscillation can only sustain in the active medium if it attains at least unit gain after a round-trip between mirrors and maintains it overcoming various losses inside the cavity. If we incorporate these losses, the effective gain coefficient reduces to $k - \Upsilon$, where Υ is the loss coefficient of the medium. If round-trip gain G were less than unity, the laser oscillation would die out, while it would grow if the G value were larger than unity. Let us consider that the laser beam of intensity I_0 passes through the active medium, homogeneously filled in the length L between the space of two mirrors M_1 and M_2 with reflectivities R_1 and R_2, respectively. The beam of intensity I_0 initiates from the surface of M_1 and attains intensity I_1 ($I_1 = I_0 \exp(k - \Upsilon)L$) after traveling a length L to reach at the surface of M_2. After reflection from M_2 and traveling back to M_1, the light intensity becomes I_2 ($I_1' = I_1 R_2$ due to reflection and $I_2 = I_1' \exp(k - \Upsilon)L$), which finally becomes I_2' after reflection from M_1 to complete a round-trip ($I_2' = I_2 R_1 = I_0 R_1 R_2 \exp2(k - \Upsilon)L$). Waves starting from the surface of mirror M_1 and those that have completed one or more round trips are in the same phase. Now, the gain $G(I_2'/I_0)$ attained in a round-trip should be at least unity to sustain the laser oscillation inside the cavity, therefore $R_1 R_2 \exp2(k - \Upsilon)L = 1$ is the threshold condition, which gives a value of $\Upsilon + (2L)^{-1} \ln(R_1 R_2)^{-1}$ for threshold gain (kth) coefficient.

1.1.6
Optical Resonator

An optical resonator is an arrangement of optical components, which allows a beam of light to circulate in a closed path so that it retraces its own path multiple times, in order to increase the effective length of the media with the aim of large light amplification analogous to the positive feedback in electronic amplifiers. Combination of optical resonator with active medium is known as *optical oscillator*. A set of two parallel and optically flat mirrors, with one highly reflecting $M_1(R \approx 100\%)$ and another partially transmitting $M_2(R > 95\%)$, makes a simple optical oscillator as shown in Figure 1.2. Some of the pumped atoms in the excited states undergo spontaneous emission generating seed photons, which pass through the active medium and get amplified through stimulated emission. Most of the energy gets reflected from both the mirrors, passes through the active medium, and continues to get amplified until steady state level of oscillation is reached. After attaining this stage, amplification of wave amplitude within the cavity dies away and extra energy produced by stimulated emission exits as laser output from the window M_2. The gain coefficient inside the cavity should be greater than the threshold gain coefficient (*k*th) in order to start and maintain laser oscillation inside the cavity. Owing to the diffraction effects, it is practically difficult to maintain a perfectly collimated beam with the combination of two parallel plane mirrors, which causes significant amount of diffraction losses. Such losses could be reduced by using a combination of concave mirrors and other optics in different optical arrangements. The optical configurations, which are able to retain the light wave inside the cavity after several transversals, are known as *stable resonators*. Some of the stable resonators are shown in the Figure 1.3. Laser oscillators with different geometries have their own benefits and losses. For example, in an oscillator having assembly of two parallel mirrors, it is difficult to align them in a strictly parallel manner. A slight deviation from the parallel geometry of the laser beam causes its walk away from the cavity axis after few reflections. However, it is beneficial in the sense that a large fraction of the active medium (mode volume) is pumped in this geometry. Confocal resonators are very simple to align, although lesser fraction of the active medium is being pumped.

Every laser resonator is characterized by a quantity Q termed as *quality factor*, which is defined by $Q = (2\pi \times$ energy stored$)/($energy dissipated per cycle$)$. The Q value of laser cavities lies in the range of $\sim 10^5 - 10^6$. Significance of higher Q value lies in the sense of capacity to store larger energy. In terms of line width $\Delta \nu$,

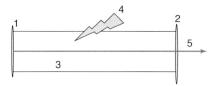

Figure 1.2 Basic geometry of laser cavity: (1) 100% and (2) 95–98% reflecting mirrors, (3) active medium, (4) pumping source, and (5) laser output.

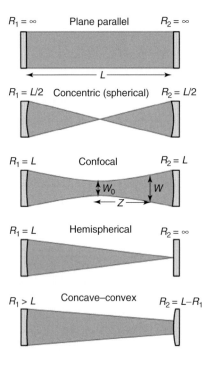

Figure 1.3 Different geometries of stable optical resonators.

and frequency ν, the quality factor can be defined as $Q = \nu/\Delta\nu$. A higher Q value associates with lower relative line width and vice versa.

A resonator that cannot maintain laser beam parallel to its axis is termed as *unstable resonator*. Such resonators suffer from high losses, but can make efficient use of the mode volume and have easy way of adjustment for the output coupling of the laser. Figure 1.4 illustrates an unstable resonator having active medium between the mirrors. Output power of the laser and inner diameter of the annular-shaped beam can be easily adjusted by varying the distance between the two reflecting

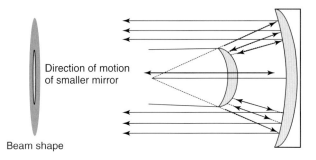

Direction of motion
of smaller mirror

Beam shape

Figure 1.4 A sketch of unstable optical resonator with annular beam shape.

mirrors. Resonators having low irradiation volume or unstable cavities require active media with a large gain coefficients, such as CO_2 gas.

1.1.7
Laser Modes

The output of laser beam actually consists of a number of closely spaced spectral lines of different frequencies in a broad frequency range. The discrete spectral components are termed as *laser modes*, and coverage range is the line width of the atomic transition responsible for the laser output. Laser modes are categorized into axial and transverse modes.

1) **Axial modes:** Let $d\phi = 2\pi/\lambda^*(2L)$ be the phase change in the laser wave after a round-trip in the cavity. In order to sustain laser oscillation inside the cavity, the phase change should be an integral multiple of 2π, that is, $2\pi/\lambda^*(2L) = 2p\pi$. In terms of frequency, this expression transforms to $v = pc/2L$; therefore separation between two adjacent p and $p+1$ modes is given by $\Delta v = c/2L$ (Figure 1.5). In the particular case of Nd: YAG (neodymium-ion-doped yttrium aluminum garnet) laser, $\lambda = 1064$ nm and $L = 25$ cm, $p = 2L/\lambda \approx 47 \times 10^4$ axial mode exists inside the laser cavity. If line width of the laser at 1064 nm is about $\Delta w = 1$ GHz, then only $\Delta w/\Delta v \approx 1$ axial mode oscillates in the cavity, while others die out. The axial modes are constructed by the light waves moving exactly parallel to the cavity axis. Light incident on a mirror and that reflected from that mirror construct a standing wave similar to a string bounded at both the ends. All the axial modes are due to the propagation of plane waves along the line joining centers of two reflecting mirrors.

2) **Transverse modes:** Unlike the plane waves propagating along the axis of the cavity in axial modes, there are some other waves traveling out of the axis that are not able to repeat their own path termed as transverse electromagnetic

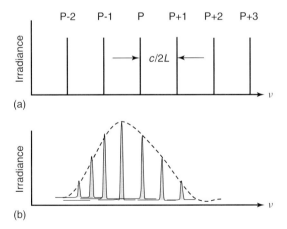

(a)

(b)

Figure 1.5 Axial laser modes (a) a simple illustration and (b) inside the laser line width, which shows that the mode at the center of the line has maximum intensity.

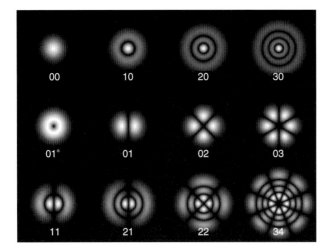

Figure 1.6 Various TEM modes of the laser.

(TEM) modes. These modes can be practically seen in the form of pattern when the laser beam falls on any surface. These modes are assigned by two integers p and q in the form of TEM_{pq}, where p and q are the number of minima in the horizontal and vertical directions, respectively, in the pattern of the laser beam. TEM_{00} means that there is no minima in the beam spot, and this is known as *uniphase mode*. On the contrary, TEM_{01} shows that there is no minima in the horizontal scanning and one minima in vertical. Laser beam spots on the screen with several TEM modes are displayed in Figure 1.6.

1.2
Types of Laser and Their Operations

Depending on the nature of the active media, lasers are classified into three main categories, namely, solid, liquid, and gas. Scientists and researchers have investigated a wide variety of laser materials as active media in each category since 1958, when lasing action was observed in ruby crystal. It is inconvenient to discuss all lasers having these materials as active media. Here, representative active medium for each of the categories and their operating principle with energy level diagram is discussed.

1.2.1
Solid Laser

1.2.1.1 Doped Insulator Laser
Solid state lasers have active media obtained by embedding transition metals (Ti^{+3}, Cr^{+3}, V^{+2}, Co^{+2}, Ni^{+2}, Fe^{+2}, etc.), rare earth ions (Ce^{+3}, Pr^{+3}, Nd^{+3}, Pm^{+3}, Sm^{+2}, $Eu^{+2,+3}$, Tb^{+3}, Dy^{+3}, Ho^{+3}, Er^{+3}, Yb^{+3}, etc.), and actinides

such as U^{+3} into insulating host lattices. Energy levels of active ions are only responsible for lasing actions, while physical properties such as thermal conductivity and thermal expansivity of the host material are important in determining the efficiency of the laser operation. Arrangement of host atoms around the doped ion modifies its energy levels. Different lasing wavelength in the active media is obtained by doping of different host materials with same active ion. $Y_3Al_5O_{12}$, $YAlO_3$, $Y_3Ga_5O_{12}$, $Y_3Fe_5O_{12}$, $YLiF_4$, Y_2SiO_5, $Y_3Sc_2Al_3O_{12}$, $Y_3Sc_2Ga_3O_{12}$, $Ti:Al_2O_3$, $MgAl_2O_4$ (spinel), $CaY_4[SiO_4]_3O$, $CaWO_4$ (Scheelite), $Cr:Al_2O_3$, NdP_5O_4, $NdAl_3[BO_3]_4$, $LiNdP_4O_{12}$, $Nd:LaMgAl_{11}O_{19}$, $LaMgAl_{11}O_{19}$, $LiCaAlF_6$, $La_3Ga_5SiO_4$, $Gd_3Sc_2Al_3O_{12}$, $Gd_3Ga_5O_{12}$, $Na_3Ga_2Li_3F_{12}$, Mg_2SiO_4 (Forsterite), CaF_2, Al_2BeO_4 (Alexandrite), and so on, are some of the important hosts. Active atom replaces an atom in the host crystal lattice. Nd:YAG is one of the best lasing material and is representative of solid state lasing materials.

1.2.1.1.1 Dopant Energy Levels in the Host Matrices

Transition metal and rare earth ions have partially filled and unfilled 3d and 4f subshells, respectively. For example, the electronic configurations of trivalent Cr and Nd ions are as follows:

$$Cr^{+3}: 1s^2 2s^2 2p^6 3s^2 3p^6 3d^3$$

$$Nd^{+3}: 1s^2 2s^2 2p^6 3s^2 3p^6 3d^{10} 4s^2 4p^6 4d^{10} 4f^3 5s^2 5p^6$$

There are unshielded partially filled d electrons in the transition metal ions, while partially filled 4f electrons of the rare earth ions are shielded by 5p and 5s sub shells. Owing to the electronic shielding of inner subshells in rare earth ions, crystal field effect on the energy levels of transition metal ions are pronounced as compared to that on energy levels of rare earth ions. When one of these ions is doped into a host lattice, three main types of interactions occur: (i) columbic interaction between electrons in the unfilled shell, (ii) the crystal field, and (iii) spin–orbit interactions. The columbic interaction between electrons causes splitting of energy levels of a single electron configuration into several levels denoted by pair of values of L and S (L and S are vector sum of angular, l, and spin, s, momenta, respectively, of electrons). Crystal field splitting dominates for transition metal, while spin–orbit interaction is the major contributor for rare earth ions in the modification of energy level of isolated host atom. The energy level diagram for $Cr^{+3}:Al_2O_3$ (ruby) and $Nd^{+3}:YAG$ are displayed in Figure 1.7.

1.2.1.1.2 Pumping Techniques in Solid State Lasers

Pumping of electrons from the ground state to the excited state to achieve population inversion condition is an essential requirement for lasing. Optical pumping is the best and most efficient pumping method for solid state active media due to their broad optical absorption bands. A significant fraction of incident optical energy can be easily used for the pumping of ground state electrons using pulsed as well as continuous light sources. Excess light energy raises temperature of the laser materials; therefore pulsed light sources are more suitable for dissipation of heat

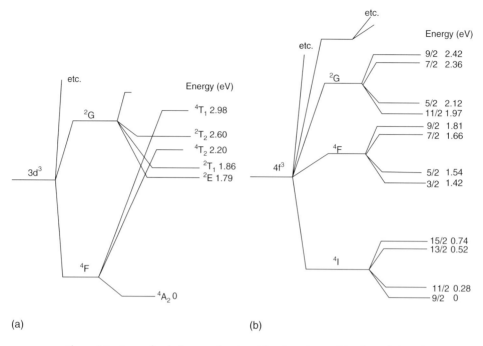

(a) (b)

Figure 1.7 Energy level diagrams for doped insulator lasers: (a) ruby and (b) Nd:YAG lasers.

through circulating water jackets. Low-pressure quartz/glass-sealed krypton/xenon lamps are mostly used for pulsed pumping light sources, while tungsten halogen lamps and high-pressure mercury discharge lamps are utilized for continuous optical pumping. An inductive, capacitive, and resistive (LCR) circuit and trigger unit as shown in Figure 1.8 is basically used for operating the flashtube. The detail circuit diagram of the power supply is presented in Ref. [3]. High-voltage pulse of the trigger coil ionizes some gas in the tube and makes it conductive. This causes rapid discharge of the capacitor through the tube and generation of intense optical radiation for almost few milliseconds. A small inductor in the series protects damage of the tube due to high capacitor discharge current. Light source and active medium should be arranged in such a way that maximum pumping radiation falls on the active medium. Active media in solid state lasers are cylindrical and rod shaped with few millimeters diameter and few centimeter lengths. Several arrangements of cylindrical flash lamp and rod-shaped active media are used for optical pumping to get laser radiation. The flash lamp and active medium assembly are placed inside gold-plated reflectors of circular or elliptical cross section. In the first practical operating laser, ruby rod was pumped by helical flash lamp inside the cylindrical reflecting cavity. Such arrangement has significant uniformity of irradiation inside the rod but exhibits poor optical coupling. Side-by-side arrangement of flash lamp and laser rod inside the cylindrical gold-plated reflector or wrapping both together with a metal foil are simpler approaches having good optical coupling

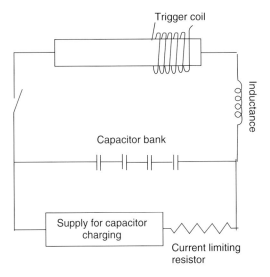

Figure 1.8 Trigger unit and LCR circuit diagram for solid state lasers.

but poor uniformity of irradiation. An elliptical reflector having flash lamp at one focus and laser rod at the other focus is the most popular and best way of optical pumping in the solid state lasers. Light radiation leaving from the first focus gets focused close to the axis of laser rod placed at the second focus to make uniform energy distribution. Combination of a number of elliptical reflectors having laser rod at the common foci and several flash lamps at the other foci is used for better optical pumping with more uniform energy distribution. Various geometries for the arrangement of laser rod and flash lamps are illustrated in Figure 1.9. Nd:YAG laser is widely used in the processing of materials and various characterizations. Here we discuss energy level diagram and operating principles of Nd:YAG lasers.

1.2.1.1.3 Nd:YAG Laser Construction and Operation

The schematic diagram of Nd:YAG laser head as shown in Figure 1.10, consists of oscillator section, rear mirror, quarter-wave plate, Pockel cells, polarizer, pump chambers, injection seeder, output coupler, D-Lok monitor, fold mirrors, amplifier section, harmonic generator (HG), temperature controller, dichroic mirrors, and Beam Lock pointing sensor. It may have single or multipump chambers, and each chamber consists of single or multiple flash lamps depending on the power of laser. The laser head end panel contains coolant, output connector, coolant input connector, neutral/ground connector, control cable connector, high-voltage connector, Q-switch input connector, and nitrogen purge input connector. The HGs have potassium di-hydrogen phosphate (KDP) and beta barium borate (BBO) crystals for frequency doubling and tripling, respectively. It can be operated in long pulse and Q-switch modes. Long pulse mode has light pulses of almost $200\,\mu s$ duration and separated from each other by $2-4\,\mu s$. The total energy of the pulse

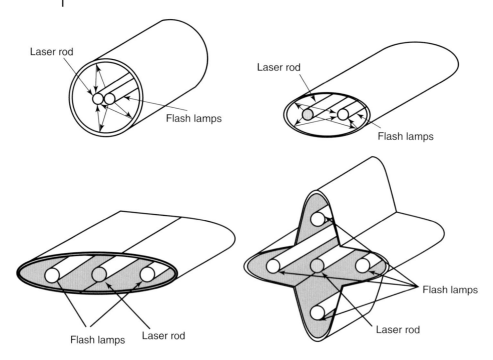

Figure 1.9 Different geometries for the arrangement of flash lamp and laser rod in solid states lasers.

train is similar to that of a single Q-switched pulse. During Q-switched operation, the pulse width is less than 10 ns and the peak optical power is tens of megawatts.

The properties of Nd:YAG are the most widely studied and best understood of all solid state laser media. Its energy level diagram, optical arrangements for Q-switching and stable and unstable resonators are depicted in Figure 1.11. The active medium is triply ionized neodymium, which is optically pumped by a flash lamp whose output matches principle absorption bands in the red and near infrared (NIR). Excited electrons quickly drop to the $F_{3/2}$ level, the upper level of the lasing transition, where they remain for a relatively longer time ($\sim$230 μs). The strongest transition is $F_{3/2} \rightarrow I_{11/2}$, emitting a photon in NIR region (1064 nm). Electrons in the $I_{11/2}$ state quickly relax to the ground state, which makes its population low. Therefore, it is easy to build up a population inversion for this pair of states with high emission cross section and low lasing threshold at room temperature. There are also some other competing transitions at 1319, 1338, and 946 nm from the same upper state, but having lower gain and a higher threshold than the 1064 nm wavelength. In normal operation, these factors and wavelength-selective optics limit oscillation to 1064 nm. A laser comprising just an active medium and resonator will emit a pulse of laser light each time the flash lamp fires. However, the pulse duration will be long, about the same as the flash lamp and its peak power will be low. When a Q-switch is added to the resonator to shorten the pulse, output peak power is raised dramatically. Owing to the long lifetime of $F_{3/2}$, a large

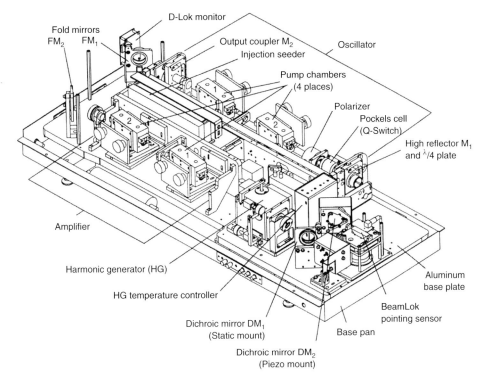

Figure 1.10 Assembly of various components in the head of an Nd:YAG laser system with four pump chambers.

population of excited neodymium ions can build up in the YAG rod in a way similar to which a capacitor stores electrical energy. When oscillation is prevented for some time to build up high level of population inversion by electro-optical Q-switching and after that if the stored energy gets quickly released, the laser will emit a short pulse of high-intensity radiation.

1.2.1.2 Semiconductor Laser

Semiconductor lasers also known as *quantum well lasers* are smallest, cheapest, can be produced in mass, and are easily scalable. They are basically p-n junction diode, which produces light of certain wavelength by recombination of charge carrier when forward biased, very similar to the light-emitting diodes (LEDs). LEDs possess spontaneous emission, while laser diodes emit radiation by stimulated emission. Operational current should be higher than the threshold value in order to attain the condition of population inversion. The active medium in a semiconductor diode laser is in the form of junction region of 2 two-dimensional layers. No external mirror is required for optical feedback in order to sustain laser oscillation. The reflectivity due to the refractive index differences between two layers or total internal reflection to the active media is sufficient for this purpose. The diodes end faces are cleaved, and parallelism of reflecting surfaces is assured. Junction

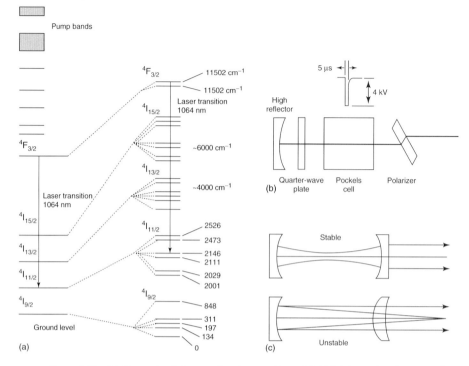

Figure 1.11 (a) Energy level diagram for the transition of Nd:YAG laser (b) The Q-switch comprises a polarizer, a quarter-wave plate, high quality reflector, and pockels cell, and (c) stable and unstable resonator configurations.

made from a single type of semiconductor material is known as *homojunction*, while that obtained from two different semiconductors is termed as *heterojunction*. Semiconductors of p and n type with high carrier density are brought together for constructing p-n junction with very thin ($\approx 1\,\mu$m) depletion layer. Figure 1.12 illustrates GaAs homojunction semiconductor diode laser. Lasing occurs in the confined narrow region, and optical feedback is done by reflections between cleaved end faces. For GaAs $n = 3.6$, therefore reflectivity R from the material–air interface is $R = (n-1)^2/(n+1)^2 = 0.32$, which is small but sufficient for lasing.

When the operating current is small, the population inversion built compensates losses in the system and no lasing action is done. Increase of the current above a critical value named as *threshold current* commences lasing action, and the intensity of laser radiation increases rapidly with further increase in the operating current. Semiconductor lasers have large divergence compared to any other laser systems, which is due to their small cross section of active region. Actual dimension (d) of the active medium is of the order of light wavelength (λ), which causes diffraction and hence divergence by an angle of $\theta \approx \lambda/d$. Homojunction semiconductor lasers have some disadvantages over heterojunction lasers. Both of the laser systems should have confinement of injected electrons and emitted light in the junction region in order to initiate efficient stimulated emission process. In

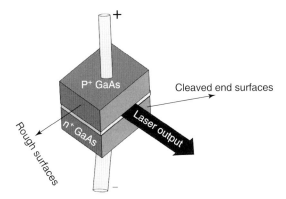

Figure 1.12 Basic geometry of semiconductor laser system.

the homojunction laser, confinement of light is the consequence of the presence of hole and electrons close to the junction. Homojunction lasers operate under such confinement mechanism but have high threshold current density and low efficiency. Electrons have to travel different distances before they recombine with the holes. In contrast, heterojunction lasers exhibit much higher lasing efficiency and low threshold current density compared to their homojunction counterparts. Another difficulty with homojunction laser is to prevent the radiation from spreading out sideways from the gain region, which causes loss instead of gain. Therefore they can only be used in the pulsed mode.

Heterojunction lasers are constructed by sandwiching a thin layer of GaAs between two layers of ternary semiconductor compound $Ga_{1-x}Al_xAs$ with comparatively lower refractive indices and higher band gap energy. Lower refractive indices of surrounding layers causes confinement of laser radiation inside the active medium by the mechanism of total internal reflection, which makes laser oscillation to sustain in the medium. Higher band gap energy of the surrounding media creates potential barrier to prevent charge carriers to diffuse from the junction region, that is, provides a way for the confinement of charge carriers in the junction region, which enhances the condition of population inversion and hence stimulated emission. The electrical circuit for pumping the semiconductor diode lasers is similar to the doped insulator lasers.

1.2.2
Gas Laser

Gas lasers are widely available in almost all power (milliwatts to megawatts) and wavelengths (UV-IR) and can be operated in pulsed and continuous modes. Based on the nature of active media, there are three types of gas lasers viz atomic, ionic, and molecular. Most of the gas lasers are pumped by electrical discharge. Electrons in the discharge tube are accelerated by electric field between the electrodes. These accelerated electrons collide with atoms, ions, or molecules in the active media and

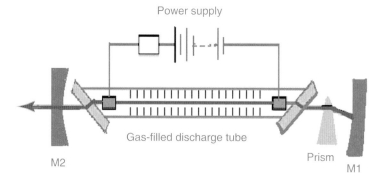

Figure 1.13 Construction of gas laser system (argon ion laser with prism-based wavelength tuning).

induce transition to higher energy levels to achieve the condition of population inversion and stimulated emission. An example of gas laser system is shown in Figure 1.12.

1.2.2.1 Atomic Gas Laser; He:Ne Laser

He−Ne laser is the simplest and representative of atomic gas lasers. The active medium is a 10 : 1 mixture of He and Ne gases filled in a narrow tube of few millimeter diameters and 0.1−1 m long at a pressure of about 10 Torr. Discharge tube and circuit are very similar as shown in Figure 1.13. A resistant box is used in series with power supply in order to limit the discharge current because tube resistance falls too low once discharge is initiated. Energy levels of Ne atom are directly involved in the laser transitions, and He atoms provide an efficient excitation mechanism to the Ne atoms. Helium atoms from their ground state 1^1S, are pumped to the excited atomic states 2^1S and 2^3S by impact with accelerated electrons in the discharge tube. Neon atoms have 3s and 2s atomic states, which are closer to the 2^1S and 2^3S states of the helium atoms, respectively. Collision between excited helium atoms in the 2^1S and 2^3S states and neon atoms in the ground states reinforce transfer of energy from helium to neon atoms. Helium atoms in the 2^1S and 2^3S states excite neon atoms from ground state to the 3s and 2s states, respectively, and return to the ground state. Excited states 3s and 2s of Ne atom have longer life times as compared to its lower (3p and 2p states), therefore they serve as metastable states and are used in achieving the condition of population inversion between s and p states. Transitions 3s → 3p, 3 s → 2p, and 2 s → 2p of neon atoms are consequences of lasing at 3.39 μm, 632.8 nm, and 1.15 μm wavelengths, respectively. Lifetimes of 3p and 2p atomic states are shorter; therefore Ne atoms from these states rapidly decay to the 1s state by nonradiative transitions. Neon atoms in the 1s state go to the ground state after losing energy through collision with the wall of the tube. The energy level diagram of the He−Ne laser is displayed in Figure 1.14. Another important atomic laser is copper vapor laser, but it is beyond the scope of this book.

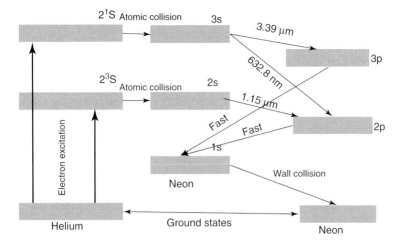

Figure 1.14 Energy level diagram for He−Ne laser system.

1.2.2.2 Ion Laser: Argon Ion Laser

1.2.2.2.1 Physical Construction
Argon ion laser is one of the widely used ion gas lasers, which typically generates several watts power of a green or blue output beam with high beam quality. The core component of an argon ion laser is an argon-filled tube made of ceramics, for example, beryllium oxide, in which an intense electrical discharge between two hollow electrodes generates a plasma with a high density of argon (Ar^+) ions. A solenoid around the tube (not shown in Figure 1.13) is used for generating a magnetic field, which increases output power of the beam by magnetic confinement of the plasma near the tube axis.

A typical device, containing a tube with a length of the order of 1 m, can generate 2.5−5 W of output power of laser beam in the green spectral region at 514.5 nm, using several tens of kilowatts of electric power. The dissipated heat is removed with a chilled water flow around the tube. The laser can be switched to other wavelengths such as 457.9 nm (blue), 488.0 nm (blue−green), or 351 nm (ultraviolet) by rotating the intracavity prism. The highest output power is achieved on the standard 514.5 nm line. Without an intracavity prism, argon ion lasers have a tendency for multiline operation with simultaneous output at various wavelengths.

1.2.2.2.2 Working of Ar Ion Laser
The argon ion laser is a four level laser, which facilitates to achieve population inversion and low threshold for lasing. The neutral argon atoms filled between two hollow electrodes inside the plasma tube (Figure 1.13) are pumped to the 4p energy level by two steps of collisions with electrons in the plasma. The first step ionizes atoms to make ions in the 3p (E_1) state, and the second one excites these ions from the ground state E_1 either directly to the $4p^4$ levels (E_3) or to the $4p^2$ levels (E_4), from which it cascades almost immediately to the $4p^2$ (E_3). The 4p ions

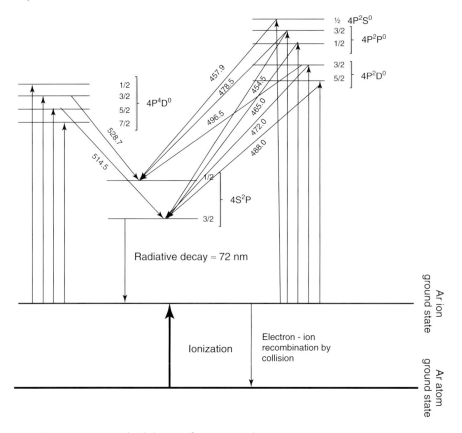

Figure 1.15 Energy level diagram for argon ion laser system.

eventually decay to 4s levels (E_2), either spontaneously or when stimulated to do so by a photon of appropriate energy. The wavelength of the photon depends on the specific energy levels involved and lies in between 400 and 600 nm. The ion decays spontaneously from 4s to the ground state, emitting a deep ultraviolet photon of about 72 nm. There are many competing emission bands as shown in Figure 1.15. These can be preferentially selected using a prism in front of one of the end mirrors. This prism selects a specific wavelength to send it back through the cavity to stimulate identical emissions, which stimulates more and more emissions and make regenerative process. This facilitates laser to operate at a single wavelength. Removal of the prism allows for broadband operation, that is, several wavelengths are kept rather than keeping only a particular wavelength. The mirrors reflect a number of lines within a maximum range of about 70 nm. Energy level diagram showing various transitions of Ar ion laser is illustrated in Figure 1.15.

1.2.2.3 Molecular Laser

Unlike isolated atoms and ions in atomic and ionic lasers, molecules have wide energy bands instead of discrete energy levels. They have electronic, vibrational, and rotational energy levels. Each electronic energy level has a large number of vibrational levels assigned as V, and each vibrational level has a number of rotational levels assigned as J. Energy separation between electronic energy levels lies in the UV and visible spectral ranges, while those of vibrational–rotational (separations between two rotational levels of the same vibrational level or a rotational level of one vibrational level to a rotational level from other lower vibrational level) levels, in the NIR and far-IR regions. Therefore, most of the molecular lasers operate in the NIR or far-IR regions.

1.2.2.3.1 Carbon Dioxide (CO₂) Laser

Carbon dioxide is the most efficient molecular gas laser material that exhibits for a high power and high efficiency gas laser at infrared wavelength. It offers maximum industrial applications including cutting, drilling, welding, and so on. It is widely used in the laser pyrolysis method of nanomaterials processing. Carbon dioxide is a symmetric molecule (O=C=O) having three (i) symmetric stretching [i00], (ii) bending [0j0], and (iii) antisymmetric stretching [00k] modes of vibrations (inset of Figure 1.16), where i, j, and k are integers. For example, energy level [002] of molecules represents that it is in the pure asymmetric stretching mode with 2 units of energy. Very similar to the role of He in He–Ne laser, N_2 is used as intermediately in CO_2 lasers. The first, $V = 1$, vibrational level of N_2 molecule lies close to the (001) vibrational level of CO_2 molecules. The energy difference between vibrational levels of N_2 and CO_2 in CO_2 laser is much smaller (0.3 eV) as compared to the difference between the energy levels of He and Ne (20 eV) in He–Ne laser; therefore comparatively larger number of electrons in the discharge tube of CO_2 laser having energies higher than 0.3 eV are present. In addition to

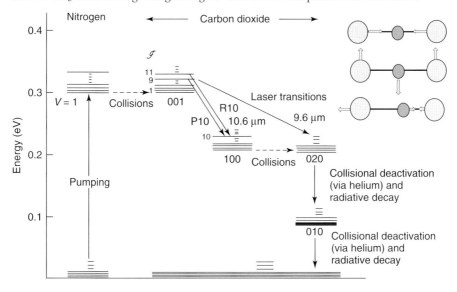

Figure 1.16 (a) Absorption, emission curves, and (b) energy level diagram of dye laser system.

this, $V = 1$ state of N_2 is metastable, which provides longer time for the collision between excited N_2 molecules and the ground state CO_2 molecules to excite them to (001) state. These two favorable conditions make it easy to attain high level of population inversion between 001 and 100, and 020 vibrational states of CO_2.

Transitions between 001 initial level to 100 and 020 final vibrational states make stimulated emissions of several IR radiations between 9.2 and 10.6 µm wavelengths. Helium gas is also mixed in the gas mixture in order to increase efficiency of lasing. Helium helps in transporting waste heats to the tube wall and de-exciting (100) and (020) energy levels by collision process. The amounts of N_2, CO_2, and He in CO_2 laser depends on the type and application of system, but usually, the amount of nitrogen and CO_2 molecules are comparable, while helium concentration is higher than either. Low pressure ($\sim$10 Torr) is generally used for CW lasers, while quite higher pressure is used for high-energy and pulsed laser applications. Depending on power level and beam quality of CO_2 lasers, various internal structures are being used, and they are called *sealed tube laser, gas flow laser*, transversally excited atmospheric (TEA) laser, and *gas dynamic laser*. Detailed discussions of these are beyond the scope of the book, although interested readers may consult Ref. [4].

In the far-IR region of 10 µm wavelength, the usual optical material has large absorbance and therefore cannot be used as windows and reflecting mirrors in the cavity. Materials such as Ge, GaAs, ZnS, ZnSe, and some alkali halides having transparency in the IR region are used.

1.2.2.3.2 Nitrogen Laser

Lasing transition in N_2 laser takes place between two electronic energy levels, therefore this laser operates in the ultraviolet region at 337 nm wavelength. Here, upper electronic level has a shorter lifetime compared to the lower one, hence CW operation cannot be achieved, but pulsed operation with narrow pulse width is possible. The pulse width is narrow because as soon as lasing starts, population of the lower state increases, while that at upper state decreases and rapidly a state at which no lasing is possible is rapidly achieved. Such a laser system is known as *self-terminating*.

1.2.2.3.3 Excimer Lasers

Excimers are molecules such as ArF, KrF, XeCl, and so on, that have repulsive or dissociating ground states and are stable in their first excited state. Usually, there are less number of molecules in the ground state; therefore direct pumping from ground state is not possible. Molecules directly form in the first excited electronic state by the combination of energetic halide and rare gas ions. Condition of population inversion can be easily achieved because the number of molecules in the ground state is too low as compared to that in the excited state. Lasing action is done by transition from bound excited electronic state to the dissociative ground state. Population in the ground state always remains low because molecules dissociate into atoms at this point. Usually a mixture of halide such as F_2 and rare gas such as Ar is filled into the discharge tube. Electrons in the discharge tube dissociate and ionize halide molecules and create negative halide ions. Positive

Ar^+ and negative F^- ions react to produce ArF^* molecules in the first excited bound state, followed by their transition to the repulsive ground state to commence lasing action. Various excimer lasers are developed in the wavelength range of 120–500 nm with 20–15% efficiency and up to 1 J peak and 200 W average powers. These lasers are widely used in materials processing and characterizations as well as for the pumping of dye lasers.

1.2.3
Liquid Laser

Liquids are more homogeneous as compared to solids and have larger density of active atoms as compared to the gasses. In addition to these, they do not offer any fabrication difficulties, offer simple circulation ways for transportation of heat from cavity, and can be easily replaced. Organic dyes such DCM (4-dicyanomethylene-2-methyl-6-*p*-dimethylaminostyryl-4H-pyran), rhodamine, styryl, LDS, coumarin, stilbene, and so on, dissolved in appropriate solvents act as gain media. When the solution of dye molecules is optically excited by a wavelength of radiation with good absorption coefficient, it emits radiation of longer wavelength, known as *fluorescence*. The energy difference between absorbed and emitted photons is mostly used by nonradiative transitions and creates heat in the system. The broader fluorescence band in dye/liquid lasers facinates them with the unique feature of wavelength tuning. Organic dye lasers, as tunable and coherent light sources, are becoming increasingly important in spectroscopy, holography, and in biomedical applications. A recent important application of dye lasers involves isotope separation. Here, the laser is used to selectively excite one of several isotopes, thereby inducing the desired isotope to undergo a chemical reaction more readily. The dye molecules have singlet (S_0, S_1, and S_2) and triplet (T_1 and T_2) group of states with fine energy levels in each of them (Figure 1.17). Singlet

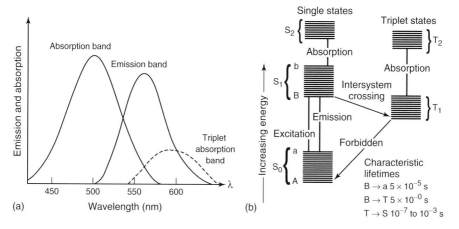

Figure 1.17 Energy level diagram for carbon dioxide (molecular) laser system. Inset shows three mode of vibration of CO_2 molecule.

and triplet states correspond to the zero and unit values of total spin momentum of electrons, respectively. According to selection rules for transitions in quantum mechanics, singlet–triplet and triplet–singlet transitions are quite less probable as compared to the transitions between two singlet or two triplet states. Optical pumping of dye molecules initially at the bottom of S_0 state transfers them to the top of S_1 state. Collisional relaxation of these molecules takes them to the bottom of S_1 state, from where they transit to the top of S_0 state with stimulated emission of radiation. Most of the states in the complex systems are usually neither pure singlet nor pure triplet. Singlet states have small contribution of triplet and vice versa. In the case of most of the dye molecules, unfortunately, T_1 state lies just below the S_1, therefore few molecules transit from S_1 to T_1 by losing some energy through nonradiative transitions. Difference between T_1 and T_2 states is almost same as the wavelength of lasing transition, therefore emitted lasing radiation gets absorbed, which reduces laser gain and may cease the laser action. Therefore, some of the dye lasers operate in the pulsed mode with the pulse duration shorter than the time required to attain a significant population in the state T_1. Some of the dyes also absorb laser radiation corresponding to the transition from S_1 to upper singlet transitions. Therefore, one should select the dye molecule so that energy differences between these states do not lie between the ranges of laser radiation.

1.3
Methods of Producing EUV/VUV, X-Ray Laser Beams

EUV and VUV coherent light sources, that is, EUV/VUV lasers, are in high demand in order to continue the validity of Moore's law in future, in the high-density data writing on disks, materials synthesis and characterizations, and spectroscopic investigations. For example, in photolithography for making the micro/nanopatterns on microelectronic chips, the width of the pattern is proportional to the wavelength of laser light used. Therefore, we would reach a limit for further miniaturization of electronic devices, which causes failure of the well-known Moore's law if shorter wavelength lasers would not be developed. Similarly, for data writing on the optical discs, if we have shorter wavelength lasers, larger amount of data can be written on the same size of disc. In addition to these, such sources are the future of 3D high-density data writing, microscopy at the atomic scale, crystallography, and medical sciences. Following are the methods for developing short wavelength lasers.

1.3.1
Free Electron Lasers (FEL)

In contrast to the other laser sources, free electron lasers (FELs) have an active medium that consists of a beam of free electrons, propagating at relativistic velocities in a spatially periodic magnet (undulator). Here, electrons experience the Lorentz force, execute transverse oscillations, and emit synchrotron radiation in the forward direction (Figure 1.18). We know that an accelerated charged particle

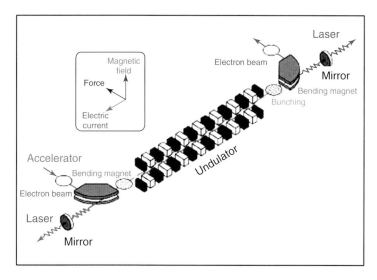

Figure 1.18 Construction geometry of free electron laser system.

moving with relativistic velocity emits radiation, which may be considered as spontaneous. The emitted spontaneous radiation interacts with the electron beam in order to stimulate laser radiation. Since electric field of light is perpendicular to the direction of its propagation, electrons cannot interact with photons unless they also have a velocity component parallel to the light electric field. Hence, electron beam is allowed to travel through wiggler magnet, which generates a spatially periodic magnetic field along the direction of propagation of electron beam. After passing through such field, electrons undergo transverse oscillations as shown in Figure 1.18. In the present case, wavelength of emitted radiation is given by the expression $\lambda = \lambda_u/2\gamma^2$ where λ_u is the undulator length and γ is the ratio of total electron energy to its rest energy. Similar to other lasers, emitted radiation oscillates between two mirrors of the laser cavity and interacts with the electron beam to enhance power of the laser radiation before leaving the cavity as laser output. Since mirrors have very low reflectivity in the X-ray region, therefore there is severe problem in achieving intense X-ray laser beam. In another mechanism known as self-amplified stimulated emission (SASE), radiation of the spontaneous emission from relativistic electron interacts with the same electron to stimulate it for stimulated emission. Following are the salient features of FELs.

1) **Tunability:** FELs generate coherent, high-power radiation that is widely tunable, currently ranging in wavelength from microwaves, through terahertz radiation and infrared, to the visible spectrum, to soft X-rays. They have the widest frequency range of any laser type. Wavelength of FELs can be easily tuned by varying electron energy and period and amplitude of magnetic field. More shorter wavelength can be produced by harmonic generation.

2) **Pulse duration:** Based on the electron beam time structure pulse duration ranges from CW to ultrashort pulsed regime (fraction of picoseconds)
3) **Coherence:** It is transverse and longitudinal for oscillators and coherent seed amplifiers, while only transverse for SASE
4) **Brilliance:** Depending on the status of the art of the electron beam technology, the FEL brilliance can be larger, in some spectral regions (in particular in VUV-X), by many orders of magnitude than the brilliance of the existing sources (lasers and synchrotron radiation).

We do not go into much details, but interested readers are referred to Refs. [6–11].

1.3.2
X-Ray Lasers

X-ray lasers provide coherent beam of electromagnetic radiation with the wavelength range from ∼30 nm down to ∼0.01 nm. The first X-ray laser beam was initiated in 1980 by underground nuclear explosion at Nevada test site, while the first laboratory demonstration of X-ray laser was made in 1984 in the form of Nova laser. Most of the experimental demonstration of light amplification in this spectral range has come from the high-density plasma produced by interaction of high-energy laser with the solid target. Like the above case of FEL, unavailability of good-quality of cavity mirrors has utilized the mechanism of SASE in X-ray lasers. In such case intensity of the laser beam depends on the amplifier length. Gain of the active medium depends on the mass and temperature of the ions and multiplicity and population of the upper/lower levels. Most of the X-ray lasers utilize transitions among L, M, N, and so on, shells of highly ionized atoms. The short life times of the excited states (picoseconds) require very large pumping rates in order to achieve and maintain the condition of population inversion. Laser produced plasmas (LPPs) have high electron temperatures (∼0.1–1.0 keV) and densities (∼10^{18-21} cm^{-3}), which is required for higher degree of ionization and excitation in X-ray lasers. Therefore LPPs are used as primary medium for X-ray lasers. In addition to these, LPPs have uniform density and temperature and thus can provide good media for amplification and propagation of X-rays. Of the various processes suggested, two main (i) collisional excitation and (ii) electron recombination are responsible for attaining the population inversion condition. In case of collisional excitation upper state of the lasing has lower probability of decay with dipole radiation, compared to the lower state, which creates the condition of population inversion between them. In contrast, the rate of population at any state by three-body recombination process is proportional to the forth power of the principle quantum number. The combination of preferential population of upper levels and fast radiative decay of lower states leads to an inversion amongst the $N = 2, 3, 4, \ldots$ states [12].

1.3.3
EUV/VUV Lasers through Higher Harmonic Generation

Higher harmonic generation (HHG) is a nonlinear optical process used for the generation of shorter wavelength laser light from the interaction of high-intensity longer wavelength lasers source with nonlinear optical medium. The obtained new frequencies are integral multiples ($n\omega$) of the fundamental (ω) frequency of original laser light. This phenomenon was first observed in 1961 by Franken *et al.* [13] with ruby laser and quartz as nonlinear medium. The first HHG result was found in 1988, which shows that intensity of the spectra decreases with the increase of order, reaches a plateau, where the intensity remains constant for several orders, and finally ends abruptly at a position called *high harmonic cutoff*. They are portable sources of EUV/soft X-rays, synchronized with the fundamental laser and operated at the same frequencies with much shorter pulse width. These are more spatially coherent compared to X-ray lasers and cheaper than FELs. The harmonic cutoff increases linearly with increasing laser intensity up to the saturation intensity I_{sat} where harmonic generation stops. The saturation intensity can be increased by changing the atomic intensities of lighter noble gases, but these have lower conversion efficiency. HHG strongly depends on the driving laser field; therefore the produced harmonics have similar spatial and temporal coherence. Owing to the phase matching and ionization conditions required for HHG, the produced new pulses are with shorter pulse duration compared to the driving laser. Mostly, harmonics are produced in very short time frame, when phase matching condition is satisfied. Instead of shorter temporal window, they emit colinearly with the driving laser pulse and have very tight angular confinement. Gaseous media and LPP on the solid surfaces are two sources used as nonlinear active media for the generation of harmonics.

The shortest wavelength producible with the harmonic generation is given by cutoff of the plateau, which is given by the maximum energy gained by ionized electron from the light electric field. The energy of cutoff is given by $E_{max} = I_P + 3.17U_P$, where U_P is the pondermotive potential from the laser field and I_P is the ionization potential. It is assumed that electron is initially produced by ionization into the continuum with zero initial velocity and is accelerated by the laser electric field. After half period of the laser electric field, direction of motion of electron is reversed and it is accelerated back to the parent nuclei. After attaining high kinetic energy of the order of hundreds of electron volts (depending on the intensity of laser field), when electron enters into the parent nuclei, it emits Bremsstrahlung-like radiation during a recombination process with the atoms as it returns back to the ground state. The three-step model of ionization, acceleration of electron, and its recombination with parent nuclei by the emission of EUV photons is shown in Figure 1.19.

Wide wavelength ranges of laser systems starting from hundreds of micrometers (molecular liquid lasers) to X-ray regions (FELs, X-ray lasers, and HHG lasers) with CW and pulsed lasers having various pulse width from milliseconds to attoseconds and repletion rates from single pulse to the megahertz are available nowadays.

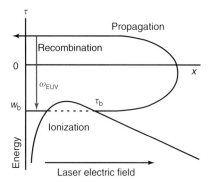

Figure 1.19 Three-step model of ionization and acceleration of electron and its recombination with parent nuclei by the emission of EUV photons.

Figure 1.20 represents the spectrum of available laser system with wavelength range and active media.

1.4
Properties of Laser Radiation

Light produced from the lasers have several valuable characteristics not shown by light obtained from other conventional light sources, which make them suitable for a variety of scientific and technological applications. Their monochromaticity, directionality, laser line width, brightness, and coherence of laser light make them highly important for various materials processing and characterization applications. These properties are discussed separately in the following subsections.

1.4.1
Monochromaticity

Theoretically, waves of light with single frequency v of vibration or single wavelength λ is termed as *single color* or *monochromatic* light source. Practically, no source of light including laser is ideally monochromatic. *Monochromaticity* is a relative term. One source of light may be more monochromatic than others. Quantitatively, degree of monochromaticity is characterized by the spread in frequency of a line by Δv, line width of the light source, or corresponding spread in wavelength $\Delta \lambda$. For small value of $\Delta \lambda$, frequency spreading, Δv, is given as $\Delta v = -(c/\lambda^2)\Delta \lambda$ and $\Delta \lambda = (c/v^2)$. The most important property of laser is its spectacular monochromaticity. Based on the type of laser media, solid, liquid, or gas and molecular, atomic, or ions, and the type of excitation, produced laser line consists of color bands that range from broad (as dye laser $\Delta \lambda \sim 200$) to narrow (for gas discharge lines, $\Delta \lambda \sim 0.01$ nm). Utilizing suitable filters one can get the monochromaticity as good as a single line of lasing transition. But such a single line also contains a set of closely spaced lines of discrete frequencies, known as *laser*

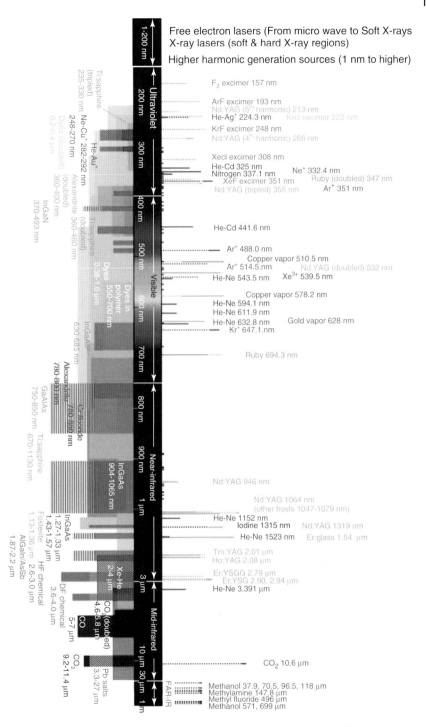

Figure 1.20 Spectrum of available laser systems, their wavelength range and active media.

modes (Figure 1.5). Suppressing all other modes excluding the central intense mode using mode locking one can increase monochromaticity of laser line. Suppressing of modes is possible by increasing the separation ($\Delta\nu_{\text{sep}} = c/2L$) between two modes, which can be done by reducing the cavity length. When the axial mode separation approaches the line width, $\Delta\nu$, of the lasing transition it is possible that only single mode oscillates. Now, the line width of the laser is equal to that of the single longitudinal mode, which is too narrow. The width of the laser line is directly related to the quality factor Q of the cavity and is given by $Q = \nu/\Delta\nu$. The quality factor Q actually defines ($Q = 2\pi \times$ energy stored in the resonator at resonance/energy dissipated per cycle) the ratio of energy stored in the cavity at resonance condition and energy dissipated per cycle. High degree of monochromaticity of laser line is required in the diagnosis of closely spaced rotational levels of molecules using selective excitation of that level. The laser line would be absolutely monochromatic if it is oscillating with single frequency, that is, width of the laser line is zero ($\Delta\nu = 0$). The single-mode laser has the highest degree of monochromaticity, but it has also not achieved the ideal monochromaticity condition.

1.4.2
Directionality

One of the most striking properties of laser is its directionality, that is, its output is in the form of an almost parallel beam. Owing to its directional nature it can carry energy and data to very long distances for remote diagnosis and communication purposes. In contrast, conventional light sources emit radiation isotropically; therefore, very small amount of energy can be collected using lens. Beam of an ideal laser is perfectly parallel, and its diameter at the exit window should be same to that after traveling very long distances, although in reality, it is impossible to achieve. Deviation in the parallelism of practical laser beam from the ideal is not due to any fault in the laser design, but due to diffraction from the edges of mirrors and windows. From the theory of diffraction, we know that circular aperture has angle of diffraction given by $\theta = \sin^{-1}(1.22\lambda/D)$. The spreading of the beam does depend on the physical nature of aperture and on the type of transverse mode oscillating inside the cavity. In the particular case of TEM_{00} mode oscillating inside the nearly confocal cavity, the value of minimum diameter at the center of cavity is given by $w_0 = (\lambda r/2\pi)^{1/2}$, where r is the radius of curvature of cavity mirrors (Figure 1.3). The beam diameter, w, varies with distance, z, from the point of minimum diameter and is given by $w = w_0(2z/r)$. The beam radius at any point inside and outside the cavity is determined by the distance from the cavity axis where intensity reduces $1/e$ times of its maximum value.

1.4.3
Coherence

Coherence is one of the striking properties of the lasers, over other conventional sources, which makes them useful for several scientific and technological applications. The basic meaning of coherence is that all the waves in the laser beam remain spatially and temporarily in the same phase. Photons generated through stimulated emission are in phase with the stimulating photons. For an ideal laser system, electric field of light waves at every point in the cross section of beam follows the same trend with time. Such a beam is called *spatially coherent*. The length of the beam up to which this statement is true is called *coherence length* (L_C) of the beam. Another type of coherence of the laser beam is temporal coherence, which defines uniformity in the rate of change in the phase of laser light wave at any point on the beam. The length of time frame up to which rate of phase change at any point on the laser beam remains constant is known as *coherence time* (t_C). Let P_1 and P_2 be two points on the same laser beam at time t and $t + \tau$. The correlation function between phases at these two points is termed as *mutual coherence function*, which is a complex number with magnitude between 0 and 1 (0 for completely incoherent beam and 1 for ideally coherent). *Coherence time* (t_C) is also defined as the time taken by the atoms/molecules in active medium to emit a light wave train of length L_C. These two coherences are thus related by $t_C = L_C/c$. The coherence time of the laser beam is almost inverse ($t_C \simeq 1/\Delta \nu$) of the width, $\Delta \nu$, of the laser transition. The lasers operating in the single mode (well-stabilized lasers) have narrow line width, therefore they exhibit higher coherence time and coherence length compared to those operating in multimode. Spatial and temporal coherences of continuous laser beams are much higher compared to those of pulsed laser systems because temporal coherence in the pulse lasers are limited by the presence of spikes within the pulse or fluctuation in the frequency of emission.

1.4.4
Brightness

Lasers are more intense and brighter sources compared to other conventional sources such as the sun. A 1 mW He–Ne laser, which is a highly directional low divergence laser source, is brighter than the sun, which is emitting radiation isotropically. Brightness is defined as power emitted per unit area per unit solid angle. In the particular case of 1 mW He–Ne laser with 3.2×10^{-5} rad beam divergence and 0.2 mm spot diameter at exit window the solid angle ($\pi (3.2 \times 10^{-5})^2$) is 3.2×10^{-9} sr and spot area ($\pi (2 \times 10^{-4})^2$) at the exit window is $1.3 \times 10^{-7} \text{m}^2$. Thus the brightness of the beam is given by ($(1 \times 10^{-3})/(1.3 \times 10^{-7} \text{ m}^2 \times 3.2 \times 10^{-9} \text{ sr})) = 2.4 \times 10^{12} \text{ W/m}^{-2} \text{ sr}^{-1}$, which is almost 10^6 times brighter than the sun ($1.3 \times 10^6 \text{ W/m}^{-2} \text{ sr}^{-1}$).

1.4.5
Focusing of Laser Beam

In practice, every laser system has some angle of divergence, which increases the spot size of laser beam and reduces its brightness. If a convergent lens of suitable focal length is inserted in the path of the laser beam, it focuses laser energy into small spot area at focal point. If w_L is the radius of the beam and f is the focal length of convergent lens, then radius of the spot at focal point is given as $r_s = \lambda f / \pi w_L$, where λ wavelength of laser radiation. If D is the lens diameter and the whole aperture is illuminated by laser radiation (i.e., $w_L = D/2$) then $r_s = 2\lambda f / \pi D$ or $r_s = 2\lambda F / \pi$, where $F = f/D$ is f number of the lens.

In the case of a particular Nd:YAG laser operating at 1064 nm wavelength and 35 mJ/pulse energy, and 10 ns pulse width. It is focused by a convex lens of $F = 5$, and whole of the lens area is illuminated by laser. The spot diameter at the focal point is given by $r_s = 2 \times 1.064 \times 10^{-6} \times 5/\pi = 3.4\,\mu m$. The irradiance of the laser beam is given by $I = E(J)/\text{pulse width (s)} \times (\pi r_s^2) = 35 \times 10^{-3}/(10 \times 10^{-9}\,s \times \pi (3.4 \times 10^{-6})^2) \approx 10^{16}\,W\,m^{-2}$.

1.5
Modification in Basic Laser Structure

Addition of some electronic, optical, or electro-optical systems between the active media and mirror to modify the pulse width, pulse shape, and energy/pulse and generation of integral multiple of laser frequency is important for advanced technological applications. Mode locking or phase locking, Q-switching, pulse shaping, pulse compression and expansion, frequency multiplication, and so on, are some commonly used methods in advanced laser technology.

1.5.1
Mode Locking

1.5.1.1 Basic Principle of Mode Locking

Mode locking is a technique in optics by which a laser can be made to produce light pulses of extremely short duration of the order of picoseconds (10^{-12} s) or femtoseconds (10^{-15} s). The basis of this technique is to induce constant phase relationship between the modes of laser cavity. Simply, same phase of δ can be chosen for all laser modes. Such a laser is called *mode-locked* or *phase-locked laser*, which produces a train of extremely narrow laser pulses separated by equal time intervals. Let N modes are oscillating simultaneously in the laser cavity with $(A_0)_n$, ω_n, and δ_n being the amplitude, angular frequency, and phase of the nth mode. All these parameters vary with time, therefore modes are incoherent. The output of such laser is a linear combination of n different modes and is given by $A(t) = \sum_{n=0}^{N} (A_0)_n e^{i(\omega_n t + \delta_n)}$ expression. For simplicity, frequency of the nth mode can be written as $\omega_n = \omega - n\Delta\omega$, where ω_n is the mode with highest frequency and $\Delta\omega = c\pi/L$ is the angular frequency separation between two modes. If all the modes have same amplitude and we force the various modes to maintain same relative phase δ to one another, that is, we mode lock the laser such that $\delta_n = \delta$, then the expression for resultant amplitude will be $A(t) = A_0 e^{i(\omega t + \delta)} \sum_{n=0}^{N} e^{-i\pi nct/L} = A_0 e^{i(\omega t + \delta)} \sin(N\phi/2)/\sin(\phi/2)$, where $\phi = \pi ct/L$. The irradiance of the laser output is given by $I(t) = A(t)A(t)^* = A_0^2 \sin^2(N\phi/2)/\sin^2(\phi/2)$, which is the periodic function of the period $\Delta\phi = 2\pi$ in the time interval $t = 2L/c$ (time of round-trip inside the cavity). The maximum value of irradiance is $N^2 A_0^2$ at $\phi = 0$ or $2p\pi$ (p is integer). Irradiance has zero value for $N\phi/2 = p\pi$, where p is an integer with values neither zero nor a multiple of N. This makes $\phi = 2p\pi/N = \pi ct/L$ or $t = (1/N)(2L/c)p$. Therefore, separation between two consecutive minima, that is, pulse width of a single laser pulse is $\Delta t = (1/N)(2L/c)$. Hence, the output of a mode-locked laser has sequence of short pulses of pulse duration $(1/N)(2L/c)$ separated in time by $2L/c$. The ratio of pulse separation to the pulse width is equal to the number of modes N, which shows that there should be a large number of modes in the cavity in order to get high-power short duration (picoseconds and femtoseconds) laser pulses.

1.5.1.2 Mode Locking Techniques

1.5.1.2.1 Active Mode Locking

We have discussed that mode locking is achieved by inducing the longitudinal mode to attain the fixed phase relationship, which may be exploited by varying the loss of the laser cavity at a frequency equal to the intermode separation $c/2L$. Let us consider a shutter between the active medium and output mirror, which is closed for most of the time and is opened after every $2L/c$ seconds (corresponding to the time of round-trip) and remains open for short duration of $(1/N)(2L/c)$ seconds. If the laser pulse train is as long as the shutter remains opened and arrives at the shutter exactly at the time of its opening, the pulse train is unaffected by the presence of shutter. The segment of the pulse, which arrives before opening and after closing of the shutter, will be clipped. Thus phase relationship of the modes is maintained by periodic oscillation of the shutter. An electro-optical or acousto-optical crystal operating on the principle of Pockels or Kerr effect respectively, may be used as a periodic shutter.

In the former case of electro-optical switching, a polarizer and an electro-optical crystal are arranged in between the active medium and laser exit mirror so that laser beam passes through the polarizer before entering into the crystal. Laser light from the active medium passes through the polarizer and gets plane polarized. When this polarized beam passes twice through the crystal (with appropriate electric field along the direction of light propagation) before returning to the polarizer, the plane of polarization is rotated by an angle of $90°$, which does not allow the light beam to enter into the active media through the polarizer. In other words, the shutter is effectively closed. If there is no field along the crystal in the direction of propagation of light, there will be no rotation of the plane of polarization of light and it can pass though the polarizer and enter into the active medium (shutter is open). Similarly in acousto-optical switching, a 3D grating pattern is established in a medium (water or glass, not active medium of lasing) by the incident and reflected sound waves created by piezoelectric transducer attached at one end of the medium. This grating diffracts a part of the laser beam and creates a high loss.

1.5.1.2.2 Passive Mode Locking

Passive mode locking method consists of placing a saturable absorber inside the cavity. Saturable absorbers are molecules having a nonlinear decrease in absorption coefficients with the increase in the irradiance of light. The saturable absorber is placed between active laser medium and mirror. If an intense pulse of light passes through the saturable absorber placed inside the laser cavity, the low-power tails (weaker modes) of the pulse are absorbed because of the absorption of dye molecules. The high-power peak of the pulse is, however, transmitted because the dye is bleached. Owing to this nonlinear absorption, the shortest and most intense fluctuation grows, while the weaker dies out.

1.5.2
Q-Switching

Q-switching, sometimes known as *giant pulse formation*, is a technique by which a laser can be made to produce a pulsed output beam. The technique allows the production of light pulses with extremely high (of the order of approximately gigawatts) peak power, much higher than would be produced by the same laser if it were operating in a CW mode. Compared to mode locking, another technique for pulse generation with lasers, Q-switching leads to much lower pulse repetition rates, much higher pulse energies, and much longer pulse durations. Both techniques are sometimes applied at once. In contrast to mode locking, where we achieve a train of pulses, Q-switching provides a single strong and short pulse of laser radiation.

Q-switching is achieved by putting a variable attenuator inside the laser's optical resonator. When the attenuator is functioning, light that leaves the gain medium does not return and lasing cannot begin. This attenuation inside the cavity corresponds to a decrease in the *Q factor* or *quality factor* of the optical resonator. A high Q factor corresponds to low resonator losses per round-trip, and vice versa. The variable attenuator is commonly called a "Q-switch," when used for this purpose. Initially the laser medium is pumped while the Q-switch is set to prevent feedback of light into the gain medium (producing an optical resonator with low Q). This produces a population inversion, but laser operation cannot yet occur since there is no feedback from the resonator. Since the rate of stimulated emission is dependent on the amount of light entering the medium, the amount of energy stored in the gain medium increases as the medium is pumped. Owing to losses from spontaneous emission and other processes, after a certain time, the stored energy will reach some maximum level; the medium is said to be *gain saturated*. At this point, the Q-switch device is quickly changed from low to high Q, allowing feedback and the process of optical amplification by stimulated emission to begin. Because of the large amount of energy already stored in the gain medium, the intensity of light in the laser resonator builds up very quickly; this also causes the energy stored in the medium to be depleted almost as quickly. The net result is a short pulse of light output from the laser, known as a *giant pulse*, which may have very high peak intensity. Similar to mode locking, active and passive techniques are used for Q-switching. Electro-optical and opto-acoustic (active Q-switching) switches, and saturable absorbers (passive Q-switching) are used to prevent feedback signal into the active media.

1.5.3
Pulse Shaping

Pulse shaping is a technique of optics, which modifies temporal profile of a pulse from laser. It may lengthen or shorten pulse duration, can generate complex pulses, or generate multiple pulses with femtosecond/picosecond separation from a single laser pulse. A pulse shaper may act as a modulator. Modulating function is applied on the input pulse to get the desired pulse. The modulating function in pulse shapers may be in time or a frequency domain (obtained by Fourier transform of time profile of pulse). In the pulse shaping methods optical signal is converted into electronic signal, where the presence and absence of pulses are designated by 1 and 0, respectively. There are two well-known pulse shaping techniques (i) direct space to time pulse shaper (DST-PS) and (ii) Fourier transform pulse shaper (FT-PS). In DST-PS, electrical analog of the desired output signal is used as a modulator with the electrical analog of input pulses. In contrast to DST-PS, the FT-PS uses a modulating function, which is a Fourier transform of the required sequence. In other words, a specific temporal sequence of pulses, the modulating function, is its Fourier transform and acts in frequency domain.

Optical design is a common experimental arrangement consisting of a combination of two sets of grating, a pair of mirrors, a pair of convex lenses, and a pulse shaper (see Figure 1.21a). By placing different types of optical components we can design various types of laser pulses. For example, employing suitable filter at focal plane can remove optical radiation of undesired frequency. A slab of transparent material with varying thickness can offer different path lengths for different frequency components and can decide which component of light will come out first. A concave lens at the focal plane provides longer path length for higher frequency component, and vice versa (Figure 1.21b). This causes the lower frequency component to come first and the higher frequency one comparatively later, resulting in positively chirped[1] output pulse (frequency increases with time). In contrast to this, a planoconvex or biconvex lens makes longer path length for lower frequency, and vice versa, which makes higher frequency to come out first and shorter frequency after that, that is, negative chirped output. Both positive and negative chirping lengthens the pulse duration, that is, stretching of the laser pulse. Removing or reducing the chirp causes compression (reduction of pulse duration) of the pulse.

---

1) Chirping is a mechanism by which different components of light frequency of a pulse from laser comes out at different time. When frequency of light components of pulse increases with time it is called *positive chirping*, decreases with time negative chirping and if constant with time no chirping (un-chirped).

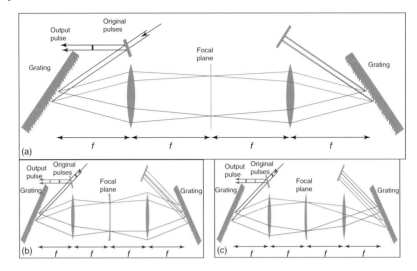

Figure 1.21 (a) Basic optical geometries for laser pulse shaping, (b) positive chirped, and (c) negative chirped pulse shaping.

References

1. Wilson, J. and Hawkes, J.F.B. (eds) (1987) *Lasers: Principles and Applications*, Prentice Hall Publications. ISBN: 013523697-5.
2. O'Shea, D.C., Callen, W.R., and Rhodes, W.T. (eds) (1977) *Introduction to Lasers and their Applications*, Addison-Wesley Publishing Company, Inc., Philippines. ISBN: 020105509-0.
3. Ifflander, R. (ed.) (2001) *Solid State Laser Material Processing: Fundamental Relations & Technical Realizations*, Springer-Verlag, Berlin, Heidelberg. ISBN: 354066980-9.
4. Weber, M.J. (ed.) (1991) *Handbook of Laser Science and Technology*, CRC Press, Boca Raton, FL, Ann Arbor, MI, Boston, MA. ISBN: 084933506-X.
5. Letokhov, V.S., Shank, C.V., Shen, Y.R., and Walther, H. (eds) (1991) *Interaction of Intense Laser Light with Free Electrons, M.V. Fedorov; Laser Science and Technology and International Handbook*, Harwood Academic Publishers GmbH, Switzerland. ISBN-3718651262.
6. Dattoli, G. and Renieri, A. (1984) Experimental and theoretical aspects of Free-Electron Lasers, in *Laser Handbook*, vol. **4** (eds M.L. Stich and M.S. Bass), North Holland, Amsterdam.
7. Marshall, J.C. (1985) *Free Electron Lasers*, Mac Millan Publishing Company, New York.
8. Luchini, P. and Motz, H. (1990) *Undulators and Free-Electron Lasers*, Claredon Press, Oxford.
9. Colson, W.B., Pellegrini, C., and Renieri, A. (eds) (1990) *Laser Handbook*, Free-Electron Lasers, vol. **6**, North Holland, Amsterdam.
10. Brau, C.A. (1990) *Free-Electron Lasers*, Academic Press, Oxford.
11. Dattoli, G., Renieri, A., and Torre, A. (1995) *Lectures in Free-Electron Laser Theory and Related Topics*, World Scientific, Singapore.
12. Dunn, J., Osterheld, A.L., Shepherd, R., White, W.E., Shlyaptsev, V.N., and Stewart, R.E. (1998) Demonstration of x-ray amplification in transient gain nickel-like palladium scheme. *Phys. Rev. Lett.*, **80** (13), 2825–2828.
13. Franken, P.A., Hill, A.E., Peters, C.W., and Weinreich, G. (1961) Generation of optical harmonics. *Phys. Rev. Lett.*, **7**, 118.

2
Introduction of Materials and Architectures at the Nanoscale

Subhash Chandra Singh, Haibo Zeng, Chunlei Guo, Ram Gopal, and Weiping Cai

Nanoscale materials have transferred science and technology into a new era, where the devices are so tiny and smart that they can swim through the veins when surgery is performed on diseased organs and cells. Materials at the nanoscale are 1000 times lighter and 1000 times stronger than those of steel, which may be the future of cheap satellite launch vehicles. They are useful in fields such as energy harvesting and storage, fabrication of intense but inexpensive light sources, nano-sized efficient electronic devices, medicine, life sciences, and surgery, as catalysts in the chemical industry, as fertilizers for plant growth stimulators, monitoring and removing of environmental pollutants, and so on. This chapter starts with the origin and historical development and introduction of nanomaterials and flows through the band theory of solids and quantum confinement; defects and imperfections of nanomaterials; metal, semiconductor, and insulator nanoparticles; the various synthesis methods and techniques of nanomaterials characterizations; self-assembly and induced assembly as well as aggregation and agglomeration; applications of lasers in the synthesis, modification, and characterization of nanomaterials; and ends with the summary and future prospects.

2.1
Origin and Historical Development

It was on 29 December 1959 at the annual meeting of the American Physical Society that noted physicist Richard Feynman, in his plenary talk entitled *"There is plenty of room at the bottom,"* invited and suggested that theoreticians and experimentalists work on the new branch of physics [1]. He suggested a way to manipulate individual atoms and molecules using some precise tools. Feynman predicted that the entire 24 volumes of the Encyclopedia Britannica could be written on the head of a pin. This would be on a scale where gravity would be meaningless, while surface tension and van der Waals interaction would become more important. The term *nanotechnology* was first employed by Norio Taniguchi in 1974 in his paper [2], where he stated that "Nanotechnology mainly consists of processing, separation, consolidation, and deformation of materials by single atom or molecule." Later in the 1980s,

Nanomaterials: Processing and Characterization with Lasers, First Edition.
Edited by Subhash Chandra Singh, Haibo Zeng, Chunlei Guo, and Weiping Cai.

Dr. K.E. Drexler promoted the technological importance of nanoscale phenomena and devices [3]. He envisioned self-replicating robots at the molecular scale in *Engines of Creation: The Coming Era of Nanotechnology* in 1986. Nanotechnology and nanoscience got a boost in the early 1980s with two major developments in the form of cluster science and invention of the scanning tunneling microscope (STM). This development led to the discovery of fullerenes in 1985 and carbon nanotubes a few years later. In the year 2000, the US president's Council of Advisors on Science and Technology founded the National Nanotechnology Initiative to coordinate with Federal nanotechnology research and development. Nanoscience and nanotechnology now top the scientific, academic, and technological research and most countries around the world are spending maximum amounts of their research grants in these fields in the hope that they will be the future of green and safe energy needs, nanotools for internal surgery and medical diagnosis, the basic need of chemical industries, and several other direct and indirect needs of mankind and society.

2.2
Introduction

Nanoscience is the study of the science of objects at the length scale of 10^{-9} m, and its deployment in the fabrication of devices for mankind and humanity is known as *nanotechnology*. Objects in the size range of 1–100 nm are called *nanomaterials*. They are made of atoms arranged in a particular geometry. For example, a spherical nanoparticle of 1 nm radius is formed by clustering of almost 1000 atoms with 1 Å atomic radius. Most of the chemical and physical properties of materials depend on the motion their electrons are allowed to perform. In isolated atoms (most confinement) having a dimension of 1 Å, electronic motion is highly confined to discrete energy levels, and atoms can absorb only a certain frequency of incident radiation. In contrast, electrons in the bulk materials have the freedom to move anywhere inside the lattice (free electrons), that is, they have continuous energy levels (least confinement) and therefore can absorb a wide spectral range of frequencies from incident radiation.

Materials in the size range of 100 nm to 10 μm have physical and chemical properties very similar to that of the bulk. In contrast, reduction in particle sizes down to 100 nm makes a considerable change in almost all the physical and chemical properties compared to that of their bulk counterparts. This phenomenon becomes more and more significant when size reduces below the quantum confinement limit, which is different for different materials, but usually lies in the size range of 1–10 nm for most of the materials. For example, particles in the nanoscale have much lower melting points and lattice constants compared to the bulk because of the increase in surface-to-volume ratio with the reduction in size. Higher surface-to-volume ratio results in surface atoms with lesser number of neighboring atoms compared to those inside the bulk, which is the cause of smaller cohesive energy and hence lower melting temperature. Decrease in the stability temperature is the cause of the decrease in ferromagnetism and ferroelectricity when materials transit from bulk to nanoscale. Increase in the surface-to-volume ratio with a higher number of dangling bonds[1] fascinates nanoparticles with high reactivity and catalytic properties. Gas sensing, surface adsorption, catalytic activity, surface functionalization and assembly, and so on, are some properties of nanomaterials stimulated by surface dangling bonds. Nanomaterials exhibit various interesting size-dependent properties, which are neither shown by bulk nor by atomic and molecular entities. For example, opaque substances become transparent (copper), stable materials turn combustible (aluminum), and insoluble materials become soluble (gold) when they turn from bulk to nanoscale. A material such as gold, which is chemically inert at normal scales, can serve as a potential chemical catalyst at the nanoscale. The density of assembly of nanomaterials is higher compared to the assembly of microparticles with a constant volume.

Nature is the best bottom-up nanotechnologist. The origin of life on earth is accomplished by the synthesis of amino acids and nucleic acids, the building blocks of proteins and hence life, from the reaction between gases CH_4, H_2, and NH_3 at high temperatures. Various amino/nucleic acids are arranged in different ways and form a number of biological molecules, such as nucleic acids, phospholipids (essential part of the cell membrane), DNA, RNA, and so on. These molecules combine to form the basic structure called the *cell* and the first single-cell living organism in the form of the protozoa. A large number of cells combine to construct various tissues, which are the bases of organs in the metazoan. Figure 2.1 illustrates the nanoarchitectural evolution of cadmium hydroxide nanoflowers that is analogous to the evolution of human life.

2.3
Band Theory of Solids

Nanocrystals are the periodic arrangement of atoms/molecules. Isolated atoms have discrete energy levels with a single degeneracy. When two atoms come close to each other, their energy levels get perturbed because of Coulombic interaction and the molecule thus produced has two molecular energy levels. Each energy level of the resultant daughter molecule has a double degeneracy. In a similar manner, when N atoms ($N \approx 10^3 - 10^5$; for 1 Å atomic radius) make a cluster of sizes in the range of 1–100 nm, the outermost (valence cell) unfilled as well as core electronic energy levels get perturbed and broadened in the form of bands with degeneracy N. For large values of N ($N > 10^4$ or particle size more than 20 nm), these N degenerate subenergy levels (sub-bands) inside the bands are so close to each other that electrons can easily transit between them even below room temperature, and therefore can be considered as a single band. If there are n electronic energy levels

---

1) Atoms on the surface of the particles have unsaturated bond termed as dangling bonds. It is an unsatisfied valence on an immobilised atom.

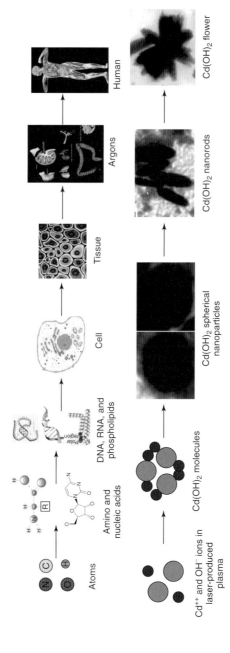

Figure 2.1 Nanoarchitectural evolution of cadmium oxide nanoflowers analogous to the evolution of human life from nonliving atoms.

in each atomic entity, and N such atoms make a single nanoparticle, the resultant particle has n energy bands and each band consists of N sub-bands. The highest occupied energy band is called the *valence band* (V_B), while the lowest unoccupied band is termed as the *conduction band* (C_B). The difference in energy between the top of a valence band and the bottom of the conduction band is termed as *band gap* (ΔE_g) energy. In metals, valence and conduction bands overlap, and a large number of free electrons in the conduction bands are therefore available for electrical and thermal transport. In contrast to the metals, insulators have energy band gap too high to transform electrons from the valence band to the conduction band.

In monoatomic systems such as particles of Si, Ag, Au, and so on, the outermost filled/partially filled (3p orbital in Si; N is the number of atoms in the nanoparticle) atomic orbitals (AOs) of N atoms get linearly combined to produce valence bands with N sublevels. Similarly, a conduction band with N sublevels is made by a linear combination of the first vacant AO (4s orbital for Si) of N constructing atoms. In diatomic semiconductor nanomaterials such as CdSe or HgS, the occupied molecular orbitals (MOs) are made of a linear combination of AOs of the negatively charged anions (e.g., Se^-, S^-), while the unoccupied MOs are made of AOs of the metallic cations (Cd^{++}, Hg^{++}, etc.). There are a number of occupied and unoccupied MOs. The highest occupied molecular orbital (HOMO) serves as the valence band, while the lowest unoccupied molecular orbital (LUMO) is the conduction band.

The bands are not flat, but have several hills and valleys on their surfaces, as shown in Figure 2.2. If one or more of the hills of the valence band lie below the valleys of the conduction band, the materials are called *direct band gap semiconductors*; otherwise, they are indirect band gaps (Figure 2.2). Oxides of zinc, titanium, copper, and so on, are direct band gap semiconductors, while Si nanoparticles, GaP, and so on, are indirect. The same material may also possess a direct as well as an indirect nature. For example, cadmium oxide has a direct band gap (3.4 eV) in the near UV region and an indirect band gap in the far infrared (IR) (1.2 eV). Evolution of bands in the Li ($1s^2$, $2s^1 2p^0$) metal is shown in Figure 2.3 where valence and conduction bands have overlapping energy levels,

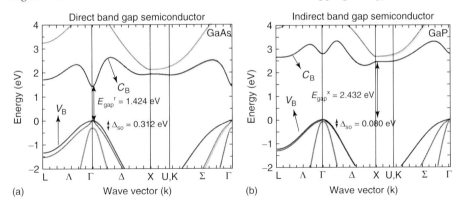

Figure 2.2 Direct and indirect band gap semiconductor materials.

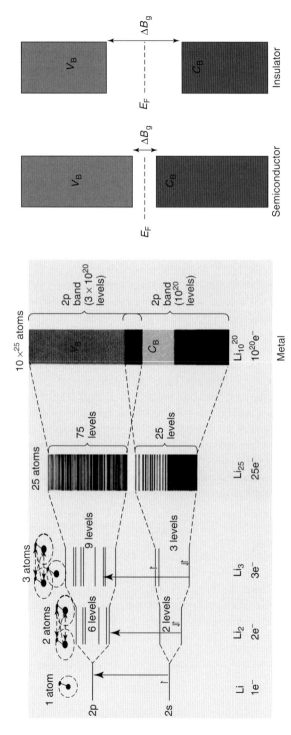

Figure 2.3 Evolution of bands of solids from discrete energy levels of individual atoms. Bands for metals, semiconductors, and insulators.

that is, no band gap energy. Semiconductors have band gap energies in between metals ($\Delta E_g = 0$) and insulators ($\Delta E_g > \approx 5$ eV). Semiconductors with band gap energies between 0.1 and ≈ 2.5 eV (such as InAS, GaAS, GaP, etc.) are assigned as narrow band gap semiconductors, while those with higher ranges than this are wide band semiconductors (ZnO, TiO_2, and CuO).

For smaller values of N (size below the quantum confinement[2] regime), sublevels are separate enough that transitions from lower to higher energy sublevels are not possible at room temperature, that is, valence and conduction bands have discrete sublevels. Increasing number of atoms, N, causes broadening of each level, which decreases the separation between the top of the valence band and the bottom of the conduction band. In other words, an increase in the size of the particle causes a decrease in the band gap energy in the regime at the nanoscale. Beyond this, any increase in the size has no effect on the band gap energy, and we can say that the material has reached to bulk form.

Nanomaterials made of single element are known as *monoatomic* (Ag, Au, Pt, Si, Ge, etc.), those with two elements are termed *diatomic* (metal-oxide, metal-carbide, metal-nitride, metal-sulfide, metal-selenide, etc.), those made of three elements are triatomic (InGaAs, InGaP, ZnCdS, etc.), and those having more than three are called *polyatomic* (biological materials, organic nanoparticles, pervoskite, etc.) nanoparticles. For more details on band theory, readers are requested to see Refs. [4−8].

2.4
Quantum Confinement

As we have stated earlier, the confinement of electrons inside the crystal determines the physical and chemical properties of materials. The size and shape of materials determine the confinement of motion of electrons, and hence are key factors of change in properties when moving from bulk to nanodimensional space. Let us consider a bulk crystal having all three dimensions larger than 100 nm. If we shrink only one dimension (1D) down to 100 nm, leaving the other two invariant, it will be in the shape of a thin film, which is known as *1D confinement or quantum well (QWl) or 2D nanostructure*. Similarly, if we reduce the size from the two perpendicular directions, it takes the form of a rod, with cross section in the nano regime. Such type of confinement is 2D confinement and 1D structure and is also known as *quantum wire (QWr)*. Finally, reduction of sizes down to the nanoscale from all the three perpendicular directions yields *quantum dots (QDs)* or 0D nanostructures and is known as *3D confinement* [9−14].

The change in the various chemical and physical properties of nanostructured materials can be explained on the basis of quantum mechanics, because when the size of particles becomes comparable to the de-Broglie wavelength

2) Size below which optical and electronic properties change rapidly.

($\lambda_B = h/P$) of electrons in the solids, the quantum nature of the material becomes relevant. Quantum confinement becomes significant when the size of the particle is comparable to the exciton's Bohr radius.[3] Employing the hydrogen atom model, the exciton Bohr radius is estimated as $r_{ex} = (\varepsilon_0 \varepsilon_r h^2)/(\pi \mu e^2)$. As ε_r and μ are different for different materials, therefore r_{ex} differs from material to material. For example, it is 11.5 nm for GaAs and 10 nm for zinc oxide. When the size of particles reduces to this value, carrier movement is restricted.

Density of states (DOSs) in solid state physics is the number of energy states (orbitals) per unit energy range per unit volume. In the bulk crystal, the conduction electrons are free to move in the entire volume; therefore, the DOS spectrum is continuous, proportional to the square root (DOS$\alpha\sqrt{E}$) of energy, and expressed as $D(E)_{bulk} = \frac{8\pi\sqrt{2}}{h^3} m^{*3/2}\sqrt{E - E_C}$ for $E \geq E_C$, where m^* is the effective mass of the electron in the solid and E_C is the bottom of the conduction band (DOSs in the conduction band). In contrast to bulk macroscopic crystals, QDs have discrete energy levels; therefore, DOS is the summation of these states and is given as $D(E)_{0D} = \sum_n 2\delta(E - E_n)$, where E_n is the nth quantum state energy. In between bulk (continuous DOSs) and QDs (discrete energy levels), QWs have a sawtooth-like function for DOS; for quantum wells, it is a step function as shown in Figure 2.4. The band gap energy of materials increases with the increase in the degree of quantum confinement (Figure 2.4). The band gap energies for $QD_{5\,nm\times5\,nm\times5\,nm} > QWr_{5\,nm\times5\,nm\times\infty} > QWl_{5\,nm\times\infty\times\infty} >$ bulk. For detailed studies, readers are suggested to consult Refs. [15–19].

The first relationship between band gap energy (E_g) and size of the particle is based on the particle in the box model [20], which is given as $E_g = E_g(\infty) + \hbar^2\pi^2/2\mu a^2$, where $E_g(\infty)$ is the band gap of bulk, μ is reduced mass ($\mu = m_e m_h/m_e + m_h$) of an electron hole (exciton) pair, and a is the size of the nanoparticle. This model is modified by Brus [21] and Kayanuma [22] by adding the terms for electron hole coulombic interaction ($-1.786e^2/\varepsilon_r a$) and correlation energy ($0.284E_a$) to the existing particle in the box model. Then the size-dependent band gap energy of nanomaterials becomes $E_g = E_g(\infty) + \hbar^2\pi^2/2\mu a^2 - 1.786e^2/\varepsilon_r a + 0.284E_a$, where $E_a = \mu e^4/2\varepsilon_r^2\varepsilon_0^2\hbar^2 = 13.56\left(\mu/\varepsilon_r^2 m_e\right)$ is the Rydberg (spatial correlation energy) for the bulk semiconductor. The effective dielectric constant ε_r and effective mass μ describe the effect of the homogeneous medium in the quantum box. This equation gives a good fit of the experimentally observed relation between E and a for the weak confinement regime but cannot explain the observations in the strong confinement regime, where the effective mass approximation with infinite barriers breaks down. Further modification and improvements in the quantum confinement theory establishes a relation between shift in band gap energy (ΔE_g) of nanoparticles and its size as $\Delta E_g = a^{-n}$ [23–25]. This expression fits better for size-dependent band gap data with different n values.

According to the theory of quantum confinement, holes in the valence band and electrons in the conduction band are confined spatially by the potential barrier of the surface or trapped by a monopotential well of the quantum box. Because

---

3) The natural separation between an electron that leaves the valence band and the hole it leaves behind results in an electron hole pair, which is called an *exciton*.

Shape/size of nanostructures   Graphical representation of DOS   Expression for DOS

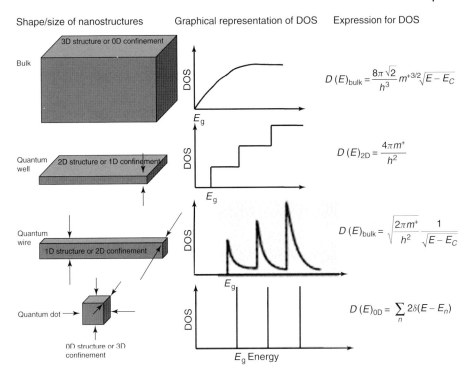

$$D(E)_{bulk} = \frac{8\pi\sqrt{2}}{h^3} m^{+3/2}\sqrt{E - E_C}$$

$$D(E)_{2D} = \frac{4\pi m^+}{h^2}$$

$$D(E)_{bulk} = \sqrt{\frac{2\pi m^+}{h^2}} \frac{1}{\sqrt{E - E_C}}$$

$$D(E)_{0D} = \sum_n 2\delta(E - E_n)$$

Figure 2.4 Confinement of dimension of bulk crystal to the quantum well, quantum wire, and quantum dots. Density of states of bulk, 10 nm thick quantum well, 10 nm × 10 nm cross-sectional quantum wire, and 10 nm × 10 nm × 10 nm quantum dot with corresponding expressions.

of the increase in confinement of both electrons and holes with decrease in size, the lowest optical energy transition from the top of the valance band to the bottom of the conduction band increases in energy, effectively increasing the band gap. In semiconductor QDs, excitons Bohr radius is determined by the extension in the wave functions of the electrons and the holes. When the size of the particle becomes smaller and reaches close to the Bohr radius, these wave functions get confined by the QD limits and the carriers' kinetic energy increases, leading to the blueshift in the absorption spectrum. Blueshift in the absorption maxima and increase in the band gap [26] energy with size reduction are depicted in Figure 2.5. In addition to a change in the absorption energy, the reduction in size gives rise to the appearance of discrete energy levels at the band edge.

The DOSs in the quasi-continuum of energy levels increases rapidly above the conduction band edge and below the valance band edge. For semiconductor QDs the situation is different. Owing to the limited number of atoms in the QD (typically $10^2 - 10^4$), discrete atomic-like energy levels appear at the edges of the bands (Figure 2.6). This quantum size effect is reflected in the absorption

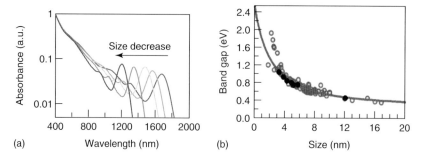

(a)   Wavelength (nm)   (b)   Size (nm)

Figure 2.5 Variation in (a) SPR absorption peak position and (b) band gap energy of PbSe QDs with size. (Source: Reprinted with permission from Ref. [26] © American Chemical Society.)

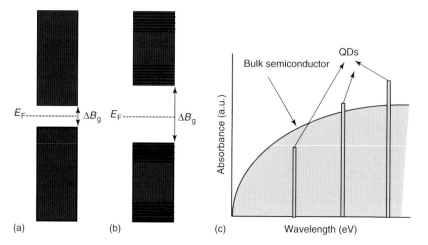

(a)   (b)   (c)   Wavelength (eV)

Figure 2.6 Bands of (a) bulk and (b) 0D quantum dot semiconductors and their absorbance spectra. Band edges of the QDs have discrete energy level of subbands (c) sketch of absorbances for bulk and quantum dots.

spectra (Figure 2.6) by the appearance of discrete absorption bands. As the particles' dimensions get smaller, this structure becomes more pronounced because the separation between energy levels increases.

2.5
Defects and Imperfections

Defects and imperfections are one of the most interesting topics of solid state physics. An ideal crystal should have a periodic arrangement of atoms and the local position of any atoms with respect to each other is repeated at the atomic scale. In other words, the local environment of each basic building entity of the crystal should be strictly the same as the other atoms of the same element. Practically, most of the micro- and nanocrystals deviate from the ideal situation and possesses several types of defects. The various types of defects are (i) point or 0D, (ii) line or 1D, (iii) planar or 2D, and (iv) volume or bulk named as 3D defects.

2.5.1
Point Defect

A few extra atoms of the host at the interstitial position (out of the plane of the regular lattice), an impurity atom in the regular lattice or interstitial position, and missing atoms (vacancy) in the regular lattice are some examples of point defects. Vacancies of anions/cations in ionic solids (NaCl, CsCl) are sometimes called *F-centers* or Schoktty defects. Interstitials are atoms that occupy a site in the crystal at which there should usually not be any atom. When an atom from the regular lattice moves at an interstitial position and creates a vacancy, then the pair of vacancy and interstitial defect is called *Frenkel defect* or *Frenkel pair*. When an impurity atom (guest atom) replaces a host atom from the regular crystal lattice, it is known as a *substitutional defect*. If the guest atom and the replaced host atom are of the same valence state, then it is called an *isovalent substitution*; otherwise, it is *aliovalent*. Various point defects are shown in Figure 2.7a. In semiconductors such as ZnO, TiO_2, GaAs, InGaAs, and so on, cation and anion vacancies and their presence at the interstitial site cause point defects in the nanomaterials and QDs. A cation at the interstitial position and an anion vacancy usually lie below the conduction band and act as a donor, while a cation vacancy and an anion at the interstitial position lie above the valence band and play the role of acceptors (Figure 2.7b).

2.5.2
Line Defects

Edge and screw dislocations are examples of line defects. Edge dislocation occurs when a plane of atoms gets terminated in the middle of a crystal (incomplete plane), which causes bending of the neighboring atomic planes around the edge of the incomplete plane so that the crystal structure is perfectly ordered on either side. The magnitude and direction of stress due to defect strain is presented by Burgers vector b and is perpendicular to the edge dislocation. The screw dislocation comprises a structure in which a helical path is traced around the linear defect by the atomic planes of atoms in the crystal lattice (Figure 2.7c).

2.5.3
Planar Defects

Twin boundaries, antiphase boundaries, and stacking faults are examples of planar defects. Grain boundaries are formed when the direction of crystal growth

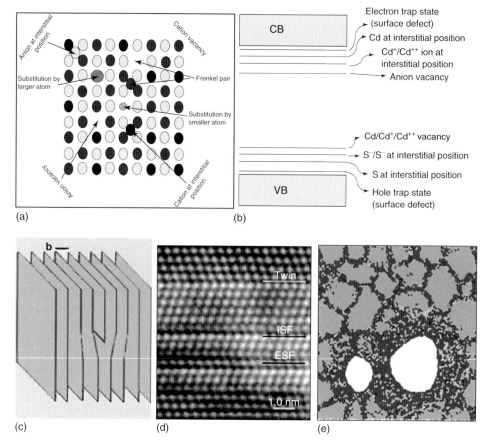

(a)

(b)

(c)

(d)

(e)

Figure 2.7 (a) Various point defects in the crystal, (b) energy level corresponding to point defects, (c) edge dislocation, (d) HRTEM image showing various planar defects (Source: adapted from website: *www.nrel.gov/measurements/trans.htm*), and (e) bulk defect as void. CB, conduction band; VB, valence band; ESF, extrinsic stacking fault.

abruptly changes. This may be the interface between two single crystal regions of different orientations. In case of antiphase ordering, the direction of growth is the same but the phase is different. For example, if ABABABAB is the usual crystal without defects, then ABABBABA is the crystal with antiphase boundary defects. A stacking fault is a one- or two-layer interruption in the stacking sequence, for example, if ABCABCABCABC … is the sequence of fcc crystal structure, then ABCCABCABC … (missing of two AB layers) is a stacking fault. Stacking is the displacement of one plane of atoms corresponding to others. When adjacent layers are shifted slightly, it is termed as intrinsic stacking fault (ISF), while the presence of an interleaving layer between two slightly shifted layers is known as a *extrinsic stacking fault* (*ESF*). Twin boundaries, ISFs, ESF are shown in Figure 2.7d.

2.5.4
Volume or Bulk Defects

If dopant or impurity atoms are insoluble in the crystal, they get clusterized and precipitated into the crystal as large atoms to form defects. Sometimes there is a vacancy of a large number of atoms at the same point in the crystal, called a *void* (Figure 2.7e). For more information, readers are requested to see Refs. [4–10, 15–19].

Any deviation from the periodic arrangement of atoms or molecules causes defects and perturbations in the band structures. Desired and undesired injection of impurity atoms may modify the band gap and create its own energy level either below the conduction band (donor level) or above the valence band (acceptor level) depending on its energy and interaction with the host. For example, in diatomic semiconductors, metal or metal ion vacancies act as acceptors, while their presence at interstitial positions acts as donors. In contrast, energy levels of the neutral, singly, or doubly charged anion lie below the conduction band and are employed as donors, while their existence at interstitial sites work as acceptors to create more holes in the valence bands.

Another type of defects that usually occurs in QDs is due to their enhanced surface effects, which lead to some surface states called *traps*. Energy levels of these trap states lie between the band gaps. The surface trap states just below the conduction band act as electron traps, while those above the valence band are known as *hole traps*. These trap states as well as other energy levels between valence and conduction bands arise due to the various types of defects in the QDs and nanomaterials; retain charges, allowing nonradiative recombination; and reduce electronic, photoconducting, and optical properties. They are beneficial in the conversion of shorter wavelength photons into longer ones, charge storage, adsorption, and so on. The surface states can be eliminated by surface passivation, which is the process of bonding surface atoms with higher band gap materials. Ideal termination of the growth of nanostructures naturally removes structural reconstruction, leaving no strain on the surfaces, and produces an atomically abrupt jump in the chemical potential for the electron or holes at the interfaces. For example, growth of Si termination by amorphous silicon dioxide and that of the $Al_{1-x}Ga_xAs$ by GaAs are examples of passivation. In the case of Si, passivation is achieved with an amorphous, disordered material that can accommodate its local bonding geometry to that of the underlying semiconductor, while $Al_{1-x}Ga_xAs$ crystal structures and bond lengths of the two materials ($Al_{1-x}Ga_xAs$ and GaAs) are matched to achieve surface passivation epitaxially. Thermal, laser, and radiation treatments of nanomaterials may reduce or eliminate defects and produce pure, dense, and crystalline products.

2.6
Metal, Semiconductor, and Insulator Nanomaterials

In Section 2.3, we presented metals, semiconductors, and insulators on the basis of the band theory of solids. As we proceed from bulk to nanoscale by decreasing the number of atoms in the cluster, the separation between valence and conduction band increases, which may transform metal into semiconductor and semiconductor into insulator. Figure 2.3 depicts the significant band gap between valence and conduction bands for the Li clusters with only 25 atoms. This means that conductor nanoparticles may become semiconductors at extremely small ($\approx$<1 nm), particularly at the subnanometer scale. Similarly, a decrease in the size of semiconductor QDs may cause its transformation into insulators. Metals, semiconductors, and insulators are assigned on the basis of electrical and thermal conductivities, which itself depend on the density of free electrons. Materials having the same size, shape, and chemical composition, but different crystal structure, may have different electrical conductivity. Different spatial arrangements of atoms/molecules into the bulk/nanocrystal have different band gap energies, DOSs for electrons as well as holes and hence different electrical, thermal, and optical conductivities. For example, different values of the chiral vector (n,m) (zig-zag ($n,0$), armchair (n,n), and chiral (n,m), etc.) of carbon atoms in carbon nanotubes provide different electrical behavior of the tube. For a given (n,m) nanotube, if $n = m$, the nanotube is metallic; if $n - m$ is a multiple of 3, then the nanotube is semiconducting, with a very small band gap; otherwise, the nanotube is a moderate semiconductor. Thus, all armchair ($n = m$) nanotubes are metallic, and nanotubes (6,4), (9,1), and so on, are semiconducting [27]. Surface treatment or capping of nanomaterials with other nanomaterials or molecules modifies their band gap energy and hence electronic properties. Dielectric constant of material increases as we move from metal to semiconductor to insulators.

2.6.1
Metal Nanoparticles and Their Size-/Shape-Dependent Properties

Nanoparticles of metals are particularly interesting systems because of their ease of synthesis as well as size, shape, and surface modifications. As already stated, in metals, electrons are highly delocalized over a large space (i.e., least confinement), which causes overlapping of valence and conduction bands, giving them their conducting properties. The quantum confinement regime for metal nanoparticles is too small ($\approx$<2 nm); therefore, optical and electronic properties of metal nanoparticles closely resembles those of their corresponding bulk counterparts. In metals, the decrease in their sizes below the electron mean free path[4] gives them the most fascinating property of their intense surface plasmon resonance (SPR) absorption bands in the visible spectral region, which gives them their mesmerizing colors reminiscent of molecular dyes. The absorption spectra of molecular systems and

---

4) Distance traveled by electrons between two scattering collisions with lattice atoms.

semiconductor particles can be understood in terms of quantum mechanics, while SPR absorption of metals is the result of coherent oscillation of free electron from one surface of the particle to the other. Such strong SPR absorption induces strong coupling of metal nanoparticles with electromagnetic radiation of light, which gives the brilliant color to colloidal metal nanoparticles, enhancement of light intensity on the nanostructured metallic surface [28], and surface-enhanced Raman scattering [29, 30].

When a beam of light having wavelength λ is incident on a metal nanoparticle with diameter much less than the light wavelength ($d \ll \lambda$), and the intensity of light is uniform across the entire particle. Then oscillation of the electromagnetic (EM) field results in collective oscillation of weakly bound and free electrons opposite to the field direction. When the incident light frequency matches the intrinsic electron oscillation frequency, the light gets absorbed resulting in SPR absorption. Displacement of electrons and cation or lattice atoms opposite to and in the direction of the field, respectively, appears as corresponding negative and positive surface charges at opposite faces of the particle (Figure 2.8). Displacement of positive and negative charges from their equilibrium positions produces a transient dipole moment, which oscillates with the frequency of incident radiation. The smaller sized particles have more tightly bound electrons resulting in a smaller dipole moment and hence higher resonance frequency or shorter wavelength of plasmon absorption. To put it another way, smaller sized particles have less distance between two surfaces, which causes a smaller time period for coherent oscillation of electrons from one surface to the other and hence a longer frequency of plasmon absorption. The spectral shape of the plasmon absorption band for a given particle was given by Gustav Mie as follows:

$$E(\lambda) = \frac{24\pi^2 N a^3 \varepsilon_{ex}^{3/2}}{\lambda \ln(10)} \left[\frac{\varepsilon_i(\lambda)}{\left[\varepsilon_r(\lambda) + \chi \varepsilon_{ex}\right]^2 + \varepsilon_i(\lambda)^2} \right]$$

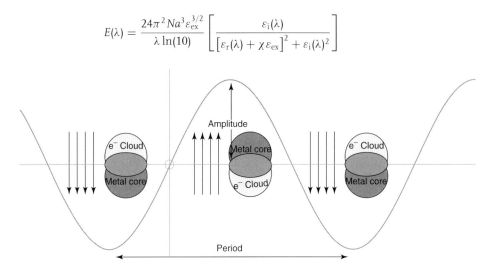

Figure 2.8 Oscillation of electrons from one surface of the particles to the other, resulting in oscillation of dipole moments under the EM field of incident radiation.

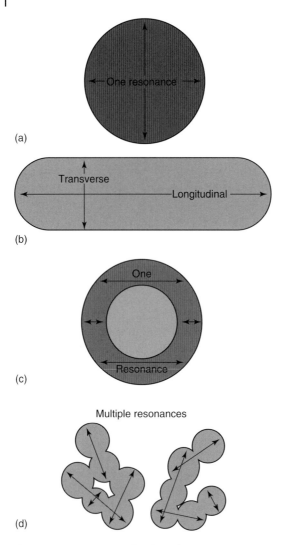

Figure 2.9 Presentation of surface plasmon resonance (SPR) modes in different shapes of particles. (Source: Represented with permission from Ref. [31] © American Chemical Society.)

where ε_r and ε_i are the real and imaginary components, respectively, of the dielectric function of the metal, ε_{ex} is the dielectric function of the matrix (surrounding of the particle), a is the radius of the particle, χ is the factor related to the eccentricity of the particle, and N is the number of atoms present in the particle. The above equation gives the position and shape of the absorption spectra of spherical and spheroidal metal nanoparticles. Figure 2.9 illustrates the schematic presentation of SPR absorption modes for different shapes of particles and aggregates [31]. Spherical particles have a completely symmetrical geometry and therefore only one mode

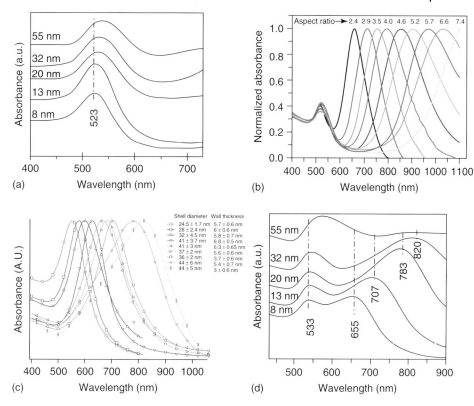

Figure 2.10 Surface plasmon resonance (SPR) spectra of (a) spherical gold nanoparticles of different sizes, (b) gold nanorods of different aspect ratios, (c) gold nanoshells with different shell diameter and wall thickness, and (d) aggregates of spherical gold nanoparticles shown in (a). (Source: Figure 2.10a,d are reprinted with permission from Ref. [33] while those of Figure 2.10a,c are reprinted with permission from Ref. [31] © American Chemical Society.)

(frequency) of SPR absorption is observed, that is, all other modes are degenerate (Figure 2.9a). With the increase in particle size, a group of electrons has to travel a longer distance between two surfaces, and therefore the smaller frequency of SPR absorption. This causes a shift of the absorption band toward longer wavelength with the increase in the size of particles (Figure 2.10a). In one-dimensional nanostructures such as metal nanorods and nanowires, there are two perpendicular modes of coherent vibrations of electrons along the longitudinal and transverse directions (Figure 2.9b). Electrons have to travel a longer distance along the length compared to the width; therefore, the longitudinal mode has a shorter frequency (longer wavelength) of SPR absorption, while the transverse mode corresponds to higher frequency (shorter wavelength) absorption. Figure 2.10b shows SPR absorption spectra of gold nanorods of different aspect (length/diameter) ratios. Hollow nanostructures such as nanorings, spherical hollow shells, core/shell nanostructures, and nanotubes have also more than one SPR band. Nanorings and spherical hollow shells have two modes of vibrations, as shown in Figure 2.9c. When shell radius is comparative to the wall thickness, the two modes are too close to observe experimentally. Increase in shell diameter and decrease in wall thickness cause separation of the higher frequency mode as peaks from the more intense lower frequency mode (Figure 2.10c). The mechanism of synthesis of larger nanostructures from smaller ones with the diffusion of interfaces between them is known as *aggregation* and the product is called an *aggregate*. Diffusion of the interface between original small-sized particles provides a longer path length for coherent oscillation of electrons in addition to the existing one, and hence the longer wavelength for SPR absorption (Figures 2.9d and 2.10d). Figure 2.10d illustrates SPR absorption of aggregates of gold nanoparticles, whose original absorption spectra are shown in Figure 2.10a. It is observed in most of the metallic nanostructures that lower frequency SPR has higher oscillator strength[5] compared to higher frequency SPRs. Noble metal nanoparticles also give weak fluorescence emission corresponding to interband d $\rightarrow$ sp transition [32]. The scattering and absorption of light by metal nanoparticles is described in Chapter 8.

Large thermal conductivity, low toxicity, easy functionalizability, and hemocompatibility of metal nanoparticles made them useful in biological applications such as photothermal treatment of cancerous cells, killing of defective cells, drug delivery at the selective sites, optical enhancement, optical sensors, and biosensors. The plasmonic properties of metal nanoparticles are mostly utilized in most of the applications.

2.6.2
Semiconductor Nanoparticles and Their Size-Dependent Properties

Materials having $0 < E_g < \approx 4.5$ eV and electrical conductivity in the range of roughly $10^3 - 10^{-8}$ S cm^{-1} are classified as semiconductors. They generally possess increase in band gap energy and redox potential, while decrease in the wavelengths of exciton absorption and emission maxima with the decrease in size. The oscillator strength of the first excitonic transition in semiconductors is inversely proportional to the third order ($f_{ij} \alpha a^{-1/3}$) of the size. Decrease in the particle size causes confinement of the electrons that reduces electron–phonon coupling in the semiconductor, which may principally increase the electron lifetime in the lattice with decrease in nonradiative emission process and temperature rise of the system. Electrons in bulk semiconductors interact with the lattice, holes, defects, and impurity atoms, if any. Two factors may affect electron lifetime and hence electronic behavior of semiconductor QDs compared to the bulk. The first is the decrease in DOSs for both electrons and phonons, which results in the decrease in electron–phonon interaction and hence the increase in electron lifetime with decrease in size. The second factor is electron–hole recombination, which increases

---

5) An atom or a molecule can absorb light and undergo a transition from one quantum state to another. The oscillator strength, f_{ij}, is a dimensionless quantity to express the strength of the transition from lower state i to the upper state j.

with decrease in size because of the less availability of space. There is a competitive approach between the two opposite processes, which yields low size-dependent electronic properties of semiconductor QDs.

As semiconductors have less density of free or weakly bound electrons compared to the metal nanoparticles, these electrons are the cause of SPR absorption in nanostructures. Therefore, SPR absorption is less pronounced in semiconductor nanostructures compared to those of metals. Figure 2.5 shows shorter wavelength shift in SPR band of PbSe semiconductor QDs with decrease in size. Dependence of SPR absorption on the size and shape of the semiconductor nanoparticles is almost the same as that for metals. Semiconductor nanomaterials have wide applications in energy harvesting and saving, fabrication of electro-optical and optoelectronic devices such as light-emitting diodes (LEDs), solar cells, light sensors, and so on, and in the fabrication of active and passive electrical components such as resistors, transistors, diodes, and field-effect transistors (FETs). Band gap energy of the semiconductor nanomaterials is highly sensitive to the surface structure and adsorbed molecules, which governs their electrical and photoelectrical conductivities. The semiconductor nanoparticles are also being used for the fabrication of sensors and transducers, photodegradation of organic molecules and water purification, drug delivery, fluorescence imaging and biotagging, and staining of cells.

2.6.3
Insulator Nanoparticles

An *insulator*, or a *dielectric*, is a material that resists the flow of electric current. An insulating material has atoms with tightly bonded valence electrons. These materials are used in parts of electrical equipment, also called *insulators* or *insulation*, and are intended to support or separate electrical conductors without passing current through themselves. Glass, paper, Teflon, plastic, and some polymers are insulating materials.

2.7
Various Synthesis Methods of Nanoscale Materials

We can get nanodimensional materials by two approaches. The first approach includes collecting/adding an atomic-sized entity, Å, in a particular manner to get nanodimensional materials. This is similar to the addition of bricks in the construction of a building, where a single brick is also made up of a collection of micron-/millimeter-sized grains of soil and ash. Making larger sized objects from a collection of smaller ones is called *bottom-up technology*. As opposed to this, the breaking of a single large-sized particle into several smaller ones is called *top-down technology*. The natural process of the evolution of life (Figure 2.1), body structuring of any living organism, construction of mountains, rocks, and so on, by small-sized dust particles come under bottom-up technology. Figure 2.11 illustrates bottom-up and top-down approaches. Most of the chemical methods

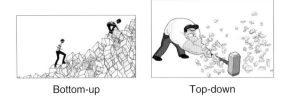

Bottom-up   Top-down

Figure 2.11 Cartoons illustrating bottom-up and top-down approaches of nanomaterials processing.

such as sol–gel, chemical precipitation, citrate reduction, micelles, hydrothermal, solvothermal, chemical pyrolysis, vapor deposition, sonochemical, electrochemical, microemulsion, combustion, and so on, are bottom-up approaches of nanomaterials synthesis. Watson–Crick base pairing of various bases to construct the helical structure of DNA, RNA, and other nucleic acids; concepts of supramolecular chemistry; and molecular recognition to automatically arrange single molecular components into some useful conformation (molecular self-assembling) are also examples of bottom-up approach of nanotechnology. Evaporating, melting, or breaking of bulk materials into atoms or smaller particles thermally or by focusing laser, electron, ion, or molecular beams are some examples of top-down approaches. Bulk target materials are either directly fragmented into smaller nanosized particles or evaporated into atoms, ions, or molecules as intermediates, which act as building blocks for the growth of nanostructures. Ball milling, magnetron, and radio frequency (RF) sputtering, various lithographic methods, atomic layer deposition (ALD), laser ablation followed by deposition or condensation, molecular beam epitaxy, and thermal evaporation are some of the methods of top-down approaches. Top-down and bottom-up approaches are also known as physical and chemical methods, respectively. Most of the methods discussed in this book for the processing of nanostructures and their modification are top-down approaches.

2.8
Various Techniques of Materials Characterization

2.8.1
Light Beam Characterization Techniques (200–1000 nm)

UV–visible absorption and photoluminescence spectroscopy are used for characterization of optical properties, calculation of size and band gap energy, and determination of defects and impurities in the crystals. Diffuse reflectance and ellipsometry analyze reflected light for the determination of thickness and refractive index of dielectric, semiconductor, and thin metal films. Dynamic light scattering technique can determine size and distribution of the particle in colloidal form. Raman spectroscopy and Raman imaging are useful in structural characterization, size determination for spherical particles, and defect measurements. Ultrafast laser spectroscopy provides information about the several dynamic properties such as electron dynamics, electron lifetime measurement, relaxation processes, and role of defects. Several light-based characterization techniques are described in the subsequent chapters.

2.8.2
Infrared (IR) Characterization (1000–200 000 nm)

The space between energy levels of molecular rotation and vibrations (symmetric and antisymmetric stretching, bending, rocking, etc.) lies in the IR region. When a beam of IR light in the range of 50–10 000 cm^{-1} falls on nanomaterials, IR energies corresponding to the vibrational and rotational energy levels of the molecules of the particles are absorbed. Assignment of these peaks provides information about the functional group, such as –OH, –C=O, –COOH, and so on, present in the nanoparticles. Molecules adsorbed on the surface of nanoparticles can be detected using IR spectroscopy. IR spectroscopy is limited only to polar molecules.

2.8.3
X-Ray-Beam-Based Characterization Methods

X-ray diffraction (XRD) is useful in structural characterization and size measurement using Scherrer's equation ($a = K\lambda/\beta \cos\theta$, where K is the shape factor, λ the wavelength of the X-ray beam, β is full width at half maximum (FWHM) at position of diffraction angle 2θ, and a is the mean size of the ordered (crystalline) domains, which may be smaller or equal to the grain size). The dimensionless shape factor, K, has a typical value of about 0.9, but varies with the actual shape of the crystallite. The Scherrer equation is limited to nanoscale particles. It is not applicable to grains larger than about 100 nm, which precludes those observed in most metallographic and ceramographic microstructures. The width of the diffracted X-ray peak would solely give particle size, if contribution from inhomogeneous strain and instrumental effect into the line width is zero; otherwise, size calculated using the Scherrer equation has a larger value than the actual particle size [34–38].

NEXAFS (near-edge X-ray absorption fine structure) is an element-specific electron spectroscopic technique that is highly sensitive to bond angles, bond lengths, and the presence of adsorbates. It is widely used in surface science and has also been used to study polymers and magnetic materials. NEXAFS is synonymous with *XANES* (X-ray absorption near-edge structure) but NEXAFS by convention is usually reserved for soft X-ray spectroscopy ($h\nu < 1$ keV). NEXAFS concentrates on fine structures within about 30 eV of the absorption edge, while extended X-ray absorption fine structure (EXAFS) considers the *extended* spectrum out to much higher electron kinetic energies.

X-ray photoelectron spectroscopy (XPS) is a quantitative spectroscopic technique that measures the elemental composition, empirical formula, and chemical and electronic states of all the elements present within a material. XPS spectra are recorded using irradiation of a sample with X-ray beam with simultaneous measurement of kinetic energy and number of photoelectrons that escape from the top 1 to 10 nm of the material being analyzed. XPS requires ultrahigh vacuum (UHV) conditions. It is a nondestructive surface chemical analysis technique of as-received or slightly surface-treated samples. It is also termed as elemental spectroscopy for chemical analysis (ESCA), as it can detect elements with atomic number $Z \geq 3$, with parts per million (ppm) detection limits.

2.8.4
Electron-Beam-Based Characterization Methods

An electron beam has 10^5 times shorter wavelength than the visible light photons; therefore, *electron microscopes* can produce an electronically magnified image with magnification up to 10^6, whereas light microscopes are limited to $10^3\times$ magnification. In *transmission electron microscope* (TEM), the electron beam from the electron gun is accelerated using $+100$ keV potential, collimated and focused by electrostatic and electromagnetic (EM) lenses, and transmitted through the sample that has some parts that are transparent for electrons, while the other scatters them out of the beam. Images of the opaque or scattering part of the specimen are recorded on the phosphor screen. *High-resolution transmission electron microscope* (HRTEM) has allowed the production of images with resolution below 0.5 Å with magnifications more than 50 million times. HRTEM can determine the position of atoms and defects within the material. In contrast to this, *scanning electron microscope* (*SEM*) does not take the complete image of the sample at a time. It produces images by scanning across a rectangular area of the sample. At each point on the sample, the incident beam loses some energy, which is converted into some other form such as heat, emission of low-energy secondary electrons, light emission (cathodoluminescence), or X-ray emission. Intensity of any of these signals coming from any position of the sample is recorded at the same time when an electron beam falls at that point. Scanning of points on the sample surface by the electron beam generates an image of the sample by emission from those points. Resolution of the SEM image is much lower than that of the TEM image, but it is able to image a bulk sample of several centimeters in size with 3D representation. Adding the scanning system of electron beam into TEM makes *scanning transmission electron microscope* (*STEM*), which can take 3D, high-resolution images of more than centimeter-sized samples. Instead of the above three microscopes, *reflection electron microscope* (*REM*) utilizes detection of elastically reflected electrons from the sample. The *low-voltage electron microscope* (*LVEM*) is a combination of SEM, TEM, and STEM in one instrument, which operates at a relatively low-electron accelerating voltage of 5 kV useful for biological samples. *Spin polarized low-energy electron microscope* (*SPLEEM*) is used for observing the microstructure of magnetic domains.

   Energy-dispersive X-ray spectroscopy (EDS or EDX) is an analytical technique used for the elemental analysis or chemical characterization of a sample. It is based on the analysis of emitted X-ray lines from the sample when exposed with the beam of charged particles. Its characterization capabilities are due in large part to the fundamental principle that each element has its unique atomic entities that allows emission of its characteristic X-ray lines. Characteristic X-ray beams from the sample are stimulated by interaction of high-energy charged particles such as electron or proton beams or beams of X-rays with the sample. In the case of proton bombardment, it is called proton-induced X-ray emission (PIXE).

Low-energy electron diffraction (LEED) is a technique for the determination of the surface structure of crystalline materials employing bombardment of a collimated beam of low-energy electrons (20–200 eV) and observation of diffracted electrons as spots on a fluorescent screen. It is used in two ways (i) qualitatively, where the diffraction pattern is recorded and analysis of the spot positions gives information on the symmetry of the surface structure and (ii) quantitatively, where the intensities of diffracted beams are recorded as a function of incident electron beam energy to generate the so-called I–V curves. By comparison with theoretical curves, these may provide accurate information on atomic positions on the surface at hand. The qualitative analysis may reveal information about the size and rotational alignment of the surface adsorbate unit cell with respect to the substrate unit cell [39, 40].

Reflection high-energy electron diffraction (RHEED) and reflection high-energy loss spectrum (RHELS) are other electron-beam-based characterization techniques of nanomaterials.

2.8.5
Nuclear Radiation and Particle-Based Spectroscopy

Positron annihilation spectroscopy (PAS), sometimes specifically referred to as positron annihilation lifetime spectroscopy (PALS) is a nondestructive spectroscopy technique to study voids and defects in solids. The technique is based on the fact that when electrons and positrons come close to each other γ rays are produced by the process of annihilation. If positrons are injected into a solid body, their lifetime will strongly depend on whether they annihilate in a region with high electron density or in a void where electrons are scarce or absent. By comparing the fraction of positrons that have a longer lifetime to those that annihilate quickly, insight can be gained into the voids or the defects of the structure.

Neutrons are electrically neutral with spin $+1/2$ particles; therefore, they do not interact with charge but interact with spin. *Neuron beam scattering spectroscopy* [41] is useful in the determination of spin DOS and defects in the magnetic materials. Wavelength and energy of thermal neutrons ideally match the interatomic distances and excitation energies, respectively in condensed matter; therefore, neutron scattering is able to directly examine the static and dynamic properties of the material.

Nuclear magnetic resonance (NMR) and *Mössbauer (γ-ray) spectroscopy* probe tiny changes in the energy levels of an atomic nucleus in response to its environment. Owing to the high energy and extremely narrow line widths of gamma rays,

Mössbauer spectroscopy is one of the most sensitive techniques in terms of energy (and hence frequency) resolution, capable of detecting changes in just a few parts per 10^{11}.

2.9
Self-Assembly and Induced Assembly, Aggregation, and Agglomeration of Nanoparticles

Self-assembly is group of mechanisms in which a disordered system of preexisting components forms an organized structure or pattern because of the local interactions among the components themselves without any external direction or forces. If forming of the assembly is induced by some chemicals or radiation, it is termed as *induced assembly*. Unlike self-organization, which is very similar to self-assembly but a nonequilibrium process, self-assembly is a spontaneous process that leads to equilibrium by reducing its free energy. Self-assembly requires components to remain essentially unchanged throughout the process. Weak interactions such as van der Waals, capillary, $\pi-\pi$, and hydrogen bonds play a key role in self-assembly rather than strong interactions such as covalent, ionic, and metallic bonds. Driving of atoms and molecules to assemble into larger structures in chemical reactions is not a self-assembly process because a strong interaction takes part in chemical reactions. It is difficult and requires advanced expensive nanolevel precession systems such as an atomic force microscope (AFM) tip for the fabrication of a single nanowire/rod/particle device. Self-assembly of nanoparticles or nanorods or nanowires into larger microlevel architectures, which can be viewed easily by a simple optical microscope, provides a cheap way of fabrication of nanomaterials-based devices.

   Aggregation and agglomeration of nanomaterials are different from that of their self-assembly and self-organization. Aggregation is direct mutual interaction of nanoparticles, atoms, or molecules via one of the several forces such as van der Waals, chemical bonding, hydrophobic effects, and magnetic/electric attraction. When there are collisions between particles in solution during Brownian motion, fluid shearing, and differential settling, they will attach to each other and the walls between them get dissolved to make larger sized particles. Aggregation plays a significant role in several industrial and technological processes such as combustion, filtration, and gas-phase synthesis. Coating nanoparticles decreases aggregation by two mechanisms: charge stabilization or steric stabilization. Particle concentration can affect the stable size of aggregates formed and the speed of aggregation. As concentration increases, aggregation is more pronounced and aggregates may become large enough to settle out via gravity. *Aggregation* is a consequence of strong attractive forces and is irreversible, while *agglomeration* is not as strong as aggregation and is more readily reversible, that is, they are easier to break apart into smaller agglomerates or individual particles. Bare particles aggregate strongly. Surface-coated particles agglomerate eventually, but can be broken up readily.

2.10
Applications of Lasers in Nanomaterial Synthesis, Modification, and Characterization

Lasers have shown their potential role in the synthesis, modification, organization, and assembling of nanoscale materials. They are also applicable as tools for medical, surgical, biological, and industrial applications of nanomaterials. Various types of laser systems are used for the fabrication of thin films, nanoparticles, wires, and rods in gaseous and vacuum environments by pulsed laser ablation. Nanoparticle-assisted pulsed laser deposition (NAPLD) provides the way for the synthesis of nanoarchitectures by laser ablation of metal, ceramic, and graphite targets in the quartz furnace. Laser-vaporization-controlled condensation (LVCC) is another novel approach utilizing laser ablation of the target at a higher temperature and nanomaterials are collected on a cold substrate. Synthesis of any type of nanoparticles such as metals, semiconductor, insulator, biological, or organic nanoparticles employing pulsed laser ablation in liquid media (PLAM) of the corresponding target is a novel, easy, quick, and green approach of nanomaterials synthesis in highly colloidal form. For example, a pellet of DNA can be used for the synthesis of DNA nanoparticles. Nanoparticles from the solution of liquid media can be transferred onto a desired substrate in the form of film using spin coating or dip coating for the fabrication of nanomaterials-based devices. Any type of nanomaterial in controlled size, shape, and distribution can be synthesized in powder form using laser pyrolysis of corresponding gaseous precursors. Nano-structures can be fabricated on the selective site of solid target surface for micro- and nanoelectronic device fabrications using laser-induced nanostructuring (LIN). Figure 2.12 presents a hierarchical ordering of various laser-based nanomaterials processing routes.

Photolithography is another application of laser in micro- and nanofabrications for micro- and nanoscale electronics (Figure 2.13d,e) [42]. Availability and future development of vacuum ultraviolet/extreme ultraviolet (VUV/EUV lasers in the range of 10–50 nm spectral range can make patterns and devices of nano sizes and is a unique way for the validation of Moore's law in future. Two-photon-polymerization (2PP) [43] is another micro-/nanofabrication technique based on interaction of fem-tosecond laser radiation, which induces a highly localized chemical reaction leading to polymerization of the photosensitive material with present resolution down to 100 nm (Figure 2.13a,b). Surface releif grating fabricated by 488 nm line of Ar ion laser is shown in Figure 2.13c. Figure 2.13d,e illustrates the 3D periodic structure constructed by a photolithographic route based on the interference of three or four beams of the visible light in continuous wave (CW) mode and laser-initiated cationic polymerization of epoxy [43]. Laser can decompose metal precursors and synthesize corresponding oxides if the process is carried out in air. Zinc oxide nanostructures are synthesized by laser-induced decomposition [44] of $Zn(AcAc)_2$ saturated solution in water/alcohol (Figure 2.13f) femtosecond laser irradiation of urea solution results growth of Urea crystal as shown in Figure 2.13g,h. Laser chemical vapor deposition (LCVD) is a well-established approach for the fabrication of metal or metal oxide thin films and nanostructures. LCVD consists of laser photolysis of gaseous

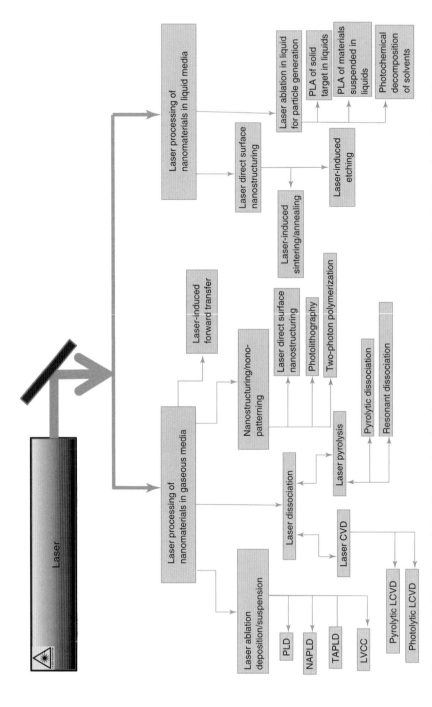

Figure 2.12 A hierarchical ordering of various laser-based nanomaterials processing routes. PLD, pulsed laser deposition; TAPLD, temperature-assisted pulsed laser deposition; PLA, pulsed laser ablation.

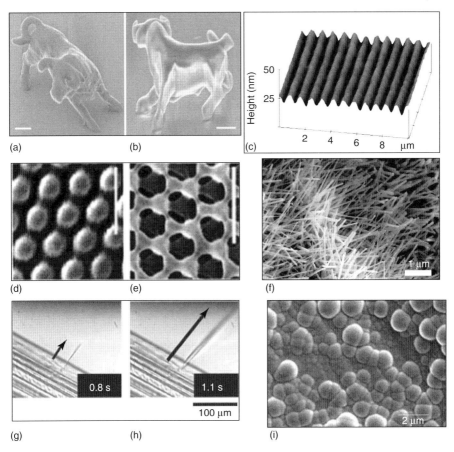

Figure 2.13 Nanostructures produced by various laser-based material synthesis methods: (a,b) nanobulls by two-photon polymerization (Source: reprinted with permission from Ref. [43]), (c) surface relief grating fabricated by 488 nm of Ar ion laser (Source: reprinted with permission from Ref. [48]), (d,e) periodic 3D nanostructures created by multibeam interference of visible laser (d) in the absence and presence of transversally excited atmospheric (TEA) laser (Source: reprinted with permission from Ref. [43]), (f) ZnO nanorods synthesized by laser-induced decomposition of An (AcAc)$_2$ (Source: reprinted with permission from Ref. [44]), (g,h) growth of urea crystals by femtosecond laser irradiation of urea solution (Source: reprinted with permission from Ref. [49]), and (i) Se nanostructured film grown by LCVD of diethyl selenium (Source: reprinted with permission from Ref. [45]).

precursors (e.g., diethyl selenium for the synthesis of Se nanostructured film, Figure 2.13i), which causes photodissociation of the precursor into the element (e.g., Se) and gets deposited on the substrate [45, 46]. Pyrolytic LCVD is a thermally driven process in which laser heating of the substrate surface raises its temperature up to a level that is required for chemical dissociation and deposition [47].

Lasers are widely used in the modification of the size, shape, distribution, phase, and surface properties of nanomaterials and can induce surface as well as bulk

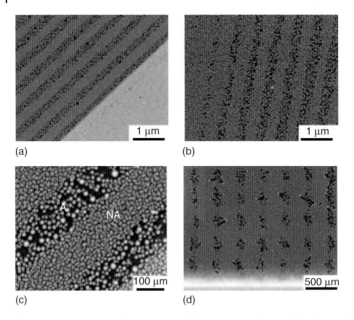

(a)

(b)

(c)

(d)

Figure 2.14 Grating and nanopattern formation using laser-induced assembly of gold nanoparticles on 40 nm thick Si3N4 membrane (a–c) single-pulse 532 nm irradiation at the normal irradiance and (d) backside irradiation of the same beam. (Source: Reprinted with permission from Ref. [52] © Nature publishing group.)

defects in the crystals for various scientific and technological applications. Lasers can induce synthesis of semiconductor or metal nanoparticles inside the bulk of transparent glass or silica matrix at any space-selective location by the photo precipitation of the corresponding precursor solution [50]. Pulsed femtosecond laser can remove defects from the nanocrystals as described by Romero *et al.* [51] and is termed as *laser nanosurgery of defects.* The grating pattern is constructed by employing single laser pulse irradiation of disordered, evaporated gold (silver) films of discrete island deposited on a 40 nm thick membrane [52] (Figure 2.14). Geometry of the patterns depends on experimental parameters such as (i) wavelength, polarization, and angle of incidence of the laser pulse (ii) the properties of the membrane, and (iii) the position of the laser spot with respect to the membrane edge. Lasers have increased application of nanomaterials in biodiagnosis, life sciences, and nanosurgery. Metal nanoparticles excited by lasers act as a molecular heater for selective killing and therapy of cancerous or infected cells, which is known as photo ablation therapy (PAT) [53]. The impulsive force generated by focusing of a laser pulse is able to inject nanoparticles [54] inside the cell for various diagnosis applications (Figure 2.15).

Lasers have potential applications in the characterization of nanoscale materials and it is continuously increasing day by day with the development of science and technology in the field of lasers. Laser Raman spectroscopy provides size,

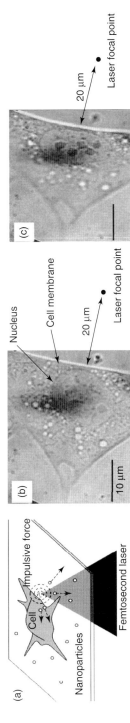

Figure 2.15 Injection of nanoparticles inside the cell using femtosecond laser-induced impulsive force (a) experimental view and TEM images of cells (b) before and (c) after particle injection. (Source: Reprinted with permission from Ref. [54].)

shape, defect, and structural characterization and its imaging provides a micro-scopic view of nanomaterials. Band gap, size, defect, and other optical properties of the nanoscale materials can be diagnosed by employing laser-excited photolu-minescence. The principle of light dynamic scattering (LDS) has added size and distribution measurement capability of lasers to the list of its uses in nanomate-rials characterization. Lasers can generate ultrasonic waves at the solid surfaces, which can diagnose defects and cracks and thermal properties of materials. Laser acoustic/photo acoustic spectroscopy is other interesting application of lasers in the characterization of thermal, relaxation, and defects inside the crystals. Multiphoton and higher harmonic generation (HHG)-produced wavelength-excited fluorescence microscopy provides a comparatively cheap system for the imaging of nanoscale particles. Resolution of such microscopic systems is continuously increasing with the availability of EUV/VUV and X-ray laser sources. Ultrashort pulsed laser excita-tion of insulator/semiconductor nanomaterials transfers electron from the valence band to the conduction band resulting in the increase in conductivity for a short time, that is, transition from insulator/semiconductor to conductor takes place for a short time and materials may be used as switch. Therefore, photoconductivity of nanomaterials and its dependence on various environmental parameters can be studied using laser excitations. Ultrafast laser spectroscopy measures lifetime and dynamics of electrons and holes, role of defects and surface adsorbates, and electron–electron, electron–phonon, and phonon–phonon interaction in the nanomaterials. High intensity of lasers is used for the study of several nonlinear optical properties of nanomaterials such as nonlinear absorption and reflection coefficients, nonlinear refractive indices, HHG efficiency, and so on. Laser-induced thermal pulses in the materials are used for the characterization of space charge distribution, electrical polarization, and their dielectric constant. Laser-induced breakdown spectroscopy (LIBS) [55] and laser-ablated inductively coupled plasma mass spectroscopy (LA-ICPMS) [56] techniques are used for determination of chemical composition of nanomaterials.

2.11
Summary and Future Prospects

Now, a wide variety of state-of-the-art laser systems having wavelengths starting from a few nanometers (EUV/VUV free electrons and X-ray lasers) to hundreds of microns (molecular laser), CW and pulsed, pulse widths starting from attosecond, femtosecond to milliseconds, and frequencies from a few hertz to megahertz are available for the processing of nanoscaled materials and their characterizations. Attosecond laser pulses are ready for probing electronic motions, can determine transition time from one state to the other, and can detect molecular evolution of pathogen of cancers, AIDS, and other diseases at their very initial stages. However, they do not have sufficient pulse energy to ablate targets, act as lithography source for the fabrication of nanoscale patterns, writing 3D dense information on optical discs, and so on. Development of intensity of short wavelength (1–50 nm), attosecond laser sources will be the future of modern science and technology and they will not only be able to solve several undiagnosed problems of materials, life, medical science but also open a new avenue in the field of nanomaterials processing and their applications.

References

1. (a) (1960) Caltech's Engineering and Science (b) *http://www.zyvex.com/nanotech/feynman.html*.
2. Taniguchi, N. (1974) On the basic concept of nano-technology. Proceedings of the International Conference on Production Engineering, London, Part II British Society of Precision Engineering.
3. Drexler, K.E. (1991) *Nanosystems: Molecular Machinery, Manufacturing, and Computation*, MIT PhD thesis, John Wiley & Sons, Inc., New York. ISBN: 0471575186.
4. Altmann, S.L. (1991) *Band theory of Solids: An Introduction from the Point of View of Symmetry*, Oxford Science Publications. ISBN: 019855866-X.
5. Singleton, J. (2001) *Band Theory and Electronic Properties of Solids*, Oxford Master Series in Condensed Matter Physics, Oxford University Press. ISBN: 019850-6449, 6457.
6. Gautreau, R. and Savin, W. (1999) *Schaum's Outlines, Modern Physics*, McGraw-Hill Professional.
7. Kittel, C. (2009) *Introduction to Solid State Physics*, Wiley India Pvt, Ltd. ISBN-8126510455, 9788126510450.
8. Ibach, H. and Luth, H. (1981) *Solid State Physics: An Introduction to Material Science*, Springer-Verlag, Berlin, Heidenberg. ISBN-9783540938033.
9. Ren, S.Y. (2006) *Electronic States in Crystals of Finite Size; Quantum Confinement of Bloch Waves*, Springer Sciences and Business Media, Inc. ISBN-100387263039.
10. Madelung, O. (1978) *Introduction to Solid State Theory*, Springer-Verlag, Berlin, Heidelberg, New York. ISBN-10:354060443X.
11. Chelikowsky, J.R. and Cohen, M.L. (1976) *Phys. Rev. B*, **19**, 556.
12. Maeda, Y. (1995) *Phys. Rev. B*, **51**, 1658.
13. Delly, B. and Steigmeier, E.F. (1993) *Phys. Rev. B*, **47**, 1397.
14. Takagahara, T. and Takeda, K. (1992) *Phys. Rev. B*, **46**, 15578.
15. Chen, G. (2005) *Nanoscale Energy Transport and Conversion*, Oxford University Press, New York.
16. Streetman, B.G. and Banerjee, S. (2000) *Solid State Electronic Devices*, Prentice Hall, Upper Saddle River.
17. Muller, R.S., and Kamins, T.I. (2003) *Device Electronics for Integrated Circuits*, John Wiley & Sons, Inc., New York.
18. Kittel, C. and Kroemer, H. (1980) *Thermal Physics*, W.H. Freeman and Company, New York.
19. Sze, S.M. (1981) *Physics of Semiconductor Devices*, John Wiley & Sons, Inc., New York.
20. Efros, A.L. and Efros, A.L. (1982) *Sov. Phys. Semicond.*, **16**, 772.
21. Brus, J.E. (1984) *J. Lumin.*, **31**, 381.
22. Kayanuma, Y. (1988) *Phys. Rev. B*, **38**, 9797.
23. Micic, O.I., Sprague, J., Lu, Z., and Nozik, A.J. (1996) *Appl. Phys. Lett.*, **68**, 3150.
24. Albe, V., Jouanin, C., and Bertho, D. (1998) *Phys. Rev. B*, **58**, 4713.
25. Fu, H. and Zunger, A. (1997) *Phys. Rev. B*, **55**, 1642.
26. Moreels, I., Lambert, K., Muynck, D.D., Vanhaecke, F., Poelman, D., Martins, J.C., Allan, G., and Hens, Z. (2007) *Chem. Mater.*, **19**, 6101.
27. Lu, X. and Chen, Z. (2005) *Chem. Rev.*, **105**, 3643.
28. Kim, S., Jin, J., Kim, Y.-J., Park, I., Kim, Y., and Kim, S.-W. (2008) *Nature*, **435**, 757.
29. Michaels, A.M., Nirmal, M., and Brus, L.E. (1999) *J. Am. Chem. Soc.*, **121**, 9933.

30. Jacson, J.B., Westcott, S.L., Hirsch, L.R., West, J.L., and Halas, N.J. (2003) *Appl. Phys. Lett.*, **82**, 257.

31. Schwartzberg, A.M. and Zhang, J.Z. (2008) *J. Phys. Chem. C*, **112**, 10333.

32. Mohamed, M.B., Volkov, V., Link, S., and El-Sayed, M.A. (2000) *Chem. Phys. Lett.*, **317**, 517.

33. Basu, S., Pande, S., Jana, S., Bolisetty, S., and Pal, T. (2008) *Langmuir*, **24**, 5562.

34. Webpage for Peak Assign- ment and Structural Analysis *http://database.iem.ac.ru/mincryst/.*

35. Cullity, B.D. and Stock, S.R. (2001) *Elements of X-Ray Diffraction*, 3rd edn, Prentice-Hall Inc., pp. 167–171. ISBN-0-201-61091-4.

36. Jenkins, R. and Snyder, R.L. (1996) *In- troduction to X-ray Powder Diffractometry*, John Wiley & Sons, Inc., pp. 89–91. ISBN-0-471-51339-3.

37. Klug, H.P. and Alexander, L.E. (1974) *X-Ray Diffraction Procedures*, 2nd edn, John Wiley & Sons, Inc., pp. 687–703. ISBN-978-0471493693.

38. Warren, B.E. (1969) *X-Ray Diffraction*, Addison-Wesley Publishing Co., pp. 251–254. ISBN-0-201-08524-0.

39. Goodman, P. (General editor) (1981) *Fifty Years of Electron Diffraction*, D. Reidel Publishing.

40. Human, D. *et al.* (2006) Low energy electron diffraction using an electronic delay-line detector. *Rev. Sci. Instrum.*, **77**, 023302.

41. Furrer, A., Mesot, J., and Strässle, T. (2009) *Neutron Scattering in Condensed Matter Physics*, World Scientific. ISBN: 978-981-02-4830-7981-02-4830-X.

42. Yang, S., Megens, M., Aizenberg, J., Wiltzius, P., Chaikin, P.M., and Russel, W.B. (2002) *Chem. Mater.*, **14**, 2831.

43. Kawata, S., Sun, H.-B., Tanaka, T., and Takada, K. (2001) *Nature*, **697**, 412.

44. Fauteux, C., Longtin, R., Pegna, J., and Therriault, D. (11036) *Inorg. Chem.*, **46**, 2007.

45. Mazumder, J. and Kar, A. (1995) in *Theory and Application of Laser Chemical Vapor Deposition in Lasers, Photonics, and Electro-Optics by in Near-field Nano-optics: from Basic Principles to Nano-Fabrication and Nanophotonics* (eds M. Ohtsu and H. Hori), Kulwar Academic/Plenum Publisher, New York. ISBN: 0-306-45897-7.

46. Ouchi, A., Bastl, Z., Boha, J., Orita, H., Miyazaki, K., Miyashita, S., Bezdicka, P., and Pola, J. (2004) *Chem. Mater.*, **16**, 3439.

47. Bondi, S.N., Lackey, W.J., Johnson, R.W., Wang, X., and Wang, Z.L. (2006) *Carbon*, **44**, 1393.

48. Gao, J., He, Y., Xu, H., Majumder, J., Kar., A., Song, B., Zhang, X., Wang, Z., and Wang, X. (2007) *Chem. Mater.*, **19**, 14.

49. Yoshikawa, H.Y., Hosokawa, Y., and Masuhara, H. (2006) *Cryst. Growth Des.*, **6**, 302.

50. Hamzaoui, H.E., Bernard, R., Chahadih, A., Chassagneux, F., Bois, L., Jegouso, D., Hay, L., Capoen, B., and Bouazaoui, M. (2010) *Nanotechnology*, **21**, 134002.

51. Romero, A.H., Garcia, M.E., Valencia, F., Terrones, H., and Terrones, M. (2005) *Nano Lett.*, **5**, 1361.

52. Eurenius, L., Hagglund, C., Olsson, E., Kasemo, B., and Chakarov, D. (2008) *Nat. Photonics*, **2**, 360.

53. Zhang, J.Z. (2010) *J. Phys. Chem. Lett.*, **1**, 686.

54. Yamaguchi, A., Hosokawa, Y., Louit, G., Asahi, T., Shukunami, C., Hiraki, Y., and Masuhara, H. (2008) *Appl. Phys. A.*, **93**, 39.

55. Sabsabi, M. and Cielo, P. (1995) *Appl. Spectrosc.*, **49**, 499.

56. (a) Chaoui, N., Millon, E., and Muller, J.F. (1998) *Chem. Mater.*, **10**, 3888 (b) Kuhn, H.-R. and Gunther, D. (2003) *Anal. Chem.*, **75**, 747.

Part I
Nanomaterials: Laser Based Processing Techniques

Nanomaterials: Processing and Characterization with Lasers, First Edition.
Edited by Subhash Chandra Singh, Haibo Zeng, Chunlei Guo, and Weiping Cai.
© 2012 Wiley-VCH Verlag GmbH & Co. KGaA. Published 2012 by Wiley-VCH Verlag GmbH & Co. KGaA.

3
Laser–Matter Interaction

3.1
High-Intensity Femtosecond Laser Interactions with Gases and Clusters

Alan M. Heins and Chunlei Guo

3.1.1
Introduction

With the advancement of femtosecond laser technologies, laser interactions with gases and gas-phase clusters have become a major research area, especially at high intensities. These interactions are ubiquitous. For example, even when the experimental intent is to study condensed matter–laser interactions, the high-intensity ultrashort laser pulses will still interact with the ambient gas. Very low density gases provide an excellent laboratory approximation to the isolated atoms or molecules commonly studied in atomic and molecular physics. On the other hand, there is a great deal of interest in studying gases at or near atmospheric density. Some atomic physics experiments, such as fluorescence measurements, are also more easily carried out at high densities. At high densities, plasma effects obscure the atomic physics somewhat, but the plasmas produced by femtosecond lasers are interesting in their own right, with some falling into the strongly coupled regime.

Laser cluster physics is an emerging area of research. It offers the promise of solid densities of atoms, but over a region sufficiently small that the intensity attenuation of the laser pulse can be neglected, a major simplification. Clusters are formed naturally in many laboratory processes, such as expansion of supersonic jets, ablation of solid materials, and condensation of saturated vapors. Although a much younger field than laser–gas interaction, laser–cluster interaction has already produced many interesting results, such as a subpicosecond source of hard X-rays and even the possibility of benchtop nuclear fusion.

3.1.2
Laser–Atom Interactions

Owing to the large value of the characteristic electric fields in atoms on a laboratory scale ($E_{atomic} = 5.14 \times 10^{11}$ V m^{-1}) – and the rapid motion of bound electrons

Nanomaterials: Processing and Characterization with Lasers, First Edition.
Edited by Subhash Chandra Singh, Haibo Zeng, Chunlei Guo, and Weiping Cai.
© 2012 Wiley-VCH Verlag GmbH & Co. KGaA. Published 2012 by Wiley-VCH Verlag GmbH & Co. KGaA.

($t_{atomic} = 24$ as) – fast, energetic systems are needed to study atomic dynamics in the nonperturbative regime. For many years, the predominant laboratory tools for studying atomic dynamics were synchrotrons [1] and ion accelerators [2, 3]; of these, only ion accelerators can produce nonperturbative electric fields. With recent advances in high-intensity femtosecond (fs: 1 fs $= 10^{-15}$ s) laser techniques [4–8], femtosecond lasers can readily generate a high electric field with a strength comparable to or even exceeding the Coulomb-binding potential that holds an atom or molecule together [9–11]. When the field becomes strong enough, the field can suppress the Coulomb potential, and the electron can tunnel through the Coulomb barrier formed by the laser field, ionizing the atom. An illustration can be seen in Figure 3.1.1. In this high-field, light-atom/molecule interactions can be significantly different from those in weak electromagnetic fields, and many interesting phenomena have been observed, such as above-threshold ionization [12], nonsequential ionization (NSI) [13], high harmonic generation (HHG) [14, 15], and attosecond pulse generation [16]. At present, amplified femtosecond laser systems can deliver pulses as short as a few femtoseconds [17] and intensities as high as 10^{22} W cm^{-2} [18]. This can be compared with the subattosecond interaction time and the (Weizsächer–Williams) equivalent intensity of 10^{19} W cm^{-2} of the most energetic ion accelerators [19]. It can be seen that femtosecond lasers can currently produce larger electric fields than the ion-collision techniques, but still lag in time resolution [20]. Various nonlinear techniques can allow subpulse, and even sublaser-cycle, time resolution in some experiments [21], which reduces the temporal gap between the two technologies.

The strength of the femtosecond laser is the wealth of laser parameters that can be adjusted to control atomic dynamics more precisely. Laser polarization [22], chirp [23–25], and carrier-envelope phase (CEP) [26] all represent degrees of freedom of the femtosecond laser system that are unavailable in ion colliders.

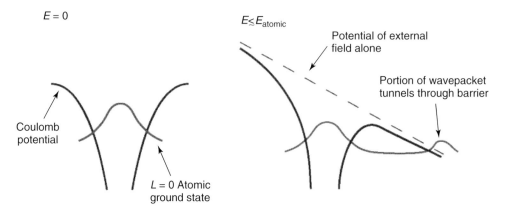

Figure 3.1.1 Tunneling ionization of an electron. On the left, the external (laser) electric field is zero, and the electron (here chosen to be in an $L = 0$ angular momentum state) resides in a Coulomb well.

When an external field – approaching but not equaling E_{atomic} – is turned on, the wave packet can tunnel through the potential barrier and ionize the atom.

The tight focusing achievable with a laser, especially when coupled with a molecular beam source, allows interaction volumes as small as a few cubic microns, a major advantage in sensitive particle momentum experiments. Lasers lend themselves very naturally to pump-probe experiments, which have produced a wealth of results in atomic and molecular dynamics. Finally, the femtosecond laser has the advantages of smaller scale and relative affordability compared to accelerators or synchrotrons.

The cyclic nature of the laser field plays an important role in its interaction with atoms and molecules. In the strong-field regime, however, the main role of the laser frequency is that it supplies the oscillation frequency of the ponderomotive potential that an ionized electron encounters. The cyclic variation of this field keeps the electron close to its parent ion. This continued close proximity of ions and electrons blurs the distinction between atomic and plasma physics [27], even in very low density gases where the interatomic interactions become negligible. The coherent control of the electrons provided by the ponderomotive field opens up several possibilities for studying the remaining atomic or molecular core. Recombination of the electron with the parent ion produces odd harmonics of the laser frequency through the process of HHG [14, 15, 23, 28]. Alternatively, electrons scattered from the ion in a single collision can produce a high-resolution image of the molecular residue, a process known as laser-induced electron diffraction (LIED) [29–32]. While the freshly ionized atom or molecule can be probed by other methods, the advantage of "recolliding" the daughter electron with its parent is the accurate synchronization with the laser pulse that this method provides [29]. The time between ionization and recollision can be adjusted by changing the laser wavelength.

Most effects in single-electron ionization and sequential multiple-electron ionization of rare gas atoms can be relatively well understood by the single-active-electron (SAE) approximation [33]. Time-dependent quantum mechanical calculations using the SAE approximation have been shown to provide accurate single and sequential multiple ionization rates, above-threshold ionization yields, high harmonic spectra, and angular distributions for rare gas atoms in strong laser fields [33–35]. In the tunneling regime [36], the SAE-based Ammosov–Delone–Krainov (ADK) tunneling model also provides an accurate fit to the single-electron ionization rate and sequential multiple-electron ionization rate of rare gas atoms [37–40]. The ADK model fails dramatically, however, for phenomena such as nonsequential double ionization (NSDI) [13, 41]. NSDI normally involves an enhancement in the double ionization yields compared to those based on sequential rates. In contrast to the sequential double ionization, in which the first and second ionizations are assumed to be independent events (and the timing between them stochastic), in NSDI the emission of the two electrons is highly correlated. Various models for NSDI have been proposed and debated. The rescattering model [27] attributes the phenomena to an (e, 2e) collision when the ponderomotive field returns the electron to its parent ion. The shake-off model proposes that NSDI occurs when the removal of one electron promotes a second electron to the continuum [41].

3.1.3
Laser–Molecule Interactions

Molecules have more degrees of freedom and a greater diversity of electronic structures than atoms and are expected to have richer dynamics. In the past, however, studies have shown that many aspects of strong-field ionization in molecules are similar to those in atoms [39, 40]. For example, argon and N_2, which have very similar ionization potentials, also have a very similar ionization behavior (Figure 3.1.2) [42]. This is not the case for molecular oxygen: while O_2 has an ionization potential similar to that of xenon, its ionization rate is an order of magnitude lower over a significant intensity range [42, 43]. It has been further verified that Xe^+ behaves normally and agrees with the prediction of the ADK model, but O_2^+ has an anomalously low single-ionization yield [42]. Various explanations have been proposed; they generally make note of the fact that O_2 has an open shell for its ground state, while most other molecules are closed shell [44, 45]. Extensive efforts have been devoted to developing suitable theoretical models to explain the anomalous ionization behavior in O_2 [43–48]. However, only a few models are able to quantitatively account for the observed suppression in O_2^+ [44, 45, 48], and these few models are all based on the original concept

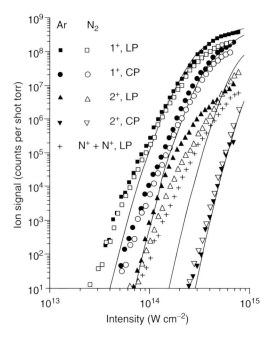

Figure 3.1.2 Atomic and molecular ionization. The figure shows that there is a close agreement in single-ion yield between argon (IP: 15.76 eV) and N_2 (IP: 15.58 eV), for both linear and circular polarizations. Furthermore, both curves are well fit by the ADK model (solid lines) at high intensities. This indicates that N_2 behaves like an atom. (Source: From Ref. [42].)

proposed in Ref. [42] that the detailed electronic structures play the dominant role in influencing electron ionization dynamics.

Similar to the case of single-electron ionization discussed above, double ionization also exhibits greater complexity when our study evolves from atoms to molecules. The greater diversity of electronic structures of molecules again allows us to gain a significantly better insight into the mechanisms of NSDI. Furthermore, interesting behaviors have been seen in molecules that cannot be explained by models used in rare gas atoms [49–51]. Recently, studies of asymmetric diatomic molecules and larger molecules also yielded interesting results [51–54].

Multiple ionization of molecules often leads to their dissociation, in cases where the Coulomb repulsion exceeds the bond strength of the remaining bonding orbitals, a process known as *Coulomb explosion* [55, 56]. Single ionization can also lead to dissociation, through creation of antibonding molecular cations or bond softening [52, 57]. Laser polarization plays an important role in molecular ionization. Molecules will generally dissociate more easily for a polarization vector along an intermolecular axis than perpendicular to it [22]. The longer the molecule, the more pronounced this effect becomes, and long, linear molecules will dissociate (often into more than two pieces) in parallel polarization at intensities far below what the atomic ADK model would predict [58, 59].

As larger molecules are considered, there is eventually an overlap with what is known as *cluster physics*. The interaction of clusters with femtosecond laser pulses is discussed in a later section.

3.1.4
High-Pressure Atomic Physics

The ultrahigh vacuum work is important for laser ionization of high-density gas because it demonstrates the ways in which the laser interacts with the individual atoms or molecules, allowing the initial state of the plasma to be determined. In general, it is more difficult to go in the other direction – to deduce the atomic physics from the study of the high-density plasma – because of the additional complication of plasma effects on the ions. However, the high-pressure experiments can provide an important complementary perspective to the low-pressure ones. Comparing the requisite conservation of energy with the fragment kinetic energies actually observed in time-of-flight experiments, many atoms are expected to be left in an excited state following the laser pulse [60, 61]. However, as the microchannel plate (MCP) detectors used in these experiments are not sensitive to the excitation levels of the incoming particles, these excitations can merely be postulated. High-density gases produce a sufficient quantity of these excited states that their decay can be observed optically. Using this method, a new excitation channel of N_2 has been discovered [62]. The suppression of the O_2-ion yield relative to xenon has also been confirmed with optical techniques in the high-pressure regime [63]. This is an important confirmation, as the MCPs have differing (and generally, unknown) sensitivities to different ions.

3.1.5
Strongly Coupled Plasmas

At the conclusion of the laser pulse, plasma effects take over. The ultrashort nature of the pulse, coupled with its very high ionization efficiency, combines to provide access to some unusual portions of plasma parameter space. In contrast to the condensed matter case, femtosecond ionization of an atmospheric pressure gas does not produce a significant number of collisions of electrons with adjacent atoms or ions [64]. These collisions do appear in gases at several atmospheres [64]. This leads to a much lesser degree of electron heating through inverse bremsstrahlung than that which occurs in solids [65, 66]. The high rate of ionization combined with a comparatively low electron temperature allows plasmas in which the Coulomb potential energy is a significant fraction of, or even exceeds, the electron kinetic energy. Such plasmas are said to be *nonideal* or *strongly coupled*, and are an active field of research [67].

The degree of nonideality in a plasma can be increased by adding dust particles or clusters to form a "dusty plasma." Because the electrons in a laser-generated plasma will initially have a higher temperature than the ions, they will collide with the dust particles first and begin to endow them with a negative charge [68]. Eventually, positive ions will also begin accretion on the dust nucleus, but the initial bombardment with electrons will ensure the cluster remains negative [68]. The large Coulomb potential of the dust particles significantly enhances the nonideality of these plasmas. At moderate coupling, these systems are known as *Coulomb liquids*; above a certain coupling value, they can undergo Wigner crystallization to become "solids" (Figure 3.1.3). However, their properties are different from that of ordinary solids and liquids, and they are best described as exotic states of matter [69, 70].

Because of their low temperatures and high densities, ultrafast laser-produced plasmas rapidly decay by recombination. The most common recombination processes at high densities – dissociative recombination and three-body recombination – typically leave the resultant atoms or molecules in highly excited states, producing a significant population inversion [71]. This effect is being explored as a method for producing an X-ray laser [72].

3.1.6
Clusters

Clusters are groupings of atoms that have properties intermediate between those of isolated atoms and condensed solids [73]. There is no well-defined lower or upper limit on the number of bound atoms that may be regarded as a cluster. Clusters with tens of thousands of atoms have been shown to still have properties different from the bulk solid, while the study of very small structures, such as C_{60}, overlaps with that of molecular physics. Very small clusters may show properties dramatically different from the bulk material; for example, mercury clusters with fewer than 17 atoms behave as nonmetals [74]. A central issue in cluster physics is

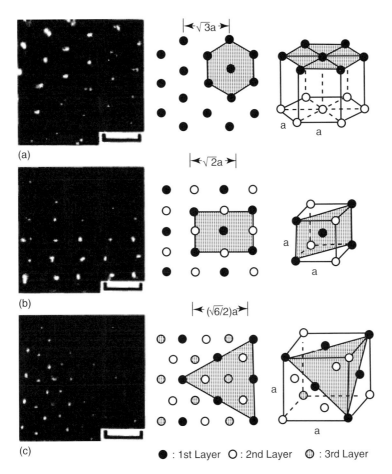

● : 1st Layer   O : 2nd Layer   ◉ : 3rd Layer

Figure 3.1.3 Wigner crystallization in an argon plasma seeded with 10 μm SiO_2 particles, (a) is hexagonal, (b) body-centered cubic, and (c) face-centered cubic. (Source: From Ref. [70].)

the number of atoms required to change the cluster symmetry from the symmetry characteristic of the icosahedra of very small clusters to that of the bulk crystal [75]. Other features, such as absorption spectra, also show major changes as the cluster size grows [76]. Many clusters are highly reactive and can exist only in low-pressure environments, but some metal and carbon clusters can persist at room temperature and atmospheric pressure [77–79] (Figure 3.1.4).

3.1.7
Laser–Cluster Production

Clusters can be produced by the laser ablation of a solid target, followed by a rapid cooling of the ejecta with a pulse of cold gas from a supersonic jet [80]. One of the most studied clusters, buckminsterfullerene, was first produced by laser ablation of

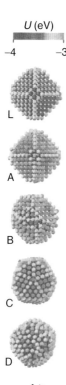

U (eV)

−4   −3

L

A

B

C

D

Figure 3.1.4 Melting of a cluster 459 gold atoms; the sequence from L to A to D corresponds with increasing energy. The cluster transitions from face-centered cubic at L, to partially icosahedral at C, and finally to disordered collection at D. The temperature plays a significant role not only in cluster formation but also in ordering. The greyscale values represent the energies of each atom. (Source: From Ref. [79].)

a graphite surface followed by rapid cooling with a helium jet [81]. Numerous metal and semiconductor clusters have been produced by the same method, including copper and aluminum [82], germanium and silicon [83], and gallium arsenide [84].

3.1.8
Laser–Cluster Interaction

In addition to forming clusters through chemical or plasma effects, lasers also show a strong interaction with preformed clusters [85]. Clusters present a laser pulse with a target having the local density of a solid, but of sufficiently small size that every atom in the cluster can experience the same strength of laser field, simplifying the analysis of effects otherwise observed in bulk solids [65]. The highest charge state attainable by ultrashort pulse ionization has been experimentally observed to be considerably higher when clusters are present than in the atomic precursor gas alone [86, 87]. This ionization enhancement can be attributed to at least two effects [65]: (i) the presence of nearby ions lowers the ionization potential of an atom by enhancing the electric field it sees (space–charge effect) and (ii) the increased local density increases the probability of dephasing electron–atom collisions, which allows the electrons to absorb more ponderomotive energy and collisionally ionize more atoms. These enhancements in ionization will only be significant for pulses shorter than the Coulomb explosion time of the cluster; that is, for subpicosecond pulses [88].

The increase in the ionization level allows for the production of kilovolt-energy X-rays in gas–cluster mixtures, which are not observed in the gas alone [65]. Charge states as high as Kr^{27+} and X-ray energies up to 2.1 keV have been observed in an experiment involving the supersonic-cooling-induced cluster formation in noble gases, irradiated with 8×10^{18} W cm^{-2} subpicosecond pulses [66]. These energetic X-ray sources offer the possibility of high spatial resolution (due to the small size of both individual clusters and the entire laser focus) and high temporal resolution, as the X-ray emission is prompt [66, 86] and cannot occur after the cluster Coulomb explodes. The hard cutoff in X-ray emission time is not shared by more traditional laser-excited X-ray spectroscopy (LEXS) setups, in which the solid target is not completely destroyed by the laser pulse [89]. More recent experiments have found ions with energies up to 1 MeV using laser pulses of 2×10^{16} W cm^{-2} on xenon clusters, opening up the possibility of tabletop fusion experiments [90].

The time dynamics of laser–cluster interaction have been studied with both nanosecond and femtosecond lasers. For nanosecond lasers, it has been found that the cluster is vaporized primarily by the plasma created by the laser, rather than by the laser itself [91]. Because nanosecond lasers generally produce much lower peak intensities than femtosecond lasers, they cannot ionize clusters as quickly to produce the intense Coulomb explosion discussed above. Furthermore, the plasmas produced by nanosecond lasers tend to be hot, as the laser can continue to heat the plasma after it forms, which makes the plasma itself much more effective in destroying solid material. Clusters of 10 μm diameter or less are generally completely evaporated by these plasmas, even though this is somewhat material dependent [92, 93].

In contrast, with ultrafast lasers, the details of cluster evaporation depend more on pulse length than cluster size [94, 95]. Because the Coulomb explosion in clusters releases a very large kinetic energy, the cluster can change size appreciably even during a subpicosecond pulse [96]. When the exploding cluster reaches a certain critical radius, it has been experimentally observed that the absorption of energy from the laser pulse increases dramatically [96]. This may be due to enhanced ionization between pairs of cluster ions, analogous to the enhanced ionization mechanism in diatomic molecules [97, 98]. Or it may be due to the entire nanoplasma passing through a plasmon resonance with the laser at the critical radius [99, 100]. In either case, the pulse will transfer maximal energy to the expanding cluster if it reaches the critical radius simultaneously with the pulse envelope maximum, and hence a sensitive dependence on pulse length. It is found, for the near-infrared pulse typically used in ultrafast studies, that clusters are fully vaporized for intensities above $10^{12} - 10^{13}$ W cm^{-2} [96].

3.1.9
Aerosol Monitoring

Because the laser pulse interacts strongly with nanoclusters, it is developed in a variety of techniques for measuring airborne aerosol concentrations. In a technique known as gas-phase laser-induced breakdown spectroscopy (LIBS)

[101, 102], a high-intensity laser is focused on a gas sample. The subsequent plasma emission is analyzed, and characteristic lines are recorded in a spectrometer for elemental identification. Compared to other methods of chemical assay of a gas sample, LIBS offers the advantages of being noncontact – meaning it can be used for gases in hostile or otherwise inaccessible environments – and rapid. The kilohertz repetition rates of modern chirped-pulse amplifier (CPA) lasers enable continuous, near-real-time monitoring of the gas composition.

Aerosols that have been studied with LIBS include many with important ramifications for environmental and human health, including heavy metal aerosols produced in industrial operations [103, 104] and bioaerosols [105, 106]. Trace quantities of explosives have also been detected [107]. Heavy metals tend to have a large number of persistent lines, which are easily identifiable in LIBS spectra [102]. The trace metal analysis of natural aerosols – such as soil-generated dust, smoke from fires and volcanoes, and salt spray from the sea – tend to be highly specific to the source [102]. LIBS aerosol analysis can therefore be used to track the progress of volcanic ash or the movement of air masses after a dust storm. Gas-phase LIBS has found additional applications in the study of combustion processes [108] and rocket engine monitoring [102].

While nanosecond lasers have traditionally been used for LIBS, femtosecond lasers offer the promise of simpler physics. As the femtosecond pulse is too short to significantly interact with the plasma it creates, processes found in nanosecond laser-produced plasmas such as laser-supported combustion, detonation, and radiation waves are not excited [102]. Furthermore, of particular importance for aerosol analysis, the strong expansion waves produced in the intensely heated nanosecond plasma can alter the particulate concentration in the probed region due to differing "slip factors" between the dust and background gas [109]. As discussed above, femtosecond lasers provide less heating because of a lower number of collisions and may mitigate this effect. Finally, femtosecond lasers have been found to produce more stable plasma signals than nanosecond lasers [110].

3.1.10
Atmospheric Effects

In addition to monitoring the atmosphere, femtosecond laser pulses can significantly modify it. High-intensity pulses have been shown to induce cloud formation [111]. While a similar effect from high-energy particles is well known in the saturated vapor of Wilson cloud chambers [112], a surprising feature of femtosecond-laser-induced condensation is that it can occur in significantly undersaturated gases. This is believed to be due primarily to the rapid formation of nucleating chemical compounds, with the shockwave accompanying ionization also possibly playing a role.

Laser filamentation is a well-studied process in the atmosphere [113–116]. The nonlinear refractive index of air causes a sufficiently intense pulse to focus as it propagates. When the pulse focuses on the breakdown intensity of air, ionization begins, causing plasma defocusing. If a pulse of power above 3 GW [111] is focused

(a)   (b)

Figure 3.1.5 (a) Spontaneous electrical discharge compared with an electrical discharge triggered with a (b) femtosecond-laser-pulse-produced filament. (Source: From Ref. [108].)

on air with a sufficiently weak lens, plasma defocusing can be made to balance nonlinear self-focusing, and one or more partially ionized "filaments" are formed. The intensity in these filaments is "clamped" in the mid-10^{13} W cm^{-2} range [116] for distances that can exceed 100 m [117], allowing high intensities to persist well outside the Rayleigh range of the focusing optic.

Numerous uses of these filaments have been proposed. The ion channel they create can reliably trigger high-voltage discharges in the laboratory [117–119] (Figure 3.1.5). They have also been shown to trigger electric discharges in thunderclouds, opening up the possibility that they could someday be used for the controlled triggering of lightning to prevent damage to structures [120]. Filaments can be used for light detection and ranging (LIDAR), and biological aerosols have been studied with two-photon fluorescence with this method [121]. Filament excitation of the atmosphere is an attractive option for long-range pollution monitoring, due to a strong backscattered signal that returns to the point of the laser beam origin. The conditions that exist in the filament also allow a comparatively background-free molecular fluorescence spectrum to be obtained in atmospheric-density molecular gases [122]. Recombination-driven lasing has also been observed inside filaments [123].

Useful nonlinear optical effects in atmospheric density gases are not limited to those produced in filaments. Self-phase modulation of an ultrashort pulse tends to redshift the leading edge of a pulse, and blueshift the trailing edge, eventually expanding its spectrum considerably [124]. This process is known as *supercontinuum generation*. While it occurs in filaments [116], it can be produced in a more stable and convenient form for laboratory use inside a hollow, noble-gas-filled fiber. The supercontinuum produced from these fibers can be compressed with chirped mirrors to a few femtoseconds [17]. The few femtosecond pulses created by this process can be carrier-envelope stabilized and in turn provide a key component for the production of even shorter attosecond pulses through HHG [125]. Third-harmonic generation through Brunel mixing has also been demonstrated in femtosecond-pulse-ionized noble gases [63].

3.1.11
Conclusion and Outlook

High-intensity femtosecond laser interaction with gases and clusters is a diverse and growing field. Research in the area spans a spectrum from the very theoretical (quantum dynamics of isolated atoms) to the very practical (airborne pollution monitoring). Tabletop high-intensity laser systems are only about a decade old, and the technology is by no means mature. Fiber-based amplifiers, in particular, are emerging as a strong competitor to traditional free-space optics-based chirped-pulse amplifiers [126–128]. The all-fiber amplifiers offer the important benefits of robustness, ease of operation, and low cost, and their development will open up the high-intensity regime to many more applied physics laboratories. On the theoretical physics side, higher intensity facilities continue to be opened and perfected. A number of laboratories now routinely conduct studies at relativistic intensities ($>10^{18}$ W cm^{-2}), where strong magnetic fields complement the strong electric fields as a method to probe matter [129–131]. Finally, the development of laser cluster physics opens up yet another route for producing extreme fields, through local field enhancement.

References

1. Schmidt, V. (1992) *Rep. Prog. Phys.*, **55**, 1483.
2. Martinson, I. and Gaupp, A. (1974) *Phys. Rep.*, **15**, 113.
3. Beyer, H.F. and Shevelko, V.P. (eds) (1999) *Atomic Physics with Heavy Ions*, Springer-Verlag, Berlin.
4. Moulton, P.F. (1986) *J. Opt. Soc. Am. B*, **3**, 125.
5. Spence, D.E., Kean, P.N., and Sibbett, W. (1991) *Opt. Lett.*, **16**, 42.
6. Keller, U. (2003) *Nature*, **424**, 831.
7. Strickland, D. and Mourou, G. (1985) *Opt. Commun.*, **56**, 219.
8. Perry, M.D. and Mourou, G. (1994) *Science*, **264**, 917.
9. DiMauro, L.F. and Agostini, P. (1995) *Adv. At. Mol. Opt. Phys.*, **35**, 79.
10. Gavrila, M. (ed.) (1992) *Atoms in Intense Laser Fields*, Academic, New York.
11. Protopapas, M., Keitel, C.H., and Knight, P.L. (1997) *Rep. Prog. Phys.*, **60**, 389.
12. Agostini, P., Fabre, F., Mainfray, G., Petite, G., and Rahman, N. (1979) *Phys. Rev. Lett.*, **42**, 1127.
13. Walker, B., Sheehy, B., DiMauro, L.F., Agostini, P., Schafer, K.J., and Kulander, K.C. (1994) *Phys. Rev. Lett.*, **73**, 1227.
14. McPherson, A., Gibson, G., Jara, H., Luk, T.S., McIntyre, I.A., Boyer, K., and Rhodes, C.K. (1987) *J. Opt. Soc. Am. B*, **4**, 595.
15. Salières, P., L'Huillier, A., Antoine, P., and Lewenstein, M. (1999) *Adv. At. Mol. Phys.*, **41**, 83.
16. Brabec, T. and Krausz, F. (2000) *Rev. Mod. Phys.*, **72**, 545.
17. Steinmeyer, G., Sutter, D.H., Gallmann, L., Matuschek, N., and Keller, U. (1999) *Science*, **286**, 1507.
18. Bahk, S.W., Rousseau, P., Planchon, T.A., Chvykov, V., Kalintchenko, G., Maksimchuk, A., Mourou, G.A., and Yanovsky, V. (2004) *Opt. Lett.*, **29**, 2837.
19. Moshammer, R., Schmitt, W., Ullrich, J., Kollmus, H., Cassimi, A., Dörner, R., Jagutzki, O., Mann, R., Olson, R.E., Prinz, H.T., Schmidt-Böcking, H., and Spielberger, L. (1997) *Phys. Rev. Lett.*, **79**, 3621.
20. Légaré, F., Lee, K.F., Litvinyuk, I.V., Dooley, P.W., Wesolowski, S.S., Bunker, P.R., Dombi, P., Krausz, F., Bandrauk, A.D., Villeneuve, D.M., and Corkum, P.B. (2005) *Phys. Rev. A*, **71**, 013415.
21. Smirnova, O., Mairesse, Y., Patchkovskii, S., Dudovich, N., Villeneuve, D., Corkum, P., and Ivanov, M. (2009) *Nature*, **460**, 972.
22. Wu, J., Zeng, H., and Guo, C. (2007) *Phys. Rev. A*, **75**, 043402.

23. Zhou, J., Peatross, J., Murnane, M.M., Kapteyn, H.C., and Christov, I.P. (1996) *Phys. Rev. Lett.*, **76**, 752.

24. Assion, A., Baumert, T., Helbing, J., Seyfried, V., and Gerber, G. (1996) *Chem. Phys. Lett.*, **259**, 488.

25. Chelkowski, S., Bandrauk, A., and Corkum, P.B. (1990) *Phys. Rev. Lett.*, **65**, 2355.

26. Kling, M.F., Siedschlag, C., Verhoef, A.J., Khan, J.I., Schultze, M., Uphues, T., Ni, Y., Uiberacker, M., Drescher, M., Krausz, F., and Vrakking, M.J.J. (2006) *Science*, **312**, 246.

27. Corkum, P.B. (1993) *Phys. Rev. Lett.*, **71**, 1994.

28. Paul, P.M., Toma, E.S., Breger, P., Mullot, G., Augé, F., Balcou, P., Muller, H.G., and Agostini, P. (2001) *Science*, **292**, 1689.

29. Zuo, T., Bandrauk, A.D., and Corkum, P.B. (1996) *Chem. Phys. Lett.*, **259**, 313.

30. Corkum, P. and Krausz, F. (2007) *Nat. Phys.*, **3**, 381.

31. Lin, C.D., Le, A.T., Chen, Z., Morishita, T., and Lucchese, R. (2010) *J. Phys. B: At. Mol. Opt. Phys.*, **43**, 122001.

32. Meckel, M., Comtois, D., Zeidler, D., Staudte, A., Pavicic, D., Bandulet, H.C., Pépin, H., Kieffer, J.C., Dörner, R., Villeneuve, D.M., and Corkum, P.B. (2008) *Science*, **320**, 1478.

33. Kulander, K.C. (1987) *Phys. Rev. A*, **35**, 445.

34. Schafer, K.J., Yang, B., DiMauro, L.F., and Kulander, K.C. (1993) *Phys. Rev. Lett.*, **70**, 1599.

35. Yang, B., Schafer, K.J., Walker, B., Kulander, K.C., Agostini, P., and DiMauro, L.F. (1993) *Phys. Rev. Lett.*, **71**, 3770.

36. Keldysh, L.V. (1964) *Zh. Eksp. Teor. Fiz.*, **47**, 1945 [(1965) *Sov. Phys. JETP* 20, 1307].

37. Ammosov, M.V., Delone, N.B., and Krainov, V.P. (1986) *Zh. Eksp. Teor. Fiz.*, **64**, 2008 [(1986) *Sov. Phys. JETP* 64, 1191].

38. Augst, S., Meyerhofer, D.D., Strickland, D., and Chin, S.L. (1991) *J. Opt. Soc. Am. B*, **8**, 858.

39. Chin, S.L., Liang, Y., Decker, J.E., Ilkov, F.A., and Ammosov, M.V. (1992) *J. Phys. B*, **25**, L249.

40. Walsh, T.D.G., Ilkov, F.A., Decker, J.E., and Chin, S.L. (1994) *J. Phys. B*, **27**, 3767.

41. Fittinghof, D.N., Bolton, P.R., Chang, B., and Kulander, K.C. (1992) *Phys. Rev. Lett.*, **69**, 2642.

42. Guo, C., Li, M., Nibarger, J.P., and Gibson, G.N. (1998) *Phys. Rev. A (Rapid Commun.)*, **58**, R4271.

43. Talebpour, A., Chien, C.Y., and Chin, S.L. (1996) *J. Phys. B*, **29**, L677.

44. Guo, C. (2000) *Phys. Rev. Lett.*, **85**, 2276.

45. Muth-Bohm, J., Becker, A., and Faisal, F.H.M. (2000) *Phys. Rev. Lett.*, **85**, 2280.

46. DeWitt, M.J. and Levis, R.J. (1998) *J. Chem. Phys.*, **108**, 7739.

47. Saenz, A. (2000) *J. Phys. B*, **33**, 4365.

48. Tong, X.M., Zhao, Z.X., and Lin, C.D. (2002) *Phys. Rev. A*, **66**, 033402.

49. Guo, C. and Gibson, G.N. (2005) *Phys. Rev. A*, **63**, 040701(R).

50. Guo, C., Li, M., and Gibson, G.N. (1999) *Phys. Rev. Lett.*, **82**, 2492.

51. Guo, C. (2005) *J. Phys. B*, **38**, L323.

52. Guo, C. (2005) *Phys. Rev. A*, **71**, 021405(R).

53. Wu, J., Zeng, H., and Guo, C. (2006) *Phys. Rev. A*, **74**, 031404(R).

54. Wu, J., Zeng, H., and Guo, C. (2007) *J. Phys. B*, **40**, 1095.

55. Codling, K., Frasinski, L.J., and Hatherly, P.A. (1989) *J. Phys. B*, **22**, L321.

56. Boyer, K., Luk, T.S., Solem, J.C., and Rhodes, C.K. (1989) *Phys. Rev. A*, **39**, 1186.

57. Bucksbaum, P.H., Zavriyev, A., Muller, H.G., and Shumacher, D.W. (1990) *Phys. Rev. Lett.*, **64**, 1883.

58. Castillejo, M., Couris, S., Koudoumas, E., and Martin, M. (1998) *Chem. Phys. Lett.*, **289**, 303.

59. Levis, R.J. and DeWit, M.J. (1999) *J. Phys. Chem. A*, **103**, 6493.

60. Wu, J., Zeng, H., Wang, J., and Guo, C. (2006) *Phys. Rev. A*, **73**, 051402(R).

61. Quaglia, L. and Cornaggia, C. (2000) *Phys. Rev. Lett.*, **84**, 4565.

62. Gibson, G., Luk, T.S., McPherson, A., Boyer, K., and Rhodes, C.K. (1989) *Phys. Rev. A*, **40**, 2378.

63. Siders, C.W., Rodriguez, G., Siders, J.L.W., Omenetto, F.G., and Taylor, A.J. (2001) *Phys. Rev. Lett.*, **87**, 263002.

64. Wood, W.M., Siders, C.W., and Downer, M.C. (1991) *Phys. Rev. Lett*, **67**, 3523.

65. Rose-Petruck, C., Schafer, K.J., Wilson, K.R., and Barty, C.P.J. (1997) *Phys. Rev. A*, **55**, 1182.

66. McPherson, A., Luk, T.S., Thompson, B.D., Borisov, A.B., Shiryaev, O.B., Chen, X., Boyer, K., and Rhodes, C.K. (1994) *Phys. Rev. Lett.*, **72**, 1810.

67. Fortov, V.E., Iakubov, I.T., and Khrapak, A.G. (2006) *Physics of Strongly Coupled Plasma*, Oxford University Press, Oxford.

68. Bellan, P.M. (2006) *Fundamentals of Plasma Physics*, Cambridge University Press, New York.

69. Morfill, G.E. (1999) *Phys. Rev. Lett.*, **83**, 1598.

70. Chu, J.H. and Lin, I. (1994) *Phys. Rev. Lett.*, **72**, 4009.

71. Hasted, J.B. (1972) *Physics of Atomic Collisions*, 2nd edn, American Elsevier, New York.

72. Glover, T.E., Crane, J.K., Perry, M.D., Lee, R.W., and Falcone, R.W. (1995) *Phys. Rev. Lett.*, **75**, 445.

73. Castleman, A.W. Jr. and Keesee, R.G. (1988) *Science*, **241**, 36.

74. Haberland, H., von Issendorff, B., Yufeng, J., Kolar, T., and Thanner, G. (1993) *Z. Phys. D.*, **26**, 8.

75. Mandich, M.L. (2006) in *Springer Handbook of Atomic, Molecular, and Optical Physics*, vol. **1** (ed. G.W.F. Drake), Springer, New York, p. 589.

76. Haberland, H. (ed.) (1994) *Clusters of Atoms and Molecules*, Springer, Berlin.

77. de Jongh, L.J. *et al.* (1992) in *Cluster Models for Surface and Bulk Phenomena* (eds G. Pacchioni, P.S. Bragus, and F. Parmigiani), Plenum: New York, pp. 151–168.

78. Buo, B.C., Kerns, K.P., and Castleman, A.W. (1992) *Science*, **255**, 1411.

79. Cleveland, C.L., Luedtke, W.D., and Landman, U. (1998) *Phys. Rev. Lett.*, **81**, 2036.

80. Smalley, R.E. (1992) *Acc. Chem. Res.*, **25**, 98.

81. (a) Rohlfing, E.A., Cox, D.M., and Kaldor, A. (1984) *J. Chem. Phys.*, **81**, 3322; (b) Kroto, H.W., Heath, J.R., O'Brien, S.C., Curl, R.F., and Smalley, R.E. (1985) *Nature*, **318**, 162.

82. Powers, D.E., Hansen, S.G., Geusic, M.E., Puiu, A.C., Hopkins, J.B., Dietz, T.G., Duncan, M.A., Langridge-Smith, P.R.R., and Smalley, R.E. (1982) *J. Phys. Chem.*, **86**, 2556.

83. Heath, J.R., Liu, Y., O'Brien, S.C., Zhang, Q.L., Curl, R.F., Tittel, F.K., and Smalley, R.E. (1985) *J. Chem. Phys.*, **83**, 5520.

84. O'Brien, S.C., Liu, Y., Zhang, Q., Heath, J.R., Tittel, F.K., Curl, R.F., and Smalley, R.E. (1986) *J. Chem. Phys.*, **84**, 4074.

85. Baumert, T. and Gerber, G. (1994) *Adv. At. Mol. Opt. Phys.*, **35**, 163–208.

86. McPherson, A., Thompson, B.D., Borisov, A.B., Boyer, K., and Rhodes, C.K. (1994) *Nature*, **370**, 631.

87. Snyder, E.M., Buzza, S.A., and Castleman, A.W. Jr. (1996) *Phys. Rev. Lett.*, **77**, 3347.

88. Ditmire, T., Donnelly, T., Falcone, R.W., and Perry, M.D. (1995) *Phys. Rev. Lett.*, **75**, 3122.

89. Kmetec, J.D., Gordon, C.L., Macklin, J.J., Lemoff, B.E., Brown, G.S., and Harris, S.E. (1992) *Phys. Rev. Lett.*, **68**, 1527.

90. Ditmire, T., Tisch, J.W.G., Springate, E., Mason, M.B., Hay, N., Smith, R.A., Marangos, J., and Hutchinson, M.H.R. (1997) *Nature*, **386**, 54.

91. Carranza, J.E. and Hahn, D.W. (2002) *Anal. Chem.*, **74**, 5450.

92. Ottesen, D.K., Wang, J.C.F., and Radziemski, L.J. (1989) *Appl. Spectrosc.*, **43**, 967.

93. Hahn, D.W., Flower, W.L., and Hencken, K.R. (1997) *Appl. Spectrosc.*, **51**, 1836.

94. Lamour, E., Prigent, C., Rozet, J.P., and Vernhet, D. (2005) *Nucl. Instrum. Methods B*, **235**, 408.

95. Köller, L., Schumacher, M., Köhn, J., Teuber, S., Tiggesbäumker, J., and Meiwes-Broer, K.H. (1999) *Phys. Rev. Lett.*, **82**, 3786.

96. Saalmann, U., Siedschlag, C., and Rost, J.M. (2006) *J. Phys. B: At. Mol. Opt. Phys.*, **39**, R39.

97. Seideman, T., Ivanov, M.Y., and Corkum, P.B. (1995) *Phys. Rev. Lett.*, **75**, 2819.

98. Zuo, T. and Bandrauk, A.D. (1995) *Phys. Rev. A*, **52**, R2511.

99. Döppner, T., Fennel, T., Diederich, T., Tiggesbäumker, J., and Meiwes-Broer, K.H. (2005) *Phys. Rev. Lett.*, **94**, 013401.

100. Ditmire, T., Donnelly, T., Rubenchik, A.M., Falcone, R.W., and Perry, M.D. (1996) *Phys. Rev. A*, **53**, 3379.

101. Radziemski, L.J. and Cremers, D.A. (eds) (1989) *Laser-Induced Plasma and Applications*, Marcel Dekker, New York.

102. Singh, J.P. and Thakur, S.N. (eds) (2007) *Laser-Induced Breakdown Spectroscopy*, Elsevier, Amsterdam.

103. Neuhauser, R.E., Panne, U., Niessner, R., and Wilbring, P. (1999) *Fresenius J. Anal. Chem.*, **364**, 720.

104. Buckley, S.G., Johnsen, H.A., Hencken, K.R., and Hahn, D.W. (2000) *Waste Manage.*, **20**, 455.

105. Morel, S., Leone, N., Adam, P., and Amouroux, J. (2003) *Appl. Opt.*, **42**, 6184.

106. Boyain-Goitia, A., Beddows, D.C.S., Griffiths, B.C., and Telle, H.H. (2003) *Appl. Opt.*, **42**, 6119.

107. De Lucia, F.C., Harmon, R.S., McNesby, K.L., Winkel, R.J., and Miziolek, A.W. (2003) *Appl. Opt.*, **42**, 6148.

108. Blevins, L.G., Shaddix, C.R., Sickafoose, S.M., and Walsh, P.M. (2003) *Appl. Opt.*, **42**, 6107.

109. Hohreiter, V. and Hahn, D.W. (2005) *Anal. Chem.*, **77**, 1118.

110. Heins, A.M. and Chunlei, G. (2012) *Opt. Lett.*, **37**, 599.

111. Rohwetter, P., Kasparian, J., Stelmaszczyk, K., Hao, Z., Henin, S., Lascoux, N., Nakaema, W., Petit, Y., Queisser, M., Salamé, R., Salmon, E., Wöste, L., and Wolf, J.P. (2010) *Nat. Photonics*, doi: 10.1038/NPHOTON.2010.115

112. Wilson, C.T.R. (1911) *Proc. R. Soc. London, Ser. A*, **85**, 285.

113. Braun, A., Korn, G., Liu, X., Du, D., Squier, J., and Mourou, G. (1995) *Opt. Lett.*, **20**, 73.

114. Lange, H.R., Grillon, G., Ripoche, J.F., Franco, M.A., Lamouroux, B., Prade, B.S., and Mysyrowicz, A. (1998) *Opt. Lett.*, **23**, 120.

115. Brodeur, A., Chien, C.Y., Illkov, F.A., Chin, S.L., Koserava, O.G., and Kandidov, V.P. (1997) *Opt. Lett.*, **22**, 304.

116. Couairon, A. and Mysyrowicz, A. (2007) *Phys. Rep.*, **441**, 47.

117. Kasparian, J., Rodriguez, M., Méjean, G., Yu, J., Salmon, E., Wille, H., Bourayou, R., Frey, S., André, Y.B., Mysyrowicz, A., Sauerbrey, R., Wolf, J.P., and Wöste, L. (2003) *Science*, **301**, 61.

118. Zhao, X.M., Diels, J.C., Wang, C.Y., and Elizondo, J.M. (1995) *IEEE J. Quantum Electron.*, **31**, 599.

119. Ackermann, R., Méchain, G., Méjean, G., Bourayou, R., Rodriguez, M., Stelmaszczyk, K., Kasparian, J., Yu, J., Salmon, E., Tzortzakis, S., André, Y.B., Bourrillon, J.F., Tamin, L., Cascelli, J.P., Campo, C., Davoise, C., Mysyrowicz, A., Sauerbrey, R., Wöste, L., and Wolf, J.P. (2005) *Appl. Phys. B*, **82**, 561.

120. Kasparian, J., Ackermann, R., André, Y.B., Méchain, G., Guillaume, P., Rohwetter, P., Salmon, E., Stelmaszczyk, K., Yu, J., Mysyrowicz, A., Sauerbrey, R., Wöste, L., and Wolf, J.P. (2008) *Opt. Express*, **16**, 5757.

121. Mejean, G., Kasparian, J., Yu, J., Frey, S., Salmon, E., and Wolf, J.P. (2004) *Appl. Phys. B*, **78**, 535.

122. Talebpour, A., Abdel-Fattah, M., Bandrauk, A.D., and Chin, S.L. (2001) *Laser Phys.*, **11**, 68.

123. Luo, Q., Liu, W., and Chin, S.L. (2003) *Appl. Phys. B*, **76**, 337.

124. Boyd, R.W. (2003) *Nonlinear Optics*, 2nd edn, Academic Press, Amsterdam, p. 357.

125. Sansone, G., Benedetti, E., Calegari, F., Vozzi, C., Avaldi, L., Flammini, R., Poletto, L., Villoresi, P., Altucci, C., Velotta, R., Stagira, S., De Silvestri, S., and Nisoli, M. (2006) *Science*, **314**, 443.

126. Limpert, J., Schreiber, T., Clausnitzer, T., Zöllner, K., Fuchs, H., Kley, E., Zellmer, H., and Tünnermann, A. (2002) *Opt. Express*, **10**, 628.

127. Röser, F., Eidam, T., Rothhardt, J., Schmidt, O., Schimpf, D.N., Limpert, J., and Tünnermann, A. (2007) *Opt. Lett.*, **32**, 3495.

128. Zaouter, Y., Boullet, J., Mottay, E., and Cormier, E. (2008) *Opt. Lett.*, **33**, 1527.

129. Mangles, S.P.D., Murphy, C.D., Najmudin, Z., Thomas, A.G.R., Collier, J.L., Dangor, A.E., Divali, E.J., Foster, P.S., Gallacher, J.G., Hooker, C.J., Jaroszynski, D.A., Langley, A.J., Mori, W.B., Norreys, P.A., Tsung, F.S., Viskup, R., Walton, B.R., and Krushelnick, K. (2004) *Nature*, **431**, 535.

130. Rousse, A., Phuoc, K.T., Shah, R., Pukhov, A., Lefebvre, E., Malka, V., Kiselev, S., Burgy, F., Rousseau, J.P., Umstadter, D., and Hulin, D. (2004) *Phys. Rev. Lett.*, **93**, 135005.

131. Gopal, A., Tatarakis, M., Beg, F.N., Clark, E.L., Dangor, A.E., Evans, R.G., Norreys, P.A., Wei, M.S., Zepf, M., and Krushelnick, K. (2008) *Phys. Plasmas*, **15**, 122701.

3.2
Laser-Matter Interaction: Plasma and Nanomaterials Processing
Subhash Chandra Singh

3.2.1
Introduction

Pulsed laser ablation (PLA) of solids in a gaseous environment was started just after the invention of the ruby laser in 1960 for the processing and characterization of materials. The pioneering work of R.M. White [1–3] on the generation of mechanical waves on the surface of elastic solids and fluids on absorption of pulsed ruby laser beam initiated research in the area of laser plasma synthesis and materials processing. Soon after that, the use of ruby laser for the synthesis of plasma [4, 5]; photoelectric effect from laser-irradiated metal target [6–10]; laser-induced emission of electrons [6–10], ions, and neutral atoms from a solid target [9]; and laser-induced thermionic emissions [8] was reported [4, 5]. PLA of a solid target in the gaseous phase is now widely used for the processing of plasma as pulsed sources of X-rays, ions, electrons [9, 10], and light [11, 12], in the nuclear fusion and fission [13], cutting, drilling, welding, and annealing [14] of materials, as well as selective fabrication of nanostructures on the surfaces [15], and deposition of thin films and nanostructures on the substrates [16]. In the context of nanomaterials, synthesis by PLA in the gaseous phase is widely used for the pulsed laser deposition (PLD) of thin films in the conventional chamber with vacuum or under background gas with dilute pressure, in nanoparticle-assisted pulsed laser deposition (NAPLD), and in temperature-assisted pulsed laser deposition (TAPLD) for the synthesis of nanoparticles (NPs), nanowires (NWs), and nanorods (NRs), and laser vaporization followed by controlled condensation (LVCC) for the fabrication of various nanoarchitectures. In addition to these, PLA in the gaseous phase is used for the fabrication of novel nanoarchitectures and nanopatterns on the surface of solid metallic, semiconductor, and polymer surfaces to modify their optical, electronic, and surface properties [15]. The processes of laser–matter interaction, plasma processing, plasma expansion, and condensation of plasma species into NPs are different for long (millisecond and nanosecond) and ultrashort (picoseconds and femtosecond) laser pulses. Nanosecond laser systems are widely used in laser-based materials processing since the invention of the ruby laser, but the employment of picoseconds and femtosecond laser systems for the PLA processing of nanoscaled materials in vacuum, gas phase, and liquid media has been surprisingly increasing in recent years.

3.2.2
Influences of Laser Irradiance on Melting and Vaporization Processes

When laser irradiance is below the ablation threshold of the target materials, no melting and vaporization occurs, but laser-induced excitations of free electrons are

Nanomaterials: Processing and Characterization with Lasers, First Edition.
Edited by Subhash Chandra Singh, Haibo Zeng, Chunlei Guo, and Weiping Cai.
© 2012 Wiley-VCH Verlag GmbH & Co. KGaA. Published 2012 by Wiley-VCH Verlag GmbH & Co. KGaA.

significant. The high density of photoexcited electrons and their lifetimes can be diagnosed using time-resolved optical absorption of the transparent material in pump-probe geometry under a femtosecond timescale. Light-induced hot electrons are produced, which are neither in equilibrium with the sample materials nor with the electron gas itself on the surface. Electronic emission can take place through the combination of thermionic and photoelectric processes from the surface of the sample if electrons attain enough energy to conquer the surface energy barrier [17]. Excited electrons having energy less than the surface potential barrier get relaxed by several relaxation processes including collision with phonons and plasmons, defects, and holes [18]. When laser irradiance rises close to the ablation threshold of the target materials, a large number of photoexcited free and bound electrons (or holes for nonmetallic samples) interact with the other steady electrons (or holes) and with the lattice through electron–electron and electron–phonon interaction processes, respectively. Interaction of excited phonons with the other steady lattice atoms through phonon–phonon interactions leads to phonon or lattice waves in the sample. When the lattice temperature or energy of phonon waves increases the melting point of the materials, an irreversible surface melting will occur at the irradiated spot (liquid front in Figure 3.2.1a). If additional laser energy is available, the liquid front at the liquid–air interface gets vaporized in the form of neutral atoms or molecules of the sample (vapor front in Figure 3.2.1a). Laser–target interaction and ejection of material from the target surface generates two shock waves that travel normal to the target surface in the air and inside the target (Figure 3.2.1a). The incoming laser irradiance should be higher than the threshold irradiance, $(I_{th} = T_b/\alpha\sqrt{K_T\rho c_p/\tau_p})$ (where T_b is the boiling temperature, α is the absorption coefficient of laser light by the target, K_T is the thermal conductivity at temperature T, ρ is the density, c_p is the specific heat at constant pressure, and τ_p is the pulse width of the laser) in order to evaporate target atoms/molecules in the time shorter than pulse duration. Since the pressure of quickly vaporizing materials is much higher than the ambient atmospheric (if vacuum-free expansion; with the increase of background-gas-confined expansion) pressure, the vaporized material leaves the surface of the target. If the laser irradiance is higher than the threshold irradiance $(I > I_{th})$, rigorous evaporation of atoms/molecules from the target surface will start. For the incident laser irradiance of I W cm^{-2}, $t_1 = K\rho c_p T_b^2/I^2\alpha^2$, nanoseconds are the minimum time required to start the evaporation process. Ablation threshold is the specific characteristic of the materials that usually depends on the melting and boiling points and the absorption coefficient at a particular wavelength. Laser ablation of the multielemental target at a particular irradiance may cause higher irradiance than the ablation threshold for some of the elements and lower than the ablation threshold for others, which will cause fractional vaporization of the target. When laser irradiance is raised higher than the ablation threshold of all the elements of the target, elements with a lower ablation threshold will expand with faster kinetic energy as compared to those of the higher ablation threshold. For the deposition of multielemental materials, thin films with the stoichiometry of the mother target, elemental composition of the laser-produced plasma should be equal to the elemental composition of the mother target. In addition to this, the

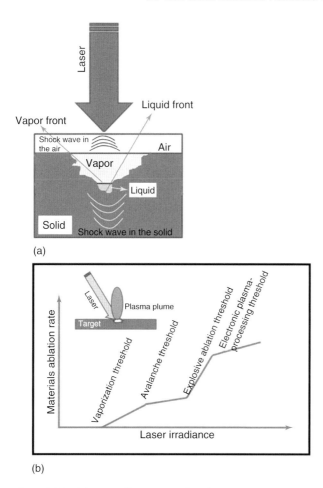

Figure 3.2.1 Schematic illustrations of (a) laser–target interaction in the gas phase and (b) variation in the material removal rate with laser irradiance and various ablation thresholds.

substrate should be placed at a distance parallel to the target, where all the plasma species attain the same kinetic energy in order to achieve smooth morphology; otherwise, deposits from one element (of the lower kinetic energy) will be ablated by the other moving with higher kinetic energies.

The vapor produced at the target-surface-containing neutral atoms is transparent to the laser radiation; therefore, the evaporation process occurs continuously. Multiphoton absorption of laser radiation by neutral atoms results in seed electrons. These electrons oscillate in the electric field of laser radiation, attain a high kinetic energy, and collide with the heavy neutral particles. At enough higher laser irradiance, I_{avth}, (*avalanche threshold*), however, the time when the electron energy becomes equal to the ionization potential of a neutral atom ($KE_e = I$), an energetic electron ionizes an atom through collision to produce one more electron. Now these two electrons gain energy from the incident laser beam until they ionize two other atoms. This process is called *avalanche ionization* or *optical vapor breakdown*, and the minimum required irradiance to start the process is called an *avalanche threshold* [19]. This process converts laser-produced neutral transparent vapor to dense hot plasma. For $I > I_{avth}$, optical breakdown or avalanche time of transparent vapor is given by $t_2 = -\alpha_1/I \ln\left(1 - \alpha_2/\lambda^2 I\right)$, where α_1 and α_2 are numerical factors [19]. This expression shows that the optical breakdown or plasma-processing time is dependent on the irradiance and wavelength of the laser used for ablation. Increase in the electron density of the plasma through the mechanism of avalanche ionization increases the absorption coefficient of the plasma at laser wavelength and hence shields the target surface against ablation. Target shielding reduces the ablation of target materials, and hence the density of the plasma in the laser path declines. This process allows the laser to reach the target for plasma processing. Processing of the plasma by laser and target shielding by the plasma behave as a valve. Under certain laser parameters, equilibrium between the target shielding and plasma processing is maintained, which allows a channel for the laser through the plasma for the continuous ablation of the target surface called *laser plasma channeling* [19, 20]. Figure 3.2.1b presents a schematic illustration of the dependence of rate of removal of material on laser irradiation along with laser ablation thresholds for the start of different ablation mechanisms.

With the further increase in the laser irradiance ($>20\,\mathrm{GW\,cm^{-2}}$ for 3 ns pulse width and 266 nm wavelength PLA of silicon) above a particular irradiance value for given materials, the material reaches a thermodynamic critical point, where it experiences fast transition from superheated liquid to a mixture of vapor and liquid droplets. Explosive ejection of mixture of vapor and liquid droplets occurs at the surface of material [20]. It might be possible that fractionation gets reduced under these conditions, because material is ablated from the surface in the form of micro-sized liquid droplets or particles. When the sample temperature reaches close to the critical temperature, which is several times larger than the vaporization temperature, more mass from the sample surface is removed with the consequence of same absorbed laser energy. The *explosive ablation threshold* is the laser irradiance at which a simple ablation mechanism changes to explosive ablation or phase explosion. Figure 3.2.2 illustrates time-resolved shadowgraphy of laser-produced silicon plasma (266 nm, 3 ns pulse width, and $8.7 \times 10^{10}\,\mathrm{W\,cm^{-2}}$), which exhibits the violent ejection of particles on a microsecond timescale [20, 21]. Early time shadowgraphs (2–40 ns) showed the propagation of shock waves from solid surface to the ambient air due to the pressure difference between the dense laser-produced plasma and ambient environment [21]. Bulk mass removal from the silicon surface was observed after 400 ns of the laser pulse, reached a maximum at 1.5 μs, and lasted for about 30 μs. The process of droplet ejection can be understood in terms of explosive boiling, established by Martynyuk [22, 23], which requires rapid heating of the target surface. PLA of Si using 3 ns pulse width, 266 nm wavelength, and $8.7 \times 10^{10}\,\mathrm{W\,cm^{-2}}$ irradiance can exceed the heating rate $10^{12}\,\mathrm{{}^\circ C\,s^{-1}}$ with thermal diffusion on the order of 10^{-11} s, which rapidly forms a melt layer that propagates into the bulk silicon during the laser pulse. At the thermodynamic critical point, density fluctuations can create vapor bubbles into the

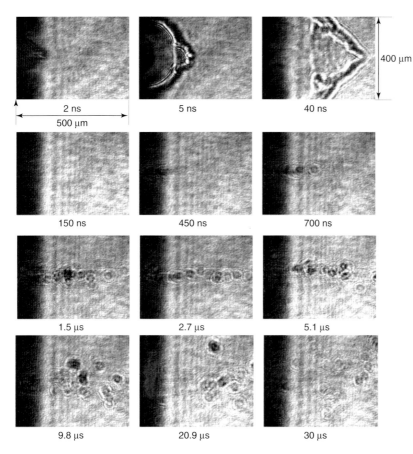

Figure 3.2.2 Shadowgraphs of temporal characteristic of mass ejection from the sample surface at different times after irradiation with 266 nm wavelength, 3 ns pulse width, and 8.7×10^{10} W cm^{-2} laser irradiation. (Source: Reprinted with permission from Russo *et al.* [21] copyright @ Springer.)

superheated liquid silicon [23]. Vapor bubbles with sizes greater than the critical radius will grow, while those shorter to this value will collapse. Vapor bubbles of the sizes equal or greater than the critical value undergo rapid transition into a mixture of vapor and liquid droplets. Rapid expansion of high-pressure bubbles in the liquid results in an explosive ejection of molten droplets from the target surface [22, 23].

Laser ablation of materials in the air or background gases causes synthesis of early stage of the plasma (on the order of picoseconds), and hence provides an ablation threshold on the order of 10^{12} W cm^{-2}. Pump-probe interferometry investigation [22] revealed $\sim 10^{20}$ cm^{-3} electron number density for this early stage plasma, which shows that plasma evolutes from the solid surface and not from the laser-induced gas ablations. Laser-induced thermionic or photoelectrons, generated at the early stage of laser–matter interaction, absorb laser energy

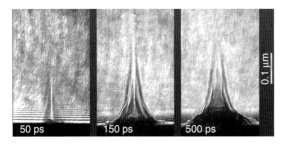

Figure 3.2.3 Temporal evolution of single-pulse picoseconds (35 ps, 1064 nm, $\sim$4 $\times$ 10^{12} W cm^{-2}) laser-produced electronic plasma of copper [20]. (Source: Reprinted with permission from Russo *et al.* [20] copyright @ American Chemical Society.)

through the mechanism of inverse bremsstrahlung to attain high kinetic energy. These energetic electrons ionize the gaseous molecules to produce *electronic plasma*, which expands promptly during the laser pulse. Since electronic plasma is produced in the picoseconds time duration (50–500 ps), it may absorb the rising part of the nanosecond laser pulse unlike the avalanche-ionized plasma that absorbs the tailing part of laser pulse. Three shadowgraphs of the electronic plasma expansion recorded at different times after the interaction of picoseconds laser pulse with the copper target (35 ps, 1064 nm, $\sim$4 $\times$ 10^{12} W cm^{-2}) are illustrated in Figure 3.2.3 [20]. The cone-shaped spatial distribution of this early stage electronic plasma is substantially different from that of the hemisphere-shaped nanosecond plasma. Dependence of plasma temperature and electron density on laser irradiance in the irradiance range of 1–60 GW cm^{-2} are presented in Figure 3.2.4a. Electron and plasma temperature as well as number density changes significantly at about 20 GW cm^{-2}, which was considered as the threshold for explosive ablation of silicon. It was reported that for laser irradiance below the explosive ablation threshold, dependence of electron number density and plasma temperature on the laser irradiance (ϕ) follows power law $n_e \propto \phi^{0.24}$ and $T_e \propto \phi^{0.54}$, respectively [21]. However, above the ablation threshold, electron number density increases much faster, $n_e \propto \phi^{1.54}$, while plasma and electron temperature increases much slower, $T_e \propto \phi^{0.25}$ with the increase in laser irradiance [21]. Dependence of crater depth formed at the target surface on the laser irradiance is shown in Figure 3.2.4b. Crater profiles are also shown for the laser irradiances of 13 and 21 GW cm^{-2}. Below the explosive ablation threshold, the crater is shallow and its depth, D, varies as $D \propto \phi^{0.47}$, while above the explosive ablation threshold, the crater is deeper and its depth follows the $D \propto \phi^{0.61}$ relationship with laser irradiance [21].

3.2.3
Influence of Laser Pulse Width and Pulse Shape

In the above section, it has been mentioned that laser irradiance = pulse energy/(focal spot area $\times$ pulse width), should be higher than the ablation threshold of the target material in order to start melting and vaporization. For a given focal spot size on the surface of the material having a particular ablation

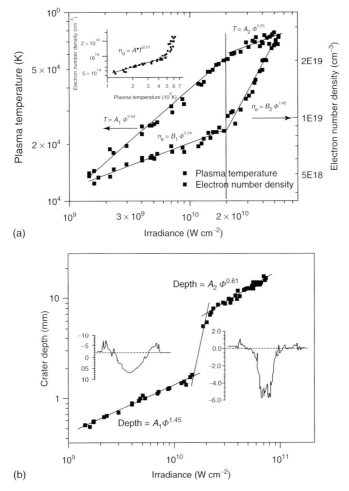

(a)

(b)

Figure 3.2.4 (a) Plasma temperature (T) and electron number density (n_e) versus laser irradiance (delay time = 30 ns). Below and above a threshold irradiance of explosive ablation ($\approx$20 GW cm^{-2}). The inset shows the relation between T and n_e on a log–log scale. (b) Crater depth versus laser irradiance below and above ablation threshold of explosive ablation. (Source: Reprinted with permission from Russo *et al.* [21] copyright @ springer.)

threshold, lasers of longer pulse duration require higher pulse energy and vice versa. Therefore, continuous wave (CW) lasers require much higher energy for the melting/evaporation of materials as compared to pulsed lasers. In the same way, femtosecond laser systems need much smaller pulse energy as compared to millisecond laser systems. Effect of CW and longer and shorter pulse width laser ablation on the plasma characteristics and hence properties of processed materials are qualitatively different because of the different mechanism associated with the laser–matter interaction, plasma processing, plasma heating, and removal of material from the irradiated area. In the case of CW laser irradiation, lateral temperature of the material surface beyond the laser spot size increases with the increase in irradiation time, which causes melting and ablation of material surface out of the irradiation region. In contrast to this, shorter pulse width picoseconds and femtosecond laser systems cause less damage threshold beyond the focal spot region as compared to the longer pulse width laser pulses.

Mechanisms related to the hydrodynamic phenomenon, thermocapillarity, back-pressure, and so on, play major roles in material removal from the surface irradiated by millisecond pulse width lasers. Temporal evolution of the plasma characteristic and melting front velocity for a given laser pulse width also depend on the shape of the laser pulse. Mazhukin *et al.* [24] simulated the effects of pulse width, pulse shape, and materials properties (thermal conductivity and heat of fusion) on the temporal evolution of surface temperature, evaporation layer thickness, melt pool thickness, and evaporation and melting front velocities for the copper and titanium targets. Simulation results showed that the dominance of one process over the other strongly depends on the shape of the laser pulse. If the intensity of the pulse is accelerating with the time (rising triangle) or a stationary process is achieved (square or rectangular pulse), the mechanism of evaporation dominates over the melting, and the relationship between the molten pool Δ_1 and the evaporated materials Δ_{ev} satisfies the condition $\Delta_1/\Delta_{ev} \ll 1$. In contrast to this, pulses with a falling temporal profile and other profiles that have no constant component (Gaussian pulse) exhibit melting pool dominance with $\Delta_1/\Delta_{ev} \gg 1$. Velocity of vaporization front and surface temperature are almost independent of the pulse duration for the pulses longer than 10^{-6} s, while they decrease sharply with the decrease in laser pulse duration below 10^{-6} s. It was also shown that the thermal phenomenon initiated by short energy pulses ($\tau \leq 10^{-6}$ s) leads to the ablation of a larger quantity of materials for metals having low heat conductivity (for example, titanium). The quantity of materials evaporated by longer pulses ($\tau \approx 10^{-4}-10^{-3}$ s) depends on the value of the specific heat of vaporization of materials instead of thermal conductivity.

Variation in pulse width in the region of nanoseconds to femtoseconds results in more changes in the ablation mechanisms as compared to the microsecond to millisecond pulse durations. For longer pulses on the order of 10 ns, there is a small heating of the solid phase and shock wave generation, while most of the laser energy is spent in heating and vaporization of the liquid layer [24]. Therefore, the evaporation process continues for a longer time after the end of the irradiation time of longer pulse lasers as compared to the shorter ones (Figure 3.2.5a). In contrast to this, maximum parts of the ultrafast laser pulses are spent for the heating of the solid materials at the solid–liquid interface and for the generation of shock waves in the solid phase and gas medium. Therefore, the velocity of the solid–liquid interface (v_{ls}) and liquid–vapor interface (v_{lv}) are higher for shorter pulse laser ablation. At a particular pulse width in the nanosecond region (1 ns), more laser energy is delivered at the liquid–vapor interface resulting in v_{lv} larger than the v_{ls} in contrast to the ablation in the picosecond and femtosecond region

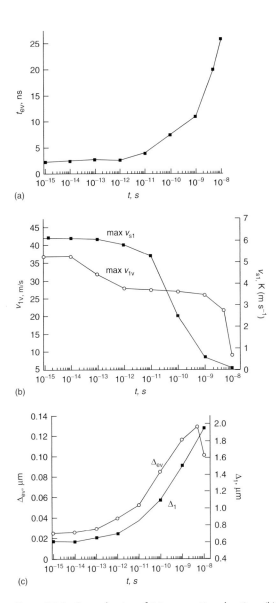

Figure 3.2.5 Dependencies of (a) evaporation duration, (b) maximum velocities of the solid–liquid interface (v_{sl}), and liquid–vapor interface (v_{lv}), and (c) thicknesses of liquid (Δ_l) and vapor (Δ_{ev}) layers, on the pulse width of the laser. (Source: Reprinted with permission from Mazhukin *et al.* [24] copyright @ Pleiades Publishing, Ltd.)

(Figure 3.2.5b). Thicknesses of the vapor and liquid layers are higher for longer pulse width laser ablation (Figure 3.2.5c).

3.2.4
Influences of Laser Wavelength on Ablation Threshold and Plasma Parameters

The wavelength of the laser beam plays a significant role in the processing of plasma, materials cutting, drilling, and welding as well as nanomaterials synthesis. Rate of ablation depends on the absorption coefficient of the laser line by the target material. Different target materials have different absorption coefficients for a particular wavelength of the light, and the absorbance coefficient of particular materials changes from one wavelength to the other. There are few reports on the investigation of the wavelength dependence ablation threshold. For InP target with a band gap energy of $1.34\,eV$ (≈ 925 nm), ablation threshold increases linearly with the increase in ablation wavelength but have different slopes above and below the band gap energy. Rate of increase in the ablation threshold with wavelength ($dI_{th}/d\lambda$) below the band gap energy is higher as compared to ($dI_{th}/d\lambda$) above the band gap energy [25]. Laser light is absorbed by its interaction with free electrons present in the plasma, which is named as inverse bremsstrahlung absorption α_{ib} and is approximated by $\alpha_{ib}(cm^{-1}) = 1.37 \times 10^{-35}\lambda^3 n_e^2 T_e^{1/2}$. Absorption of the laser light by the plasma increases its temperature. Since inverse bremsstrahlung absorption is proportional to the third power of the wavelength, the plasma produced with larger IR lasers have much higher temperatures as compared to those produced with UV lasers. In contrast to this, shorter wavelength photons have more capability to ionize neutral atoms to produce electrons, and therefore plasmas produced with shorter wavelength lasers have higher electron density. In addition, it has been observed that shorter wavelength lasers produce nanomaterials or thin films having close stoichiometry of the mother target, smoother surfaces of the films, comparatively smaller size of NPs and clusters during PLD, and smoother edges and holes in the cutting and drilling process. Absorption of the laser light by the electrons present in the plasma through the electron–ion inverse bremsstrahlung process ($\alpha_{IB,e-I}$) varies with the third order of the wavelength. Therefore, plasma shielding of the target is higher for longer wavelength lasers. Absorption of the trailing part of the higher wavelength laser pulse by the plasma increases the temperature of electrons and ions as well as their kinetic energy, which yields rough surfaces of the deposited films. Figure 3.2.6 presents the dependencies of plasma temperature, electron density, and inverse bremsstrahlung absorption on the laser irradiance for 1064, 532, and 355 nm wavelengths of the Nd:YAG laser.

3.2.5
Influences of Background Gas Pressure on the Plasma Characteristic and Morphology of Produced Materials

Laser ablation of target materials under vacuum conditions results in supersonic expansion of the plasma plume. Size of the plasma plume at a particular delay time

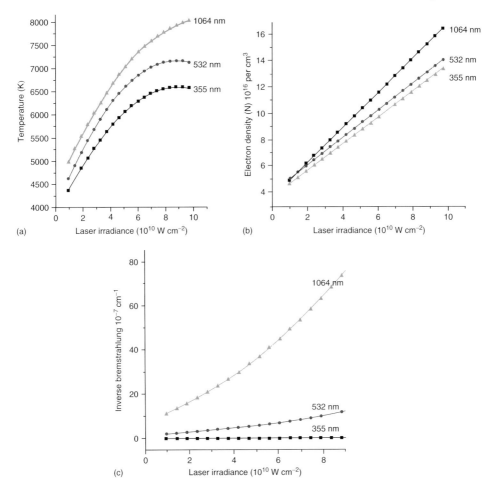

Figure 3.2.6 Dependencies of (a) plasma temperature (b) electron density, and (c) inverse bremsstrahlung absorption on laser irradiance for 1064, 532, and 355 nm wavelengths of Nd:YAG laser.

is longer, but lifetime is shorter for vacuum-produced plasma as compared to those produced in the gaseous ambience. Position of plasma plume front produced in the vacuum varies linearly with time, that is, moving with the constant velocity because there is no decelerating force against plasma expansion. Increase in the background gas pressure results in the confinement of the plasma plume, increase in collisions of plasma species with gas molecules, and in turn deceleration of plasma expansion velocity. Background gas confines electrons and ions of the plasma plume in the smaller volume and increases the lifetime of the excited species, which results in more intense and longer time emission from the plasma plume, and hence the plume is sustained for a longer time at higher pressures. Under a vacuum or a diluted gas environment, there is almost negligible deceleration, and therefore plasma expands normal to the target surface and achieves a conical or an ellipsoidal shape. While under higher background gas pressure, the normal expanding plume experiences much higher deceleration as compared to the lateral expanding plume; therefore, it achieves a spherical shape. There are fast moving components of highly ionized ions on the plasma front and slow moving components of neutral species on the base of the plume closer to the target. The fast and slow moving components get separated earlier in vacuum as compared to the higher background gas pressure where they move together for a longer time. Under vacuum and dilute gas pressures, the plasma–gas interface is smooth and continuous up to the lifetime of the plasma, while at higher pressures distortions or perturbations at the interface occurs due to the Rayleigh–Taylor instability.

Position of the plume front expanding under vacuum follow the linear position (R)–time (t) relation $R(t) = at$; where a is the constant (Figure 3.2.7A), whereas the plume expanding under the diluted gas atmosphere follows the blast wave model $R(t) = at^n$, where a and n are constants depending on the nature and pressure of the gas (Figure 3.2.7B). At high background gas pressure, confinement of the plasma is much stronger, and deceleration (dv/dt) on the expansion of the plasma plume becomes proportional to the expansion velocity, that is, $dv/dt \, \alpha - \beta v$; where v is the plume expansion velocity and β is the stopping coefficient. The solution of this equation is $R(t) = R(0) \left[1 - e^{-\beta t}\right]$, where $R(0)$ is distance at which the plume comes to rest. This equation is known as *drag model* for the plume expansion under high background gas pressure. Figure 3.2.7 shows $R - t$ curves for plume expansion under vacuum, diluted gas (0.1 and 1 torr), and high-pressure (10 and 70 Torr) background gas atmospheres. Kerdja *et al.* [26] and Geohegan [27] have reported a transition region where the initial expansion following the drag model develops into a stable shock structure at low pressures. This occurs due to the coalescence of the slow component with the fast component of the leading edge of the plasma at later times.

Pressure inside the plasma plume is much higher as compared to the background at the initial time, which decreases as plasma expands with the time. At a certain moment when the pressure inside the plasma becomes equal to the background gas pressure, expansion of plume ceases. The distance from the target surface at which the expansion of the plasma plume stops is known as *plume length* (L) and follows the equation $L = A \left[(\gamma - 1) \varepsilon E\right]^{1/3\gamma} P^{-1/3\gamma} V_i^{(\gamma-1)/3\gamma}$ [29], where γ is the specific heat ratio of vapor, ε the fraction of energy absorbed, $E = (1 - R_l) E_0$ is the laser energy absorbed; where E_0 is the incident laser energy, R_l the reflectivity of the target at laser wavelength, P the ambient gas pressure, and $V_i = v_0 \tau (\pi r^2)$ is the initial volume of the plasma; where v_0 is the initial expansion velocity of the plasma, τ the pulse width of the laser, and r is the radius of the laser spot size on the target. A is the geometrical factor that depends on the shape of the plasma plume.

As far as electron density and electron temperature of the plasma are concerned, electron temperature increases with the increase in background gas pressure, reaches a maximum value and decreases after that, while electron density decreases with the increase in the background gas pressure. Molecular and atomic weight of

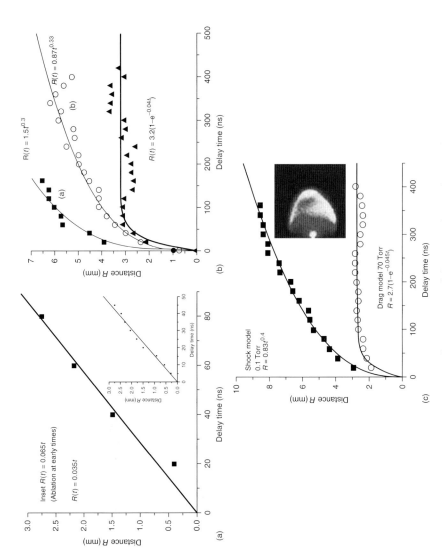

Figure 3.2.7 Position–time $(R-t)$ curves for expansion of aluminum plasma plume under (A) vacuum (inset shows $R-t$ plot under vacuum for early stage up to 40 ns), (B) (a) 0.01 torr, (b) 1 torr, and (c) 10 torr background pressures of N_2, and (C) 0.1 and 70 Torr (inset shows formation of shock wave at 0.1 Torr at 140 ns (inset shows formation of shock wave at 0.1 Torr at 140 ns (Source: Reprinted with permission from Sharma and Thareja [28] copyright @ Elsevier.)

the background gases also affect plasma characteristics, rate of ablation, size, shape, and morphology of the nanostructures produced. Under particular experimental conditions, higher atomic/molecular weight gases show faster effects on the change in the plasma characteristics and on the morphology of the nanomaterials produced. For example, argon as a background gas shows higher rate of increase in the electron temperature, reaches maximum, and also decreases at a higher rate with the increase in the background gas pressure as compared to He. In a similar way, electron density decreases more rapidly with the increase in argon pressure as compared to helium. PLA-produced crater depth on the target surface under helium atmosphere is slightly higher than that produced in the argon atmosphere and shows almost no change with the increase in the pressure up to almost 10 Torr. Rate of decrease in crater depth with the increase in background gas pressure above 10 Torr is higher for argon as compared to helium. Pressure and nature of the background gas also affect crystallinity, surface morphology, size, and composition of the obtained product. In PLD of thin films or NPs, thickness of the film or amount of NPs at particular substrate to target distance, and substrate to target distance to deposit maximum thickness of the film or maximum number of particles depend on the background gas pressure as well as the nature of the gas. At particular background gas pressure, gas of higher molecular/atomic weight exerts higher force against the expansion of the plasma plume and condense them in the form of nanomaterials. Distance of travel of the plasma plume before cooling under higher molecular/atomic weight gas is also smaller as compared to lower atomic/molecular weight gas. Therefore, at constant gas pressure, lower atomic/molecular background gas requires larger target to substrate distance to deposit maximum thickness of the film or to collect maximum amount of nanomaterials [30]. In other words, for a particular substrate to target distance, lighter molecular/atomic gases require higher background gas pressure as compared to heavier molecular/atomic gases. Lateral expansion of the plasma plume determines the length of the deposited film on the substrate placed at a particular distance from the target. At a particular background gas pressure, lighter atomic/molecular gases exert lower deceleration along normal as well as lateral flow of the plasma plume; therefore, lateral as well as normal expansion of the plasma plume is higher for lighter gases as compared to heavier ones. Therefore, longer lateral expansion of the plume length under a given pressure of a lighter background gas provides a longer film on the substrate [30]. Weight fraction of rutile phase in laser-produced TiO_2 NPs increases almost linearly with the increase in $Ar/O_2(5:5)$ background pressure in the range of 1–11 Torr [31]. Similarly, the size of Fe–Co particles increases linearly with the increase in background gas pressure [32]. Size of the PLD-produced Co NPs increases with the increase in Ar pressure in the range of 10–100 Pa, attains maximum almost at 95 Pa, and starts decreasing after that. Mean diameter of NPs produced at 100 Pa was nearly 3.3 nm, while that produced with 10^4 Pa background gas pressure has an average particle size of almost 1.2 nm [33].

Retaining elemental composition of deposited thin film or nanomaterials inherently from their mother target is a challenging task for multielemental PLD.

Stoichiometry of the product and its crystallinity depend on the background gas pressure and molecular/atomic weight of the gas. At a given fluence PLD in the background of a given molecular/atomic gas, the stoichiometry of the product is maintained up to a certain pressure and departs from the stoichiometry of the mother target with the increase in pressure beyond this. The critical pressure below which the stoichiometry of the mother target is maintained is higher for lighter molecular/atomic gases. Zbroniec *et al.* [34] investigated the effects of ambient gas pressure on the compositional changes of iron oxide particles. They plotted the XRD peak intensity ratio $R = I(Fe_2O_3)/I(FeO) + I(Fe_2O_3)$ of the PLD product where Fe_2O_3 was used as target. Increase in the laser fluence requires increase in the critical pressure of the corresponding gas to deposit the stoichiometric product (Figure 3.2.8a,b). Increase in the background gas pressure up to a particular value increases the rate of condensation of plasma species in the form of nanocondensates, and therefore increases the thickness of the film on the substrate at a given substrate to target distance. Further, increase in pressure along with maintaining other experimental parameters confines plasma plume and results in lesser number of species reaching the substrate, which causes decrease in the thickness of the film (Figure 3.2.8c). Pressure and nature of background gas also determine the morphology of PLD-produced nanostructures [35].

3.2.6
Double Pulse Laser Ablation

Aims of multiple pulse laser ablations are to increase the rate of deposition of film thickness and to deposit films at a larger surface area of substrate with more homogeneity in thickness. Rate of increase in the thickness could be easily achieved by increasing the repetition rate of the pulse with monitoring other parameters. But when two or more plasmas are produced in such a way that they are expanding parallel to each other with some overlapping/colliding regions or they are approaching each other at oblique incidence/heading, then plasma characteristics are changed. In colliding plasma, one plasma reheats the other to modify plasma characteristics [36]. Figure 3.2.9 shows spatial variation in the plasma temperature of the single- and colliding-laser-produced plasma [36].

3.2.7
Electric- and Magnetic-Field-Assisted Laser Ablation

Since laser-produced plasma have charged particles such as electrons and ions, and their motion inside the plasma produces transient electric and magnetic fields. Motion of these charged particles and hence the reaction between the plasma species inside the plasma as well as between plasma species and background gas, at plasma gas interface, or inside the gas can be controlled by external electric or magnetic fields. Optical emission intensity of the plasma plume increases with the application of the electric field in the direction of plume expansion [37]. Application of an electric field opposite to the direction of plume expansion

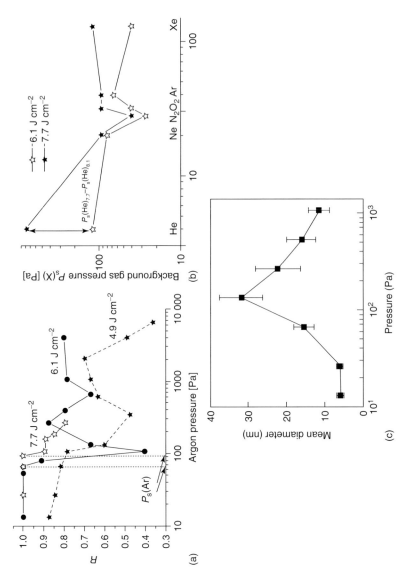

Figure 3.2.8 (a) Variation in the XRD peak ratio, R, with argon pressure, ratio R = 1 means preservation of stoichiometry of the mother target. (b) The effect of the atomic/molecular mass of the background gas on the $P_S(X)$ pressure; where P_S is the pressure required to attain ratio R = 1. (Source: Reprinted with permission from Zebroniec et al. [34] copyright @ Elsevier.) (c) Variation in the mean diameter of TiO_2 NPs with the Ar pressure. (Source: Reprinted with permission from Koshizaki et al. [35] copyright @ Elsevier.)

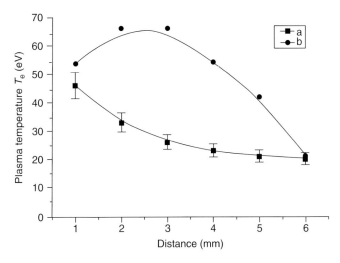

Figure 3.2.9 Temperature as a function of distance from the target (a) in the case of a single plasma plume and (b) in the case of colliding plasmas. (Source: Reprinted with permission from Atwee and Kunje [36] copyright @ Institute of Physics.)

produces single crystalline particles with comparatively larger size, while field in the direction of plume flow results in polycrystalline materials with smaller grain sizes [37]. Application of magnetic field on laser-produced plasma results its spatial confinement. Application of magnetic field in the direction of plume expansion latterly confines plasma expansion, which makes the plume more cylindrical. Effects of magnetic field on the plume dynamics and plasma properties appear for longer (>300 ns) delay for the copper plasma. At higher delay, flare emission in front of the target surface and near the magnets appears, which shows that the plume front moves out from the target with comparatively higher, 3×10^6 cm s^{-1}, expansion velocity [38]. Employment of magnetic field results in increase in the optical emission intensity of spectral lines and decrease in the temperature near target after 500 ns delay. Since application of magnetic field latterly confines the plasma plume and gives it cylindrical shape, the thickness of the deposited film increases [38] (Figure 3.2.10).

3.2.8
Effect of Laser Polarization

Interaction of polarized laser light during PLA of materials induces transient magnetic fields [39]. This transient magnetic field can guide charge particles inside the plasma to control the morphology of nanostructured materials. In addition to this, plane polarized light having an electric field vector oscillating at a different angle from the direction of plume expansion can control motion of charge particles inside the plasma, and therefore has the capability to control morphology and other characteristics of the plasma. Chen *et al.* [39] studied the effects of laser

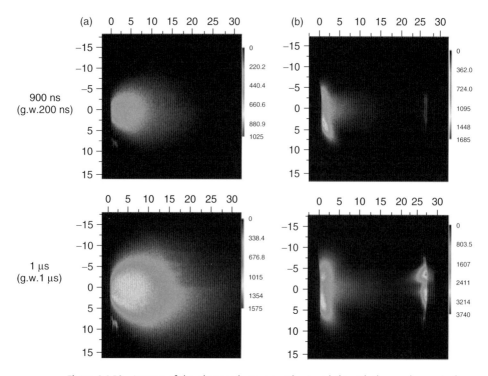

Figure 3.2.10 Images of the plasma plume (a) without and (b) with the employment of magnetic field. Delay time and gate width are given on the left. (Source: Reprinted with permission from Pagano and Lunney [38] copyright @ Institute of Physics.)

polarization on the fast electron emission from an aluminum target irradiated by ultrashort laser pulses. It is observed that the jet emission of outgoing fast electrons is collimated in the polarization direction for s-polarized laser irradiation, whereas highly directional emission of outgoing fast electrons is found in the direction close to the normal of the target when p-polarized laser was used. Plasma parameters depend strongly on laser polarization. For instance, interaction of p-polarized ultrafast laser pulses with copper target shows emission of γ-rays of energy ∼400 keV, which corresponds to electron temperatures of 66 and 52 keV in normal and reflective directions, respectively of the solid surface, and hot electrons were emitted from the plasma in the normal directions. Unlike this, no γ-ray emission was observed, which led to the deduction that 26 keV temperatures for the hot electron and emission of hot electrons was parallel to the laser field [40].

3.2.9
Conclusions

Mechanisms of laser–target interaction, processing of plasma, its spatial and temporal diagnosis, and processes of ejection of material from the target have been discussed. Effects of various laser parameters such as laser irradiance, wavelength, pulse width, and polarization on the plasma properties are described. Influences of background gas pressure on the expansion dynamics and plasma parameters have been also described. Rate of materials removal and size and morphology of product nanomaterials are correlated with the plasma parameters.

Acknowledgments

Dr. Singh is thankful to the Irish Research Centre for Science, Engineering and Technology (IRCSET) for providing EMPOWER postdoctoral fellowship for financial assistantances during the compilation of this chapter.

References

1. White, R.M. (1962) *IRE Trans. Instrum.*, **I–II**, 294.
2. White, R.M. (1963) *J. Appl. Phys.*, **34**, 2123.
3. White, R.M. (1963) *J. Appl. Phys.*, **34**, 3559.
4. Weichel, H., David, C.D., and Avizonis, P.V. (1968) *Appl. Phys. Lett.*, **13**, 376.
5. Cunningham, P.F., Campbell, R.N., and Michaelis, M.M. (1986) *J. Phys. E: Sci. Instrum.*, **19**, 957.
6. Logothetis, E.M. and Hartman, P.L. (1967) *Phys. Rev. Lett.*, **18**, 581.
7. Ahmad, S.R. and Walsh, D. (1972) *J. Phys. D: Appl. Phys.*, **5**, 1157.
8. Logothetis, E.M. and Hartman, P.L. (1969) *Phys. Rev.*, **187**, 460.
9. Cobb, J.K. and Murray, J.J. (1965) *Br. J. Appl. Phys.*, **16**, 271.
10. Langer, P., Tonon, G., Floux, F., and Ducauze, A. (1966) *IEEE J. Quantum Electron.*, **2**, 499.
11. Yeates, P., White, J., and Kennedy, E.T. (2010) *J. Appl. Phys.*, **108**, 093302.
12. Dardis, J. and Costello, J.T. (2010) *Spectrochim. Acta B*, **65**, 627.
13. Schwoerer, H., Ewald, F., Sauerbrey, R., Galy, J., Magill, J., Rondinella, V., Schenkel, R., and Butz, T. (2003) *Europhys. Lett.*, **61**, 47–52.
14. Steen, W.M. (2003) *Laser Material Processing*, 3rd edn, Springer, ISBN-10: 1852336986; ISBN-13: 978-1852336981.
15. Vorobyev, A.Y. and Guo, C. Direct femtosecond laser nanostructuring and nanopatterning on metals, in (2012) *Nanomaterials Processing and Characterization with Lasers*, Chapter 5 (eds S.C. Singh *et al.*), Wiley-VCH Publication, pp. 69–80.
16. Singh, S.C. *et al.* (eds) (2012) *Nanomaterials Processing and Characterization with Lasers*, Chapter 4.1 to 4.4, Wiley-VCH Publication.
17. Miller, J. C. and Haglund, R. F. (eds) (1998) *Laser Ablation and Desorption*, Academic, New York.
18. Mao, S.S., Mao, X.L., Greif, R., and Russo, R.E. (1998) *Appl. Phys. Lett.*, **73**, 1331.
19. Andrev, A.A., Mak, A.A., and Solovyev, N.A. (2000) *An Introduction to Hot Laser Plasma Physics*, Nova Science Publishers, Inc., Hutington, New York, ISBN-1-56072-8035.
20. Russo, R.E., Mao, X., and Mao, S.S. (2002) *Anal. Chem.*, **74**, 70A–77A.
21. Russo, R.E., Mao, X.L., Liu, H.C., Yoo, J.H., and Mao, S.S. (1999) *Appl. Phys. A: Mater. Sci. Process.*, **69**, S887–S894.
22. Martynyuk, M.M. (1974) *Sov. Phys. Tech. Phys.*, **19**, 793.
23. Martynyuk, M.M. (1976) *Sov. Phys. Tech. Phys.*, **21**, 430.
24. Mazhukin, V.I., Mazhukin, A.V., and Lobok, M.G. (2009) *Laser Phys.*, **19**, 1169–1178.
25. Borowiec, A. and Haugen, H.K. (2003) Proceedings of the CLEO.
26. Kerdja, T., Abdelli, S., Ghobrini, D., and Malek, S. (1996) *J. Appl. Phys.*, **80**, 5365.
27. Geohegan, D.B. (1992) *Appl. Phys. Lett.*, **60**, 2732.

28. Sharma, A.K. and Thareja, R.K. (2005) *Appl. Surf. Sci.*, **243**, 68–75.

29. Singh, R.K., Holland, O.W., and Narayan, J. (1990) *J. Appl. Phys.*, **68**, 233.

30. Zbroniec, L., Sasaki, T., and Koshizaki, N. (2002) *Appl. Surf. Sci.*, **197–198**, 883–886.

31. Matsubara, M., Yamaki, T., Itoh, H., Abe, H., and Asai, K. (2003) *Jpn. J. Appl. Phys.*, **42**, L479–L481.

32. Happy, Mohanty, S.R., Lee, P., Tan, T.L., Springham, S.V., Patran, A., Ramanujan, R.V., and Rawat, R.S. (2006) *Appl. Surf. Sci.*, **252**, 2806–2816.

33. Li, Q., Sasaki, T., and Koshizaki, N. (1999) *Appl. Phys. A*, **69**, 115–118.

34. Zbroniec, L., Sasaki, T., and Koshizaki, N. (2004) *Appl. Phys. A*, **79**, 1783–1787.

35. Koshizaki, N., Narazaki, A., and Sasaki, T. (2002) *Appl. Surf. Sci.*, **197–198**, 624–627.

36. Atwee, T. and Kunze, H.-J. (2002) *J. Phys. D: Appl. Phys.*, **35**, 524–528.

37. Park, H.S., Nam, S.H., and Park, S.M. (2007) *J. Phys.: Conf. Ser.*, **59**, 384–387.

38. Pagano, C. and Lunney, J.G. (2010) *J. Phys. D: Appl. Phys.*, **43**, 305202.

39. Chen, L.M., Zhang, J., Li, Y.T., Teng, H., Liang, T.J., Sheng, Z.M., Dong, Q.L., Zhao, L.Z., Wei, Z.Y., and Tang, X.W. (2001) *Phys. Rev. Lett.*, **87**, 225001.

40. Ping, Z., Tian-Jiao, L., Li-Ming, C., Yu-Tong, L., Duan-Bao, C., Zu-Hao, L., Jing-Tang, H., Zhi-Yi, W., Long, W., Xiao-Wei, T., and Jie, Z. (2001) *Chin. Phys. Lett.*, **18**, 1374.

4
Nanomaterials: Laser-Based Processing in Gas Phase

4.1
Synthesis and Analysis of Nanostructured Thin Films Prepared by Laser Ablation of Metals in Vacuum

Rashid Ashirovich Ganeev

4.1.1
Introduction

Nanostructures of different materials are of special interest because of their potential applications in optoelectronics and nonlinear optics. The structural, optical, and nonlinear optical parameters of nanoparticles are known to differ from those of the bulk materials because of the quantum confinement effect. Silver [1, 2], copper [2–4], and gold [2, 5] are among the most useful metals suited for nanoparticle preparation in optoelectronics and nonlinear optics. Further search of prospective materials in nanoparticle form, their preparation, and application are of considerable importance nowadays.

Studies in the past on nanoparticles prepared using metal vapor deposition [6], reduction of some salts by alkalides [7], and solution dispersion method [8] have revealed many interesting structural and optical properties of these materials. The laser ablation of metals in vacuum and liquids is among the perspective techniques that can also be successfully applied to the preparation of nanoparticle-containing media. Recently, the application of laser ablation in liquids for the preparation of semiconductor [9, 10] and metal [11] colloids has been demonstrated.

It is known that the metal ablation in air is significantly less efficient than that in vacuum because of redeposition of the ablated material. The ablation rates in vacuum can be calculated using a thermal model that also allows estimating the ablation rates of other metals from basic optical and thermal properties [12]. A comparison of the morphology of the deposition by nanosecond and picosecond ablation shows unequivocally the advantages of short-pulse ablation for the preparation of nanoparticles [13].

To prove the generality of the vacuum deposition method and its potential use for preparing nanoparticles, we considered various metals and analyzed this process at different focusing conditions of the laser radiation. The most interesting and new

Nanomaterials: Processing and Characterization with Lasers, First Edition.
Edited by Subhash Chandra Singh, Haibo Zeng, Chunlei Guo, and Weiping Cai.
© 2012 Wiley-VCH Verlag GmbH & Co. KGaA. Published 2012 by Wiley-VCH Verlag GmbH & Co. KGaA.

features of laser ablation and nanoparticle formation during irradiation of the solid targets have been recently observed in the case of short laser pulses (100 fs to 1 ps). In this section, we study the formation of nanoparticles of silver, chromium, stainless steel, and indium in vacuum using much longer (subnanosecond) laser pulses. The structural properties of the nanoparticles deposited on different substrates are reported. The effect of the focusing condition of the laser (tight or weak focusing) on the films prepared by the laser ablation of bulk targets in the vacuum is analyzed. Our results show that the nanoparticles can be formed using long laser pulses under tight focusing conditions.

4.1.2
Experimental Details

The ablation of various materials was carried out in vacuum using uncompressed pulses (of 300 ps duration) from a chirped-pulse amplification Ti : sapphire laser system (Thales Lasers S.A.). The samples were placed inside a vacuum chamber. Uncompressed pulses from the Ti:sapphire laser ($\lambda = 793$ nm, $\tau = 300$ ps, $E = 30$ mJ, 10 Hz pulse repetition rate) were focused on a bulk target at two regimes of focusing. In the first case (referred to as *tight focusing*), the intensity of laser radiation was in the range of 2×10^{12} W cm^{-2}, and in the second case (referred to as *weak focusing*), the intensity was considerably lower (4×10^{10} W cm^{-2}). Silver, indium, stainless steel, and chromium were used as targets. Float glass, silicon wafer, and various metal strips (silver, copper, and aluminum) were used as the substrates and were placed at a distance of 50 mm from the targets. The deposition was carried out in oil-free vacuum ($\approx 1 \times 10^{-4}$ mbar) at room temperature.

The structure of the deposited films of ablated metals was analyzed using different techniques. For this purpose, the nanoparticle-containing films were deposited on different substrates. The nanoparticle formation is governed by the thermodynamic conditions at the target surface. The presence of nanoparticles was inferred by analyzing the spatial characteristics and the spectral absorption of the deposited material.

The absorption spectra of the deposited films were analyzed using a fiber optic spectrometer (USB2000). The analysis of the sizes of deposited nanoparticles was carried out using the total reflection X-ray fluorescence (TXRF). Details of the TXRF are described in Ref. [14]. The structural properties of the deposited films were analyzed using a scanning electron microscope (SEM, Philips XL30CP), an atomic force microscope (AFM, SOLVER PRO, NT-MDT), and a transmission electron microscope (TEM, Philips CM200).

4.1.3
Results and Discussion

The absorption spectra of the materials deposited on transparent substrates (float glass) were used to determine the presence or absence of nanoparticles. The presence of nanoparticles was inferred from the appearance of strong absorption

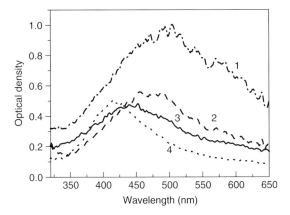

Figure 4.1.1 Absorption spectra (curves 1–3) of the silver films deposited at different tight focusing conditions and the absorption spectrum of silver nanoparticles (curve 4) implanted inside the silica glass plate by ion bombardment.

bands associated with surface plasmon resonance (SPR). Figure 4.1.1 presents the absorption spectra of silver films deposited on float glass substrates. Earlier work [11] has shown that the SPR of spherical silver nanoparticles induces a strong absorption in the range of 410–480 nm depending on the preparation technique. In this study, we observed a variation in the position of the absorption of Ag deposition, which depended on the conditions of excitation and evaporation of bulk target by the 300 ps pulses, which interacted with the surface at different tight focusing conditions. However, in all these cases, the peaks of SPR were centered in the range of 440–490 nm (Figure 4.1.1, curves 1–3). In the case of the deposition of a silver film at weak focusing conditions, no absorption peaks were observed in this region, indicating the absence of nanoparticles.

In Figure 4.1.1, curve 4 shows analogous measurements made on a sample of Ag nanoparticles embedded in silica glasses using the ion bombardment, reported in Ref. [15]. In this work, the thickness of the implanted layer was 60 nm, and the size of the silver nanoparticles was reported to vary from 4 to 8 nm. It is observed that the absorption curve of this sample is quite similar to those of silver deposited on the glass surfaces in this case (Figure 4.1.1, curves 1–3). The only difference is that the position of the peak of SPR (415 nm) was on the shorter wavelength side (Figure 4.1.1, curve 4).

Much attention has been devoted during the past few years to precisely determine the spatial arrangement in two- and three-dimensional structures of metals. However, ordering and using the nanomaterials necessitate synthesis of monodispersed individual nanoparticles, for which no general method is presently available. We describe, later in the text, the analysis of synthesized nanoparticles by laser ablation of bulk targets at two different conditions of ablation.

The structure of ablated silver debris was analyzed by studying the films deposited on silicon substrates. One of the aims of this study was to investigate whether the plumes contain nanoclusters. The presence of the latter could be

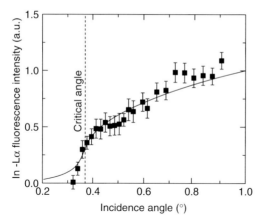

Figure 4.1.2 Recorded X-ray fluorescence profile for an indium deposition prepared by ablation under weak focusing conditions. The profile shows that the indium is deposited in the form of a continuous monoatomic film instead of nanoparticles. The solid line shows a fitted profile. The critical angle of indium is ~0.36° for 8.50 keV X-ray energy.

responsible for the enhancement of nonlinear optical characteristics. In particular, high-order harmonic conversion efficiency may be affected because of the quantum confinement effect during propagation of the femtosecond laser pulse through the nanoparticle-containing plasma. Harmonic generation using single atoms and multiparticle aggregates has been reported in Ref. [16] for argon atoms and clusters.

TXRF measurements were performed using a TXRF technique [17] for the analysis of the structural properties of the deposited material. The angular dependence of the fluorescence intensity in the total reflection region [18, 19] can be successfully used to identify the presence of nanoparticles on a flat surface. This was done for the deposition in the tight focusing condition of the laser. In the case of weak focusing, it showed a thick filmlike deposition of metal, without any inclusion of nanoparticles (see the TXRF image in the case of the ablation of indium at the weak focusing conditions, Figure 4.1.2).

It can be seen from this figure that the fluorescence intensity of In-Lα decreased abruptly below the critical angle for indium (~0.36° at 8.5 keV). For incident angles larger than the critical angle, In-Lα fluorescence intensity increased monotonously. It showed a behavior similar to that for a thick film of atomic indium deposited on the substrate.

The fluorescence measurements were carried out using a Peltier-cooled solid-state detector (EurisusMesures EPXR 10–300), a spectroscopy amplifier AMP 6300, and a multichannel pulse height analyzer card installed in a personal computer. The solid-state detector had an energy resolution of 250 eV at 5.9 keV. A well-collimated primary beam, from a line focus Cu X-ray tube, was used as an excitation source. Figure 4.1.3 shows the X-ray fluorescence trace recorded in the case of silver nanoparticles deposited on a glass substrate at tight focusing conditions. It can be seen from this figure that the angle-dependent fluorescence

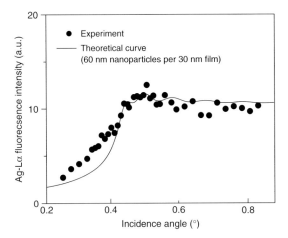

Figure 4.1.3 Recorded X-ray fluorescence profile of silver nanoparticles prepared by laser ablation in vacuum and deposited on a float glass substrate. The dots show experimental data, while the solid line shows a fitted profile.

profile of the silver film shows the presence of nanostructure on the flat surface, as well as indicates a monoatomic layer. The average size of the silver nanoparticles was determined by fitting the recorded fluorescence profile using the CATGIXRF program [20]. The solid line presents the best theoretical fit, from which the average size of the nanoparticles was estimated to be 60 nm, while the thickness of the layer of monoatomic silver particles was estimated to be 30 nm. The TEM measurements also confirmed the presence of silver nanoparticles in these deposited films.

Our SEM studies of the structural properties of the deposited films showed that, in the tight focusing condition, these films contained a lot of nanoparticles with variable sizes. In the weak focusing condition, the concentration of nanoparticles was considerably smaller compared to that in the tight focusing condition. Figure 4.1.4a shows the SEM images of the deposited chromium nanoparticles on the surface of a silicon wafer. In the case of weak focusing, the deposited film was almost homogeneous with a few nanoparticles appearing in the SEM images (see Figure 4.1.4a showing the deposition of chromium), while in tight focusing, plenty of nanoparticles ranging from 30 to 100 nm appeared in the SEM images (see Figure 4.1.4b showing the deposition of stainless steel). The average size of these nanostructures was estimated to be 60 nm. An enlarged SEM image of the silver nanoparticles prepared in tight focusing conditions is presented in Figure 4.1.4c. The average size of these spherical clusters was also measured to be 60 nm. The same behavior was observed in the case of other targets. These studies showed that the material of the target does not play a significant role in the formation of nanoparticles in the case of laser ablation using 300 ps laser pulses in tight focusing conditions. Hence, further studies on the influence of substrate material on the nanoparticle deposition were carried out using silver target.

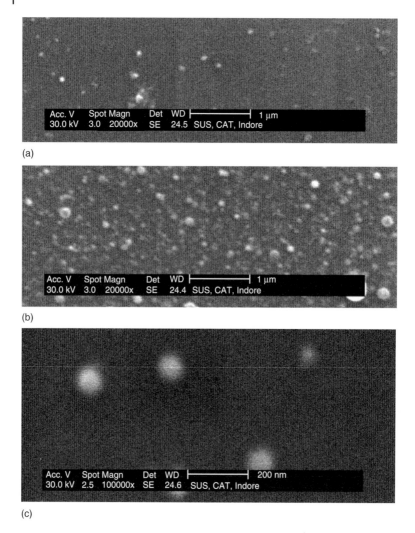

Figure 4.1.4 SEM images of the chromium, stainless steel, and silver nanoparticles deposited on silicon wafer as substrate. These images were obtained under (a) weak focusing (chromium deposition), (b) tight focusing (stainless steel deposition), and (c) tight focusing (silver deposition) conditions. (Note the different magnifications used.) The average size of the spherical clusters in all the three cases was measured to be 60 nm.

Figure 4.1.5 shows the SEM images of silver deposited on a copper substrate. One can see a considerable difference in the concentrations of nanoparticles in the cases of weak and tight focusing. It may be pointed out that there is a special interest in silver nanoparticles because of their potential applications. In general, highly dispersed metals have a much higher surface area for a given volume, and hence they can be useful for efficient catalytic conversion. Silver nanoparticles are widely used for surface-enhanced Raman scattering. Silver nanoparticles have an

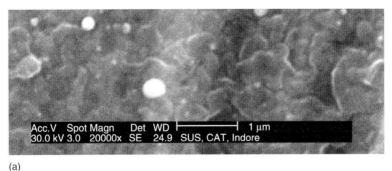

(a)

(b)

Figure 4.1.5 SEM images of the silver nanoparticles deposited on a copper substrate under (a) weak focusing and (b) tight focusing conditions.

advantage over other metal nanoparticles (e.g., gold and copper) from the point of view of the position of the SPR of silver, which is far from the interband transitions. This enables one to investigate the optical and nonlinear optical effects in the silver nanoparticles by focusing on the surface plasmon contribution.

Further studies on the characteristics of nanosized structures of the deposited materials were carried out using AFM. AFM measurements were carried out in noncontact mode in an ambient environment. Silicon cantilever tips of resonant frequency $>180\,kHz$ and spring constant $5.5\ N\ m^{-1}$ were employed. Figure 4.1.6a shows the AFM image of the silver nanoparticles deposited on the surface of copper strip. The average size of silver nanoparticles was 65 nm. In contrast to this, the AFM images obtained from the deposited films prepared under the weak focusing condition showed considerably smaller number of nanoparticles. Figure 4.1.6b shows an AFM picture of silver deposition prepared under these conditions. The image indicates the presence of very few nanoparticles. The same difference in AFM pictures was observed in the case of indium deposition under the two focusing conditions.

To characterize the ablation process, the temporal and spectral characteristics of the radiation emitted by the plume were also studied. The oscilloscope traces showed a considerable increase in the duration of the plasma emission in the case of tight focusing that could be expected considering the excitation conditions. A combined analysis of the spectra and oscilloscope traces in the cases of two different regimes

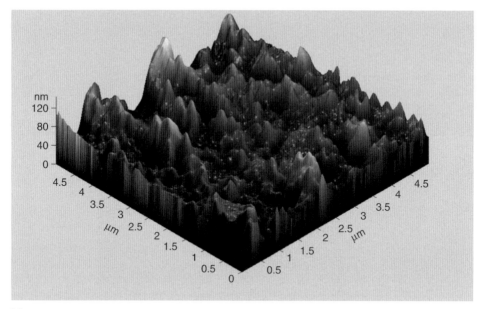

(a)

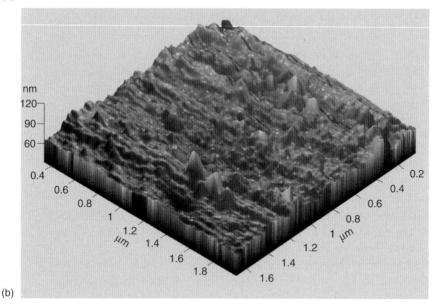

(b)

Figure 4.1.6 AFM images of (a) the silver nanoparticles deposited on a copper strip in the tight focusing condition and (b) the silver nanoparticles deposited on an aluminum strip in the weak focusing condition. Very few nanoparticles are seen in the case (b) of weak focusing condition compared to the case (a) of tight focusing condition.

of the excitation of plasma revealed that the structureless continuum appearing in the spectrum in the case of tight focusing is due to the emission from hot nanoparticles produced during laser ablation. Such hot nanoparticles behave like black body radiators emitting for a longer time till they get cooled down.

It is well accepted that when a solid target is ablated by the laser radiation, the ablating material is in the form of atoms, ions (and electrons), and clusters. These atoms and clusters tend to aggregate during the laser pulse or soon afterwards, leading to the formation of larger clusters. The reported results (Ref. [21]) also indicate that the ablation processes in the picosecond and femtosecond timescales are very different compared to the nanosecond one. In addition to the early experimental observations, several theoretical studies have suggested that rapid expansion and cooling of the solid density matter heated by a short laser pulse may result in nanoparticle synthesis via different mechanisms. Heterogeneous decomposition, liquid-phase ejection and fragmentation, homogeneous nucleation and decomposition, and photomechanical ejection are among those processes that can lead to nanoparticle production [22–24]. Short pulses, contrary to nanosecond pulses, do not interact with the ejected material and thus avoid complicated secondary laser interactions. Further, short pulses heat a solid to higher temperature and pressure than do longer pulses of comparable fluence, since the energy is delivered before significant thermal conduction can take place.

The model developed in Ref. [25] for aluminum predicts that for short laser pulses at intensities in the range 10^{12}–10^{13} W cm^{-2}, the adiabatic cooling drives the system into a metastable region of its phase diagram, resulting in the production of a relatively large fraction of nanoparticles. At larger intensities ($\geq 10^{14}$ W cm^{-2}), the system can never reach the metastable region, resulting in an almost atomized plume.

Pulsed laser deposition using short pulses has gained some interest because of a number of advantages over other processes, such as the possibility of producing materials with a complex stoichiometry and a narrowed distribution of nanoparticle sizes with reduced porosity. Typically, laser deposition is carried out in an ambient gas, which quenches the ablated plume, thus controlling the mean particle size [26]. However, some previously reported studies [13, 27], as well as this study, suggest that nanoparticles are generated as a result of some relaxation processes of the extreme material state reached by the irradiated target surface. This stands in stark contrast to the formation of nanoparticles during nanosecond laser ablation in a background gas, where vapor condensation is considered to be an important mechanism.

4.1.4
Conclusions

A study of the metal deposition by laser ablation in vacuum using subnanosecond laser pulses under different focusing conditions has been performed. It is shown that nanoparticle-contained films can be formed using long laser pulses under tight focusing conditions. The average size of the nanoparticles was ∼60 nm for the various targets (silver, indium, chromium, and stainless steel) used in the study.

Acknowledgments

The support from Raja Ramanna Centre for Advanced Technology, Indore, India, to carry out this work is gratefully acknowledged. The author also acknowledges the fruitful collaboration with P. D. Gupta, P. A. Naik, U. Chakravarty, H. Srivastava, C.Mukherjee, M. K. Tiwari, and R. V. Nandedkar.

References

1. Ganeev, R.A., Ryasnyanskiy, A.I., Stepanov, A.L., and Usmanov, T. (2004) *Opt. Quantum Electron.*, **36**, 949.

2. Ganeev, R.A., Ryasnyansky, A.I., Stepanov, A.L., Marques, C., da Silva, R.C., and Alves, E. (2005) *Opt. Commun.*, **253**, 205.

3. Ryasnyansky, I., Palpant, B., Debrus, S., Ganeev, R.A., Stepanov, A.L., Can, N., Buchal, C., and Uysal, S. (2005) *Appl. Opt.*, **44**, 2839.

4. Falconieri, M., Salvetti, G., Cattaruza, E., Gonella, F., Mattei, G., Mazzoldi, P., Piovesan, M., Battaglin, G., and Polloni, R. (1998) *Appl. Phys. Lett.*, **73**, 288.

5. Debrus, S., Lafait, J., May, M., Pinçon, N., Prot, D., Sella, C., and Venturini, J. (2000) *J. Appl. Phys.*, **88**, 4469.

6. Yan, X.M., Ni, J., Robbins, M., Park, H.J., Zhao, W., and White, J.M.(2002) *J. Nanopart. Res.*, **4**, 525.

7. Tsai, K.L. and Dye, J.L. (1991) *J. Am. Chem. Soc.*, **113**, 1650.

8. Zhao, Y., Zhang, Zh., and Dang, H. (2003) *J. Phys. Chem. B*, **107**, 7574.

9. Ganeev, R.A., Baba, M., Ryasnyansky, A.I., Suzuki, M., and Kuroda, H. (2005) *Appl. Phys.*, **80**, 595.

10. Ganeev, R.A. and Ryasnyansky, A.I. (2005) *Opt. Commun.*, **246**, 163.

11. Ganeev, R.A., Baba, M., Ryasnyansky, A.I., Suzuki, M., and Kuroda, H. (2004) *Opt. Commun.*, **240**, 437.

12. Anisimov, S.I., Imas, Y.A., Romanov, G.S., and Khodyko, Y.V. (1970) *High Power Radiation Effect in Metals*, Nauka, Moscow.

13. Amoruso, S., Ausanio, G., Bruzzese, R., Vitiello, M., and Wang, X. (2005) *Phys. Rev. B*, **71**, 033406.

14. Tiwari, M.K., Sawhney, K.J.S., Gowri Sankar, B., Raghuvanshi, V.K., and Nandedkar, R.V. (2004) *Spectrochim. Acta B*, **59**, 1141.

15. Ganeev, R.A., Ryasnyansky, A.I., Stepanov, A.L., and Usmanov, T. (2004) *Phys. Status Solidi B*, **241**, R1.

16. Donnelly, T.D., Ditmire, T., Neuman, K., Perry, M.D., and Falcone, R.W. (1996) *Phys. Rev. Lett.*, **76**, 2472.

17. Tiwari, M.K., Gowrishankar, B., Raghuvanshi, V.K., Nandedkar, R.V., and Sawhney, K.J.S. (2002) *Bull. Mater. Sci.*, **25**, 435.

18. Bedzyk, M.J., Bommarito, G.M., and Schildkraut, J.S. (1989) *Phys. Rev. Lett.*, **62**, 1376.

19. de Boer, D.K.G. (1991) *Phys Rev. B*, **44**, 498.

20. Tiwari, M.K. (2006) Calculation of Grazing Incidence X-Ray Fluorescence Intensities from a Layer Samples. RRCAT Internal Report 2006, RRCAT, Indore, India.

21. Teghil, R., D'Alessio, L., Santagata, A., Zaccagnino, M., Ferro, D., and Sordelet, D.J. (2003) *Appl. Surf. Sci*, **210**, 307.

22. Glover, T.E. (2003) *J. Opt. Soc. Am. B*, **20**, 125.

23. Jeschke, H.O., Garsia, M.E., and Bennemann, K.H. (2001) *Phys. Rev. Lett.*, **87**, 015003.

24. Perez, D. and Lewis, L.J. (2003) *Phys. Rev. B*, **67**, 184102.

25. Amoruso, S., Bruzzese, R., Spinelli, N., Velotta, R., Wang, X., and Ferdeghini, C. (2002) *Appl. Phys. Lett.*, **80**, 4315.

26. Sturm, K., Fähler, S., and Krebs, H.U. (2003) *Appl. Surf. Sci.*, **154-155**, 462.

27. Amoruso, S., Bruzzese, R., Spinelli, N., Velotta, R., Vitiello, M., Wang, X., Ausanio, G., Iannotti, V., and Lanotte, L. (2004) *Appl. Phys. Lett.*, **84**, 4502.

4.2
Synthesis of Nanostructures with Pulsed Laser Ablation in a Furnace

Rusen Yang and Jung-Il Hong

4.2.1
General Consideration for Pulsed Laser Deposition: an Introduction

With a pulsed laser deposition (PLD) system equipped with a high vacuum pump and substrate heater, materials with various nanostructures can be deposited onto a substrate. Deposition of two-dimensional structures such as thin films with PLD has been considered to be one of the common standard techniques. Nowadays, its capability to grow nanowires (NWs) is being recognized and its application is expanding beyond the synthesis of thin films [1]. A diagram of the common PLD system is shown in Figure 4.2.1. The power of the laser pulse should be high enough to ablate the target material (typically on the order of ~ 1 J cm^{-2}). As the laser pulse hits the surface of the target material, atoms are ablated from the surface and they fly in the direction perpendicular to the target surface. The velocity of the atoms can easily reach up to a few tens of kilometers per second depending on the species of the target atoms and the energy of the laser pulses [2]. As these high-energy atoms fly toward the substrate, they collide with the gas molecules within the chamber, which results in the change of the energy as well as the chemical reactions before they reach the surface of the substrate. After the atoms reach the surface of the substrate, they diffuse around the surface before finally settling down at the energetically stable or metastable sites. Depending on the dynamic balances between these procedures, the shape of the final (nano)structures is decided.

Compared to the widely used vacuum oven system for the physical vapor deposition of NWs [3], the PLD system in Figure 4.2.1 has several advantages such as separate control of substrate temperatures, capability to mount multiple targets, wider variation of the pressure, etc. Therefore, combinations of different physical structures as well as chemical phases can be achieved in a controlled manner.

4.2.1.1   One-Dimensional Nanostructure

In general, thin films grow in a low-pressure environment. On the other hand, the growth of NWs or nanorods requires the pressure of a few torr or above and a relatively high growth temperature (typically a few hundred degrees Celsius). Therefore, by varying the pressure and temperature, the grown nanostructure may exhibit various morphologies. In this chapter, we focus on one-dimensional NW structures since a vast amount of literature is available for thin-film synthesis and processing with PLD.

Iron oxide NWs shown in Figure 4.2.2a and films with different surface morphologies as in Figure 4.2.2b,c demonstrate how sensitively the structure changes with the synthesis conditions. NWs in Figure 4.2.2a were grown at the substrate temperature of 820 °C in an Ar atmosphere of ~ 0.8 torr pressure. Growth in a slightly lower pressure results in the film with surface morphology as in Figure 4.2.2b,

Nanomaterials: Processing and Characterization with Lasers, First Edition.
Edited by Subhash Chandra Singh, Haibo Zeng, Chunlei Guo, and Weiping Cai.

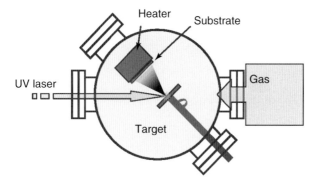

Figure 4.2.1 Basic structure of a PLD system. Multiple targets can be mounted usually, and the chamber pressure should be adjustable from UHV to up to ~1 atm to synthesize nanostructured materials. (Reproduced with permission from Dr. Jung-Il Hong.)

which appears to be formed by piling up nanoparticles. In the meantime, the growth in a higher temperature yields the structure shown in Figure 4.2.2c, which was grown at 950 °C, showing the crystalline facets that formed as neighboring grains fused together with the help of a high substrate temperature.

In Figure 4.2.3a, ZnO NWs were first grown on Si wafer with PLD system and iron oxide was subsequently deposited on top of ZnO NWs, resulting in the rough coating of iron oxide over the surface of ZnO NWs. Two steps were carried out consecutively within a system without breaking the vacuum of the chamber but by changing the substrate temperature and chamber pressure of the PLD system loaded with multiple targets. Because the grain size of the coating is in the same order as the NWs on which the film grows, the coating cannot stay smooth unless the crystalline structure of the coating and NW matches perfectly.

The process conditions for the overcoating can be further varied for yet another type of NW modification such as bending of the NWs as shown in Figure 4.2.3b. The bending of ZnO in Figure 4.2.3b was achieved by depositing energetic Zn and O atoms from the ZnO target. Those atoms were directed to hit the NWs from only one side of the ZnO NWs [4]. In order to keep the directionality and high kinetic energy of Zn and O atoms, the deposition was performed in an ultrahigh vacuum (UHV) atmosphere ($\sim 10^{-6}$ torr range) at room temperature. Similar effects were observed when the NWs were coated with different materials but the efficiency of bending varied depending on the deposited materials. When the laser-ablated energetic atoms hit one side of the NWs, their lateral impact created a horizontal force on the NWs, causing them to bend. Increased elasticity of NWs has been reported [5]; hence, severe bending of NWs as shown in Figure 4.2.3b can be accommodated for thin enough NWs. When the NW thickness was over ~500 nm, it was observed that the NWs cracked as they were bent. As more material is deposited on the side of the deposition, a hit-stick-hold process of the deposited atoms may create and hold the shape of the NWs. Alternatively, the fastest atoms may penetrate the NW and thus be implanted below the surface, which may develop strain on the stretched side of the NW [6].

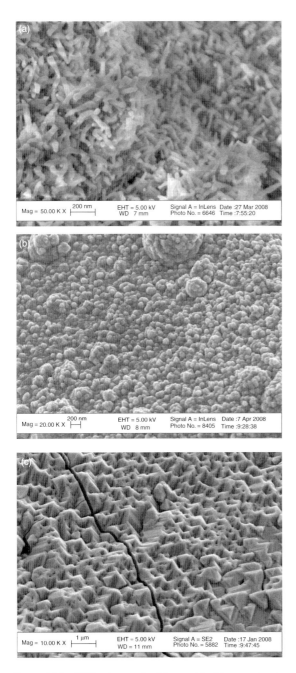

Figure 4.2.2 SEM micrograph of iron oxide (a) nanowires/nanorods and (b,c) films with different morphologies, all grown by PLD on the substrates of alumina or silicon. (Reproduced with permission from Dr. Jung-Il Hong.)

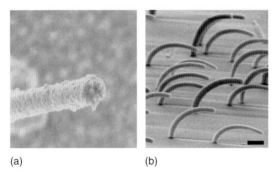

(a)   (b)

Figure 4.2.3 (a) ZnO nanowire was coated with iron oxide. Diameter of the nanowire is ~300 nm. (b) Deposition of energetic atoms on the side of nanowires may bend the nanowires [4]. The scale bar represents 1 µm. (Reproduced with permission from Dr. Jung-Il Hong.)

An example of the chemical modification is alloying or doping of NWs. For semiconductors such as Si, ZnO, or GaN, the doping determines the conduction type (p- or n-type) whose control is crucial for the implementation of these nanostructures in real devices applications. However, as the size of the nanostructure decreases, the homogeneous distribution of the dopant atoms becomes more difficult to achieve [7], although it may not impossible. In spite of the known difficulties, doped NWs began to be synthesized recently and PLD was demonstrated to be one of the useful tools for fabricating doped NW structures [8, 9]. With a systematic study by Lin *et al.* [9], the relationship between the detailed NW shape and growth conditions was also explored with Li-doped $Zn_{0.95}Mg_{0.05}O$ and it was reconfirmed that there exists a relatively small window of process conditions within which the growth of NWs can be initiated (Figure 4.2.4). Without doping, ZnO and ZnMgO NWs are n-type semiconductors. By doping with Li, synthesis of p-type NWs was shown. Furthermore, alloying Mg with ZnO allows the synthesis of NWs with controlled band gaps. Much attention is necessary for the control of Mg concentration as the synthesis conditions in this case also change the Mg concentration, x, in the $Zn_{1-x}Mg_xO$ NWs, which is proportional to the band gap increase.

4.2.2
Thermal-Assisted Pulsed Laser Deposition

The solid vapor deposition process in a furnace is widely adopted for the creation of different nanostructures [10]. Tube furnace can achieve control over the growth temperature at the substrate, flowing gas type and speed, and pressure. Tuning those growth parameters and the source materials, different nanostructures have been successfully created, such as nanobelts (ZnO, SnO_2, In_2O_3, CdO, or Ga_2O_3) [11], nanorings [12], nanohelixes [13–15], nanocombs [16], nano tetralegs [17], nanopropellers [18], and so on. In addition, pulsed lasers can achieve an extremely high temperature within an extremely short time and have also been employed for the synthesis of nanostructures, such as fullerenes [19], carbon nanotubes [20],

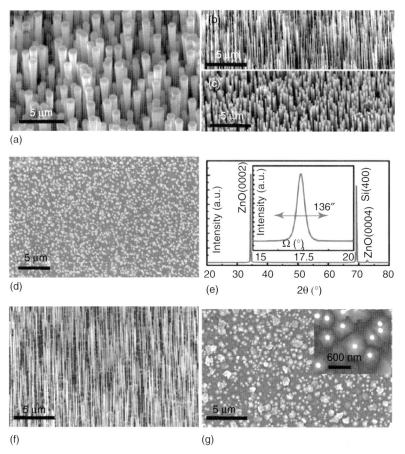

Figure 4.2.4 (a) NWs prepared at 5.0 torr with an Ar/O2 flow ratio of 6 for 20 min followed by at 4.5 torr for 20 min. These NWs show a larger diameter at the top compared with that at the bottom. NWs grown under (b) 5.0 torr of pure argon. (c) 5.0 torr at Ar/O2 flow ratio of 6. The comparison between (b) and (c) reveals that the aspect ratio becomes smaller when the oxygen partial pressure is higher. (d) Top view of the NWs of (c), revealing that the NWs are perfectly vertical to the silicon substrate. (e) The corresponding XRD spectra. Inset, rocking curve of (0002) peak shows a small full width at half maximum (FWHM) value of 136″. From (a–e), the target is Zn0.95Mg0.05O: Li ceramic (0.5 at.% doping concentration of Li). (f) NWs produced from another target containing 1.5 at.% Na under 5.0 torr with argon/oxygen flow ratio of 6. (g) Top view of NWs in (f). Inset, magnified top view of NWs, which appear as bright white spots and grow from hexagonal bases. (Ref. [9] Reproduced with permission from the American Chemical Society.)

and NWs (ZnO [21] and SnO₂ [22]). Integration of two systems, tube furnace and pulsed laser ablation, can provide effective control over the growth environment during the growth of nanostructures through the solid vapor deposition process and create different nanostructures, which cannot be easily achieved otherwise.

We discuss here the growth of nanostructures with a thermal-assisted pulsed laser deposition (TAPLD) process combining furnace and pulsed laser deposition. First, the general experimental setup, synthesis procedure, and growth mechanism are introduced followed by the detailed investigation of zinc phosphide [23] and iron oxide [24] nanostructures from this new growth method.

4.2.2.1 Furnace System

Figure 4.2.5 schematically illustrates a typical furnace system for the synthesis of nanostructures. A furnace system usually consists of three main components: the furnace, the carrier gas module, and the pressure control module. Different furnaces can be used according to the growth temperature and other requirements. An alumina tube or a quartz tube is generally placed inside the tube furnace to serve as a growth chamber. Heating elements of the furnace are equally spaced to provide uniform heat within the tube furnace. Cooling water is circulated inside the cover caps, which are sealed tightly to both ends of the tube. As a result, a desired temperature gradient in the tube can be maintained during the growth of nanostructures. One end cap is fitted with a gas inlet connection for introducing carrier gas from the carrier gas module. The other end cap is connected to the pressure control module, which maintains the desired pressure for the synthesis and exhausts the waste gases into a hood. The temperature controller of the furnace can precisely control the temperature of the tube furnace during the growth process [25].

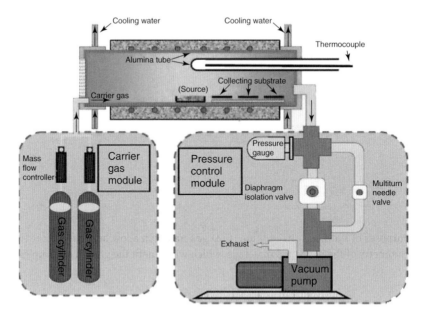

Figure 4.2.5 Schematic diagram of a furnace system, which includes furnace, pressure control module, and carrier gas module [25]. (Reproduced with permission from Dr. Rusen Yang.)

In most cases, source powders were placed in the middle of the furnace and the nanostructures were deposited on collecting substrates, which is placed at a lower temperature region in the downstream direction of the carrier gas. Temperature control is one of the most important parameters for nanostructure synthesis. The furnace usually has a thermocouple placed at its center but outside the alumina tube. This thermocouple is used to monitor the heating progress of an experiment. The temperature gradient along the tube and the local deposition temperature are critical for controlling the nanostructures. The thermocouple inside the tube can be used for the calibration purpose.

Besides the tube furnace, this synthesis setup also contains two more important parts, the carrier gas module and the pressure control module, which are used to control the gas type and flow speed and the pressure level. The carrier gas module consists of gas cylinders, regulators, and mass flow controllers. Different gases or gas mixtures, such as N_2, Ar, and O_2, are introduced and regulated by the mass flow controller. The pressure control module is connected to the outlet of the furnace system. The vacuum level is monitored with pressure gauges. The pressure within the synthesis system can be controlled by two valves. One valve is for coarse adjustment and the other is for fine adjustment. Mechanical pumps are used for the growth of most nanostructures although other pumps can also be incorporated to improve the vacuum.

4.2.2.2 Laser Ablation Setup

PLD is a conventional method for high-quality thin-film deposition, where a high-power pulsed laser beam is focused inside a vacuum chamber to strike a target of the desired composition. Material is then vaporized from the target and deposited as a thin film on a substrate facing the target. Inspired by the fast development of nanoscience and nanotechnology, the laser ablation technique has also been used to fabricate nanostructures and achieved great success with different nanomaterials, including carbon nanotubes [19, 26, 27].

Laser ablation is the process of removing materials from a solid surface by irradiating it with a laser beam. The materials are heated by the absorbed laser energy and evaporate or sublime at low laser flux. In contrast, the materials are typically converted to plasma at high laser flux. Pulsed lasers are usually used for this process, while a continuous wave laser beam can also do the job as long as the laser intensity is high enough.

Pulsed laser ablation has many advantages in fabricating nanostructures. First, due to the highly intense energy of the laser spot, almost any material can be ablated for purposes of synthesis. In addition, PLD generally can allow better control over stoichiometry of the deposited materials, which will benefit the growth of complex functional materials.

The furnace and the laser ablation can also be integrated together to allow the creation of nanostructure using the TAPLD process, as shown in Figure 4.2.6. The reaction chamber, temperature control, carrier gas module, and pressure control in the furnace system are linked to a pulsed laser system. The pulse energy and the repetition rate can be adjusted for the optimum synthesis condition. The pressure,

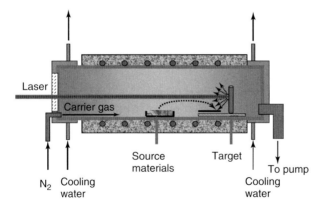

Figure 4.2.6 Schematic setup for laser ablation system [25]. (Reproduced with permission from Dr. Rusen Yang.)

temperature, and carrier gas can also be adjusted with the furnace system, as mentioned in the previous section. Such an integrated system can be tuned for the growth of novel nanostructures with excellent controllability.

4.2.2.3 Experimental Procedure

TAPLD shares a common furnace system with the thermal evaporation experiments. Accordingly, we discuss the thermal evaporation first and add a few notes for laser ablation. There are several processing parameters such as temperature, pressure, carrier gas (including gas species and its flow rate), evaporation time period, and collecting substrates, which can be controlled and need to be selected properly before and/or during the thermal vaporization. Other parameters related to the laser system also need to be considered when PLD is employed, which are discussed separately. The source temperature selection mainly depends on the volatility of the source material. Usually, it is slightly lower than the melting point of the source when pure source materials are used. When a mixture source, such as ZnO with graphite, is used as source material, this temperature can be significantly lowered. The evaporation rate or vapor pressure of the source material(s) can help to determine the pressure. However, the thermodynamic data are not always available for certain materials. In most cases, preliminary testing experiments are necessary to find a reasonable starting point. The carrier gases used are either nitrogen (N_2) or argon (Ar). O_2 can also be used when oxides are desired nanomaterials. In comparison, ammonia is usually used for the growth of nitride nanomaterials. The substrates used to grow nanostructures can be classified into two categories: polycrystalline and single-crystalline substrates. The use of a particular type of substrate over another is determined by the experiment design.

The first step in the synthesis process is the preparation of the alumina tube, source materials, and collecting substrate. The clean tube is placed inside the furnace before loading the source materials and collecting substrates. The source materials are weighed, placed in an alumina crucible, and then transferred to the center of the tube. Collecting substrates, either single crystal or polycrystalline, are cut and placed in the tube and positioned a certain distance away from the source materials in the middle. The substrates were also placed above the substrate in some experiments [28]. Once the substrates and source materials are in position, the alumina tube is sealed with two water-cooling end caps before the next evacuation step.

The thermal evaporation process is very sensitive to the concentration of oxygen in the growth system. Oxygen influences not only the volatility of the source material, and the stoichiometry of the vapor phase, but also the formation of the final products. As such, the synthesis process involves an evacuation process to reduce the initial oxygen content in the system before the synthesis. The system is held at vacuum for a minimum of 60 min, after which the synthesis process begins. A designated pressure and temperature set point need to be chosen before the synthesis process. The carrier gas is introduced into the chamber first. By adjusting the coarse and fine valves on the vacuum system, the rate of evacuation for the chamber can be manipulated and the system pressure can be maintained at a specific set point before evaporation of the source material. The furnace is heated up and maintained at the elevated temperature for a certain time before cooling down. The system is held at constant pressure during the whole process. Once the cooling process is completed, the chamber can be vented and the nanostructures collected and analyzed.

TAPLD uses the same furnace system and follows the aforementioned procedure of preparation, evacuation, thermal evaporation, and cooling off. However, some modification is also needed because of the characteristics of lasers. Laser ablation requires a solid target instead of loose powders. The designed source materials were pressed to form a cylindrical target for laser ablation. The alumina tube was replaced with a quartz tube, which allowed us to precisely focus the laser onto the target.

The target was placed vertically in the chamber, and collecting substrates were positioned close to the target and at the upstream side, as shown in the schematic diagram in Figure 4.2.6. More source materials were also loaded in the middle of the tube. After that, the chamber was evacuated for at least 60 min before introducing the carrier gas. The tube chamber was then maintained at a constant pressure. The excimer laser was turned on with very low pulse energy and repetition rate for alignment and focusing at the target. The laser was turned off, the top of the furnace was closed, and the furnace began to heat up. When the furnace reached, or almost reached, the designed synthesis temperature, the excimer was turned on again with a preset pulse energy and repetition rate. The synthesis would last for a preset time, and once the furnace began to cool off the laser was turned off, in sequence.

4.2.3
Single-Crystalline Branched Zinc Phosphide Nanostructures with TAPLD

4.2.3.1 Properties of Zn_3P_2
In recent years, increasing attention has been paid to new materials for solar cell and optoelectronic devices. Among such compounds, Zn_3P_2 is a relatively new

and promising material. Zn_3P_2 is an $A_3(II)B_2(V)$-type semiconductor with natural p-type conductivity in conventionally grown crystals because of the additional phosphorus atoms in the interstitial positions.

The fundamental absorption measurement for undoped single crystals of Zn_3P_2 presented an exponential edge within 1.4–1.6 eV [29]. Such a location of the absorption edge is exactly at the optimum range demanded for solar energy conversion [30]. In addition, the distinct changes in the absorption curve slopes of polarized light denote the energies of allowed direct transition. The difference between the edges of the photoresponse spectra for polarized light implies that Zn_3P_2 can also be used for detecting a particular type of light polarization. The large optical absorption coefficient ($>10^4$ cm^{-1}) and a long minority diffusion length ($\sim$13 μm) of Zn_3P_2 permits high current collection efficiency [31]. In addition, the constituent materials are abundant and inexpensive and would allow the large-scale deployment of devices such as solar cells, infrared (IR) and ultraviolet (UV) sensors, lasers, and light polarization step indicators [29, 32].

4.2.3.2 Zn_3P_2 Nanostructures

In order to investigate the potential applications of Zn_3P_2, several kinds of hetero-junctions have been designed, such as InP/Zn_3P_2 [33], Mg/Zn_3P_2 [34], $Zn_3P_2/ZnSe$ [35, 36], ITO/Zn_3P_2 [37], and ZnO/Zn_3P_2 [38]. However, the majority of research on Zn_3P_2 has been limited to thin films, and very little work has been done in the nanoscale range except very few reports on the synthesis of Zn_3P_2 nanoparticles [39–41] and on the synthesis of nanotrumpets with an unavoidable ZnO layer coated on the surface [42]. Because of the large excitonic radii, Zn_3P_2 is expected to exhibit a pronounced quantum size effect, which has been observed for Zn_3P_2 nanoparticles [40]. However, the electric property and photoresponse of Zn_3P_2 nanostructures, and heterojunctions made from Zn_3P_2 nanostructures, have not been reported until the success of the integration of the furnace and the pulsed laser ablation.

This section discusses the TAPLD process for the creation of single-crystalline tree-shaped Zn_3P_2 nanostructure arrays, NWs, and nanobelts. The morphology, crystal structure, and optical properties are also determined using electron microscopy and photoluminescence (PL). Optoelectronic measurements of single Zn_3P_2 NWs and the crossed heterojunction indicate that Zn_3P_2 is very sensitive to light and the heterojunction exhibits enhanced performance, which implies the future applications of Zn_3P_2 nanostructures from TAPLD.

4.2.3.2.1 Zn_3P_2 Nanostructure Fabrication through the TAPLD Process

The setup for the fabrication of Zn_3P_2 nanostructures is shown schematically in Figure 4.2.5. The two ends of the tube were closed and the water cooled to ensure a stable synthesis environment inside. A mixture of 0.8 g ZnO and 0.12 g graphite powder was placed in the middle of the tube. In order to transport the source vapor and reduce possible oxidization during the synthesis, N_2 was introduced from one end of the tube at a flow rate of 25 sccm during the entire synthesis process. A cylindrical target containing Zn_3P_2, ZnO, and Zn (in ratios of 2 : 1 : 1

by weight) was placed downstream. A single-crystalline (111) Si wafer was placed in front of the target, serving as a collecting substrate. In order to minimize the residual oxygen, the tube was first evacuated to 10^{-2} mbar and held for 2–3 h before the introduction of N_2 and maintained at a constant pressure of 200 mbar thereafter. The furnace was then heated up at the rate of 50 °C per min to the desired peak temperature of 1100 °C and held for another 20 min before slowly cooling down. Once the temperature exceeded 900 °C, a Compex Series Excimer 102 Laser (248 nm, 10 Hz, 30 kV, ~300 mJ) started to generate pulsed laser energy and kept shining a laser spot, 1 mm wide and 5 mm long, on the target during the synthesis until the furnace was cooled down to 900 °C.

When the synthesis process was completed, dark yellowish fuzzy materials were found around the laser-ablated pit on the target and on the surface of the substrate. The morphology of the as-synthesized materials was examined using a LEO 1530 field emission scanning electron microscope (SEM), and the composition of Zn and P was first confirmed with energy-dispersive X-ray spectrum (EDS) attached to the SEM.

4.2.3.2.2 Structural Characterization

Figure 4.2.7 shows typical SEM images of as-synthesized Zn_3P_2 nanostructures. A top view of the as-synthesized nanostructures on a Si substrate is present in Figure 4.2.7a and its inset, which shows that those hierarchical nanostructures take a sixfold symmetry. The tree shape is further revealed in the side view of the Zn_3P_2 nanostructures in Figure 4.2.7b, which is the as-synthesized product from the target lying on a conductive carbon tape for SEM imaging. The constituent branches of those tree-shaped structures can be several tens of micrometers in length. The diameter of those branches is in the range of 10 nm to a few hundred nanometers.

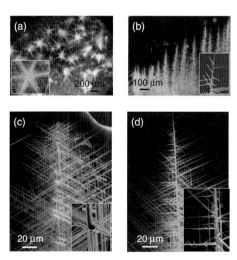

Figure 4.2.7 SEM images of tree-shaped nanostructures of Zn_3P_2 [25]. (Reproduced with permission from the American Chemical Society and Dr. Rusen Yang.)

Generally, the branches at the root of the central trunk are larger than those at the top part. A single branch is uniform and gradually gets smaller toward its tip. The tree-shaped nanostructure can get very complex with the growth of secondary branches, as shown in Figure 4.2.7c,d. Nanostructures seem to have only four sets of orientations: three sets of branches have a sixfold symmetry with the fourth set having branches in the middle and perpendicular to the aforementioned three sets. Such regularity indicates a direct correlation between the morphology and its crystallographic structures. This correlation is also confirmed in transmission electron microscopy (TEM) analysis and is discussed later on.

Zn_3P_2 nanobelts are also formed from this TAPLD process, as shown in Figure 4.2.8. The thickness is in the range of 100 nm to several hundred nanometers, but the same nanobelt has a very uniform thickness throughout the entire belt. In comparison, the widths vary dramatically, even within a single nanobelt. As can be seen from Figure 4.2.8a, some nanobelts can be as wide as 10 μm at the root and get noticeably narrower toward the tip. Smaller nanobelts have also been observed, as shown in the inset in Figure 4.2.8b with a width around 100 nm. Big belts are also observed (Figure 4.2.8c). No matter how wide the belt is, the thickness is generally well below 100 nm. Considering their super long length, those nanobelts are still overall very uniform. Besides those hierarchical nanostructures and derivative nanobelts, numerous Zn_3P_2 NWs with diameters up to 100 nm have also been found on the Si substrates.

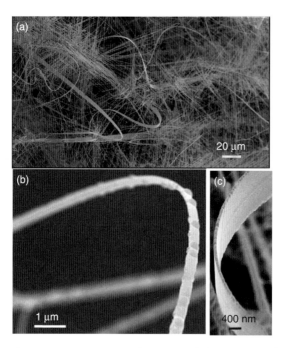

Figure 4.2.8 SEM images of Zn_3P_2 nanobelts [25]. (Reproduced with permission from the American Chemical Society and Dr. R.S. Yang.)

EDS analysis confirmed the composition of tree-shaped nanostructures, NWs, and nanobelts. A typical spectrum in Figure 4.2.9a, which was taken from a Zn_3P_2 branch during TEM analysis, indicates the presence of Zn and P from the nanostructure, with Cu and C peaks originating from the sample grid, and O peaks from contamination during the sample preparation or from partial oxidization of the nanostructure.

X-ray diffraction (XRD) characterization in Figure 4.2.9b, performed using a PANalytical X-Pert Pro MRD with copper K-alpha radiation, confirms that the phase of the nanostructures is tetragonal Zn_3P_2 (JCPDS Card No. 65-2854) with lattice constants $a = 8.095$ Å and $c = 11.47$ Å. XRD analysis indicates that the synthesized nanostructures are single-phased Zn_3P_2 with very good crystallinity. No Zn, ZnO, or any other phases are detectable.

Zn_3P_2 nanostructures were further examined using a Hitachi HF-2000 (field emission gun (FEG)) TEM, as shown in Figure 4.2.10. Figure 4.2.10a presents a TEM image of a part of a Zn_3P_2 nanostructure with selected area electron diffraction (SAED) patterns recorded from the circled areas labeled (b–g) and shown in Figure 4.2.10b–g, respectively. All the branches in Figure 4.2.10a are in the same plane, and the angle between them is about 60°. As expected, all the SAED patterns from different locations are along the same zone axis, [021]. In addition, in-plane rotations are found for some of the SAED patterns, indicating the existence of twin structures when the branches are formed. Figure 4.2.10h presents

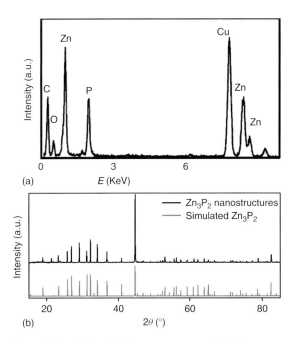

Figure 4.2.9 EDS and XRD measurements of Zn_3P_2 nanostructures [23]. (Reproduced with permission from the American Chemical Society.)

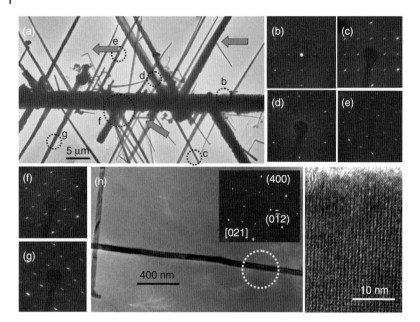

Figure 4.2.10 A detailed crystal structure analyses with TEM. (a) Low-magnification TEM image of Zn_3P_2 structure with SAED patterns from circled areas labeled (b–g) and shown in (b–g), respectively. The arrowheads indicate secondary branch growth. (h) TEM image of Zn_3P_2 nanowires with SAED from the circled area and (i) a high-resolution TEM image [23]. (Reproduced with permission from the American Chemical Society and Dr. Rusen Yang.)

a TEM image of a Zn_3P_2 nanobelt with the same SAED as that of nanostructures in Figure 4.2.10a, which is reasonable because nanobelts always grow from those nanostructures, based on SEM observation. According to those SAED patterns, we conclude that the central wire in Figure 4.2.10a and the nanobelts in Figure 4.2.10h have a growth front of $(0\bar{1}2)$. The top and bottom branches in Figure 4.2.10a have growth fronts of $(3\bar{1}2)$, $(31\bar{2})$, or the opposite directions. Secondary branches can also grow from grown branches, as indicated by arrowheads in Figure 4.2.10a. The high-resolution TEM image in Figure 4.2.10i taken from the circled area in Figure 4.2.10h illustrates the nearly perfect single-crystalline arrangement of atoms in the NWs. Amorphous materials are scarcely seen on the surface. In addition, no ZnO layer has ever been seen on the surface of those Zn_3P_2 nanostructures [42]. Electron microscopy and XRD data demonstrate that perfectly crystalline Zn_3P_2 nanostructures free from an oxidization layer can be obtained.

4.2.3.3 Properties and Devices Fabrication

4.2.3.3.1 Optical Properties
The optical performance of Zn_3P_2 nanostructures is very interesting. Figure 4.2.11 shows reflection spectrum from 1 to 2 eV and PL spectrum under the excitation

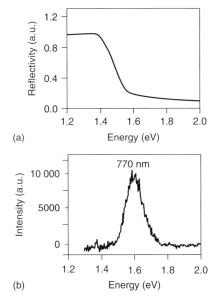

(a)

(b)

Figure 4.2.11 (a) Reflection spectrum of Zn_3P_2 nanostructures. (b) The corresponding photoluminescence spectrum [23]. (Reproduced with permission from the American Chemical Society.)

of 326 nm at room temperature. From the reflectance spectrum it can be seen that energies below 1.4 eV are almost totally reflected by the Zn_3P_2 nanostructures without absorption, while a distinct edge was found at ~1.4 eV, indicating a strong absorption of higher energies [43]. The corresponding PL spectrum provides direct evidence of this distinct absorption edge, at which a broad peak of 1.4–1.7 eV can be found. The broad spectrum found in PL likely results from the free exciton emission near the band edge, namely, acceptor-bound excitons [44].

4.2.3.3.2 Device Design and Fabrication Techniques

Without going into too much of detail, we present the schematic structure of the devices in the top inset in Figure 4.2.12a. The whole process includes four steps, which includes bottom electrode fabrication with photolithography followed by metallization and liftoff, nanostructure manipulation with dielectrophoresis technique, top electrode deposition with focus ion beam lithography, and finally device characterization.

We used a light-emitting diode (LED) as a light source, green (532 nm, <5 mW), red (680 nm, <5 mW), and white lamp (continuous wavelength) to test the photoresponse of single Zn_3P_2 NWs and a heterojunction structure with ZnO/Zn_3P_2 NWs. Measurement in the dark verified the ohmic contact between the NW and the electrode. Utilization of a different light source revealed the response of the tested device to different wavelengths.

The performance of a single Zn_3P_2 nanostructure is shown in Figure 4.2.12a. The representative nanodevice is schematically illustrated in the upper inset of

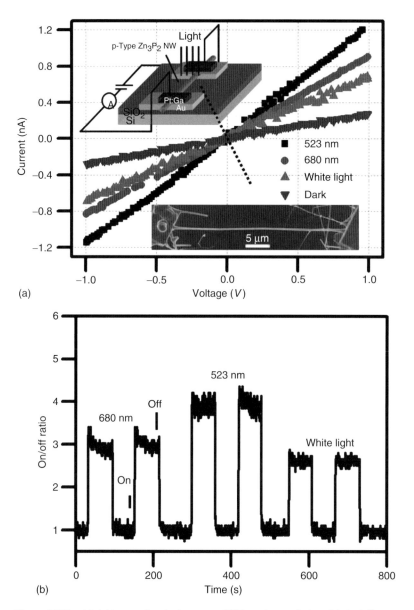

Figure 4.2.12 (a) I–V curve for single Zn₃P₂ NW in dark or under the illumination of different wavelength light. Upper inset shows the device configuration under the illumination of light. Bottom inset shows the corresponding SEM image of the Zn₃P₂ NW-based nanodevice. (b) On/off ratio as a function of the time under red (680 nm), green (523 nm), or white light illumination, respectively [23]. (Reproduced with permission from the American Chemical Society.)

Figure 4.2.12a, and the corresponding SEM image of the nanodevice is inset in Figure 4.2.12a. The linear I–V characteristic without illumination of light (blue line) confirms the ohmic contact between the electrodes and Zn_3P_2 NW, and a resistance of about $3.63 \times 109 \, \Omega$ is derived. The resistivity of Zn_3P_2 NW was calculated to be about $5.6 \times 102 \, \Omega$ cm with the length and the diameter as 35 μm and 232 nm, respectively, based on the SEM characterization. The dark resistance here is higher than that of its thin-film counterpart $(2-3 \times 10^2) \, \Omega$ cm [45], possibly due to the contact resistance.

On the other hand, under the illumination of light at a wavelength of 532 nm (green color with power <5 mW), 680 nm (red color with power <5 mW), or continuous wavelength from a white LED lamp, the resistance decreased significantly due to excess electron–hole pairs (EHPs) excited by the illuminating light, which has an energy larger than the band gap of Zn_3P_2. In addition, the desorption of the contaminants from the surface due to the illumination increased the conductivity as well. It is reasonable to expect that the shorter wavelength results in higher photoconductivity because the light with higher energy can create more EHPs inside the NW. In addition, the excess energy ($h\nu$-Eg) created phonons in the lattice and raised the temperature, resulting in larger electron and hole mobility and contributing a higher conductivity. The on/off ratio, which is defined as the current under the illumination over the dark current, is presented in Figure 4.2.12b as a function of the illuminated time, indicating the high sensitivity of Zn_3P_2 NWs to the illumination of different wavelengths of light. The response time for all of the light is considerably less than 1 s with the on/off ratio being about 3 for red light (680 nm), 4–5 for green light (532 nm), and 2–3 for white light from an LED lamp.

Having a shorter response time and a higher on/off ratio than that of a single NW [46], the p–n photodiode is an alternative form of the photoconductor and can be constructed with crossed NWs [47, 48]. Two-step dielectrophoresis with ZnO NWs and then Zn_3P_2 NWs can form a crossed structure with the heterojunction formed at the interface. The final structure and schematic diagram are illustrated in the insets of Figure 4.2.13a.

The ZnO/Zn_3P_2 nano heterojunction can be considered as an abrupt N^+P junction, in which the depletion region is almost located at the Zn_3P_2 side [38]. The N^+P junction characteristic is revealed by the rectification behavior of the I–V measurement in Figure 4.2.13a. The I–V curves for the photodiode under reverse bias and in the dark and under the illumination of light with wavelengths of 532 or 680 nm are presented in Figure 4.2.13b, showing an apparent current enhancement by the light. The increase in the current results from the generation of EHPs inside the depletion region and nearby under the excitation of the light.

The lights have a greater effect on a reverse-biased photodiode than on a forward-biased photodiode, consistent with the fact that the photoexcited EHPs can significantly influence the concentration of minority carriers, which dominate the current through a reverse-biased diode, as shown in Figure 4.2.13b. Interestingly, all of the current from the illumination of light shows a similar trend: significantly greater than dark current under low reverse bias and more and more comparable to dark current under higher reverse bias. At low reverse bias, photon-generated

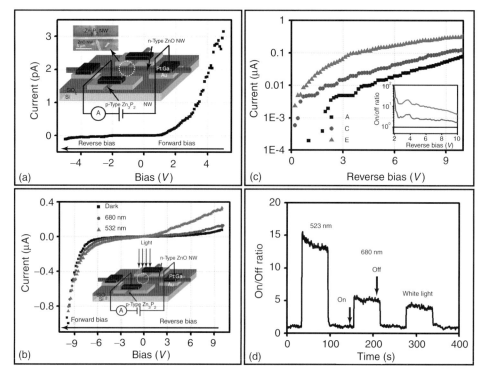

Figure 4.2.13 (a) I–V curve for ZnO/Zn₃P₂ nanoscale heterojunction at reverse and forward bias. Inset shows the prototype of the nanodevice, in which a ZnO NW was placed on top of a Zn₃P₂ NW. (b) I–V curve of ZnO/Zn₃P₂ heterojunction under illumination of different wavelengths as displayed in logarithmic scale under reverse bias. (c) The on/off ratio as a function of the time under red (680 nm), green (523 nm), or white light illumination, respectively. The inset is the schematic of the device [23]. (Reproduced with permission from the American Chemical Society.)

carriers are overwhelming compared to thermal-generated carriers and result in a surprisingly high on/off ratio (~102), as shown in the inset of Figure 4.2.13c. However, thermal-generated EHPs increase with increased bias and contribute more and more to the reverse current, as can be seen in the dark current curve in Figure 4.2.13a. Accordingly, the on/off ratio should decrease with increased reverse bias, which confidently agrees with our result in the inset of Figure 4.2.13c. Despite the decrease at a high reverse bias, the on/off ratio can still reach 9–10, which is superior to that of other single NWs as photoconductors [49–52]. In addition, the photon response of the photodiode to different lights is shown in Figure 4.2.13d, in which we applied a fixed reverse bias of 5 V and turned on and off the green light (523 nm), red light (680 nm), and the white light lamp serially. The high sensitivity, quick response, and nanoscale size can benefit the ZnO/Zn₃P₂ nanoscale heterojunction as a candidate for a highly efficient and spatially resolved photon detector.

4.2.3.4 Summary of the Zn_3P_2 Nanostructures

In summary, tree-shaped Zn_3P_2 nanostructures, nanobelts, and NWs were synthesized in a TAPLD process. The PL spectrum of Zn_3P_2 nanostructures shows a broad emission centered at 770 nm with a slight blueshift with regard to that of the bulk Zn_3P_2. A strong absorption from UV to near-IR is apparent from the reflectivity measurement of those nanostructures. Taking advantage of the direct band gap of 1.5 eV, high absorption coefficient, large minority diffusion length, and high crystallinity free from oxidization, those nanostructures have potential applications in solar cells, broad range photodetectors, lasers, and so on. Preliminary investigation into their application has been performed by the optoelectric measurement of a single Zn_3P_2 NW and a nanoscale photodiode. Considering their small size, we expect that Zn_3P_2 nanostructures grown from the TAPLD process will play an important role in nano-optoelectronics. In addition, the TAPLD process can also be used for the creation of other nanomaterials. The next section gives another example, iron oxide nanostructures from the TAPLD process.

4.2.4
Aligned Ferrite Nanorods, NWs, and Nanobelts with the TAPLD Process

The TAPLD process has also successfully created aligned α-Fe_2O_3 (hematite), ε-Fe_2O_3, and Fe_3O_4 (magnetite) nanorods, nanobelts, and NWs on alumina substrates. Gold has been used as a catalyst during the growth, and the presence of spherical gold catalyst particles at the tips of the nanostructures indicates the vapor–liquid–solid (VLS) growth mechanism. A series of experiments revealed a primitive "phase diagram" for growing these structures based on several designed pressure and temperature parameters. TEM analysis has shown that the rods, wires, and belts are single crystalline and grow along $<111>_m$ or $<110>_h$ directions. XRD measurements confirm phase and structural analysis. Interesting magnetic behavior, particularly at room temperature, has also been found from superconducting quantum interference device (SQUID) measurements. Magnetite 1D nanostructure was created for the first time without using a template. In addition, long ε-Fe_2O_3 nanobelts and NWs were also observed for the first time.

4.2.4.1 Introduction

The advances in growth and characterization techniques have led to the production of modern magnetic materials, which reveal a range of fascinating phenomena and applications, such as ultrahigh-density memory storage [53, 54] and advanced communications devices [55] with ferrite nanoparticles. ε-Fe_2O_3 is an ideal candidate for such applications because of its high room temperature coercivity (maximum $H_c = 20$ kOe) [56]. Iron oxide NWs and nanorods have been fabricated through templating [57], wet chemical growth [58], and thermal decomposition [59]. However, these methods often suffered from undesired contaminants; lack of control over morphology, size, and orientation; grainy structures; and the need for annealing treatments post synthesis [57, 60]. The integration of PLD and thermal vapor deposition proves to be a promising method for simple synthesis of hematite and magnetite 1D nanostructure arrays and long ε-Fe$_2$O$_3$ nanobelts, overcoming many of the problems encountered with other methods, while still granting spatial and size control. The first known synthesis of ε-Fe$_2$O$_3$ long nanobelts also resulted from the TAPLD method. This section discusses a primitive "phase diagram" for growing these structures based on several designed pressure and temperature parameters. The structure of the NWs was revealed by TEM and XRD, and the magnetic property was investigated by SQUID.

4.2.4.2 Experimental Method

The experimental setup is very similar to the setup in Figure 4.2.6, while the target is placed in the middle of the furnace, as shown in Figure 4.2.14. The cylindrical pallet in the middle was made of ~5 g of Fe$_3$O$_4$ powder (Alfa Aesar 98%). The polycrystalline alumina wafer substrates coated with 2 nm of Au was placed next to the target. Argon (99.999%) flowing at 50 sccm facilitated reduction of adsorbed oxygen and then stopped to allow the system to return to a low pressure of around 0.02 mbar. The furnace was then allowed to heat up at the rate of 20 °C per min to the desired maximum temperature under a well-maintained ambient pressure prescribed according to the chart shown in Figure 4.2.15a. Argon flow gas was simultaneously introduced into the system at 50 sccm and was maintained during the synthesis.

Once the temperature and pressure had stabilized, pulsed laser ablation of the target was started and it lasted for 60 min with Compendex Series Excimer 102 Laser (20 Hz, 30 kV, ~300 mJ). At the end of the ablation, the laser, furnace, and flow gas were turned off, and the system was evacuated back down to 0.02 mbar and cooled to room temperature. The substrate with the deposited sample was carefully collected and loaded onto an SEM stub for analysis. Most samples appeared red/orange to gray in color with rainbow-like fringes on the substrate radiating away from the leading edge.

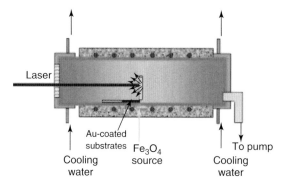

Figure 4.2.14 Schematic showing a PLD synthesis apparatus. Laser energy is directed through a treated glass window to a pressed magnetite target. Vapors are released and re-condensed as 1D nanostructures nucleated on gold catalyst particles [24]. (Reproduced with permission from the American Chemical Society.)

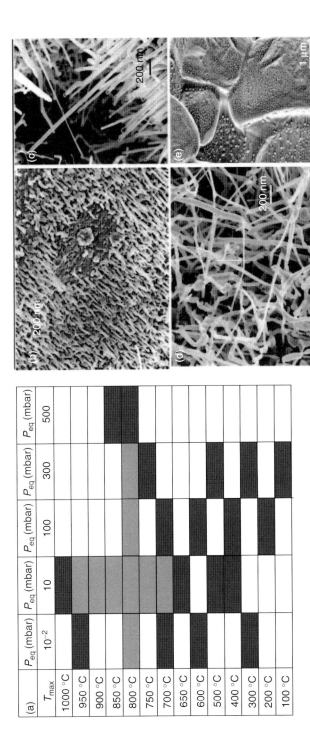

Figure 4.2.15 (a) Table showing tested pressure/temperature combinations and observed outcomes. Each cell represents a unique experiment. Gray cells indicate combinations that produced high-density 1D nanowire growth. Black cells indicate no or very low growth. White cells indicate untested parameter combinations or those in which results were inconclusive. All other parameters were kept constant in these experiments. (b) SEM image showing short rods synthesized at 700 °C, 10 mbar. (c) SEM image showing long belts synthesized at 900 °C, 10 mbar. (d) SEM image displaying greater secondary growth after 120 min of laser energy exposure compared to the usual 60 min growth time at 900 °C, 10 mbar. (e) SEM image showing no 1D growth at 900 °C, 10 mbar without laser energy [24]. (Reproduced with permission from the American Chemical Society.)

4.2.4.3 Results and Discussion

4.2.4.3.1 Systematic Synthesis Study

The conditions necessary for 1D nanostructure growth using the TAPLD process were revealed with a series of experiments in which nanostructure growth was correlated with the systematic variations in experimental parameters, including the maximum ambient temperature (T_{max}) inside the tube and the controlled equilibrium pressure (P_{eq}) during laser ablation, as shown in Figure 4.2.15a. All other variables, such as substrate type, catalyst density, target species, Ar flow rate, temperature ramp rate, ablation time, laser energy, and laser frequency, were unchanged. SEM characterization determines the outcome from each experiment. Black cells in Figure 4.2.15 correspond to experiments without NW growth, gray cells correspond to parameters yielding successful growth, and white cells correspond to untested parameter combinations, or those in which results were inconclusive.

A successful experiment in this study was defined as one showing uniform high-density growth of nanostructures with length dimensions at least three times that of the catalyst particle diameter. The best 1D ferrite nanostructure growth consistently occurred from 700 to 950 °C at 10 mbar pressure. Figures 4.2.15b and 4.2.2c show the length variation of nanostructures according to chamber temperature (T_{max}). Nanorods grown at 700 °C in Figure 4.2.15b were aligned, and single crystalline just like those in Figure 4.2.15c grown at 900 °C, with the only difference being nanostructure length. This result illustrates that ferrite nanostructure size can be tuned with the reaction chamber temperature.

Laser ablation time over 60 min did not significantly increase the mean nanostructure length from the experiments carried out at 900 °C and 10 mbar. However, increased branching and secondary growth were observed in the prolonged growth. Figure 4.2.15d demonstrates this finding with an SEM image taken from a sample ablated for 120 min with all the other parameters the same as that in Figure 4.2.15c. The critical role of laser ablation was also proved with an experiment in the absence of a laser. This experiment was conducted at 10 mbar, 900 °C, a parameter combination producing very high density ferrite nanostructures in the presence of pulsed lasers. The result in Figure 4.2.15e shows no growth at all on the substrate, confirming that the laser irradiation is the principle energy source responsible for target vaporization and the resulting nanostructure growth.

4.2.4.3.2 Electron Microscope Characterization of Magnetite and Hematite Nanostructures

Impurity doping can also be realized with the TAPLD growth method. Mg-doped ferrite NWs were shown in Figure 4.2.16a,b and pure iron oxide nanostructures in Figure 4.2.16c,d. As these images demonstrate, both pure and Mg-doped 1D ferrites are of high-quality single-crystalline nanostructures. The NWs and nanobelts in Figure 4.2.16 were grown at a maximum ambient temperature of 900 °C. Although nanostructure dimensions varied greatly within samples, NWs and nanobelts grown at 900 °C were generally highly dense, and on the order of 1 µm long × 30 nm wide

Figure 4.2.16 SEM images of Mg-doped and pure iron oxide 1D nanostructures. (a) Low-magnification SEM image of aligned iron oxide nanowires doped with Mg showing local alignment along alumina crystallites. Inset shows EDS chemical signature. (b) Higher magnification SEM image showing aligned Mg-doped iron oxide nanorods. (c) SEM image showing long nanobelt structure most likely having ε-Fe$_2$O$_3$ microstructure, without the presence of Mg. (d) SEM image showing a group of aligned iron oxide nanowires with Mg absent [24]. (Reproduced with permission from the American Chemical Society.)

with the longest NWs growing up to ~5 μm. Most long structures appeared to possess a rectangular cross section, giving them a beltlike morphology; however, those with a more circular cross section were observed as well, especially at the higher temperatures. In comparison, nanorods were the prevalent morphology, with typical dimensions 500 nm long × 30 nm wide under the 700 °C condition. The density of these structures increases as temperature increases.

Each grain of the substrate in Figure 4.2.16a promotes aligned growth in a specific direction relative to the plane of the substrate. In other words, some grain faces are more favorable and resulted in the growth of longer NWs. Later synthesis on c-plane and a-plane sapphire (single-crystalline Al_2O_3) produced high-density arrays with perpendicular and in-plane growth, respectively, which further confirmed the influence of the crystallographic structure on the NW growth. In addition, a spherical gold particle was always observed at the tip of each NW, as shown in Figure 4.2.16b, which indicates that the NW was always initiated by the gold catalyst particle. As a result, researchers should be able to achieve the control of the angle of growing nanostructures relative to the substrate and the spatial location by carefully engineering surface orientation on single-crystalline substrates and the precise placement of the catalyst materials.

Figure 4.2.16c is a typical image of the frequently observed long nanobelts from TAPLD growth. Electron diffraction analysis, as well as XRD data, indicates that these long nanobelts exhibit a ε-Fe_2O_3 phase of iron oxide. It has been reported that addition of column II cations such as Sr_2^+ or Ba_2^+ to ε-Fe_2O_3 stabilized the phase and allowed for growth of larger nanocrystals [56]. EDS analysis did not reveal Mg in all long ε-Fe_2O_3 nanobelts from the TAPLD process, indicating that its presence is not necessary for ε-Fe_2O_3 nanobelt formation using this method.

TEM characterization revealed both hematite and magnetite nanostructures from the same sample, as shown in Figure 4.2.17a–e. Adding a piece of iron foil to the chamber to act as an oxygen "getter" helped increase the relative magnetite yield. This finding indicates that a phase can be controlled through careful variation of oxygen partial pressure within the tube. The bright field image of a magnetite NW in Figure 4.2.17a has a spherical gold catalyst particle visible at the tip, implying that the NW follows a VLS growth mechanism. The high-resolution TEM image in Figure 4.2.17b illustrates the near-perfect single-crystalline arrangement of atoms in the NW. Very little amorphous material is present on the surface. The electron diffraction pattern in the inset provides further evidence of the single-crystalline nature of this NW and shows the growth direction along [$1\bar{1}1$]. However, magnetite NWs along the <110> directions have also been observed during TEM characterization.

Hematite nanorods with Au catalyst particles at the tip were also shown in Figure 4.2.17c. The high-resolution image in Figure 4.2.17e clearly shows the ordered arrangement of atoms and the lack of a significant amorphous layer at the surface. The nanorod in this figure grew along the [$01\bar{1}$] direction, as seen in the electron diffraction pattern shown in Figure 4.2.17d.

4.2.4.4 Summary of the Iron Oxide Nanostructures

TAPLD was successfully employed for the growth of single-crystalline magnetite and hematite NWs and nanobelts. This same method also produces the first known long ε-Fe_2O_3 and Mg stabilized ε-Fe_2O_3 NWs and nanobelts. This TAPLD growth method offers many benefits over previous synthesis methods, including higher quality structures, higher density structures, greater spatial and size control, ability to grow complex structures, and obviating the need for a template. The systematic

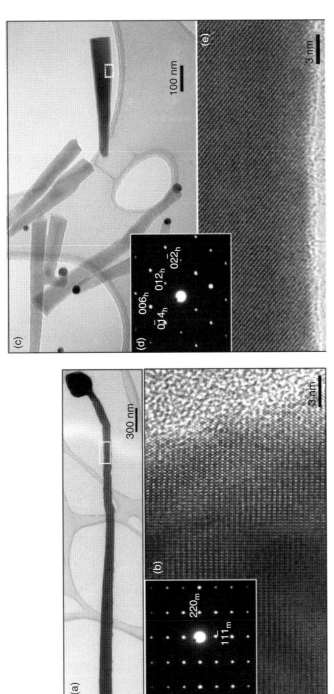

Figure 4.2.17 TEM data on magnetite and hematite nanostructures. (a) Low-magnification bright-field TEM image of a magnetite nanowire. (b) High-resolution TEM image of a nanowire taken from the side edge of the wire in (a), illustrating high-quality single-crystalline nature of the wire. (c) Low-magnification bright-field TEM image of hematite nanorods with Au particle tips. (d) Electron diffraction image taken from the dark rod in the previous image shows the single-crystalline nature of the rod. (e) High-magnification TEM image of the section outlined in (c), showing well-ordered structure and very little amorphous material on nanorod surface [24]. (Reproduced with permission from the American Chemical Society.)

study of the effect of growth parameters on the size, morphology, and phase of the grown product should be helpful for those interested in utilizing this technique as well as those hoping to further elucidate the underlying mechanisms behind VLS NW growth.

References

1. (a) Sun, Y., Fuge, G.M., and Ashfold, M.N.R. (2004) Growth of aligned ZnO nanorod arrays by catalysis-free pulsed laser deposition methods. *Chem. Phys. Lett.*, **396**, 21–26. (b) Morales, A.M. and Lieber, C.M. (1998) A laser ablation method for the synthesis of crystalline semiconductor nanowires. *Science*, **279**, 208–211.

2. Perriere, J., Millon, E., Seiler, W., Boulmer-Leborgne, C., Craciun, V., Albert, O., Loulergue, J.C., and Etchepare, J. (2002) Comparison between ZnO films grown by femtosecond and nanosecond laser ablation. *J. Appl. Phys.*, **91**, 690–696.

3. Dai, Z.R., Pan, Z.W., and Wang, Z.L. (2002) Growth and structure evolution of novel tin oxide diskettes. *J. Am. Chem. Soc.*, **124**, 8673–8680.

4. Shen, Y., Hong, J.I., Xu, S., Fang, H., Zhang, S., Ding, Y., Snyder, R.L., and Wang, Z.L. (2010) A general approach for fabricating arc-shaped composite nanowire arrays by pulsed laser deposition. *Adv. Funct. Mater.*, **20**, 703–707.

5. Han, X.D., Zhang, Z., and Wang, Z.L. (2007) Experimental nanomechanics of one-dimensional materials by in-situ microscopy. *Nano*, **2**, 249–271.

6. Norton, D.P., Park, C., Budai, J.D., Pennycook, S.J., and Prouteau, C. (1999) Plume-induced stress in pulsed laser deposited CeO_2 films. *Appl. Phys. Lett.*, **74**, 2134–2136.

7. Erwin, S.C., Zu, L., Haftel, M.I., Efros, A.L., Kennedy, T.A., and Norris, D.J. (2005) Doping semiconductor nanocrystals. *Nature*, **436**, 91–94, and references therein.

8. Lin, S.S., Song, J.H., Lu, Y.F., and Wang, Z.L. (2009) Identifying individual n- and p-type ZnO nanowires by

the output voltage sign of piezoelectric nanogenerator. *Nanotechnology*, **20**, 365703.

9. Lin, S.S., Hong, J.I., Song, J.H., Zhu, Y., He, H.P., Xu, Z., Wei, Y.G., Ding, Y., Snyder, R.L., and Wang, Z.L. (2009) Phosphorus doped $Zn(1-x)Mg(x)O$ nanowire arrays. *Nano Lett.*, **9**, 3877–3882.

10. Wang, Z.L. (2004) Zinc oxide nanostructures: growth, properties and applications. *J. Phys.: Condens. Matter.*, **16** (25), R829–R858.

11. Pan, Z.W., Dai, Z.R., and Wang, Z.L. (2001) Nanobelts of semiconducting oxides. *Science*, **291** (5510), 1947–1949.

12. Kong, X.Y. *et al.* (2004) Single-crystal nanorings formed by epitaxial self-coiling of polar nanobelts. *Science*, **303** (5662), 1348–1351.

13. Kong, X.Y. and Wang, Z.L. (2003) Spontaneous polarization-induced nanohelixes, nanosprings, and nanorings of piezoelectric nanobelts. *Nano Lett.*, **3** (12), 1625–1631.

14. Yang, R.S., Ding, Y., and Wang, Z.L. (2004) Deformation-free single-crystal nanohelixes of polar nanowires. *Nano Lett.*, **4** (7), 1309–1312.

15. Gao, P.M. *et al.* (2005) Conversion of zinc oxide nanobelts into superlattice-structured nanohelices. *Science*, **309** (5741), 1700–1704.

16. Lao, C.S. *et al.* (2006) Formation of double-side teethed nanocombs of ZnO and self-catalysis of Zn-terminated polar surface. *Chem. Phys. Lett.*, **417** (4–6), 358–362.

17. Dai, Y. *et al.* (2002) Synthesis and optical properties of tetrapod-like zinc oxide nanorods. *Chem. Phys. Lett.*, **358** (1–2), 83–86.

18. Gao, P.X. and Wang, Z.L. (2004) Nanopropeller arrays of zinc oxide. *Appl. Phys. Lett.*, **84** (15), 2883–2885.

19. Kroto, H.W. *et al.* (1985) C60: buckminsterfullerence. *Nature*, **318**, 162–163.

20. Thess, A. *et al.* (1996) Crystalline ropes of metallic carbon nanotubes. *Science*, **273**, 483–487.

21. Ganesan, P.G. *et al.* (2005) ZnO nanowires by pulsed laser vaporization: synthesis and properties. *J. Nanosci. Nanotechnol.*, **5** (7), 1125–1129.

22. Liu, Z.Q. *et al.* (2003) Laser ablation synthesis and electron transport studies of tin oxide nanowires. *Adv. Mater.*, **15** (20), 1754–1757.

23. Yang, R.S. *et al.* (2007) Single-crystalline branched zinc phosphide nanostructures: synthesis, properties, and optoelectronic devices. *Nano Lett.*, **7**, 269–275.

24. Morber, J.R. *et al.* (2006) PLD-assisted VLS growth of aligned ferrite nanorods, nanowires, and nanobelts-synthesis, and properties. *J. Phys. Chem. B*, **110** (43), 21672–21679.

25. Yang, R. (2007) Oxide Nanomaterials: synthesis, structure, properties and novel devices, PhD Thesis, Georgia Institute of Technology.

26. Guo, T. *et al.* (1995) Catalytic growth of single-walled nanotubes by laser vaporization. *Chem. Phys. Lett.*, **243** (1–2), 49–54.

27. Guo, T. *et al.* (1995) Self-assembly of tubular fullerenes. *J. Phys. Chem.*, **99** (27), 10694–10697.

28. Kuo, T.J., Lin, C.N., Kuo, C.L., and Huang, M.H. (2007) Growth of ultralong ZnO nanowires on silicon substrate by vapor transport and their use as recyclable photocatalysts. *Chem. Mater.*, **19**, 5143–5147.

29. Misiewicz, J. *et al.* (1994) Zn3P2 – a new material for optoelectronic devices. *Microelectron. J.*, **25** (5), R23–R28.

30. Loferski, J.J. (1956) Theoretical considerations governing the choice of the optimum semiconductor for photovoltaic solar energy conversion. *J. Appl. Phys.*, **27** (7), 777–784.

31. Fessenden, R.W., Sobhanadri, J., and Subramanian, V. (1995) Minority-carrier lifetime in thin-films of Zn3P2 using microwave and optical transient measurements. *Thin Solid Films*, **266** (2), 176–181.

32. Kakishita, K., Aihara, K., and Suda, T. (1994) Zinc phosphide epitaxial-growth by photo-mocvd. *Appl. Surf. Sci.*, **80**, 281–286.

33. Park, M.H. *et al.* (1996) Low resistance Zn3P2/InP heterostructure Ohmic contact to p-InP. *Appl. Phys. Lett.*, **68** (7), 952–954.

34. Hava, S. (1995) Polycrystalline Zn_3P_2 Schottky photodiode – vacuum surface effects. *J. Appl. Phys.*, **78** (4), 2808–2810.

35. Kakishita, K., Aihara, K., and Suda, T. (1994) Zinc phosphide epitaxial-growth by photo-MOCVD. *Appl. Surf. Sci.*, **80**, 281–286.

36. Kakishita, K., Aihara, K., and Suda, T. (1994) Zn3P2 photovoltaic film growth for Zn3P2/Znse solar-cell. *Solar Energy Mater. Solar Cells*, **35** (1–4), 333–340.

37. Suda, T. (1990) Zinc phosphide thin-films grown by plasma assisted vapor-phase deposition. *J. Cryst. Growth*, **99** (1–4), 625–629.

38. Nayar, P.S. (1982) Properties of zinc phosphide zinc-oxide heterojunctions. *J. Appl. Phys.*, **53** (2), 1069–1075.

39. Weller, H., Fojtik, A., and Henglein, A. (1985) Photochemistry of semiconductor colloids – properties of extremely small particles of Cd3P2 and Zn3P2. *Chem. Phys. Lett.*, **117** (5), 485–488.

40. Green, M. and O'Brien, P. (2001) A novel metalorganic route to nanocrystallites of zinc phosphide. *Chem. Mater.*, **13** (12), 4500–4505.

41. Buhro, W.E. (1994) Metalloorganic routes to phosphide semiconductors. *Polyhedron*, **13** (8), 1131–1148.

42. Shen, G.Z. *et al.* (2006) Single-crystalline trumpetlike zinc phosphide nanostructures. *Appl. Phys. Lett.*, **88** (14), 143105.

43. Bryja, L., Jezierski, K., and Misiewicz, J. (1993) Optical-properties of Zn_3P_2 thin-films. *Thin Solid Films*, **229** (1), 11–13.

44. Fuke, S. *et al.* (1986) Some properties of Zn_3P_2 polycrystalline films prepared by hot-wall deposition. *J. Appl. Phys.*, **60** (7), 2368–2371.

45. Kakishita, K., Ikeda, S., and Suda, T. (1991) Zn3P2 epitaxial-growth by

MOCVD. *J. Cryst. Growth*, **115** (1–4), 793–797.

46. Sze, S.M. (1981) *Physics of Semiconductor Devices*, John Wiley & Sons, Inc., New York.

47. Duan, X.F. *et al.* (2001) Indium phosphide nanowires as building blocks for nanoscale electronic and optoelectronic devices. *Nature*, **409** (6816), 66–69.

48. Zhong, Z.H. *et al.* (2003) Synthesis of p-type gallium nitride nanowires for electronic and photonic nanodevices. *Nano Lett.*, **3** (3), 343–346.

49. Philipose, U. *et al.* (2006) Conductivity and photoconductivity in undoped ZnSe nanowire array. *J. Appl. Phys.*, **99** (6), 066106.

50. Law, J.B.K. and Thong, J.T.L. (2006) Simple fabrication of a ZnO nanowire photodetector with a fast photoresponse time. *Appl. Phys. Lett.*, **88** (13), 133114.

51. Mathur, S. *et al.* (2005) Size-dependent photoconductance in SnO_2 nanowires. *Small*, **1** (7), 713–717.

52. Kim, K. *et al.* (2005) Photoconductivity of single-bilayer nanotubes consisting of poly(p-phenylenevinylene) (PPV) and carbonized-PPV layers. *Adv. Mater.*, **17** (4), 464–468.

53. Fried, T., Shemer, G., and Markovich, G. (2001) Ordered two-dimensional arrays of ferrite nanoparticles. *Adv. Mater.*, **13**, 1158–1161.

54. Gubin, S.P., Spichkin, Y.I., Yurkov, G.Y., and Tishin, A.M. (2002) Nanomaterial for high-density magnetic data storage. *Russ. J. Inorg. Chem.*, **47**, S32–S67.

55. Chen, R.S., Yung, E.K.N., Ji, F., and Dou, W.B. (2003) Development of the microstrip circulator with a magnetized ferrite sphere in millimeter waveband. *Int. J. Infrared Millimeter Waves*, **24**, 813–828.

56. Ohkoshi, S., Sakurai, S., Jin, J., and Hashimoto, K. (2005) The addition effects of alkaline earth ions in the chemical synthesis of epsilon-Fe_2O_3 nanocrystals that exhibit a huge coercive field. *J. Appl. Phys.*, **97**, 10K312.

57. Zhang, D., Liu, Z., Han, S., Li, C., Lei, B., Stewart, M.P., Tour, J.M., and Zhou, C. (2005) Magnetite (Fe_3O_4) core-shell nanowires: Synthesis and magnetoresistance. *Nano Lett.*, **4**, 2151–2155.

58. Liao, Z.-M., Li, Y.-D., Xu, J., Zhang, J.-M., Xia, K., and Yu, D.-P. (2006) Spin filter effect in magnetite nanowire. *Nano Lett.*, **6**, 1087–1091.

59. Kelm, K. and Mader, W. (2005) Synthesis and structural analysis of epsilon-Fe_2O_3. *Z. Anorg. Allg. Chem.*, **631**, 2383–2389.

60. Yang, J.B., Xu, H., You, S.X., Zhou, X.D., Wang, C.S., Yelon, W.B., and James, W.J. (2006) Large scale growth and magnetic properties of Fe and Fe_3O_4 nanowires. *J. Appl. Phys.*, **99**, 08Q507.

4.3
ZnO Nanowire and Its Heterostructures Grown with Nanoparticle-Assisted Pulsed Laser Deposition

Bingqiang Cao, Ruiqian Guo, and Tatsuo Okada

Semiconductor nanowires have attracted considerable recent research interest because of their unique physical properties and possible applications as building blocks for functional nanodevices. However, the fundamental research and application development were greatly hampered by the limited rational growth of semiconductor nanowires and their heterostructures. The aim of this chapter is to provide an updated review of the progress in pulsed laser deposition (PLD) growth of zinc oxide (ZnO) nanowires. A proposed nanoparticle-assisted pulsed laser deposition (NAPLD) growth mechanism is discussed in detail, which is different from the traditional vapor–liquid–solid (VLS) nanowire growth mechanism. Various experimental conditions that influence the NAPLD growth of ZnO nanowires are discussed. This is followed by a discussion of the two-step growth of homogeneous ZnO-based core–shell nanowire heterostructures using the low-density nanowire arrays as physical templates. The final part is a summary and a personal outlook.

4.3.1
Introduction

One-dimensional (1D) nanowires, including nanobelts and nanorods, represent an important kind of nanostructures at the forefront of nanoscience and nanotechnology. Nanowires are typically single-crystalline, highly anisotropic, semiconducting, insulating, and/or metallic nanostructures that result from rapid growth along one direction [1]. They usually have big aspect ratio (length/diameter) and uniform cross section. It is generally accepted that nanowires provide a good system to investigate the dependence of electrical and thermal transport or mechanical properties on dimensionality and size. Nanowires are also expected to play an important role as both interconnects and functional units for optoelectric nanodevices as they represent one of the best defined and controlled classes of nanoscale building blocks [2]. The controlled growth has been the focus of much attention not only in terms of understanding the fundamental growth mechanism but also for the applications in nanoscale optoelectronics. One of the interesting materials is zinc oxide (ZnO), a wide band gap II–VI semiconductor that has a direct band gap of about 3.37 eV and a big exciton binding energy of 60 meV at room temperature [3–6]. It is a very promising functional material suitable for optoelectronic device applications. Growth of nanowires with precisely controlled chemical composition, diameter, length, doping, growth direction, and surface properties is highly desired now. The controlled synthesis of ZnO nanowires and their heterostructures is the first step toward their functional device applications.

Generally speaking, single-crystal nanowires can be synthesized by promoting the fast crystallization of solid structures along one direction through controlling

Nanomaterials: Processing and Characterization with Lasers, First Edition.
Edited by Subhash Chandra Singh, Haibo Zeng, Chunlei Guo, and Weiping Cai.
© 2012 Wiley-VCH Verlag GmbH & Co. KGaA. Published 2012 by Wiley-VCH Verlag GmbH & Co. KGaA.

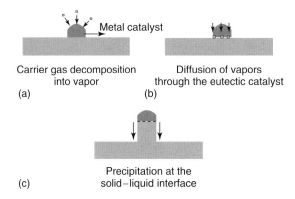

Carrier gas decomposition into vapor

(a)

Diffusion of vapors through the eutectic catalyst

(b)

Precipitation at the solid–liquid interface

(c)

Figure 4.3.1 (a–c) Scheme of the metal catalyst VSL growth mechanism of nanowires.

the experimental parameters during growth. Vapor-phase methods are probably the most widely explored strategies for growth of one-dimensional nanowires. Numerous techniques, such as thermal deposition, chemical vapor deposition (CVD), metal-organic chemical vapor deposition (MOCVD), and molecular beam epitaxy (MBE) have been developed to prepare precursors to the gas phase for the nanowire growth. Among all vapor-phase methods, the so-called vapor–liquid–solid (VLS) mechanism seems to be very successful in growing single-crystalline nanowires in large quantities [7–9].

The VLS mechanism can be divided into three main stages: nucleation, precipitation, and deposition, as demonstrated in Figure 4.3.1 [10–12]. In the nucleation stage, metallic nanoparticle catalysts are formed on a substrate and their diameter is typically around 10 nm. Usually, the background pressure can influence the catalyst size, and the growth temperature needs to be adjusted in order to maintain the catalyst in the liquid state. Then, the source material carrier gas is introduced into a growth chamber maintained above the eutectic temperature (Figure 4.3.1a). The carrier gas reacts in the chamber to form liquid eutectic particles and the source materials diffuse through the catalyst droplets (Figure 4.3.1b). When the eutectic alloy becomes saturated, reacted source materials precipitate at the liquid–solid interface. This is the precipitation process (Figure 4.3.1c). Therefore, anisotropic growth goes on while the source gas flow is maintained. This process finally results in the 1D nanowire growth. At the end of the process, nanowires of high purity are obtained except at one tip, which contains the solidified metallic catalyst.

For ZnO nanowires grown with the VLS method, gold is usually selected as the catalyst for many vapor-phase growth routes [13–15]. However, many of the experimentally achieved growth of ZnO nanowires have not yielded a satisfactory elucidation of the underling mechanism. For example, in most cases, the gold catalysts were not clearly observed at the tips of ZnO nanowires. Another important application for the metal-catalyzed VLS growth is to use catalyst pattern to control the nanowire density. According to the VLS growth mechanism, the ZnO nanowires were expected to show the same density as that of the catalyst pattern. However, the patterned metal catalyst array that is supposed to function as a nucleation template for the subsequent growth of ZnO nanowires [16–19] via the VLS model does not work as well as other semiconductor nanowires [20, 21]. Moreover, as high temperatures in the range of 800–1000 °C are usually needed for the growth of ZnO nanowires, high mobility and diffusion of the metals can destroy the predefined catalyst pattern. Until now, the density control of ZnO nanowire growth is still a challenge. So, when metal catalysts were applied, the exact growth mechanism for ZnO nanowires is still not clear. A clear understanding of the present VLS process and development of a new synthesis mechanism can help design the nanowire growth more rationally and predictably.

Pulsed laser deposition (PLD) is generally recognized as a growth method for thin films, with excellent electronic and optical properties by condensation of a laser plume ablated from a single target, excited by the high-energy laser pulses far from equilibrium [22]. High-quality ZnO films can heteroepitaxy on GaN, sapphire, and even silicon substrates [23]. PLD can also be adopted to grow nanowires. Morales and Lieber [24] first used laser ablation to synthesize silicon and germanium nanowires through the VLS mechanism with iron as catalyst. Few reports of ZnO nanorod synthesis by PLD methods have appeared thus far. Yan *et al.* [25] demonstrated growth of oriented Ga-doped ZnO nanorod arrays on GaN/sapphire substrates, but only at certain specific Ga-doping levels. Lorenz *et al.* [26] have reported growth of ZnO nanowires on sapphire in the presence of gold colloid catalysts.

In this chapter, we review the progress that we have made in growth of ZnO nanowires and their heterostructures through a new growth mechanism, nanoparticle-assisted pulsed laser deposition (NAPLD). After the introduction, in Section 4.3.2, the growth evolution from a two-dimensional (2D) ZnO nanowall to a 1D nanowire by PLD is discussed. Section 4.3.3 presents the nanowire growth process with NAPLD. Controlled growth of ZnO nanowires is highlighted, and influence of different experimental conditions on nanowire growth is discussed in Section 4.3.4. New results of growth of ZnO nanowire heterostructures based on the controlled NAPLD growth are illustrated. Finally, this chapter ends with some concluding remarks.

4.3.2
From 2D Nanowall to 1D Nanowire with PLD

Conventional PLD has been widely adopted to grow ZnO films with relative flat surfaces under high vacuum conditions from 2 to 10^{-5} mtorr [27]. The surface morphology of the ZnO film grown with PLD is important for their surface and interface qualities, such as low-loss surface acoustic wave (SAW), thin quantum wells (QWs), laterally homogeneous metal-semiconductor Schottky contacts, and optical emission properties. Therefore, the control and optimization of surface properties is essential for the successful applications of ZnO thin films in related

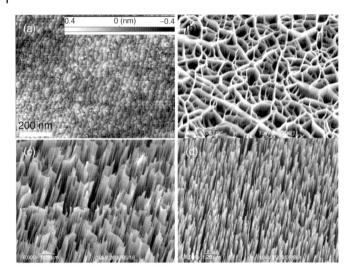

Figure 4.3.2 Morphology evolution from 2D film, nanowall to 1D nanowire grown with PLD under different pressures, (a) 10^{-3} torr, (b) 20 torr, (c) 70 torr, and (d) 140 torr.

device configurations. It has been found that the surface roughness is dependent on the oxygen partial pressure during PLD. For example, the optimized growth pressure for ZnMgO films is about 1×10^{-3} mbar with a minimum average roughness of 0.5 nm. Generally speaking, the average roughness of the ZnO PLD film increases with the growth pressure.

Figure 4.3.2 shows the morphology evolution of the ZnO PLD samples with the growth pressure from 10^{-3} to 140 torr. At high vacuum conditions, the ZnO film has a rather flat surface and the average roughness is only about 1 nm. When the background pressure increases to 20 torr, the surface of the film becomes very rough and exhibits a 2D interconnected nanowall structure. With the growth pressure further increasing, the interconnected nanowalls break and show a longer nanobelt structure. Finally, when the pressure is higher than 100 torr, 1D ZnO nanowire arrays grow on the sapphire substrate. It is clearly illustrated that the growth of ZnO nanowires with PLD was mainly determined by the growth pressure and catalysts were not needed. It is expected that the growth mechanism of ZnO nanowires with PLD is not the traditional VLS process, as discussed in more detail in the following section.

4.3.3
NAPLD Nanowire Growth Mechanism

In order to study the growth mechanism of the ZnO nanowire growth by PLD without a catalyst, the influence of the chamber pressure on the laser plume dynamics was measured with UV Rayleigh scattering during the nanowire growth [28]. Figure 4.3.3 shows the nanoparticle spatial distribution at 1 ms after laser ablation in the O_2 background at different pressures. The ZnO target was fixed

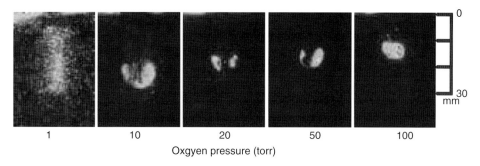

Figure 4.3.3 Rayleigh scattering visualization of ZnO nanoparticles in ablation plume at different growth pressures; the right scale is the distance from the target. (From Ref. [28].)

at the top of the image and the ablation laser beam reached the target from the bottom. The scale at the right-hand side indicates the distance from the target. In O_2, the formation of nanoparticles was clearly observed by Rayleigh scattering even at the low pressure of 1 torr. After the formation of the nanoparticles in the gas phase, they were transported onto the substrates by the carrier gas. Even at temperatures as low as $500\,°C$, the ablated nanoparticles can crystallize in a highly c-axis oriented manner. Therefore, it indicates that the surface reaction of the nanoparticles transported on a heated substrate plays an important role in the growth of nanorods.

To disclose the growth process of the nanorods synthesized by the high-pressure PLD method, it is important to know the initial stage of the nanowire growth [29]. Figure 4.3.4a–f shows SEM images of the obtained ZnO nanostructures after laser ablation of 10, 30, 60, 180, and 1800 s, respectively. After 10 s laser ablation, ZnO nanoparticles with an average diameter of 100 nm were deposited on the sapphire substrate at the initial growth stage. These nanoparticles were formed in the gas phase because of the high background pressure, as observed with Rayleigh scattering. When the particles were transported onto the substrate at high temperature, they melted and migrated on the substrate. During this process, the nanoparticles aggregated each other. As a result of the melting temperature getting higher with the size of aggregated particles, they precipitated at some places on the substrate. When the ablation time increased to 30 s, most of the ZnO particles merged into a continuous network and larger clusters. Some of the isolated nanocrystals still existed and could serve as nucleation centers to initiate the growth of ZnO nanorods, which can be seen in Figure 4.3.4b. Meanwhile, the larger clusters and the continuous network led to the growth of ZnO films/junks on the substrate. From Figure 4.3.4c, it can be seen that after ablation for 1 min, bigger nanoparticles that formed on the ZnO buffer layer began to grow into nanorods. Then, 3 min later, nanowires with big nanoparticles as roots were formed, as shown in Figure 4.3.4d. Along the growth direction, the nanowire slightly tapered. Therefore, such nanowires grown with high-pressure PLD consisted of bigger nanoparticles as roots and thinner nanowires as stems, which was significantly different from that of the axial growth nanowires previously reported [30, 31]. For

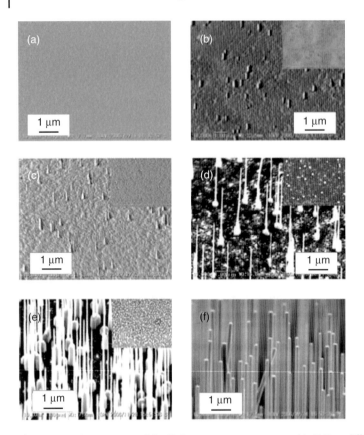

Figure 4.3.4 SEM images of the ZnO nanostructures grown with PLD at different growth times, (a) 10 s, (b) 30 s, (c) 1 min, (d) 3 min, (e) 20 min, (f) 60 min. (From Ref. [29].)

such nanowires obtained at the ablation time of 180 s, the average diameter of the nanowire stem was about 150 nm and its average length was about 4 µm, while the average size and length of the root were both about 500 nm. From this stage, the crystal growth was then dominated by vertical growth and the nanorods grew longer. As shown in Figure 4.3.4e, both the root and stem became longer and bigger as the ablation time increased to 1200 s. Meanwhile, the root exhibited more clearly a hexagonal rod with prismatic morphology on the top. After a long time growth over 1800 s, the nanowires with tapered tips gradually became uniform hexagonal cylinders with flat tips.

From the above studies, it is concluded that extrinsic catalyst particles are not necessary for ZnO nanowire growth.

However, large hexagonal ZnO particles usually exist between the nanowires and the substrate. Therefore, the growth mechanism of ZnO nanowires synthesized with high-pressure PLD is different from the traditional metal-catalyzed VLS mechanism. However, there is some resemblance. That is, the ZnO nanoparticles play an important and possibly similar but self-catalyzed role in directing the

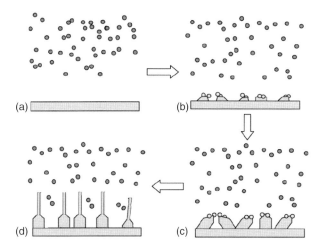

Figure 4.3.5 Growth model of nanoparticle-assisted pulsed laser deposition (NAPLD). (a) Nanoparticles were formed because of laser ablation. (b) Nanoparticles were transported onto the substrate, where they melt and relatively large crystals grow. (c) At the tips of the nanoparticles, nanowires started to grow via a self-catalyzed process. (d) Nanowires with bigger nanoparticles as roots were finally grown. (From Ref. [32].)

nanowire growth. A proposed growth mechanism is schematically illustrated in Figure 4.3.5 [32]. On laser ablation, ZnO nanoparticles were formed in the gas phase at relatively high pressure. Then, such nanoparticles were transported onto the heated substrates by carrier gas, where the nanoparticles melted and grew into hexagonal nanocrystals bounded with six equivalent crystallographic {01-10} planes, as shown in Figure 4.3.5a,b. Basically, the morphology of the crystals is related to the relative growth rates of various crystal facets bounding the crystal [33]. The ionic and polar structure of ZnO can be described as hexagonal close packing of oxygen and zinc atoms. They exhibit several crystal planes: two basal polar {0001} faces and six low-index nonpolar {01-10} planes parallel to the c-axis. The nonpolar planes are the most stable ones, and the polar ones are metastable with the highest surface energy [34]. Therefore, the growth rate of the polar plane {0001} is faster than that of the nonpolar plane. Moreover, surface diffusion is the most important rate-limiting process of the ZnO nanowire growth. Therefore, the (0001) plane easily disappears and instead it is capped with the lower surface energy facet of {01-11} surfaces. After the growth of {01-11} facets, the Zn-terminated (0001) plane is the most likely remaining facet that tends to have tiny ZnO particles at the growth front of the root. It provides an active site for the nanowire growth [35], as shown in Figure 4.3.5c.

Owing to the preferential faster growth along the c-axis, it is expected that the nanowires show tips at the ends. However, the growth velocity along [0001] was suppressed in higher vapor pressure. Therefore, the incoming atoms were driven to migrate and form flat surfaces [36–38]. Finally, both the root and nanowire became thicker and flattened with the gradual increase in ablation time, as shown in Figure 4.3.5d.

4.3.4
Controlled Nanowire Growth with NAPLD

4.3.4.1 Influence of Substrate–Target Distance

To investigate the influence of substrate–target (S–T) distance on the ZnO nanowire NAPLD growth, a series of ZnO nanowires were grown at various S–T distances [39], while other growth conditions were kept the same. All SEM images shown in Figure 4.3.6 were taken at similar center areas of the substrates with the same magnification. Clearly, the growth density of the nanowires shows S–T distance dependence. The nanowires grown at an S–T distance of 12 mm show the lowest density of vertically aligned ZnO nanowires, as shown in Figure 4.3.6a. Bigger S–T distances result in higher densities, as demonstrated in Figure 4.3.6b–d. When the S–T distance increases to 22 mm, the ZnO nanowires

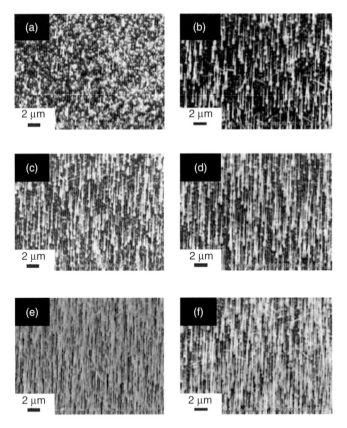

Figure 4.3.6 SEM images of ZnO nanowires grown at different substrate–target distances, (a) 12 mm, (b) 15 mm, (c) 18 mm, (d) 20 mm, (e) 22 mm, and (f) 25 mm. (From Ref. [39].)

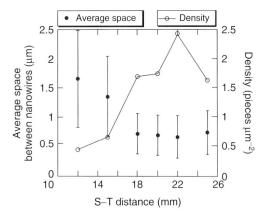

Figure 4.3.7 Dependence of nanowire growth density (circle) and average space (dot) among nanowires on substrate–target (S–T) distances. (From Ref. [39].)

show the highest growth density, as shown in Figure 4.3.6e. However, when the S–T distance was increased to 25 mm, the density of nanowires slightly decreased, as shown in Figure 4.3.6f. Figure 4.3.7 shows the dependence of nanowire density and average space among nanowires on S–T distance, which were calculated from the SEM data shown in Figure 4.3.6.

Besides the large difference in growth density, the ZnO nanowires also exhibit different diameters. The nanowires with a diameter of 130 nm and a length of 1.5 mm were grown at an S–T distance of 12 mm, exhibiting the minimum size, as shown in Figure 4.3.6a. When the S–T distance increased to 15 mm, the diameter of nanowires increased to 200 nm and the length to 2.2 mm, as shown in Figure 4.3.6b. With a further increase in the S–T distance, the lengths of nanowires also increased. The nanowires grown at an S–T distance of 22 mm exhibit the maximum length of about 5 mm, that is, the higher the density, the larger the nanowires. It was possibly caused by the laser plume position that was closely related to the S–T distance. When the substrate was located close to the ablated plume, more nanoparticles that formed in the gas phase were transported onto the substrate, which can serve as nucleation centers for initiating the nanowire growth to achieve a higher density. In addition, during the nanowire growth, more nanoparticles were deposited at the growth fronts of the nanowires, which may promote further growth of the nanowires along the c-axis, resulting in longer nanowires with bigger diameters.

4.3.4.2 Influence of Laser Energy

In the NAPLD growth process, the ablated species from the ZnO target were confined in a small volume, the quartz tube, with the stoichiometric composition and condensed into nanoparticles in the background gas. Then, the nanoparticles were transported onto the substrate as a precursor to form the nanowires. It is expected that the growth of ZnO nanowires can be controlled by varying the flux of the nanoparticles, which was determined by laser operation conditions.

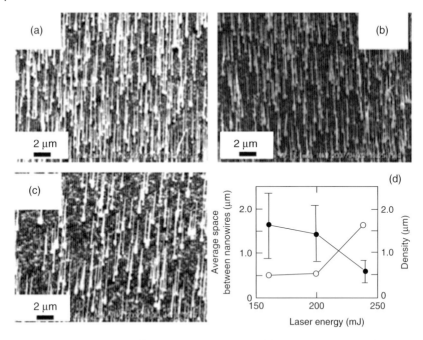

Figure 4.3.8 SEM images of ZnO nanowires grown at different laser energies of (a) 240, (b) 200, and (c) 160 mJ. (d) Average spacing distances among nanowires (circle) and the corresponding density (dot) as a function of laser energy. (From Ref. [39].)

Figure 4.3.8 shows the SEM images of the nanowires grown on sapphire substrates with different laser energies of 240, 200, and 160 mJ. Figure 4.3.8d shows the average distances among nanowires and the corresponding densities of nanowires, measured from Figure 4.3.8a–c as the function of laser energy density.

It was found that the density of nanowires increased with the increasing laser energy densities. By increasing laser energy, more species were ablated from the ZnO target and more nanoparticles are formed in the quartz tube. The nanoparticles arrived on the substrate and migrated on the surface. These nanoparticles aggregated to form the nucleation center for the nanowire growth. The probability of the aggregation became higher when more particles were transported onto the substrate surface. This assumption also explains other experimental observations that the surface density of nanowires increases with the increasing laser frequency and the heating temperature of the furnace. A higher laser frequency also increased the density of nanoparticles, and a high temperature promoted the migration of nanoparticles on the substrate.

4.3.4.3 Influence of Substrate Annealing
The selection of single-crystal substrates is critical for achieving deterministic control of nanowire growth behavior. A close match of both symmetry and lattice constant between the substrate and the grown nanowire is essential for

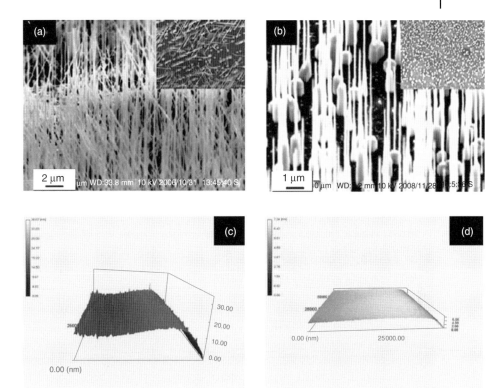

Figure 4.3.9 (a,b) SEM images of ZnO nanowires grown with NAPLD on as-received and annealed sapphire substrates, respectively. (c,d) AFM images of the as-received and annealed sapphire substrate surfaces. (From Ref. [41].)

successful heteroepitaxy and is anticipated to influence strongly the nanowire growth direction [40].

To study the morphology dependence on the substrate conditions, the synthesis of ZnO nanowire was performed on as-received and annealed sapphire substrates [41].

The SEM image shown in Figure 4.3.9a reveals that ZnO nanowires grown on the as-received sapphire substrates are mixtures of high-density vertically aligned and tilted nanowires with a length of about 10 μm or even longer. From Figure 4.3.9b, it can be seen that the ZnO wires grown on the sapphire substrates annealed at 1000 °C for 1 h are nanonails. Each nanonail consists of a hexagonal rod-shaped root with prismatic morphology and a tapered stem with a sharp tip. The average diameter of the stem is about 160 nm, and its length is about 5 mm. The average size of the root is about 600 nm, and its length is up to 500 nm or longer. The top view of the nanowires shown in the inset of Figure 4.3.9b exhibits many bright dots, indicating that the nanowires are vertically grown on the annealed sapphire substrate. Therefore, substrate annealing shows a clear influence on the morphologies of the ZnO nanowires.

To study the surface changes of the sapphire substrates after annealing, atomic force microscopy (AFM) was applied. Figure 4.3.9c,d shows the AFM images of sapphire substrates before and after annealing. The AFM image shown in Figure 4.3.9c reveals the irregular corrugation and crystallographic defects of the as-received sapphire substrate. After thermal annealing, the substrate surface becomes flatter, which can be clearly seen in Figure 4.3.9d. The root mean square (RMS) roughness of sapphire substrates decreases obviously from 4.3 to 0.3 nm after the thermal annealing process. Correspondingly, the ZnO nanowires on annealed sapphire substrates show different growth behavior. It can be concluded that the flatter substrate surface induced the vertical growth of ZnO nanowires by heteroepitaxy [42]. However, the irregularly rough surface leads to random direction growth, which is consistent with the results of SEM analysis shown in Figure 4.3.9a,b. High-resolution transmission electron microscopy (TEM) was expected to check the interface properties between the nanowires and the applied substrates.

4.3.4.4 Influence of Wetting Layer

From the above discussions, it can be seen that the nanowire density shows some change by optimization of the ZnO PLD growth conditions through NAPLD. In addition, according to the traditional VLS mechanism proposed for nanowire growth, a general route to control the growth density is to use patterned catalytic metal nanoparticles [43]. Many methods, such as phase-shift lithography, electron beam lithography, and nanosphere lithography, have been adopted to pattern the catalytic particles [16, 17, 44]. Then, semiconductor nanowires with an expected density similar to that of the catalyst pattern can be grown via the VLS mechanism. However, the metal particle patterning process mentioned above is usually complicated and of low yield. Moreover, the high temperature for nanowire growth usually results in the high mobility and diffusion of the metal catalysts, which can destroy the predefined catalyst pattern. From the former discussion, we also know that the ZnO nanowire growth mechanism shows different characteristics from that of the traditional VLS nanowire mechanism. Therefore, for density control, the prepatterned metal catalyst array that is supposed to function as a template for the subsequent growth of ZnO nanowires via the VLS model does not work as well as other semiconductor nanowires. Until now, the density control of ZnO nanowire growth is still a challenge.

Figure 4.3.10a shows the SEM image of ZnO nanowire arrays grown with NAPLD on pure sapphire substrate. The typical nanowire density is $1–10 \, \mu m^{-2}$. However, by introducing a ZnO wetting layer ($\sim$200 nm) on the sapphire substrate, the density of ZnO nanowires decreases clearly. Figure 4.3.10b is the SEM image of the ZnO nanowires grown at the cross-sectional area from pure sapphire substrate to sapphire with the ZnO wetting layer. The decreasing growth density is clearly demonstrated. Figure 4.3.10c shows a typical SEM image of such sparse ZnO nanowires with density smaller than $0.1 \, \mu m^{-2}$ grown on a sapphire substrate with ZnO wetting layer.

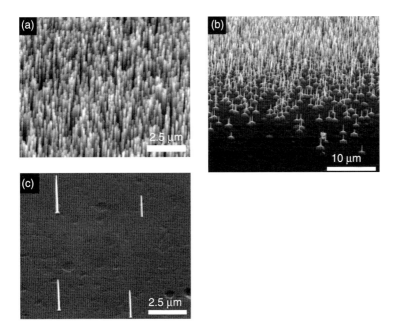

Figure 4.3.10 (a) SEM images of high-density ZnO nanowire arrays grown with PLD on pure sapphire substrates. (b) The nanowire density decreases in the cross-sectional area. (c) Typical SEM image of a low-density ZnO nanowire array grown with the same PLD on a sapphire substrate with a ZnO wetting layer.

Moreover, the nanowire density on ZnO wetting layer also shows S–T distance dependence. Figure 4.3.11 shows the SEM images of ZnO nanowires grown on different parts of a sapphire substrate (1×1 cm^2) with ZnO wetting layer. Figure 4.3.11a,c,e shows the general view of the nanowire arrays under smaller SEM magnification, which corresponds to the S–T distances of 12, 9, and 6 mm, respectively. Clearly, the nanowire density increases when the S–T distance increases, which is along the ablated plume expansion direction. This S-T-dependent behavior is similar to the case discussed in Chapter 4.1. However, the nanowire density is obviously smaller. Figure 4.3.11b,d,f shows the enlarged SEM images in Figure 4.3.11a,c,e respectively. The majority of the nanowires shows two clearly distinct regions, with a narrower upper wire section grown over a micrometer-sized base section. Therefore, it indicates that the nanowire growth on the wetting layer can also be described with the NAPLD mechanism. Such low-density ZnO nanowires are perpendicular to the substrate and very uniform along the whole length. The typical diameter is around 200 nm. The average nanowire densities grown on the wetting layer in Figure 4.3.11 are much smaller than that of nanowires grown on pure sapphire. The typical growth densities at different S–T distances were estimated from the SEM images and plotted in Figure 4.3.12. The growth density of ZnO nanowires without using a wetting layer was also included for comparison. In general, by using a wetting layer and adjusting the S–T distances,

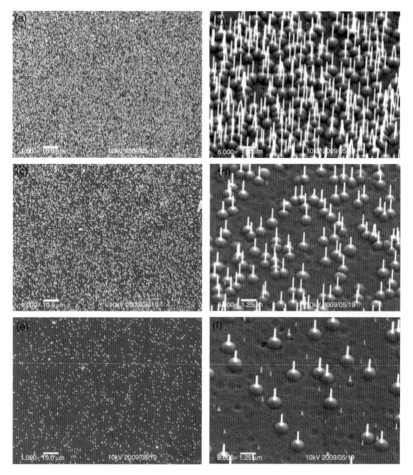

Figure 4.3.11 SEM images of low-density ZnO nanowire arrays grown on sapphire substrates with a 200 nm ZnO wetting layer. The substrate–target distance for (a,b), (c,d), and (e,f) is 12, 9, and 6 mm, respectively. (a,c,e) Clearly show the decrease in the nanowire density. (b,d,f) With same scale bars of 1.25 µm are the enlarged SEM images of (a,c,e), respectively.

the density of the ZnO nanowire array grown with NAPLD can be controlled from 10 to 10^{-2} µm^{-2}.

In case of sapphire substrates with ZnO wetting layers, when the ablated source materials were transported onto the substrate, because of the exact lattice match, some of them were consumed for film epitaxial growth. The wetting layer thickness before and after the nanowire growth were measured with ellipsometry. Owing to the ZnO film incorporation, the increase in thickness of the wetting layer varied from 500 to 1000 nm depending on the measurement points. Therefore, only part of the transported source materials were provided to grow nanowires. Accordingly, the nucleation sites will reduce. In comparison with high-density nanowires,

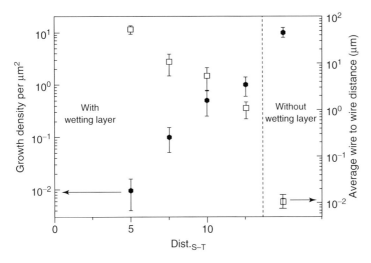

Figure 4.3.12 Dependence of the ZnO nanowire growth density (dot) and average space among wires (circle) on the substrate–target distance (Dist.$_{S-T}$) when a wetting layer was used. The data of nanowires grown on sapphire substrates without wetting layer was also shown for comparison.

the reduction in nucleation sites is definitely determinative for the low-density nanowire growth. By careful observation of such low-density nanowires, two kinds of nucleation sites were found. Some ZnO begins to nucleate on the buffer layer and form pyramidlike particles. Then, nanowires grow on such nanoparticles, which is the typical NAPLD growth behavior. Other nanowires grow directly on the ZnO wetting layer. These sites should be some sparse defects formed during the film growth, which have smaller activation energy. These defects can also act as effective collecting sites for incoming atoms and growth proceeds preferably therein. Such defects cannot be as effective as the ZnO nanoparticles in guiding the nanowire growth. So, in general, it is not surprising that the former case is important.

4.3.5
Growth of Nanowire Heterostructures Based on Low-Density Nanowires

Analogous to planar semiconductor technology where impressive advances have been made toward controlling heterostructures for various applications [45], NW heterostructures hold tremendous promise for nanoelectronics and photonics applications [46, 47]. Modulated nanostructures in which the composition and/or doping are varied on the nanometer scale represent important targets of synthesis since they could enable a new and unique function and potential for integration in functional nanodevices. The growth of crystalline overlayers on nanowires and the formation of core–shell nanowire heterostructures are important for controlling surface properties and enabling new function [48–50]. Taking advantage of the

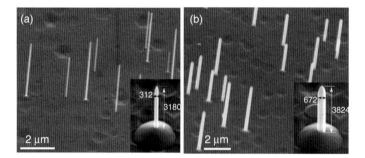

Figure 4.3.13 (a) SEM image of low-density ZnO nanowires grown on sapphire substrate with a ZnO wetting layer. (b) SEM image of ZnO/ZnMgO core–shell nanowires using the nanowires shown in (a) as cores. Insets of (a) and (b) compare the diameter and length of the core and core–shell nanowires by selecting exactly the same nanowire.

nanowires, low-density and growth flexibility of PLD, ZnO/ZnMgO core–shell nanowire heterostructures, and ZnO/ZnMgO multishells, QW nanowires were demonstrated using the low-density ZnO nanowires as physical templates.

Figure 4.3.13a,b shows the same ZnO nanowire before and after a growth turn of 3000 laser pulse ablation of an ZnMgO target in a conventional low-pressure PLD chamber. The insets are exactly the same ZnO core and the corresponding ZnO/ZnMgO core–shell nanowire. It can be seen clearly that the nanowires were homogeneously coated on both tips and side facets because of the low density and geometry of the core nanowires. By careful measurements of the length and diameter of ZnO cores and the corresponding core–shell nanowires with SEM, the average axial and radial growth rates are estimated to be 0.19 and 0.035 nm per pulse, respectively, under the applied PLD growth conditions.

Figure 4.3.14a shows the energy-dispersive X-ray spectrum (EDS) taken from the middle part (red line) and sidewall surface (green line) of the nanowire, where the magnesium signal from ZnMgO barrier layer is clearly observed. As the nanowire is a cylinder, the EDS signal detected from the side shell comes from both the top and bottom parts of the projected nanowire, while the signal from the middle part is only from the top surface of the nanowire. Therefore, the magnesium signal detected from the side surface is about 2 at.% higher than that of the middle part, which indicates that a ZnMgO layer was coated on the whole nanowire. A uniform and clear magnesium intensity mapping is detected as shown in the inset, which further proves the formation of a uniform ZnMgO shell.

Figure 4.3.14b shows the cathodoluminescence (CL) spectra measured from a single ZnO/ZnMgO core–shell nanowire at room temperature with a scanning electron microscope. The CL of the ZnO core shows the typical near-band-gap (NBG) emission at 3.26 eV. Besides this peak, a clear peak at 3.52 eV was also detected from the core–shell wire, which can be attributed to the exciton recombination in ZnMgO shell. The magnesium concentration is estimated to be 10 wt% according to the experience data on planar ZnO/ZnMgO QW growth [51]. Between these peaks from ZnO and ZnMgO, no other CL peaks were detected, indicating that

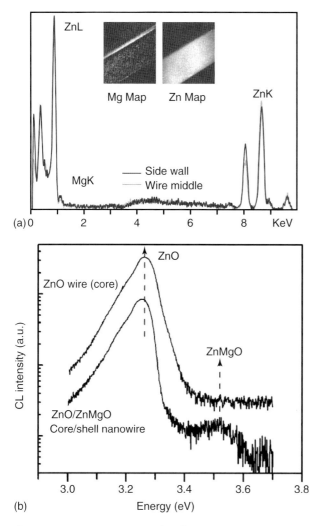

Figure 4.3.14 (a) EDS spectra taken from the main body (light gray) and sidewall (dark gray) of a ZnO/ZnMgO core–shell nanowire. Magnesium signal from ZnMgO shell was clearly detected. Insets are the zinc and magnesium intensity maps. (b) Room temperature CL spectra of single ZnO nanowire (up) and ZnO/ZnMgO core–shell nanowire (down).

there is no significant intermediate layer formation between the ZnO core and the ZnMgO shell [52].

The ZnMgO/ZnO/ZnMgO multishell structure was also demonstrated by *in situ* changing the targets. To confirm the formation of a radial ZnO QW along the nanowire, TEM was applied to check the microstructure of the nanowires with multishells. Figure 4.3.15 shows the general cross-sectional view of the ZnO/ZnMgO nanowire taken along the *c*-axis. The nanowires are either defect free

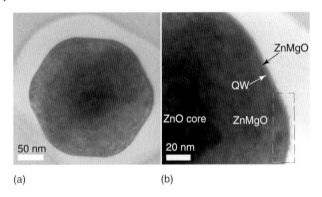

Figure 4.3.15 (a) A general view of the hexagonal cross-sectional TEM image of an ZnO/ZnMgO QW nanowire taken with an electron beam parallel to its longitudinal axis. (b) A QW with dark intensity is clearly shown.

or contain, in some cases, basal stacking faults as the TEM contrast is sensitive to the atomic number of the imaged materials. The dark contrast region along the nanowire perimeter is a ZnO middle layer; consequently, the outer shell and the inner core are the ZnMgO barrier layers, which is consistent with our targeted QW structure. Moreover, the ZnO QW occurs homogeneously around the whole nanowire perimeter and the QW is even continuous at the $60°$ corners in passing from one {10−10}-type sidewall to the next one, as indicated by the dashed quadrangle. This continuity of the QW in passing from one {10−10}-plane to the next one is facilitated by the smearing out of the $60°$ corner due to the deposition of the thicker ZnMgO barrier layer and is important for the growth of the homogeneous QW structure.

Figure 4.3.16 shows the CL spectrum measured at different parts from a single core–shell nanowire as illustrated at 10 K. The CL detected along the nanowire exhibits similar spectra characters, indicating that the core–shell structure is uniform. The emissions at 3.361 eV and lower energy are the donor-bound excitons (D^0X) and related phonon replicas. The peak at 3.521 eV is from the ZnMgO barrier. The new peak at 3.456 eV was detected from the nanowire QW [53]. It is the exciton recombination confined in the radial nanowire QW with an estimated thickness of 2.5 nm according to its emission energy (Figure 4.3.15). Therefore, homogeneous radial QW nanowires were grown using the low-density nanowires as templates. The usually encountered shadowing effect in the case of high-density nanowires was effectively eliminated.

4.3.6
Conclusions

It has been demonstrated that the PLD is an elegant and straightforward method with a huge potential for different material synthesis tasks. For nanowire growth, one principal problem that has to be solved is how to control the material nuclei and growth in one dimension. The physical parameters, such as temperature,

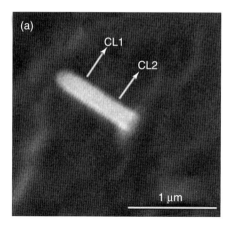

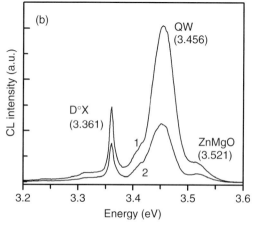

Figure 4.3.16 (a) An SEM image of the ZnO/ZnMgO QW nanowire taken before CL measurements. (b) CL spectra at $T = 10$ K measured at different parts of a single ZnO/ZnMgO QW nanowire shown in (a).

pressure, substrates, and carrier gas, have to be tuned in such a way to induce and promote one-dimensional crystal growth. In this chapter, a new ZnO nanowire growth mechanism, NAPLD, was proposed and discussed in detail. This growth process is different from the traditional VLS nanowire growth mechanism, which uses the extrinsic metal catalysts to realize the 1D crystal growth. Therefore, it is expected to eliminate the structural and physical influences caused by the extrinsic metal particles. Different experimental factors that influence the nanowire growth through NAPLD were discussed in detail. The NAPLD has great flexibility in nanowire growth, especially for tuning the nanowire composition and growth density. On the basis of these controlled growth achievements, a two-step combinational PLD strategy was developed to grow ZnO-/ZnMgO-based core–shell nanowire heterostructures. Structural and optical characterizations indicated that the ZnO QW nanowires had homogeneous properties along the radial direction.

At present, significant new results involving the growth of 1D semiconductor oxide nanostructure materials, especially for nanowires, experimental characterizations, and theoretical predictions are being reported almost every week. The wider dissemination and the use of growth methods for controlled material growth with selected composition and heterostructures undoubtedly will enable the advancement of material technologies and facilitate wider adoption of semiconductor nanowires. Significant challenges involving the controlled growth, processing, and assembly of 1D nanowires remain to be addressed before these materials displace the prevalence of top-down fabricated devices. PLD synthesis of inorganic nanomaterials is a technique with a relatively simple basic mechanism. Although the physical process parameters are comparatively easy to control, little detail is available about the chemical and physical processes taking place. The aim will be a modeling of the parameters in order to make predictions about the nanowire growth with expected composition, orientation, density, and heterostructure.

Acknowledgments

This work is supported by NSFC (51002065, 11174112) and Outstanding Youth Project from Shandong provincial science foundation. BC thanks the Taishan Scholar Professorship (TSHW20091007) tenured at University of Jinan. The Program for New Century Excellent Talents in University (NCET-11-1027) and the Research Foundation for Returned Oversea Students, Ministry of Education, China, are also acknowledged.

References

1. Burke, P.J. (2007) *Nanotubes and Nanowires*, World Scientific, New Jersey.
2. Lieber, C.M. and Wang, Z.L. (2007) *MRS Bull.*, **32**, 99. (review).
3. Klingshirn, C. (2007) *Phys. Status Solidi B*, **244**, 3027. (review).
4. Wang, Z.L. (2009) *Mater. Sci. Eng., R*, **64**, 33. (review).
5. Yi, G.C., Wang, C.R., and Park, W.I. (2005) *Semicond. Sci. Technol.*, **20**, S22. (review).
6. Jagadish, C. and Pearton, S. (eds) (2006) *Zinc Oxide Bulk, Thin Films and Nanostructures: Processing, Properties and Applications*, Elsevier, Amsterdam.
7. Huang, M.H., Wu, Y.Y., Feick, H., Tran, N., Weber, E., and Yang, P.D. (2001) *Adv. Mater.*, **13**, 113.
8. Wang, Z.L. (2004) *J. Phys. Condens. Matter*, **16**, R829. (review).
9. Norton, D.P., Heo, Y.W., Ivill, M.P., Ip, K., Pearton, S.J., Chisholm, M.F., and Steiner, T. (2004) *Mater. Today*, **6**, 35. (review).
10. Harmand, J.C., Patriarche, G., Péré-Laperne, N., Mérat-Combes, M.N., Travers, L., and Glas, F. (2005) *Appl. Phys. Lett.*, **87**, 203101.
11. Kodambaka, S., Tersoff, J., Reuter, M.C., and Ross, F.M. (2007) *Science*, **316**, 729.
12. Noor Mohammad, S. (2008) *Nano Lett.*, **8**, 1532.
13. Yao, B.D., Chan, Y.F., and Wang, N. (2002) *Appl. Phys. Lett.*, **81**, 757.
14. Park, W.I., Yi, G.C., Kim, M.Y., and Pennycook, S.J. (2002) *Adv. Mater.*, **14**, 1814.
15. Li, S.Y., Lee, C.Y., and Tseng, T.Y. (2003) *J. Cryst. Growth*, **247**, 357.
16. Greyson, E.C., Babayan, Y., and Odom, T.W. (2004) *Adv. Mater.*, **16**, 1348.
17. Ng, H.T., Han, J., Yamada, T., Nguyen, P., Chen, Y.P., and Meyyappan, M. (2004) *Nano Lett.*, **4**, 1247.

18. Wang, X., Summers, C.J., and Wang, Z.L. (2004) *Nano Lett.*, **4**, 423.

19. Fan, H.J., Lee, W., Scholz, R., Dadgar, A., Krost, A., Nielsch, K., and Zacharias, M. (2005) *Nanotechnology*, **16**, 913.

20. Jensen, L.E., Bjök, M.T., Jeppesen, S., Persson, A.I., Ohlsson, B.J., and Samuelson, L. (2004) *Nano Lett.*, **4**, 1961.

21. Mohan, P., Motohisa, J., and Fukui, T. (2005) *Nanotechnology*, **16**, 2903.

22. Willmott, P.R. and Huber, J.R. (2000) *Rev. Mod. Phys.*, **72**, 315. (review).

23. Eason, R. (ed.) (2007) *Pulsed Laser Deposition of Thin Films*, John Wiley & Sons, Inc., New Jersey.

24. Morales, A.M. and Lieber, M. (1998) *Science*, **279**, 208.

25. Yan, M., Zhang, H.T., Widjaja, E.J., and Chang, R.P.H. (2003) *J. Appl. Phys.*, **94**, 5240.

26. Lorenz, M., Kaidashev, E.M., Rahm, A., Nobis, T., Lenzner, J., Wagner, G., Spemann, D., Hochmuth, H., and Grundmann, M. (2005) *Appl. Phys. Lett.*, **86**, 143113.

27. Lorenz, M. (2008) Pulsed laser deposition of ZnO-based films, in *Transparent Conductive ZnO* (eds K. Ellmer, A. Klein, and B. Rech), Springer: Berlin, Chapter 7, pp. 305–360.

28. Kawakami, M., Hartanto, A.B., Nakata, Y., and Okada, T. (2003) *Jpn. J. Appl. Phys.*, **42**, L33.

29. Guo, R.Q., Nishimura, J., Ueda, M., Higshihata, M., Nakamura, D., and Okada, T. (2007) *Appl. Phys. A*, **89**, 141.

30. Liu, X., Wu, X.H., Cao, H., and Chang, R.P. (2004) *J. Appl. Phys.*, **95**, 3141.

31. Pan, Z.W., Dai, S., Rouleau, C.M., and Lowndes, D.H. (2005) *Angew. Chem. Int. Ed.*, **44**, 274.

32. Okada, T. and Suehiro, J. (2007) *Appl. Surf. Sci.*, **253**, 7840.

33. Mullin, J.W. (2001) *Crystallization*, Butterworth-Heinemann, Oxford.

34. Kong, X.Y., Ding, Y., Yang, R.S., and Wang, Z.L. (2004) *Science*, **303**, 1348.

35. Wang, Z.L., Kong, X.Y., and Zou, J.M. (2003) *Phys. Rev. Lett.*, **91**, 185502.

36. Xu, C.X., Sun, X.W., Dong, Z.L., and Yu, M.B. (2004) *Appl. Phys. Lett.*, **85**, 3878.

37. Leung, Y.H., Djurisic, A.B., Gao, J., Xie, M.H., and Chan, W.K. (2004) *Chem. Phys. Lett.*, **385**, 155.

38. Chen, S., Fan, Z., and Carroll, D.L. (2002) *J. Phys. Chem. B*, **106**, 10777.

39. Guo, R.Q., Nishimura, J., Matsumoto, M., Higashihata, M., Nakamura, D., and Okada, T. (2008) *Jpn. J. Appl. Phys.*, **47**, 741.

40. Kuykendall, T., Pauzauskie, P.J., Zhang, Y.F., Goldberger, J., Sirbuly, D., Denlinger, J., and Yang, P.D. (2004) *Nat. Mater.*, **3**, 524.

41. Guo, R.Q., Nishimura, J., Higashihata, M., Nakamura, D., and Okada, T. (2008) *Appl. Surf. Sci.*, **254**, 3100.

42. Pant, P., Budai, J.D., Aggarwal, R., Narayan, R.J., and Narayan, J. (2009) *J. Phys. D: Appl. Phys.*, **42**, 105409.

43. Fan, H.J., Werner, P., and Zacharias, M. (2006) *Small*, **6**, 700.

44. Wang, X.D., Summers, C.J., and Wang, Z.L. (2004) *Nano Lett.*, **4**, 423.

45. Manasreh, O. (2005) *Semiconductor Heterostructures and Nanostructures*, McGraw-Hill, New York.

46. Lieber, C.M. (2002) *Nano Lett.*, **2**, 81.

47. Lauhon, L.J., Gudiksen, M.S., and Lieber, C.M. (2004) *Philos. Trans. R. Soc. London. Ser. A*, **362**, 1247–1260.

48. Wu, Y., Xiang, J., Yang, C., and Lieber, C.M. (2004) *Nature*, **430**, 61.

49. Hua, B., Motohisa, J., Kobayashi, Y., Hara, S., and Fukui, T. (2009) *Nano Lett.*, **9**, 112.

50. Lee, C.H., Yoo, J.K., Doh, Y.J., and Yi, G.C. (2009) *Appl. Phys. Lett.*, **94**, 043504.

51. Choopun, S., Vispute, R.D., Yang, W., Sharma, R.P., Venkatesan, T., and Shen, H. (2002) *Appl. Phys. Lett.*, **80**, 1529.

52. Makino, T. (2005) *Superlattices Microstruct.*, **38**, 231. (review).

53. Makino, T., Chia, C.H., Tuan, N.T., Sun, H.D., Segawa, Y., Kawasaki, M., Ohtomo, A., Tamura, K., and Koinuma, H. (2000) *Appl. Phys. Lett.*, **77**, 975.

4.4
Laser-Vaporization-Controlled Condensation for the Synthesis of Semiconductor, Metallic, and Bimetallic Nanocrystals and Nanoparticle Catalysts

M. Samy El-Shall

4.4.1
Introduction

It is now well established that nanoparticles (1–100 nm) exhibit unique chemical and physical properties that differ from those of the corresponding bulk materials [1–5]. Semiconductor nanoparticles from ~1 to ~20 nm in diameter are often called *quantum dots, nanocrystals,* or *Q-particles*. In this size regime, the particles possess short-range structures that are essentially the same as the bulk semiconductors, yet have optical and/or electronic properties that are dramatically different from the bulk. The confinement of electrons within a semiconductor nanocrystal results in the shift of the band gap to higher energy with smaller crystalline size. This effect is known as the *quantum size effect* [1–5]. In the strong confinement regime, the actual size of the semiconductor particle determines the allowed energy levels and thus the optical and electronic properties of the material.

Owing to their finite small size and the high surface-to-volume ratio, nanoparticles often exhibit novel and sometimes unique properties. Furthermore, bimetallic alloy nanoparticles can display new properties that arise from the combination of different compositions of metals on the nanoscale. The characterization of these properties can ultimately lead to identification of many potential uses and applications ranging from catalysis, ceramics, microelectronics, sensors, pigments, and magnetic storage to drug delivery and biomedical applications. The applications of nanoparticles are thus expected to enhance many fields of advanced technology particularly in the areas of catalysis, chemical and biological sensors, optoelectronics, drug delivery, and media storage [1–5].

A wide range of scientifically interesting and technologically important nanoparticles have been produced by both chemical and physical methods [1–5]. The synthesis of nanocrystals by colloidal methods involves nucleation (the initial formation of the appropriate semiconductor bond), growth (the formation of a highly crystalline core), and passivation of the nanocrystal surface [6–8]. The passivation step is important in stabilizing the colloid and controlling the growth of the nanoparticles, preventing the agglomeration and fusing of the particles, and allowing the solubility of the nanoparticles in common solvents [6–8].

The vapor-phase synthesis of nanoparticles involves the generation of the vapor of the material of interest, followed by the condensation of clusters and nanoparticles from the vapor phase [4, 9]. The vapor may be generated by thermal, laser, and electron beam evaporation. Compound precursors can also be used, and different sources of energy may be used to decompose the precursor such as microwave plasma, laser pyrolysis, laser photolysis, or combustion flame. The size of the nanoparticle is determined by the particle residence time, temperature of the

Nanomaterials: Processing and Characterization with Lasers, First Edition.
Edited by Subhash Chandra Singh, Haibo Zeng, Chunlei Guo, and Weiping Cai.
© 2012 Wiley-VCH Verlag GmbH & Co. KGaA. Published 2012 by Wiley-VCH Verlag GmbH & Co. KGaA.

chamber, precursor composition, and pressure. Low-temperature flames can also be used to supply the energy to decompose the precursors. Flame synthesis is most common for the production of oxides [4, 5]. Pure metal particles are best produced by gas condensation [5, 10]. In spite of the success of this method, there are, however, some problems and limitations such as possible reactions between metal vapors and oven materials, inhomogeneous heating that can limit the control of particle size distribution, limited success with refractory metals due to low vapor pressures, and difficulties in controlling the composition of the mixed metal particles because of the difference in composition between the alloys and the mixed vapors.

Laser vaporization provides several advantages over other heating methods including the production of a high-density vapor of any metal, the generation of a directional high-speed metal vapor from the solid target that can be useful for directional deposition of the particles, the control of the evaporation from specific spots on the target, and the simultaneous or sequential evaporation of several different targets [9, 11]. In 1994, we introduced a novel technique to synthesize nanoparticles of controlled size and composition [12]. The technique combines the advantages of pulsed laser vaporization with controlled condensation (LVCC) in a diffusion cloud chamber under well-defined conditions of temperature and pressure. It allows the synthesis of a wide variety of nanoparticles of metal oxides, carbides, and nitrides [13–50].

In this chapter, we present the synthesis of nanoscale particles by the LVCC method. The chapter consists of three major sections. The first is a brief review of the processes of nucleation and growth in supersaturated vapors for the formation of clusters and nanoparticles. The second section deals with the application of laser vaporization for the synthesis of nanoparticles and the description of the LVCC process. In the third section, we present some examples of nanoparticles synthesized using this approach and discuss some of their selected properties and applications.

4.4.2
Brief Overview of Nucleation and Growth from the Vapor Phase

Nucleation of liquid droplets from the vapor phase can occur homogeneously or heterogeneously. Homogeneous nucleation occurs in the absence of any foreign particles or surfaces when the vapor molecules themselves cluster to nuclei within the supersaturated vapor. According to the classical nucleation theory (CNT), embryonic clusters of the new phase can be described as spherical liquid droplets with the bulk liquid density inside and the vapor density outside [51, 52]. The free energy of these clusters relative to the vapor is the sum of two terms: a positive contribution from the surface free energy and a negative contribution from the bulk free energy difference between the supersaturated vapor and the liquid. The surface free energy results from the reversible work used in forming the interface between the liquid droplet and the vapor. For a cluster containing n atoms or molecules, the interface energy is given by

$$\sigma A(n) = 4\pi\sigma \left(\frac{3v}{4}\right)^{2/3} n^{2/3} \tag{4.4.1}$$

where σ is the interfacial tension or surface energy per unit area, $A(n)$ is the surface area of the clusters, and v is the volume per molecule in the bulk liquid. Since n molecules are transferred from the vapor to the liquid, the bulk contribution to the free energy is $n(\mu_l - \mu_v)$ where μ_l and μ_v are the chemical potentials per molecule in the bulk liquid and vapor, respectively. Assuming *ideal* vapor, it can be shown that

$$(\mu_l - \mu_v) = -n\, k_B T \ln S \tag{4.4.2}$$

where k_B is the Boltzmann constant, T the temperature, and S the vapor supersaturation ratio defined as $S = P/P_e$, where P is the pressure of the vapor and P_e is the equilibrium or "saturation" vapor pressure at the temperature of the vapor (T).

The sum of the contributions in Eqs. (4.4.1) and (4.4.2) is the reversible work (free energy) $W(n)$ done in forming a cluster containing n atoms or molecules. This free energy is given by

$$W(n) = nk_B T \ln S + 4\pi\sigma \left(\frac{3v}{4\pi}\right)^{2/3} n^{2/3} \tag{4.4.3}$$

Because of the positive contribution of the surface free energy, there is a free energy barrier that impedes nucleation. Equation (4.4.3) expresses the competition between "bulk" and "surface" behavior in determining cluster stability and, ultimately, cluster concentration in the supersaturated vapor. For a saturated vapor where $S = 1$, the first term (bulk contribution) vanishes and $W(n)$ is proportional to $n^{2/3}$. For $S > 1$, however, the first term provides a negative contribution to $W(n)$. Increasing S reduces the barrier $W(n)$ and, therefore, enhances the probability that fluctuation processes will allow some clusters to grow large enough to overcome the barrier and grow into stable droplets. The smallest cluster n^* (critical size or nucleus) that can grow with a decrease in free energy is determined to be

$$n* = \frac{32\pi\sigma^3 v^2}{3(k_B T \ln S)^3} \tag{4.4.4}$$

Substituting n^* into Eq. (4.4.3) yields the barrier height $W(n^*)$, given by

$$W(n*) = \frac{16\pi\sigma^3 v^2}{3(k_B T \ln S)^2} \tag{4.4.5}$$

It is clear from Eq. (4.4.5) that increasing the supersaturation (S) reduces the barrier height and the critical cluster size (n^*). It is important to emphasize that S can be increased either by increasing P or by decreasing P_e. The pressure P can be increased by increasing the rate at which atoms are placed in the vapor or decreasing the rate at which they leave the region where nucleation is occurring. The equilibrium vapor pressure can be decreased by decreasing the temperature since P_e is approximately given by

$$P_e \approx \exp\left(\frac{-L(0)}{RT}\right) \tag{4.4.6}$$

where $L(0)$ is the latent heat of evaporation per mole at zero temperature and R is the gas constant. It should be noted that a more exact expression for P_e would include the temperature dependence of the latent heat of evaporation.

The rate of homogeneous nucleation J in a supersaturated vapor, usually defined as the *number of drops nucleated per cubic centimeter per second*, is given by

$$J = K \exp\left(\frac{-W(n^*)}{k_B T}\right) \tag{4.4.7}$$

The pre-exponential factor K incorporates both the effective collision rate of vapor molecules from a nucleus of size n^* and the departure of the cluster distribution from equilibrium [51, 52]. The rate of nucleation J is clearly very sensitive to the height of the free energy barrier W^*. Thus, if the vapor becomes sufficiently supersaturated, the barrier is reduced to the point where nucleation occurs at a high rate. Preexisting surfaces (e.g., aerosol or dust particles), ions, or large polymer molecules greatly accelerate the rate of nucleation by lowering W^*, defined by Eq. (4.4.5). Such surfaces accomplish this by reducing the amount of work required to provide the interface in nucleation (since a surface already exists), while ions accomplish this by dielectric polarization so that the barrier W^* can be lowered to the point where a single ion can induce the formation of a macroscopic liquid drop. This represents almost the ultimate in amplification and detection [53, 54].

4.4.3
The LVCC Method

The LVCC method is based on the formation of nanoparticles by condensation from the vapor phase. Previous work was based on using thermal evaporation or sputtering to produce supersaturated metal vapors [9, 11, 17, 29, 48]. In the LVCC method, the process consists of pulsed laser vaporization of a metal target into a selected gas mixture under controlled temperature and pressure. The laser vaporization produces a high-density vapor within a very short time, typically 10^{-8} s, in a directional jet that allows directed deposition. Desorption is possible from several targets simultaneously, yielding mixed particles. An important feature is the use of an upward diffusion cloud chamber at well-defined temperatures and pressures. A temperature differential between the end plates produces a convection current into which the metal is evaporated. A sketch of the chamber with the relevant components for the production of nanoparticles is shown in Figure 4.4.1a.

The chamber consists of two horizontal stainless steel plates separated by a glass ring. A metal target of interest is set on the lower plate, and the chamber is filled with a pure carrier gas such as helium or Ar (99.99% pure) or a mixture containing a known composition of a reactant gas (e.g., O_2 in case of oxides; NH_3 for nitrides, CH_4 or C_2H_2 for carbides, etc.). The metal target and the lower plate are maintained at a temperature higher than that of the upper one (temperatures are controlled by circulating fluids). The top plate can be cooled to less than 150 K by circulating liquid nitrogen. The large temperature gradient between the bottom and top plates

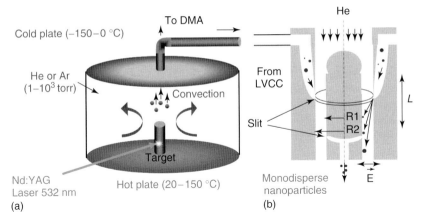

Figure 4.4.1 (a) Experimental setup for the synthesis of nanoparticles using the LVCC method. (b) Experimental setup for the LVCC method coupled with a differential mobility analyzer (DMA).

results in a steady convection current that can be enhanced by using a heavy carrier gas such as Ar under high-pressure conditions (10^3 torr).

The metal vapor is generated by pulsed laser vaporization using the second harmonic (532 nm) of Nd:YAG laser (15–100 mJ per pulse depending on the target used, 10^{-8} s pulse). The laser beam is moved on the metal surface in order to expose a new surface to the beam and assure good reproducibility of the amount of metal vapor produced. Following the laser pulse, the ejection of the metal atoms and their eventual interaction with the ambient atmosphere take place. Since the target surface where evaporation occurs is located near the middle of the chamber (about 0.5 reduced height) and the ambient temperature rapidly decreases near the top plate, it is likely that maximum supersaturation develops within the upper half of the chamber above the surface target (closer to the target than to the top plate). Since the equilibrium vapor pressure (P_e) is approximately an exponential function of temperature, the metal vapor reaches maximum supersaturation in the condensation zone near the top plate (note that the metal vapor is supersaturated in any place in the chamber but its maximum supersaturation is close to the top plate). This supersaturation can be made as large as desired by increasing the temperature gradient between the chamber plates. (The higher the supersaturation, the smaller the size of the nucleus required for condensation.) However, the most important role of increasing the temperature gradient between the chamber plates is the creation of a strong convection current within the chamber. The role of convection is to remove the small particles away from the nucleation zone (once condensed out of the vapor phase) before they can grow into larger particles. The rate of convection increases with the temperature gradient in the chamber. Therefore, by controlling the temperature gradient, the total pressure, and the laser power (which determines the number density of the metal atoms released in the vapor phase), it is possible to control the size of the condensing particles.

Nichrome heater wires wrapped around the glass ring provide sufficient heat to prevent condensation on the ring and to maintain a constant temperature gradient between the metal target and the top plate. In a typical run, the laser operates at 20 Hz for about 1–2 h. Then, the chamber is brought to room temperature and the particles are collected under atmospheric conditions. Glass slides or metal wafers can be attached to the top plate when it is desired to examine the morphology of as-deposited particles. No particles are found on any other place in the chamber except on the top plate; this supports the assumption that nucleation takes place in the upper half of the chamber and that convection carries the particles to the top plate where deposition occurs.

For the synthesis of bimetallic nanoparticles, a mixture of the appropriate elemental powder (micrometer-sized particles) is prepared in a specific molar ratio using a mortar and a pestle. The mixture is then pressed at 500 MPa using a hydraulic press in order to shape it into a cylindrical disk target. Pure metal nanoparticles are also prepared from bulk metals under the same experimental conditions for comparison with the bimetallic alloy nanoparticles.

Since laser vaporization of metallic and semiconductor targets produces a significant fraction of ions and charged species (about 10^6 ions and 10^{14} atoms per laser pulse), the charged nanoparticles produced through the LVCC process can be classified based on their electrical mobility in a dilute inert gas flow using a low-pressure differential mobility analyzer (LP-DMA) [42, 48, 55, 56]. To prepare size-selected nanoparticles and measure the size distribution produced under specific experimental conditions, the LVCC chamber is coupled to a LP-DMA as shown in Figure 4.4.1b. The principles of operation of the LVCC–DMA system are briefly discussed below [42, 48].

By applying an electric field between two plates of the LVCC chamber, the charged particles are attracted to one of the two plates by an electrostatic force depending on their polarities, and they deposit on the plates. Here, the electrostatic velocity of the particles depends on their mobility. When the thermal gradient is applied between two plates of the LVCC chamber, the particles are drifted by the thermophoretic force. Therefore, the thermophoretic force is available for collecting all size ranges of particles on the cold plate. On the other hand, one can utilize the size dependence of electrostatic force for "sizing" nanoparticles.

In order to control the transport of nanoparticles, the LVCC technique was modified by introducing a helium gas flow as the carrier gas instead of the static system. Two mesh plates were located at the top and bottom of the chamber to produce a vertical uniform flow. The generated particles were transported by the gas flow from the bottom to the top of the chamber with balancing external forces [42]. The electric field was generated by applying positive DC voltage to a bottom plate and by electrically grounding a top plate. Since part of the particles generated by laser vaporization and condensation was originally charged because of the attachment of electrons or positive ions in the laser-induced plasma, they were transported by the electrostatic force. It should be noted that the negatively charged particles move toward the gas flow when positive voltage is applied to the bottom plate. As a result, the positively charged particles accelerate toward the outlet (top plate) but the negatively charged particles might balance with gas flow under the electrostatic field. Similarly, the particles were transported by the thermophoretic force when a temperature gradient was produced by feeding hot water (up to 90 °C) to the bottom and by cooling a top plate with liquid nitrogen. Consequently, the particles were accelerated or decelerated depending on the balance between gas flow and external force.

The size and/or amount of the nanoparticles deposited on the plate can be controlled by tuning the strength of the applied field. In order to analyze the effect of the external field on the morphology and the size distribution of the deposited particles, the particles were collected on the glass plate or microgrid for observation using scanning and transmission electron microscopes (SEM and TEM). The rest of the particles, which did not deposit on the plates, were transported out of the LVCC chamber with a carrier gas [flow rate; 1 SLM (standard liter per minute)] and then introduced into a low-pressure differential mobility analyzer (LP-DMA) [42]. In the LP-DMA, the particles are differentially classified utilizing the balance of electrical mobility and gas flow. The mobility of the classified particles is related to the applied voltage to the LP-DMA. The mobility equivalent diameter of the nanoparticles can be controlled by the applied voltage to the LP-DMA. After the classification by the LP-DMA, all the monodispersed particles are collected by the impaction onto the solid substrate for TEM and SEM measurements. By measuring the electric current, from the charged particles, the number density of the particles can be obtained. The mobility size distribution of the gas-phase particles can be observed on line by scanning the DMA-applied voltage and measuring electric current.

4.4.4
Silicon Nanocrystals

The discovery that porous Si nanocrystals emit visible light with a high quantum yield has raised hopes for new photonic Si-based devices [57, 58]. This discovery has also stimulated interest in the synthesis of Si nanocrystals, which are believed to be the luminescent centers in porous silicon [59–63].

The physical, optical, and photoluminescence (PL) properties of Si nanocrystals prepared by the LVCC technique were systematically studied [20, 22, 27, 29, 32]. Detailed characterizations showed that the Si nanocrystals have a 1–2 nm surface oxide layer over a crystalline Si core, whose lattice constant is similar to that of bulk crystalline Si. A size distribution analysis showed an average particle size of 5–6 nm, with a relatively broad size distribution [22, 32]. Selected-area electron diffraction patterns of the Si nanocrystals confirm the expected lattice spacings for Si in the diamond cubic phase. Figure 4.4.2 shows typical TEM images of the Si nanocrystals and the associated electron diffraction (ED) pattern.

Figure 4.4.3a shows the UV–vis absorbance of the Si nanoparticles suspended in methanol. The spectrum shows the absorption features associated with indirect band gap transitions, particularly the yellow absorption tail, which extends from 400 nm across the visible and the stronger absorption from 370 to 240 nm. From the plot of $(\alpha h\nu)^{1/2}$ versus $h\nu$ (where α is the reciprocal absorption length and $h\nu$ is

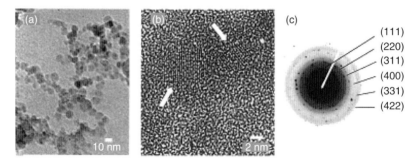

Figure 4.4.2 (a) TEM and (b) HRTEM (high-resolution transmission electron microscopic) images and (c) electron diffraction pattern from Si nanocrystals prepared by the LVCC method.

the photon energy), the band gap of the Si nanocrystals is calculated as 1.78 eV as shown in Figure 4.4.3b. For comparison, the indirect band gap of bulk Si is 1.1 eV.

Dispersed PL spectra obtained with 355 nm pulsed laser excitations are shown in Figure 4.4.4. The spectra differ by the position of the boxcar gate, which ranges from a 0.1 μs delay with respect to the laser excitation pulse to a 8 μs delay. The short delay spectrum enhances the blue emission component, since the lifetime associated with it is short (<20 ns) compared to the lifetimes of the red emission. The shape of the blue emission appears similar to that from SiO_2 nanoparticles, and it is probably due to the oxidized surface layer of the Si nanoparticles [14].

The incorporation of Si nanocrystals in polymer films may lead to the development of novel materials that combine several properties such as the visible PL and elasticity. Block copolymers are interesting materials that exhibit several different morphologies depending on the composition and structure of the copolymer. Figure 4.4.5 displays the PL data obtained from a sulfonated styrene-ethylene/butylene-styrene triblock copolymer containing the Si nanocrystals. The PL from the copolymer film alone and from the suspended Si nanocrystals in methanol are also shown for comparison. It is clear that the copolymer film containing the Si nanocrystals exhibits the PL characteristic of the pure copolymer and of the suspended Si nanocrystals. This indicates that the PL properties of Si nanocrystals can be retained after the incorporation of the particles in the polymer film. This may have interesting applications in the design of new materials for optical display.

4.4.5
Laser Alloying of Nanoparticles in the Vapor Phase

Recently, bimetallic alloy nanoparticles have gained much interest because of the additional new properties that arise from the combination of different compositions of metals on the nanoscale [3, 48]. At a fundamental level, information on the evolution of electronic structures of bimetallic nanoparticles as a function of size and composition and their effects on the optical absorption spectra continue to

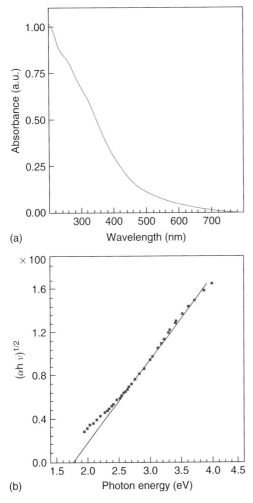

Figure 4.4.3 (a) UV–vis absorbance of Si nanocrystals suspended in methanol. (b) $(\alpha h \nu)^{1/2}$ versus $h\nu$ for the absorption tail in the visible region.

be a major goal of research in nanostructured materials. On a practical level, the unique optical properties of metallic and bimetallic nanoparticles are exploited for a variety of applications including optical markers for biomolecules, biological sensors, optical filters, surface enhancement in Raman spectroscopy, and ultrafast nonlinear optical devices.

The presence of a surface plasmon resonance (SPR) band in the visible region of the absorption spectrum of noble metallic nanoparticles was recognized a long time ago, and its origin is attributed to the collective oscillation of the free conduction electrons induced by an interacting electromagnetic field [64, 65]. The oscillation frequency is determined by the metal electron density, the effective electron mass, and the shape and size of the charge distribution [66]. As the particle

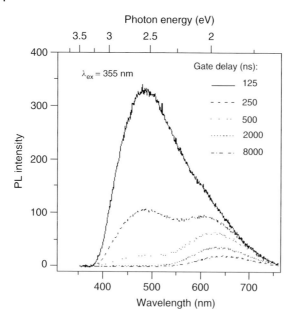

Figure 4.4.4 Dispersed emission of Si nanocrystals (solid) obtained at different gate delays from the 355 nm excitation laser pulse.

size becomes smaller than the mean free path of the free electrons, the plasmon band broadens until it disappears. For example, gold particles less than 1 nm in diameter had no plasmon absorption band. Nanocomposites, that is, nanoalloys and core–shell nanoparticles are expected to exhibit different SPR characteristics. Indeed, core–shell nanoparticles of gold–silver exhibit two distinct SPR bands [65]. Nevertheless, gold–silver nanoalloys exhibit one single plasmon band and its absorption wavelength depends on the alloy composition [67].

Gold and silver nanoparticles have been intensively studied because their SPR frequencies lie in the visible region of the electromagnetic spectrum, usually centered around ∼520–540 and ∼400–420 nm, respectively [64–72]. Much of the interest in studying gold and silver nanoparticles as well as their alloys is due to a variety of biomedical applications [68–71]. Because of the efficient conversion of the optical excitation energies of the plasmon oscillation into heat, the excited gold and silver nanoparticles generate a tremendous amount of heat following the plasmon absorption. These hot nanoparticles have been proposed for critical applications in medicine including, for example, the selective destruction of cancer cells following the irradiation of target cells. Silver nanoparticles have been used as antibacterial agents or bacteria sensors [72]. The alloying of gold and silver nanoparticles to form complexes with DNA ligands may promise new treatments for cell diseases and for cancer diagnosis and treatment [68, 73].

Several synthesis methods have been developed for the design of bimetallic nanocomposites (i.e., alloy and/or core–shell structure) of gold–silver nanostructures [64–72]. Gold–silver alloy nanoparticles have been prepared chemically by

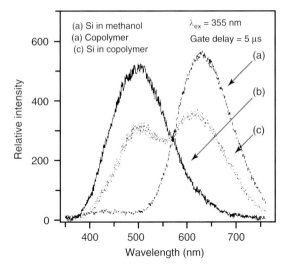

Figure 4.4.5 PL spectra of (a) Si nanocrystals in methanol, (b) triblock copolymer film, and (c) the triblock copolymer film containing the Si nanocrystals.

co-reduction of silver and gold compounds in a one-phase liquid system. Laser ablation methods have been developed to prepare bimetallic and monometallic nanoparticles in solution [74–79]. The LVCC method has been used for the vapor-phase synthesis of bimetallic alloy nanoparticles such as gold–silver, gold–palladium, and gold–platinum [40, 48]. The advantages of the vapor-phase synthesis are contamination-free products (as compared to chemical reductions in solutions), elimination of the chemical precursors and solvents, and, in most cases, production of highly crystalline nanoparticles. Furthermore, by coupling of the LVCC method with a DMA, size-selected nanoparticles can be prepared from the vapor phase [40, 43, 48].

4.4.5.1 Gold–Silver Alloy Nanoparticles

The DMA size distributions of the Au, Ag, and Au–Ag alloy nanoparticles generated by the 532 nm laser vaporization are shown in Figure 4.4.6. The average particle size for Ag (12 nm) appears to be significantly smaller than that of Au (20 nm) and of the Au–Ag alloy (22 nm). The stoichiometric coefficients given for the alloy nanoparticles represent the atomic ratio of the gold and silver in the nanoparticles as determined from the ICP-MS (inductively coupled plasma mass spectrometry) analysis. The particles generated by 1064 nm have a smaller size as compared to those generated by the 532 nm. However, in all cases, the size distributions exhibit significant broadening and tailing, characteristic of aggregated nanoparticles grown from the vapor phase.

Figure 4.4.7 displays a TEM image of the 25 nm selected Au–Ag (47 at.% Au) alloy nanoparticles prepared using the LVCC–DMA system. Although the nanoparticles are still aggregated, the primary particles appear to exhibit identical sizes. Individual

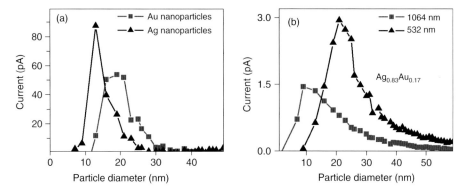

Figure 4.4.6 (a) Size distributions of Au and Ag nanoparticles and (b) $Ag_{0.83}Au_{0.17}$ alloy nanoparticles prepared by laser vaporization of bulk targets using 532 and 1064 nm lasers.

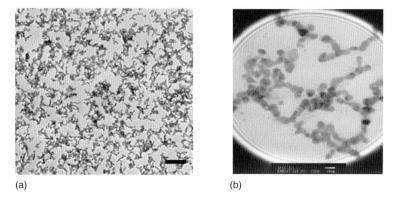

Figure 4.4.7 (a) TEM of 25 nm selected Au–Ag alloy nanoparticles, scale bar = 100 nm, and (b) high-resolution TEM of $Au_{0.28}Ag_{0.72}$ alloy nanoparticles, scale bar = 10 nm.

monodisperse particles can be deposited only if the number density of the particles is kept very low to avoid the aggregation of the particles.

The UV–vis absorption spectra of dispersed Au, Ag, and Au–Ag alloy nanoparticles in water are shown in Figure 4.4.8. A single plasmon band at ∼533 and ∼420 nm was observed for the Au and Ag nanoparticles, respectively. The broadening in the absorption plasmon bands of Au and Ag observed in Figure 4.4.8 is attributed to the broad particle size distributions shown in Figure 4.4.6 and to the aggregation of the nanoparticles, where small particles have a high tendency to aggregate faster than big particles because of their high surface energy. Therefore, the SPR bands are inhomogeneously broadened by the different particle sizes and aggregates of different sizes and shapes.

In Figure 4.4.8a, the absorption spectrum of the alloy $Au_{0.53}Ag_{0.47}$ nanoparticles exhibits a distinct peak with a maximum observed at 487 nm. This plasmon peak is located at an intermediate position between the Au and the Ag surface plasmon bands. The single plasmon band implies that the particles are spherical rather than rods or triangles, which would have two or three plasmon peaks, respectively

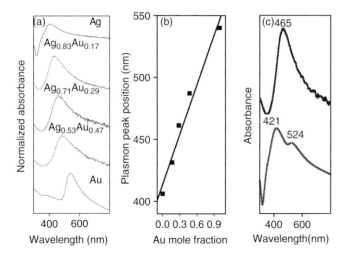

Figure 4.4.8 (a) UV–vis absorption spectra of Ag–Au nanoparticles prepared by the LVCC method, (b) plasmon peak position as a function of the Au content in the alloy nanoparticles, and (c) UV–vis absorption spectra of $Au_{0.29}Ag_{0.71}$ alloy nanoparticles (upper graph) and a physical mixture of Au and Ag nanoparticles with the same molar ratio found in the alloy (lower graph).

[67, 80, 81]. It also implies the formation of an alloy between Au and Ag nanoparticles rather than of a mixture, which would simply exhibit two distinct plasmon bands corresponding to Au and Ag. Figure 4.4.8c displays the UV–vis spectra of the alloy and the mixture having the same molar ratio of Au (71)/Ag (29). The mixture had two distinct peaks at 421 and 524 nm corresponding to the absorption plasmon of the silver and the gold nanoparticles, respectively, while the alloy nanoparticles show only a single plasmon peak at 465 nm. The plasmon peak depends on the composition of the alloy prepared. It shifts linearly to higher energy with increased silver content in the nanocomposite alloy, as shown in Figure 4.4.8b. The linear relationship between silver content in gold–silver alloy and its surface plasmon energy has been extensively observed for many alloys prepared by different chemical reduction routes in liquids; for example, in single-phase systems [67, 82], in binary phase systems [83, 84], in microemulsion systems [85], and also by radiation chemistry [86] and laser ablation in solution [78, 87].

4.4.5.2 Size Control by Laser Irradiation of Nanoparticles in Solutions

The gold and silver nanoparticles dispersed in water were irradiated with the fundamental (1064 nm) as well as the second harmonic (532 nm) of a pulsed Nd:YAG laser at (power density $4.3–7.6 \times 10^7$ W cm^{-2}) for 20 min [48]. The UV–vis absorption spectra, as well as the TEM micrographs, showed clearly that after irradiation with the 532 nm light, the particle sizes decreased and nearly monodispersed particles were formed, while the nonirradiated particles showed a broad size distribution with significant amounts of nonspherical particles. For the Au nanoparticles, the plasmon peak was shifted from ∼533 to ∼520 nm. Similar results have been

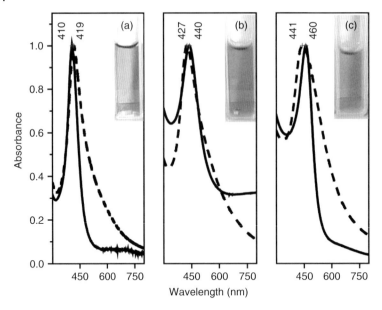

Figure 4.4.9 UV–vis spectra of (a) $Au_{0.17}Ag_{0.83}$, (b) $Au_{0.29}Ag_{0.71}$, and (c) $Au_{0.47}Ag_{0.53}$ after irradiation with the 532 nm (solid line) and the 1064 nm (dotted line) lasers for 20 min.

reported by Koda and coworkers [76] where the 532 nm irradiation of a chemically prepared gold colloid resulted in a significant reduction in the average particle size. On the other hand, the decrease in the average particle size was less pronounced when the 1064 nm laser beam was used. Accordingly, the plasmon band was slightly blueshifted from ~533 to 528 nm. The 532 nm radiation is more effective in reducing the particle size due to the strong plasmon absorption by gold, which results in both melting and vaporization, whereas in the case of 1064 nm radiation only vaporization takes place. The melting of the nanoparticles leads to the shape change from nonspherical to spherical particles, and the size reduction is due to the vaporization of the particles.

A comparison between the absorption spectra for the $Au_{0.17}Ag_{0.83}$, $Au_{0.29}Ag_{0.71}$, and $Au_{0.53}Ag_{0.47}$ nanoparticles irradiated with 532 and 1064 nm is shown in Figure 4.4.9. The gradual change of color with the change in the Au–Ag compositions is clear in the photographs of the alloy solutions shown in Figure 4.4.9. The surface plasmon bands for $Ag_{0.83}Au_{0.17}$, $Au_{0.71}Ag_{0.29}$, and $Au_{0.53}Ag_{0.47}$ nanoparticles were blueshifted from 431, 465, and 487 nm to 410, 440, and 460 nm, respectively, after 20 min irradiation with the 532 nm laser. On the other hand, the plasmon bands were shifted to 419, 427, and 441 nm, respectively, after 20 min irradiation with 1064 nm. The broadening in the UV–vis absorption spectra for Ag–Au particles decreased after irradiation, indicating a decrease in the size reduction of the nanoparticles, which could lead to a narrow size distribution.

The effect of irradiation wavelength on the alloy nanoparticles dispersed in water was also examined using TEM. The TEM micrographs for $Ag_{0.71}Au_{0.29}$ nanoparticles

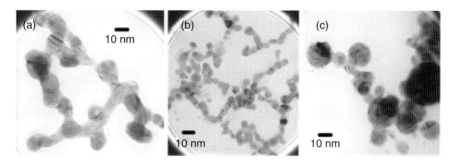

Figure 4.4.10 TEM images of the $Ag_{0.71}Au_{0.29}$ nanoparticles (a) as-prepared, (b) after 20 min irradiation with 532 nm, and (c) after 20 min irradiation with 1064 nm. Scale bar = 10 nm.

are shown in Figure 4.4.10a with and without irradiation in water medium [48]. The as-prepared particles from the LVCC were nonspherical with facets and a broad size distribution. After being irradiated with 532 nm for 20 min (12 000 laser pulses) in water, size reduction was observed from the TEM results, where melting, reshaping, and vaporization took place to yield almost monodispersed spherical particles, as shown in Figure 4.4.10b. However, with the 1064 nm irradiation, only a small size reduction occurred in addition to reshaping to spherical particles as shown in Figure 4.4.10c. It was also observed that a broad size distribution still remained, even after the 1064 nm irradiation.

4.4.5.3 Gold–Palladium Alloy Nanoparticles

Au–Pd nanoparticles were prepared from a 50–50 mol% Au–Pd target. The TEM images shown in Figure 4.4.11 indicate that the average size of the alloyed nanoparticles is 8($\pm$3) nm [40].

The XRD of the alloy Au–Pd nanoparticles, shown in Figure 4.4.12, matches well the diffraction pattern for the AuPd alloy [88]. It is significant to note that the observed peaks at 38.94, 45.33, and 66.19 (2θ), which correspond to reflections

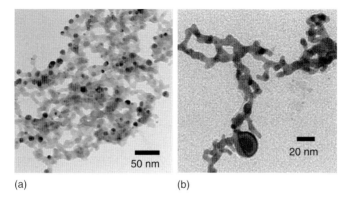

(a)   (b)

Figure 4.4.11 (a,b) TEM of Au–Pd alloy nanoparticles.

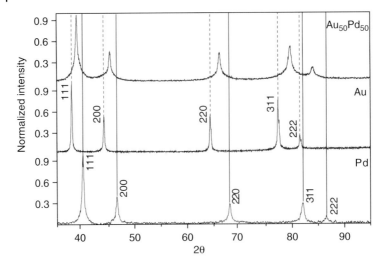

Figure 4.4.12 XRD data of the Au, Pd, and AuPd alloy nanoparticles.

from the 111, 200, and 220 planes, respectively, are significantly shifted from those of pure Au and Pd as shown in Figure 4.4.12. It is also significant to note that while the XRD of the LVCC target indicates a physical mixture of Au and Pd, the XRD diffraction pattern of the Au–Pd nanoparticles confirms the formation of the AuPd alloy in the vapor phase without any indication of the presence of its starting materials. In fact, there is a wide composition range of the Au/Pd target (30–70 mol% Au) where the AuPd alloy nanoparticles are exclusively produced. Outside this range, a mixture of the AuPd alloy and Au or Pd nanoparticles are obtained.

4.4.6
Intermetallic Nanoparticles

Intermetallic materials based on the aluminides of transition metals such as iron, nickel, titanium, and cobalt are potential candidates for a variety of high-temperature structural applications and are known to operate well beyond the operating temperatures of conventional materials. This is primarily due to their excellent oxidation and corrosion resistances that are attributed to the formation of a continuous, fully adherent alumina layer on the surface when exposed to air or oxygen atmospheres even at temperatures as high as $1000\,^{\circ}C$ or higher. For example, iron aluminide alloys have tensile strengths comparable to most ferritic and austenitic alloys and in parallel possess a density lower than that of stainless steel [89, 90]. They have excellent resistance to corrosion in many aqueous environments and to sulfidation in H_2S and SO_2 gases. In addition, their high electrical resistivity that increases with temperature and low cost of production have made these intermetallics excellent candidates for various applications such as heating elements, insulation wrapping, tooling, catalytic converters, automotive exhaust manifolds, fabric and paper cutting, piping, and hot gas filters [91, 92]. However, the bulk materials of iron aluminide suffer from the lack of room temperature ductility, low high-temperature strength, and severe hydrogen embrittlement in air by reaction with water vapor [89]. These problems impede the use of iron aluminides at higher temperatures where their corrosion resistance capability could be strongly exploited. To overcome these problems, efforts have been focused to synthesize ultrafine iron aluminide alloy powder, which provides a good combination of strength and ductility because of the fine grain size of the product [93, 94].

4.4.6.1 FeAl and NiAl Intermetallic Nanoparticles

The FeAl and NiAl nanoparticle powders prepared by the LVCC method appear to consist of flakes with lateral dimensions of about some fractions of a millimeter. The surface is slightly corrugated and the flakes appear to have a low density, since the free fall in air is quite slow. All the nanoparticle powders display a black color. The SEM images of the as-deposited Ni and NiAl nanoparticles show the typical weblike structure morphology, similar to the morphology observed for other nanoparticles prepared by the LVCC method [9, 11, 17, 31, 33–35]. The web network is characterized by well-defined pores with diameters of 1–2 μm. The high porosity could make these nanoparticles suitable for catalytic applications.

Figure 4.4.13 displays typical TEM images of the FeAl and NiAl nanoparticles. Most of the primary particles have average diameters in the range of 5–7 nm with a small percentage (∼2–4%) of larger particles that are ∼20–30 nm in diameter. Figure 4.4.14 displays several TEM images of 30 and 14 nm selected FeAl nanoparticles prepared using the LVCC–DMA system [42, 43]. Although it is difficult to distinguish individual particles because of the overlapping of the particles on the grid (Figure 4.4.14a,c,e), most of particles are spherical "droplet-like" particles with identical sizes. Individual monodisperse particles can be deposited only if the number density of the particles is kept very low to avoid the aggregation of the particles as shown in Figure 4.4.14b,d.

The XRD patterns for Ni and Al nanoparticles have four strong lines assigned to reflections from the 111, 200, 220, and 311 planes at scattering angles (2θ) of 44.49,

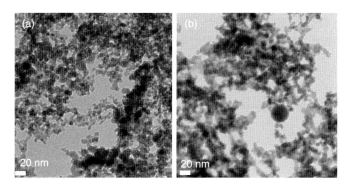

Figure 4.4.13 TEM of (a) FeAl and (b) NiAl nanoparticles, respectively. Scale bar is 20 nm.

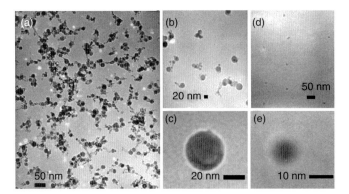

Figure 4.4.14 TEM of selected FeAl nanoparticles (a–c) 30 nm and (d,e) 14 nm.

51.81, 76.41, and 92.89, respectively, for Ni and 38.47, 44.71, 65.09, and 78.21, respectively, for Al. The lattice constants, assuming a cubic unit cell, for Ni and Al nanoparticles were calculated to be 3.5242 and 4.0495, compared to 3.523 and 4.0494 for the bulk materials, respectively [95]. The XRD pattern for nanoparticle sample obtained following the ablation of the Ni–Al powder mixture (51.8 at.% Ni) is shown in Figure 4.4.15 along with the XRD patterns for pure Ni and Al nanoparticles. It is clear that the XRD pattern of the NiAl sample does not match any of the XRD peaks of pure Ni or Al nanoparticles. However, it matches the diffraction pattern of bulk $Ni_{0.58}Al_{0.42}$ alloy [95], which means that NiAl intermetallic nanoparticles were formed in the vapor phase using the LVCC method. The strong diffraction lines at the scattering angles (2θ) of 31.11, 44.63, 64.97, and 82.27 were assigned to the diffraction from 100, 110, 200, and 211 planes, respectively, of the B2 crystal structure of NiAl. The calculated lattice parameter for NiAl nanoparticles (2.868) is in good agreement with the bulk lattice parameter (2.871) [95].

A mixture of Ni and Al nanoparticles, as the major product, was produced from the vaporization of a powder mixture with (75 at.% Ni), as identified from the XRD data. When the molar percentage of Ni was decreased to 50 at.% in the target, alloy nanoparticles NiAl were obtained, and the absence of pure Ni and/or Al nanoparticles was evident. The alloy nanoparticles were assigned to $Ni_{0.9}Al_{1.1}$ as identified from the XRD pattern, which matched the pattern of the $Ni_{0.9}Al_{1.1}$ bulk alloy [95]. A mixture of pure Al and intermetallic $Ni_{0.9}Al_{1.1}$ nanoparticles was obtained when a target with 25.0 and 10 at.% Ni was vaporized in the LVCC chamber. A mixture with more Al nanoparticles was obtained from 10.0 at.% Ni target as compared to that prepared from the 25.0 at.% Ni target. It is evident that only one intermetallic phase (NiAl) is obtained in the preparation of Ni/Al nanoparticles using targets consisting of mixtures of Ni and Al bulk powders. However, by increasing the molar ratio of Al or Ni content, alloy nanoparticles, in addition to the pure components (depending on which metal is in excess), are obtained. This result can be explained by the greater stability of the NiAl phase compared to other nickel aluminide phases such as $NiAl_3$, Ni_2Al_3, and Ni_3Al, since the NiAl phase has the highest heat of formation [96, 97].

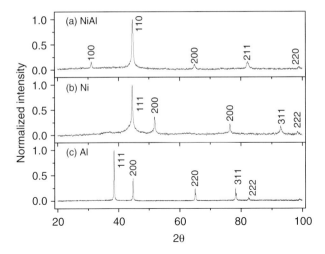

Figure 4.4.15 XRD of (a) NiAl, (b) Ni, and (c) Al nanoparticles prepared by the LVCC method.

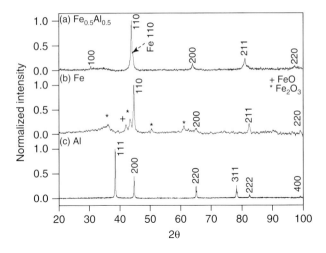

Figure 4.4.16 XRD of (a) $Fe_{0.5}Al_{0.5}$, (b) Fe, and (c) Al nanoparticles prepared by the LVCC method.

The XRD pattern of the nanocrystalline sample prepared from (50:50 at.%) of Fe and Al metallic powder mixture is shown in Figure 4.4.16, along with the diffraction patterns of pure Fe and Al nanoparticles. Neither the Fe nor Al diffraction pattern matched the nanopowder sample patterns prepared from the Fe–Al (50:50) powder mixture. By comparing the XRD data to the database, good agreement was found between the XRD patterns of prepared nanopowder and bulk $Fe_{0.5}Al_{0.5}$ intermetallic alloy [95]. The observed prominent peaks at scattering angles (2θ) of 30.91, 44.15, 64.23, and 81.27 are assigned to scattering from the 100, 110, 200, and 211 planes, respectively, of the FeAl crystal lattice [95].

4.4.7
Growth of Filaments and Treelike Assembly by Electric Field

Charged nanoparticles are of immediate interest to the disciplines of materials, environmental, health, atmospheric, and astrophysical sciences. In materials science, applications to semiconductor manufacturing and surface coatings could be important. Also, consolidation of nanostructured assemblies is expected to significantly enhance the plastic behavior and the ductility of metallic, intermetallic, and ceramic materials. In environmental and health sciences, the control of charged airborne particulates and drug delivery could be of significant benefit. Furthermore, the knowledge and the behavior of nanoparticles under the influence of electric and magnetic fields is important in understanding and predicting the behavior of interstellar dust particles and other granular materials on earth.

Since laser vaporization of metals and semiconductors produces a significant portion of charged (by electrons and ions) nanoparticles, the assembly of these nanoparticles into filaments and long chain fibers can be achieved by applying an electric field during the LVCC synthesis. Application of an electric field between the bottom and top plates of the LVCC chamber during the preparation of the metallic and intermetallic nanoparticles results in the formation of filaments and fibers as shown in Figure 4.4.17 [37]. The influence of an electric field during the vaporization and condensation of several metallic (Fe, Ti, Ni, Cu, and Zn), intermetallic (FeAl, Ti_3Al, NiAl, and CuZn), and semiconductor (Si) nanoparticles has been investigated [35–37, 42, 43]. In the presence of an electric field across the LVCC chamber, the nanoparticles aggregate as little chains and stack end to end. Generally, the chains grow perpendicular on the top and bottom plates of the LVCC chamber as shown in Figure 4.4.17a. Eventually, the chains bridge the top and bottom metal plates (electrodes), indicating that both negatively and positively charged particles are involved. The filamentlike chains grow into 3D treelike structures with the increasing amount of nanoparticles produced at longer vaporization times as shown in Figure 4.4.17b. As with magnets, two chain aggregates would repel one another when their like charges face each other. The repulsion is unstable, however, and the chains would then swing around very rapidly and stick together. In other words, the chain rotates to orient the dipoles so that the positive end would face the negative end, thus attracting one another. It is interesting to note that these aggregates stretch at a higher field strength and contract when the field is turned off.

The filament growth depends strongly on the type of the nanoparticles assembled. The growth of Si filaments requires higher field strength (200 V cm^{-1}) than what is typically used for the metallic and intermetallic nanoparticles investigated (50–60 V cm^{-1}). In addition, the Si filaments (Figure 4.4.17c) are noticeably shorter than the metallic and intermetallic filaments, and they exhibit a fractal morphology. Interestingly, nanoparticles formed by laser vaporization of an Si–Pt mixed target produces long filaments, which aggregate to form complex dense treelike assemblies as shown in Figure 4.4.17d. The XRD and ED of the nanoparticles of the chain aggregates are similar to those obtained from the nanoparticles prepared in the

(a) FeAl nanoparticles (60 V cm^{-1})

(b) FeAl nanoparticles (60 V cm^{-1})

(c) Si nanoparticles (200 V cm^{-1})

(d) Si–Pt nanoalloy (200 V cm^{-1})

Figure 4.4.17 (a–c) Digital photographs and (d) SEM image of nanoparticle filaments grown in the presence of an electric field across the LVCC chamber plates.

absence of the field, thus indicating that the field has no effect on the composition of the particles.

Because of dipole forces, γ-Cu$_5$Zn$_8$ intermetallic nanoparticles showed a much higher tendency in forming filaments as compared to pure Cu and Zn nanoparticles [36]. Growth of filaments is caused by particle aggregation at one of the chamber plates, thereby extending them toward the opposite plate. When the metal target is placed in contact with the bottom plate of the chamber, it creates a gradient in the electric field and causes the filaments to grow toward the target as shown in the photographs of the Cu–Zn nanoparticle fibers displayed in Figure 4.4.18 [36]. The filament growth starts as thin small wires extending from the top plate to the center of the CuZn target as shown in Figure 4.4.18a. This pattern continues until the surface of the target is completely covered (Figure 4.4.18b–f) and then the filaments start to bridge the top and bottom plates of the chamber (Figure 4.4.18g–i). Quite dense fibers can be produced by increasing the amount of nanoparticles formed at extended laser vaporization times as shown in Figure 4.4.18i. The XRD pattern

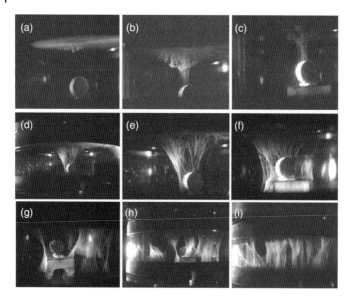

Figure 4.4.18 Digital photographs of γ-Cu_5Zn_8 nanoparticle filaments assembled by the application of 60 V/cm electric field across the LVCC chamber plates. The distance between the chamber plates is 5 cm. The photographs were taken at 10 minutes intervals.

of the nanoparticle filaments is consistent with the γ-Cu_5Zn_8 phase similar to the nanoparticles formed in the absence of an electric field. The TEM analysis indicates that the primary particles are 10–15 nm in diameter.

Figure 4.4.19a,b displays SEM images of the as-deposited Fe and Ni filaments on glass slides placed on the top plate of the LVCC chamber. It is clear that the filament and fiber morphology is quite different from the weblike morphology observed with no electric field applied during the experiment. In the presence of a field, the filaments condense as a bundle of wires as in a wool yarn. The filaments can be a few centimeters long, and they tangle together with neighboring wires to form bundles.

The effect of the electric field on the formation of the chain aggregates acts through the polarization of the charges on the nanoparticle's surface [37]. For larger particles, the effect of the electrostatic charge is overpowered by the gravity effect, but for nanoparticles, the electrostatic forces can be predominant. Since mixtures of positive and negative charges are present on the nanoparticles surfaces, some of the nanoparticles can have net charges. There are two effects that when combined may lead to the sticking of particles of the same net charge. The dipole force is very strong near the surface of the particle; further away from the surface, the net charge or monopole force becomes evident. When two individual particles are separated by a certain distance, they will respond to the monopole charge between them, and if the monopoles are the same, the particles will repel. However, by orienting their dipoles so that they are attracting one another, and since at this short distance the dipole dominates over the net charge, the two particles will stick together.

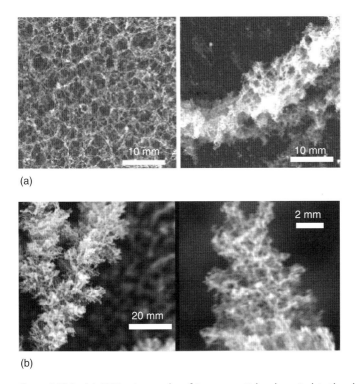

Figure 4.4.19 (a) SEM micrographs of Fe nanoparticles deposited in the absence (left image) and the presence of electric field (200 V cm^{-1}) (right image). (b) SEM micrographs of Ni filament nanoparticles prepared under the influence of an electric field (60 V cm^{-1}).

The filamentlike and treelike aggregates may have some special applications as fillers (additives) to increase the elastic modulus and tensile strength of polymers such as low-strength rubbers. Consolidated materials based on the nanoparticle filaments may show enhanced plasticity, that is, they may exhibit significantly better elongations as compared to cast and powder-processed components. This may lead to new applications as additives and reinforcing agents for low-viscosity polymers and oils [37].

Another important implication of the growth of charged nanoparticles is related to the role of nanoparticles in the formation of interstellar dust and in catalyzing reaction processes [98, 99]. The study of the effects of electric and magnetic fields on the behavior of iron-containing nanoparticles might serve as a model to simulate the interactions of charged particles in space in preparation for the human exploration program [100]. Detailed knowledge about charged particles is essential for the human mission to Mars, where a super arid environment, radiation effects, and the charging up of dust particles by friction as they whip around in the wind amplify electrostatic charges. Because the Martian wind systems are global and there is nothing to interrupt the motion of the dust, the particles are expected to be everywhere. Investigations of electrostatic charge in microgravity predict that *"when astronauts land on Mars, it's going to be very, very dusty, and the dust will probably be charged electrostatically wherever they go!"* [100].

4.4.8
Upconverting Doped Nanocrystals by the LVCC Method

The conversion of near-infrared (NIR) photons to higher energy UV or visible light via multiple absorptions or energy transfer, known as upconversion (UC) phosphors, is an area of extensive research because of its fundamental interest and tremendous potential for photonic and biophotonic applications [101]. The efficient generation of color tunable and white light through the UC process is an important research goal in this area that offers a variety of applications such as displays, back light, sensors, biolabeling, and photodynamic therapy [101]. Nanocrystals can provide several advantages including high efficiency due to the minimum energy loss by scattering and the possibility of biological applications, which require particle size less than 50 nm. Rare-earth-doped nanocrystals provide several attractive features for efficient UC processes [102–105]. These include high thermal stability, chemical durability, stable photocycles, and convenient excitations with inexpensive commercial CW (continuous wave) lasers. The doping with Yb^{3+} as a sensitizer makes it possible to excite with the commercially available NIR 980 nm laser diode to obtain red, green, and blue emissions [106–108]. Various methods for the synthesis of doped rare earth nanocrystals have been reported, including, for example, coprecipitation, solution combustion, hydrothermal processes, sol–gel, colloidal solutions, and flame synthesis [102–111]. Most of the reported methods require high temperatures and postannealing processes to induce crystallinity and increase the quantum yield. The LVCC method was successfully used for the vapor-phase synthesis of upconverting Y_2O_3 nanocrystals doped with Yb^{3+}, Er^{3+}, and Tm^{3+} to generate red, green, and blue emissions that can be adjusted to produce white light [49, 50]. The solid nanophosphors are produced in high purity directly from the vapor phase, which is desirable for specific applications involving remote sensing and identification.

In the experiments, commercial micrometer-sized powders of rare earth oxides 99.9% RE_2O_3 where RE = Yb, Er, Ho, and Tm and Y_2O_3 (99.99%) were compressed into targets using a pellet press. The target of interest was placed on the lower plate of the LVCC chamber, and the chamber was filled with O_2. The target was vaporized using the second-harmonic generation (532 nm) of Nd:YAG laser (30–50 mJ pulse^{-1}, 10^{-8} s pulse). The temperature gradient between the lower and top plates produces a steady convective current that helps to rapidly remove the nanoparticles from the vaporization zone before they can grow to larger particles. Figure 4.4.20 displays a typical TEM image of an as-prepared sample containing 1% (wt) Er_2O_3, 4% Yb_2O_3, and 95% Y_2O_3. The nanocrystals form a weblike structure with an average particle size of 10–15 nm.

The XRD pattern of the Y_2O_3 nanocrystals prepared by the LVCC method matches perfectly with the monoclinic structure rather than the cubic structure of the original pellet as evident from the XRD data shown in Figure 4.4.21a. It is clear

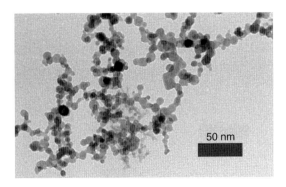

Figure 4.4.20 TEM of Yb^{3+}/Er^{3+}-codoped Y_2O_3 nanocrystals.

that the laser vaporization process transforms the cubic Y_2O_3 bulk powder into a monoclinic structure in the synthesized nanocrystals. This is consistent with the recent flame synthesis study, which indicates that this transformation occurs when the flame temperature exceeds 2900 K [111]. The phase transformation induced by the laser vaporization appears to be efficient since the XRD of the nanocrystals can be assigned predominantly to the monoclinic phase with a minor component left from the cubic phase as shown in Figure 4.4.21a. It should be noted that both pure Yb_2O_3 and Er_2O_3 nanoparticles retained their cubic crystal structures in the individual samples after the LVCC process.

Figure 4.4.21b (bottom) displays the XRD pattern of the mixed bulk powder of 1% Er_2O_3, 4% Yb_2O_3, and 95% Y_2O_3, which is used to form the pellet for the LVCC experiment. The pattern shows predominantly the cubic phase of bulk Y_2O_3 with additional small peaks assigned to Yb_2O_3 and no peaks that can be assigned to Er_2O_3, consistent with the 4 and 1% concentrations of Er_2O_3 and Yb_2O_3, respectively. Interestingly, after the LVCC process, the Er_2O_3/Yb_2O_3-doped Y_2O_3 nanocrystals produced exhibit the monoclinic Y_2O_3 lattice (Figure 4.4.21b, top) with no evidence for the presence of either Yb_2O_3, indicating that the Yb^{3+} are well dispersed throughout the Y_2O_3 lattice. The monoclinic lattice is expected to enhance the UC emission intensity of the Er_2O_3/Yb_2O_3-doped Y_2O_3 nanocrystals due to the enhancement of phonon isolation following absorption of the 980 nm photons by the Yb^{3+} ions [111, 112]. It is significant to note that the LVCC process results in efficient cubic to monoclinic phase transformation of the Y_2O_3 lattice without the detrimental effect of increasing particle size, which is unavoidable at the high temperatures (>2900 K) necessary for the phase transformation. This is clearly one of the important advantages of the LVCC method in the synthesis of doped nanocrystals as compared to other methods.

Figure 4.4.22 compares the UC spectra of bulk Y_2O_3 powders doped with (8%) $Yb^{3+}/(6\%)$ Er^{3+}, (10%) $Yb^{3+}/(2\%)$ Ho^{3+}, and (6%) $Yb^{3+}/(2\%)$ Tm^{3+} with the corresponding Y_2O_3-doped nanocrystals prepared by the LVCC method. While there is not much of an apparent difference in the intensity of Er^{3+}/Yb^{3+}-codoped Y_2O_3 nanoparticles when compared to its bulk counterpart, as shown in Figure 4.4.22a,

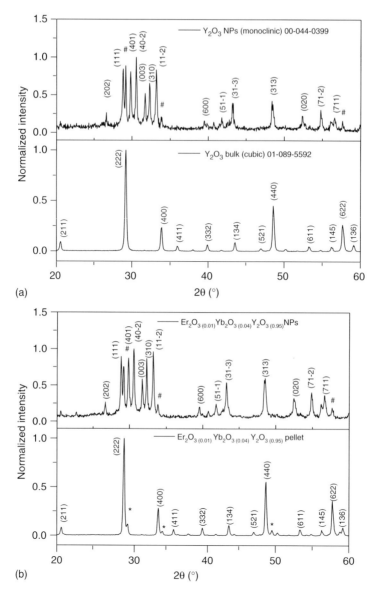

Figure 4.4.21 (a) XRD data of (bottom) bulk Y₂O₃ target showing the cubic phase and (top) Y₂O₃ nanocrystals prepared using the LVCC method showing predominantly the monoclinic phase. Peaks marked # are due to minor component of the cubic phase within the monoclinic phase. (b) XRD data of (bottom) bulk Y₂O₃ target containing 1% Er₂O₃ and 4% Yb₂O₃ powder showing the cubic phase and the presence of Yb₂O₃ as indicated by the peaks marked (*), and (top) 1% Er₂O₃, 4% Yb₂O₃-doped Y₂O₃ nanocrystals prepared using the LVCC method showing predominantly the monoclinic phase and the disappearance of the Yb₂O₃ peaks. Peaks marked # are due to minor component of the Y₂O₃ cubic phase within the monoclinic phase.

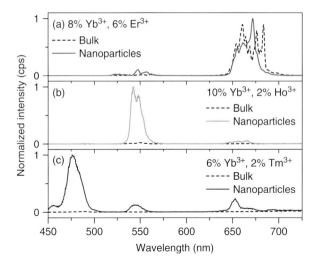

Figure 4.4.22 Upconversion emission spectra of bulk Y_2O_3 powder and Y_2O_3 nanocrystals doped with (a) (8%) Yb^{3+}/(6%) Er^{3+}, (b) (10%) Yb^{3+}/(2%) Ho^{3+}, and (c) (6%) Yb^{3+}/(2%) Tm^{3+}.

the Yb^{3+}/Ho^{3+} and Yb^{3+}/Tm^{3+}-doped Y_2O_3 nanoparticles are significantly brighter than their bulk counterparts as shown in Figure 4.4.22b,c, respectively.

The flexibility and control of the LVCC synthesis approach in tuning the color of the UC emissions is demonstrated by the preparation of Y_2O_3 nanocrystal samples doped with 10% Yb^{3+}/2% Er^{3+}/2% Ho^{3+} and 16% Yb^{3+}/1% Er^{3+}/1% Tm^{3+}. The UC emission spectra of these samples are shown in Figure 4.4.23.

The incorporation of Ho^{3+} within the Yb^{3+}/Er^{3+}-codoped Y_2O_3 nanocrystals results in the enhancement of the green emissions as shown in Figure 4.4.23a

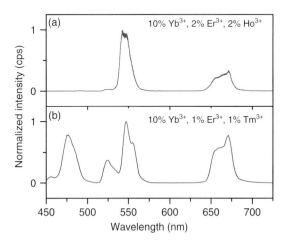

Figure 4.4.23 Upconversion emission spectra of Y_2O_3 nanocrystals doped with Yb^{3+}/Er^{3+}/Ho^{3+} and Yb^{3+}/Er^{3+}/Tm^{3+}.

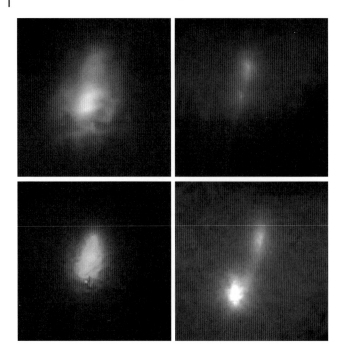

Figure 4.4.24 Photographs of the emitted light showing (a) red, (b) green, (c) blue and (d) white light produced upon the 980 nm laser excitation of (16%) Yb^{3+}/(6%) Er^{3+}, (10%) Yb^{3+}/(2%) Ho^{3+}, (6%) Yb^{3+}/(2%) Tm^{3+}, and (16%) Yb^{3+}/(1%) Er^{3+}/(1%) Tm^{3+}-doped Y_2O_3 nanocrystals, respectively. The colored photographs appear in the printed Figure as light and dark grey colors with different degree of darkness representing the actual observed colors.

in comparison with the UC emission spectrum of the Yb^{3+}-/Er^{3+}-codoped Y_2O_3 nanocrystals shown in Figure 4.4.23b. The nearly equal intensity of the red and green emissions from the 10% Yb^{3+}, 2% Er^{3+}, and 2% Ho^{3+}-codoped Y_2O_3 nanocrystals results in an overall greenish yellow color. Figure 4.4.24 displays representative images of UC emissions from the Ln^{3+}-doped Y_2O_3 nanocrystals taken with a digital camera in dim light. These colors can easily be seen with the naked eye under dimmed room light. It is clear that a variety of doped nanocrystals with control dopant concentrations can be designed and tested using the LVCC method.

4.4.9
Supported Nanoparticle Catalysts by the LVCC Method

Nanophase metal and metal oxide catalysts, with controlled particle size, high surface area, and more densely populated unsaturated surface coordination sites, could potentially provide significantly improved catalytic performance over conventional catalysts [113, 114]. These nanoparticle oxides enable catalytic activation at significantly lower temperatures for the reduction of sulfur dioxide and the oxidation of carbon monoxide [114–117]. Research in this area is motivated by the possibility of designing nanostructured catalysts that possess novel catalytic properties such as low-temperature activity, selectivity, stability, and resistance to poisoning and degradation [113–117]. Such catalysts are essential for technological advances in environmental protection, improving indoor air quality, and in chemical synthesis and processing.

The discovery of highly effective supported coinage metal (Cu, Au) on metal oxide nanocatalysts, and the subsequent attempts to elucidate their essential features, exemplifies well the new opportunities in nanocatalysis research [115, 118, 119]. There seems to be no doubt, however, that activity in these systems is derived from the presence of nanometer-scale metal structures. The gold-based nanocatalysts are distinct technologically in that they function at (or below) normal temperature and humidity, catalyzing air purification via complete combustion of noxious waste gases using ambient oxygen as an oxidant.

The low-temperature oxidation of carbon monoxide is one of the current important environmental issues since small exposure (parts per million) to this odorless invisible gas can be lethal [120]. Therefore, the discovery that Au nanoparticles between 2 and 5 nm are exceptionally active for low-temperature CO oxidation has stimulated extensive research to develop highly active catalysts to remove even a small amount of CO from the local environment [116, 119, 121–123]. There is also a strong incentive to develop active supported catalysts that utilize small amounts of the noble metals such as gold and palladium. The catalytic activity of the Au-based catalysts depends on various factors such as the type of the support, the Au precursor, the preparation conditions, the pretreatment conditions, and the catalytic reaction conditions [116, 119, 121–123]. Among these factors, the nature and shape of the support are expected to play major roles given the strong tendency of Au nanoparticles to efficiently adsorb CO molecules but the inability to activate oxygen molecules [116, 119, 121–123].

The LVCC method has been successful for the preparation of a variety of supported and unsupported metallic and bimetallic nanoparticle catalysts [39, 40, 43, 46, 47]. To prepare Au nanoparticle catalysts dispersed on different supports such as SiO_2, Al_2O_3, ZrO_2, and CeO_2, commercial micrometer-sized powders were used to prepare pellet targets of selected compositions of the catalyst/support system. It should be noted that the LVCC method is uniquely suited for the preparation of supported nanoparticle catalysts where the active metal nanoparticles such as Au are homogeneously dispersed on the surfaces of the oxide support [39, 40]. If the weight percentage of the Au in the target is kept low with respect to the oxide support ($<5\%$), laser vaporization from such targets within an Ar atmosphere results in the formation of small Au nanoparticles (5–10 nm) deposited on the surface of larger (50–100 nm) support particles as shown in the SEM and TEM images of the Au/CeO_2 nanoparticles in Figure 4.4.25.

Figure 4.4.26a compares the catalytic activity for CO oxidation of the supported Au/CeO_2 nanoparticles prepared by the LVCC method with the activity of the 5% Au in a CeO_2 micrometer-sized powder mixture used to prepare the evaporated target. It is clear that the supported Au/CeO_2 nanoparticles exhibit much higher activity

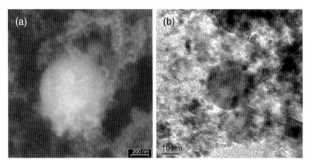

Figure 4.4.25 SEM and TEM of Au/CeO$_2$ nanoparticles.

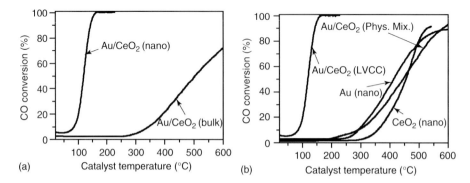

Figure 4.4.26 (a,b) Catalytic activity for the CO oxidation on different catalysts. Phys. mix., physical mixture.

compared to the bulk powder under similar conditions. The 5% supported Au/CeO$_2$ nanoparticles show a 3% CO to CO$_2$ conversion at $<20\,°$C, a 50% conversion at 118 °C, and a 100% conversion at 163 °C. For comparison, the bulk mixture reaches a maximum conversion of only 60% at temperatures near 600 °C. The activities of the individual Au and CeO$_2$ nanoparticles prepared under similar conditions as well as 5% physical mixture of Au nanoparticles in CeO$_2$ nanoparticles are compared with the activity of the 5% supported Au/CeO$_2$ nanoparticles as shown in Figure 4.4.26b.

Although the individual nanoparticles showed higher activities than the corresponding micrometer-sized particles, the physical mixture of nanoparticles did not exhibit any enhanced activity over the individual components as shown in Figure 4.4.26b. The very different activity of the physical mixture of a 5% Au and 95% CeO$_2$ nanoparticles from that of the 5% supported Au/CeO$_2$ nanoparticles clearly indicates that laser vaporization of the mixed Au–CeO$_2$ target under the appropriate LVCC conditions can produce Au nanoparticles supported on CeO$_2$ nanoparticles with significant metal–support interaction.

The effects of the oxide support on the catalytic activity of Au nanoparticles prepared by chemical methods have been extensively investigated [124–127]. Reducible oxides such as CeO$_2$, TiO$_2$, and ZrO$_2$ provide suitable active sites that enhance the

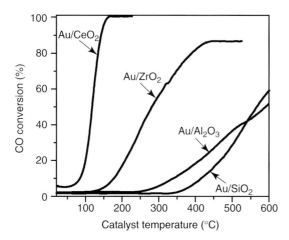

Figure 4.4.27 Catalytic activity for the CO oxidation on 5% Au on different oxide supports.

catalytic activity. The results show that Au nanoparticles supported on different oxides prepared by the LVCC method exhibit a variety of conversion efficiency and light-off temperatures depending on the type of support. Figure 4.4.27 displays the catalytic activities of catalysts containing 5% Au nanoparticles supported on CeO_2, ZrO_2, Al_2O_3, and SiO_2 nanoparticles prepared using the LVCC method. It is clear that the Au/CeO_2 catalyst has the highest activity followed by Au/ZrO_2, Au/Al_2O_3, and finally Au/SiO_2. The high activity of the Au/CeO_2 nanoparticle catalyst is attributed to the strong interaction between Au and CeO_2 and to the oxygen storage capacity and redox properties of CeO_2 nanoparticles [124–127].

The results also indicate that the Au/CeO_2 nanoparticle catalyst exhibits excellent stability and reproducibility under typical reaction conditions. For example, at 163 °C, the 100% conversion was constant over a test period of 20 h under the flow of the reaction mixture. This stability appears to be one of the most important advantages of using vapor-phase synthesis of nanoparticle catalysts because of the elimination of chemical precursors and organic solvents.

4.4.10
Conclusion

The LVCC technique combines the advantages of pulsed LVCC from the vapor phase under well-defined conditions of temperature and pressure. It allows the synthesis of a wide variety of nanoparticles including semiconductors, metals, metal oxides, nanoalloys, intermetallics, doped nanocrystals, and supported nanocatalysts.

Acknowledgments

We gratefully acknowledge NASA for the partial support of this research through grant NNX07AU16G. We also thank NSF (OISE-0938520) for the support of the

"US-Egypt Advanced Studies Institute on Nanomaterials and Nanocatalysis for Energy, Petrochemicals and Environmental Applications," which facilitated the completion of this work.

References

1. Liz-Marzan, L.M. and Kamat, P.V. (eds) (2003) *Nanoscale Materials*, Kluwer Academic Publishers, Boston, Dordrecht, London.

2. Ozin, G.A. and Arsenault, A.C. (2005) *Nanochemistry: A Chemical Approach to Nanomaterials*, Royal Society of Chemistry, RSC Publishing, Cambridge.

3. Johnston, R.L. and Ferrando, R. (eds) (2008) *Nanoalloys: From Theory to Application*, Royal Society of Chemistry, Cambridge.

4. Edelstein, A.S. and Cammarata, R.C. (eds) (1996) *Nanomaterials: Synthesis, Properties and Applications*, Institute of Physics, Bristol and Philadelphia.

5. Hadijipanyis, G.C. and Siegel, R.W. (eds) (1994) *Nanophase Materials: Synthesis, Properties, Applications*, Kluwer Academic Publications, London.

6. Carbone, L. and Cozzoli, P.D. (2010) *Nano Today*, **5**, 449–493.

7. Nagarajan, R. and Hatton, T.A. (eds) (2008) *Nanoparticles: Synthesis, Stabilization, Passivation and Functionalization*, American Chemical Society, Washington, DC.

8. Talapin, D.V., Lee, J.-S., Kovalenko, M.V., and Shevchenko, E.V. (2010) *Chem. Rev.*, **110**, 389–458.

9. El-Shall, M.S. and Li, S. (1998) in *Advances in Metal and Semiconductor Clusters* (ed. M.A. Duncan), Jai Press, Ltd, London, p. 115.

10. Siegel, R.W. (1994) *Nanostruct. Mater.*, **4**, 121.

11. El-Shall, M.S. and Edelstein, A.S. (1996) in *Nanomaterials: Synthesis, Properties and Applications*, Chapter 2 (eds A.S. Edelstein and R.C. Cammarata), Institute of Physics Publishing, Bristol and Philadelphia, pp. 13–54.

12. El-Shall, M.S., Slack, W., Vann, W., Kane, D., and Hanley, D. (1994) *J. Phys. Chem.*, **52**, 3067–3070.

13. (a) El-Shall, M.S., Graiver, D., and Pernisz, U.C. (1996) Silica Nanoparticles. US Patent 5, 580, 655 (b) Graiver, D., Pernisz, U.C., and El-Shall, M.S. (1997) Silicon Nanoparticles. US Patent 5, 695, 617.

14. El-Shall, M.S., Li, S., Turkki, T., Graiver, D., Pernisz, U.C., and Baraton, M.E. (1995) *J. Phys. Chem.*, **99** 17805.

15. El-Shall, M.S., Graiver, D., Pernisz, U.C., and Baraton, M.I. (1995) *Nanostruct. Mater.*, **6**, 297.

16. Baraton, M.I. and El-Shall, M.S. (1995) *Nanostruct. Mater.*, **6**, 301.

17. El-Shall, M.S., Li, S., Graiver, D., and Pernisz, U. (1996) in *Nanotechnology: Molecularly Designed Materials*, ACS Symposium Series, Vol. **622**, Chapter 5 (eds G. Chow and K.E. Gonsalves), p. 79.

18. El-Shall, M.S. (1996) *Appl. Surf. Sci.*, **106**, 347.

19. Jonsson, B.J., Turkki, T., Strom, V., El-Shall, M.S., and Rao, K.V. (1996) *J. Appl. Phys.*, **79**, 5063.

20. Li, S., Silvers, S., and El-Shall, M.S. (1997) in *Advances in Microcrystalline and Nanocrystalline Semiconductors*, Materials Research Society Symposium Proceedings Series, Vol. **452** (eds P.M. Fauchet, R.W. Collins, P.A. Alivisatos, I. Shimizu, T. Shimada, and J.C. Vial), Materials Research Society, Pittsburgh, Pennsylvania, p. 141.

21. Li, S., Silvers, S., and El-Shall, M.S. (1997) in *Advances in Microcrystalline and Nanocrystalline Semiconductors*, Materials Research Society Symposium Proceedings Series, Vol. **452** (eds P.M. Fauchet, R.W Collins, P.A. Alivisatos, I. Shimizu, T. Shimada, and J.C. Vial), Materials Research Society, Pittsburgh, Pennsylvania, p. 389–394.

22. Li, S., Silvers, S., and El-Shall, M.S. (1997) *J. Phys. Chem.*, **101**, 1794.

23. El-Shall, M.S. and Li, S. (1997) *SPIE*, **3123**, 98.
24. Li, S., Germanenko, I.N., and El-Shall, M.S. (1998) *J. Phys. Chem.*, **102**, 7319.
25. Li, S. and El-Shall, M.S. (1998) *Appl. Surf. Sci.*, **127**, 330.
26. El-Shall, M.S. and Li, S. (1999) *Nanostruct. Mater.*, **12**, 215.
27. Germanenko, I.N., Li, S., Silvers, S.J., and El-Shall, M.S. (1999) *Nanostruct. Mater.*, **12**, 731.
28. Li, S., Germanenko, I.N., and El-Shall, M.S. (1999) *J. Cluster Sci.*, **10**, 533.
29. Germanenko, I.N., Dongol, M., Pithawalla, Y.B., Carlisle, J.A., and El-Shall, M.S. (2000) *Pure Appl. Chem.*, **72**, 245–255.
30. Pithawalla, Y.B., Deevi, S.C., and El-Shall, M.S. (2000) *Intermetallics*, **8–9**, 1225.
31. Pithawalla, Y.B., El-Shall, M.S., Deevi, S.C., and Rao, K.V. (2001) *J. Phys. Chem. B*, **105**, 2085–2090.
32. Germanenko, I.N., Li, S., and El-Shall, M.S. (2001) *J. Phys. Chem.*, **105**, 59–66.
33. Pithawalla, Y.B., Deevi, S.C., and El-Shall, M.S. (2001) *Adv. Powder Metall. Part. Mater.*, **9**, 99–109.
34. Pithawalla, Y.P., Deevi, S.C., and El-Shall, M.S. (2002) *Mater. Sci. Eng. A*, **329**, 92–98.
35. El-Shall, M.S. (2003) *McGraw-Hill Year book of Science and Technology*, McGraw-Hill, pp. 268–272.
36. Pithawalla, Y.B., El-Shall, M.S., and Deevi, S. (2003) *Scr. Mater.*, **48**, 671–676.
37. El-Shall, M.S., Abdelsayed, V., Pithawalla, Y.B., Alsharach, E., and Deevi, S.C. (2003) *J. Phys. Chem. B*, **107**, 2882–2886.
38. Glaspell, G.P., Jagodzinski, P.W., and Manivannan, A. (2004) *J. Phys. Chem. B*, **108**, 9604–9607.
39. Yang, Y., Saoud, K.M., Abdelsayed, V., Glaspell, G., Deevi, S., and El-Shall, M.S. (2006) *Catal. Commun.*, **7**, 281–284.
40. Abdelsayed, V., Saoud, K.M., and El-Shall, M.S. (2006) *J. Nanopart. Res.*, **8**, 519–531.
41. Abdelsayed, V., Glaspell, G., Saoud, K., Meot-Ner, M., and El-Shall, M.S. (2006) in *Astrochemistry: From Laboratory Studies to Astronomical Observations*, AIP Conference Proceedings, Vol. **855** (eds R.I. Kaiser, P. Bermath, A.M. Mebel, Y. Osamura, and S. Petrie), American Institute of Physics, Melville, New York, pp. 76–85.
42. Abdelsayed, V., El-Shall, M.S., and Seto, T. (2006) *J. Nanopart. Res.*, **8**, 361–369.
43. Glaspell, G., Abdelsayed, V., Saoud, K.M., and El-Shall, M.S. (2006) *Pure Appl. Chem.*, **78**, 1671–1693.
44. Mautner, M.N., Abdelsayed, V., Thrower, J.D., Green, S.D., Collings, M.P., McCoustra, M.R.S., and El-Shall, M.S. (2006) *Faraday Discuss.*, **133**, 103–112.
45. Abdelsayed, V. and El-Shall, M.S. (2007) *J. Chem. Phys.*, **126**, 024706.
46. Radwan, N.R.E., El-Shobaky, G.A., Hassan, H.M.A., and El-Shall, M.S. (2007) *Appl. Catal. A*, **331**, 8–18.
47. Glaspell, G., Hassan, H.M.A., Elzatahry, A., Abdelsayed, V., and El-Shall, M.S. (2008) *Top. Catal.*, **47**, 22–31.
48. Abdelsayed, V., Glaspell, G., Nguyen, M., Howe, J., and El-Shall, M.S. (2008) *Faraday Discuss.*, **138**, 163–180.
49. Glaspell, G., Wilkins, J., Anderson, J., and El-Shall, M.S. (2008) *Proceedings of the SPIE*, Infrared Technology and Applications XXXIV, Vol. **6940** (eds B.F. Andresen, G.F. Fulop, and P.R. Norton), International Society for Optics and Photonics, San Diego, pp. 69403B/1–69403B/7.
50. Glaspell, G., Anderson, J., Wilkins, J., and El-Shall, M.S. (2008) *J. Phys. Chem. C*, **112**, 11527–11531.
51. Kashchiev, D. (2000) *Nucleation, Basic Theory with Applications*, Butterworth Heinemann, Burlington.
52. Abraham, F.F. (1974) *Homogeneous Nucleation Theory*. Academic Press, New York.
53. (a) McGraw, R. and Reiss, H. (1979) *J. Colloid Interface Sci.*, **72** 172 (b) Reiss, H. (1987) *Science*, **238**, 1368.
54. (a) Kane, D., Daly, G.M., and El-Shall, M.S. (1995) *J. Phys. Chem.*, **99** 7867 (b) Kane, D. and El-Shall, M.S. (1996) *Chem. Phys. Lett.*, **259**, 482.
55. Makino, T., Suzuki, N., Yamada, Y., Yoshida, T., Seto, T., and Aya, N. (1999) *Appl. Phys. A.*, **69**, S243–S247.

56. Seto, T., Kawakami, Y., Suzuki, N., Hirasawa, M., and Aya, N. (2001) *Nano Lett.*, **1**, 315–318.

57. Cullis, A.G. and Canham, L.T. (1991) *Nature*, **353**, 335.

58. Canham, L.T., Cullis, A.G., Pickering, C., Dosser, O.D., Cox, T.I., and Lynch, T.P. (1994) *Nature*, **368**, 133.

59. Littau, K.A., Szajowski, P.J., Muller, A.J., Kortan, A.R., and Brus, L.E. (1993) *J. Phys. Chem.*, **97**, 1224.

60. Wilson, W.L., Szajowski, P.F., and Brus, L.E. (1993) *Science*, **262**, 1242.

61. Kanemitsu, Y., Ogawa, T., Shiraishi, K., and Takeda, K. (1993) *Phys. Rev. B*, **48**, 4883.

62. Brus, L.J., Szajowski, P.F., Wilson, W.L., Harris, T.D., Schuppler, S., and Citrin, P.H. (1995) *J. Am. Chem. Soc.*, **117**, 2915.

63. Kanemitsu, Y. (1995) *Phys. Rep.*, **263**, 1–91.

64. Link, S. and El-Sayed, M.A. (2000) *Int. Rev. Phys. Chem.*, **19**, 409.

65. Hu, M., Chen, J., Li, Z.-Y., Au, L., Hartland, G.V., X., Li, Marquez, M., and Xia, Y. (2006) *Chem. Soc. Rev.*, **35**, 1084.

66. Kelly, K.L., Coronado, E., Zhao, L.L., and Schatz, G.C. (2003) *J. Phys. Chem. B*, **107**, 668.

67. Link, S., Wang, Z.L., and El-Sayed, M.A. (1999) *J. Phys. Chem. B*, **103**, 3529.

68. Jain, P.K., Qian, W., and El-Sayed, M.A. (2006) *J. Am. Chem. Soc.*, **128**, 2426.

69. Jain, P.K., Lee, K.S., El-Sayed, I.H., and El-Sayed, M.A. (2006) *J. Phys. Chem. B*, **110**, 7238.

70. Huang, X., El-Sayed, I.H., Qian, W., and El-Sayed, M.A. (2006) *J. Am. Chem. Soc.*, **128**, 2115.

71. Loo, C., Lowery, A., Halas, N., West, J., and Drezek, R. (2005) *Nano Lett.*, **5**, 709.

72. Panacek, A., Kvitek, L., Prucek, R., Kolar, M., Vecerova, R., Pizurova, N., Sharma, V.K., Nevecna, T.J., and Zboril, R. (2006) *J. Phys. Chem. B*, **110**, 16248.

73. El-Sayed, I.H., Huang, X., and El-Sayed, M.A. (2006) *Cancer Lett.*, **239**, 129.

74. Takeuchi, Y., Ida, T., and Kimura, K. (1997) *J. Phys. Chem. B*, **101**, 1322.

75. Mafune, F., Kohno, J., Takeda, Y., and Kondow, T. (2002) *J. Phys. Chem. B*, **106**, 7575.

76. Takami, A., Kurita, H., and Koda, S. (1999) *J. Phys. Chem. B*, **103**, 1226.

77. Hodak, J.H., Henglein, A., Giersig, M., and Hartland, G.V. (2000) *J. Phys. Chem. B*, **104**, 11708.

78. Zhang, J., Worley, J., Denommee, S., Kingston, C., Jakubek, Z.J., Deslandes, Y., Post, M., Simard, B., Braidy, N., and Botton, G.A. (2003) *J. Phys. Chem. B*, **107**, 6920.

79. Chen, Y.H. and Yeh, C.S. (2001) *Chem. Commun.*, 371.

80. Link, S., Burda, C., Nikoobakht, B., and El-Sayed, M.A. (1999) *Chem. Phys. Lett.*, **315**, 12.

81. Sun, Y. and Xia, Y. (2002) *Science*, **298**, 2176.

82. Mallin, M.P. and Murphy, C.J. (2002) *Nano Lett.*, **2**, 1235.

83. He, S.T., Xie, S.S., Yao, J.N.H., Gao, H.J., and Pang, S.J. (2002) *App. Phys. Lett.*, **81**, 150.

84. Han, S.W., Kim, Y., and Kim, K. (1998) *J. Colloid Interface Sci.*, **208**, 272.

85. Chen, D.H. and Chen, C.J. (2002) *J. Mater. Chem.*, **12**, 1557.

86. Treguer, M., de Cointet, C., Remita, H., Khatouri, J., Mostafavi, M., Amblard, J., Belloni, J., and de Keyzer, R. (1998) *J. Phys. Chem. B*, **102**, 4310.

87. Lee, I., Han, S.W., and Kim, K. (2001) *Chem. Commun.*, 1782.

88. Huang, C.-Y., Chiang, H.-J., Huang, J.-C., and Sheen, S.-R. (1999) *Nanostruct. Mater.*, **10**, 1393–1400.

89. McKamey, C.G. (1996) in *Physical Metallurgy and Processing of Intermetallic Compounds* (eds N.S. Stoloff and V.K. Sikka), Chapman and Hall, Thomson Publishing, New York, p. 351.

90. Stoloff, N.S. (1998) in *Iron Aluminides: Alloy Design, Processing, Properties and Applications*, Materials Science and Engineering A, Vol. **258** (eds S.C. Deevi, J.H. Schneibel, D. Morris, and V. Sikka), Elsevier Science S. A., p. 1.

91. Deevi, S.C. and Sikka, V.K. (1996) *Intermetallics*, **4**, 357.

92. Lilly, A.C., Deevi, S.C., and Gibbs, Z.P. (1998) in *Iron Aluminides: Alloy Design, Processing, properties and Applications*, Materials Science and Engineering A, Vol. **258** (eds S.C. Deevi, J.H. Schneibel,

D. Morris, and V. Sikka), Elsevier Science S. A., p. 42.

93. Deevi, S.C. and Sikka, V.K. (1997) in *Nickel and Iron Aluminides: Processing, Properties, and Applications* (eds S.C. Deevi, V.K. Sikka, P.J. Maziasz, and R.W. Cahn), ASM International, Materials Park, p. 157.

94. Amilis, X., Nogues, J., Surinach, J.S., Lutterotti, L., and Baro, M.D. (1999) *Nanostruct. Mater.*, **12**, 801.

95. X'Pert High Score Plus, Version 3.0. (2009) PANanalytical: Almelo, The Netherlands.

96. Dong, S., Hou, P., Yang, H., and Zou, G. (2002) *Intermetallics*, **10**, 217–223.

97. Chrifi-Alaoui, F.Z., Nassik, M., Mahdouk, K., and Gachon, J.C. (2004) *J. Alloys Compd.*, **364**, 121–126.

98. Henning, T. (1998) *Chem. Soc. Rev.*, **27**, 315.

99. Bernstein, M.P., Sandford, S.A., and Allamandola, L.J. (1999) *Sci. Am.*, **281**, 42.

100. Marshall, J. (2000) Researching Mars Dust. NASA Microgravity News, Vol. 7.

101. Prasad, P.N. (2004) *Nanophotonics*, John Wiley & Sons, Inc., New York.

102. Silver, J., Martinez-Rubio, M., Ireland, T., Fern, G., and Withnall, R. (2001) *J. Phys. Chem. B*, **105**, 948.

103. Capobianco, J., Boyer, J., Vetrone, F., Speghini, A., and Bettinelli, M. (2002) *Chem. Mater.*, **14**, 2915.

104. Vetrone, F., Boyer, C., Capobianco, J., Speghini, A., and Bettinelli, M. (2003) *J. Phys. Chem. B*, **107**, 1107.

105. Patra, A., Ghosh, P., Chowdhury, P.S., Alencar, M.A.R., Lozano, W., Rakov, N., and Maciel, G.S. (2005) *J. Phys. Chem. B*, **109**, 10142.

106. Matsuura, D., Hattori, H., and Takano, A. (2005) *J. Electron. Chem. Soc.*, **152**, H39.

107. Sivakumar, S., van Veggel, F.C.J.M., and Raudsepp, M. (2005) *J. Am. Chem. Soc.*, **127**, 12464.

108. Boyer, J., Vetrone, F., Cuccia, L.A., and Capobianco, J.A. (2006) *J. Am. Chem. Soc.*, **128**, 7444.

109. Heer, S., Lehmann, O., Haase, M., and Gudel, H.U. (2003) *Angew. Chem. Int. Ed.*, **42**, 3179.

110. Sun, Y., Liu, H., Wang, X., Kong, X., and Zhang, H. (2006) *Chem. Mater.*, **18**, 2726.

111. Qin, Z., Yokomori, T., and Ju, Y. (2007) *Appl. Phys. Lett.*, **90**, 073104.

112. Patra, A., Friend, C.S., Kapoor, R., and Prasad, P.N. (2003) *Appl. Phys. Lett.*, **83**, 284.

113. Somorjai, G.A. (1994) *Introduction to Surface Chemistry and Catalysis*, Wiley-Interscience, New York.

114. Moser, W.R. (ed.) (1996) *Advanced Catalysts and Nanostructured Materials*, Academic Press.

115. Haruta, M., Tsubota, S., Kobayashi, T., Kageyama, H., Genet, M.J., and Delmon, B. (1993) *J. Catal.*, **144**, 175.

116. Chen, M.S. and Goodman, D.W. (2004) *Science*, **306** 252–255.

117. Chou, K.C., Markovic, N.M., Kim, J., Ross, P.N., and Somorjai, G.A. (2003) *J. Phys. Chem. B*, **107**, 1840.

118. Hayashi, T., Tanaka, K., and Haruta, M. (1998) *J. Catal.*, **178**, 566–575.

119. Moreau, F. and Bond, G.C. (2006) *Catal. Today*, **114**, 362.

120. World Health Organization (1999) Carbon Monoxide: Environmental Health Criteria, World Health Organization, Geneva, p. 213.

121. Bond, G.C. and Thompson, D.T. (1999) *Catal. Rev. Sci. Eng.*, **41**, 319–388.

122. Haruta, M. (1997) *Catal. Today*, **36**, 153–166.

123. Hutchings, G.J. (1996) *Gold Bull.*, **29**, 123–130.

124. Travarelli, A., Leitenburg, C., Dolcetti, G., and Boaro, M. (1999) *Catal. Today*, **50**, 353–367.

125. Fernandez-Garcia, M., Martinez-Arias, A., Salamanca, L.N., Coronado, J.M., Anderson, J.A., Conesa, J.C., and Soria, J. (1999) *J. Catal.*, **187**, 474–486.

126. Fu, Q., Weber, A., and Flytzani-Stephanopoulos, M. (2001) *Catal. Lett.*, **77**, 1–3.

127. Luengnaruemitchai, A., Ouswan, S., and Gulari, E. (2003) *Catal. Commun.*, **4**, 215–221.

5
Nanomaterials: Laser-Induced Nano/Microfabrications

5.1
Direct Femtosecond Laser Nanostructuring and Nanopatterning on Metals

Anatoliy Vorobyev and Chunlei Guo

5.1.1
Introduction

Nanostructured materials have found a wide range of applications in optics, opto-electronics, photonics, plasmonics, optical biosensing, nanofluidics, optofluidics, biomedicine, and other areas. In many of these applications, surface nanostructuring is used as an effective way for modifying optical, mechanical, chemical, wetting, and other properties of materials. Therefore, further advances in nanotechnology intrinsically depend on the development of more efficient techniques for controlled surface nanostructuring and nanopatterning. With the advent of the femtosecond laser, the potential for the femtosecond laser in ablation-based nanoprocessing of materials [1, 2] has been demonstrated. Further studies of both femtosecond laser ablation and its applications have given rise to a new field of controlled nanostructuring and nanopatterning of a large variety of materials including metals, semiconductors, glasses, ceramics, and polymers. Several femtosecond laser techniques based on laser ablation have been developed for surface nanostructuring and nanopatterning of solids. Near-field nanomachining techniques have been developed in previous studies [3, 4]. Wu *et al.* [5] have developed a laser-assisted chemical etching technique for nano- and microstructuring of silicon. Koch *et al.* [6] have demonstrated nanotexturing of metal films by femtosecond laser-induced melt dynamics. Deposits from femtosecond laser-ablated plume material have also produced nanostructures [7–9]. In a number of studies, it was found that laser-induced periodic surface structures (LIPSSs) that had previously been generated using long-pulsed lasers can be also produced by femtosecond laser pulses on semiconductors [10–12], dielectrics [13], and metals [14–16]. It has been also demonstrated that a large variety of nano- and microstructures can be produced by direct femtosecond laser processing of metals [14, 17]. Moreover, it has been found that these surface structures significantly modify optical properties of the treated surface [14, 18]. In this chapter, we focus on a review of surface nanostructuring of

Nanomaterials: Processing and Characterization with Lasers, First Edition.
Edited by Subhash Chandra Singh, Haibo Zeng, Chunlei Guo, and Weiping Cai.
© 2012 Wiley-VCH Verlag GmbH & Co. KGaA. Published 2012 by Wiley-VCH Verlag GmbH & Co. KGaA.

metals using the direct femtosecond laser processing technique and the changes in metals' properties caused by the laser-induced surface structural modifications. Femtosecond laser micromachining of other materials has also been actively studied [19].

5.1.2
Basic Principles of Surface Nanostructuring by Direct Femtosecond Laser Ablation

When a femtosecond laser beam is focused tightly on a metal surface, it causes material ablation in the irradiated spot. As a result of ablation, the surface of the metal undergoes a structural modification. Physical processes underlying the femtosecond laser ablation of metals can be found in a number of studies (see, for example, Refs. [20–24]). A typical experimental setup for surface structuring of materials using the direct femtosecond ablation technique is shown in Figure 5.1.1 [25]. Femtosecond laser pulses are focused by a lens onto a sample mounted on an XY-translation stage. To monitor the energy of femtosecond laser pulses, a beam splitter and joulemeter are used. This experimental setup configuration allows structuring of a single spot on the sample when the translation stage sits still, a single line when sample is translated along the X- or Y-axis, or a large area when the sample is raster scanned. Other laser beam focusing geometries have been used, for example, through a cylindrical lens [26].

Nanoscale surface structures produced by direct femtosecond laser ablation can be classified into the following categories: (i) nanostructures (nanocavities, nanospheres, nanoprotrusions, and nanowires), (ii) laser-induced periodic structures (periodic nanogrooves), (iii) extended microgroove structures covered with nanostructures, and (iv) single nanoholes and arrays of nanoholes. Below, we discuss these structures in detail.

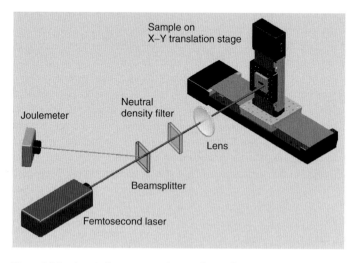

Figure 5.1.1 A typical experimental setup for surface nanostructuring.

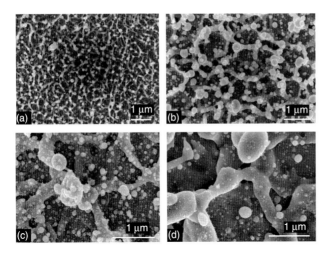

Figure 5.1.2 SEM images of the central part of the irradiated spot on copper following ablation at $F = 1.52$ J cm^{-2} (a) surface after 1 shot featuring only random nanostructures in the form of nanoprotrusions and nanocavities, (b) surface after 2 shots featuring only random nanostructures in the form of spherical nanoprotrusions and nanocavities, (c) surface after 10 shots featuring both nano- and microstructures, and (d) surface after 1000 shots showing microstructures being dominating.

5.1.3
Nanostructures

The nanostructures generated by direct femtosecond laser ablation have been studied for ablation both in air [14, 17, 27–30] and in liquids [31]. One systematic study of laser-induced surface modifications was reported in Ref. [17]. It studied copper processed by femtosecond laser pulses. The evolution of surface nanostructuring with increasing number of laser shots is shown by the scanning electron microscopic (SEM) images in Figure 5.1.2 [17], where one can see that the laser-induced surface nanostructures consist of nanoprotrusions, nanospheres, nanocavities, and nanopores that vary in size.

Figure 5.1.3b shows the surface structures produced by a single-pulse irradiation [17], where the characteristic types of initial nanostructures are labeled. For comparison, Figure 5.1.3a shows an undamaged area of the sample using the same scale as in Figure 5.1.3a. It is seen in Figure 5.1.3b that surface structuring is initiated on random highly localized nanoscale sites. The typical structures include circular nanopores with a diameter in the range of 40–100 nm, nanoprotrusions with a diameter in the range of 20–70 nm and a length of 20–80 nm, nanocavities of arbitrary form, and nanorims around nanocavities. Therefore, under these femtosecond laser processing conditions, nanoscale features down to a size of 20 nm are produced. One can see from Figure 5.1.3b that a nanopore or nanocavity is immediately accompanied by a nanorim or nanoprotrusion, indicating a nanoscale material relocation to an adjacent site. This one-to-one nanoscale dips and protrusions occur randomly over the laser spot, suggesting an initial

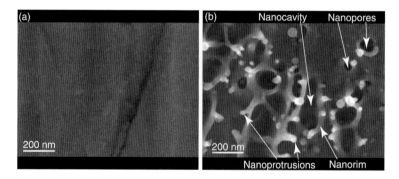

Figure 5.1.3 (a) A typical image of a sample surface before irradiation and (b) nascent nanostructures induced on copper by ablation at $F = 0.35$ J cm^{-2} and $N = 1$. Note that (a) does not show exactly the same spot on the sample as in (b).

nonuniform laser energy deposition. When the incident laser fluence is close to the laser ablation threshold, the spatial variations in deposited laser energy can produce a melt at localized nanoscale sites within the irradiated spot. When laser fluence is sufficiently high to produce ablation, the atoms ejected from the nanomelts produce a recoil pressure that squirts liquid metal outside of the nanomelt. Studies on nanostructuring of other metals have been also performed for gold [14, 29], titanium [28], silver [29], and aluminum [27, 30, 31].

The following mechanisms have been suggested to explain the formation of nanostructure and fine microstructures: (i) femtosecond laser-induced melt dynamics [17, 29, 31], (ii) cavitation nanospallation [32, 33], and (iii) redeposition of ablated species back to the irradiated sample [7, 14]. The first mechanism of nanostructuring occurs because of localized nanoscale melts, where a high radial temperature gradient in a nanomelt induces a radial surface tension gradient (Marangoni force) that expels the liquid to the periphery of the nanomelt. This melt flow leads to the formation of nanocavities, nanoprotrusions, and nanorims due to fast freezing of the expelled liquid. This regime of nanostructuring occurs at low laser fluence and low number of laser shots. With the increase in laser fluence or the number of laser shots, the surface nanomelts grow and then merge in a large melt pool, and ejection of liquid nanoparticles from the melt pool begins to occur. In this regime, those liquid nanoparticles that are frozen at a moment of their detachment process will form surface nanoprotrusions of various shapes [31]. The second mechanism explains the formation of surface nanostructures by deformation of the spallation layer by cavitation bubbles. The superfast cooling of the surface melt "roughened" by the cavitation bubbles results in surface nanocavities. The third mechanism of the surface structure formation depends on ambient gas pressure. It is known that ablated species are intensively deposited back onto the irradiated sample and form aggregates of nanoparticles when ablation occurs in an ambient gas of atmospheric pressure [7, 14]. However, the back deposition significantly diminishes when the ambient gas pressure decreases [34]. Since femtosecond laser nanostructuring is observed both in air and in vacuum, the redeposition mechanism is not dominant. However, it may significantly contribute to surface nanostructuring when the sample is treated using raster scanning. Although most studies on laser nanostructuring were carried out in air, some promising results were also obtained when sample ablation was performed in liquids [31].

5.1.4
Femtosecond Laser-Induced Periodic Structures (Periodic Nanogrooves) on Metals

LIPSSs on solids were first observed in 1965 [35] and since then were extensively studied using relatively long-pulsed lasers. Usually, these LIPSSs show a regular groove structure with a period on the incident laser wavelength scale and oriented perpendicularly to the polarization of the incident light. The formation of these LIPSSs on metals is often explained by the interference of the incident laser light with the excited surface plasmon (SP) polaritons that results in spatial periodic energy distribution on the surface. For a linearly polarized laser light, the period d of LIPSSs is given by [16]

$$d = \lambda/(\eta \pm \sin\theta) \text{ with } \mathbf{g} \parallel \mathbf{E} \tag{5.1.1}$$

where λ is the incident light wavelength, θ is the angle of the incident light, $\eta = \mathrm{Re}[\varepsilon/(\varepsilon + 1)]^{1/2}$ is the real part of the effective refractive index of the air–metal interface for SPs, ε is the dielectric constant of the metal, $\mathbf{g}$ is the grating vector, and $\mathbf{E}$ is the electrical field vector of the incident laser light. With the advent of the femtosecond laser, the formation of LIPSS on semiconductors [10–12, 36], dielectrics [13], and metals [14–16, 28] has also been observed. In several recent studies [16, 18], it was shown that femtosecond laser-induced periodic surface structure (FLIPSS) are densely covered by nanostructures, in contrast to smooth LIPSS produced by long-pulsed lasers. Another distinctive feature of these nanostructure-covered FLIPSS is that its period is appreciably less than that of the regular LIPSS whose period is approximately equal to the laser wavelength at normal incident laser light. The reduced period of the nanostructure-covered FLIPSS is due to a change in the effective refractive index of the air–metal interface when nanostructures develop on the metal surface that affects the propagation of

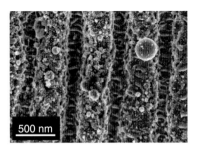

Figure 5.1.4 SEM image of a femtosecond laser-induced periodic surface structure on a metal. Both ridges and valleys of FLIPSS are extensively textured with nanostructures.

excited SP polaritons [16]. An example of FLIPSS is shown in Figure 5.1.4. The depth of FLIPSS is typically on the nanoscale [16]; therefore, FLIPSS can be also termed as a *periodic nanogroove pattern* or *nanogroove grating*. Various aspects of FLIPSS generation on metals and alloys have been extensively studied in a number of recent works [37–42].

5.1.5
Nanostructure-Textured Microstructures

5.1.5.1 Nanostructure-Textured Microgroove Structures
When the treated sample is scanned across a tightly focused laser beam, a single extended microgroove or an array of parallel extended microgrooves can be produced. It has been found that under certain conditions of direct femtosecond laser processing, one can produce microgrooves that are extensively covered with nanostructures [27, 43]. An example of such nanostructure-textured microgrooves on platinum is reproduced from Ref. [43] in Figure 5.1.5.

Figure 5.1.5a shows an overall view of periodic parallel microgrooves with a period equal to the vertical step between the adjacent horizontal scanning lines (about 100 μm). Both ridges and valleys of microgrooves are extensively textured with a rich variety of surface structures, including nanocavities, nanoprotrusions, and microscale aggregates formed by fused nanoparticles, as seen in Figure 5.1.5b,c,d.

Another approach for fabrication of the surface microgrooves is to use a mask projection technique. As reported in Ref. [44], this technique allows production of high-quality microgroove arrays at a high processing speed.

5.1.5.2 Nanostructure-Textured Columnar Microstructures
Figure 5.1.6 shows surface morphology of titanium evolving from nanostructures to columnar microstructures textured with nanostructures [28]. Owing to potential biomedical and optical applications, ultrafast laser generation of columnar microstructures has been studied in a number of works [28, 45–48]. Under

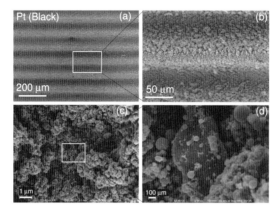

Figure 5.1.5 SEM images showing structural features of black platinum [43].

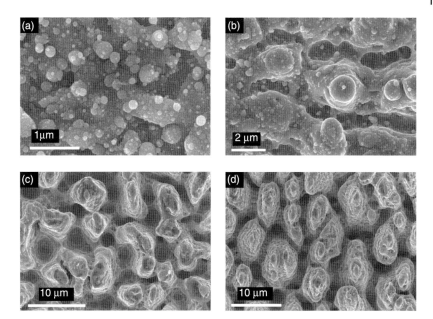

Figure 5.1.6 Surface nano- and microtopography of titanium following femtosecond laser treatment at $F = 0.35$ J cm^{-2}. (a) Nano- and microroughness after a 1-shot laser treatment, (b) typical random microroughness covered with nanostructures after a 40-shot treatment, (c) typical columnar microstructures after a 100-shot treatment, and (d) typical columnar microstructures after a 200-shot treatment.

different laser processing conditions, a variety of shapes and sizes of columnar structures were produced, such as conical columnar microstructures textured with nanostructures [46, 47].

The nanoscale texture on the columnar structures can be fabricated in the form of irregular nanostructures [28] or regular nanoripples [47]. Fabrication of columnar structures on Ti plates using femtosecond laser ablation in various liquids has been recently reported by Yang *et al.* [48].

5.1.6
Single Nanoholes and Arrays of Nanoholes

In 1995, Pronko *et al.* [1] have demonstrated the ability of femtosecond laser with a Gaussian beam profile to drill single nanoholes in metal films through ablation at the laser fluence slightly above the damage threshold. Under these ablation conditions, only a central part of the focused laser beam can ablate the metal, resulting in a nanohole. Using a femtosecond laser beam tightly focused to a spot size of 3000 nm, Pronko *et al.* have produced nanoholes with a diameter of 300 nm and a depth of 52 nm in a silver film. This technique has been further studied for producing subdiffraction structures on transition metals (Cr, Mo, W, and Fe) [2], where it has been found that by imaging a small aperture onto the metal surface, nanoholes can be produced with a better quality compared to beam focusing. Nanomilling metal surfaces using near-threshold femtosecond laser pulses has been studied in Ref. [49], where craters with a depth of about 10 nm were produced. Vestentoft and Balling [50] have produced arrays of nanoholes using a self-assembled microlens array formed by deposition of quartz microspheres on a metal surface. The quartz microspheres were deposited using a suspension of the quartz microspheres in water or ethanol. When the solute evaporates, the microspheres form a pattern that depends on their concentration in the suspension. Ablation by a single laser pulse results in both nanoholes in the metal and the removal of the microspheres. A nanohole-patterning technique based on interferometric femtosecond laser ablation has been studied in Refs. [51–54]. With this technique, a four-beam interference pattern on a metal surface is used for ablation that allows production of an array of nanoholes on the surface. It should also be noted that depending on the material and the ambient medium, femtosecond laser surface structuring can involve surface chemical compositioning [30, 55]. However, few studies have been performed in addressing this aspect.

5.1.7
Applications of Femtosecond Laser-Induced Surface Structures on Metals

5.1.7.1 Modification of Optical Properties
In one study [14], it was shown that surface nano- and microstructures produced by direct femtosecond laser structuring can significantly alter optical properties of metals. The absorptance of a metal can be controllably enhanced from its intrinsic value of a small percentage to almost 100%, resulting in a black appearance of the metal surface [14]. Following this study, black [27, 43] and color [27, 31, 56] metals have been created using femtosecond laser surface structuring. As an example, Figure 5.1.7 shows photographs of black and blue [56] titanium samples, respectively.

Total hemispherical reflectances of blue and black Ti plotted as a function of wavelength are shown in Figure 5.1.8, where reflectance of a titanium sample before laser processing is also shown for a comparison. One can see from the spectral reflectance of the blue titanium that it has a greater absorption in the

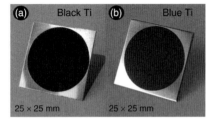

Figure 5.1.7 (Color online) Photographs of black (a) and (b) blue titanium (shown in gray color).

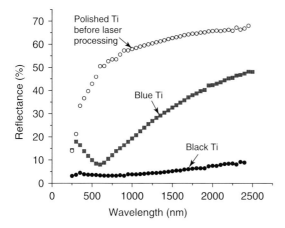

Figure 5.1.8 (Color online) Spectral reflectance of the black and blue Ti samples as a function of wavelength. For a comparison, the reflectance of the polished sample before femtosecond laser processing is also shown.

green and red wavelengths, resulting in a blue color. On the other hand, the total reflectance of the black Ti drops to below 10% in the entire wavelength range between 250 and 2500 nm, rendering the treated surface pitch black.

The SEM study shows that unique patterns of surface structures at nano- to microscales are formed in the black and colored metals, and these structures drastically modify the optical responses of metals. The metal blackening technique has been demonstrated in a variety of metals, such as Al [27], Pt [43], W [57], Au [25], and Ti90/Al6/V4 alloy [25]. Similar effects have also been attempted in other studies [53, 54, 58, 59]. Furthermore, the metal blackening technique has also produced highly absorptive metals over an ultrabroad electromagnetic spectrum, ranging from ultraviolet to terahertz [60].

FLIPSSs can also affect optical properties of metals significantly [18, 27, 57, 61, 62]. In contrast to the colored metal discussed above that shows a solid color at all viewing angles, metals with FLIPSSs exhibit different colors at different viewing angles [27, 61]. The viewing angle dependence of FLIPSS colors can produce iridescent colors [27] and even color images [63]. Furthermore, metals with FLIPSSs also have a clear polarization effect depending on whether the incident light polarization is parallel or perpendicular to the FLIPSS grating vector.

5.1.7.2 Modification of Wetting Properties

In a recent study, a unique pattern of surface structures, which shows dramatic wetting effects, was produced. Figure 5.1.9 shows that a metal structured with nanotextured microgrooves can pump liquids uphill [64]. Snapshots of methanol spreading on a vertically standing Pt sample against the gravity are shown in Figure 5.1.9a–d. The self-propelled directional motion of the liquid is explained by the capillary effect in open microgrooves [65–67]. Another interesting wetting property of the structured platinum is that if the bottom part of a vertically standing

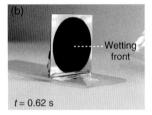

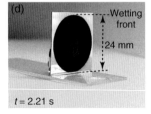

Figure 5.1.9 Snapshots showing methanol running uphill on a vertically standing platinum sample. (a–d) Dynamics of methanol running uphill.

sample is submerged in a methanol reservoir, the methanol spreads only along the microgrooves that are submerged in the methanol but does not expand laterally to grooves that are not initially in contact with the methanol in reservoir. However, when this experiment is repeated with the sample enclosed in a transparent container to suppress evaporation, similar vertical wetting strip is formed rapidly but the liquid also spreads slowly in the lateral directions until the entire structured black surface becomes wet. In this case, the methanol spreading speed in the lateral direction is about an order of magnitude lower than that along the vertical grooves. This observation shows that, besides the capillary effect, evaporation also plays a role in the unidirectional spreading of methanol in open air. Therefore, in addition to a physical wetting boundary formed by the outmost groove walls that are in contact with the methanol reservoir, the evaporation actually creates a virtual wall leading to a significantly enhanced anisotropic (essentially unidirectional) wetting behavior. Studies on modifying wetting properties on silicon [68–70] and glass [71] were carried out and reported.

5.1.7.3 Biomedical Applications

Platinum, titanium, gold, and biograde stainless steel are materials of choice in various biomedical applications, such as implants, implantable devices, biosensors, and cell-handling devices. Numerous *in vitro* and *in vivo* studies have demonstrated that surface topography of a metal is an important factor affecting biological response. For example, it was found that both microscale and nanoscale surface textures enhance the performance of titanium implants [72]. It was also shown that both irregular (roughness) and regular (extended parallel grooves) surface structures influence cell proliferation [72]. Laser processing of biomaterials has been studied in the past using long laser pulses. It was found that laser treatment provides both suitable surface texture and less surface contamination as compared with other methods. However, femtosecond laser surface texturing of metal biomaterials still remains a poorly studied area [28, 44, 48, 73, 74]. A systematic study of femtosecond laser-induced surface structures on platinum and their effects on both hydrophobicity and fibroblast cell proliferation has been reported in Ref. [73], where it was found that the femtosecond laser-induced surface structures suppress the fibroblast cell proliferation. Moreover, it was also found that there is a specific correlation between the femtosecond laser-induced hydrophobicity and cell growth: the higher the hydrophobicity, the lower the fibroblast proliferation. This observation provides the ability to suppress fibroblast proliferation by producing superhydrophobic surface structures and predict cell response using simple wetting tests on the structured surfaces. Femtosecond laser-induced surface structures on titanium have also been studied in the past [28, 44, 48]. Femtosecond laser-generated microgrooves and ''lotus-like'' structures on Ti were studied in Ref. [44] from a perspective of surface functionalization of orthopedic titanium implants. Effects of these surface structures on the behavior of both human fibroblasts and MG-63 osteoblasts have been studied *in vitro*. It was found that the lotus-like structures are superhydrophobic and promote the proliferation of osteoblasts but suppress the proliferation of the fibroblasts that is beneficial for the orthopedic titanium implants. Fabrication of columnar structures on Ti plates using femtosecond laser ablation in various liquids has been recently reported by Yang *et al.* [48]. It was demonstrated that through laser ablation of titanium in supersaturated hydroxyapatite aqueous solution, one can improve biocompatabilty of the structured surface due to incorporation of Ca and P into the surface structure.

5.1.7.4 Other Applications

5.1.7.4.1 Generation of X-Ray Bursts

Volkov *et al.* [75] reported that FLIPSSs can be used for enhancing X-ray emission. In that study, the authors first produced FLIPSSs on the surface of Fe and then irradiated the FLIPSSs by a femtosecond laser pulse at an intensity of 10^{16} W cm^{-2}. It was found that X-ray emission from the surface with FLIPSSs is significantly enhanced compared with a smooth surface.

5.1.7.4.2 Enhanced Photoelectron Emission from FLIPSS

A fundamental mechanism of FLIPSS formation on metals is a modulation of the absorbed laser energy by the interference between the incident laser light and the excited SPs. Owing to this plasmonic nature, FLIPSSs can efficiently support the excitation and propagation of SPs. In other words, FLIPSSs can effectively couple the incident laser light to SPs. Hwang *et al.* [76] have studied the effects of FLIPSSs on photoelectron emission. It was found that the excitation of SPs via FLIPSSs significantly enhances the photoelectron emission from a FLIPSS metal surface. Furthermore, it was also reported that nanostructures on FLIPSSs relax the phase-matching conditions for SPs excitation and significantly broaden its angular coupling range, causing an enhanced photoelectron emission over a broad range of the incident angles [76].

5.1.7.4.3 Modifying Emission of Thermal Radiation from Incandescent Light Sources

Vorobyev *et al.* [57] have explored the effects of femtosecond laser surface structuring of a tungsten incandescent lamp filament on its radiation. In the study, the authors applied a surface structuring technique to enhance the absorptance of the tungsten filament surface. As a result, the lamp emission enhances significantly. Furthermore, the study showed that one can also obtain partially polarized light as well as control the spectral range of the thermal radiation emission from the femtosecond laser-structured tungsten lamp.

5.1.8
Summary

Direct femtosecond laser material nanostructuring emerged recently as a new research area, and significant progresses have been made in studying the surface structures and their applications. In terms of applications, for example, the black and colored metals produced by direct femtosecond laser structuring may find applications in areas such as photonics, plasmonics, optoelectronics, stealth technology, airborne/spaceborne devices, solar energy absorbers, and thermophotovoltaics. The unique wetting and wicking properties of the processed materials may find applications in nano/microfluidics, optofluidics, lab-on-chip technology, fluidic microreactors, chemical and biological sensors, biomedicine, and heat transfer devices. The studies show that the femtosecond laser can produce unique surface structures that lead to controllable material properties that may not be easily obtainable otherwise.

References

1. Pronko, P.P., Dutta, S.K., Squier, J., Rudd, J.V., Du, D., and Mourou, G. (1995) Machining of submicron holes using a femtosecond laser at 800-nm. *Opt. Commun.*, **114**, 106–110.
2. Korte, F., Serbin, J., Koch, J., Egbert, A., Fallnich, C., Ostendorf, A., and Chichkov, B.N. (2003) Towards nanostructuring with femtosecond laser pulses. *Appl. Phys.*, **A77**, 229–235.
3. Nolte, S., Chichkov, B.N., Welling, H., Shani, Y., Liebermann, K., and Terkel, H. (1999) Nanostructuring with spatially localized femtosecond laser pulses. *Opt. Lett.*, **24**, 914–916.
4. Chimmalgi, A., Choi, T.Y., Grigoropoulos, C.P., and Komvopoulos, K. (2003) Femtosecond laser apertureless near-field nanomachining of metals assisted by scanning probe microscopy. *Appl. Phys. Lett.*, **82**, 1146–1148.
5. Wu, C., Crouch, C.H., Zhao, L., Carey, J.E., Younkin, R., Levinson, J.A., Mazur, E., Farrell, R.M., Gothoskar, P., and Karger, A. (2001) Near-unity below-band-gap absorption by microstructured silicon. *Appl. Phys. Lett.*, **78**, 1850–1852.
6. Koch, J., Korte, F., Bauer, T., Fallnich, C., Ostendorf, A., and Chichkov, B.N. (2005) Nanotexturing of gold films by femtosecond laser-induced melt dynamics. *Appl. Phys.*, **A81**, 325–328.
7. Pereira, A., Cros, A., Delaporte, P., Georgiou, S., Manousaki, A., Marine, W., and Sentis, M. (2004) Surface nanostructuring of metals by laser irradiation: effects of pulse duration, wavelength and gas atmosphere. *Appl. Phys.*, **A79**, 1433–1437.
8. Amoruso, S., Ausanio, G., Bruzzese, R., Vitello, M., and Wang, X. (2005)

Femtosecond laser pulse irradiation of solid targets as a general route to nanoparticle formation in a vacuum. *Phys. Rev.*, **B71**, 033406.

9. Eliezer, S., Eliaz, N., Grossman, E., Fisher, D., Gouzman, I., Henis, Z., Horovitz, Y., Frankel, M., Maman, S., and Lereah, Y. (2004) Synthesis of nanoparticles with femtosecond laser pulses. *Phys. Rev.*, **B69**, 144119.

10. Borowiec, A. and Haugen, H.K. (2003) Subwavelength ripple formation on the surfaces of compound semiconductors irradiated with femtosecond laser pulses. *Appl. Phys. Lett.*, **82**, 4462–4464.

11. Costache, F., Kouteva-Arguirova, S., and Reif, J. (2004) Sub--damage--threshold femtosecond laser ablation from crystalline Si: surface nanostructures and phase transformation. *Appl. Phys.*, **A79**, 1429–1432.

12. Bonse, J., Munz, M., and Sturm, H. (2005) Structure formation on the surface of indium phosphide irradiated by femtosecond laser pulses. *J. Appl. Phys.*, **97**, 013538.

13. Hnatovsky, C., Taylor, J.R., Rajeev, P.P., Simova, E., Bhardwaj, V.R., Rayner, D.M., and Corkum, P.B. (2005) Pulse duration dependence of femtosecond-laser-fabricated nanogratings in fused silica. *Appl. Phys. Lett.*, **87**, 14104–14106.

14. Vorobyev, A.Y. and Guo, C. (2005) Enhanced absorptance of gold following multi-pulse femtosecond laser ablation. *Phys. Rev.*, **B72**, 195422.

15. Wang, J. and Guo, C. (2005) Ultrafast dynamics of femtosecond laser-induced periodic surface pattern formation on metals. *Appl. Phys. Lett.*, **87**, 251914.

16. Vorobyev, A.Y., Makin, V.S., and Guo, C. (2007) Periodic ordering of random surface nanostructures induced by femtosecond laser pulses on metals. *J. Appl. Phys.*, **101**, 034903.

17. Vorobyev, A.Y. and Guo, C. (2006) Femtosecond laser nanostructuring of metals. *Opt. Express*, **14**, 2164–2169.

18. Vorobyev, A.Y. and Guo, C. (2007) Effects of nanostructure-covered femtosecond laser-induced periodic surface structures on optical absorptance of metals. *Appl. Phys.*, **A86**, 321–324.

19. Gattass, R.R. and Mazur, E. (2008) Femtosecond laser micromachining in transparent materials. *Nat. Photonics*, **2**, 219–224.

20. Anisimov, S.I. and Luk'yanchuk, B.S. (2002) Selected problems of laser ablation theory. *Phys. Usp.*, **45**, 293–324.

21. Gamaly, E.G., Rode, A.V., Luther-Davies, B., and Tikhonchuk, V.T. (2002) Ablation of solids by femtosecond lasers: ablation mechanism and ablation thresholds for metals and dielectrics. *Phys. Plasmas*, **9**, 949–957.

22. Stoian, R., Rosenfeld, A., Ashkenasi, D., Hertel, I.V., Bulgakova, N.M., and Campbell, E.E.B. (2002) Surface charging and impulsive ion ejection during ultrashort pulsed laser ablation. *Phys. Rev. Lett.*, **88**, 097603.

23. Zhigilei, L.V., Lin, Z., and Ivanov, D.S. (2009) Atomistic modeling of short pulse laser ablation of metals: connections between melting, spallation, and phase explosion. *J. Phys. Chem.*, **C113**, 11892–11906.

24. Demaske, B.J., Zhakhovsky, V.V., Inogamov, N.A., and Oleynik, I.I. (2010) Ablation and spallation of gold films irradiated by ultrashort laser pulses. *Phys. Rev.*, **B82**, 064113.

25. Vorobyev, A.Y. and Guo, C. (2010) Metallic light absorbers produced by femtosecond laser pulses. *Adv. Mech. Eng.*, **2010** (Art. ID 452749), pp. 1–4.

26. Zuhlke, C.A., Alexander, D.R., Bruce, J.C. III, Ianno, N.J., Kamler, C.A., and Yang, W. (2010) Self assembled nanoparticle aggregates from line focused femtosecond laser ablation *Opt. Express*, **18**, 4329–4339.

27. Vorobyev, A.Y. and Guo, C. (2008) Colorizing metals with femtosecond laser pulses. *Appl. Phys. Lett.*, **92**, 041914.

28. Vorobyev, A.Y. and Guo, C. (2007) Femtosecond laser structuring of titanium implants. *Appl. Surf. Sci.*, **253**, 7272–7280.

29. Hwang, T.Y., Vorobyev, A.Y., and Guo, C. (2009) Ultrafast dynamics of femtosecond laser-induced nanostructure formation on metals. *Appl. Phys. Lett.*, **95**, 123111.

30. Li, X., Yuan, C., Yang, H., Li, J., Huang, W., Tang, D., and Xu, Q. (2010) Morphology and composition on Al surface irradiated by femtosecond laser pulses. *Appl. Surf. Sci.*, **256**, 4344–4349.

31. Stratakis, E., Zorba, V., Barberoglou, M., Fotakis, C., and Shafeev, G. (2009) Laser writing of nanostructures on bulk Al via its ablation in liquids. *Nanotechnology*, **20**, 105303.

32. Inogamov, N.A., Zhakhovskii, V.V., Ashitkov, S.I., Petrov, Y.V., Agranat, M.B., Anisimov, S.I., Nishihara, K., and Fortov, V.E. (2008) Nanospallation induced by an ultrashort laser pulse. *JETP*, **107**, 1–19.

33. Zhakhovskii, V.V., Inogamov, N.A., and Nishihara, K. (2008) New mechanism of the formation of the nanorelief on a surface irradiated by a femtosecond laser pulse. *JETP Lett.*, **87**, 423–427.

34. Vorobyev, A.Y. and Guo, C. (2007) Shot-to-shot correlation of residual energy and optical absorptance in femtosecond laser ablation. *Appl. Phys.*, **A86**, 235–241.

35. Birnbaum, M. (1965) Semiconductor surface damage produced by ruby lasers. *J. Appl. Phys.*, **36**, 3688–3689.

36. Ouyang, H., Deng, Y., Knox, W.H., and Fauchet, P.M. (2007) Photochemical etching of silicon by two photon absorption. *Phys. Status Solidi A*, **204**, 1255–1259.

37. Yang, Y., Yang, J., Xue, L., and Guo, Y. (2010) Surface patterning on periodicity of femtosecond laser-induced ripples. *Appl. Phys. Lett.*, **97**, 141101.

38. Qi, L., Nishii, K., and Namba, Y. (2009) Regular subwavelength surface structures induced by femtosecond laser pulses on stainless steel. *Opt. Lett.*, **34**, 1846–1848.

39. Golosov, E.V., Emel'yanov, V.I., Ionin, A.A., Kolobov, Y.R., Kudryashov, S.I., Ligachev, A.E., Novoselov, Y.N., Seleznev, L.V., and Sinitsyn, D.V. (2009) Femtosecond laser writing of subwave one_dimensional quasiperiodic nanostructures on a titanium surface. *JETP Lett.*, **90**, 107–110.

40. Huang, Y., Liu, S., Li, W., Liu, Y., and Yang, W. (2009) Two-dimensional periodic structure induced by single-beam femtosecond laser pulses irradiating titanium. *Opt. Express*, **17**, 20756–20761.

41. Li, Z., Li, P., Fan, J., Fang, R., and Zhang, D. (2010) Energy accumulation effect and parameter optimization for fabricating of high-uniform and large-area period surface structures induced by femtosecond pulsed laser. *Opt. Lasers Eng.*, **48**, 64–68.

42. Okamuro, K., Hashida, M., Miyasaka, Y., Ikuta, Y., Tokita, S., and Sakabe, S. (2010) Laser fluence dependence of periodic grating structures formed on metal surfaces under femtosecond laser pulse irradiation. *Phys. Rev.*, **B 82**, 165417.

43. Vorobyev, A.Y. and Guo, C. (2008) Femtosecond laser blackening of platinum. *J. Appl. Phys.*, **104**, 053516.

44. Fadeeva, E., Schlie, S., Koch, J., and Chichkov, B.N. (2010) Selective cell control by surface structuring for orthopedic applications. *J. Adhes. Sci. Technol.*, **24**, 2257–2270.

45. Tsukamoto, M., Kayahara, T., Nakano, H., Hashida, M., Katto, M., Fujita, M., Tanaka, M., and Abe, N. (2007) Microstructures formation on titanium plate by femtosecond laser ablation. *J. Phys. Conf. Ser.*, **59**, 666–669.

46. Oliveira, V., Ausset, S., and Vilar, R. (2009) Surface micro/nanostructuring of titanium under stationary and non-stationary femtosecond laser irradiation. *Appl. Surf. Sci.*, **255**, 7556–7560.

47. Nayak, B.K. and Gupta, M.C. (2010) Self-organizedmicro/nanostructures in metal surfaces by ultrafast laser irradiation. *Opt. Lasers Eng.*, **48**, 940–949.

48. Yang, Y., Yang, J., Liang, C., Wang, H., Zhu, X., and Zhang, N. (2009) Surface microstructuring of Ti plates by femtosecond lasers in liquid ambiences: a new approach to improving biocompatibility. *Opt. Express*, **17**, 21124–21133.

49. Kirkwood, S.E., Taschuk, M.T., Tsui, Y.Y., and Fedoseyevs, R. (2007) Nanomilling surfaces using near-threshold femtosecond laser pulses. *J. Phys. Conf. Ser.*, **59**, 591–594.

50. Vestentoft, K. and Balling, P. (2006) Formation of an extended nanostructured metal surface by ultra-short laser pulses: single-pulse ablation in the high-fluence limit. *Appl. Phys.*, **A84**, 207–213.

51. Nakata, Y., Okada, T., and Maeda, M. (2004) Lithographical laser ablation using femtosecond laser. *Appl. Phys.*, **A79**, 1481–1483.

52. Bekesi, J., Klein-Wiele, J.H., and Simon, P. (2003) Efficient submicron processing of metals with femtosecond UV pulses. *Appl. Phys.*, **A76**, 355–357.

53. Paivasaari, K., Kaakkunen, J.J.J., Kuittinen, M., and Jaaskelainen, T. (2007) Enhanced optical absorptance of metals using interferometric femtosecond ablation. *Opt. Express*, **15**, 13838–13843.

54. Kaakkunen, J.J.J., Paivasaari, K., Kuittinen, M., and Jaaskelainen, T. (2009) Morphology studies of the metal surfaces with enhanced absorption fabricated using interferometric femtosecond ablation. *Appl. Phys.*, **A94**, 215–220.

55. Dou, K., Knobbe, E.T., Parkhill, R.L., Irwin, B., Matthews, L., and Church, K.H. (2003) Femtosecond study of surface structure and composition and time-resolved spectroscopy in metals. *Appl. Phys.*, **A76**, 303–307.

56. Vorobyev, A.Y. and Guo, C. (2008) Metal colorization with femtosecond laser pulses. *Proc. SPIE*, **7005**, 70051T.

57. Vorobyev, A.Y., Makin, V.S., and Guo, C. (2009) Brighter light sources from black metal: significant increase in emission efficiency of incandescent light sources. *Phys. Rev. Lett.*, **102**, 234301.

58. Iyengar, V.V., Nayak, B.K., and Gupta, M.C. (2010) Ultralow reflectance metal surfaces by ultrafast laser texturing. *Appl. Opt.*, **49**, 5983–5988.

59. Yang, Y., Yang, J., Liang, C., and Wang, H. (2008) Ultra-broadband enhanced absorption of metal surfaces structured by femtosecond laser pulses. *Opt. Express*, **16**, 11259–11265.

60. Vorobyev, A.Y., Topkov, A.N., Gurin, O.V., Svich, V.A., and Guo, C. (2009) Enhanced absorption of metals over ultrabroad electromagnetic spectrum. *Appl. Phys. Lett.*, **95**, 121106.

61. Vorobyev, A.Y. and Guo, C. (2008) Spectral optical responses of femtosecond laser-induced periodic surface structures on metals. *J. Appl. Phys.*, **103**, 043513.

62. Vorobyev, A.Y., Makin, V.S., and Guo, C. (2009) Optical properties of femtosecond laser-induced periodic surface structures on metals. Conference Proceedings of 52nd IEEE International Midwest Symposium on Circuits and Systems, pp. 905–908 (IEEE Catalog Number: CFP09MID-CDR; ISBN: 978-1-4244-4480-9).

63. Dusser, B., Sagan, Z., Soder, H., Faure, N., Colombier, J.P., Jourlin, M., and Audouard, E. (2010) Controlled nanostructrures formation by ultra fast laser pulses for color marking. *Opt. Express*, **18**, 2913–2924.

64. Vorobyev, A.Y. and Guo, C. (2009) Metal pumps liquid uphill. *Appl. Phys. Lett.*, **94**, 224102.

65. Romero, L.A. and Yost, F.G. (1996) Flow in an open channel capillary. *J. Fluid Mech.*, **322**, 109–129.

66. Ha, J.M. and Peterson, G.P. (1998) Capillary performance of evaporating flow in microgrooves: an analytical approach for very small tilt angles. *ASME J. Heat Transfer*, **120**, 452–457.

67. Nilson, R.H., Tchikanda, S.W., Griffiths, S.K., and Martinez, M.J. (2006) Steady evaporating flow in rectangular microchannels. *Int. J. Heat Mass Transfer*, **49**, 1603–1618.

68. Zorba, V., Persano, L., Pisignano, D., Athanassiou, A., Stratakis, E., Cingolani, R., Tzanetakis, P., and Fotakis, C. (2006) Making silicon hydrophobic: wettability control by two-lengthscale simultaneous patterning with femtosecond laser irradiation. *Nanotechnology*, **17**, 3234–3238.

69. Baldacchini, T., Carey, J.E., Zhou, M., and Mazur, E. (2006) Superhydrophobic surfaces prepared by microstructuring of silicon using a femtosecond laser. *Langmuir*, **22**, 4917–4919.

70. Vorobyev, A.Y. and Guo, C. (2010) Laser turns silicon superwicking. *Opt. Express*, **18**, 6455–6460.

71. Vorobyev, A.Y. and Guo, C. (2010) Water sprints uphill on glass. *J. Appl. Phys.*, **108**, 123512.

72. Bruncttc, D.M., Tcngvall, P., Textor, M., and Thomsen, P. (eds) (2001) *Titanium*

in Medicine: Material Science, Surface Science, Engineering, Biological Responses and Medical Applications, Springer, Berlin.

73. Fadeeva, E., Schlie, S., Koch, J., Chichkov, B.N., Vorobyev, A.Y., and Guo, C. (2009) in *Contact Angle, Wettability and Adhesion*, vol. 6 (ed. K.L. Mittal), VSP/Brill, Leiden, pp. 163–171.

74. Vorobyev, A.Y. and Guo, C. (2009) Femtosecond laser surface structuring of biocompatible metals. *Proc. SPIE*, **7203**, 720300.

75. Volkov, R.V., Golishnikov, D.M., Gordienko, V.M., and Savel'ev, A.B. (2003) Overheated plasma at the surface of a target with a periodic structure induced by femtosecond laser radiation. *JETP Lett.*, **77**, 473–476.

76. Hwang, T.Y., Vorobyev, A.Y., and Guo, C. (2009) Surface-plasmon-enhanced photoelectron emission from nanostructure-covered periodic grooves on metals. *Phys. Rev.*, **B79**, 085425.

5.2
Laser-Induced Forward Transfer: an Approach to Direct Write of Patterns in Film Form

Hironobu Sakata and Moriaki Wakaki

5.2.1
Introduction

Laser-induced forward transfer (LIFT) is defined as the *laser ablation of materials* in a film form predeposited on a transparent substrate forwarded to another substrate placed in close proximity (normally $\sim$1 µm to several hundred micrometers) to the donor (film plus substrate) [1]. This film transfer can take place in air or in vacuum at the focused point of the laser when the donor is irradiated by a single-pulse or a multipulse beam of different kinds of lasers with the wavelength ranging from vacuum ultraviolet (VUV) to near-infrared (NIR) range. A review work of this technique including applications was described recently [2]. It is important to note that LIFT occurs at laser fluences more than the threshold fluence at the focused point of the laser, which is normally the interface between the film and the substrate, and this threshold depends on the kind of film and its thickness. Recently, the kinds of films have extended from metals, oxides, and composites to biomaterials such as cells and proteins and DNA. The typical objectives of LIFT are the microfabrication of electronic devices, protection coatings, and diagnostic systems of biomaterials and biosensing devices. Film materials and conditions of the LIFT technique applied to metallic and oxide films and some biomaterials are summarized in Table 5.2.1.

It is noted that the LIFT process is the "dry process," different from that of photolithography. The advantage of this technique is the easy setup of the apparatus for experiments. However, it is unsuitable for mass production engineering because of the two-step handling of film materials.

In this chapter, characteristics and features of various films transferred by LIFT are reviewed, and conditions of laser irradiations for further development of nanoprocessing and nanostructuring of different materials to obtain deposits of good quality are also summarized.

5.2.2
Principle and Method

The first attempt of LIFT was performed by Bohandy *et al.* [1] for Cu films on fused silica substrates in vacuum using an ArF excimer laser (193 nm). Adrian *et al.* [3] analyzed the experimental results using a one-dimensional thermal diffusion equation with a properly penetrating heat source from the laser and a moving boundary between solid and molten phases of the Cu film. The finite element method was applied for the calculation. The following sequence of events was suggested by this simulation: the laser pulse heats the back surface of the film to reach melting, then the melted interface propagates through the film until it

Table 5.2.1 Laser-induced forward transferred materials and lasers used for the studies.

Transferred material	Substrate	Deposit feature size^a (μm)	Laser (λ, pulse width)	Reference number
Cu	Si	50	ArF (193 nm, ~15 ns)	[1]
Cu	SiO$_2$	50	ArF (193 nm, ~15 ns)	[3]
Cu, Ag	SiO$_2$	15	2ωNd:YAG (532 nm, 10–15 ns)	[4]
Al	Si, glazed alumina	200	Nd:YAG (1.06 μm, ~120 ns) and Nd:glass (1.06 μm, 40 ns)	[5]
Au	Quartz, glass	–	KrF (248 nm, ~25 nm) and 2ωNd : YAG (532 nm, ~15 ns)	[6]
Ti, Cr	Glass	10	Ruby (694 nm, 20 ns)	[7]
W, Cr	Glass	10	Ar+ (515 nm, 1 μs) and Nd:YAG (1.06 μm, 100 μs)	[8]
YBaCuO, BiSrCaCuO	MgO, Si	100	ArF (193 nm, 20 ns) and Nd:YAG (1.06 μm, 5 ns)	[9, 10]
Au, Al	Si, quartz	7–10	ArF (193 nm, 18 ns)	[11]
Pd	Quartz, ceramics, polymers	150	ArF (193 nm), KrF (222 nm), KrCl (248 nm), XeCl (308 nm), and XeF (351 nm)	[12]
Diamond	Si	10	Cu laser (510 nm, 20 ns) and KrF (248 nm, 15 ns)	[13]
Cr, Pt, Mo, In, In$_2$O$_3$	Glass, Si	1	KrF (248 nm, 500 fs)	[14]
Pt, Cr, In$_2$O$_3$	Glass	3	KrF (248 nm, 500 fs)	[15]
Ag, Au, Ni–Cr, BaTiO$_3$	Glass, Al$_2$O$_3$	25	KrF (248 nm,30 ns) and XeCl (308 nm)	[16]
Au–Sn	Si	30	Ti: sapphire (775 nm, 100–800 fs)	[17]
Au	SiO$_2$	–	Ti: sapphire (775 nm, 150 fs)	[18]
TiO$_2$–Au	Glass	–	2ωNd : YAG (532 nm, 10 ns)	[19]
V$_2$O$_5$	Glass	–	2ωNd : YAG (532 nm, 10 ns)	[20]
Phosphor (Y$_2$O$_3$: Eu/Zn$_2$SO$_4$: Mn)	Quartz	~120	KrF (248 nm, 25 ns)	[21]
DNA (lambda phage)	Quartz	~120	KrF (248 nm, ~500 fs)	[22]
cDNA (human)	Glass	40	3ωNd : YAG (355 nm, 10 ns)	[23]
Antigen (*Treponema pallidum*)	Glass	80	3ωNd : YAG (355 nm, 10 ns)	[24]
Fungi (conidia)	Quartz	~350	KrF (248 nm, 30 ns)	[25]

^{a}Diameter of the deposited area by the LIFT.

reaches the front surface, and shortly before the melting occurs through the film the back surface reaches the boiling point; thus, the metal vapor pressure at the back propels the molten Cu to the receiving target. This vapor-driven propulsion model explains the fact that a transfer deposit for a 1.2 μm thick film was yielded by the 60 mJ laser pulse energy but not by the 30 mJ pulse energy.

Kántor *et al.* [7] performed LIFT depositions of high melting point material (200 nm thick) Ti or Cr films on a glass by using a ruby laser (694 nm, and 20 ns pulse width, single-pulse irradiation). They calculated the temperature profile for different laser fluences by the finite element method using a one-dimensional heat conduction equation. The calculated maximum temperature of the Ti film reached 1941 K (melting point of Ti) for a 240 mJ cm^{-2} fluence, which explained the complete removal of Ti films. However, the experiment revealed that poorly adherent 200 nm thick Ti films showed the transfer at ~100 mJ cm^{-2} fluence, which corresponded to the calculated temperature of ~1500 K and was lower than the melting point of Ti. As a result, the LIFT occurred at the solid phase of Ti, which means that the above-described melt propulsion model is not valid for Ti films. In the case of 173 nm thick Cr films with high adherence to a glass substrate initially, the experiment started to give LIFT at a fluence of ~250 mJ cm^{-2}, which corresponded to the calculated temperature of 2073 K (melting point of Cr). In these cases, the film removal is affected by solid–solid adherence between the substrate and the film and is a thermomechanical blow off in the solid phase for poor adherence on irradiation. Similar results were also reported for W films [26]. Clear contours of both residual and transfer-deposited W films suggest the above removal mechanism.

The basic setup of the LIFT experiment is shown in Figure 5.2.1. A transparent substrate precoated with a film, defined as a *donor*, is placed in close proximity to a receiving substrate, defined as a *receiver*, with the spacing between ~1 μm and several hundred micrometers. Larger spacing normally produces spread of the deposit feature size (diameter of the deposited area by the LIFT) [27]. Laser can be suitably chosen so that the light absorption of the film to be transferred is large and the donor substrate is transparent at the wavelength of the laser. The fluence of the laser is an important parameter to obtain transferred deposits with good quality. From various experimental results, a laser fluence slightly higher than the threshold fluence of the target material is preferable to obtain uniform, smooth, and noncoalescent deposits. The laser beam should be focused at the interface of the donor film.

5.2.3
LIFT of Materials

5.2.3.1 Metals and Single Element

5.2.3.1.1 Copper
Attempts at LIFT started in the mid-1980s. Bohandy *et al.* [1] deposited 0.41 μm thick Cu films precoated on silica plates by LIFT on silicon substrates by a

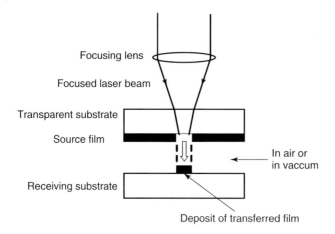

Focusing lens

Focused laser beam

Transparent substrate

Source film

In air or
in vaccum

Receiving substrate

Deposit of transferred film

Figure 5.2.1 Basic setup of LIFT experiment.

single-pulse irradiation of an ArF excimer laser (193 nm) with ~15 ns pulse width. They used a quartz cylindrical focusing lens to obtain the transferred film with a linear profile by the LIFT. The donor and receiver substrates were set in a vacuum chamber under 10^{-4} Pa. The gap between the source and the receiver substrate was estimated to be less than 10 µm. The Cu deposition was observed for the pulse energies from 60 to 139 mJ. For the 110 mJ pulse energy, a deposited copper line with a width of 40 µm was observed, just the same as the width of the removed copper source film. At the 139 mJ pulse energy, the width of the deposited copper line became larger (~60 µm), indicating that a lot of copper splattered away from the edges of the deposited copper line.

The analysis of the LIFT process [3] using a one-dimensional heat transfer model showed that the back of the source film, that is, the metal at the donor–substrate interface, reaches the boiling point of the metal before the melted interface propagates through the film to reach the front surface. As a result, the vapor-driven propulsion of the metal occurs from the source film to the target. This was consistent with the observation [4] of single-pulse experiments of copper (and silver) on silica substrates, showing the copper deposition with a ~50 µm diameter (Figure 5.2.2), where they used a frequency-doubled Nd:YAG laser (532 nm) with 10–15 ns pulse width and the maximum pulse energy of 50 mJ.

5.2.3.1.2 Gold

Baseman *et al.* [6] measured the minimum fluence of the LIFT for Au films (400 nm thick) on quartz or glass donor substrates. In air experiments, the minimum fluence of ~100 mJ cm^{-2} was obtained for the KrF excimer laser (248 nm, about 25 ns pulse width). The threshold fluence was the same for 248 and 532 nm (frequency-doubled Nd:YAG laser, about 15 ns pulse width). The amount of the threshold increased with increasing Au film thickness but independent of the laser wavelength.

Substrate dependence with respect to the Au film transfer was investigated using an ArF excimer laser (193 nm, 18 ns pulse width) [11]. Teflon-AF,

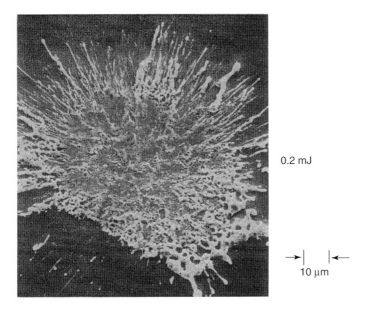

0.2 mJ

10 µm

Figure 5.2.2 SEM image of Cu deposit transferred by LIFT from 0.6 µm thick Cu film using a single pulse of 532 nm radiation and 0.2 mJ pulse energy [4].

polymethacrylonitrile (PMAN), and quartz, on which Au films (23 nm thick) have been precoated, were used as donor substrates. Quartz and silicon were used as receiving substrates. The threshold laser fluences for the LIFT were 7.9 and 9.7 mJ cm^{-2} for Teflon-AF and PMAN substrates, respectively, while that for the quartz substrate was 17.8 mJ cm^{-2}. For a 350 nm thick Au film on the quartz, the fluence of about 300 mJ cm^{-2} was necessary to transfer a pattern to the quartz substrate, but a splashed aspect of the transferred Au was observed by scanning electron microscopy (SEM). On the other hand, for the Au film (350 nm thick) on the Teflon-AF film (0.1–0.2 µm thick) precoated on the quartz donor substrate (i.e., double-layered films), the LIFT was observed at the fluence of 166 mJ cm^{-2}, and the Au film transferred on a Si receiver substrate exhibited a clear pattern without splashing.

That degradation of the polymer donor substrate occurred during the backside laser irradiation was confirmed by X-ray photoelectron spectroscopy (XPS) and static secondary ion mass spectroscopy (SSIMS) analyses.

The LIFT study at shorter pulses was recently performed for Au films (200 nm thick) using a frequency-doubled Ti:sapphire femtosecond pulsed laser (400 nm, 150 fs pulse width) [18]. The quartz donor substrate was in near contact with the receiving quartz substrate in air. The Au film transfer occurred when the laser energy exceeded 150 nJ/pulse at a pulse repetition rate of 1 kHz at ambient temperature. Transferred Au depicted ~3 µm lines in width, which were the same as the removed size in the donor film. A Au–Sn alloy disk was also used for such experiments with a Ti:sapphire femtosecond pulsed laser (775 nm, 100 fs to 3.5 ps pulse width, and 0.05–0.15 mJ pulse energy) in a vacuum chamber [17], and a 1.8 μm thick Au-Sn pad was obtained on a Si receiver substrate by a single-pulse shot.

5.2.3.1.3 Aluminum

Microexplosive bonding of Al films to Cu and Si has been attempted by the LIFT technique [28]. An Al film (500 nm thick) was deposited in vacuum onto a glass slide and served as an exploding film, on which a bonding Al film (1 μm thick) was deposited again as a flyer film with the diameter of 3 mm. Bonding of Al exploding and flyer films was observed on both Cu and Si receiver substrates after the LIFT using a Nd:YAG laser (1.06 μm, 1 ns pulse width) for an intensity larger than $\sim 1.0 \times 10^9$ W cm^{-2}, and distinct spots (200 μm in size) were obtained in a rough vacuum (3.3–9.3 Pa). Adhesion of the Al on Cu was satisfactory. In a SEM picture, a continuous film bonding was found in the Al-Si shots, while an intermixing of the two metals was observed in the Al-Cu shots.

The LIFT of Al onto glazed alumina ceramics and Si using Nd:YAG laser (1.06 μm, $\sim$120 ns pulse width) and a Nd:glass laser (40 ns pulse width) was investigated [5], where the Al film (0.8–5.7 μm thick) was deposited on a 1.5 mm thick glass plate by the magnetron sputtering. The receiver substrate was a sliced alumina ceramic (630 μm thick) coated with about 75 μm thick glaze. The glaze consisted of different oxides such as SiO_2, Al_2O_3, BaO, PbO, and CaO.

A Si wafer with a thermally grown $\sim$1 μm thick silica was used. The LIFT experiment conditions were in air at ambient temperature. The results showed that for high laser intensity 2.1×10^7 W cm^{-2} on the Al film target (0.8 μm thick), Al deposits showed "soft" removal without severe modifications, whereas for a higher laser intensity (1.5×10^8 W cm^{-2}), a droplet of Al from the target (2.7 μm thick) scattered on the same glazed alumina receiver substrate (Figure 5.2.3), suggesting the effect of surface tension for molten Al during transfer. Bullock *et al.* [8] reported the laser-generated vapor plume in irradiating Al films by Nd:YAG laser in vacuum. Time-integrated reflectivity was observed, indicating a sharp ablation threshold at 50 mJ pulse energy.

5.2.3.1.4 Tungsten

Tungsten is effective as one of the model materials for the LIFT experiments to elucidate the mechanism of laser-induced transfer, since it has a high melting point (3653 K). This attempt [29] was carried out using 100 nm thick vacuum-evaporated tungsten films on a glass donor substrate in contact with a receiver silicon substrate using an Ar$^+$ laser (515 nm, 1 μs to 1 ms pulse width) as a laser source. The onset of the ablation of W films started at the laser power of 40–50 mW. Several patterns of tungsten film were observed on a silicon substrate through the LIFT experiments with various powers (50–476 mW) and pulse widths (20–200 μs). Model calculations revealed that no melting of tungsten film could take place below $\sim$130 mW laser power because the possible maximum temperature in the experiments was below the melting point $T_m = 3653$ K at these laser powers.

The start of the ablation for a W film (200 nm thick) deposited on a glass donor substrate by the electron beam evaporation using a diode-pumped YAG

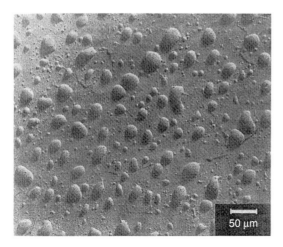

Figure 5.2.3 SEM image of Al deposit on a glazed alumina ceramic receiver transferred by LIFT at higher laser energy density of 1.5×10^8 W cm^{-2} [5].

laser (1.06 µm, 100 µs to 1 ms pulse widths) with the power 0–200 mW was nearly the same for the W films irradiated by the Ar$^+$ laser, since the onset of ablation transfer began at 50–55 mW laser power [26]. From temperature calculation, it was also shown that no melting of the W films could take place up to 135 mW since all the temperatures were below the melting point $T_m = 3653$ K, and the ablation was thought to start in solid phase in the experimental power regime. Figure 5.2.4 shows a good printing but with rather irregular contours at the laser irradiation of the power 110 mW.

Kántor *et al.* [7] investigated in detail the transfer of W films with 100 nm thickness on a glass plate using the diode-pumped Nd:YAG laser (1.06 µm, 20–230 mW peak power), where the receiver substrate–film distance was less than 1 µm. They varied the pulse width from 100 to 1000 µs, and a triangular pulse was used. When a 100 nm thick W film was irradiated with 500 µs width triangular pulses with 130 mW peak power, a complete transfer of the W film was produced in a circular form about 500 µm in diameter.

5.2.3.1.5 Titanium

The LIFT of Ti, a metal with a high melting point (1941 K) and stable at ambient temperature, was reported [26] using glass donor substrates coated with 200 nm thick Ti films deposited by vacuum evaporation. As samples for the comparative study, 173 nm thick Cr films from commercial products (Hoya Co. Ltd) were also used for the LIFT experiments. Initial check of adherence showed that the Ti films were poorly adhered to the donor glass substrate, while Cr films were well-adhered on the donor glass substrates. A ruby laser (694 nm, 20 ns pulse width, 0–1.1 J cm^{-2} fluence) was used for single-pulse experiments. The film removal from the donor glass substrate and the deposition on the receiver glass were characterized by monitoring the transmittance at the wavelength of 633 nm

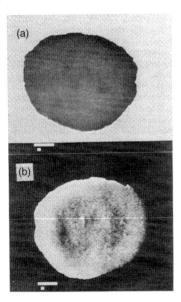

Figure 5.2.4 (a) SEM images of the ablated area on a glass donor substrate and (b) the transferred print on the receiving glass substrate of a tungsten film with a thickness of 200 nm at a single-pulse irradiation with a diode-pumped YAG laser (1063 nm, 65 mW peak power, and 1 ms pulse width) (scale : 1 µm) [14].

for both substrates. The transfer of the titanium films was found to occur at the fluence of $\sim$100 mJ cm^{-2}, and a complete removal of Ti films from the donor substrate took place at the fluence of 200–240 mJ cm^{-2} as shown in Figure 5.2.5. However, an abrupt ablation threshold was observed for poorly adhered Ti films. At the fluence just above this threshold, the whole irradiated area of the film was removed, resulting in complete ablation without a remnant of the removed film. The transmittance of the transferred printed area started to decrease above 240 mJ cm^{-2}, indicating the onset of deposition, where the prints consisted of resolidified splashes. The ablated areas became darker because of the formation of a hazy layer on the target with increasing fluence up to $\sim$550 mJ cm^{-2}. Above the fluence of 550 mJ cm^{-2}, the formation of the vapor phase induced the enhanced efficiency of removal and transfer. These results were compared with those of the model calculations of the temperature as a function of laser fluence.

5.2.3.1.6 Chromium
Chromium is frequently used for reflecting mirrors in the visible region by depositing it onto the donor glass substrate, and the adherence of the film is remarkable compared with that of Au and Ag films. The LIFT characteristics of Cr films were investigated in parallel with the study of Ti films. Kántor *et al.* [26] also studied the single-pulse shot LIFT for Cr films (173 nm thick), a commercial product of optical chrome mask (HOYA Co. Ltd.) and well adhered initially to a donor glass, onto a receiver glass substrate using a ruby laser (694 nm, 20 ns pulse

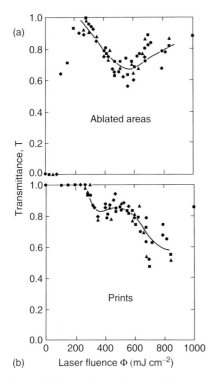

Figure 5.2.5 (a) Transmittance of the ablated Ti films and (b) the transfer prints of the films on a glass receiver substrate at 633 nm (initial Ti film thickness: 200 nm), under a single shot in air from a Q-switched ruby laser (694 nm) with 20 ns pulse width. Kinds of the data points correspond to the different spacings of 15–55 μm between the donor and the receiver [26].

width) in air at ambient temperature. The removal of the film and the corresponding transferred print were characterized by monitoring the optical transmittances of the donor and the receiver glasses as a function of the laser fluence. The ablated area increased approximately linearly with the increase in the fluence from 350 to 600 mJ cm^{-2}. Unlike poorly adherent Ti films [26], the ablation was incomplete in this fluence range: locally removed and transferred films showed isolated holes and islands. Material depositions began at $\sim$300 mJ cm^{-2} and the transmittance of the print deposited by the LIFT gradually decreased with increasing fluence from 300 to 600 mJ cm^{-2}.

Later, the LIFT study of sputtered Cr films with various film thicknesses (40–200 nm) on quartz wafers was performed by a single-pulse irradiation using a KrF excimer laser (wavelength of 248 nm, pulse width of 500 fs, and pulse energy of 13 mJ) [15]. A SEM image of Cr dot deposits transferred from a 40 nm thick Cr film on the donor glass substrate is shown in Figure 5.2.6, where the laser-illuminated area is 4×4 μm. The LIFT ablation threshold fluences for 40 and 80 nm thick Cr films were obtained from the fluence dependence of the deposit feature size

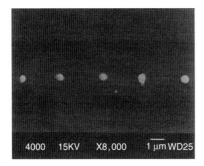

Figure 5.2.6 SEM image of isolated Cr dots deposited on a glass receiver substrate in 10 Pa vacuum by a single-pulse shot of a KrF excimer laser (248 nm, 500 fs pulse width, and 100 mJ cm^{-2} fluence) [15].

(the diameter of the deposited area) as 100 and 150 mJ cm^{-2}, respectively. The spread of the ablated Cr film was measured by varying the spacing between the donor and receiver surfaces from near-contact to 500 μm, and the results showed that the feature size of the deposits increased nearly linearly from 4 to 40 μm with the increase in the spacing from 0 to 300 μm, respectively, for the fluences of 156 and 260 mJ cm^{-2}.

The laser-induced "backward" transfer of thin Cr films was also carried out using a KrF excimer laser (248 nm, 500 fs pulse width, and 10 Hz repetition rate) by Papakonstantinou *et al.* [27]. In this configuration, a laser beam was incident through a transparent receiver (without film) on the film precoated on the donor. The film material was ablated and transferred backward to the receiver substrate. By this transfer scheme, 30–40 nm thick Cr films predeposited by the pulsed laser deposition on a donor quartz substrate could be transferred onto a glass plate, where the laser fluences were 150 and 200 mJ cm^{-2}. The latter case with higher fluence presented a crater in the center and splattered Cr dots around it on the receiver glass, and the aspect was similar to the Al prints on the glazed alumina ceramics [5].

5.2.3.1.7 Palladium

The LIFT process applied to the thin film deposition of Pd [12] was not the direct method for the Pd film transfer but that for the surface nucleation of Pd nanocrystals, followed by the post electroless copper plating on various receiver substrates. The LIFT technique was performed on the donor quartz substrate with Pd-acetate films (100–500 nm thick) precoated by the spin-on technique. The excimer lasers of ArF (193 nm), KrF (248 nm), and KrCl (222 nm) were used with the fluences between 5 and 300 mJ cm^{-2}. Quartz, Al$_2$O$_3$, AlN, PI (polyimide), PMMA (polymethyl methacrylate), and PET (polyester) were used as the receiving substrates. These substrates were placed against the donor quartz with a 60 μm air gap, which was irradiated in air at ambient temperature with either a single pulse or a few pulses. In SEM images, Pd nuclei with an average diameter of ~100 nm were observed in the deposited Pd films on quartz and Al$_2$O$_3$ substrates, which were generated during the LIFT at the fluence of 80 mJ cm^{-2} and by 10 pulse irradiations.

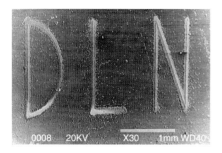

Figure 5.2.7 SEM image of patterned diamond on a Si receiver substrate after prenucleation of ultrafine diamond powders within a 2–3 μm thick photoresist layer using the LIFT process and subsequent diamond film deposition by the CVD process. A Cu vapor laser (510 nm) was used with 20 ns pulse width, 30 mW average power, 40 × 40 μm spot size, and 30 μm s^{-1} scanning velocity [13].

After the LIFT application of Pd, the prenucleated surfaces were electroless coated with copper using a commercially available electroless copper bath. The transferred Pd could be plated on ceramics and different plastics without problems.

5.2.3.1.8 Diamond

The LIFT process was applied to the prenucleation of diamond nanoparticles on a silicon wafer, followed by the CVD (chemical vapor deposition) diamond film coating [13]. The donor quartz substrate was first coated with the mixture of ultrafine diamond powders (mean particle size of 5 nm) with a photoresist of 2–3 μm thickness. The donor substrate was placed in close proximity to the Si receiver substrate. A Cu vapor laser (510 nm, 20 ns pulse width, and 10 kHz repetition rate) and a KrF excimer laser (248 nm, 15 ns pulse width) were applied for the LIFT experiments. The best transfer of diamond onto the Si wafer was obtained after the KrF laser irradiation of a single pulse at fluences of 150–200 mJ cm^{-2}. The irradiation of the laser pulse through the quartz substrate caused photoresist deposits containing diamond onto the Si wafer. The accuracy of the lateral displacement of the computer-driven X–Y stage for the laser scanning to draw the pattern was 2 μm. The irradiation was carried out in air at ambient temperature.

After the LIFT, the Si substrate was put into an auxiliary CVD reactor, which grew a diamond layer with the deposition rate $\sim$10 μm h^{-1}. The SEM image in Figure 5.2.7 exhibits the diamond deposit grown on a Si receiver substrate after patterning by the transferred diamond powders, where the nucleation on the receiver substrate was performed by the beam scanning of the Cu vapor laser.

5.2.3.2 Oxides

5.2.3.2.1 YBCO, BiSrCaCuO

Amorphous films (100–800 nm thick) of YBCO and BiSrCaCuO compounds could be transferred by the LIFT process using an ArF excimer laser (193 nm, 20 μs

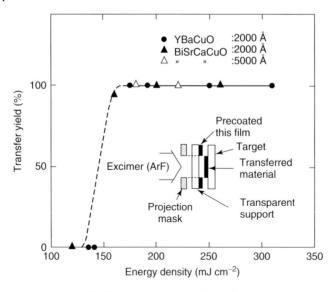

Figure 5.2.8 Yield of laser-induced forward transfer for YBCO and BiSrCaCuO films by a single-pulse irradiation of an ArF excimer laser (193 nm) with the pulse width of 20 ns as a function of laser energy density/fluence [9].

pulse width, and $0.1-0.5 \, J \, cm^{-2}$ fluence) onto Si and MgO single-crystal receiver wafers [9, 10]. These donor films were deposited initially by DC sputtering or excimer laser ablation in amorphous form on suprasil quartz or simple donor glass substrates. As a next step, these donor substrates were irradiated by a single pulse through the quartz support (backside) using (i) an ArF excimer laser (193 nm) or (ii) a Nd:YAG laser ($1.06 \, \mu m$, 5 ns pulse width, and $0.1-1 \, J \, cm^{-2}$ fluence) under a vacuum pressure (10^{-2} Pa) at ambient temperature.

As shown in Figure 5.2.8, the threshold fluence of the LIFT was found close to $150 \, mJ \, cm^{-2}$ for both components, and the transition from 0 to 100% removal was achieved over a relatively narrow range of laser fluence ($\approx 50 \, mJ \, cm^{-2}$). At the wavelength of 193 nm of the ArF excimer laser, this threshold did not seem to be significantly influenced by the film thickness (thicker than 200 nm) of the donor film. This result contrasts with the case of Au film transfer onto a quartz receiver substrate using pulsed lasers (248 and 532 nm) [6], where the threshold fluence depended on the film thickness.

The same experiments were repeated using a Nd:YAG laser ($1.06 \, \mu m$) for BiSrCaCuO films. In this case, the threshold fluence of the laser shifted to a higher value (larger than $500 \, mJ \, cm^{-2}$), which appeared to be dependent on the thickness of the precoated film. For the laser fluence largely in excess ($>300 \, mJ \, cm^{-2}$ at 193 nm) over the minimum required for the transfer ($\approx 150 \, mJ \, cm^{-2}$), the surface of the transferred layer became very rough. Under these conditions, a poor adhesion of the deposit on the receiver substrate was also observed. It is noted that holding the receiver substrate (Si or MgO) in close contact with the precoated films during the LIFT processing is required to realize the high-quality transfer.

The superconducting transition was observed for the BiSrCaCuO film, followed by a subsequent post-thermal annealing of the transferred layer on the MgO receiver substrate in O_2 atmosphere at about 865 °C for 10 min. The onset and zero-resistance temperatures were observed at around 90 and 80 K, respectively, which are typical for the Bi compounds with a stoichiometry close to 2, 2, 1, 2.

5.2.3.2.2 In$_2$O$_3$

An In_2O_3 film, a typical material for the optically transmitting electrode film, was studied as a target for the LIFT study [15]. The In_2O_3 film (200 nm thick), after initial deposition by the reactive pulsed laser deposition on a quartz donor substrate, was transferred onto a glass receiver substrate using a KrF excimer laser (248 nm, 13 mJ pulse energy, and 500 fs pulse width) at the fluence of 150 mJ cm^{-2} as microdepositions with a pixel size of 4×4 μm, which was used for demonstration of a holographic recording. XRD peaks due to the In_2O_3 structure were observed for the films deposited by the LIFT. These peaks were the same as those of the target film on the quartz. The threshold fluence of 45 mJ cm^{-2} was found for the LIFT of the In_2O_3 films with a thickness of 150–200 nm. When an InOx source film (~50 nm thick) placed a few micrometers apart against the receiver glass plate was irradiated by the same KrF excimer laser (248 nm, 500 fs pulse width, and 10 Hz repetition rate), the LIFT started at the fluence of 60 ± 20 J cm^{-2} as shown in Figure 5.2.9 [27]. In this figure, the transferred spot (LIFT focused area of 4×4 μm) from the InOx source film consists of a smooth uniform thin layer in the central region of ~1 μm in diameter, but a number of coalesced grains and some isolated ones are observed around the outer edge.

In the same study [27], the threshold fluences for the transfer of Pt and Cr films (30–40 nm thick) were around 150 ± 20 J cm^{-2}, which were considerably larger than that for In_2O_3 films. The difference in the fluence was explained by estimating the penetration depth ($l_s = \alpha^{-1}$, where α is the absorption coefficient) and the thermal diffusion length given as $l_t = (\pi D \tau)^{1/2}/2$, where D is the thermal diffusivity written as $D = k/\rho C$ and k, ρ, C, and τ are the thermal conductivity, the density of the material, the heat capacity, and the laser pulse width, respectively. The thermal diffusion length l_t for a laser pulse width of 500 fs was estimated by applying these physical parameters to Pt, Cr, and In_2O_3. The calculated results were that $l_s(45.5$ nm$) > l_d(3.0$ nm$)$ for In_2O_3 but $l_s(8.2$ nm Pt, 7.6 nm Cr$) \leq l_t$ (10.0 nm Pt, 10.5 nm Cr) for Pt and Cr, respectively. These results mean that the heated material volume by the laser irradiation of 500 fs pulse width is determined by the skin depth and the In_2O_3 with larger l_s explains its lower threshold value due to a deeper penetration of the laser beam into the film.

In this study, the transfer by the laser backward transfer scheme was also carried out for the In_2O_3 films [27]. It is worth noting that the aspect of the backward transferred In_2O_3 prints was similar to that observed in the forward transfer scheme.

5.2.3.2.3 V$_2$O$_5$

Vacuum-deposited amorphous V_2O_5 films are n-type semiconducting and weakly absorbing in the visible range. Using a sheet beam emitted through a cylindrical

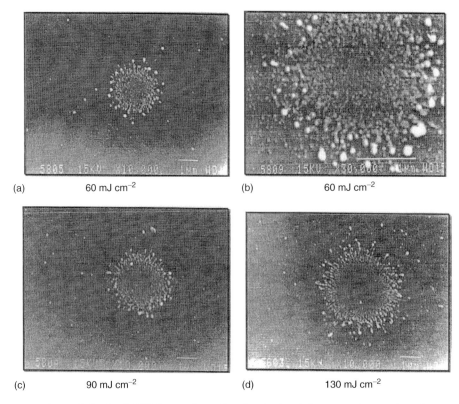

(a)   60 mJ cm^{-2} (b)   60 mJ cm^{-2}

(c)   90 mJ cm^{-2} (d)   130 mJ cm^{-2}

Figure 5.2.9 (a–d) SEM image of transferred and printed dots on a glass receiver plate transferred from a 50 nm thick In_2O_3 film on the quartz donor plate at various fluences of a KrF excimer laser (248 nm, 500 fs pulse width, and 15 mJ pulse energy) [27].

lens from a frequency-doubled Nd:YAG laser (532 nm, 10 ns pulse width, 4 J cm^{-2} fluence, and 10 Hz repletion rate), the LIFT process was applied to the amorphous V_2O_5 film (180 nm thick) on a glass donor slide in air, and a parallel stripe pattern was formed in the film on a receiving glass slide (Figure 5.2.10) [20]. XRD of the LIFT transferred film also showed the amorphous nature of the initial V_2O_5 film.

A clear stripe pattern was obtained by irradiating the sheet beam on the pair of donor and receiver substrates moving at a constant speed of 5 mm s^{-1} in the direction normal to the beam. The V_2O_5 deposit showed an irregular edge shape on the boundary of the stripe for the sample with a 0.14 mm air gap. The atomic force microscopy (AFM) observation revealed that the edge profile of the stripe was less sharp compared with the scheme in direct mechanical contact between the donor and the receiver during the irradiation. The sample with 0.14 mm air gap also showed a larger surface roughness (24.8 nm Rrms) than that of the direct contact sample (22.6 nm Rrms). SEM images showed that the V_2O_5 donor film was not completely removed from the donor substrate under this beam condition, probably due to the large adherence of the film to the donor substrate, and the

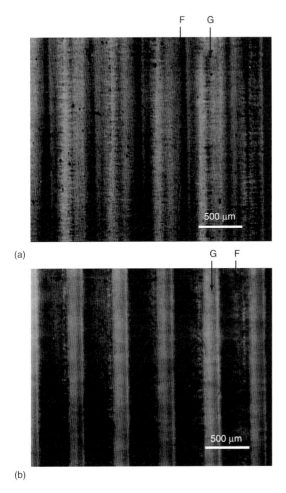

(a)

(b)

Figure 5.2.10 Optical micrograph of V_2O_5 films (180 nm thick) transferred by laser-induced forward transfer to a receiving glass substrate using a frequency-doubled Nd:YAG laser, patterned by sheet beam irradiations of the laser in (a) direct contact and in (b) 0.14 mm air gap between the donor and the receiver, where striped regions G and F correspond to the glass substrate and the deposited film, respectively [20].

residual donor film presented many large pores in the surface from which V_2O_5 was removed.

5.2.3.2.4 TiO₂–Au

TiO_2 is optically transparent in the visible range with a large band gap energy. Hence, it is rather difficult to transfer TiO_2 films by the LIFT process using the laser with the wavelength in the visible range. However, when Au nanoparticles are incorporated in TiO_2 films, the optical absorption due to the surface plasmon makes the film absorbing. Such TiO_2–Au nanocomposite films were prepared by the sol–gel process [30, 31] and targeted for LIFT experiments [19].

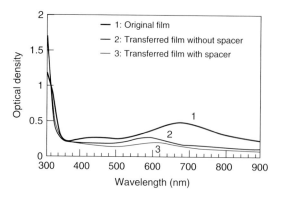

Figure 5.2.11 (1) Optical absorption spectra of 89TiO$_2$–11Au (mol%) films [31] for 100 nm thick original film, (2) forward transferred film in direct contact with receiving glass slide substrate, and (3) forward transferred film with a 0.14 mm air gap on the receiver glass slide [31].

Attempt of maskless patterning of 89TiO$_2$–11Au films (100 nm thick) coated on an optical glass donor slide by the sol–gel method was performed in air using a sheet beam of a frequency-doubled Nd:YAG laser (532 nm, 10 ns pulse width, and 3 J cm^{-2} fluence) emitted through a cylindrical lens. The sheet beam was irradiated on a pair of donor–receiver glass slides moving at the speed of 5 mm s^{-1}. The donor and the receiver glass slides were placed in direct (mechanical) contact or with 0.14 mm air gap. In both cases, a deposit pattern with linear parallel tracks from the TiO$_2$–Au film was observed on the glass slide receiver surface. SEM pictures of the transferred films showed that the grains of the transferred TiO$_2$–Au films were more coalesced for the films with the 0.14 mm gap than those in direct contact. It was found that the maximum optical absorption of the initial TiO$_2$–Au film in the visible region due to the surface plasmon shifted to shorter wavelength for the transferred and deposited film, (Figure 5.2.11) and the result was explained, using the Mie scattering model, by the porous character of the transferred TiO$_2$–Au films [31].

5.2.3.3 Other Compounds Including Biomaterials

The materials targeted by the LIFT studies have recently extended to polymers and biomaterials including biocells and DNA. These materials are normally transparent in the visible region because of wide band gaps. Then, excimer lasers with far ultraviolet (FUV) wavelength or third harmonic generation (THG) (3ω) radiation from the Nd:YAG laser are usually used for the material transfer under laser irradiation. Some typical applications of these materials using the LIFT are presented.

5.2.3.3.1 Photopolymers

Photoabrasive decomposition during the LIFT was investigated for Alyltriazene photopolymer TP-6-CH$_3$ using a XeCl excimer laser (308 nm, 30 ns pulse width, and 12–105 mJ cm^{-2} fluence) [32]. The photopolymer was irradiated in air at ambient temperature. Transparent films ($\sim$20 to over 500 nm thick) of the TP-6 were prepared onto the fused silica donor substrates. The donor and the receiver substrates were placed in close contact ($<$1 µm). Ablation transfer experiments were performed using the XeCl excimer laser at a single pulse, and the ablated spots defined by a rectangular mask 500 $\times$ 500 µm in size were observed by an optical microscope. The range of the fluence for the transfer processing of the TP-6-CH$_3$ thin layer with a thickness of 150 nm was found to be at $\sim$80–100 mJ cm^{-2} for an intermediate energy-absorbing and sacrificial DRL (dynamic release layer) model system. It was revealed from the thickness measurements of the donor and transferred films that some amount of the photopolymer was consumed and decomposed during the transfer process from the thickness measurements of the donor and transferred films, and some amount was missing (e.g., thickness of 90 nm was missing for the initial film thickness of $\sim$150 nm) during the single-pulse laser ablation transfer. As a result, about 10–20 nm thick intact polymer pixels were obtained at the fluence of about 350 mJ cm^{-2} by this LIFT process.

5.2.3.3.2 Antigen and DNA

The technique of the LIFT was investigated to spot an array of a purified bacterial antigen for protein detection by protein-based biosensors [24]. A THG (3ω) Nd:YAG laser (355 nm, 10 ns pulse width, and 10 µJ pulse energy) was used for the irradiation of a single-pulse shot in air. The donor liquid film containing the *Treponerma pallidum* 17 kDa antigen mixed with phosphate buffer solution and glycerol was spread onto the glass slide precoated with a 60 nm thick Ti film. The Ti film was coated to increase the optical absorption of the sample. The receiving substrate was a microscope slide or a nylon-coated slide, and the donor–receiver spacing was set at about 100 µm. The microarrays obtained after the LIFT processes were observed by an optical microscope, where almost all the droplets formed by the transfer had diameters of about 80 µm. For the immobilization of the antigen onto the substrate, the nylon-coated substrate was shown to be adequate. A similar approach was performed for DNA microarrays of human MAP K3 cDNA using a 3ωNd : YAG laser [23].

5.2.4
Applications

The LIFT technique was applied to the direct writing and microprinting of the donor film material onto a receiver substrate. X–Y dot patterns due to the LIFT process were easily obtainable by using a computer-driven X–Y digital moving stage. Zergioti *et al.* [15] produced a computer-generated holographic pattern by the microdeposition (microdot deposition) of In$_2$O$_3$ or Cr films as shown in Figure 5.2.12. Such a computer-generated holographic pattern using an InOx film can also be produced on a quartz receiver substrate by the laser-induced backward transfer mode [27]. A reconstructed image of the microprinted Cr computer-generated holographic pattern has been reported [33].

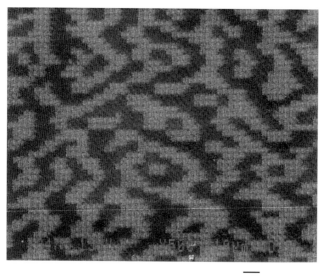

10 μm

Figure 5.2.12 SEM image of a computer-generated holographic pattern produced by the microdeposition of a Cr film on a silicon (100) receiver substrate with a pixel size 3 × 3 μm. The donor Cr film on a glass substrate had a thickness of 40 nm, and the fluence of the KrF excimer laser (248 nm, 500 fs pulse width) was 339 mJ cm^{-2} [15].

The direct microprinting/micropatterning of transparent materials by the LIFT technique was applied recently to biomaterials such as biocell, DNA [22, 23], and conidia [34]. Since DNA is normally transparent in the visible region, excimer lasers with several hundreds of nanometer wavelength [22] or THG (3ω) Nd:YAG laser (355 nm) [23] were used. Lambda phage DNA arrays printed with an X–Y–Z translation stage onto a glass receiver substrate using a KrF excimer laser (248 nm, ~500 fs pulse width, and 10 mJ pulse energy) are shown in Figure 5.2.13. The SEM image shows the transferred matrix print of the phage (~150 × 150 μm in size). Further, a scanning confocal fluorescence microscopic image was presented for the two-dimensional array of the Alexa 594-labeled lambda phage DNA [22]. A microarray preparation of a human cDNA through the LIFT process was also reported [23].

The LIFT process was also applied to the thermal transfer digital image printing system in electronics [35]. In this case, an ink sheet was in direct contact with the paper and both were wound on a motor-driven roll irradiated by a focused laser-diode-pumped Nd:YAG laser (3 μs pulse width, maximum power of 16 W); a digital high-definition image was obtained. Such a color printing technique using infrared-diode-pumped laser has been reported [16].

The LIFT method, basically a direct writing process, was recently extended to the MAPLE-DW (matrix-assisted pulsed laser evaporation-direct write) proposed by Chrisey *et al.* [36] with some experimental results. This process is characterized as follows: (i) the composite film incorporating a target material and a laser-absorbing

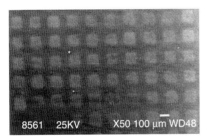

Figure 5.2.13 SEM image of the printed features of lambda phage DNA on a glass receiver substrate deposited by the LIFT process using a KrF laser (248 nm, 500 fs pulse width) [22].

matrix material for assisting the LIFT process are transferred to the receiving substrate. So, inorganic plus inorganic, organic plus inorganic, or organic plus organic composite laser-absorbing materials can be utilized as the target material system. (ii) A "ribbon" of the composite target film precoated onto a transparent donor glass or a polymer substrate is placed in close proximity (5–100 μm) to the receiver. This ribbon makes an easy and constant transfer of the composite to the receiving substrate, and by its X–Y stepwise movement during the laser transfer, any dot pattern of the target material can be generated on the receiving substrate.

The MAPLE-DW technique was initially applied to the printing in mesoscopic electronic devices, that is, Au-conductor lines, Nichrome resistor, and $BaTiO_3$ capacitor [36], by using KrF (248 nm) and XeCl (308 nm) excimer lasers at 200 mJ cm^{-2} fluence. The process was also applicable to phosphor pattern printing [21]. In this case, a phosphor ribbon was fabricated first by a 100 nm thick Au film sputtered on a quartz donor substrate, followed by the electrophoretic deposition of three kinds of phosphor powders (15 μm in size) suspended in glycerin/isopropanol solution with nitrate salts. The cathode luminescence emission spectra for the initial phosphor-coated ribbon and that of the transferred phosphor were found to be identical. Details and applications of MAPLE-DW for organic and biomaterials are reported [37].

The AFA-LIFT (absorbing film-assisted laser-induced forward transfer) is another extension of LIFT, applicable to biomaterials and macromolecules, for example, fungi [25], enzyme, cells, and so on.

In the AFA-LIFT of fungi [25], a fungus (a biomaterial) was placed, in suspension, on a transparent silica donor substrate coated with an absorbing film (vacuum-evaporated 50 nm thick Ag film) for laser-induced transfer. This carrier was set on the glass receiver with an air gap of ~1 mm. The KrF excimer laser (248 nm, 30 ns pulse width) was irradiated on the samples at fluences of 35–200 mJ cm^{-2} in a single-pulse irradiation. After the LIFT process in air at room temperature followed by an incubation of the transferred fungi for 20 h at 25 °C, islands of the transferred fungi were clearly observed by an optical microscope. It is noted that an ejecting speed of the fungi from the target attained up to 1150 m s^{-1} on the laser irradiation at the fluence of 355 mJ cm^{-2} [34].

5.2.5
Summary and Conclusion

The LIFT is a novel technique of removing and transferring a thin film predeposited on a transparent donor substrate to another receiving substrate by the irradiation of a pulsed laser on the donor film in air or in vacuum. It is a direct writing and dry processing technique, but a two-step process with regard to film handling, that is, film predeposition onto the donor substrate by any method of deposition and subsequent transfer to the receiver by laser irradiation. According to these features of the LIFT process, this technique is applicable to the experimental stage of researches because of the easy setup of the apparatus but not to the mass production stage because of multistep handling of transferring films.

Owing to the direct printing, the LIFT has useful advantages of depositing dots or 1D–2D patterns by transferring the film in a relatively short time (several tens of nanoseconds to picoseconds) of a single-pulse or multipulse irradiation. To obtain deposits with good quality, that is, well-defined deposits by the LIFT, the following factors proposed from the previous works have to be taken into account:

1) The wavelength of the laser for the LIFT should be adequate so that the laser beam can be absorbed strongly by the source film for transfer. So, the excimer lasers or 3ω to 4ω Nd:YAG laser are preferable depending on film materials.
2) The laser fluence for transfer that yields a well-defined deposition should be slightly larger than the threshold fluence for the films to be transferred, which depends on the thickness and kinds of film; normally, these are from several tens to some hundred millijoules per square centimeter. The transfer at the lower fluence that causes a soft detachment of the films, which is preferable to prepare the transferred dots with good quality.
3) Film thickness of about 200 nm is suitable for metals and oxides. Larger thickness requires a higher fluence of the laser and produces splashing and coalescence of the transferred films; thus, the contour of the deposition becomes ambiguous.
4) Spacing between the donor and the receiver influences largely the shape of the deposition. Larger spacing produces larger but ambiguous shape of the deposition. A spacing of several micrometers is normally adequate for inorganic films.
5) The adherence between the source film and the initial substrate (donor substrate) influences the threshold of the LIFT; lower adherence of the film to substrate causes a transfer with lower threshold fluence, as seen in the case of plastic substrates.

A variety of LIFT studies were performed using the films of metals, oxides, composites, and organic and biomaterials in the electronics or bioscience area. The extension of LIFT can be expected as a useful tool for micro- to nanoprocessing in patterning accompanied by micro- to nanostructuring of simple and complex material in the film form. Hence, in the next generation, 2D–3D multilayered patterned structures should be realized by the extended LIFTs in order to generate deposits exhibiting new properties.

References

1. Bohandy, J., Kim, B.F., and Adrian, F.J. (1986) Metal deposition from a supported metal film using an excimer laser. *J. Appl. Phys.*, **60** (4), 1538–1539.
2. Zergiotti, I., Koudourakis, G., Vainos, N.A., and Fotakis, C. (2002) in *Direct Write Technologies for Rapid Prototyping Applications* (eds A. Piqué and D.B. Chrisey), Academic Press, New York, pp. 493–516.
3. Adrian, F.J., Bohandy, J., Kim, B.F., Jette, A.N., and Thomson, P. (1987) A study of the mechanism of metal deposition by the laser-induced forward transfer process. *J. Vac. Sci. Technol.*, **B5** (5), 1490–1494.
4. Bohandy, J., Kim, B.F., Adrian, F.J., and Jette, A.N. (1988) Metal deposition at 532nm using a laser transfer technique. *J. Appl. Phys.*, **63** (4), 1158–1162.
5. Schultze, V. and Wagner, M. (1991) Laser-induced transfer of aluminium. *Appl. Surf. Sci.*, **52**, 303–309.
6. Baseman, R.J., Froberg, N.M., Andreshak, J.C., and Schlesinger, Z. (1990) Minimum fluence for laser blow-off of thin gold films at 248 and 532nm. *Appl. Phys. Lett.*, **56** (15), 1412–1414.
7. Kántor, Z., Thóth, Z., and Szörényi, T. (1995) Metal pattern deposition by laser-induced forward transfer. *Appl. Surf. Sci.*, **86**, 196–201.
8. Bullock, A.B., Borton, P.R., and Mayer, F.J. (1997) Time-integrated reflectivity of laser-induced back-ablated aluminum thin film targets. *J. Appl. Phys.*, **82** (4), 1828–1831.
9. Fogarassy, E., Fuchs, C., Kerherve, F., Hauchecorne, G., and Perriere, J. (1989) Laser-induced forward transfer of high-Tc YBaCuO and BiSrCaCuO superconducting thin films. *J. Appl. Phys.*, **66** (1), 457–459.
10. Fogarassy, E., Fuchs, C., Kerherve, F., Hauchecorne, G., and Parrière, J. (1989) Laser-induced forward transfer: a new approach for the deposition of high Tc superconducting thin films. *J. Mater. Res.*, **4** (5), 1082–1086.
11. Lätsch, S., Hiraoka, H., Nieveen, W., and Bargon, J. (1994) Interface study on laser-induced material transfer from polymer and quartz surfaces. *Appl. Surf. Sci.*, **81**, 183–194.
12. Esrom, H., Zhan, J.-Y., Kolgelschatz, U., and Pedraza, A.J. (1995) New approach of a laser-induced forward transfer for deposition of patterned thin metal films. *Appl. Surf. Sci.*, **86**, 202–207.
13. Pimenov, S.M., Shafeev, G.A., Smolin, A.A., Konov, V.I., and Vodolaga, B. (1995) Laser-induced forward transfer of ultra-fine diamond particles for selective deposition of diamond films. *Appl. Surf. Sci.*, **86**, 208–212.
14. Kántor, Z., Tóth, Z., Szörényi, T., and Tóth, A.L. (1994) Deposition of micrometer-sized tungsten pattern by laser transfer technique. *Appl. Phys. Lett.*, **64** (25), 3506–3508.
15. Zergioti, I., Mailis, S., Vainos, N.A., Papakonstantinou, P., Kalpouzos, C., Grigoropoulous, C.P., and Fotakis, C. (1998) Microdeposition of metal and oxide structures using ultrashort laser pulses. *Appl. Phys.*, **A66**, 579–582.
16. Blanchet, G.B., Loo, Y.-L., Rogers, J.A., Gao, F., and Fincher, C.R. (2003) Large area, high resolution, dry printing of conducting polymers for organic electronics. *Appl. Phys. Lett.*, **82** (3), 463–465.
17. Bähnisch, R., Groß, W., and Menschig, A. (2000) Single-shot, high repetition rate metallic pattern transfer. *Microelecron. Eng.*, **50**, 541–546.
18. Tan, B., Venkatakrishnan, K., and Tok, K.G. (2003) Selective surface texturing using femtosecond pulsed laser induced forward transfer. *Appl. Surf. Sci.*, **207**, 365–371.
19. Sakata, H., Chakraborty, S., Yokoyama, E., Wakaki, M., and Chakravorty, D., (2005) Laser-induced forward transfer of TiO$_2$-Au nanocomposite films for maskless patterning. *Appl. Phys. Lett.*, **86** (11), 114104-1–114104-3.
20. Chakraborty, S., Sakata, H., Yokoyama, E., Wakaki, M., and Chakravorty, D. (2007) Laser-induced forward transfer technique for maskless patterning of amorphous V$_2$O$_5$ thin films. *Appl. Surf. Sci.*, **254**, 638–643.

21. Fitz-Gerald, J.M., Piqué, A., Chrisey, D.B., Rack, P.D., Zeleznik, M., Auyeung, R.C.Y., and Lakeou, S. (2004) Laser direct writing of phosphor screens for high-resolution displays. *Appl. Phys. Lett.*, **96** (11), 3478–3481.

22. Karaikou, A., Zergioti, I., Fotakis, C., Kapsetaki, M., and Kafetzopoulos, D. (2003) Microfabrication of biomaterials by the sub-ps laser-induced forward transfer process. *Appl. Surf. Sci.*, **208–209**, 245–249.

23. Serra, P., Colina, M., Fernández-Pradas, J.M., Sevilla, L., and Morenza, J.L. (2004) Preparation of functional DNA microarrays through laser-induced forward transfer. *Appl. Phys. Lett.*, **85** (9), 1639–1641.

24. Serra, P., Fernández-Pradas, J.M., Berthet, F.X., Colina, M., Elvira, J., and Morenza, J.L. (2004) Laser direct writing of biomolecule microarrays. *Appl. Phys.*, **A79**, 949–952.

25. Hopp, B., Smausz, T., Antal, Zs., Kresz, N., Bor, Zs., and Chrisey, D. (2004) Absorbing film assisted laser induced forward transfer of fungi (Trichoderma conidia). *J. Appl. Phys.*, **96** (6), 3478–3481.

26. Kántor, Z., Thóth, Z., and Szörényi, T. (1992) Laser-induced forward transfer: the effect of support-film interface and film-to-substrate distance on transfer. *Appl. Phys.*, **A54**, 170–175.

27. Papakonstantinou, P., Vainos, N.A., and Fotakis, C. (1999) Microfabrication by UV femtosecond laser ablation of Pt, Cr and indium oxide thin films. *Appl. Surf. Sci.*, **151**, 159–170.

28. Alexander, D.E., Was, G.S., and Mayer, F.J. (1988) Laser-driven micro-explosive bonding of aluminium films to copper and silicon. *J. Mater. Sci.*, **23**, 2181–2186.

29. Tóth, Z., Szörényi, T., and Tóth, A.L. (1993) Ar+ laser-induced forward transfer: a novel method for micrometer-size surface patterning. *Appl. Surf. Sci.*, **69**, 317–320.

30. Matsuoka, J., Yoshida, H., Nasu, H., and Kamiya, K. (1997) Preparation of gold microcrystal-doped TiO_2, ZrO_2 and Al_2O_3 films through sol-gel process. *J. Sol-Gel Sci. Technol.*, **9**, 145–155.

31. Sakka, S., and Kozuka, H., (1998) Sol-gel preparation of coating films containing noble metal colloids. *J. Sol-Gel Sci. Technol.*, **13**, 701–705.

32. Nagel, M., Fardel, R., Feurer, P., Häberli, M., Nüesch, F.A., Lippert, T., and Wokaun, A. (2008) Aryltriazene photopolymer thin films as sacrificial release layers for laser-assisted forward transfer systems: study of photoablative decomposition and transfer behavior. *Appl. Phys.*, **A92**, 781–789.

33. Mailis, S., Zergioti, I., Koundourakis, G., Ikiades, A., Patentalaki, A., Papakonstantinou, P., Vainos, N.A., and Fotakis, C. (1999) Etching and printing of diffractive optical microstructures by a femtosecond excimer laser. *Appl. Opt.*, **38** (11), 2301–2308.

34. Hopp, B., Smausz, T., Barna, N., Vass, Cs., Antal, Zs., Kredics, L., and Chrisey, D. (2005) Time-resolved study of absorbing film assisted laser induced forward transfer of *Trichoderma longibrachiatum* conidia. *J. Phys. D: Appl. Phys.*, **38**, 833–837.

35. Irie, M. (2003) 1.0W high-power small-sized Nd: Yttrium Garnet (YAG) laser optical head for a laser-induced printing system. *Jpn. J. Appl. Phys.*, **42**, 1633–1636.

36. Chrisey, D.B., Piqué, A., Fitz-Gerald, J., McGill, R.C.Y., Wu, H.D., and Duinigan, M. (2000) New approach to laser direct writing active and passive mesoscopic circuit elements. *Appl. Surf. Sci.*, **154–155**, 593–600.

37. Fitz-Gerald, J.M., Rack, P.D., Ringeisen, B., Young, D., Modi, R., Auyeung, R., and Wu, H.-D. (2002) in *Direct Write Technologies for Rapid Prototyping Applications* (eds A. Piqué and D.B. Chrisey), Academic Press, New York, pp. 517–553.

5.3
Laser-Induced Forward Transfer: Transfer of Micro-Nanomaterials on Substrate

Qing Wang, Vahit Sametoglu, and Ying Yin Tsui

5.3.1
Introduction of Laser-Induced Forward Transfer (LIFT)

In the microelectronics industry, lithography is the mature technique for micro-/nanofabrication. However, for many new applications that require modification or repair of existing MEMS (microelectromechanical systems) devices, lithography is probably not the best technique. Laser direct-write (LDW) processes, which can modify, add, and subtract materials, have been developed to be an important complementary technology in the micro-/nanofabrication area. As pointed out in Ref. [1], LDW processes can be categorized into three main classes: laser direct-write subtraction (LDW−), laser direct-write modification (LDWM), and laser direct-write addition (LDW+). LDW−, the most common LDW technique, applies laser to interact with materials, leading to the photochemical, photothermal, or photophysical ablation on the substrate. It can be used in laser cutting, drilling, etching, and laser cleaning [2]. LDWM utilizes a moderate laser energy, which is not high enough to result in ablation, to cause a permanent structural or chemical change in the material. A typical application is the rewritable compact disk, whose phase is transited between crystalline and amorphous material [3] when illuminated by a diode laser. LDW+ uses a laser-induced process to add material onto the substrate; a typical example is the laser chemical vapor deposition (LCVD), utilizing pyrolytic or photolytic decomposition in gases to produce patterns on the substrate [4]. Because of the need for an organometallic precursor, and operation requirement of a vacuum environment, LCVD has limited applications [5]. Laser-induced forward transfer (LIFT) is another LDW+ process, in which a portion of a thin film is transferred in the form of microdots (or microdisks) from a transparent donor substrate to an acceptor substrate near the pulsed lasers. This technique was first demonstrated by Bohandy *et al.* [6] to produce direct writing of 50 μm wide Cu lines on Si and fused silica substrates by using single nanosecond (ns) excimer laser pulses (193 nm) under high vacuum (10^{-6}mbar). It was then demonstrated that this process can be carried out under atmospheric conditions without the requirement of vacuum by the same group in 1988 [7]. Compared with the LCVD technique, LIFT is simpler and does not need a vacuum system and chemical decomposition. It can be potentially used in various areas requiring the addition of a certain material.

Figure 5.3.1 shows the typical LIFT process. A continuous thin film is first deposited onto the substrate, which is transparent to the laser. The film is usually between a few tens of nanometers and up to about a micrometer thick, and this coated substrate is called the *donor substrate*. An acceptor substrate is placed underneath, in close proximity to the film on the donor substrate. A laser pulse (typically, a hundred femtoseconds to tens of nanoseconds) is focused onto the film

Nanomaterials: Processing and Characterization with Lasers, First Edition.
Edited by Subhash Chandra Singh, Haibo Zeng, Chunlei Guo, and Weiping Cai.
© 2012 Wiley-VCH Verlag GmbH & Co. KGaA. Published 2012 by Wiley-VCH Verlag GmbH & Co. KGaA.

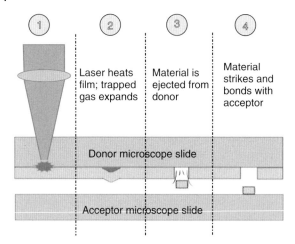

Figure 5.3.1 A schematic diagram showing the LIFT process.

through the donor substrate, causing the formation of vapor at the interface. The force tears out a portion of the film causing it to eject from the donor substrate at high speed. The ejected film travels through the gap between the donor and acceptor substrates. It subsequently strikes and bonds with the acceptor substrate.

The ejected film is slowed down by the air resistance and thus limits how far the film can travel. In one study [8], it was found that LIFT transfer is possible across gaps as wide as around 100 μm. The observation seems to agree with a simple analysis examining how far a blast wave can propagate. A 130 fs, 800 nm laser pulse with several microjoules of energy is focused onto a spot of 2.5 μm radius at e^{-2} intensity to irradiate the donor substrate coated with 80 nm thick Au. After the laser energy is absorbed by the Au film, the plasma formed expands as a blast wave into the ambient air. The blast wave is expected to come to a stop at a distance R_{stop}, which is on the order of $R_L = (E/P)^{1/3}$ [9], where E is the absorbed energy and P is the ambient air pressure. This radius represents the point where the energy of the air displaced by the blast wave is equal to the deposited energy. A previous result with femtosecond pulses [10] indicates that the expansion occurs up to around 0.5 R_L, where the drag of background gas brings the expansion to a rest. In our case with $E \sim 2$ μJ, R_{stop} is estimated to be around 130 μm, which is consistent with the experimental observations. It was reported previously [11] that the best LIFT results can be obtained with the donor and acceptor substrates in proximity to each other.

Metallic materials transferred by using nanosecond pulses (ns-LIFT) have several shortcomings, such as poor film quality due to the energy deposition leading to crack and debris around the transferred spot, and oxidation of metal and delamination of the transferred layers because of the melting and solidification during the transfer process. Typically, these shortcomings can be overcome by using ultrashort picosecond or femtosecond laser pulses (ps-LIFT or fs-LIFT). For a nanosecond pulse, the absorbed laser energy will diffuse during the laser pulse duration and thus the molten region will be over a much larger volume than the laser-absorbed region defined by the laser focal spot and laser penetration depth. On the other hand, for an ultrashort laser pulse, heat diffusion is very small during the pulse duration [12, 13].

Since the LIFT process was first proposed in 1986, a variety of metals have been demonstrated to be transferred by LIFT, including Al [14], Au [8, 15, 16], W [17, 18], Cr [19], Cu [6], and Ag [7, 20]. Other inorganic and organic materials such as In_2O_3 [19], Al_2O_3 [21], high-temperature superconductors (YBCO and BiSrCaCuO) [22], ZnO [23], carbon nanotubes (CNTs) [24], peptides [25], proteins [26], living cells [27, 28], and lambda phage DNA microarrays [29] were reported to be transferred by LIFT. The LIFT process is ideal for the applications in prototyping and custom device fabrication as well as in the modification and repair of existing devices or surfaces whose topography or chemistry makes traditional micropatterning techniques impossible. Using a LIFT system with an XYZ micropositioning system, material could be added to or removed from microdevices programmatically as needed. Various applications have been realized by directly transferring metals and other inorganic and organic materials. Complicated patterns can be printed to substrates by building up LIFT microdots by moving the acceptor substrate with the computer-controlled translation stage. For example, LIFT has been used to print holographic diffraction patterns on glass [19] and to print electrodes for microfluidic devices [8]. Sensors, microbatteries, interconnects, antennae, and solar cells were also fabricated by LIFT [30–33]. By combining the LIFT technique with other LDW techniques such as laser micromachining, even embedded electronic devices and circuits were reported to be fabricated [34, 35]. The LIFT technique can be combined with laser micromachining for the fine-tuning of MEMS. Many electromechanical components on a MEMS device require that their mechanical resonant frequencies be matched to system requirements or to other on-chip frequencies. It has been suggested that laser micromachining be used to tune electromechanical components, increasing its operational frequency by material removal [36]. The LIFT technique, on the other hand, can be a complementary process and can be potentially used to add a small amount of material to the component, thereby decreasing its operation frequency.

5.3.2
Spatial Resolution of the LIFT Process

The spatial resolution of the LIFT process depends on the size of the LIFT spot. Typically, a laser with a Gaussian spatial profile is used in LIFT. The intensity of a Gaussian laser beam profile can be described by

$$I = I_0 \exp\left[-2\frac{R^2}{w^2}\right]$$

where I_0 is the peak intensity, R the radial distance from the laser beam axis, and ω the beam waist radius.

Since laser ablation typically has a sharp energy threshold, the region of the film that is exposed to the laser above the ablation threshold will be removed. For the

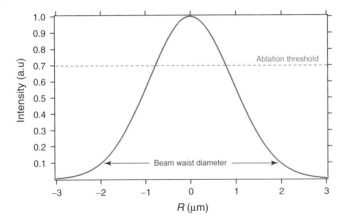

Figure 5.3.2 Intensity profile of a Gaussian laser beam.

region of the film below this threshold, little permanent change occurs in the film. Heat diffusion blurs the energy threshold by allowing the transfer of heat through the film. Thus, ultrashort pulses permit ablation spots smaller than the laser beam waist. If the ablation threshold is close to the peak intensity of the laser spot, as in Figure 5.3.2, the deposition spot may be significantly smaller than the e^{-2} point of the laser profile (the beam waist).

The sizes of the LIFT spots depend on the fluence (as shown in Figure 5.3.3). In the experiment, a donor substrate coated with 80 nm thick Cr is used. The 800 nm, 130 fs laser has a beam waist diameter of 3.7 μm. The diameter of the LIFT spots rapidly increases from ~0.2 to ~2 J cm^{-2}. The increase in the size of the LIFT spots is relatively small for fluences from ~2 to ~20 J cm^{-2}. However, the increase in the size of the LIFT spots is large for fluences from ~0.2 to ~2 J cm^{-2}. The

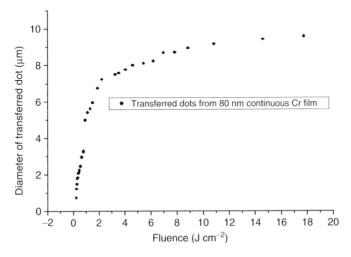

Figure 5.3.3 Diameters of transferred Cr dots versus fluences [37].

diameters of the LIFT spots are ∼1 μm or less, ∼7, and ∼10 μm for laser fluencies of ∼0.2, ∼2, and ∼20 J cm^{-2}, respectively. The smallest Cr LIFT spots are around 700 nm. This happens when the fluence is kept around the transfer threshold at ∼0.2 J cm^{-2}.

The morphology of the LIFT spots is different at different fluences (Figure 5.3.4). Roughly speaking, there are three types of LIFT spot morphology. First, at fluences near the transfer threshold, the LIFT spots consist of a main large piece and several smaller pieces of films (Figure 5.3.4a). Second, at fluences several times above the threshold fluence, the LIFT spots consist of many patches that consist of droplets with sizes on the order of hundreds of nanometers; also, significant amount of debris is observed (Figure 5.3.4b). As the fluence continues to increase, a ringlike transferred pattern is observed and the amount of debris increases further (Figure 5.3.4c). It was suggested that this was due to the center of the pattern being too hot to bond effectively to the acceptor substrate [38]. Similar

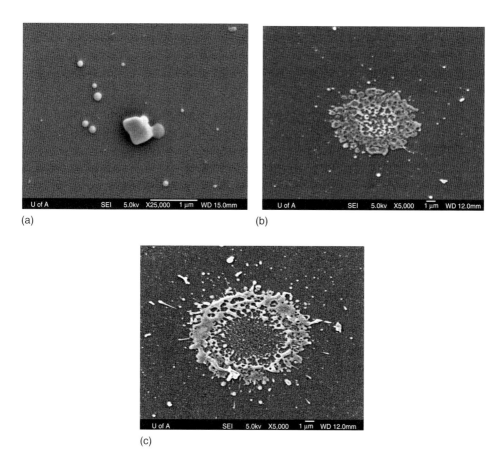

(a)

(b)

(c)

Figure 5.3.4 Morphology of LIFT spots at various laser fluences ((a) $F - 0.25$ J cm^{-2}, (b) $F = 2.2$ J cm^{-2}, and (c) $F = 17.7$ J cm^{-2}) [37].

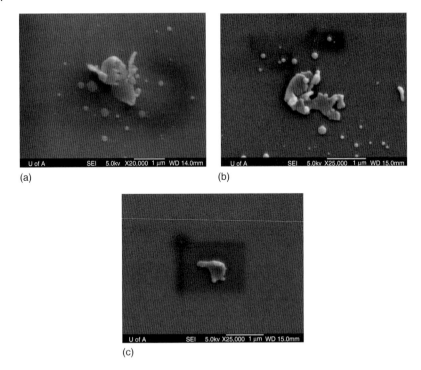

Figure 5.3.5 SEM images of transferred spots from 80 nm continuous Cr film ((a) $F = 0.28$ J cm^{-2}, (b) $F = 0.26$ J cm^{-2}, and (c) $F = 0.24$ J cm^{-2}) [37].

observations for morphology and minimum feature sizes have been reported for fs-LIFT using donor substrates with a 100 nm thick Al [39]. The authors reported that under optimum conditions the smallest feature size achievable with a 60 × (0.65 NA (numerical aperture)) microscope objective is 700 ± 100 nm.

A small fluctuation of the laser fluence could lead to very different results, as shown in Figure 5.3.5. Thus, to transfer submicrometer LIFT spots, the laser fluence must be precisely controlled.

The thickness of the film on the donor substrate is an important parameter. It has been reported that the minimum achievable spatial resolution of transferred Cr material by fs-LIFT is around 330 nm [40]. In that study, the film on the donor substrate is very thin (30 nm) so that the laser is expected to melt through the entire thickness of the film and would cause a molten Cr droplet to deposit on the acceptor substrate. Numerical simulations confirm [37] that the melting depth is ∼40 nm for experimental conditions as in Ref. [40]. Figure 5.3.6 illustrates the two different mechanisms when the donor substrate with thin (thinner than the melting depth) and thick (thicker than the melting depth) films are used. The technique that uses a film on the donor substrate with the thickness of the film less than the melting depth is called *nanodroplet LIFT*. In such a technique, the size of the LIFT droplet is expected to be dictated by the thickness of the film and the material properties

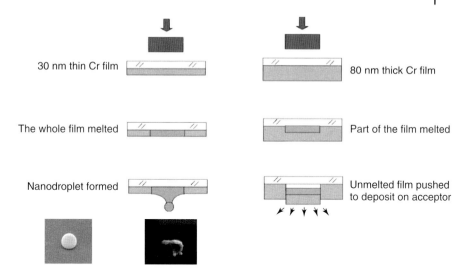

Figure 5.3.6 A schematic diagram of mechanisms for LIFT from (a) a 30 nm thin Cr film and (b) an 80 nm thick Cr film.

(in particular its liquid surface tension). A similar technique (called *microdroplet LIFT*) using a nanosecond laser pulse instead of femtosecond laser pulses has been reported [41] to produce Ni droplets with diameters of ∼3 μm. In the experiment, a laser beam from a Q-switch laser system was focused to a laser spot diameter of ∼25 μm, which is much larger than the 3 μm Ni droplets. Transfer of materials in the molten phase has been demonstrated for Cr [40], Ni [41], Cu [42], $FeSi_2$ [43], and Au [44].

The size of a nanodroplet depends on the molten volume, which is defined by the laser focal spot size and the film thickness. As shown in Figure 5.3.7, the diameter of the Au droplets decreases when smaller focal spots or thinner films are used. A 220 nm Au nanodroplet was obtained using a 10 nm Au film.

Another technique to transfer smaller submicrometer dots is illustrated schematically in Figure 5.3.8. Submicrometer-sized LIFT spots that are much smaller than the focal spot diameter are fabricated on the donor substrate by using photolithography or e-beam lithography. The laser pulses are then focused on the top of these prepatterned spots and transferred onto the acceptor substrate. Since these small prepatterned spots are isolated from each other, they will not be affected by the surrounding material during the transfer process. The size of the transferred spot is thus expected to be similar to the one on the donor substrate.

It is found that the LIFT spots transferred with laser fluences near the threshold are mostly debris-free, remain intact in one piece, and keep the same profile as the prepatterned spots on the donor substrate. A slight variation of fluence does not result in any significant change. Figure 5.3.9 shows an SEM image of a LIFT spot, which is transferred from a donor substrate consisting of an array of 80 nm thick and 550 nm diameter Cr spots prepatterned using the e-beam lithography.

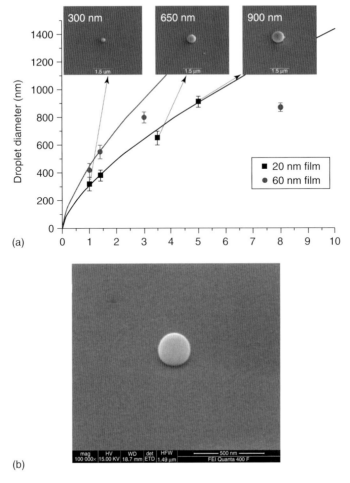

(a)

(b)

Figure 5.3.7 (a) Size dependence of the transferred droplets on the focusing conditions and Au film thickness; (b) a 220 nm Au droplet transferred using a 10 nm thick film. (Reproduced with permission from Kuznetsov *et al.* [44].)

The energy fluence used is 0.29 J cm^{-2}, and a large laser spot with a beam waist diameter of ~12 μm produced is used in the experiment by defocusing.

When a small laser spot with a beam waist diameter of 3.7 μm is used, the fraction of successful transfer is low. This is probably because the laser beam misses or partly misses the nanodot on target substrate due to laser pointing errors.

5.3.3
Transfer of Thermally and Mechanically Sensitive Materials

An extreme condition of high temperature and high pressure exists during the LIFT process. Many delicate materials such as functional materials and biological

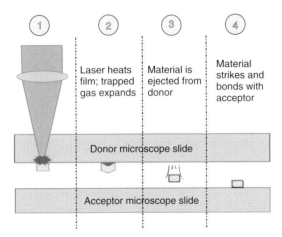

Figure 5.3.8 LIFT process with the prepatterned donor substrate.

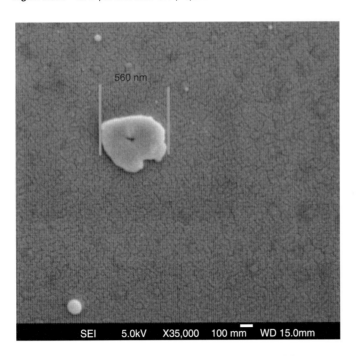

Figure 5.3.9 Transferred Cr nanodots from the prepatterned donor (50 μm off focus) with $F = 0.29\ \mathrm{J\,cm^{-2}}$ [37].

cells may not survive such a condition. Thus, ways of mitigating the situation are required. One way is to use more sophisticatedly designed LIFT donor substrates. Employing multilayered films for LIFT was first applied by Tolbert, originally called *laser ablation transfer* [46]. In this method, a thin intermediate layer consisting of laser absorption materials is first deposited on the transparent donor substrate. The target material that is to be transferred is then deposited on top of this laser absorption material. During the LIFT process, the incident laser pulse interacts with the absorption layer, resulting in vaporization, which will then force the target material that is in contact with the absorption layer to be removed and transferred toward the acceptor substrate. This approach will reduce the damage of the target material, since laser energy mainly reacts with the intermediate layer and make the transfer of materials with weak absorption of the laser radiation possible. Using this approach, laser dye [47], phosphor powders (Y_2O_3:Eu and Zn_2SiO_4:Mn) [45], polymers [48], organic conducting polymer [49], biomolecule microarrays [50–52], peptides [25], proteins [26], and living cells [27] were successfully transferred with application of metals (Au, Cr, or Ti) as the absorption layer (several tens of nanometers thick). Organic light-emitting pixels were successfully transferred by LIFT using 30–40 nm Ag nanoparticles as an absorption material on the donor substrate. Polymers [28, 53–55], instead of metal materials, were also used as absorption materials on donor substrates. Stem cells were successfully transferred using commercial polymer as an absorption material [28]. Cells [53], organic light-emitting diode (OLED) pixels [54], and semiconductor nanocrystal quantum dots (NQDs) [55] were successfully transferred using LIFT with photosensitive triazene polymer (TP) as a sacrificial layer. For the transfer of OLED pixels and semiconductor NQDs, the donor substrate used consists of a layer of TP, a layer of metal, and a layer of LIFT material (OLED pixels or CdSe NQDs). The TP sacrificial layer is vaporized by the UV laser pulse and the rapidly expanding organic vapor pushes the metal-LIFT material bilayer toward the acceptor substrate. The metal layer serves as an electrode, and it also prevents the exposure of the OLED pixels or CdSe NQDs to the UV radiation (Figure 5.3.10).

Pique's group from the Naval Research Laboratory (NRL) has developed the MAPLE DW (matrix-assisted pulsed laser evaporation direct-write) technique,

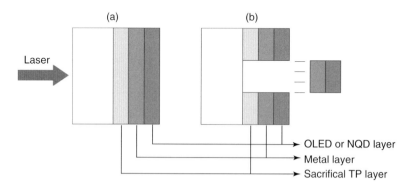

Figure 5.3.10 A schematic diagram for LIFT of OLED pixels or NQDs. (a) The laser pulse propagates from left to right. The donor substrate consisting of three layers: the first layer from the left is the sacrificial TP layer, the second layer is the metal layer, and the third layer is the OLED or NQD material. (b) The zone affected by the laser will be transferred forward without the sacrificial layer.

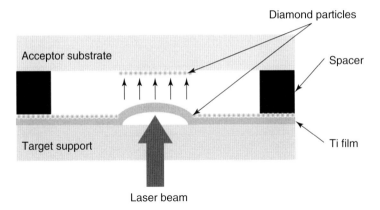

Figure 5.3.11 A schematic diagram for BB-LIFT of diamond nanopowder.

which combined the LIFT and matrix-assisted pulsed laser evaporation (MAPLE) [56]. This approach employs the mixture of soluble materials to be deposited and a solvent phase that is usually precooled to low temperature as the target material on the donor substrate. When the target is irradiated with laser pulse, the solvent is rapidly vaporized and pumped away, propelling the solute toward the acceptor substrate to form a uniform thin film with minimal decomposition. This approach has been applied to transfer metals, ceramics, and polymeric materials such as Ag, Au, BaTiO$_3$, and BTO [57, 58]. Various types of physical and chemical sensor devices, microbatteries and biosensors have been also prototyped by this approach [30]. More recently, an approach named laser decal transfer was proposed by the same NRL research group [5, 59]. This approach uses viscous nanoparticle suspensions (Ag in their case) as inks for the target material. It was found that the viscosity of the suspension plays an important role in the ability to perform the decal transfers. When the viscosity was low (<100 cP), the transferred material showed a high degree of spattering with significant surrounding debris. However, when high viscosity (≈100 000 cP) was applied, the breakup of the transferred material was prevented. Patterns with sharp shape but little debris, while maintaining excellent conductivity and adhesion, were achieved on the acceptor substrate [5].

Another technique called blister-based laser-induced forward transfer (BB-LIFT) was used to transfer diamond nanopowder [60]. In this study, the diamond nanopowder is spread over the metal surface on a Ti-coated donor substrate and a 50 ps visible laser pulse is used to heat the metal film on the donor substrate (Figure 5.3.11). The laser-heated metal film works like a flexible membrane; as it moves outward, it pushes the nanopowder toward the donor substrate.

Not all functional materials require complex donor substrates. CNT field emission cathodes have been successfully transferred by ns-LIFT using a simple CNT-coated glass substrate [24]. The field emission properties of the transferred CNT are not degraded.

References

1. Arnold, C.B. and Piqué, A. (2007) Laser direct-write processing. *MRS Bull.*, **32**, 32.
2. Stratoudaki, K.S. *et al.* (2000) Comparative study of different wavelengths from IR to UV applied to clean sandstone. *Appl. Surf. Sci.*, **157**, 1.
3. Feinleib, J. *et al.* (1971) Properties of (Se,S)-based chalcogenide glass films, and an application to a holographic supermicrofiche. *Appl. Phys. Lett.*, **18**, 254.
4. Tsui, Y.Y. (2001) Thin film deposition, in *LIA Handbook of Laser Materials Processing* (ed. J.F. Ready), Magnolia Publishing Inc., p. 691 (ISBN: 0-941463-02-8).
5. Piqué, A. *et al.* (2008) Digital microfabrication by laser decal transfer. *J. Laser Micro/Nanoeng.*, **3**, 163.
6. Bohandy, J. *et al.* (1986) Metal deposition from a supported metal film using an excimer laser. *J. Appl. Phys.*, **60**, 1538.
7. Bohandy, J. *et al.* (1988) Metal deposition at 532 nm using a laser transfer technique. *J. Appl. Phys.*, **63**, 1158.
8. Germain, C. *et al.* (2007) Electrodes for microfluidic devices produced by laser induced forward transfer. *Appl. Surf. Sci.*, **253**, 8328.
9. Taylor, G. (1950) The formation of a blast wave by a very intense explosion. I. Theoretical discussion. *Proc. R. Soc. London, Ser. A*, **201**, 159.
10. Yalcin, S. *et al.* (2005) Images of femtosecond laser plasma plume expansion into background air. *IEEE Trans. Plasma Sci.*, **33**, 482.
11. Kantor, Z., Toth, Z., and Szorenyi, T. (1992) Laser induced forward transfer: the effect of support-film interface and film-to-substrate distance on transfer. *Appl. Phys. A*, **54**, 170.
12. von der Linde, D. *et al.* (1997) Laser-solid interaction in the femtosecond time regime. *Appl. Surf. Sci.*, **109-110**, 1.
13. Gamaly, E.G. *et al.* (1999) Ultrafast ablation with high-pulse-rate laser. Part I: theoretical consideration. *J. Appl. Phys.*, **85**, 4213.
14. Alexander, D.E. *et al.* (1988) Laser-driven micro-explosive bonding of aluminum films to copper and silicon. *J. Mater. Sci.*, **23**, 2181.
15. Baseman, R.J. *et al.* (1990) Minimum fluence for laser blow-off of thin gold films at 248 and 532 nm. *Appl. Phys. Lett.*, **56**, 1412.
16. Latsch, S. *et al.* (1994) Interface study on laser-induced material transfer from polymer and quartz surfaces. *Appl. Surf. Sci.*, **81**, 183.
17. Toth, Z. *et al.* (1993) Ar laser induced forward transfer (LIFT): a novel method for micrometer size surface patterning. *Appl. Surf. Sci.*, **69**, 317.
18. Kantor, Z. *et al.* (1995) Metal pattern deposition by laser-induced forward transfer. *Appl. Surf. Sci.*, **86**, 196.
19. Zergioti, I. *et al.* (1998) Microdeposition of metal and oxide structures using ultrashort laser pulses. *Appl. Phys. A*, **66**, 579.
20. Adrian, F.J. *et al.* (1987) A study of the mechanism of metal deposition by laser induced forward transfer process. *J. Vac. Sci. Technol., B*, **5**, 1490.
21. Greer, J.A. and Parker, T.E. (1988) Laser-induced forward transfer of metal oxides to trim the frequency of surface acoustic wave resonator devices. *Proc. SPIE*, **998**, 113.
22. Fogarassy, E. *et al.* (1989) Laser-induced forward transfer of high-Tc YBaCuO and BiSrCaCuO superconducting thin films. *J. Appl. Phys.*, **66**, 457.
23. Kini, A. *et al.* (2007) ZnO nanorod micropatterning via laser-induced forward transfer. *Appl. Phys. A*, **87**, 17.
24. Chang-Jian, S.K. *et al.* (2006) Fabrication of carbon nanotube field emission cathodes in patterns by a laser transfer method. *Nanotechnology*, **17**, 1184.
25. Dinca, V. *et al.* (2007) Development of peptide-based patterns by laser transfer. *Appl. Surf. Sci.*, **254**, 1160.
26. Barron, J.A. *et al.* (2005) Printing of protein microarrays via a capillary-free fluid jetting mechanism. *Proteomics*, **5**, 4138.

27. Barron, J.A. *et al.* (2004) Biological laser printing of genetically modified Escherichia coli for biosensor applications. *Biosens. Bioelectron.*, **20**, 246.

28. Doraiswamy, A. *et al.* (2006) Excimer laser forward transfer of mammalian cells using a novel triazene absorbing layer. *Appl. Suf. Sci.*, **252**, 4743.

29. Karaiskou, A. *et al.* (2003) Microfabrication of biomaterials by the sub-ps laser-induced forward transfer process. *Appl. Surf. Sci.*, **208-209**, 245.

30. Piqué, A. *et al.* (2003) Laser direct-write of miniature sensor and microbattery systems. *RIKEN Rev.*, **50**, 57.

31. Arnold, C.B. *et al.* (2004) Direct-write laser processing creates tiny electrochemical systems. *Laser Focus World*, **5**, S9.

32. Auyeung, R.C.Y. *et al.* (2004) Laser fabrication of GPS conformal antennas. *SPIE Proc.*, **5339**, 292.

33. Kim, H. *et al.* (2004) Dye-sensitized solar cells using laser processing techniques. *SPIE Proc.*, **5339**, 348.

34. Piqué, A. *et al.* (2005) Laser direct-write of embedded electronic components and circuits. *SPIE Proc.*, **5713**, 223.

35. Piqué, A. *et al.* (2006) Embedding electronic circuits by laser direct-write. *J. Microelectron. Eng.*, **83**, 2527.

36. Kirkwood, S.E. *et al.* (2007) Nanomilling surfaces using near-threshold femtosecond laser pulses. *J. Phys. Conf. Ser.*, **59**, 591.

37. Wang, Q. (2009) Lateral resolution in laser induced forward transfer, MSc Thesis. University of Alberta.

38. Papakonstantinou, P. *et al.* (1999) Microfabrication by UV femtosecond laser ablation of Pt, Cr andindium oxide thin films. *Appl. Surf. Sci.*, **151**, 159.

39. Bera, S. *et al.* (2007) Optimization study of the femtosecond laser-induced forward-transfer process with thin aluminum films. *Appl. Opt.*, **46**, 4650.

40. Banks, D.P. *et al.* (2006) Nanodroplets deposited in microarrays by femtosecond Ti:sapphire laser-induced forward transfer. *Appl. Phys. Lett.*, **89**, 192107.

41. Willis, D.A. and Grosu, V. (2005) Microdroplet deposition by laser-induced forward transfer. *Appl. Phys. Lett.*, **86**, 244103.

42. Yang, L. *et al.* (2006) Microdroplet deposition of copper film by femtosecond laser-induced forward transfer. *Appl. Phys. Lett.*, **89**, 16110.

43. Narazaki, A. *et al.* (2009) Nano- and microdot array formation by laser-induced dot transfer. *Appl. Surf. Sci.*, **255**, 9703.

44. Kuznetsov, A.I. *et al.* (2009) Laser-induced backward transfer of gold nanodroplets. *Opt. Express*, **17**, 18820.

45. Fitz-Gerald, J.M. *et al.* (2000) Laser direct writing of phosphor screens for high-definition displays. *Appl. Phys. Lett.*, **76**, 1386.

46. Tolbert, W.A. *et al.* (1993) High-speed color imaging by laser ablation transfer with a dynamic release layer: fundamental mechanisms. *J. Imaging Sci. Technol.*, **37**, 411.

47. Nakate, Y. *et al.* (2002) Transfer of laser dye by laser-induced forward transfer. *Jpn. J. Appl. Phys.*, **41**, L839.

48. Boutopoules, C. *et al.* (2008) Liquid phase direct laser printing of polymers for chemical sensing applications. *Appl. Phys. Lett.*, **93**, 191109.

49. Rapp, L. *et al.* (2009) Characterization of organic material micro-structures transferred by laser in nanosecond and picoseconds regimes. *Appl. Surf. Sci.*, **255**, 5439.

50. Fernández-Pradas, J.M. *et al.* (2004) Laser-induced forward transfer of biomolecules. *Thin Solid Films*, **453-454**, 27.

51. Serra, P. *et al.* (2004) Laser direct writing of biomolecule microarrays. *Appl. Phys. A*, **79**, 949.

52. Duocastella, M. *et al.* (2008) Laser-induced forward transfer of liquids for miniaturized biosensors preparation. *J. Laser Micro/Nanoeng.*, **3**, 1.

53. Kattamis, N.T. *et al.* (2007) Thick film laser induced forward transfer for deposition of thermally and mechanically sensitive materials. *Appl. Phys. Lett.*, **91**, 171120.

54. Fadel, R. *et al.* (2007) Fabrication of original light-emitting diode pixels by laser-assisted forward transfer. *Appl. Phys. Lett.*, **91**, 061103.

55. Xu, J. *et al.* (2007) Laser-assisted forward transfer of mult-spectral nanocrystal

quantum dot emitters. *Nanotechnology*, **18**, 025403.

56. Chrisey, D.B. *et al.* (2000) New approach to laser direct writing active and passive mesoscopic circuit elements. *Appl. Surf. Sci.*, **154-155**, 593.

57. Chrisey, D.B. *et al.* (2000) Direct writing of conformal mesoscopic electronic devices by MAPLE DW. *Appl. Surf. Sci.*, **168**, 345.

58. Piqué, A. *et al.* (2000) Direct writing of electronic and sensor materials using a laser transfer technique. *J. Mater. Res.*, **15**, 1872.

59. Piqué, A. *et al.* (2008) Laser decal transfer of electronic materials with thin film characteristics. *SPIE Proc.*, **6879**, 687911–687911.

60. Kononenko, T.V. *et al.* (2009) Laser transfer of diamond nanopowder induced by metal film blistering. *Appl. Phys. A*, **94**, 531.

5.4
Laser-Induced Forward Transfer for the Fabrication of Devices

Matthias Nagel and Thomas Lippert

5.4.1
Introduction

In conjunction with the increasing availability of cost-efficient laser units during the recent years, laser-based micromachining techniques have been developed as an indispensable industrial instrument of "tool-free" high-precision manufacturing techniques for the production of miniaturized devices made of nearly every type of materials. Laser cutting and drilling, as well as surface etching, have grown meanwhile to mature standard methods in laser micromachining applications where a well-defined laser beam is used to remove material by laser ablation. As an accurately triggerable nonmechanical tool, the ablating laser beam directly allows a subtractive direct-write engraving of precise microscopic structure patterns on surfaces, such as microchannels, grooves, and well arrays, as well as for security features. Therefore, laser direct-write (LDW) techniques imply originally a controlled material ablation to create a patterned surface with spatially resolved three-dimensional structures, and gained importance as an alternative to complementary photolithographic wet-etch processes. However, with more extended setups, LDW techniques can also be utilized to deposit laterally resolved micropatterns on surfaces, which allows, in a general sense, for the laser-assisted "printing" of materials [1, 2].

As outlined in Figure 5.4.1, the basic setup for an additive direct-write deposition of materials is the laser-induced forward transfer (LIFT), where the laser photons are used as the triggering driving force to catapult a small volume of material from a source film toward an acceptor substrate. In contrast to the classic laser-ablation micropatterning setup where material is removed from the top surface (Figure 5.4.1a), for LIFT applications the laser interacts from the inverse side of a source film, which is typically coated onto a nonabsorbing carrier substrate. The incident laser beam propagates through the transparent carrier before the photons are absorbed by the back surface of the film. Above a specific threshold of the incoming laser energy, material is ejected from the target source and catapulted toward a receiving surface that is placed either in close proximity to or even in contact with the donor film. With adequate tuning of the applied laser energy the thrust for the forward propulsion is generated within the irradiated film volume. The absorbed laser photons cause a partial ablation of the film material, which induces a sharply triggered volume expansion coupled with a pressure jump that catapults the overlying solid material away. However, the energy conversion processes as well as the phase transitions involved in the LIFT process are complex and affected by a large number of diverse parameters. Therefore, the highly dynamic interactions between the laser and the transferred material are not easy to describe in fundamental models. Apart from laser parameters such as emission wavelength,

Nanomaterials: Processing and Characterization with Lasers, First Edition.
Edited by Subhash Chandra Singh, Haibo Zeng, Chunlei Guo, and Weiping Cai.
© 2012 Wiley-VCH Verlag GmbH & Co. KGaA. Published 2012 by Wiley-VCH Verlag GmbH & Co. KGaA.

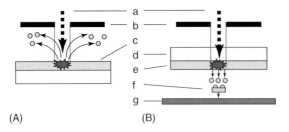

(A)   (B)

Figure 5.4.1 Conventional laser ablation (A) and laser-induced forward transfer (LIFT) setup (B), with (a) incoming laser pulse, (b) projection optics and mask, (c) surface layer of material to be ablated, (d) laser-transparent carrier substrate, (e) donor layer for transfer, (f) catapulted flyer to be deposited, and (g) receiver substrate.

pulse duration, pulse shape, focal spot size, and fluence, the properties of the transfer material system like thickness and composition, optical absorptivity, heat conductivity, thermal diffusion, viscosity, phase change behavior, and so on play an important role too. Further parameters such as the geometry of the transfer setup including the distance between the donor and receiver substrates, the properties of the carrier materials, and their surface morphology have also to be taken into consideration. For successful LIFT applications, this wide parameter space has to be specifically adapted and optimized for individual material systems to reach a smooth and controlled deposition with high resolution and conservation of integrity and functionality also of sensitive materials.

The fabrication of device structures requires the formation of complex high-resolution patterns in three dimensions with a reasonable speed. At present, commercially available, computer-controlled translation stages as well as galvanometric scanning modules enable a precise and rapid motion of either the substrates or the laser beam, and therefore the tailored direct-write deposition of small volumes of individual materials. During the past years, LIFT-based techniques have gained fundamental academic interest as complex photodynamic systems. A broad variety of device applications and demonstrators have been described in the literature, involving inorganic, organic, and macromolecular materials, and even biological systems. Since the LIFT process allows deposition of complex and multicomponent materials in one transfer step directly from a separately prepared donor substrate and without further wet-developing processes, this technique appears preferentially suitable for the fabrication of microdevices where sensitive and functional materials have to be arranged in a controlled manner without degrading their desirable properties. The following sections focus on the LIFT process for fabrication of various devices, ranging from simple metal thin-film patterns to demanding biological applications where living cells and enzymes are deposited for biosensor applications. The great versatility of the LIFT direct-write process allows the transfer of a wide range of materials and opens up new possibilities compared to conventional solvent-based deposition techniques such as screen or inkjet printing.

For the fabrication of microstructured surfaces, high-resolution photolithography techniques are far developed and well established [3], but this patterning method requires a sequence of numerous chemical processing steps, normally based on a wet-etching development of a photoresist layer irradiated by a projection mask. However, the processing conditions require chemicals and solvents that are not always compatible with sensitive materials. Therefore, further alternative nonlithographic techniques for the deposition of patterned materials on surfaces have been brought into action for the fabrication of devices, most of them based on contact or jet printing procedures.

Microcontact printing (µCP), also called *soft lithography*, uses a relief pattern on a prestructured stamp of elastomeric silicone polymers for the transfer of patterned structures of soft materials through conformal direct contact [4]. The material to be transferred has to be formulated as an appropriate ink in which the stamp is immersed before the micropattern on the receiver surface is formed by stamping [5]. This allows a precise repetitive pattern transfer for a wide range of materials that are not compatible with lithographic resist processes, for example, biomaterials and polymer precursors. Further benefits of µCP processes are comparably low fabrication and material costs, a high throughput and accuracy by automation, as well as the possibility of pattern deposition even on curved surfaces. However, µCP is a wet deposition technique, since the materials to be stamped need to be dissolved or dispersed in a suitable rheological system [6].

Screen printing (serigraphy) is a well-established method for production of graphical artwork, and has been adapted as a further wet process to deposit material patterns directly by a stencil mask technique. A fine-woven mesh stretched on a frame supports a layer of the predesigned stencil (the screen), where the stencil openings determine the 1 : 1 image that will thus be imprinted by the passing of a highly viscous ink or any other paste medium through the open areas of the ink-permeable layer onto a substrate. The ink paste is deposited and patterned in one step, but the image resolution is rather limited, with single feature sizes in the submillimeter range (down to about 100 µm). Owing to the possibility of preparing thick layers from pastelike materials, screen printing has found applications for the fabrication of electric and electronic devices on an industrial scale [7]. It is used for printing semiconducting materials and metal pastes, for example, for conducting lines on wafer-based solar photovoltaic (PV) cells, for circuit boards, and further printed electronic devices where a relatively high layer thickness, but not a high resolution, is important [8]. Nevertheless, dielectric or passivating layers for organic semiconductor devices, and even complete organic field-effect transistors (OFETs) can be screen printed, by using families of electrically functional electronic or optical inks to create active or passive devices [9]. However, the process is not very flexible since precisely structured printing masks have to be individually fabricated.

Compared to the screen printing technology, slightly higher resolutions can be achieved by inkjet printing, where small droplets of much less viscous functional inks are catapulted onto the receiver surface. Inkjet printing enables a noncontact material transfer and is a digitally data-driven direct-write deposition method that can be set up with relatively low effort (no masks or screens are needed) and also at laboratory scale. Inkjet technology can be applied to deposit materials on substrates, but the material to be jetted must be compatible with the printhead used and must be formulated with a viscosity within a specific range [10]. Normally, the image pattern is generated by the computer-addressed two-dimensional translation of the printer head in coordination with the trigger of the droplet ejection through the nozzle orifices. Droplet deposition is well suited for low-viscosity, soluble materials like organic semiconductors, but with high-viscosity materials like organic dielectrics and dispersed particles, such as inorganic metal inks, difficulties due to clogging of the nozzles repeatedly occur. Owing to the drop-wise deposition of layers, their homogeneity is limited, and the resolution (about 50 μm) depends on the minimum droplet volume that can be deposited. The throughput is determined by the available droplet dispense rate, which typically does not exceed 25 kHz. Simultaneous usage of many nozzles as well as prestructuring of the substrate can improve the productivity and resolution, respectively. Inkjet printing is preferably used for the fabrication of organic semiconductors in organic light-emitting diodes (OLEDs) [11], but OFETs have also been completely prepared using this method [12–14]. Furthermore, biological microarrays of nucleic acids and proteins, as well as other sensing and detector devices, can be prepared by means of inkjet printing. A layer-by-layer deposition of polymer precursors allows for rapid 3D prototyping determined by digital CAD files. Three-dimensional objects and freestanding structures in the shape of each cross section can be created by successive deposition of material layers, for example, by jetting liquid photopolymers that are immediately cured after deposition by UV light. As an alternative, layers of a fine powder (plaster, corn starch, or resins) are selectively bonded by codepositing a solidifying adhesive from the inkjet nozzle head.

As a common issue, all mentioned microprinting methods require viscous inks or rheological suspension systems, and can therefore be classified as wet or solvent-based deposition procedures. At least a drying step for the solvent removal is necessary to fix the printed structure on top of the receiver surface, if not more complex annealing or even curing processes by chemical postdeposition modifications. In contrast to conventional ink-dye-based printing techniques of graphical art images, the need for a versatile deposition technology for free-forming materials [15, 16] and for multilayer devices raises a number of materials problems. Higher resolutions will be needed if organic transistor architectures are to be printed. Also, it must be possible to print bubble- and pinhole-free layers to avoid shorting of devices. Further, printing on dense rather than porous substrates requires that the ink interacts constructively with the substrate for optimum adhesion. For stacked device architectures, multiple layers must be printed such that they form discrete unmixed layers with clearly defined interfaces. Since multilayered stacks are the most favored design in device fabrication, either a strict orthogonality in the solubility of different materials or a subsequent selective cross-linking of previously solution-deposited layers is required. This solvent issue is one of the main reasons why applications of the so-called direct-write transfer methods, where solid layers are transferred from a carrier to the substrate, are of practical interest.

Laser-based methods as a preparative tool for thin-film deposition, such as *pulsed laser deposition* (PLD) [17–22] and *matrix-assisted pulsed laser evaporation* (MAPLE) [18, 23] have been developed to grow thin films of inorganic and organic materials. Both methods are based on laser ablation of a target material under vacuum conditions. On the laser interaction, inorganic materials get vaporized to form a plasma, which transports the components via the gas phase toward a receiver substrate where the redeposition process takes place. Even when such laser-based deposition methods enable the controlled growth of homogeneous thin films, they have the disadvantage that only complete layers with no lateral resolution are formed. In addition, these vapor-phase deposition processes are restricted in that they are applicable only to polymers with inherently high molecular masses, and only a few polymers have been deposited successfully by PLD with UV lasers [24–26]. This is not really surprising since UV photons will induce reactions so that such compounds tend to degrade or decompose if directly exposed to UV laser irradiation. Only polymers that depolymerize, that is, form the monomer on UV laser irradiation by a defined photochemical cleavage mechanism, may be used for this approach [27]. The deposited monomers then react on the substrate to form a new polymer film [28]. Therefore, the deposited polymer films will most probably have a different molecular weight and weight distribution than the starting material, and may also contain decomposition products. Attempts to deposit films of the biopolymer lysozyme by PLD from a solid pressed target revealed that the target crumbled up and broke into small pieces during direct mulitpulse laser exposition [29]. One possible approach to deposit thin polymer films by PLD is the application of mid-infrared (IR) radiation, which is tuned to certain absorption bands of the polymer (resonant infrared pulsed laser deposition, RIR-PLD) [30–32]. A modified approach to PLD that is more gentle is MAPLE [18, 23], which uses, in principle, the same setup as PLD, with the main difference that the target consists of a frozen solution of the polymer. The laser is then used to vaporize the solvent that is removed by the vacuum pumps while the polymer chains are propelled and deposited on the substrate. However, this approach works only for polymers that can be dissolved or dispersed in an appropriate solvent, and it should also work best when the laser is preferentially absorbed by the solvent matrix. The formation of high-quality thin films is possible, although problems with the homogeneity of the films and trapped solvents exist. Thin-film deposition of the conducting polymer PEDOT:PSS (poly(3,4-ethylenedioxythiophene)-polystyrene sulfonate) was studied with a combination of *resonant infrared laser vapor deposition* (RIR-LVD) and the MAPLE approach. The PEDOT:PSS was frozen in various matrix solutions and deposited using a tunable, mid-IR free electron laser. When adequately absorbing solvents (e.g., isopropanol and *N*-methyl pyrrolidinon) were added to the aqueous matrix, and the laser wavelength was properly adjusted to selectively excite the specific co-matrix, deposited films were smooth and exhibited electrical conductivity [33].

In the context of device fabrication, thermal evaporation processes also play an important role. Metals as well as sufficiently volatile and thermally stable organic compounds can be deposited from the gas phase within high-vacuum chambers in the form of thin-layered two-dimensional patterns by using appropriate shadow masks. However, such vapor deposition techniques via masks suffer from the disadvantage that normally most of the evaporated material recondenses on the shadow mask area itself, and only a small part is deposited by passing the stencil apertures on the receiver surface. Therefore, a large part of the evaporated material is unproductive or even lost if not recovered. Further, small openings in the mask (as, e.g., needed for the fabrication of arrays of small pixels for display screens) tend to become overgrown by the evaporated material, and have therefore to be cleaned frequently. Printing larger areas with such delicate masks proved also to be rather difficult because of shape distortions and their weak mechanical stability. In addition, the method is essentially limited for low volatile compounds such as ionic compounds, ceramics, and polymers.

For device microfabrication, many and diverse material classes have to be integrated with a high local accuracy. In order to circumvent the above-summarized problems concerning the solvent- and vapor-based microdeposition of materials, novel approaches based on LIFT processes that enable a solvent-free (or "dry") direct-write deposition of solid thin films, necessary for the fabrication of devices with sensitive functional materials, have been developed. Since LIFT-based direct-write techniques are versatile with respect to the materials that can be microdeposited, such methods were developed and adapted for a variety of different device-relevant fabrication steps. For example, the deposition of micron-sized metal patterns by LIFT allows not only the direct-write of pixel masks but also the application of electrical conducting paths for electronic circuits that can simultaneously serve as metal cathodes or anodes. Mainly, inorganic compounds are used for the LIFT-based fabrication of electrochemical microcells, which are designed as small-scale energy storage and power delivery devices, such as microbatteries, microsupercapacitors, or even PV power-generating systems for remote sensors [2]. Further LIFT-fabricated microdevices include electronic on-chip circuits where semiconductor components, resistors, capacitors, and thin-film transistors (TFTs) were transferred together with metal-inks for electrical contacts. A promising extension for LIFT applications is the fabrication of organic optoelectronic devices such as color screen displays based on electroluminescent materials, or "plastic electronic" devices where components such as TFTs are built up with semiconducting organic and polymeric materials. Finally, a recent field of LIFT applications has been established for the fabrication of chemosensors and biosensors where biomaterials such as nucleic acids, proteins, enzymes, cells, and microorganisms have to be accurately deposited for detecting microdevice systems. Envisioned further applications include energy harvesting with thermoelectric and piezoelectric materials and radiofrequency identification (RFID) tag systems with added functionalities.

The following sections present a compilation of examples of LIFT techniques for the microdeposition of various material systems and with regard to the potential for microdevice fabrication applications.

5.4.2
LIFT Techniques for Direct-Write Applications

5.4.2.1 Traditional LIFT

As outlined in Figure 5.4.1, the donor system for basic LIFT setups consists usually of a single source material layer that has been separately precoated as a thin plane film on top of a laser-transparent carrier substrate, typically fused silica supports or polymer films, depending on the laser wavelength. A small area of the transfer material corresponding to the beam spot dimensions of a projection-mask-shaped laser is directly exposed to the incoming laser radiation, mostly applied as individual single pulses. Patterns can be generated by scanning and modulating the laser or moving the donor–acceptor stack on a computer-controlled translation stage. For an optimum laser–material interaction, the source material layer has to provide a high absorption coefficient at the laser emission wavelength, and has to be heat resistant, as well as insensitive to phase transitions and thermomechanical stress.

As far as documented in the literature, already in the early 1970s, the first demonstration of the LIFT principle was reported [34] for laser-writing image recording with common polyethylene-backed typewriter ribbons as the donor, and paper or polyester (PET) foils as the accepting substrates (Figure 5.4.2A) [35]. Compared to conventional typewriter machines or dot matrix printers, the laser-triggered scanned material transfer recording technique features an inherently noncontact, nonimpact printing system with instant viewability, and requires furthermore no process chemicals. Graphical line patterns with widths from less than 30 µm to more than 130 µm have been recorded with a 1064 nm Nd:YAG laser, depending on the focus spot size and the air gap between donor and accepting substrate, and with energy densities in the range of 1–3 J cm^{-2}. The experiments were extended further to PET-foil-supported dyes and ink materials, proving that the laser-assisted material transfer recording is not restricted to a black carbon ribbon system. Systematic studies were performed with regard to the influence of the laser

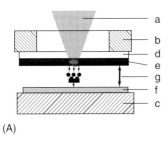

(A)

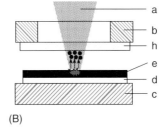

(B)

Figure 5.4.2 Two material transfer recording setups as analogously published in Ref. [35]. (A) represents the basic LIFT setup, whereas (B) shows a reversed transfer setup for a "backscattering" process that corresponds to a vertical pulsed laser deposition (VPLD) arrangement. The components are (a) focused and outline-shaped laser pulse, (b) front retainer, (c) rear retainer, (d) transparent polyethylene typewriter ribbon carrier film, (e) black carbon coating, (f) accepting surface of the recording medium (g) spacer gap, and (h) top receiver material transparent to laser beam.

fluence and the spacer gap distance on the quality of deposited patterns. The best resolution with smooth contours of the printed patterns was found when the donor ribbon and the accepting sheet were in direct contact.

A mechanistic model for the material transfer process involves light-to-heat conversion (LTHC) within the opaque film of dye pigments dispersed in a translucent binder matrix. Depending on the concentration of the absorbing particles, the incoming IR laser radiation propagates into the coating layer volume where it is absorbed within a certain penetration depth. The absorbed laser energy causes an ultrafast strong heating of the irradiated film volume close to the support interface, which immediately induces vapor generation by phase transitions together with shock and thermal stress waves. The vapor emission leads to an attendant buildup in pressure and induces a volume expansion together with a thermal strain. If the opaque layer is thicker than the optical penetration depth, the vaporization-built pressure will first blast a small cavity in the material layer. Finally, it breaks open the surface and delivers the momentum that catapults the spot volume toward the receiver surface. At this point, a hole in the opaque dye coating is formed, whereas the ejected material is accelerated along trajectories that depend on the pressure gradient and momentum of the expanding gas. Laser ablation itself is always accompanied by shock waves generated by the intense, nearly instantaneous thermal expansion of a point area at the surface of the illuminated material. With an air gap of about 100 μm the deposited structures showed less sharp contours accompanied by some randomly distributed particulate debris, which stems from the mechanical burst-open process of the donor film surface when the detached coating shell spot (the catapulted "flyer"; see component (f) in Figure 5.4.1) is forward propelled. In an air gap, the propagation and even reflection of shockwaves at the receiver surface may also interfere with the ejected material, as is discussed in a later section.

Control experiments were performed with a reverse transfer setup (Figure 5.4.2B) [35] to test the differences between the two kinds of material deposition by ablation transfer. For the catapulting in the forward direction (Figure 5.4.2A) and for a given coating thickness, the transferable material was either completely removed above a certain threshold laser fluence or not at all. This observation reflects the fact that the thermal ablation process and therefore the vapor bubble formation necessary for the forward catapulting starts at the carrier–coating interface (i.e., the "backside" of the film). If the formed pressure is high enough, the overlying shell of the coating will fracture and be completely split off as the flyer. However, in the reverse transfer setup (Figure 5.4.2B) which works more along the classic laser ablation, small amounts of material could be removed gradually from the top surface depending on the applied laser fluence. In cases where not all of the material was removed, ablated structures in the transferable coating were observed.

5.4.2.1.1 Enhanced Donor Systems for Printing Applications

In a US patent filed in 1971 [36], a slight extension of the donor system in the printing setup is disclosed for a laser imaging application similar to that described above. The donor film coated on a transparent PET substrate comprises heat-absorbing carbon black particles dispersed in a self-oxidizing binder consisting of nitrocellulose. The incoming IR laser heats the energy-absorbing particles and induces a blow-off or explosion of the energetic binder matrix. This causes the removal of the coating, leaving a clear area on the supporting substrate that corresponds to a positive image. If a receiver substrate is mounted as a backing in contact with the donor film, the deposited material forms an image recording. The laser power required at the focus point is sufficient only to heat the self-oxidizing nitrocellulose binder to initiate combustion. The blow-off provides the thrust to carry the carbon black particles toward the receiver sheet. Even when no details concerning the resolution and pattern accuracy are given, this example demonstrates that the laser-triggered decomposition of energetic materials can, in principle, be advantageous for the material transfer.

A further improvement of the laser-driven transfer method was disclosed in 1976 in a further patent for the preparation of planographic printing plates [37]. The formulation of the donor substrate coating contains, in addition to the above-mentioned carbon black particles and the self-oxidizing nitrocellulose binder, a cross-linkable prepolymer resin together with a thermotriggered cross-linker agent. After a laser-responsive ignition of the blow-off that causes the transfer of the prepolymer resin coating together with the cross-linking agent, the cross-linking reaction is initiated by the heat generated during the laser irradiation and the exothermic decomposition of the binder. Owing to the cross-linking reaction, the deposited polymer features are fixed after thermal curing on the receiving plate, which could directly be used for the copy printing process, whereas the donor film shows again the opposite image recording. This example demonstrates that by using appropriate material systems, in principle, laterally structured functional units such as printing plates for graphic arts can be directly fabricated by laser-assisted material transfer techniques.

5.4.2.1.2 LIFT of Metal Patterns

The transfer of small metal features from thin metal donor films coated on fused silica carriers by the LIFT process was first investigated by Bohandy *et al.* [38] and published in a paper in 1986 entitled "Metal deposition from a supported metal film using an excimer laser." In contrast to the previous studies of graphical print applications where argon-ion and Nd:YAG lasers with emission lines in the visible and IR range, respectively, were used, metal films were irradiated with single pulses of a UV excimer laser (ArF, 193 nm emission wavelength, 15 ns pulse length), and 50 and 15 µm wide lines of copper and silver were transferred in a high-vacuum chamber from fused silica carriers precoated with metal donor films ~400 nm in thickness onto silicon receiver substrates. The experimental apparatus corresponds to the conventional LIFT setup outlined in Figure 5.4.1, which was initially placed in a vacuum chamber. With a donor–receptor distance in the range of some micrometers, deposition of metal patterns was observed above a threshold pulse energy of around 60 mJ. With increasing pulse energies, the width of the transferred line got broader, and with energies around 140 mJ more metal "splatter" and surrounding molten debris deposits have been observed away from the edges of the line. As already described above, analogous control experiments have also been performed, where the metal film was irradiated through a transparent receiver substrate located in front of the film, according to the vertical pulsed laser deposition (VPLD) in Figure 5.4.2B. The metal deposits produced by that backscattering ablation experiment were found to be much worse compared to the forward transferred ones.

This study was the beginning of systematic investigations concerning the transfer mechanism and applications of the LIFT technology for the direct-write printing of conducting lines as interconnects in electronic circuit devices. One benefit of the LIFT direct-write process is the fact that the size of the written metal deposits corresponds closely to the size and shape of the laser focal spot. Mechanistic models of the laser-induced forward propulsion of metal features involve heating, melting, and vaporization of the metal film by the laser. The LIFT mechanism is triggered by the absorption of the incident laser photons at the carrier–metal interface. Depending on the surface properties and the optical constants (e.g., reflectivity R and absorption coefficient α) of the metal layer, a part of the incoming laser pulse will be reflected or even scattered, whereas the absorbed laser energy induces an ultrafast heating of the irradiated area. Owing to the high absorption coefficient α of metals, the typical penetration depths of UV photons in metal layers are typically below 20 nm, that is, the pulse energy is deposited in a very thin volume of the metal film where it gets converted into heat and leads to melting and local vaporization. Depending on the pulse length and the individual thermal parameters of both the metal layer and the underlying substrate, such as thermal conductivity, specific heat, and density, a rapid propagating phase change takes place within the film. Above a certain energy threshold, the pressure of the metal vapor that is trapped between the source film and its carrier support is high enough for a rupture of the film accompanied by the vapor-driven propulsion and lift-off of the remaining film shell [39]. For these reasons, each metal shows individual LTHC properties, depending on the combination of its inherent material properties with respect to the specific interaction with the laser and its parameters, such as the emission wavelength, fluence, pulse duration, and shape. Owing to the high thermal conductivity of metals, thermal diffusion has a strong influence on the ablation behavior, since the local heating gradient is relatively small compared to materials with a lower thermal conductivity (as, e.g., the carrier substrate). Typically, for metals and pulse lengths in the nanosecond regime the thermal diffusion length is roughly on the order of two times the optical penetration depths. For pico- and, particularly, femtosecond pulsed lasers, thermal diffusion becomes less important, and the same incident laser energy stays with ultrashort pulses significantly more localized (i.e., "more concentrated") in the irradiated focal point. Therefore, pulse length and beam homogeneity also play an essential role in the thermal ablation process, since the morphology and cleanliness of the resulting features and deposits is crucially influenced by the combination of all these parameters, which finally define the processing window for a successful LIFT deposition. A further problem that can be observed with the LIFT deposition of metals concerns the wettability of the receiver surface because of the surface tension of molten metals, which often shows blurred and splashed droplet structures.

In Chapter 5.2, Sakata and Wakaki presented a compilation on general LIFT studies of various metals. In the context of device fabrication, metal deposition by LDW methods can be useful for creating conductive connections and contacts in electronic circuits by "laser placing." Also, a repair of burned or interrupted interconnecting lines can be envisioned. However, as already mentioned above, the pattern quality and homogeneity of metal patterns deposited by LIFT is often unsatisfactory because of the distinct splattering and spallation behavior of transferred metal features that often appear substantially blurred by precipitation of small droplets along the rough microfeature boundaries. Recently, it was reported that for the LIFT deposition of zinc with subpicosecond pulses of a 248 nm excimer laser over a transfer gap of 100 µm the obtained structures were found to be much sharper and less blurred when the transfer was carried out in a moderate vacuum [40]. Obviously, the ultrashort pulsed UV LIFT provides a highly forward-directed material plume with only a small angular divergence in vacuum, so that only small amounts of metal is deposited outside the main track of deposited material. Nevertheless, scanning electron microscopic (SEM) images reveal at a higher magnification that the line structures are composed of many individual submicronsized droplet fragments and features, indicating that the metal film is not transferred as an intact unit.

At IBM Laboratories *laser-induced chemical vapor deposition* (LCVD) of high-purity metal films has been applied to a wide variety of microelectronics and device applications. The ability to deposit conducting materials has enabled specific applications for the laser-induced repair of "open" circuit defects, laser-induced deposition of interconnects for customization of circuits and the laser-based repair of high-resolution photolithographic projection masks [41]. Such masks consist of a precisely patterned thin metal layer usually coated on fused silica carriers. For the repair of clear and opaque defects in the metal mask layer material addition is needed. Besides focused ion beam (FIB) metal deposition processes, laser-induced deposition staining has been used [42]. Controlled quality thin metallic films were deposited on the defect sites by locally defined UV laser-induced decomposition of volatile tungsten, gold, molybdenum, and chromium coordination complexes as metal precursors. The optical transmission and phase-shifting properties of the deposits can be matched well to the films commonly used in phase shifting masks. But the LCVD thin-film direct-write technique is limited by the slow deposition rate, narrow choice of materials, and process complexity, including the need for operating in vacuum. For the repair of further pattern defects and trimming of microsturctures, locally resolved material removal has been performed with femtosecond laser ablation and FIB bombardment techniques.

A recent study on the creation of conductive lines by LIFT was carried out using 110 fs pulses of an 800 nm laser focused through microscopy objectives [43]. From gold donor films with a metal layer thickness between 40 and 80 nm, conductive lines with a width of 15–30 µm were written between contact electrodes. The same deposition was used to demonstrate the repair of a damaged Pt microfluidic heater element with a small Au metal pad. Owing to the rather nanoparticulate appearance of the deposited metal, 15–40 subsequent overwrites were necessary to create lines with a peak line high of 0.6–3 μm, which showed a resistance between 100 and 400 Ω for a 1 mm long line. This is rather a high value since the porous nature of the deposited metal greatly increases the resistivity of the lines. The conductivity could be improved by a factor of 2 by gentle electrical heating by which many microdroplets are fused together, reducing conductivity bottlenecks. Thus, lines with conductivities in the range of 10^{-6} Ω m, approximately 100 times the value for bulk gold, can be produced. In an earlier study, LIFT deposition of conducting Au lines transferred with nanosecond pulses of an excimer laser from a 1.6 μm thick Au donor over a 25 μm gap resulted in 40 μm wide lines with a height of about 10 μm after 100 laser pulses with multiple stepwise overwrites [44]. While LCVD-deposited Au from metal complex precursors gave homogeneous gold lines with a resistivity about 10 times that of bulk Au, a higher resistivity was obtained from gold deposited by LIFT [41].

Since ultrafast laser pulses in the subpicosecond regime are known to have precise breakdown thresholds and minimal thermal diffusion in metals, uncontrollable melting processes are prevented, and therefore they offer advantages in precision micromachining and microdeposition applications [45]. The higher quality of the femtosecond lasers in terms of better resolution and more clearly defined morphology of the LIFT deposits has been demonstrated for the creation of diffractive optic thin-film devices in the form of binary-amplitude computer-generated hologram masks [46, 47]. With a modified 248 nm femtosecond excimer laser system, complex pixelated patterns were laser transferred pixel by pixel from a 400 nm thick chromium donor layer with a high lateral resolution of the individual square Cr deposit features of about 3 μm × 3 μm. It was also demonstrated that more than one layer of metal can be written on top of each other by creating multilevel holographic structures, comprising three layers of Cr pixels subsequently deposited. As an experimental extension, temporally shaped femtosecond laser pulses have been studied for controlling the size and morphology of LIFT-deposited micron-sized metallic structures of Au, Zn, and Cr as a function of the pulse separation time of double pulses [48]. Depending on the electron–phonon coupling in the metals, tailored pulse shaping has an influence on the ultrafast early-stage excitation processes, and this has effects on the dynamics of the ejection and deposition process.

LIFT of gold fringe patterns was achieved by projection of a laser interference pattern onto an Au-coated donor target [49]. The inherent energy distribution of the incoming interference beam has been projected as finely structured gold line patterns on the receiver substrate. This interference-based LIFT allows deposition of well-resolved structures without the need for a corresponding mask.

By using appropriate projection masks, arrays of more than one individual pixel can be deposited by one laser pulse. Simultaneous deposition of more than one small single pixel per pulse allows faster laser printing of repetitive-patterned structured and dot arrays, which, in principle, is useful for device series fabrication. A simultaneous deposition of gold multidot arrays was demonstrated by using a target Au film that is evaporated onto a regular two-dimensional (2D) lattice of microspheres formed by self-assembly on a transparent support [50]. The Au-coated SiO_2 microspheres had a diameter of 6 μm and were irradiated with defocused single pulses of a nanosecond KrF excimer laser through the quartz carrier. Each microsphere represents a small microlens focusing the laser beam onto a small spot on the top radius where the metal film gets locally ablated. The size and index of refraction of the microspheres determine the degree of focusing and thereby the area of the illuminated gold film on the calotte of a particular sphere. Arrays of hexagonal Au dots with an average diameter of ∼2.5 μm have been transferred sucessfully onto the receiver substrate. A similar setup with polystyrene microbeads as the focusing elements has been used to deposit arrays of submicron metal features [51]. A femtosecond laser emitting at 800 nm was used to deposit dotted Ti, Cr, and Au patterns. With appropriate optical parameters, deposition of submicron dot arrays could be achieved by this microbead-assisted parallel LIFT approach. However, the size of the projection mask has to fit with the laser beam diameter, and the energy distribution within the beam profile has to be taken into account. Owing to their more homogeneous beam profile, rare gas excimer lasers have therefore some advantages compared to lasers with a Gaussian-shaped energy profile.

This was recently shown with the fabrication of micron and submicron dot arrays of semiconducting β-FeSi$_2$ by LIFT with a grid-pattern projection mask [52]. Demagnification of a pinhole photomask in a UV-laser-compatible projection system down to 1 mm × 1 mm allowed the simultaneous microdeposition of highly regular dot patterns with high site control and downsized feature dimensions below 1 μm. Again, the pattern quality was found to depend strongly on the applied fluence of the 248 nm excimer laser (20 nm pulse length), and for the given system the optimum processing window had to be determined. Within the fluence range closely above the threshold fluence of transfer where the smallest regular droplets were deposited on the receiver, no corresponding apertures or holes were found in the donor layer; only bumplike curved arches have been observed on the layer surface. As an interesting contribution to the elucidation of the involved LIFT mechanism, the authors investigated the morphology of the backside interface between the source film and transparent support after irradiation. The used part of the donor film was peeled off from the support and regular concave hollows with the shape of the projected laser beam were found at the irradiated sites. The formation of these hollow gaps can be attributed to the fast phase transition processes during the laser-induced local superheating where liquefaction and gas bubble formation take place. Only a small droplet is ejected from the molten surface, whereas the remaining liquid layer freezes out again immediately and entraps the formed vapor bubble. A similar phenomenon with "frozen" remaining structures for laser fluences close to the ablation threshold was documented in a study on the transfer dynamics during the LIFT process of aluminum with an IR laser [53].

In order to enhance the pattern quality of conducting metal deposits and to avoid the splashing and droplet-like debris formation during the LIFT process, the effect of an additional 1 μm thick polymer coating layer on top of the 400 nm thick metal donor film was tested [54]. Unfortunately, no information has been disclosed on the type of polymer used in this study and which laser was applied; only pulse repetition rates up to 80 kHz have been mentioned and the influence of the scan speed (obviously in the range of some tens of micrometers per second) at a fluence around 8 J cm^{-2}. What happens to this overlying polymer coating during the laser-induced ejection process under these conditions has not been clearly described; however, the edge quality of the deposited patterns appears to be enhanced. It can be imagined that the polymer top coating has an influence on the metal evaporation dynamics, and modifies by that the droplet formation during the forward ejection. Also, a certain "taping effect" of the polymer layer to the ejected metal features can be imagined. Obviously, the polymer then gets thermally decomposed since, allegedly, by energy-dispersive X-ray spectroscopy (EDX) analyses no carbon components could be detected in the deposited metal patterns. Conducting test microelectrodes from Cr, Cu, and Al with a high of 5 μm were LIFT deposited by multiple layer-by-layer overwrites. The microstructures consist of small densely packed metal particles, showing a specific resistance of about 100 times that of the bulk metals, but also 2 times less compared to metal electrodes deposited without the auxiliary polymer top coating.

A 2 μm thick polymer coating on top of the receiver substrate was used to investigate the LIFT deposition of conducting metal interconnects into a passivation layer of electronic devices [55]. A conventional LIFT setup with silver-coated (300 nm) donor substrates and a Nd:YAG laser ($\lambda = 1064$ nm, 7 ns) was used to catapult Ag clusters into the polymeric receiver layer at an applied fluence of 1 J cm^{-2}. Usually three to five laser pulses are required to make the embedded silver particles well interconnected, forming a conducting structure in the poly(vinyl alcohol) polymer (PVA) coating. This process has been named *laser-induced implantation*. However, this technique has some differences compared to the *laser molecular implantation* (LMI) discussed in Section 5.4.3.2. The accelerated metal ejections penetrate the polymer layer surface and get embedded. Further laser pulses push the LIFT-embedded metal clusters deeper into the receiver matrix. A PVA-encapsulated TFT and a polymer light-emitting diode (LED) device have been successfully contacted via LIFT-implanted Ag interconnects. Even when the performance of the LIFT-contacted devices was somehow reduced, this approach could bear the potential for embedding electric interconnects in already encapsulated organic electronic devices during plastic substrate processing.

As an inorganic nonmetallic material, the LIFT deposition of silicon nanoparticles was recently studied with respect to potential applications in device fabrication [56]. The reported process of deposition consists of two steps each, whereby a Nd:YAG laser (1064 nm, 5–7 ns pulse width) was used: for the preparation of the donor substrates, Si nanosized particles were deposited on glass substrates by VPLD from native silicon wavers as the target material at a high-power output of the laser, according to Figure 5.4.2B. Depending on the process gas atmosphere, strong morphological variations of the deposited Si structures were observed. While in an argon atmosphere porous films of mutually agglomerated silicon nanoparticles are formed, the same procedure performed in ambient air resulted in films consisting of networks of hyperbranched nanowires of crystalline Si interconnected by amorphous silicon oxide deposits. Both types of prefabricated donor substrates were used in the subsequent LIFT process to deposit Si-based patterns, and their morphology changes were investigated. As a test for the conservation of material functionality, fluorescence microscopy imaging was performed in order to detect luminescence emission of Si nanocrystal centers. From the analysis of the photoluminescence (PL) spectra, a core–shell structure of the deposited Si nanoparticles was derived where excitons are trapped in a confinement at the Si/SiO$_2$ interface.

LIFT of Metal Nanoparticle Inks   LDW of conductive metallic interconnection lines is of great interest for the tailored fabrication of electronic microdevices and printed electronics. During the past years, inkjet technology of metal nanoparticles was essentially developed, including the delicate thermal postprocessing steps for the fixation and functionalization of the inkjetted structures by curing and annealing. However, for inks with high contents of nanoparticles indispensable for the creation of well-conducting lines, the inherently high viscosity of such inks makes it more difficult to be applied by small-orifice nozzles in inkjet printers because of their tendency to clog and contaminate [57]. Here, the LIFT of viscous pastes can offer a promising technical improvement. The opaque ink has a high absorption coefficient, and the metal nanoparticles provide an efficient interaction with the laser so that only a portion of an ink donor layer would be vaporized to catapult the remaining coating. While the binder matrix of the paste serves only as an auxiliary compound that has anyway to be modified or removed later by a separate curing step, a chemical alteration by the LIFT process might not be so relevant. Depending on the composition of the ink, the binder matrix can be more or less volatile and adapted to an appropriate viscosity of the paste so that the fluid dynamics of the ejected features can be finely adapted toward a proper jetting and deposition behavior. In addition, tuning of the rheologic properties of the ink paste enables the deposited single-droplet features to flow and coalesce, and form by that continuous pattern structures for conducting lines. Highly viscous nanoparticle pastes enable a defined LIFT catapulting process of elastic soft patterns, which are transferred homogenously without morphological changes by droplet formation and deposited without splashing.

Silver nanoinks containing 20–50 wt% of Ag particles with a size below 100 nm were used for corresponding LIFT direct-write printing experiments [58]. A frequency-tripled Nd : YVO$_4$ laser with an emission at 355 nm and about 30 ns pulses was used to transfer the nanoink coating from a glass support toward Si wavers or polyimid (Kapton$^{\circledR}$) foils. Postdeposition curing treatment was carried out either by thermal curing in an oven for 40 min at 250 °C or with laser baking. Continuous lines, ∼20 μm in width and ∼0.5 μm in height, with good particle agglomeration could be obtained by that procedure. Lines of 470 μm in length printed between two gold electrodes on Kapton exhibited good electrical conductivity with a resistivity of less than 10 times that of bulk Ag.

Conducting lines were also LIFT printed with a pulsed IR laser using a commercially available Ag/Pt-nanoparticle-based ink (QS 300 manufactured by DuPont)

[59]. This ink has been developed for industrial microelectronic screen printing applications for producing fine lines down to 75 µm with contact masks. The rheology of such screen printing inks is specifically designed to vary with the shear forces applied to the ink. In the absence of shear forces, the ink is very viscous (thixotropy) so that deposited patterns can resist distortions once they are deposited. For the preparation of donor substrates, the ink was mixed with α-terpineol as a less-volatile solvent to get a modified matrix that was used earlier for a matrix-assisted pulsed laser evaporation direct-write (MAPLE-DW) study with inorganic powders and a 355 nm UV laser [60]. A complex fluid dynamic behavior has been observed on pulsed irradiation with a neodymium-doped yttrium lithium fluoride (Nd:YLF) laser at 1047 nm emission wavelength, including a bubble protrusion, jetting, and plume regime, depending on the surface tension of the film. Only within the bubble protrusion regime could a sufficient transfer be achieved, since considerable instability of the jet was observed in the more energetic regime, and the ejected ink splattered on contact with the receiver substrate. Without the additional solvent, the ink proved to be too viscous for thin-film fabrication by spin coating or wire coating. Therefore, the donor films were thicker and did not form a jet anymore. However, line features as small as about 20 µm could be deposited, which showed conductivity after high-temperature curing of about 75% of the specified value for conventional screen printing.

A further enhancement of the transfer pattern quality was obtained by using designed silver nanoparticle (Ag NP) inks with significantly increased viscosity [61, 62]. In the LIFT process, since the ink need not pass a small nozzle channel to be jetted on the receiver, ink pastes with much higher viscosities and a higher nanoparticle content can be employed, similar to that used for screen printing. A paste of small colloidal Ag nanoparticles (size 3–7 nm) with a viscosity of ~100 000 cP was used for printing pixel patterns and line structures with the third harmonic emission at 355 nm of a Nd : YVO_4 laser at fluences of 8–40 mJ cm^{-2}. The donor layer was coated with a 100–300 nm thick layer of the ink suspension and the transfer gap was adjusted between 10 and 50 µm. Unfortunately, no information is disclosed on the composition of the binder matrix of that Ag nanoparticle paste, so that no conclusions can be drawn about the laser–ink interaction and the energy conversion that results in well-defined pixel deposits. It may be assumed that the UV laser excites the quantum-size-effect-dependent absorption of the Ag nanoparticles, inducing an LTHC that causes the thrust for the transfer by decomposition of the thermocurable binder matrix. A thermal postdeposition curing of the transferred patterns was carried out either by *in situ* laser treatment (scanning with a 532 nm continuous wave (CW) laser) or in a convection oven at 200 °C for 30 min. Since the deposited pixels showed clearly defined edges without debris and with a homogeneous shape morphology in correspondence with the remaining ablated spots on the donor substrate, the method was called *laser decal transfer* (derived from "decal," which is a plastic, cloth, paper, or ceramic substrate that has printed on it a pattern that can be moved to another surface on contact, usually with the aid of heat or water). Conducting line structures were built either by continuous deposition of individual pixels close to each other or by stacking two layers to bridge gaps between pixels. The resistivity of such structures after thermocuring was found to be up to about two times that of bulk silver. Even freestanding, 25 µm long and 5 µm width bridging silver structures could recently be deposited over microchannels by this approach [63].

However, since such nanoinks consist of more than only one component, that is, the metal nanoparticle dispersion in the "wet" binder matrix one may note that the mechanism of such a transfer process differs in principle from the above-discussed basic LIFT technique where normally coatings of only one material are used for transfer. Owing to the role of the thermodegradable binder system, this application has some close analogy with the so-called MAPLE-DW process discussed in Section 5.4.3.1. A suspension of Ag NPs was also used as a laser-absorbing dynamic release layer (DRL) for the transfer of organic electronic materials, as described later in Section 5.4.3.3.5) [64].

5.4.2.1.3 **LIFT of Organic Materials**

Since the entire transfer material itself is directly exposed to the laser radiation in the basic LIFT process, the transfer conditions for the material to be deposited are rather harsh with respect to thermal load and photochemical impact that may cause structural alterations. As shown for metals, thermally induced phase transitions may have a strong and often detrimental influence on the quality and homogeneity of the transferred features, even when ultrashort laser pulse lengths are applied. Therefore, mostly robust and heat-tolerant compounds, such as inorganic oxides, ceramic, and dielectric materials were used for transfer studies. Nevertheless, examples can be found in the literature where organic materials were deposited successfully, but preferably by using ultrashort laser pulses in order to circumvent detrimental heat diffusion within the donor material coating. For applications where bulk properties of the transferred material are important, the decomposition of a more or less "thin" top layer of the sensitive material may be acceptable, but this is not the case for all applications, such as sensors, where the surface properties of the layers are decisive.

LIFT of Small Organic Molecules　The LIFT deposition of the pink-colored laser dye Rhodamine 610 was investigated with the second harmonic emission of a Nd:YAG laser at 532 nm [65]. When the transfer was carried out at low fluences and in vacuum over a gap of ~15 µm, corresponding pink deposits were found on the SiO_2 receiver surface, accompanied by black debris particles. As a check of the chemical integrity of the transferred molecules, fluorescence measurements were performed. It was found that for a transfer performed in air the fluorescence intensity was only quite weak, whereas under vacuum conditions a moderate intensity could be detected. Obviously, the thermal impact in an oxidizing atmosphere is rather destructive to the chemical structure of the dye, but at low pressure and in an oxygen-free environment a transfer seems feasible. Possibly, due to the reduced pressure the thermal sublimation of dye molecules is facilitated, in analogy with other thermal deposition techniques, such as chemical vapor deposition (CVD) and also PLD, which are known to be applicable to the fabrication of thin-film coatings of small organic molecules [66–68]. However, the fluorescence intensities after LIFT of the same dye component was significantly higher when the transfer was carried out with a modified donor substrate where the dye is coated on top of an additional intermediate thin gold metal film that absorbs the incoming laser pulse. This modified metal-film-assisted LIFT technique is discussed in Section 5.4.3.3.3.

A LIFT-fabricated organic TFT device could be fabricated by depositing patterns of the organic p-type semiconductor copper phthalocyanine (CuPc), which is known as a chemically and photochemically robust metalorganic material [69]. The UV–vis absorption spectrum of CuPc shows a band around 355 nm, giving rise to absorption depth of about 85 nm of the corresponding emission wavelength of a pulsed Nd:YAG picosecond laser. The LIFT deposition was achieved from 100 nm thick CuPc donor films on fused silica (Suprasil) at a fluence of 100 mJ cm^{-2}, resulting in laterally well-resolved square deposits. Source and drain electrodes have also been LIFT printed using a commercially available Ag NPs ink. The electrical characteristics and parameters of devices fabricated by LIFT with CuPc as active layer were not so far from devices made by conventional evaporation of CuPc and silver electrodes.

LIFT of Carbon Nanotubes In close analogy with the very first reported LIFT deposition of black carbon ink recordings, carbon nanotubes (CNTs) as one further carbon modification were transferred to form functional patterned arrays that work as field emission cathode devices [70]. A precoated 20 μm thick film of CNTs with diameters ranging from 20 to 40 nm was used as the donor layer, which was exposed to 10 ns pulses of a 1064 nm Nd:YAG laser. Arrays of 27 × 27 dots were transferred to a conductive indium–tin oxide (ITO)-coated glass substrate by using a 80 μm thick sieve-type through-hole contact mask with a grid pattern of small holes. Donor, hole mask, and receiver were held in close contact during the single-pulse laser irradiation at a fluence of about 320 J cm^{-2} and in ambient environment. After peel-off of the patterning mask a CNT dot array with sizes of the single bump features of ~10 μm was obtained. Such dot arrays with different thicknesses of the deposited dots were tested as field emission cathode devices where a light emission could be demonstrated if a DC voltage up to 1100 V was applied. Obviously, the field emission properties were not suppressed by the laser interaction of the CNTs during the transfer.

LIFT of multiwall carbon nanotubes (MWCNTs) dispersed in various water-soluble polymer host matrices was studied recently with single pulses of the frequency-quadrupled emission at 266 nm of a Nd:YAG nanosecond laser [71]. Polymer composites were prepared by dispersing two types of MWCNTs in either poly(acrylic acid) or poly(vinylpyrrolidone). Owing to the inherent hydrophobic properties of MWCNTs, chemically functionalized MWCNTs that bear carboxyl groups on their surface were used to enhance the solubility in aqueous media, and to avoid the well-known agglomeration and aggregation behavior of MWCNTs in hydrophilic solvents. About 1.5 μm thick films of the polymer/MWCNT composites were fabricated by spin coating, which show a strong and broad

UV absorption around 250 nm because of the electronic structure of the CNTs. Only LIFT composite films with the functionalized MWCNTs that are much more homogeneously dispersed in the donor target matrix gave reproducible deposits, whereas agglomerates of unmodified MWCNTs in the composite donor films resulted in an inhomogeneous local absorption distribution of the laser energy, and therefore to discontinuous LIFT deposition results. Deposited composite pixels with an MWCNT content up to 10% showed sufficient electrical conductivity. With the versatile variability of functional polymeric host materials, laser-based direct-write deposition of conducting metal-free microstructures based on designed composite materials has a broad potential for applications in organic electronics and sensing device fabrication.

LIFT of Polymer Films   Since polymers are macromolecular compounds that have inherently high molecular masses they normally cannot easily be thermally evaporated or sublimed without substantial decomposition, even under ultrahigh vacuum conditions. For photochemical, photothermal, or photophysical ablation with laser photons the polymers need to have corresponding radiation-absorbing chromophore moieties incorporated in the repeating units. Most of the common commercial polymers have only a weak absorbance in the IR and visible domain. Absorption in the UV range induces electronically excited molecular states that often induce the cleavage of chemical bonds, and lead therefore to polymer photodecomposition and degradation into various small fragments. Since the first reports on polymer ablation with UV lasers in 1982 [72], laser polymer processing has become an important field for various microfabrication applications, mainly for ablative structuring and pinhole drilling. The main parameters that describe polymer ablation are the ablation rate $d(F)$, that is, a certain ablation depth d obtained with defined applied laser fluence F, and the ablation threshold fluence F_{th}, which is defined as the minimum fluence where the onset of ablation can be observed [73]. The linear absorption coefficient $\alpha_{lin}(\lambda)$ represents the wavelength-dependent spectroscopic absorption properties for a given film thickness, determined under static irradiation conditions with light intensities far below the ablation threshold. Many single-pulse ablation processes can be approximately described by the following equation:

$$d(F) = \alpha_{eff}^{-1} \ln(F/F_{th})$$

where α_{eff} represents the so-called effective absorption coefficient for laser ablation, which links as a wavelength-, material-, and process-dependent index factor the observed ablation rate $d(F)$ with the applied laser fluence, and depends further inherently on the pulse length. All these parameters have to be taken into account when a polymer film is catapulted and deposited according to the LIFT method. However, without further modifications, at least a part of the entire polymer layer has to be sacrificed for the generation of the pressure necessary for the LIFT process. If an appropriate amount of small molecular fragments are generated by ablative laser-induced polymer decomposition close to the carrier substrate, the overlying part of the film can be forward ejected by the pressure buildup within the ablated volume. However, the top surface of the LIFT-deposited film will be modified or even damaged since it is the laser-exposed side of the donor target that gets partly ablated. Owing to the restricted parameter space and the sensitivity of polymeric materials, it proved rather difficult to find optimum conditions for an ablative LIFT deposition of polymer films without chemical degradation or significant deterioration of their physical properties.

The direct LIFT transfer of entire thin films of the conjugated polymer PEDOT:PSS, which is used as a conductive polymer coating in stacked electronic thin-film device architectures, was exemplarily studied with different laser wavelength and pulse durations [74]. About 300 nm thick polymer films on glass or fused silica carriers were used as donor targets and irradiated with 8 ns pulses of an 1064 nm Nd:YAG solid-state laser. Owing to a certain absorbance in the IR, the donor film could be ablated from the donor carrier, but a pixel deposition onto the receiver substrate could only be reached at higher laser fluences that cause severe traces of thermal stress and morphological surface alterations, and even damage to the flexible plastic receiver surface was observed. This indicates that only a part of the laser radiation was absorbed by the film, and the whole film volume was penetrated (and therefore heated) by the laser photons. With the second harmonic of the same laser at 532 nm, the transfer was much worse because of the much lower absorption of the film in the visible radiation domain. Only 25% of the incoming light is absorbed, and by that ablation of the film from the donor was only possible with quite high laser fluences that destroyed the film without getting acceptable deposits. PEDOT:PSS is strongly absorbing at the 248 nm emission of a UV excimer laser, and the LIFT deposition of well-defined coherent pixels can be performed at comparably low laser fluences, however, only in a very small fluence range, and with loss of the electrical properties of the conjugated polymer due to photochemically induced alterations. Trapped bubbles and pinholes in the deposited film can be seen as an indicator of the formation of small volatile fragments during laser ablation, which form a pressure pocket during irradiation of the film and diffuse into a thin polymer layer before it is forward propelled by the generated thrust. In summary, direct LIFT of thin polymer films proved difficult to obtain without substantially destructive side effects, and examples in the literature are hitherto rare.

LIFT of Biogenic Materials   Subpicosecond laser pulses of a modified 248 nm KrF excimer laser were used to study the direct-write LIFT printing of microarrays of entire biomaterials, such as DNA [75] and enzymes [76]. Since thermal effects detrimental to the irradiated material can be kept much smaller with smaller laser pulses compared to conventional nanosecond laser pulse durations, destructive effects on the absorbing material are less prominent during the LIFT process. Therefore, even delicate organic materials can be taken into account for LIFT deposition. Dynamic multiphoton absorption effects may additionally contribute to enhanced absorption properties of the irradiated material, and therefore to limit the propagation depth of the incident laser pulse within the transfer material. With such a reduced penetration depth and a much more locally restricted heat load, a higher fraction of sensitive transfer materials suffers less intense effects caused by the laser impact, enhancing the probability that delicate material systems can also be forward transferred without functional damage to their desirable properties. For microdevice fabrication, not only the quality of transferred patterns with respect to local resolution, feature morphology, and homogeneity play an essential role but also, in particular, the conservation of desired functional properties of the deposited material systems. Therefore, a check of material integrity, physical functionality, or activity after laser transfer is crucial to judge or even quantify whether a successful transfer was obtained. For biogenic materials systems, sophisticated immunochemical indicators and highly specific probe reactions were developed based, for example, on labeled antigen–antibody reactions or enzyme activity measurements that can be performed with very low concentrations and at microscale dimensions.

Horseradish peroxidiase (HRP) was used as a model enzyme for LIFT deposition [76]. In conjunction with hydrogen peroxide (H_2O_2) as the chemically active oxidizing agent, HRP efficiently catalyzes oxidation reactions. The presence and enzyme activity of HRP can be made easily visible and detectable by spectrophotometric methods when, by such oxidation reactions, a colorless compound (a leukodye) is converted into visible chromogenic product molecules that can then be physically detected. In addition, in the presence of luminol as cosubstrate the HRP-catalyzed oxidation results in a direct light emission by chemoluminescence (428 nm). With appropriate experimental conditions, this allows a quantitative and locally resolved measurement of the enzyme activity. By a corresponding staining reaction after LIFT printing of HRP micropixel patterns with subpicosecond UV pulses, it was demonstrated by confocal laser fluorescent scanning that the enzyme remained active after the LIFT process. However, in the referred paper no quantitative data are presented about the relation of enzyme activity before and post transfer.

Transferring DNA material was also demonstrated with the application of subpicosecond pulses of a UV laser [75]. Double-strand DNA molecules extracted from lambda phage particles have been used as the test material. Before coating on the donor support, the DNA strands were labeled with a fluorescent dye (Alexa 594) for detection by scanning laser confocal fluorescence microscopy after two-dimensional arrays have been LIFT printed. It could be shown that sufficient fluorescence intensity could be detected from the deposited arrays. Electrophoretic blot analysis of the DNA after an enzymatic restriction process showed a distribution pattern quite similar to the single fractions without a significant enrichment of oligomeric compounds.

The LIFT printing was also studied for the enzyme luciferase as a further biogenic material [77]. The enzymatic activity of luciferase is part of the chemical cascade of the *in vivo* bioluminescence of fireflies. A visible luminescence is emitted when luciferase reacts with the substrate luciferin in the presence of oxygen and adenosine triphosphate (ATP). A simple microsensor device for ATP microanalytical detection was fabricated, where, via the LIFT printing process, luciferase pads were deposited in prefabricated 40 μm deep cavities of a polydimethylsiloxane

(PDMS) chip on top of an integrated microphotodiode that can detect quantitatively the emitted luminescence. The third or fourth harmonic emission of a Gaussian Nd:YAG laser (355 and 266 nm, respectively, with 10 ns pulse length) were used with a focus spot of 100–300 µm at 2J cm^{-2} laser fluence. For the deposited luciferase spots, a good adhesion was found on the PDMS surface. The bioactivity of the luciferace patterns transferred on the PDMS chip was investigated by reacting the enzyme with ATP and luciferin. Patterns transferred at 266 nm showed no luminescence activity, indicating that the enzyme functionality was not preserved under these conditions. However, spots transferred at 355 nm exhibited a yellow chemoluminescence, which showed almost identical spectrophotometric plots as not-transferred luciferase films. Therefore, it can be concluded that bioactivity is not seriously diminished during the transfer process. Measurements of the photocurrent produced in the on-chip photodiode by the luminescence reaction with ATP/luciferin test mixtures allowed the derivation of curves fitting to the classic quasi-stationary-state enzyme kinetic equations according to the Michaelis–Menten model, and therefore a quantitative correlation necessary for a potential ATP-sensing microdevice.

Since the traditional LIFT process involves a direct exposition of the entire transfer material to the applied laser radiation, the method bears some crucial intrinsic restrictions concerning the properties of appropriate materials that have to tolerate a strong heat load. Thermally induced effects such as phase transitions and thermomechanical strain, as well as photochemical processes, have to be considered. For the LIFT deposition of metals, their thermoconductive properties, melting behavior, and surface tension of formed droplets have an influence on the morphology of deposits. The absorption coefficient at a given wavelength defines the permeation depth of the laser radiation within the material. Normally, smooth metal surfaces have a high degree of reflectivity, and therefore only a minor part of the applied laser photons is finally absorbed in a coating layer, whereas other materials have only a weak absorbance or are even nearly transparent to the incoming laser, so that thick layers are necessary for an efficient interaction with the laser. While evaporated metals can recondense again to atomic lattices, materials with high molecular weights, such as polymers or biogenic materials, may melt on laser irradiation, but a vapor formation necessary for an effective pressure buildup for the forward ejection can only be reached by fragmentation and degradation reactions either by thermolysis or photolytical cleavage of bonds, that is, only with loss of the initial material composition and properties. Neither variations in the emission wavelengths of the lasers from the far IR to the vacuum ultraviolet below 180 nm nor the application of ultrashort pulses down to the femtosecond regime could circumvent the limitations associated with the direct laser exposure of the transfer material coating. For that reason, various modifications of the original LIFT process were developed. Most of the extensions concern setup alterations of the donor system, either by specifically adding laser-absorbing intermediate release layers or by embedding the transfer materials into an easily evaporable host matrix.

5.4.3
Modified LIFT Methods

5.4.3.1 Matrix-Assisted Pulsed Laser Evaporation Direct-Write (MAPLE-DW)

For MAPLE-DW, the transfer material is embedded in a laser-evaporable matrix system, for example, frozen solvents or organic binders, and the laser is used to vaporize the matrix. The donor substrate consists of a transparent carrier coated with a suspension or dispersion of the particulate transfer material that can be either powders or small particles, or even a suspension of biomaterials including viable cells. The working principle of MAPLE-DW is outlined in Figure 5.4.3. By choosing appropriately absorbing and volatile components for the matrix, the laser irradiation (usually in the UV range) interacts preferably with the matrix components, which are then vaporized to form the pressure necessary to propel the embedded transfer material over a gap to the receiver. The liquid–vapor phase transition behavior and ejection of thin superheated solvent films have been investigated in detail with a time-resolved laser imaging technique [78]. This approach allowed for determining the generated pressures, the achievable superheating, and the relevant timescales of the vaporization process [79].

However, there may be problems involved with a certain UV load to the matrix-embedded transfer material due to absorption selectivity issues. Depending on the grain size of particulate systems dispersed randomly within the matrix, a high local resolution for depositing accurate patterns might also be restricted. Further, compared to the metal donor films in classic LIFT setups, MAPLE-DW donor films have typically a thickness of some micrometers, and if components of the thick matrix are co-deposited there may appear problems with contamination of the transferred systems. As a photodegradable polymeric matrix material, poly(butyl methacrylate) was used, which tends to decompose into its monomer units when exposed to UV irradiation below ~250 nm wavelength.

MAPLE-DW was originally developed as a method to rapidly prototype meso-scopic passive electronic device components, such as interconnects, resistors [80], and capacitors [1, 2, 81–84]. However, to obtain the functionality of the particulate materials after deposition, often posttransfer processing steps are necessary to obtain device compatibility, such as laser sintering, thermal annealing for the bonding

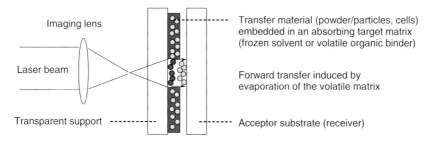

Imaging lens

Laser beam

Transparent support

Transfer material (powder/particles, cells) embedded in an absorbing target matrix (frozen solvent or volatile organic binder)

Forward transfer induced by evaporation of the volatile matrix

Acceptor substrate (receiver)

Figure 5.4.3 Working principle of matrix-assisted pulsed laser evaporation direct-write (MAPLE-DW).

of conductive powders, and especially laser trimming of blurred deposit borderlines in order to get clear-cut structure edges [85]. A series of microcircuit demonstrator devices were fabricated to study the electronic properties of such laser-deposited submillimeter-sized passive device components, such as spiral inductors [86], fractal antenna, oscillators made of resistive elements, and microcapacitors [2, 44, 87]. Two types of chemical sensors, surface acoustic wave (SAW) mass transducer and chemiresistor devices were also fabricated by MAPLE-DW of a chemoselective polymer (SFXA, a fluoroalcoholpolysiloxane), and polyepichlorhydrin (PECH), respectively [88]. Frequency response of MAPLE-DW and spray-coated SFXA SAW resonators were quite similar when exposed in on–off cycles to the nerve agent simulant dimethyl methyl phosphonate (DMPP) as test compound.

The use of a "liquid transfer vehicle" as the matrix for inorganic transfer material powders was applied to add micropower storage devices onto printed circuit boards (PCBs). Planar alkaline as well as lithium-ion microbatteries were fabricated by applying correspondingly composed inks in the MAPLE-DW process [89, 90]. In a similar manner, colloidal TiO_2 pastes were used in donor substrates to create on-chip PV microdevices based on dye-sensitized solar cell technology [91].

For the deposition of biogenic materials, MAPLE-DW contributed to the progress in the fabrication of biodevices and sensors [92]. Solutions or suspensions of active proteins, including enzymes and antibodies, could be laser deposited by this method, and evidence for the conservation of the biologic activity of these materials after transfer has been demonstrated [93]. Test assays consisting of microarrays with different immobilized proteins were fabricated and their specific antibody–protein binding capacity was checked by staining with fluorescently tagged complementary proteins.

For the fabrication of more demanding devices with delicate biomaterials such as proteins and living cells, MAPLE-DW has been proved a promising approach for LIFT deposition [94]. For the fabrication of the donor substrates, the biologic materials were trapped at low temperatures in a frozen matrix of appropriate solvents [95]. A series of studies were carried out with 193 nm ArF excimer lasers since most organic compounds absorb strongly in the short UV range below 200 nm [92, 93]. Therefore, the propagation depths (i.e., the optical length) of the laser can be kept small, and only a small layer of the coated entire donor target is destroyed by the laser radiation to deliver the propellant for the rest of the overlying donor biocoating. However, most of the biologic materials do not tolerate corresponding laser-absorptive organic solvents, but water as a completely biocompatible solvent has no strong enough absorption coefficient at 193 nm, so that the UV light can pass through the frozen biomatrix layer and damage the embedded UV-sensitive species. In order to circumvent the problem of low absorption of a frozen water matrix, UV-absorptive water-based biocompatible matrix media systems were tested (e.g., Dulbecco's Matrigel^{TM}), which contribute to the volatilization of the biolayer on rapid laser heating and form a vapor pocket that provides the thrust for the catapulting process [96]. This approach allowed transferring, for example, living rat cardiac cell arrays and stacks into a biopolymer matrix with near-single-cell resolution. In addition, patterns and even three-dimensional structures of living

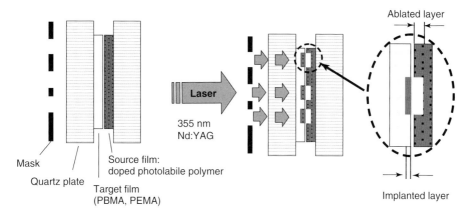

Figure 5.4.4 Scheme for laser molecular implantation (LMI) PBMA, poly(butyl methacrylate); PEMA, poly(ethyl methacrylate).

cells have been formed by MAPLE-DW. Various eukaryotic as well as prokaryotic cells grown in a corresponding culture medium matrix were transferred with the pulsed laser onto the receiver surface. Depending on the size of the focus area small cell groups from the culture could be optically selected and accurately deposited as patterns of dissected individual cell conglomerates. Bioassays and fluorescent protein marking demonstrated a high degree of viability of the cells after laser transfer into appropriate growth media. Tailored codeposition of various cell types allows not only applications for tissue engineering but also the integration of combined native cells for the fabrication of biosensor and detector devices.

5.4.3.2 Laser Molecular Implantation (LMI)

A further technique for the laser transfer of molecular compounds is LMI. The scheme of LMI is shown in Figure 5.4.4, and illustrates a certain correspondence to the setup for VPLD as depicted in Figure 5.4.2B. For LMI, a laser beam, which may be shaped by a mask, passes through a transparent substrate coated with only a weakly absorbing polymer layer (the target film as e.g. poly(butyl methacrylate), PBMA, or poly(ethyl methacrylate), PEMA) that is in contact with an absorbing polymer film containing the molecules that should be implanted, for example, fluorescent probes such as pyrene [97–99] or photochromic molecules [100].

The laser then decomposes preferentially the absorbing polymer matrix into small fragments, resulting in a pressure jump that assists the implantation of the embedded probe molecules into the nonabsorbing polymer film. A highly UV absorbing and distinctly photolabile triazene polymer used as the absorbing polymer matrix of the pyrene-doped source film on laser irradiation at 355 nm decomposes in a defined manner into small volatile fragments [101]. It was possible with this approach to implant pyrene with a resolution given by the mask into the target polymer with a depth of some few tens of nanometres. The disadvantage of this method is the limited number of available polymers with a high transparency and a quite low glass transition temperature T_g. A low T_g

makes the implantation of the transferred molecules more efficient because of a controlled softening of the target film by a heat-induced phase transition triggered by the pulsed laser irradiation. However, implantation was only possible with a maximum depth <100 nm. In later studies, phthalocyanine [102] and coumarin C545 and C6 were implanted in an analogous manner, and in the process now called *laser-induced molecular implantation technique* (LIMIT) [103, 104].

5.4.3.3 Layered Donor Systems with Intermediate Absorbing Films

Again, progress in the development of LIFT-based material deposition techniques was stimulated by process optimization in printing applications. The very first LDW image recordings used mostly black pigments and dyes for printing purposes, which were directly compatible with the use of IR solid-state and diode lasers because of their corresponding inherent strong absorption. However, laser-absorbing black pigments, such as carbon black, graphite, and metal oxides, which work as black body absorbers in conjunction with common near-infrared (NIR) lasers, are not compatible for printing other colors. To enhance the interaction of the laser with the imaging medium layer on a color donor sheet, "colorless" IR-absorbing dyes with substantially no absorption in the visible area were used as colorless sensitizers. Such IR absorbers convert the laser light to thermal energy and transfer the heat to the colorant imaging layer, which then gets ablated and transferred to the receiving element. The laser-induced ablation-transfer imaging process entails both complex nonequilibrium physical and chemical mechanisms. The ablation transfer is triggered by the rapid and transient accumulation of pressure within or beneath the mass transfer layer, initiated by laser irradiation. Various factors contribute to such a laser-triggered pressure accumulation, such as rapid gas formation via chemical decomposition combined with rapid heating of trapped gaseous products, evaporation and thermal expansion, and, in addition, propagation of shockwaves. The force produced by the immediate release of such pressure has to be sufficient for a complete transfer of the exposed area of an entire layer rather than the partial or selective transfer of components thereof. This is the particular difference of laser ablation transfer (LAT) from other material transfer imaging techniques based on equilibrium physical changes in the material, such as thermal melt transfer and dye sublimation-diffusion thermal transfer, where laser-addressed local heating is also employed. For all techniques, the formulation of the ink with an appropriate binder system plays an essential role. For thermal melt transfer applications the composition must contain low melting binder materials, whereas for laser-ablation transfer imaging the binder needs to consist of easily heat-decomposable and thermolabile components that form volatile degradation products when heated.

Former laser imaging recording processes that use self-oxidizing binders in conjunction with black absorber pigments as light-to-heat converting initiator components that trigger the blow-off reaction have already been mentioned in Section 5.4.2.1.1. As a further development, in 1992 a patent disclosed the combination of efficiently thermocleavable polymer binders doped with colorless IR ablation sensitizers as thermotriggered release system for LAT color imaging [105].

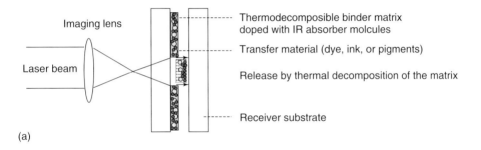

(a)

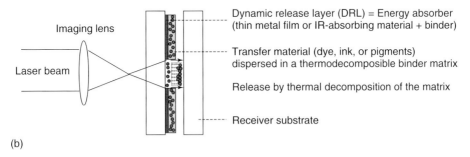

(b)

Figure 5.4.5 Scheme for laser ablation transfer imaging. In setup (a), the infrared absorbing compounds are formulated together with the coloring agents into the same binder matrix, forming a single-donor coating layer. The IR ablation sensitizers convert the near-infrared laser radiation into heat within the entire film. For setup (b), a thin intermediate metal layer is added between the carrier and the imaging coating that partially absorbs the incoming NIR radiation. Owing to its higher absorption the metal layer acts as an efficient thin-film heat converter that deposits the thermal stimulus for the decomposition of the overlying color coating in a much smaller volume close to the substrate interface, enabling a more defined ablation transfer behavior already at lower threshold fluences.

Tailor-made, thermocleavable polycarbonates and polyurethanes with low equilibrium decomposition temperatures around 200 °C were used as the binder, which was doped with Cyasorb™ IR 165 dye as the IR-absorbing sensitizer. To this functional binder matrix, commercial gravure color ink formulations containing, in addition, nitrocellulose as a self-oxidizing component, were added. Absorption of IR laser radiation by the ablation sensitizers causes a local heat accumulation that initiates the decomposition of the thermolabile binder matrix together with the ignition of the blow-off reaction of the self-oxidizing binder compound. The principle of LAT imaging is outlined in Figure 5.4.5a. Both the imaging dye composition and the IR ablation sensitizers are homogeneously formulated within the heat-degradable polymeric binder matrix, forming a single donor layer. On IR laser exposure, the binder matrix decomposes into small evaporable fragments that build up the local transient pressure to propel the imbued colored agents toward the receiver sheet.

The mechanism of LAT was investigated in detail by time-resolved ultrafast microscopy, where propelling speeds of the ejected coating of at least Mach 0.75 were observed [106]. With an optical absorption coefficient α of about 12 000 cm^{-1} at 1064 nm for such a typical color coating, the optical propagation length $1/\alpha$ of the IR photons within the absorbing layer is in the range of ~850 nm, which is in the same dimension as the donor layer thickness of around 1 μm useful for LAT imaging applications. A model calculation for such an absorber-sensitized system resulted in estimated heating rates dT/dt as high as $10^9\,°C\,s^{-1}$ and an ablation temperature above 600 °C for transient nonequilibrium superheating conditions during laser irradiation. When the coating material was irradiated above a well-defined threshold fluence of around 60 mJ cm^{-2}, essentially complete material transfer occurred provided that the pulse duration is shorter than ~200 ns, the time interval required for substantial heat transfer into the substrate. LAT made it possible to produce high-resolution full-color graphic images at comparably high printing speeds, demonstrating its potential as a versatile dry mass transfer method for also printing functional materials in direct-to-circuit-board device fabrication applications.

5.4.3.3.1 Laser Ablation Transfer (LAT) with a Metallic Release Layer

A significant improvement in the LAT process could be achieved by a further modification of the donor substrates, according to Figure 5.4.5b. The original LAT setup for color imaging applications uses organic NIR absorber dyes as ablation sensitizer dopants for the heat generation within the irradiated coating (corresponding to Figure 5.4.5a). Since only a certain maximum dopant concentration can be added to the binder matrix beside all other essential components, the NIR absorption of the whole coating stays rather small, even when the IR dyes themselves have a high specific absorption. As a consequence, the incoming NIR photons can propagate far into the coating until they are quantitatively absorbed, so that comparably thick coatings are necessary for a complete absorption of the laser radiation, and the LTHC process runs distributed within the whole film thickness. As already mentioned in the section on LIFT of metal films, metal absorption properties are in general superior to other material classes because of their characteristic electronic structure that allows optical excitation with high yields and in a wide spectral range. However, the absorption coefficient α is wavelength dependent and usually smaller for NIR radiation compared to UV excitation. While near-UV radiation is completely absorbed usually in a thin skin layer of some tens of nanometres (e.g., ~10 nm for Al at 308 nm, corresponding to an absorption coefficient of ~10^6 cm^{-1}) [107], NIR radiation has a slightly higher penetration depth in the range up to ~50 nm. But even when metals have a high absorption, they have at the same time a high degree of inherent reflectivity that causes crucial losses by back reflection, which is a disadvantage at least with respect to the high cost of laser photons. Nevertheless, thin metal layers can act as efficient light-to-heat converters, which can also be easily ablated and vaporized with a high relative increase in volume.

The improved LAT donor films include a thin metal film as a DRL, as outlined in Figure 5.4.5b. An ultrathin vapor-deposited aluminum film layer with a thickness of only ∼3 nm was used in the first study on the effect of such a metallic interlayer on the LAT imaging process [106]. Such an ultrathin Al layer represents a partially transmitting and partially absorbing interlayer (apart from the inherent reflection losses mentioned above). With stationary low-intensity irradiation at 1064 nm, such films show a reflectivity of ∼27% of the incident light, a transmission of ∼33%, and therefore a fraction of absorbed light of about 40%, which corresponds to an absorption coefficient α of about $3 \times 10^6 \text{cm}^{-1}$ at 1064 nm. However, all these values may be affected dramatically during pulsed laser irradiation due to temperature effects and phase transitions induced by ultrafast superheating, but with such a thin DRL, the threshold fluence for ablation is reduced by nearly 50%, and the ablated spots show a better shape. The ablation mechanisms involved in the DRL-assisted LAT process are complex, since the thin metal film itself absorbs only a part of the incident radiation, whereas the remaining part gets absorbed by the coating layer containing an IR sensitizer dye. Therefore, LTHC takes place within both layers, resulting in rapid heating of the metal DRL, which helps to reach the decomposition temperature necessary for the ablation of the thermolabile polymer binder matrix within a much more confined part of that layer close to the carrier interface. Hence, the transient pressure jump is generated more efficiently and more localized in a substantially smaller volume, which leads to a more controllable propulsion, and, finally, to a more homogeneous ejection of the overlying coating. Since the ultrathin DRL has a much higher absorption coefficient α compared to the sensitizer dye-doped coating, the LTHC process on a laser pulse results in a completely different temperature profile along the thickness of the overlying coating if the DRL is present. The threshold temperature for ablative decomposition is reached earlier and in a more confined zone so that the ejection process starts before the thermal diffusion during the pulse length reaches the top surface of the overlying coating. Therefore, a part of the coating stays "cold" (i.e., preserved from phase changes) and even shielded from the incident laser irradiation and can be propelled off without severe modifications.

Time-resolved ultrafast optical temperature measurements were performed for the DRL-free LAT imaging film exploiting the "molecular thermometer" of the IR absorber dye Cyasorb IR-165, which changes its absorbance coefficient in the visible range linearly with increasing temperatures [108]. The time-dependent transmission at 633 nm was measured at the onset of ablation triggered with the 1064 nm laser, and the peak temperature of the doped binder just before LAT starts was found to reach about 600 °C [109]. This is somewhat less than the melting temperature of aluminum (660 °C), and, indeed, electron microscopy inspection of ablated spots of donor films with an intermediate aluminum DRL showed that the Al metal film itself was not completely ablated, even when the metal layer obviously was temporarily molten, but not substantially vaporized. This can be attributed to the higher absorption coefficient of the thin metal film, which then acts more efficiently as a confined LTHC layer that creates a local hot spot that overshoots the thermochemical decomposition temperature for ablation of the thermolabile binder. A later study by secondary ion mass spectrometry (SIMS) analyses of the transferred pixels on the receptor surface revealed that aluminum residues are codeposited with the imaging inks. This finding is consistent with the observed phase transition of the remaining spots of the aluminum DRL on the donor after irradiation.

Also, already in 1993 the same authors reported a study on the effects of laser pulse length in the LAT process. Instead of 150 ns pulses of a common solid-state laser, they used 23 ps duration pulses generated by a Kerr lens mode-locked (KLM) 1064 nm laser [110]. It was found that the threshold fluence of ablation for donor film systems with a DRL of aluminum was reduced to 1/10th of that observed with nanosecond pulses (i.e., from 80 to 8 mJ cm^{-2}). Using the immense instantaneous peak powers of picosecond duration pulses, ablation can be induced at a lower fluence through vaporization of the thin Al interlayer. Vaporization of the metallic layer is intrinsically a more effective ablation method compared to the ablation of polymeric binders, because the relative increase in volume on vaporization is considerably larger (due to "atomization" of the metal layer) than for organic coating materials. The heating rates dT/dt of the Al layer observed with the picosecond laser were on the order of 10^{14} °C s^{-1} (i.e., about 5 orders of magnitude higher than for nanosecond pulse irradiation). Therefore, with respect to the ultrashort timescale of optical heating, heat diffusion effects via thermal conduction to the adjacent carrier and coating material within the irradiated spot are much less relevant, and the Al layer can reach and exceed the boiling point of Al, 2467 °C, already at low fluences just below the ablation threshold. The spatial distribution of heat in the dimension along the axis of propagation of the laser pulse differs essentially from the heat profile generated by about 1000 times longer nanosecond laser pulses with the same power density, where temperature losses by heat diffusion prevent the Al layer from immediate vaporization.

A similar color printing technique called *laser thermal transfer* was described, which is based on laser heating performed with 100 mW diode lasers with an emission wavelength of 825 nm and used in a modulated pulsed manner with tunable pulse durations in the range up to tens of microseconds [111]. For the donor ink sheets, a 1.8 μm thick light-absorbing layer consisting of an NIR absorber dye (a commercial bisaryl nickel complex, PA-1006 by Mitsui Chemicals Inc.) with a λ_{max} at 870 nm dispersed in a polycarbonate binder matrix was coated on top of a transparent base film, in close analogy with Figure 5.4.5b. As the coloring layer, sublimation dyes with melting points around 150 °C were coated by vacuum evaporation. Depending on the exposure time defined by the individually modulated pulse duration, increasing amounts of the sublimation dyes were deposited, when the donor film and receiver medium are in contact. Laser beam spots with a diameter of 25 and 3 μm were used to print image structures, and a resolution of 2540 dpi (corresponding to a single dot size of about 10 μm) has been reported. However, with a transfer gap of 25 μm between the donor and receiving sheet, a threshold pulse length of more than 30 μs was necessary for a transfer of the ink layer. Obviously, it needs a certain time with a given heating rate to obtain the critical temperature level for an "explosive" release mechanism similar to the LAT imaging systems with thermolabile binders, even if the timescale of laser exposure is in the multimicrosecond order, and both mechanisms, dye vaporization and ablation, may contribute to the transfer.

An evaporated black aluminum DRL in combination with an additional layer of an energetic polymer is claimed in a US patent for printing applications, and called *laser propulsion transfer* [112]. Mixed oxides of aluminum were vapor coated onto a carrier film as the DRL. On laser-induced heating, the black aluminum is exothermically oxidized to Al_2O_3, which is colorless, and serves at the same time as the initiator for the decomposition of the energetic polymer coating. The use of thermolabile gas-generating decomposing materials in conjunction with a laser-absorbing DRL allowed improving the efficiencies and speed of the printing process. Gas-producing polymers are known, for example, as energetic components of solid rocket propellants. Such polymers are composed of repeating units with nitrogen-containing precursor groups (such as azido, nitro, or triazole moieties), which on heat exposure can split off molecular nitrogen gas (N_2) during thermal decomposition, resulting in a significant volume expansion by the formed gaseous products. If such gas-producing polymers have a thermally available nitrogen content greater than about 10 wt%, they can serve as excellent propellants for thermal mass transfer operations. Examples of such energetic polymers, which exothermally decompose and form a large amount of gaseous products when heated above a certain threshold temperature on the millisecond or microsecond scale, are glycidyl azide polymer (GAP), poly[bis(azidomethyl)]oxetane (poly-BAMO), and polyvinylnitrate (PVN). Chemical structures of some nitrogen-rich polymers are shown in Figure 5.4.7. Such polymers were also studied in designed solid fuel tapes for diode laser-driven microthrusting devices for the propulsion and steering of ultralightweight minisatellites [113–115]. Owing to the involved exothermal decomposition processes, only low laser power densities are necessary to "ignite" the blow-off reaction, and therefore comparably high writing speeds of more than 50 m s^{-1} are possible with focused spots of common IR lasers on appropriate drum printing systems with galvanometric scanners. However, since most printing processes are performed in a direct-contact mode, it is difficult to differentiate whether the heat-induced mass transfer process corresponds to a LIFT mechanism or rather to a melt-type thermal transfer printing method with the laser as the fast-performing point heat source. Nevertheless, with the upcoming high-power semiconductor laser sources, new optical printing heads for miniaturized high-definition digital imaging applications based on laser-induced thermal transfer systems have been developed with printing resolutions above 3000 dpi [116–118].

5.4.3.3.2 Inorganic Materials as Absorbing Release Layers

A different concept of a laser-absorbing layer that contains volatile atoms that can be immediately released on laser heating was used to transfer 100 nm thin Al films. Hydrogenated amorphous silicon (*a*-Si:H) films formed on UV transparent quartz substrates were irradiated with 308 nm pulses of a XeCl excimer laser [119]. Films with a thickness of 50 nm and with a concentration of hydrogen atoms of 10 at.% were used as laser-responsive DRL, which transferred the top-coated Al layer at a laser fluence of 200 mJ cm^{-2}. When the *a*-Si:H-film is rapidly heated during the laser irradiation, the internally trapped hydrogen is liberated by explosive effusion and provides by adiabatic expansion the pressure for the propulsion of the Al film. With respect to device fabrication applications, microstructured arrays of Al metal lines were deposited by preparing corresponding mask-patterned targets with a target–receptor separation of 125 μm. However, similar to direct LIFT of aluminum and other metals, the transferred Al structures also show some irregularities by splashing and splattering when transferred by the hydrogen-assisted LIFT method. Cleaner transferred features were observed when this method was used to deposit Si structures directly obtained by fully laser melting the entire *a*-Si:H propellant films at higher fluences [120–122].

5.4.3.3.3 Metal Films as Absorbing Release Layers

The interaction of short-pulse laser radiation with metals causes a complex cascade of dynamic nonequilibrium excitation processes that finally lead to the thermalization of the absorbed laser light. Absorbed laser photons excite the conducting band electrons with subsequent energy relaxation, that is, the energy transfer from the hot electrons to the lattice vibrations due to the electron–phonon interactions and the electron heat conduction from the irradiated surface to the bulk of the target [123]. The propagation depth of the incoming laser photons in metals is small only because of the high (wavelength-dependent) absorption coefficients $\alpha(\lambda)$ and covers generally only a skin depth of some tens of nanometers or less. Depending on the electronic structure of the individual metals, the thermophyiscal properties, including their temperature-dependent changes, differ crucially, which has to be carefully taken into account for the design of sophisticated laser ablation applications. A dominating effect has the strength of the electron-lattice energy exchange, that is, the electron–phonon coupling that defines the temperature-dependent electron relaxation times. An electron–phonon coupling factor G has been derived from theoretical calculations and pump-probe experiments that reflect some aspects of the different ablation behavior of various metals in addition to their different physical properties, such as density, melting and boiling temperatures, heat capacity, as well as thermal and electrical conductivity. A high value of the electron–phonon coupling factor reflects an efficient energy transfer from the excited electrons in a nonequilibrium thermal distribution to the lattice. The finite time for the energy equilibration with the metal lattice is called the *electron–phonon relaxation time* τ_c. For metals with high values of the electron–phonon coupling factor G, the corresponding relaxation (or thermalization) time τ_c is shorter than for metals with low G values, with a typical range of τ_c between 0.8 ps for gold and 0.05 ps for titanium. For laser pulse durations shorter than the thermalization time τ_c, the electron distribution in the irradiated metal layer is a nonequilibrium state with a transient electron temperature much higher than the lattice temperature. Therefore, with appropriate femtosecond pulse durations below the corresponding τ_c, metals can be ablated in a much more controlled manner since side effects caused by thermal conduction are much less prominent compared to longer pulse durations where a heat transfer to the metal lattice takes place. Table 5.4.1 compiles characteristic thermophysical properties of some metals typically used as absorbing DRLs.

Table 5.4.1 Relevant thermophysical properties of some representative metals used as absorbing DRL [51, 123].

	Aluminum (Al)	Gold (Au)	Chromium (Cr)	Titanium (Ti)
Electron configuration	$3s^2 3p^1$	$5d^{10} 6s^1$	$3d^5 4s^1$	$3d^2 4s^2$
Melting temperature ($^\circ$C)	660	1063	1875	1668
Boiling temperature ($^\circ$C)	2450	2970	2665	3260
Thermalization time τ_c (ps)	0.55	0.8	0.1	0.05
Skin depth (nm)[a]	15	4	9	19
Electron–phonon coupling factor G (10^{17} W m^{-3}K^{-1})[b]	2.45	0.3	4.2	$\sim$130

[a] For 800 nm irradiation.
[b] At room temperature.

The electronic structure of the metals contributes to the thermophysical properties. It has obviously an influence, whether the electron configuration corresponds to a free electron gas model or if the d bands of transition metals are almost filled or less than half filled. The strength of the electron–phonon coupling depends on the electron temperature, and can either increase (Al, Au, Ag), decrease (e.g., for Ni, Pt), or exhibit nonmontonic changes (Ti) with increasing electron temperature [123]. Therefore, the strong temperature dependencies of the electron–phonon coupling have to be taken into account in the interpretation of experimental data obtained under conditions when the transient values of the electron temperature undergo significant variations, as this is undoubtedly the case during laser ablation and LIFT conditions.

Gold films with a thickness up to 500 nm were tested in a LIFT study for the transfer of the laser dye Rhodamine 610 on SiO$_2$ substrates with the second harmonic emission at 532 nm of a Nd:YAG laser [65]. In the same study, the laser dye was directly transferred without an intermediate metal layer, as already mentioned in Section 5.4.2.1.3. Measurements of the fluorescence activity were used as an indicator for conservation of the dye functionality. Best results with respect to the fluorescence intensity after LIFT deposition of dye spots with a diameter of about 200 μm were obtained with a thickness of the base gold film of 50 nm and in vacuum and with a rather low laser fluence of $\sim$50 mJ cm^{-2}. Under these conditions, the gold film is about two to three times thicker than the estimated optical absorption length (the so-called skin depth) of around 20 nm for 532 nm irradiation. Thicker gold films needed higher fluences for ablation, and the transferred laser dye on the acceptor substrate was found to have a less intense fluorescence response and was covered by evaporated gold. When the transfer was carried out in vacuum, the results were significantly better compared to experiments in ambient atmosphere, and the shape and morphology of the deposited spots were much more defined.

In combination with fast-imaging techniques, the laser-induced fluorescence of gold atoms was also used to visualize the time-resolved dynamics of the ejection process during LIFT propulsion [124]. Thin gold films (20, 100, and 500 nm in thickness) evaporated on fused silica substrates were irradiated by a pulsed dye laser at 440 nm (~9 ns pulse width), and the temporal and spatial distribution of the ablated species from the film were observed by imaging of the fluorescence emission at 627.8 nm after excitation by a probe laser beam (268 nm). In vacuum, the velocity of an ejected cloud of vaporized gold atoms exceeds 2×10^3 m s^{-1}, whereas emissive particles or droplets reached a speed of around 100 m s^{-1}. When an acceptor substrate was placed in front of the donor film at a distance of either 70 or 500 μm, a reflection of the ejected species was observed. This indicates that the conditions of the mechanical impact also play an essential role in the LIFT deposition process, and the velocity of the collision of the catapulted flyer with the receiving surface has to be optimized in order to get sufficient adhesion. Owing to the gas resistance under atmospheric conditions, the ejection speed is slower, but always accompanied by a shock wave in the gas phase, which may cause similar detrimental effects when reflected at the receiver surface (Section 5.4.3.3.6).

Red and green phosphorescent powders were deposited by the DRL-assisted LIFT method with a KrF excimer laser (248 nm emission, 25 ns pulses) using an intermediate 100 nm thick sputtered gold film [125]. In order to form a donor layer of the phosphorescent powders on top of the basic gold release layer, they were coated according to the concept of MAPLE-DW mentioned in Section 5.4.3.1 as a suspension (with glycerol and isopropanol as the solvents) by electrophoretic deposition, resulting in a wet viscous powder layer with a thickness of ~10 μm. On irradiation, the absorbing gold layer gets vaporized and the powder suspension with the high-boiling glycerol is catapulted over a gap of 25 μm onto the receiver. Interdigitating pattern arrays of green and red fluorescent powder deposits with a diameter of ~50 μm each were deposited and their emission spectroscopically characterized by cathodoluminescence (CL) measurements. The emission spectra of the transferred powder were identical to the corresponding CL measurements of the donor layer. The conservation of the functionality and the precise deposition of light-emissive materials as small pixels are of general interest for color screen and display device applications. However, there might be some technical restrictions for this approach for potential device integration since the transferred spots are obviously contaminated with codeposited small gold droplets that might impair the device performance.

In order to circumvent the codeposition of the metallic release layer, a modification of the MAPLE-DW process has been reported for the deposition of phosphor pastes by laser heating [126]. Instead of a thin metal film, a 200 μm thick stainless steel sheet is used, on which arrays of hemispherical blind holes with a diameter of 300 μm were etched. These cavities were filled with a viscous paste of the phosphorescent powders and served as locally resolved donor dots for the LIFT process. The binder matrix consisted of ethyl cellulose and organic solvents with boiling points between ~120 and >200 °C. On laser exposure of the back of the steel sheet, the powder mixture gets ejected because of the fast evaporation and bubble formation of the superheated solvents that form a jet. With the appropriate paste composition and laser heating conditions the stream of the phosphor paste is ejected as a spray, which is deposited as a regular plaque onto the receiver. This specific LIFT modification has been named *laser-induced thermal spray printing* (LITSP).

Thin metal films were also introduced as laser-absorbing layers in donor systems for the transfer of biomaterials. The requirement of strictly biocompatible matrix systems and the weak absorption of water-based matrix media has been mentioned in Section 5.4.3.1. With respect to the short propagation depths of UV laser photons in a metal layer (the so-called optical skin depth), only ultrathin absorbing interlayers of 35 nm for gold and 75 nm for titanium were found to be necessary as efficient energy conversion layers for the gentle transfer of various biomaterials embedded in common water-based matrix media with a frequency-quadrupled Nd:YAG laser (266 nm) [127]. Also, 85 nm thick films of biocompatible TiO_2 as UV-absorbing compound had the same effect of preserving the biomaterial from direct laser interaction and of providing a rapid LTHC that contributes to the fast vaporization of the adjacent part of the overlying water-based biomatrix to build up the catapulting pressure. By choosing the appropriate combination of laser fluence, pulse duration, and film thickness the propagation of the laser photons stays restricted to the optical length within the metal layer, whereas the generated heat can propagate (i.e., the thermal penetration) without further significant ablation into the much thicker aqueous biofilm causing local vaporization. By that, the ejection process for the transfer includes a different forward-ablation mechanism, which is more in analogy with the IR-absorbing light-to-heat converting DRL systems discussed in Section 5.4.3.3.1. Instead of the ablation of the entire donor film, a forward jetting of the viscous biomatrix is achieved, which allows a more precise deposition of droplet-like features. In a series of papers, this method was referred to as *"biological laser printing"* (BioLP™) [128] and the adjacent deposition of multiple cell types and accurate large-scale cell arrays are demonstrated, useful for potential applications in biodevice and minisensor fabrication as well as for tissue engineering [129].

Radiation-absorbing titanium layers of about 50–60 nm thickness have been also used in a LIFT setup to successfully deposit droplet-like microbumps of viscous DNA dispersions with the 355 nm emission of a frequency-tripled Nd:YAG laser [130]. The same setup was then also used for the precise spotting of microdroplet arrays of a purified bacterial antigen, which conserved its immunological reactivity after having been transferred [131]. For biodevice fabrication, where tiny amounts of delicate materials have to be manipulated, this modified LIFT approach opens up a promising orifice-free and noncontact deposition technique for the gentle and accurate direct-write of "bioinks" [132–134] with appropriate rheological properties for droplet formation during the laser-triggered jetting process [135–141]. The experimental results with an intermediate metallic absorption layer are clearly superior with respect to droplet morphology and resolution to an earlier approach called *light-hydraulic effect* (LHE), where viscous black printing inks were similarly forward ejected by direct IR laser exposure [142].

A series of bio-LIFT experiments were also reported with ~50 to 100 nm thick silver layers as radiation absorbers in conjunction with 248 nm pulses (30 ns duration) of a KrF excimer laser, and referred to as *absorbing film-assisted laser-induced forward transfer* (AFA-LIFT) [143]. Without an embedding matrix, dried conidia of the fungi *Trichoderma* were transferred over a gap of 1 mm in a laser fluence range between 35 and 355 mJ cm^{-2} [144]. Various mammalian cell types were tested with this LIFT approach, which survived the transfer process even when acceleration of the cells in the jetted droplets reached approximatively $10^7 \times g$ [145, 146]. It was observed that for all fluences above the transfer threshold the silver layer was completely ablated, leaving a transparent hole on the donor substrate. Therefore, it has to be assumed that silver residues are codeposited together with the biomaterial. Even when first microscopic inspections revealed obviously no visible metal droplet contamination of the deposited biospecies, submicrometer-sized Ag particles (250–700 nm) have been detected later in a more detailed study on the amount, size, and local distribution of such metal precipitations [147]. Since it is well known that dispersed silver particles have distinctly biocidic effects, such codeposited residues of the metallic absorber layer may restrict the viability or functionality of transferred biomaterials, for example, in biosensor devices.

Ejection of biomaterial-containing microdroplets onto bioassay chips was recently studied with absorbing ultrathin (10 nm) gold layers with 15 ns and 500 fs pulses of a 248 nm excimer laser system [148–150]. With the same transfer layer thickness of a viscous glycerol/phosphate-saline-buffer solution and with the same incident laser fluence, smaller droplets were deposited over a gap of 20 to 1000 µm with the femtosecond pulses. Although the threshold energy was lower for the femtosecond laser pulse transfer, both laser pulse regimes could be used for the pattern transfer with a model protein (Avidin) without significant deterioration of its functionality, as was shown by a key–lock reaction with Biotin as the complementary ligand. However, with nanosecond pulses the distance for transfer is longer than in the case of the femtosecond laser system, obviously due to a different jetting behavior. However, the size of the deposited droplets depends also on the dimensions of the laser beam, its fluence, and further, the composition of the transfer layer (e.g., the glycerol content) that influences the rheological behavior as well as the wettability of the accepting surface play an important role.

A very similar setup with 40 nm thick Cr films as the radiation-absorbing layer has recently been reported for the LIFT deposition of biotin dispersion microarrays with a 266 nm solid-state nanosecond laser system on low-temperature oxide layers on Si [151]. The influence of the chemical functionalization of the receiver surface was studied to obtain optimum immobilization of the deposited biotin droplets. Compatibilization by chemical modification of electronic semiconductor surfaces is a necessary prerequisite for the fabrication of biosensor devices. The LIFT deposition process of chemical sensing polymers onto prefabricated capacitive micromechanical sensor arrays was studied with respect to the integration of the LIFT technique for the production of real biosensor systems [152–154].

A combination of the two-photon polymerization (2PP) technique and LIFT with 55–60 nm thick gold films as energy-absorbing release layers allowed the combined deposition of 3D scaffolds as artificial extracellular matrix together with viable cells [155]. This represents an integrated laser-based approach for the tailored fabrication of cellular tissue constructs with a microengineered polymer scaffold matrix with controlled pore design for cell deposition. Prefabricated scaffolds were then seeded with endothelial and muscle cells by the LIFT process. Cell suspensions were dispersed as $\sim$50 µm thick layers on top of the Au-coated donor slides, and the LIFT deposition performed with single nanosecond pulses of a solid-state Nd:YAG laser at 1064 nm focused to a 45 µm diameter spot. Single droplets, each containing 20–30 cells, were ejected at a pulse energy of 66 µJ and deposited over a gap between the donor and prepared collector slide of $350 - 500$ µm. The combination of well-defined laser-fabricated porous microscaled structures with accurate LIFT microdeposition methods of sensitive biomaterials might be relevant not only for fundamental studies of cell–matrix interactions but also for future tailored applications in cell-based biosensor and device engineering.

Direct LIFT deposition of thin polymer films has been proved to be unsatisfactory because of substantial chemical alteration of the transferred macromolecular materials. However, with a metallic DRL, clearly better results are reported with respect to the pixel cohesion, homogeneity, and surface roughness, even when co-deposition of metal residues could not be avoided. Effects of the thickness of gold and silver metallic DRLs and the influence of the pulse duration on the transfer performance has recently been studied again [156] for the transfer of $\sim$300 nm thick films of the conducting polymer PEDOT:PSS, which is frequently used as an electrode-coating material in the fabrication of electronic thin-film devices. As already mentioned, direct LIFT deposition of PEDOT:PSS was tested without an intermediate absorbing DRL, and best pixel depositions were obtained only by using a UV laser. However, the conductivity of the transferred PEDOT:PSS was lost because of expected photochemical degradation reactions on direct exposure of the polymer to UV radiation. The LIFT process of films of the same conducting polymer was carried out with either 50 or 25 nm thick intermediate metallic absorber layers, and two different UV laser systems, a 248 nm KrF excimer laser with 35 ns pulse duration and a frequency-tripled Nd:YAG with 50 ps pulses at 355 nm were compared to study the influence of the different pulse durations. For the laser wavelengths 248 and 355 nm, the absorption lengths are for gold 13 and 15 nm, respectively, and for silver 15 and 51 nm. With nanosecond pulses and 50 nm Ag films, well-defined deposits are obtained. In this case, the film thickness of the Ag layer has almost the same value as the absorption length of the beam for this radiation, whereas best results for a gold release layer were obtained with 25 nm films, that is, twice the absorption length. This implies that the laser energy is converted into different metal volumes and thermophysical parameters have to be also considered. However, with the picosecond regime and gold layers, a huge amount of metallic debris has been observed all around the deposits, but they could be easily removed with a soft air blow. But SEM images and X-ray dispersive energy analyses (EDAX) show that the transferred polymer pixels are contaminated on their surface with residual nano- and microparticulate metallic debris. These metal contaminations might be a potential drawback to the integration of metal DRL-based LIFT in electronic microdevice fabrication.

Metallic absorber layers were also used for the fabrication of small electroluminescent devices by LIFT sublimation of small light-emitting molecules in a designed vacuum chamber with a CW Nd:YAG laser [157]. The donor consists of a chromium layer sputtered onto a carrier substrate, which is then coated with the organic electroluminescent materials by high-vacuum deposition. The optical system included a highly dynamic and high-precision galvanometric scanner and allowed to achieve a focus diameter of about 35 μm on a working area of 16 cm × 16 cm. Small lines of the organic emitter materials were transferred when the focused laser beam was scanned along the back of the donor substrate, heating up locally the absorber layer. The spatially selective rapid laser heats and induces the flash sublimation of the organic molecules from the donor substrate onto the receiver substrate over a spacer-defined gap. This modified LIFT technique has been named *laser-induced local transfer* (LILT). Stripe patterns of OLEDs with a line width of ~300 μm were deposited according to a multilayered thin-film OLED device architecture consisting of a hole-blocking, an electron-transporting, and an emissive layer that was laser deposited. By optimization of the processing parameters, scanning speeds between 1 and 2 m s^{-1} were reported to achieve minimum line widths of 50–70 μm. In principle, if the deposition process is performed for red, green, and blue (RGB) emitting materials, full-color red, green, and blue organic light-emitting diode (RGB-OLED) display devices could be printed with fast processing speeds and on larger areas. However, the organic materials used need to be thermally robust to tolerate the laser-induced flash sublimation step.

5.4.3.3.4 Laser-Induced Thermal Imaging (LITI)

The principle of thermal imaging has been already presented in Section 5.4.3.3.1, which has been applied mainly for digital color printing. Thermal imaging enables the printing of multiple, successive layers via a dry additive deposition process at high speed and with micron-size resolution. With the ablative transfer of solid layers, the solvent compatibility issues, which are always encountered when printing sequential layers from solution, are entirely avoided. This technique has also been extended by DuPont to fabricate organic electronic devices [6], also with polymeric material sets [158]. The flexible donor sheet comprises essentially a LTHC layer (together with an optional ejection layer), which is coated with the printable electronic materials. Donor and receiver sheets are pressed together in close contact during the transfer process, which is performed by focused IR laser irradiation either from output-modulated diode lasers or scanned cw solid-state lasers. With appropriate tuning of the laser fluence and the local irradiation duration, the release layer system is thermally decomposed into gaseous products. Their expansion thus propels the top layer of the donor film onto the receiver. However, owing to inherent heat conduction processes with transient laser irradiation in the microsecond regime, the transfer material layer also gets noticeably heated, and therefore printable materials have to be reasonably heat resistant. Such a thermoimagible organic conductor was developed (a blend of single-wall CNTs and doped polyaniline) and printed in small patterns as source and drain electrodes for TFTs. The conductivity of the deposited features depends on the applied fluence of the 780 nm IR diode laser, and decreases with increasing fluences after a maximum closely above the transfer threshold fluence. This decrease at higher fluences reflects the increasing structural degradation of the conducting polymer at elevated thermal stress. Even when this example demonstrated that components of a functioning organic electronic device can be dry printed by laser transfer, the obtained process performance proved not to be easily applicable to other material systems without refinement of the individual processing parameters and especially the material properties. From a mechanistic point of view, thermal imaging may comprise both components, fast nonequilibrium ablation processes in combination with slower equilibrium melt transfer by laser heating. DuPont has also launched a digital printing technology for manufacturing liquid crystal display (LCD) color filters that eliminates the photomasks and liquid process chemicals used in traditional photolithography. The Optilon™ TCF system [159] uses donor films for the three filter colors RGB, which are transferred and patterned on the prefabricated LCD matrix by direct laser writing. Digital files of color filter patterns are directly image transferred from the dry Optilon RGB donor films on glass substrates.

Adapted from the laser imaging and graphics applications mentioned above, 3M has developed over the past decade as an extended digital transfer method for laser thermal patterning of electronic devices, such as flat panel displays and OLED emitters. Patterning of electroactive organic materials by dry mass transfer via thermal printing techniques is intrinsically difficult because both molecular and bulk electronic properties must be preserved, even when the materials are nonvolatile or thermally labile. The 3M technique has been called *laser-induced thermal imaging* (LITI) and uses a focused CW IR laser system as a heating device in a designed exposure system with fast oscillating elliptical laser beams [160].

As outlined in Figure 5.4.6, the donor sheet consists of a flexible 75 μm thick PET support that carries the functional nontransferring layers and a top-coated ultrathin transfer layer of electronic materials. The nontransfer layers are a 1.6 μm thick absorptive LTHC layer and an additional 1 μm thick shielding interlayer. The LTHC layer is a carbon-black-based black body absorber for the applied IR radiation. The optical density and thickness of the LTHC layer are important parameters in adjusting the donor heating profile, which affects imaging quality, performance, and device efficiency. With appropriate tuning of the process parameters, heating rates up to more than $10^7 \,^{\circ}\text{C s}^{-1}$, and peak temperatures within the LTHC layer reaching from 650 to 700 °C were observed. The peak temperature within the organic transfer layer is about 350 °C, but remains above 100 °C for less than 1.5 ms. The second coated layer consists of a photocured polymer that acts as an interlayer to protect the transfer layer from chemical, mechanical, and thermal damage. The interlayer also moderates other physical defects known to arise from warping or distortion of the LTHC so that the transfer layer surface remains smooth. In addition, fine tuning of the interlayer (which may consist of more than one layer depending on the transfer layer material requirements) affects crucially

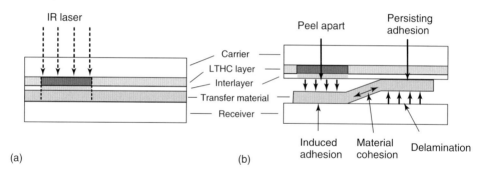

(a)

(b)

Figure 5.4.6 Scheme of the laser-induced thermal imaging (LITI) process. Donor and receiver sheet need to be in intimate contact during the mass transfer process by a thermal release/adhesion mechanism. (a) Laser irradiation of the light-to-heat conversion (LTHC) layer induces a controlled phase transition of the transfer material that leads to a local change in adhesion to the receiver surface. (b) The imaged pattern is mechanically developed by careful separation of the donor and receiver sheet.

the transfer layer release functions. Nevertheless, thermal transfer defects were observed during the LITI process of vacuum-coated OLED materials [161].

For OLED fabrication, the final coated transfer layer is commonly a thin doped evaporated emissive layer, typically some tens of nanometers in thickness. An optimal construction of phosphorescent OLED multilayer systems comprises usually a well-balanced system of charge transport and blocking layers for exciton confinement. With evaporation techniques, such stacks of multiple discrete organic layers can be deposited in reverse order on top of the interlayer on the donor sheet before the patterning step. During the LITI process, the entire multilayered stack will be transferred on the prepared receptor. A large-area laser imager with a twofold galvanoscanner system has been developed for the LITI mass transfer process with two superimposed Nd:YAG lasers focused to a Gaussian spot of 30 μm by 330 μm, delivering a laser power on the film plane up to 16 W. The optical design with fast oscillation of both elliptical beams transverse to the scan direction improves the line edge quality and precision by increasing the slope of the laser intensity profile closely above the transfer fluence threshold. Continuous variation of the scan speed and oscillation parameters such as frequency and beam dithering allow the fluence profile to be tailored to the materials and to achieve high cross-scan accuracy for writing precise line structures. Essential for good transfer results is the controlling of the thermal processes, the heat flow, and the temperature profiles from the LTHC layer to the transfer layer during the LITI transfer. A phase transition of the transfer material, for example, by exceeding the glass transition temperature T_g, facilitates the adhesion to the receptor surface. For a successful transfer, the transfer layer has to become detached from the interlayer at the same time, as outlined in Figure 5.4.6. A critical step is the separation of the irradiated donor and receiver sheet. Since there is no sharp cut or perforation along the edge of the transferred stripe, that is, along the border of exposed and unexposed donor film, the micromechanical properties of the film during removal of the donor have a strong influence on the formation of the line edges. This is the critical region of the donor–receptor laminate where a delicate balance of adhesion and release must be achieved in order to create acceptable line structures with smooth edges. Materials with high cohesion and film strength, such as certain commercial high-molecular mass light-emitting polymers (LEPs), tend to overwhelm the tearing forces at the line edge, leading to poor image quality. Therefore, industrial applications are rather focused to the patterning of vapor-deposited conventional small molecular phosphorescent emitting materials with less intermolecular forces, which have high electroluminescent efficiencies and long OLED device lifetimes. However, carefully tailored formulation of such polymeric materials as blends with electrically inert host polymers and further hole-transporting materials allowed the deposition of satisfactory line patterns with promising emission characteristics [162, 163]. Receptor substrates are typically glass coated with ITO as the transparent electrode layer and a hole-conducting layer (80 nm thick PEDOT:PSS) spin coated on top of it. In order to improve the adhesion of the transfer layer to the PEDOT-coated receiver, an additional thin siloxane layer was added in order to cover the hydrophilic surface with a hydrophobic ultrathin film to enhance the compatibility and wettability at the interface with the transferred hydrophobic polymer blend [164]. About 80 μm wide stripes of the LEP blend were transferred from donor films precoated with red-, green-, and blue-emitting materials to fabricate OLEDs. In 2003, Samsung SDI presented the first full-color active-matrix OLED display prototypes fabricated with an LITI process step [162]. From Samsung's extensive patent application activities concerning development and refining of technical process equipment compounds during the past years, it might be concluded that the LITI process is already close to industrial integration in full-color OLED flat-panel production lines.

5.4.3.3.5 Metal Nanoparticle Absorbers

The size-dependent quantum dot absorbance of colloid Ag NPs has been exploited to build tailored absorptive LTHC layers with an absorption in the visible range [64]. About 100–200 nm thick films of 30–40 nm sized Ag NPs show a strong absorption around 530 nm, which fits the emission of the 532 nm emission of frequency-doubled Nd:YAG lasers. In order to prevent the Ag NPs from aggregation and coalescence into a bulk film during spin coating, they were protected by a self-assembled alkanethiol monolayer (SAM). The absorbing nanomaterial layer is only loosely connected and can therefore be easily vaporized by the laser. The green-emitting OLED material tris-(8-hacroxyquinoline) aluminum (Alq_3) was deposited by an RIR-PLD technique to form 200–300 nm thick layers. The receptor substrate was coated with a siloxane layer as adhesion promoter and placed with a small transfer gap (<100 μm). Pattern transfer of Alq_3 pixels (0.9 mm × 0.9 mm) was obtained in a fluence range of 50–150 mJ cm^{-2}. As an indicator of the conservation of the materials properties the fluorescence of Alq_3 was measured before and after transfer, and the spectra showed no significant alterations. The method has been named *nanomaterial-enabled laser transfer* (NELT). However, it is

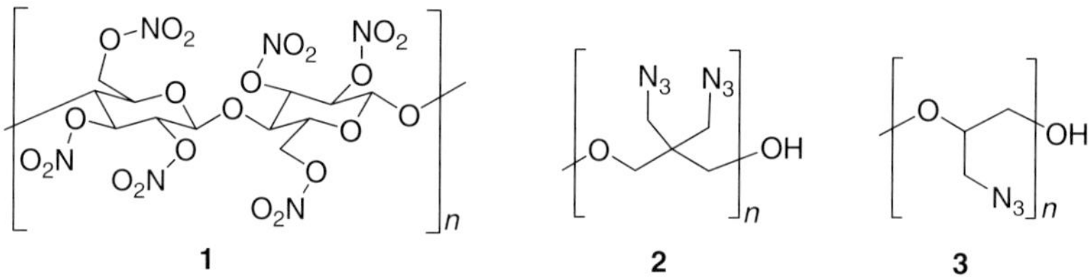

Figure 5.4.7 Chemical structures of nitrogen-rich polymers: nitrocellulose (1), poly-BAMO (2), and glycidyl azide polymer GAP (3).

not clear which amount of the ablated Ag nanoparticle release layer is co-deposited with the molecular OLED material.

5.4.3.3.6 Absorbing Polymer Release Layer Systems

The laser-triggered decomposition of energetic gas-generating polymer release layers has been presented in the section on LAT with metal absorbing layers, as well as composites of thermodecomposable polymers doped with black carbon or dispersed IR dyes as radiation absorber. Similarly, LIFT transfer of polymer composites doped with a CNT content of 5–10% (wt), which provides sufficient absorption to interact with a UV laser, has been mentioned above. Chemical structures of some representative nitrogen-rich polymers are shown in Figure 5.4.7. Nitrocellulose (or cellulose nitrate) is a self-oxidizing polymer and its energetic performance depends on the nitrogen content. With a nitrogen content higher than 13% it has been used as "gun cotton" and rocket propellant. Poly-BAMO (2) and GAP (3) are energetic polymers with a polyether backbone that bears per repeating unit two, and one azide (N_3) groups, respectively, in the side chains, that can each split off one molecule of elemental nitrogen (N_2) on thermal ignition or appropriate UV irradiation. Immediate decomposition of the labile nitrogen moieties leads to a significant volume expansion by violent gas generation that can be used as a propellant thrust.

Both nitrite ester as well as azide moieties attached as side chains to the polymer main chain show an absorption in the UV range between ∼300 and 250 nm. They can therefore be electronically excited by UV light within that range, and a phototriggered decomposition of the nitrogen-containing functional groups can be initiated on irradiation. In another type of tailor-made nitrogen-containing polymers, UV-photocleavable aryltriazene chromophores (Ar−N=N−N−) are covalently incorporated within the polymer backbone [165]. By that, UV laser irradiation of the polymer induces a photofragmentation reaction, and the main chain is cleaved into nitrogen gas and small volatile organic fragments, as outlined in Figure 5.4.8.

Films of such aryltriazene polymers show a high absorption in the range of about 250–350 nm with an absorption coefficient (at 308 nm $\alpha_{lin} \sim 100\,000$ cm^{-1}) and can be efficiently decomposed on UV exposure [166]. Therefore, films of these photosensitive polymers proved to be excellently suitable for laser ablation applications. The laser fluence threshold for ablation is very low ($\sim$20 mJ cm^{-2}), the ablation rates per laser pulse are quite high, the ablated surface patterns are sharp

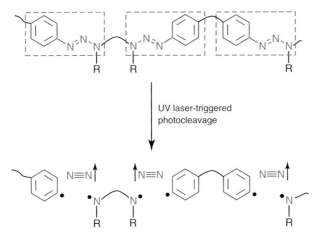

Figure 5.4.8 Schematic general structure of the UV-cleavable polymers. The polymer main chain contains aryl triazene chromophore units (dashed frames), which decompose on UV light irradiation, leading to an irreversible fragmentation of the polymer backbone into small volatile organic fragments and nitrogen gas.

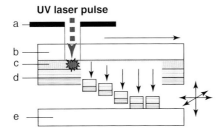

Figure 5.4.9 Principle of the modified laser-induced forward transfer (LIFT) process with an absorbing sacrificial polymer film: (a) mask, (b) carrier substrate, (c) sacrificial thin film of "exploding" aryltriazene photopolymer serving as a pressure generator on laser irradiation, (d) thin layers of transfer material, and (e) receiver substrate with transferred pixels.

and clean, and no debris by redeposition of ablated material or carbonization was observed [73, 167, 168]. The laser-triggered photofragmentation process results in an abrupt volume expansion jet of "ablation" products and a transient high-pressure jump. This effect can be utilized for LIFT applications, when the aryltriazene polymer films are used as absorbing sacrificial DRLs. The overlying film of solid and pure transfer materials is punched out and catapulted integrally toward the receiver surface as outlined in Figure 5.4.9. Various transfer materials were investigated as model systems (e.g., pyrene-doped polymethyl methacrylate (PMMA) films) [169] to define the best process conditions, such as optimum film thickness for the individual material layers with relation to the applied laser fluence [170]. As examples, the transfer of well-defined pixels of thin polymer films was demonstrated with aqueous-based films of the biopolymer gelatin and a methylated cellulose derivative [170]. The accurate LIFT deposition of thin polymer films has recently been demonstrated for microsensor applications. Homogeneous and hole-free polyethyleneimine (PEI) pixels with a low surface roughness were deposited in the same way with a triazene photopolymer DRL after process optimization [171]. Meanwhile, LIFT of PEI thin films to sensing chip units allowed measuring of chemoresponsive signals [165].

Use of the polymeric DRL also enables the accurate LIFT deposition of metal films. Transferred 80 nm thin aluminum film pixels (500 μm × 500 μm) showed clear-cut edges and a homogeneous layer morphology without traces of splashing or codeposited debris outside the pixel area. It has been found that the mechanical as well as the thermophysical material properties of the transfer layer and carrier substrate have significant effects on the transfer results. Owing to the influence of the thermal diffusion toward the carrier substrate during nanosecond laser pulse duration, the fluence threshold for ablation was found to increase substantially for ultrathin films (<50 nm), depending on the material properties of the carrier material [172].

The spatial shape of the focused laser spot directly defines the outline of the catapulted pixel. Since the sacrificial polymer release layer protects the transfer layer from the incident UV irradiation, even highly sensitive materials can be gently transferred and deposited. Only small and volatile organic fragments are formed by the laser-triggered photopolymer decomposition, and therefore only a minimum of contamination of the transferred deposits may be expected. This is a substantial advantage compared to metal absorbing layers, which are known to be codeposited with the transferred material after laser-induced volatilization during the LIFT process [147]. Compared with previous LTHC layers frequently used with IR lasers, the UV-light-triggered photodecomposition process reduces significantly the heat load to the transfer layer, which is detrimental to the properties or functionality of sensitive materials.

The aryltriazene photopolymers, used as the catapulting layer, were tailored to have an absorption peak that fits the XeCl excimer laser wavelength at 308 nm [173]. The flatter beam energy profile of the less coherent excimer laser is its principal advantage over Gaussian solid-state UV lasers, as this results in a much cleaner transfer. Other lasers that have shown how differences in pulse duration and the beam energy profile affected the quality of the transferred material have been studied [174]. Despite improving the catapulting efficiency, there was no clear advantage of shorter pulse lengths over the standard 30 ns pulse length of the XeCl laser.

The potential of aryltriazene photopolymers as catapulting sacrificial layers in LIFT applications was demonstrated with various highly sensitive materials: assisted by a ~100 nm thick sacrificial triazene polymer, DRL living mammalian neuroblast cells were transferred and gently deposited on a receiver substrate using a 193 nm ArF nanosecond excimer laser [175]. With focused laser spots of about 20 μm in diameter, plaques of cell bundles embedded in an extracellular matrix were punched out from a donor substrate and transferred to a bioreceiver over a gap of 150 μm above the threshold fluence for cell transfer of 50 mJ cm^{-2}, gentle enough that the functionality was not impaired at all, and the cells started reproducing instantly. After 48 h of the transfer, the cells showed intact preservation of nuclei with well-developed axonal extensions, demonstrating that the technique is competent in creating patterns of viable cells at low fluences. This opens up new possibilities for the manufacture of biosensors in which living cells should be precisely deposited onto microchips. By developing more automated computer-controlled LIFT setups with *in situ* monitoring by charged-coupled detector (CCD) cameras more complex cell arrays can be deposited in a reproducible way, as recently reported [176].

In the same way, semiconducting multispectral nanocrystal quantum dots (NCQDs) were successfully transferred by the photopolymer-assisted LIFT setup into laterally patterned arrays [177]. A patterned metal mask was used for the LIFT deposition of two different sizes of CdSe(CdS) nanoquantum dots (5 and 6 nm in diameter which show yellow and red color) coated on a bilayer of triazene photopolymer DRL and a thin aluminum film as conducting electrode. Interdigitating 6 × 6 matrix arrays with pixel sizes of 800 μm were printed onto ITO-coated glass substrates. PL of the NCQDs before and after transfer was found to be nearly unchanged, indicating the conservation of materials functionality during the transfer.

To prove the validity of the photopolymer-based LIFT approach, a working miniaturized model OLED device has been fabricated [178]. For that, micropixel stacks consisting of a bilayer of the electroluminescent polyparaphenylene vinylene derivative MEH-PPV (poly[2-methoxy-5-(2′-ethylhexyloxy)-p-phenylene vinylene]), together with an aluminum cathode, was transferred onto a receiver substrate coated with a prestructured ITO anode (as shows in Figure 5.4.10). On the fused silica donor carrier three individual thin layers had to be deposited one after

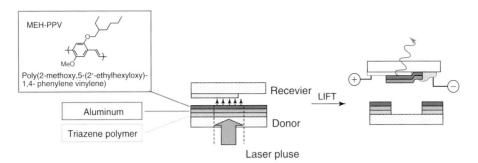

Figure 5.4.10 Layer architecture of the donor and receiver substrate for the pixel transfer with a thin layer (40–90 nm) of the electroluminescent polymer MEH-PPV, the aluminum cathode (70 nm), and the aryltriazene photopolymer film (100 nm). The incident laser pulse decomposes the sacrificial photopolymer layer and catapults the Al/MEH-PPV bilayer system in one step toward the prestructured transparent indium–tin oxide (ITO) anode on top of the receiver surface.

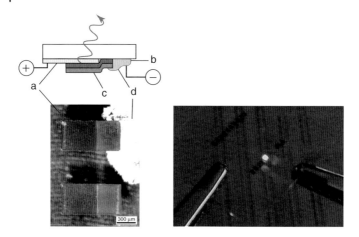

Figure 5.4.11 Resulting OLED pixel device after LIFT: the MEH-PPV layer (b) and cathode (c) are printed onto the ITO anode (a). The photograph shows two adjacent pixels seen through the ITO-coated substrate. After contacting with silver paste (d) the pixel shows an orange–red light emission.

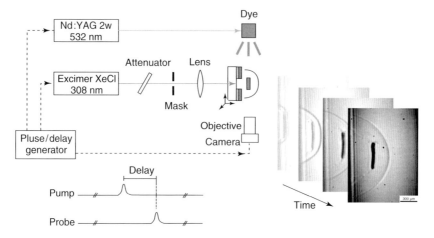

Figure 5.4.12 Setup for the two laser "pump-probe" shadowgraphy experiments for time-resolved imaging of the forward transfer processes.

another, corresponding to the inverse stack of the later OLED device: a spin-coated basic triazene photopolymer layer (100 nm) followed by a thermally evaporated Al layer (70 nm), and as the top layer, a spin-coated MEH-PPV film (90 nm) (shown in Figure 5.4.10).

With single nanosecond pulses of a 308 nm XeCl excimer laser, MEH-PPV and the aluminum cathode were directly printed onto the ITO anode in one step. The transferred pixel required that contact be made only with a DC voltage for the emission of orange light (Figure 5.4.11). The functionality of operating devices was characterized by current–voltage and electroluminescence measurements, which prove that the integrity of the transferred materials has been fully preserved during the improved LIFT deposition process. The transfer process has been further extended, by shaping the laser beam using an appropriate mask, so that a pattern of multiple pixels may be deposited with a single pulse, providing a promising method for the production of full-colored displays on flexible substrates.

To get more insight into the dynamic mechanism of the LIFT process, a time-resolved pump-probe microimaging technique known as *shadowgraphy* was adapted to visualize the ejection of micropixels and the transfer process that also involves the occurrence of shockwaves [179]. The experimental setup for such investigations is outlined in Figure 5.4.12.

Such shockwaves are generated by laser–material interactions, in this case by the laser-triggered "microexplosion" of the catapulting layer, and their strengths depend on the pressure of the ambient atmosphere. As an example, the shadowgraphy images in Figure 5.4.13 show the ejection process of a model pixel flyer (80 nm thick film of aluminum on top of 350 nm of absorber polymer) in ambient atmosphere, where the evolving shockwave together with the catapulted pixel can be clearly seen. The upper row of images shows the time-resolved development of the shockwave and the flyer at a constant fluence of 360 mJ cm^{-2} with increasing time intervals (200 ns steps) after the LIFT pump pulse. A bilayer system of 80 nm aluminum on top of a triazene DRL of 350 nm was irradiated at 308 nm with 360 mJ cm^{-2} per pulse. After 200 ns the evolving shockwave together with the ejected flyer appears with a piston-type flat surface. At 400 ns the shockwave appears more hemispherical and propagates with a higher speed than the flyer. The shape and morphology of the catapulted pixel stays intact over a distance of more than 300 μm, as these images show. The second row shows the influence of the applied fluence after a fixed time delay of 800 ns each: the propagation velocity of both the shockwave and the pixel flyer depends on the energy of the incident laser pulse, which allows a fine tuning of the transfer conditions at low fluences. However, at a fluence of already around 1 J cm^{-2} the blast power is high enough to accelerate the flyer more than the shockwave, so that it gets destroyed on interaction. Such time-resolved mechanistic investigations are helpful to define optimized conditions for the LIFT process of various transfer material layer systems. From such shadowgraphy image series, the individual propagation speed of the shockwave and the flyer can be derived depending on the laser fluence and the thickness of the sacrificial DRL. The analysis of these values allowed for deriving the first simple model for the energy balance in order to estimate the conversion efficiency of the incident laser into the kinetic energy of the ejected flyer [180].

For example, solid ceramic thin films of gadolinium gallium oxide and ytterbium-doped yttrium aluminum oxide have been deposited by the sacrificial polymer-assisted LIFT [181]. The dependence of the ablation dynamics and quality of the ejected donor material on the laser fluence and thickness of the sacrificial and donor layers were investigated by means of this shadowgraphy technique. From detailed analyses of such studies, dynamic process parameters and information on the energy conversion can be derived, such as propagation velocities of catapulted flyers and correlated shockwaves that have to be carefully tuned for

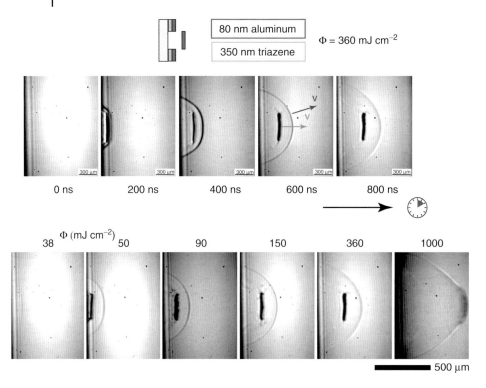

Figure 5.4.13 Shadowgraphy microimages of the time-resolved development of the shockwave and flyer ejection for a laser fluence Φ of 360mJ cm^{-2} (upper row). The flyer consists of a layer of 80 nm aluminum, which was coated on top of a 350 nm thick triazene photopolymer. The flyer stays stable over quite a long distance of more than 0.3 mm. The image sequence shows the different propagation speeds v of the flyer and the shockwave. Second row: The forward ejection of the same model system was studied to investigate the fluence dependence of the generated thrust. Images are taken each at a constant delay time of 800 ns after the laser pulse. Flyer velocity and shockwave shape depend on the applied laser fluence.

successful high-quality LIFT results necessary for reliable device fabrication [180]. Shadowgraphy studies revealed also that a smooth deposition of micropixels over a narrow transfer gap can be crucially disturbed by the interaction of the catapulted flyer with reflected shockwaves [182]. An example of the interaction of the reflected shockwave echo with a thin metal film flyer within a transfer gap of 500 μm is shown in Figure 5.4.14. The evolving shockwave front reaches the receiver substrate after 400 ns and starts to be reflected. As a consequence, the catapulted flyer collides with the shockwave echo. This interaction reduces the speed of the flyer and causes its disintegration before being deposited onto the receiver. Here, proper choice of the pressure conditions and gap distance are further variables in the multiparameter space of the LIFT process.

The functionality of triazene photopolymers as DRL has also been studied with femtosecond laser pulses at 800 nm [183] for the transfer of 10 μm ceramic disc

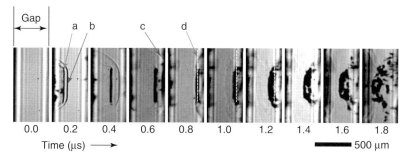

Figure 5.4.14 Shadowgraphy analysis of forward transfer with a gap distance of 0.5 mm. Sequence of images taken with a sample as shown in Figure 5.4.13 (80 nm Al on top of 350 nm triazene polymer DRL) at a fluence of 160 mJ cm^{-2}. After 200 ns time delay, the catapulted flyer (a) and the shockwave front (b) can be seen. When the shockwave has reached the receiver surface, the echo wave starts to be reflected (c), and the velocity of the incoming flyer is slowed down after interaction with the shockwave echo (see the constant position of the flyer front indicated by the dotted line (d) during following 400 ns). Finally, the flyer gets disintegrated into small fragments that are scattered backwards.

pellets [184]. The threshold transfer fluence with the photopolymer DRL was found to be only ~20% of the polymer ablation threshold at the laser wavelength. This behavior is in stark contrast to nanosecond DRL-LIFT using the same polymer, where the transfer threshold is reproducibly found to be slightly greater than the ablation threshold. This decrease has been attributed to ultrafast shockwave generation in the constrained polymer layer under femtosecond irradiation being the driving force for the femtosecond LIFT with the polymer DRL. However, further effects may also play a role here, such as white-light continuum generation in the carrier substrate, as well as standing optical waves [185].

Special Applications Using Femtosecond Lasers The use of femtosecond pulsed lasers has certain advantages, as described above, and can be used for the transfer of very small pixels. Deposition of nanoscale droplets of Cr was achieved using femtosecond Ti:sapphire LIFT, and deposits around 300 nm in diameter are obtained from a 30 nm thick source film [186]. Femtosecond lasers also have the advantage of high repetition rates, and the possibility of beam shaping (temporal and lateral profile). Temporal beam shaping can be used to optimize with feedback loops, while lateral beam shaping can be used to precut (structure) the transfer pixel before transfer. Multiple, subthreshold fluence femtosecond pulses have been used to lessen the adhesion of a donor film to a support substrate to facilitate the subsequent forward transfer step of solid "pellets." To define the area for transfer in "hard" solid donor films, such as glasses, crystals, ceramics, and so on, more precisely, and to allow for more reproducible pellet shapes, an outer ring with relatively higher intensity was added to the transfer laser pulses by means of the near-field diffraction pattern of a circular aperture. This approach was called *ballistic laser-assisted solid transfer* (BLAST) [187]. Since BLAST does not require a

DRL material, it is potentially applicable to the forward transfer deposition of any "hard" donor. A further laser-based microstructuring technique for the etching and deposition of solid materials, called *laser-induced solid etching* (LISE), utilizes also the absorption of femtosecond-duration laser pulses in a constrained metal film between two bulk substrates [188]. At least one of the two substrates has to be transparent to the laser wavelength. The very rapid pressure increase in the "sandwiched" metal film following irradiation is believed to initiate crack propagation in one or the other of the bulk substrates. By spatially shaping the laser beam, the cracking process can be controlled to etch small solid flakes of material from the substrates, and smooth, micron-scale pits and trenches in silicon and silica could be etched using LISE. As a unique feature of LISE, the material etched from the bulk substrate is removed as a single solid piece and is not shattered, melted, or vaporized by the process. Hence, the etched material can be collected on the other bulk substrate used in the process. In this way, micron-scale dots and lines of silica have been deposited onto silicon and vice versa. Minimal evidence of melting during the process has been observed, suggesting that LISE may be a useful technique for the forward transfer direct-write of intact solid materials.

Variation of the DRL LIFT Approach   A UV-absorbing poly(ethylene naphthalate) (PEN) foil (1.35 μm, showing the maximum UV–vis absorption λ_{max} around 360 nm) has been applied in a tailored laser-based setup for the contact- and contamination-free collection of histologic material for proteomic and genomic analyses. The sample of interest is prepared by a laser microdissection step followed by laser-induced transport (laser pressure catapulting) toward a collector microvessel [189, 190]. The tissue section is prepared on top of the PEN carrier membrane and the cell transfer achieved with nanosecond pulses of a N_2 laser at 337 nm. Time-resolved ejection dynamics were investigated by an ultrafast imaging system.

5.4.3.3.7   Blistering Lift-Off Methods

A 4 μm thick film absorbing layer of polyimide was used for the LIFT deposition of sensitive embryonic stem cells [191]. The film thickness is significantly larger than the laser absorption depth of about 750 nm at the 355 nm laser emission wavelength. In this case, the laser causes vaporization inside the sacrificial layer below the transfer material interface. The polymer expands because of the generated pressure with plastic deformation and builds up a curved blister. The hot gases cool and stay trapped inside the blister cavity, thereby remaining isolated from the transfer material. With appropriate tuning of the laser fluence, viscous materials are ejected from the fast bulging polyimide surface on top of the expanding cavity, and the jetted droplet size increases with increasing laser fluences. At higher laser fluences the polyimide membrane will be more and more ablated and susceptible to rupture. When the film ruptures, hot vapor from the ablation process is able to escape into the transfer material layer and blow it locally away, leading to a more energetic explosive ejection and deposition process that was found to lead to splashes. Transfer of viable mouse embryonic stem cells spread as a glycerol/cell growth media suspension on the polyimide sacrificial layer was successfully achieved over a 100 μm gap.

Titanium metal coatings with a thickness far above the absorption length at 532 nm emission were used in a setup similar to the blistering membrane [192]. The metal layer was coated with a ~1 μm thick layer of ultradispersed diamond (particle size below 200 nm) and irradiated with single 50 ps pulses. Transfer of the diamond nanopowder over a defined gap was observed between a minimum threshold fluence and the fluence for total ablation of the Ti layer, whereby the deposited diamond nanopowder was accompanied by deposition of metal droplets on the acceptor substrate (ablative transfer). In an intermediate fluence regime between these critical values, the Ti film stays optically intact without perforations and cracks. Ejection of the nanopowder can be attributed to a fast movement of the transient blister at the metal-quartz-carrier interface and the shock waves. With these transfer conditions, the transfer layer stays completely isolated from the incoming radiation and ablation products of the sacrificial layer, but the transfer efficiency is limited by the adhesion of the transfer material to the blistering layer. Since the ejection of the particles starts on top of an expanding membrane-like convexity, a fluence-dependent transverse scattering of the deposited particles was observed. Even when the method of blister-based LIFT seems to be limited to powders or viscous suspensions, it offers an interesting potential for the transfer of wet biomaterials that can be used for biodevice fabrication with complete shielding of the laser radiation.

5.4.3.3.8 Other Methods

With a polyimide-absorbing sacrificial polymer release layer, a method was reported for the laser-driven release on demand of prefabricated microstructures [193]. This DRL-LIFT-based approach was used for the batch assembly of hybrid microelectromechanical systems (MEMS) built from parts fabricated on different substrates. A classic photolithographic process has been applied to build up about 100 μm high microstructured components, but on top of a polyimide basis layer that is later used as the absorbing DRL. For the assembly step, prefabricated parts are aligned and fitted together by LIFT catapulting of the prepared microstructure to the chip with a KrF excimer laser. The same procedure was adapted later for the LIFT deposition of more complex bar die device units and Si substrates, including commercial LM555 timer chips [194]. As a proof of principle, a blinker circuit was assembled by laser transferring each of its submillimeter-sized components to a common chip board. An excimer laser was preferentially used with respect to its larger area beam capable of uniformly irradiating a bigger surface of the device, allowing for the transfer to occur with a single laser pulse. The release and transfer of single unpackaged semiconductor devices into a pocket or recess in a substrate enables the LIFT technique to perform the same way as pick-and-place machines used in circuit board assembly to fabricate embedded microelectronic circuits. As a demonstration example, a 300 μm × 300 μm sized InGaN LED was laser deposited and it emitted light when powered [195].

5.4.4
Conclusions and Future Aspects

LIFT has matured since the early reports in the early 1970s and 1986 from a curiosity to a process with many variations, as exemplified by the large number of acronyms (LAT, LITI, LILT, MAPLE-DW, AFA-LIFT, BLAST, LISE, etc.) for the process. The range of transferred materials has also steadily increased from single inorganic compounds such as metals, to imaging dyes and inks, and further to functionally defined multilayers and highly complex and sensitive biological systems. Optimization of the process for individual materials and applications by more insight in the involved mechanisms of the dynamic processes have finally led to LIFT being applied for the fabrication of working devices, such as miniaturized OLEDs or chemosensors. In summary, the recent developments suggest that variations of the LIFT technology rather than the LITI process will be used as industrial fabrication methods.

Acknowledgments

The authors acknowledge the Swiss National Science Foundation (SNSF), the Paul Scherrer Institut (PSI), and Swiss Federal Laboratories for Materials Science and Technology (EMPA) for financial support of their research presented herein. A part of their present work has been financially supported by the FP 7 European project e-LIFT(project n° 247868 −call FP7-ICT-2009-4).

References

1. Kyrkis, K.D., Andreadaki, A.A., Papazoglou, D.G., and Zergioti, I. (2006) in *Recent Advances in Laser Processing of Materials* (eds J. Perrière, E. Millon, and E. Fogarassy), Elsevier, pp. 213–241.

2. Arnold, C.B., Serra, P., and Piqué, A. (2007) Laser direct-write techniques for printing of complex materials. *MRS Bull.*, **32**, 23–31.

3. Madou, M.J. (1997) *Fundamentals of Microfabrication*, CRC Press LLC.

4. Quist, A.P., Pavlovic, E., and Oscarsson, S. (2005) Recent advances in microcontact printing. *Anal. Bioanal. Chem.*, **381**, 591–600.

5. Wilbur, J.L., Kumar, A., Biebuyck, H.A., Kim, E., and Whitesides, G.M. (1996) Microcontact printing of self-assembled monolayers: applications in microfabrication. *Nanotechnology*, **7**, 452.

6. Blanchet, G. and Rogers, J. (2003) Printing techniques for plastic electronics. *J. Imaging Sci. Technol.*, **47**(4), 296–303.

7. Hobby, A. (1990) Fundamentals of screens for electronics screen printing. *Circuit World*, **16**(4), 16–28.

8. Carr, B.P. (1994) Introduction to thick film fine line printing screens. *Microelectron. Int.*, **11**(1), 4–7.

9. Pardo, D.A., Jabbour, G.E., and Peyghambarian, N. (2000) Application of screen printing in the fabrication of organic light-emitting devices. *Adv. Mater.*, **12**(17), 1249–1252.

10. Calvert, P. (2001) Inkjet printing for materials and devices. *Chem. Mater.*, **13**(10), 3299–3305.

11. de Gans, B.-J., and Schubert, U.S. (2004) Inkjet printing of well-defined polymer dots and arrays. *Langmuir*, **20**(18), 7789–7793.

12. Gamerith, S., Klug, A., Scheiber, H., Scherf, U., Moderegger, E., and List, E.J.W. (2007) Direct ink-jet printing of Ag-Cu nanoparticle and Ag-precursor based electrodes for OFET applications. *Adv. Funct. Mater.*, **17**(16), 3111–3118.

13. Ko, S.H., Chung, J., Pan, H., Grigoropoulos, C.P., and Poulikakos, D. (2007) Fabrication of multilayer passive and active electric components on polymer using inkjet printing and low temperature laser processing. *Sens. Actuator A*, **134**(1), 161–168.

14. Plötner, M., Wegener, T., Richter, S., Howitz, S., and Fischer, W.-J. (2004) Investigation of ink-jet printing of poly-3-octylthiophene for organic field-effect transistors from different solutions. *Synth. Met.*, **147**(1–3), 299–303.

15. Lewis, J.A. and Gratson, G.M. (2004) Direct writing in three dimensions. *Mater. Today*, (July/August) 32–39.

16. Therriault, D., White, S.R., and Lewus, J.A. (2003) Chaotic mixing in three-dimensional microvascular networks fabricated by direct-write assembly. *Nat. Mater.*, **2**, 265–271.

17. Eason, R.W. (2007) *Pulsed Laser Deposition of Thin Films Applications-Led Growth of Functional Materials*, John Wiley & Sons, Inc., Hoboken, NJ.

18. Chrisey, D.B. and Hubler, G.K. (1994) *Pulsed Laser Deposition of Thin Films*, John Wiley & Sons, Inc., New York.

19. Blanchet, G.B. (1995) Deposition of poly(methyl methacrylate) films by UV laser ablation. *Macromolecules*, **28**, 4603–4607.

20. Blanchet, G.B. (1996) Prepare fluoropolymer films using laser ablation. *Chemtech*, **26**(6), 31–35.

21. Chrisey, D.B., Pique, A., McGill, R.A., Horwitz, J.S., Ringeisen, B.R., Bubb, D.M., and Wu, P.K. (2003) Laser deposition of polymer and biomaterial films. *Chem. Rev.*, **103**, 553–576.

22. Yang, X., Tang, Y., Yu, M., and Qin, Q. (2000) Pulsed laser deposition of aluminum tris-8-hydroxyquinline thin films. *Thin Solid Films*, **358**(1–2), 187–190.

23. Chrisey, D.B., Pique, A., Fitz-Gerald, J., Auyeung, R.C.Y., McGill, R.A., Wu, H.D., and Duignan, M. (2000) New approach to laser direct writing active and passive mesoscopic circuit elements. *Appl. Surf. Sci.*, **154**, 593–600.

24. Losekrug, B., Meschede, A., and Krebs, H.U. (2007) Pulsed laser deposition of smooth poly(methyl methacrylate) films at 248 nm. *Appl. Surf. Sci.*, **254**(4), 1312–1315.

25. Suske, E., Scharf, T., Krebs, H.U., Junkers, T., Buback, M., Suske, E., Scharf, T., Krebs, H.U., Junkers, T., and Buback, M. (2006) Mechanism of poly(methyl methacrylate) film formation by pulsed laser deposition. *J. Appl. Phys.*, **100**(1), 014906.

26. Kecskemeti, G., Smausz, T., Kresz, N., Toth, Zs., Hopp, B., Chrisey, D., and Berkesi, O. (2006) Pulsed laser deposition of polyhydroxybutyrate biodegradable polymer thin films using ArF excimer laser. *Appl. Surf. Sci.*, **253**, 1185–1189.

27. Blanchet, G.B. and Fincher, C.R. (1994) Laser-induced unzipping–a thermal route to polymer ablation. *Appl. Phys. Lett.*, **65**(10), 1311–1313.

28. Blanchet, G.B. and Fincher, C.R. (1994) Thin-film fabrication by laser-ablation of addition polymers. *Adv. Mater.*, **6**(11), 881–887.

29. Purice, A., Schou, J., Kingshott, P., Pryds, N., and Dinecu, M. (2007) Characterization of lysozyme films produced by matrix assisted pulsed laser evaporation (MAPLE). *Appl. Surf. Sci.*, **253**, 6451–6455.

30. Bubb, D.M., Toftman, B., Haglund, R.F. Jr., Horwitz, J.S., Papantonakis, M.R., McGill, R.A., Wu, P.W., and Chrisey, D.B. (2002) Resonant infrared pulsed laser deposition of thin biodegradable polymer films. *Appl. Phys. A*, **74**, 123–125.

31. Toftmann, B., Papantonakis, M.R., Auyeung, R.C.Y., Kim, W., O'Malley, S.M., Bubb, D.M., Horwitz, J.S., Schou, J., Johansen, P.M., and Haglund, R.F. Jr. (2004) UV and RIR matrix assisted pulsed laser deposition of organic MEH-PPV films. *Thin Solid Films*, **453–454**, 177–181.

32. Lantz, K.R., Pate, R., Stiff-Roberts, A.D., Duffell, A.G., and Smith, E.R.

(2009) Comparison of conjugated polymer deposition techniques by photoluminescence spectroscopy. *J. Vac. Sci. Technol. B*, **27**(5), 2227–2231.

33. Johnson, S.L., Park, H.K., and Haglund, R.F. Jr. (2007) Properties of conductive polymer films deposited by infrared laser ablation. *Appl. Surf. Sci.*, **253**, 6430–6434.

34. Braudy, R.S. (1969) Laser writing. *Proc. IEEE*, **57**, 1771–1772.

35. Levene, M.L., Scott, R.D., and Siryj, B.W. (1970) Material transfer recording. *Appl. Opt.*, **9**(10), 2260–2265.

36. Roberts, D.L. (1971) Laser Recording technique using combustible blow-off. US Patent 3, 787, 210, The National cash Register Comp., Dayton, filed Sept. 30, 1971.

37. Peterson, J.O.H. (1974) Printing plate by laser transfer. US Patent 3, 964, 398, Scott Paper Comp., Philadelphia, filed Jan. 17, 1974.

38. Bohandy, J., Kim, B., and Adrian, F.J. (1986) Metal deposition from a supported metal film using an excimer laser. *J. Appl. Phys.*, **60**(4), 1538–1539.

39. Veiko, V.P., Shakhno, E.A., Smirnov, V.N., Mioskovski, A.M., and Nikishin, G.D. (2006) Laser-induced film deposition by LIFT: physical mechanisms and applications. *Laser Part. Beams*, **24**, 203–209.

40. Claeyssens, F., Klini, A., Mourka, A., and Fotakis, C. (2007) Laser patterning of Zn for ZnO nanostructure growth: comparison between laser induced forward transfer in air and in vacuum. *Thin Solid Films*, **515**, 8529–8533.

41. Baum, H.H. and Comita, P.B. (1992) Laser-induced chemical vapor deposition of metals for microelectronics technology. *Thin Solid Films*, **218**, 80–94.

42. Grenon, B.J. *et al.* (1997) Methods for repair of photomasks. US Patent 6, 165, 649, IBM Corporation, Inv., filed Jan. 21, 1997.

43. Germain, C., Charron, L., Lilge, L., and Tsui, Y.Y. (2007) Electrodes for microfluidic devices produced by laser-induced forward transfer. *Appl. Surf. Sci.*, **253**, 8328–8333.

44. Chrisey, D.B., Pique, A., Fitz-Gerald, J., Auyeung, R.C.Y., McGill, R.A., Wu, H.D., and Duignan, M. (2000) New approach to laser direct writing active and passive mesoscopic circuit elements. *Appl. Surf. Sci.*, **154–155**, 593–600.

45. Zergioti, I., Mailis, S., Vainos, N.A., Papakonstantinou, P., Kalpouzos, C., Grigoropoulos, C.P., and Fotakis, C. (1998) Microdeposition of metal and oxide structures using ultrashort laser pulses. *Appl. Phys. A*, **66**, 579–582.

46. Zergioti, I., Mailis, S., Vainos, N.A., Fotakis, C., Chen, S., and Grigropoulos, C.P. (1998) Microdeposition of metals by femtosecond excimer laser. *Appl. Surf. Sci.*, **127–129**, 601–605.

47. Papakonstantinou, P., Vainos, N.A., and Fotakis, C. (1999) Microfabrication by UV femtosecond laser ablation of Pt, Cr and indium oxide thin films. *Appl. Surf. Sci.*, **151**, 159–170.

48. Klini, A., Loukakos, P.A., Gra, D., Manousaki, A., and Fotakis, C. (2008) Laser induced forward transfer of metals by temporally shaped femtosecond laser pulses. *Opt. Express*, **16**(15), 11300–11309.

49. Shah, A., Parashar, A., Singh Mann, J., and Sivakumar, N. (2009) Interference assisted laser induced forward transfer for structured patterning. *Open Appl. Phys. J.*, **2**, 49–52.

50. Landström, L., Klimstein, J., Schrems, G., Piglmayer, K., and Bäuerle, D. (2004) Single-step patterning and the fabrication of contact masks by laser-induced forward transfer. *Appl. Phys. A*, **78**, 537–538.

51. Othon, C.M., Laracuente, A., Ladouceur, H.D., and Ringeisen, B.R. (2008) Sub-micron parallel laser direct-write. *Appl. Surf. Sci.*, **255**, 3407–3413.

52. Narazaki, A., Sato, T., Kurosaki, R., Kawaguchi, Y., and Niino, H. (2008) Nano- and microdot array formation of FeSi_2 by nanosecond excimer laser-induced forward transfer. *Appl. Phys. Express*, **1**, 057001-1–057001-3.

53. Willis, D.A. and Grosu, V. (2005) Microdroplet deposition by laser-induced

forward transfer. *Appl. Phys. Lett.*, **86**, 244103.

54. Lee, B.-G. (2009) Improvement of conductive micropattern in a LIFT process with a polymer coating layer. *Electron. Mater. Lett.*, **5**(1), 29–33.

55. Chen, K.T., Lin, Y.H., Ho, J.-R., Cheng, J.-W.J., Liu, S.-H., Liao, J.-L., and Yan, J.-Y. (2010) Laser-induced implantation of silver particles into poly(vinyl alcohol) films and its application to electronic-circuit fabrication on ecapsulated organic electronics. *Microelectron. Eng.*, **87**, 543–547.

56. Rigout, M.L.A., Niu, H., Qin, C., Zhang, L., Li, C., Bai, X., and Fan, N. (1998) Fabrication and photoluminescence of hyperbranched silicon nanowire networks on silicon substrates by laser-induced forward transfer. *Nanotechnology*, **19**, 245303 (5pp).

57. Fuller, S.B., Wilhelm, E.J., and Jacobson, J.M. (2002) Ink-jet printed nanoparticle microelectromechanical systems. *J. Microelchtromech. Syst.*, **11**(1), 54–60.

58. Auyeung, R.C.Y., Kim, H., Mathews, S.A., and Piqué, A. (2007) Laser direct-write of metallic nanoparticle inks. *J. Laser Micro/Nanoeng.*, **2**(1), 21–25.

59. Kinzel, E.C., Su, S., Lewis, B.R., Laurendeau, N.M., and Lucht, R.P. (2006) Direct writing ov conventional thick film inks using MAPLE-DW process. *J. Laser Micro/Nanoeng.*, **1**(1), 74–78.

60. Young, D., Auyeung, R.C.Y., Piqué, A., Chrisey, D.B., and Dlott, D.D. (2002) Plume and jetting regimes in a laser based forward transfer process as observed by time-resolved optical microscopy. *Appl. Surf. Sci.*, **197–198**, 181–187.

61. Piqué, A., Auyeung, R.C.Y., Metkus, K.M., Kim, H., Mathews, S., Bailey, T., Chen, X., and Young, L.J. (2008) Laser decal transfer of electronic materials with thin film characteristics. *Proc. SPIE*, **6879**, 687911.

62. Piqué, A., Auyeung, R.C.Y., Kim, H., Metkus, K.M., and Mathews, S.A. (2008) Digital microfabrication by laser

decal transfer. *J. Laser Micro/Nanoeng.*, **3**(3), 163–168.

63. Auyeung, R.C.Y., Kim, H., Birnbaum, A.J., Zalalutidinov, M., Mathews, S.A., and Piqué, A. (2009) Laser decal transfer of free-standing microcantilevers and microbridges. *Appl. Phys. A*, **97**, 513–519.

64. Ko, S.H., Pan, H., Ryu, S.G., Misra, N., Grigoropoulos, C.P., and Park, H.K. (2008) Nanomaterial enabled laser transfer for organic light emitting material direct writing. *Appl. Phys. Lett.*, **93**, 151110.

65. Nakata, Y., Okada, T., and Maeda, M. (2002) Transfer of laser dye by laser-induced forward transfer. *Jpn. J. Appl. Phys.*, **41**(7B, 2), L839–L841.

66. Gupta, R.K., Gosh, K., and Kahol, P.K. (2009) Fabrication and electrical characterization of Schottky diode based on 2-amino-4, 5-imidazoledicarbonitrile (AIDCN). *Phys. E*, **41**(10), 1832–1834.

67. Blanchet, G.B., Fincher, C.R., and Malajovich, I. (2003) Laser evaporation and the production of pentacene films. *J. Appl. Phys.*, **94**(9), 6181–6184.

68. Guy, S., Guy, L., Bensalah-Ledoux, A., Pereira, A., Grenard, V., Cossoa, O., and Vauteya, T. (2009) Pure chiral organic thin films with high isotropic optical activity synthesized by UV pulsed laser deposition. *J. Mater. Chem.*, **19**, 7093–7097.

69. Rapp, L., Diallo, A.K.K., Alloncle, A.P., Videlot-ckermann, C., Fages, F., and Delaporte, P. (2009) Pulsed-laser printing of organic thin-film transistors. *Appl. Phys. Lett.*, **95**, 171109-1–171109-3.

70. Chang-Jian, S.-K., Ho, J.-R., Cheng, J.-W.J., and Sung, C.-K. (2006) Fabrication of carbon nanotube field emission cathodes in patterns by a laser transfer method. *Nanotechnology*, **17**, 1184–1187.

71. Boutopoulos, C., Pandis, C., Giannakopoulos, K., Pissis, P., and Zergioti, I. (2010) Polymer/carbon nanotube composite patterns via laser induced forward transfer. *Appl. Phys. Lett.*, **96**, 041104.

72. Srinivasan, R. and Mayne-Banton, V. (1982) Self-developing photoetching of poly(ethylene terephthalate) films by far-ultraviolet excimer laser radiation. *Appl. Phys. Lett.*, **41**(6), 576–578.

73. Lippert, T. and Dickinson, J.T. (2003) Chemical and spectroscopic aspects of polymer ablation: special features and novel directions. *Chem. Rev.*, **103**, 453–485.

74. Thomas, B., Alloncle, A.P., Delaporte, P., Sentis, M., Sanaur, S., Barret, M., and Collot, P. (2007) Experimental investigations of laser-induced forward transfer process of organic thin films. *Appl. Surf. Sci.*, **254**, 1206–1210.

75. Karaiskou, A., Zergioti, I., Fotakis, C., Kapsetaki, M., and Kafetzoupoulos, D. (2003) Microfabrication of bio-materials by the su-ps laser-induced forward transfer process. *Appl. Surf. Sci.*, **208–209**, 245–249.

76. Zergioti, I., Karaiskou, A., Papazolou, D.G., Fotakis, C., Kapsetaki, M., and Kafetzopoulou, D. (2005) Time-resolved schlieren study of sub-picosecond and nanosecond laser transfer of biomaterials. *Appl. Surf. Sci.*, **247**, 584–589.

77. Tsuboi, Y., Furuhat, Y., and Kitamura, N. (2007) A sensor for adenosine triphosphate fabricated by laser-induced forward transfer of luciferase onto poly(diemthylsiloxane) microchip. *Appl. Surf. Sci.*, **253**, 8422–8427.

78. Lang, F., Leiderer, P., and Georgiou, S. (2004) Phase transition dynamics measurements in superheated liquids by monitoring the ejection of nanometer-thick films. *Appl. Phys. Lett.*, **85**(14), 2759–2761.

79. Lang, F. and Leiderer, P. (2006) Liquid-vapour phase transitions at interfaces: sub-nanosecond investigations by monitoring the ejection of thin liquid films. *New J. Phys.*, **8**, 14 (10 pp).

80. Modi, R., Wu, H.D., Auyeung, R.C.Y., Gilmore, C.M., and Chrisey, D.B. (2001) Direct writing of polymer thick film resistors using a novel laser transfer technique. *J. Mater. Res.*, **16**(11), 3214–3222.

81. Piqué, A., Chrisey, D.B., Auyeung, R.C.Y., Fitz-Gerald, J., Wu, H.D., McGill, R.A., Lakeou, S., Wu, P.K., Nyuyen, V., and Duignan, M. (1999)

A novel laser transfer process for direct writing of electronic and sensor materials. *Appl. Phys. A*, **69** (Suppl.), S279–S284.

82. Fitz-Gerald, J.M., Wu, H.D., Piqué, A., Horwitz, J.S., Auyeung, R.C.Y., Chang, W., Kim, W.J., and Chrisey, D.B. (2000) MAPLE Direct Write: a new approach to fabricate ferroelectric thin film devices in air at room temperature. *Integr. Ferroelectr.*, **29**(1–2), 13–28.

83. Piqué, A., Chrisey, D.B., Fitz-Gerald, J.M., McGill, R.A., Auyeung, R.C.Y., Wu, H.D., Kakeou, S., Nyuyen, V., Chung, R., and Duignan, M. (2000) Direct writing of electronic and sensor materials using a laser transfer technique. *J. Mater. Res.*, **15**(9), 1872–1875.

84. Young, D., Wu, H.D., Auyeung, R.C.Y., Modi, R., Fitz-Gerald, J., Piqué, A., and Chrisey, D.B. (2001) Dielectric properties of oxide structures by a laser-based direct-writing method. *J. Mater. Res.*, **16**(6), 1720–1725.

85. Chrisey, D.B., Piqué, A., Modi, R., Wu, H.D., Auyeung, R.C.Y., and Young, H.D. (2000) Direct writing of conformal mesoscopic electronic devices by MAPLE-DW. *Appl. Surf. Sci.*, **168**, 345–352.

86. Zang, C., Liu, D., Mathews, S.A., Graves, J., Schaefer, T.M., Gilbert, B.K., Modi, R., Wu, H.-D., and Chrisey, D.B. (2003) Laser direct-write and ist application in low temperature Co-fired ceramic (LTCC) technology. *Microelectron. Eng.*, **70**, 41–49.

87. Modi, R., Wu, H.D., Auyeung, R.C.Y., Vollmers, J.E.S., and Chrisey, D.B. (2002) Ferroelectric capacitors made by a laser forward transfer technique. *Integr. Ferroelectr.*, **42**, 79–95.

88. Pique, A., Auyeung, R.C.Y., Stepnowiski, J.L., Weir, D.W., Arnold, C.B., McGill, R.A., and Chrisey, D.B. (2003) Laser processing of polymer thin films for chemical sensor applications. *Surf. Coat. Technol.*, **163–164**, 293–299.

89. Arnold, C.B., Kim, H., and Piqué, A. (2004) Laser direct write of planar alkaline microbatteries. *Appl. Phys. A*, **79**, 417–420.

90. Wartena, R., Curtright, A.E., Arnold, C.B., Piqué, A., and Swider-Lyons, K.E. (2004) Li-ion microbatteries generated by a laser direct-write method. *J. Power Sources*, **126**, 193–202.

91. Kim, H., Kushto, G.P., Arnold, C.B., Kafafi, Z.H., and Piqué, A. (2004) Laser processing of nanocrystalline TiO$_2$ films for dye-sensitized solar cells. *Appl. Phys. Lett.*, **85**, 464.

92. Chrisey, D.B., Piqué, A., Mc Gill, R.A., Horwitz, J.S., Ringeisen, B.R., Bubb, D.M., and Wu, P.K. (2003) Laser deposition of polymer and biomaterial films. *Chem. Rev.*, **103**, 553–576.

93. Wu, P.K., Ringeisen, B.R., Callahan, J., Brooks, M., Bubb, D.M., Wu, H.D., Piqué, A., Spargo, B., Mc Gill, R.A., and Chrisey, D.B. (2001) The deposition, structure, pattern deposition of biomaterial thin films by matrix-assisted pulsed-laser evaporation (MAPLE) and MAPLE direct write. *Thin Solid Films*, **398–399**, 607–614.

94. Ringeisen, B.R., Chrisey, D.B., Piqué, A., Young, H.D., Modi, R., Bucaro, M., Jones-Meehan, J., and Spargo, B.J. (2002) Generation of mesoscopic patterns of viable Escherichia doli by ambient laser transfer. *Biomaterials*, **23**, 161–166.

95. Wu, P.K., Ringeisen, B.R., Krizman, D.B., Frondoza, C.G., Brooks, M., Bubb, D.M., Auyeung, R.C.Y., Piqué, A., Spargo, B., McGill, R.A., and Chrisey, D.B. (2003) Laser transfer of biomaterials: matrix-assisted pulsed laser evaporation (MAPLE) and MAPLE direct write. *Rev. Sci. Instrum.*, **74**(4), 2546–2557.

96. Barron, J.A., Ringeisen, B.R., Kim, H., Spargo, B.J., and Chrisey, D.B. (2004) Application of laser printing to mammalian cells. *Thin Solid Films*, **453–454**, 383–387.

97. Fukumura, H., Kohji, Y., Nagasawa, K., and Masuhara, H. (1994) Laser implantation of pyrene molecules into poly(methyl methacrylate) films. *J. Am. Chem. Soc.*, **116**, 10304–10305.

98. Fukumura, H. (1997) Laser molecular implantation into polymer solids induced by irradiation below ablation threshold. *J. Photochem. Photobiol. A*, **106**, 3–8.

99. Karnakis, D.M., Goto, M., Ichinose, N., Kawanishi, S., and Fukumura, H. (1998) Forward-transfer laser implantation of pyrene molecules in a solid polymer. *Appl. Phys. Lett.*, **73**(10), 1439–1441.

100. Fukumura, H., Uji-i, H., Banjo, H., Masuhara, H., Karnakis, D.M., Ichinose, N., Kawanishi, S., Uchida, K., and Irie, M. (1998) Laser implantation of photochromic molecules into polymer films: a new approach towards molecular device fabrication. *Appl. Surf. Sci.*, **127–129**, 761–766.

101. Karnakis, D.M., Lippert, T., Ichinose, N., Kawanishi, S., and Fukumura, H. (1998) Laser induced molecular transfer using ablation of a triazeno-polymer. *Appl. Surf. Sci.*, **127–129**, 781–786.

102. Asahi, T., Yoshikawa, H.Y., Yashiro, M., and Masuhara, H. (2002) Femtosecond laser ablation transfer and phase transition of phthalocyanine solids. *Appl. Surf. Sci.*, **197–198**, 777–778.

103. Kishimoto, M., Hobley, J., Goto, M., and Fukumura, H. (2001) Microscopic laser patterning of functional organic molecules. *Adv. Mater.*, **13**(15), 1155–1158.

104. Pihosh, Y., Goto, M., Oishi, T., Kasahara, A., and Tosa, M. (2006) Process during laser implantation and ablation of Copumarin 6 in poly(butyl methacrylate) films. *J. Photochem. Photobiol. A*, **183**, 292–296.

105. Foley, D.M., Bennett, E.W., and Slifkin, S.C. (1991) Ablation-transfer imaging/recording. US Patent 5, 156, 938, (Graphics Technology Int.), filed May 29, 1991.

106. Sandy Lee, I.Y., Tolbert, W.A., Dlott, D.D., Doxtader, M.M., Foley, D.M., Arnold, D.R., and Willis, E.W. (1992) Dynamics of laser ablation transfer imaging investigated by ultrafast microscopy. *J. Imaging Sci. Technol.*, **36**(2), 180–187.

107. Seyfang, B.C., Fardel, R., Lippert, T., Scherer, G.G., and Wokaun, A. (2009) Micro-patterning for polymer electrolyte fuel cells: single pulse laser ablation of aluminium films from glassy carbon. *Appl. Surf. Sci.*, **255**, 5471–5475.

108. Tolbert, W.A., Sandy Lee, I.-Y., Doxtader, M.M., Ellis, E.W., and Dlott, D.D. (1993) High-speed color imaging by laser ablation transfer with a dynamic release layer: Fundamental Mechanisms. *J. Imaging Sci. Technol.*, **37**(4), 411–422.

109. Sandy Lee, I.Y., Wen, X., Tolbert, W.A., Dlott, D.D., Doxtader, M.M., and Arnold, D.R. (1992) Direct measurement of polymer temperature during laser ablation using a molecular thermometer. *J. Appl. Phys.*, **72**(6), 2440–2448.

110. Tolbert, W.A., Sandy Lee, I.Y., Wen, X., and Dlott, D.D. (1993) Laser ablation transfer imaging using picosecond optical pulses: ultra-high speed, lower threshold and high resolution. *J. Imaging Sci. Technol.*, **37**(5), 485–489.

111. Kinoshita, M., Hasihino, K., and Kitamura, T. (2000) Mechanism of dye thermal transfer from ink donor layer to receiving sheet by laser heating. *J. Imaging Sci. Technol.*, **44**(2), 105–110.

112. Bills, R.E., Chou, H.-H., Dower, W.V., and Wolk, M.B. (1993) Laser propulsion transfer using black metal coated substrates. US Patent 5, 308, 737, Minnesota Mining and Manufacturing Comp., filed Mar. 18, 1993.

113. Phipps, C.R., Luke, J.R., McDuff, G.G., and Lippert, T. (2003) Laser-driven micro-rocket. *Appl. Phys. A*, **77**, 193–201.

114. Urech, L., Lippert, T., Phipps, C.R., and Wokaun, A. (2007) Polymers as fuel for laser-based microthrusters: An investigation of thrust, material, plasma and shockwave properties. *Appl. Surf. Sci.*, **253**, 7646–7650.

115. Lippert, T., Urech, L., Fardel, R., Nagel, M., Phipps, C.R., and Wokaun, A. (2008) Materials for laser propulsion: ''liquid'' polymers. *Proc. SPIE*, **7005**, 700512.

116. Irie, M. and Kitamura, T. (1993) High-definition thermal transfer printing using laser heating. *J. Imaging Sci. Technol.*, **37**(3), 231–235.

117. Irie, M., Kato, M., and Kitamura, T. (1993) Thermal transfer color printing using laser heating. *J. Imaging Sci. Technol.*, **37**(3), 235–238.

118. Irie, M. (2003) 1.0 W high–power small-sized Nd: Yttrium Aluminium Garnet (YAG) laser optical Head for a laser-induced printing system. *Jpn. J. Appl. Phys.*, **42** (Part I, 4A), 1633–1636.

119. Sameshima, T. (1996) Laser beam application to thin film transistors. *Appl. Surf. Sci.*, **96–98**, 352–358.

120. Toet, D., Thompson, M.O., Smith, P.M., and Sigmon, T.W. (1999) Laser assisted transfer of silicon by explosive hydrogen release. *Appl. Phys. Lett.*, **74**(15), 2170–2172.

121. Toet, D., Smith, P.M., Sigmon, T.W., and Thompson, M.O. (2000) Spatially selective materials deposition by hydrogen-assisted laser-induced transfer. *Appl. Phys. Lett.*, **77**(2), 307–309.

122. Toet, D., Smith, P.M., Sigmon, T.W., and Thompson, M.O. (2000) Experimental and numerical investigations of hydrogen-assisted laser-induced materials transfer procedure. *J. Appl. Phys.*, **87**(7), 3537–3546.

123. Lin, Z., Zhigilei, L.V., and Celli, V. (2008) Electron-phonon coupling and electron heat capacity of metals under conditions of strong electron-phonon nonequilibrium. *Phys. Rev. B*, **77**, 075133.

124. Nakata, Y. and Okada, T. (1999) Time-resolved microscopic imaging of the laser-induced forward transfer process. *Appl. Phys. A*, **69** (Suppl.), S275–S278.

125. Fitz-Gerald, J.M., Piqué, A., Chrisey, D.B., Rack, P.D., Zelenznik, M., Auyeung, R.C.Y., and Lakeou, S. (2000) Laser direct writing of phosphor screens for high-definition displays. *Appl. Phys. Lett.*, **76**(11), 1386–1388.

126. Lee, J.H., Yoo, C.D., and Kim, Y.-S. (2007) A laser-induced thermal spray printing process for phosphor layer deposition of PDP. *J. Micromech. Microelectron.*, **17**, 258–264.

127. Barron, J.A., Wu, P., Ladouceur, H.D., and Ringeisen, B.R. (2004) Biological laser printing: A novel technique for

creating heterogeneous 3-dimensional cell patterns. *Biomed. Microdevices*, **6**(2), 139–147.

128. Barron, J.A., Krizman, D.B., and Ringeisen, B.R. (2005) Laser printing of single cells: statistical analysis, cell Viability, and stress. *Ann. Biomed. Eng.*, **33**(2), 121–130.

129. Ringeisen, B.R., Othon, C.M., Barron, J.A., Young, D., and Spargo, B.J. (2006) Jet-based methods to print living cells. *Biotechnol. J.*, **1**, 930–948.

130. Fernandez-Pradas, J.M., Colina, M., Serra, P., Dominguez, J., and Morenza, J.L. (2004) Laser-induced forward transfer of biomolecules. *Thin Solid Films*, **453–454**, 27–30.

131. Serra, P., Fernandez-Pradas, J.M., Berthet, F.X., Colina, M., Elvira, J., and Morenza, J.L. (2004) Laser direct writing of biomolecule microarrays. *Appl. Phys. A*, **79**, 949–952.

132. Serra, P., Colina, M., Fernandez-Pradas, J.M., Sevilla, L., and Morenza, J.L. (2004) Preparation of functional DNA microarrays through laser-induced forward transfer. *Appl. Phys. Lett.*, **85**(9), 1639–1641.

133. Colina, M., Serra, P., Fernandez-Pradas, J.M., Sevilla, L., and Morenza, J.L. (2005) DNA deposition through laser induced forward transfer. *Biosens. Bioelectron.*, **20**, 1638–1642.

134. Serra, P., Fernandez-Pradas, J.M., Colina, M., Duocastella, M., Dominguez, J., and Morenza, J.L. (2006) Laser-induced forward transfer: a direct- writing technique for biosensors preparation. *J. Laser Micro/Nanoeng.*, **1**(3), 236–242.

135. Colina, M., Duocastella, M., Fernandez-Pradas, J.M., Serra, P., and Morenza, J.L. (2006) Laser-induced forward transfer of liquids: study of the droplet ejection process. *J. Appl. Phys.*, **99**, 084909-1–084909-7.

136. Duocastella, M., Colina, M., Fernandez-Pradas, J.M., Serra, P., and Morenza, J.L. (2007) Study of the laser-induced forward transfer of liquids for laser bioprinting. *Appl. Surf. Sci.*, **253**, 7855–7859.

137. Duocastella, M., Fernandez-Pradas, J.M., Serra, P., and Morenza, J.L. (2008) Laser-induced forward transfer of liquids for miniaturized biosensors preparation. *J. Laser Micro/Nanoeng.*, **3**(1), 1–4.

138. Duocastella, M., Fernandez-Pradas, J.M., Serra, P., and Morenza, J.L. (2008) Jet formation in the laser forward transfer of liquids. *Appl. Phys. A*, **93**, 453–456.

139. Duocastella, M., Fernández-Pradas, J.M., Domínguez, J., Serra, P., and Morenza, J.L. (2008) Printing biological solutions through laser-induced forward transfer. *Appl. Phys. A*, **93**, 941.

140. Serra, P., Duocastella, M., Fernández-Pradas, J.M., and Morenza, J.L. (2008) Liquids microprinting through laser-induced forward transfer. *Appl. Surf. Sci.*, **255**, 5342.

141. Duocastella, M., Fernández-Pradas, J.M., Morenza, J.L., and Serra, P. (2009) Time-resolved imaging of the laser forward transfer of liquids. *J. Appl. Phys.*, **106**, 084907.

142. Vodp'yanov, L.K., Kozlov, P.S., Kucherenko, I.V., Maksimoviskii, S.N., and Radutskii, G.A. (2003) Studying the possibility of applying the light-hydraulic effect to digital printing. *Instrum. Exp. Tech.*, **46**(4), 549–553.

143. Hopp, B., Smausz, T., Antal, Z., Kresz, N., Bor, Z., and Chrisey, D. (2004) Absorbing film assisted laser induced forward transfer of fungi (Trichoderma conidia). *J. Appl. Phys.*, **96**(6), 3478–3481.

144. Hopp, B., Smausz, T., Barna, N., Vass, C., Antal, Z., Kredics, L., and Chrisey, D. (2005) Time-resolved study of absorbing film assisted laser induced forward transfer of Trichoderma longibrachiatum conidia. *J. Phys. D: Appl. Phys.*, **38**, 833–837.

145. Hopp, B., Smausz, T., Kresz, N., Barna, N., Bor, Z., Kolozsvari, L., Chrisey, D.B., Szabo, A., and Nogradi, A. (2005) Survival and proliferative ability of various living cell types after laser-induced forward transfer. *Tissue Eng.*, **11**(11/12), 1817–1823.

146. Nogradi, A., Hopp, B., Smausz, T., Kecskemeti, G., Bor, Z., Kolozsvari, L., Szabo, A., Klini, A., and Fotakis, C.

(2008) Directed cell growth on laser-transferred 2D biomaterial matrices. *Open Tissue Eng. Regenerative Med. J.*, **1**, 1–7.

147. Smausz, T., Hopp, B., Kecskemeti, G., and Bor, Z. (2006) Study on metal microparticle content of the material transferred with absorbing film assisted laser induced forward transfer when using silver absorbing layer. *Appl. Surf. Sci.*, **252**, 4738–4742.

148. Dinca, V., Kasotakis, E., Catherine, J., Mourka, A., Mitraki, A., Popescu, A., Dinescu, M., Farsari, M., and Fotakis, C. (2007) Development of peptide-based patterns by laser transfer. *Appl. Surf. Sci.*, **254**, 1160–1163.

149. Dinca, V., Farsari, M., Kafetzopoulos, D., Popscu, A., Dinescu, M., and Fotakis, D. (2008) Patterning parameters for biomolecules microarrays constructed with nanosecond and femtosecond UV lasers. *Thin Solid Films*, **516**, 6504–6511.

150. Dinca, V., Kasotakis, E., Mourka, A., Ranella, A., Farsari, M., Mitraki, A., and Fotakis, C. (2008) Fabrication of amyloid peptide micro-arrays using laser-induced forward transfer and avidin-biotin mediated assembly. *Phys. Stat. Sol. C*, **5**(12), 3576–3579.

151. Boutopoulos, C., Andreakou, P., Kafetzopoulos, D., Chatzandroulis, C., and Zergioti, I. (2008) Direct laser printing of biotin microarrays on low temperature oxide on Si substrates. *Phys. Stat. Sol. A*, **205**(11), 2505–2508.

152. Boutopoulos, C., Tsouti, V., Goustouridis, D., Chatzandroulis, S., and Zergioti, I. (2008) Liquid phase direct laser printing of polymers for chemical sensing applications. *Appl. Phys. Lett.*, **93**, 191109-1–191109-4.

153. Tsouti, V., Goustouridis, D., Chatzandroulis, S., Normand, P., Andreakou, P., Ioannou, M., Kafetzopoulos, D., Boutopoulos, C., Zergioti, I., Tsoukalas, D., Hue, J., and Rousier R. (2008) A capacitive biosensor based on ultrathin Si membranes. Reports of the 2008 IEEE Sensors Conference, IEEE Conference Publications, pp. 223–226, doi: 10.1109/ICSENS.2008.4716421

154. Tsouti, V., Boutopoulos, C., Andreakou, P., Ioannou, M., Zergioti, I., Goustouridis, D., Kafetzopoulos, D., Tsoukalas, D., Normand, P., and Chatzandroulis, S. (2009) Detection of the biotin-streptavidin interaction by exploiting surface stress changes on ultrathin Si membranes. *Microelectron. Eng.*, **86**, 1495–1498.

155. Ovsianikov, A., Gruene, M., Pflaum, M., Koch, L., Maiorana, F., Wilhelmi, M., Haverich, A., and Chichkov, B. (2010) Laser printing of cells into 3D scaffolds. *Biofabrication*, **2**, 014104.

156. Rapp, L., Cibert, C., Alloncle, A.P., and Delaporte, P. (2009) Characterization of organic material micro-structures transferred by laser in nanosecond and picosecond regimes. *Appl. Surf. Sci.*, **255**, 5439–5443.

157. Kröger, M., Hüske, M., Dobbertin, T., Meyer, J., Krautwald, H., Riedl, T., Johannes, H.-H., and Kowalsky, W. (2005) A novel patterning technique for high-resolution RGB-OLED displays: laser induced local transfer (LILT). *Mater. Res. Soc. Symp. Proc.*, **807E**, 77–82.

158. Blanchet, G.B., Loo, Y.L., Rogers, J.A., Gao, F., and Fincher, C.R. (2003) Large area, high resolution, dry printing of conducting polymers for organic electronics. *Appl. Phys. Lett.*, **82**(3), 463.

159. DuPont (2008) Displays Brochure *http://www2.dupont.com/Displays/ en_US/assets/downloads/pdf/ DisplaysBrochure.pdf* (accessed September 2010).

160. Wolk, M.B., Baetzold, J., Bellmann, E., Hoffend, T.R., Lamansky, S., Li, Y., Roberts, R.R., Savvateev, V., Staral, J.S., and Tolbert, W.A. (2004) Laser thermal patterning of OLED materials. *Proc. SPIE*, **5519**, 12–23.

161. Lamansky, S., Hoffend, T.R., Le, H., Jones, V., Wolk, M.B., and Tolbert, W.A. (2005) Laser induced thermal imaging of vacuum-coated OLED materials. *Proc. SPIE*, **5937**, 593702.

162. Chin, B.D., Suh, M.C., Kim, M.H., Kang, T.M., Yang, N.C., Song, M.W., Lee, S.T., Kwon, H.H., and Chung, H.K. (2003) High efficiency AMOLED

using hybrid of small molecule and polymer materials pattered by laser transfer. *J. Inform. Display*, **4**(3), 1–5.

163. Suh, M.C., Chin, B.D., Kim, M.-H., Kang, T.M., and Lee, S.T. (2003) Enhanced luminance of blue light-emitting polymers by blending with hole-transporting materials. *Adv. Mater.*, **15**(15), 1254–1258.

164. Lee, J.Y. and Lee, S.T. (2004) Laser-induced thermal imaging of polymer light-emitting materials on poly(3,4-ethylenedioxythiophene): silane hole-transport layer. *Adv. Mater.*, **16**(1), 51–54.

165. Dinca, V., Fardel, R., Shaw-Stewart, J., Di Pietrantonio, F., Cannatà, D., Benetti, M., Verona, E., Palla-Papavlu, A., Dinescu, M., and Lippert, T. (2010) Laser-induced forward transfer: an approach to single-step polymer microsensor fabrication. *Sens. Lett.*, **8**, 436–440.

166. Nagel, M., Hany, R., Lippert, T., Molberg, M., Nüesch, F.A., and Rentsch, D. (2007) Aryltriazene photopolymers for UV-laser applications: improved Synthesis and photodecomposition study. *Macromol. Chem. Phys.*, **208**, 277–286.

167. Lippert, T. (2004) Laser applications of polymers in polymers and light. *Adv. Polym. Sci.*, **168**, 51–246, and references therein.

168. Fardel, R., Feurer, P., Lippert, T., Nagel, M., Nüesch, F.A., and Wokaun, A. (1887) Laser ablation of aryltriazene photopolymer films: effects of polymer structure on ablation properties. *Appl. Surf. Sci.*, **254**, 1332–1337.

169. Mito, T., Tsujita, T., Masuhara, H., Hayashi, N., and Suzuki, K. (2001) Hollowing and transfer of polymethyl methacrylate film propelled by laser ablation of triazeno polymer film. *Jpn. J. Appl. Phys.*, **40** (Part 2, 8A), L805–L806.

170. Fardel, R., Nagel, M., Nüesch, F., Lippert, T., and Wokaun, A. (2007) Laser forward transfer using a sacrificial layer: Influence of the material properties. *Appl. Surf. Sci.*, **254**, 1322–1326.

171. Dinca, V., Fardel, R., Di Pietrantonia, F., Cannata, D., Benetti, M., Verona, E., Palla-Papavlu, A., Dinescu, M., and Lippert, T. (2010) Laser-induced forward transfer: An approach to single-step polymer microsensor fabrication. *Sens. Lett.*, **8**, 1–5.

172. Fardel, R., Nagel, M., Lippert, T., Nüesch, F., Wokaun, A., and Luk'yanchuk, B.S. (2008) Influence of thermal diffusion on ablation of thin polymer films. *Appl. Phys. A*, **90**, 661–667.

173. Nagel, M., Fardel, R., Feurer, P., Häberli, M., Nüesch, F.A., Lippert, T., and Wokaun, A. (2008) Aryltriazene photopolymer thin films as sacrificial release layers for laser-assisted forward transfer systems: study of photoablative decomposition and transfer behaviour. *Appl. Phys. A*, **92**, 781–789.

174. Shaw Stewart, J., Fardel, R., Nagel, M., Delaporte, P., Rapp, L., Cibert, C., Alloncle, A.-P., Nüesch, F., Lippert, T., and Wokaun, A. (2010) The effect of laser pulse length upon laser-induced forward transfer using a trazene polymer as dynamic release layer. *J. Optoelectron. Adv. Mater.*, **12**(3), 605–609.

175. Doraiswamy, A., Narayan, R.J., Lippert, T., Urech, L., Wokaun, A., Nagel, M., Hopp, B., Dinescu, M., Modi, R., Auyeung, R.C.Y., and Chrisey, D.B. (2006) Excimer laser forward transfer of mammalian cells using a novel triazene absorbing layer. *Appl. Surf. Sci.*, **252**, 4743–4747.

176. Schiele, N.R., Koppes, R.A., Corr, D.T., Ellison, K.S., Thompson, D.M., Ligon, L.A., Lippert, T.K.M., and Chrisey, D.B. (2009) Laser direct writing of combinatorial libraries of idealized cellular constructs: Biomedical applications. *Appl. Surf. Sci.*, **255**, 5444–5447.

177. Xu, J., Liu, J., Cui, D., Gerhold, M., Wang, A.Y., Nagel, M., and Lippert, T.K. (2007) Laser-assisted forward transfer of multi-spectral nanocrystal quantum dot emitters. *Nanotechnology*, **18**, 025403 (6pp).

178. Fardel, R., Nagel, M., Nüesch, F., Lippert, T., and Wokaun, A. (2007) Fabrication of organic light-emitting

diode pixels by laser-assisted forward transfer. *Appl. Phys. Lett.*, **91**, 061103.

179. Fardel, R., Nagel, M., Nüesch, F., Lippert, T., and Wokaun, A. (2009) Shadowgraphy investigation of laser-induced forward transfer: front side and back side ablation of the triazene polymer sacrificial layer. *Appl. Surf. Sci.*, **255**, 5430–5434.

180. Fardel, R., Nagel, M., Nüesch, F., Lippert, T., and Wokaun, A. (2009) Energy balance in a laser-induced forward transfer process studied by shadowgraphy. *J. Phys. Chem. C*, **113**, 11628–11633.

181. Kaur, K., Fardel, R., May-Smith, T.C., Nagel, M., Banks, D.P., Grivas, C., Lippert, T., and Eason, R.W. (2009) Shadowgraphic studies of triazene assisted laser-induced forward transfer of ceramic thin films. *J. Appl. Phys.*, **105**(11), 113119.

182. Fardel, R., Nagel, M., Nüesch, F., Lippert, T., and Wokaun, A. (2010) Laser-induced forward transfer of organic LED building blocks studied by time-resolved shadowgraphy. *J. Phys. Chem. C*, **114**, 5617–5636.

183. Bonse, J., Solis, J., Urech, L., Lippert, T., and Wokaun, A. (2007) Femtosecond and nanosecond laser damage therhods of doped and undoped trazenepolymer thin films. *Appl. Surf. Sci.*, **253**, 7787–7791.

184. Banks, D.P., Kaur, K., Gazia, R., Fardel, R., Nagel, M., Lippert, T., and Eason, R.W. (2008) Triazene photopolymer dynamic release layer-assisted femtosecond laser-induced forward transfer with an active carrier substrate. *Europhys. Lett.*, **83**, 38003.

185. Banks, D.P., Kaur, K.S., and Eason, R.W. (2009) Influence of optical standing waves on the femtosecond laser-induced forward transfer of transparent thin films. *Appl. Opt.*, **48**(11), 2058–2066.

186. Banks, D.P., Grivas, C., Mills, J.D., Eason, R.W., and Zergioti, I. (2006) Nano-droplets deposited in microarrays by femtosecond Ti:sapphire laser-induced forward transfer. *Appl. Phys. Lett.*, **89**, 193107.

187. Banks, D.P., Grivas, C., Zergioti, I., and Eason, R.W. (2008) Ballistic laser-assisted solid transfer (BLAST) of intact material from a thin film precursor. *Opt. Express*, **16**(5), 3249–3254.

188. Banks, D.P., Kaur, K.S., and Eason, R.W. (2009) Etching and forward transfer of fused silica in solid-phase by femtosecond laser-induced solid etching (LISE). *Appl. Surf. Sci.*, **255**(20), 8343–8351.

189. Vogel, A., Lorenz, K., Herneffer, V., Hüttmann, G., Smolinski, D., and Gebert, A. (2007) Mechanisms of laser-induced dissection and transport of histologic specimens. *Biophys. J.*, **93**, 4481–4500.

190. Horneffer, V., Linz, N., and Vogel, A. (2007) Principles of laser-induced separation and transport of living cells. *J. Biomed. Opt.*, **12**(5), 054016-1–054016-13.

191. Kattamis, N.T., Purnick, P.E., Weiss, R., and Arnold, C.B. (2007) Thick film laser induced forward transfer for deposition of thermally and mechanically sensitive materials. *Appl. Phys. Lett.*, **91**, 171120.

192. Kononenko, T.V., Alloncle, P., Konov, V.I., and Sentis, M. (2009) Laser transfer of diamond nanopowder induced by metal film blistering. *Appl. Phys. A*, **94**, 531–536.

193. Holmes, A.S. and Saidam, S.M. (1998) Sacrificial layer process with laser-driven release for batch assembly operations. *J. Micromech. Syst.*, **7**(4), 416–422.

194. Piqué, A., Charipar, N.A., Auyeung, R.C.Y., Kim, H., and Matthews, S.A. (2007) Assembly and Integration of thin bare die using laser direct write. *Proc. SPIE*, **6458**, 645802.

195. Mathews, S.A., Auyeung, R.C.Y., and Piqué, A. (2007) Use of laser direct-write in microelectronics assembly. *J. Laser Micro/Nanoeng.*, **2**(1), 103–107.

6
Nanomaterials: Laser-Based Processing in Liquid Media

6.1
Liquid-Assisted Pulsed Laser Ablation/Irradiation for Generation of Nanoparticles

Subhash Chandra Singh

6.1.1
Introduction

Lasers have proved their potential application in every field of science and technology including, but not limited to, medicine and biology, environment and space, defense and homeland security, information technology and communication, spectroscopy and imaging, energy, marine life, and materials processing and its characterization. Applications of lasers are increasing in these fields day by day, with the emergence of new fields for them. Experiments in laser irradiation for removal of materials from surfaces started just after the invention of the ruby laser in 1960 at the Hughes Research Laboratories [1]. Since then, pulsed laser ablation (PLA) of a solid target has been extensively employed for high-temperature and high-pressure plasma synthesis to diagnose atomic and molecular states and their lifetime, deposition of thin films and clusters, cleaning of surfaces, cutting and drilling of materials, compositional analysis, and device fabrication. In the previous chapters, laser–matter interaction for plasma processing; applications of lasers in the gas-phase synthesis of nanomaterials using conventional pulsed laser deposition (PLD), nanoparticle-assisted pulsed laser deposition (NAPLD), temperature-assisted pulsed laser deposition (TAPLD), laser-vaporization-controlled condensation (LVCC); and gas-phase fabrication of nano/micropatterns on the substrates have been extensively studied.

As discussed in Chapter 3.1, laser irradiation of a solid target creates a plasma plume that expands normal to the target surface. The rate and mechanism of removal of the target material depend on the laser irradiance falling on the surface. The plasma produced in the vacuum expands freely and follows a linear relationship for the position of the plume front with time $-R(t) = at$, where a is a constant [2]. The plasma plume becomes confined under the ambient gas, and the degree of confinement depends on the background gas pressure. For example, the expansion of the plume follows the shock wave model, $R(t) = at^b$, in low-pressure ambience

Nanomaterials: Processing and Characterization with Lasers, First Edition.
Edited by Subhash Chandra Singh, Haibo Zeng, Chunlei Guo, and Weiping Cai.
© 2012 Wiley-VCH Verlag GmbH & Co. KGaA. Published 2012 by Wiley-VCH Verlag GmbH & Co. KGaA.

and the drag model, $R(t) = a(1 - e^{-bt})$, where a and b are constants with $a > 0$ and $0 < b < 1$, in high-pressure ambience [2]. An increase in the ambient pressure causes an increase in the degree of space confinement of the plasma plume and, therefore, better reactions among the plasma species and between the plasma species and ambient environment at the plasma–ambience interface. For example, PLA in vacuum generates highly expanding plasma where plasma species fly from each other faster than the time of their clustering, and they can be deposited on the substrate in the form of an atomic layer, that is, a thin film [3, 4] (as described in Chapter 4.1). In contrast to this, PLA in a high-pressure environment causes confinement of the plasma plume and its condensation to particles that can be collected by nanoscale servers or deposited on the substrate in the form of nanostructured films [5–8] (as described in Chapter 4.3). Confinement of the plasma plume transfers it to the thermodynamic state of high temperature, high pressure, and high density (HTHPHD), where special reactions can be carried out to generate unique metastable phases of nanostructures that cannot be achieved under normal conditions. Liquids have higher densities as compared to gases; therefore, PLA in liquids generates a plasma plume under a higher degree of confinement and hence a much higher temperature, higher pressure, and higher density.

PLA of a solid target at the solid–gas interface is widely exploited by researchers for the processing of materials because of the simple mechanism associated with the processes of laser–matter interaction, plasma processing, and condensation of plasma species on the substrate in the form of nanomaterials or suspensions in the gaseous media. PLA of a solid target at the solid–liquid interface involves comparatively simple experimental arrangements, but more complex reaction mechanisms between plasma species and liquid media, which is one of the reasons for theoretical modeling and simulation of liquid-phase pulsed laser ablation (LP-PLA) being less exploited in experimental understanding than in theoretical assistance provided to vacuum or gaseous-phase PLA. In spite of some theoretical and experimental works on laser ablation of a solid target under liquid confinement, understanding of the mechanisms of laser–matter interaction, plasma processing, expansion of the plasma under confinement, reaction between plasma species and ambient liquid media, and finally, condensation of product molecules to form nanomaterials of various sizes, shapes, and compositions is still in the developmental stage. Spatial and temporal spectroscopic diagnosis of an expanding plasma plume under liquid confinement is required to determine the temporal evolution of temperature and pressure of the plasma plume due to confinement, various ionic species in the plasma plume and their reaction with the molecules of the liquid, effects of external pressure (pressure in front of the plume) and other laser parameters on the plasma characteristics, as well as influence of liquid media parameters (surface tension, viscosity, dielectric constant, dipole moment, melting and boiling points, etc.) on the rate of ablation and rate of reaction between plasma species and molecules of liquid, in order to understand the basic physical and chemical aspects involved in the LP-PLA processes.

6.1.2
Advantages of Liquid-Phase Laser Ablation over Gas Phase

Liquid-assisted pulsed laser ablation (LA-PLA), which has several names and abbreviations in the literature, namely, liquid-phase pulsed laser ablation (LP-PLA), pulsed laser ablation in liquids (PLALs), pulsed laser ablation in aqueous media (PLAAM), and so on, exhibits much simpler and cheaper experimental arrangements as compared to gas-phase laser ablations. Here, one does not require complex experimental arrangements and costly vacuum systems. Higher availability of cheaper liquids, their transportation, and storage, as compared to gases, also facilitates research on LP-PLA. There is also no need for any additional nanoparticle (NP) collectors such as substrates in PLD, NAPLD, and TAPLD experiments and nanoscale filters in gas-phase PLA. It provides a highly colloidal solution of NPs with *in situ* surface functionalization [9, 10] and *in situ* use of surfactants as growth terminators and stabilizers [11–20]. Liquid-assisted laser ablation is also used in the removal of particles from surfaces, shock wave processing, cutting, drilling, welding, and surface machining with comparatively sharper edges and corners without deposition of debris. LP-PLA-produced NPs are charged and thus possess extremely high colloidal stability. Unlike the dry nanopowders produced in gas-phase PLA, NP colloids generated by LP-PLA are not inhalable and thus lead to improved occupational safety. Since no chemical precursors are used, LP-PLA-produced colloids are of high (100%) purity and surfaces of NPs are free from any type of chemical contamination. In other words, more number of NP sites/surface atoms is available for functionalization, drug loading, and environmental sensing. LP-PLA can be applied universally to an unlimited number of laser and liquid media parameters to synthesize a wide variety of elemental and compound nanostructures. Very recently, it has been proved that the advantages associated with LP-PLA-produced NPs have more valuable applications as compared to conventionally produced particles. For example, (i) almost three to five times the number of biomolecules can be conjugated on the surface of LP-PLA-produced NPs as compared to their chemically produced counterparts [9, 10]. Higher yields of LP-PLA-produced gold NP aptamer conjugates are beneficial especially if costly functional biomolecules are conjugated [9, 10]. (ii) PLA in polymer solution allows embedding of NPs into polymer matrices for rapid nanomaterial prototyping [21–23]. (iii) LP-PLA-generated pure NPs and aggregates provide comparatively higher signal-to-noise ratio when used as substrates in surface-enhanced Raman scattering (SERS) [24, 25].

6.1.3
Classification of Liquid-Phase Laser Ablation on the Basis of Target Characteristics

The following are the target characteristics on the basis of which liquid-phase laser ablation is classified.

1) **Liquid-Phase Laser Ablation of Solid Bulk Target Materials**

Here, a solid target immersed in a liquid is irradiated/ablated with a focused laser beam at the solid–liquid interface through the transparent liquid. The laser induces plasma at the solid–liquid interface, and this laser-produced plasma heats and ionizes the liquid layer at the plasma–liquid interface to generate plasma-induced plasma. Laser-induced plasma has atoms/ions or molecules from the target material, while plasma-induced plasma contains species from the liquid. Intermixing of these two plasmas and the reaction between them produces compound nanomaterials of active metals and elemental NPs of noble metals. This approach of liquid-phase laser processing of nanomaterials has been maximally exploited in the past two decades.

2) **Liquid-Phase Laser Ablation of Suspended Nano/Microparticles**

In this case, bulk or micro/nanosized particles are dispersed in the liquids to form a solution, which is irradiated with a focused or an unfocused beam of laser by continuous stirring. Laser irradiation melts or vaporizes larger-sized particles into smaller-sized liquid drops or atoms or explosively fragments them into smaller-sized clusters. There are two pathways: thermal and nonthermal. Drops or atoms produced by melting or vaporization in the thermal pathway can re-form smaller-sized particles after cooling or condensation, respectively. This process can change the size, shape, and phase of NPs synthesized by chemical or physical methods. In the nonthermal pathway, irradiation of particles causes ejection of electrons from the particle surface and leaves behind positive charges on the surface of the particle. Positive charges on the surface of particles result in Coulombic repulsion among different portions of the particle and consequently, its fragmentation. Sizes of NPs obtained through thermal or nonthermal pathways depend on the laser and liquid medium parameters and time of ablation/irradiation. The size of the product NPs may be smaller or larger depending on the experimental conditions initial size of raw particle and the time of ablation.

3) **Liquid-Phase Laser Ablation of Metal Salts/Organic Precursors**

A solution of metal salts or liquid organic precursors of metals is laser dissociated into neutral atoms, and these atoms get clusterized to form NPs.

6.1.3.1 Liquid-Phase Laser Ablation of Solid Bulk Target Materials

LP-PLA of solid targets was pioneered in 1987 by Patil *et al.* [26], who used a pulsed laser beam to ablate an iron target in water to produce iron oxide coating on the surface. Soon after, Ogale *et al.* [27, 28] used LP-PLA as a promising method for surface modification, such as oxidation, nitriding, and carbiding. They also produced several metastable phases such as diamond particulate using LP-PLA of a graphitic target in liquid benzene. LP-PLA has numerous applications in materials processing including liquid-phase wet etching for surface patterning or nano/microfabrication, laser-induced shock-wave-based surface cleaning, surface coating of cutting tools, and generation of colloidal solution of NPs. The pioneering work by Fabbro and coworkers [29–36] on the diagnosis of thermodynamic parameters of a liquid-confined laser-produced plasma plume using optical emission spectroscopy and shock wave methods shows much higher temperature, pressure, and density of the liquid-confined plasma as compared to the freely expanding one. The shock wave induced after laser–target interaction and plasma processing propagates in the liquid and the target in the direction normal to the target surface. It creates dynamic and static stress and permanent damage to the target surface, which may be called as *shock-wave-affected zone*. Increase in the plasma pressure generates a plane longitudinal wave at the solid–liquid interface that propagates into the target and induces plastic deformation. Peyer *et al.* [35] reported that the plasma pressure generated in the water-confined regime (WCR) varies almost linearly with the laser irradiance, and at constant laser irradiance it shows higher value for shorter pulse width lasers. They also reported influence of peak power density on the amount of surface stress and free velocity of water-confined plasma using VISAR (velocity interferometer system for any reflector) measurements. Mathematical models and codes were developed for the study of thermodynamic and kinetic aspects of laser-produced plasma under liquid confinement by Fairand and Clauer [37] in 1979, Griffin *et al.* [38] in 1986, Fabbro *et al.* [39] in 1990, Sollier *et al.* [40] in 2001, Colvin *et al.* [41] in 2003, Zhang *et al.* [42] in 2004, and Wu and Shin [43] in 2005. These models are described in short in Section 6.1.3.1.3.

6.1.3.1.1 Basic Characteristics of Laser Ablation under Liquid Confinement

Laser irradiates the surface of target material at the solid–liquid interface through the transparent liquid. Energy delivered by the laser to the skin layer of target material up to the skin depth excites free electrons, which oscillate rapidly in the strong optical field of laser. Skin depth depends on the wavelength and refractive indices of target and liquid media. Since the refractive index of liquids lies in between the indices of solid and air, therefore liquids provide better phase matching for coupling of the laser light with the target. Interaction of fast moving electrons with the stationary atoms through electron–phonon interaction causes energy vibrations in the lattice and consequently, phonon–phonon interaction that creates lattice waves in the target material. Propagation of lattice wave rapidly heats the target surface higher than its melting/boiling point. If laser irradiance is above the melting threshold of the target, but below its vaporization threshold, surface melting of the target and formation of liquid droplets of target material occurs, which may be called as *molten globules*. On the contrary, if laser irradiance is higher than the vaporization threshold, it generates vapors of the target material at the solid–liquid interface. The trailing part of laser pulse ionizes this vapor to create hot laser plasma. Ablated material in the form of hot plasma/molten globule has high temperature and pressure, which expands with supersonic velocity under liquid confinement. Plasma boils the surrounding liquid layer at the plasma–liquid interface and creates high vapor pressure because of its confinement by forward liquid layers. The liquid vapor, sandwiched between plasma and forward liquid, exerts high pressure on the expanding plasma plume, which results in explosive ejection of atomic/ionic species from plasma into the liquid as well as their mixing and reaction with vapor and liquid medium in a penetrating way (penetration reaction) [44]. These penetrated plasma species get clusterized into particles either before or after reaction with molecules of liquid media. Former is the case of synthesis of elemental NPs, while the latter, depending on the nature of liquid media, leads to the formation of oxide, nitride, carbide, sulfide, and so on of the target compound [44].

6.1.3.1.2 Influences of Confinement on the Thermodynamics of Plasma: Electron Density, Plasma Temperature, and Pressure

Degree of confinement of the expanding plasma plume for PLA in liquids is much higher than the PLA in vacuum and gaseous environments. The high degree of confinement of the plasma plume imparts it to the thermodynamic state of high pressure, high temperature, and high density, which facilitates generation of metastable phases and nanostructures. A shock wave is generated when plasma expands supersonically after the absorption of the trailing part of laser pulse and continuous supply of the target material. This shock wave feels pressure due to the confinement by liquid layers in front of it and therefore exerts extra pressure on the laser-produced plasma, which may be called *plasma-induced pressure*. This extra pressure causes extra rise in temperature of the plasma plume. Confinement of the plasma plume by liquid and extra pressure on it by shock wave causes contact between plasma and target surface for longer times than gas-phase ablations, which causes increased rate of etching of the target material.

According to the work of Fabbro *et al.* [39], when a pulsed laser of $1-30$ ns pulse width is focused ($1-20$ GW cm^{-2}) on a solid target, a high pressure (>1 GPa) plasma is produced. If the surface of target is covered with a water layer of $1-5$ mm thickness or a glass overlay, the plasma plume is confined with delayed expansion. The confinement process is described in the following three steps: (i) in the pulse duration of laser, the pressure generated by the plasma induces a shock wave that propagates into the target and in the confining medium, (ii) at the time of switch off of the laser pulse, plasma maintains its pressure, which decreases during its adiabatic expansion, and (iii) at last, after the complete disappearance of the plasma, the canonball-like expansion of the heated gas inside the interface adds additional momentum to the target [39]. First, they proposed an analytical model for the confined ablation and reported the effect of laser intensity, wavelength, and pulse width on the plasma absorption, impulse, coupling coefficient, and plasma pressure for confined and direct ablation. It was observed that for particular laser intensity plasma pressure, coupling coefficient and plasma absorption are higher for confined ablation as compared to direct ablation. Longer pulse width lasers cause higher plasma absorption, impulse, and plasma pressure but smaller coupling coefficient [39].

Berthe *et al.* [45] investigated the influence of laser power density on the value of peak pressure inside the liquid-confined plasma for different laser wavelengths using VISAR technique and simulated it theoretically. Pressure induced by this confinement is four times higher and lifetime of shock wave is two to three times longer than that of the unconfined plasma. This WCR was used for surface

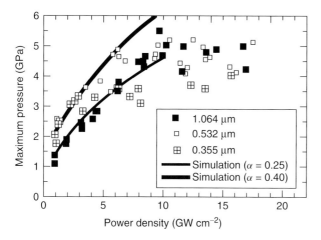

Figure 6.1.1 Peak pressures obtained in the water-confined regime from back-free velocity measurements with the VISAR technique as a function of laser power density at 1.064 µm and 0.532 and 0.355 mm laser wavelengths. Comparison with SHYLAC simulation with different interaction efficiencies $\alpha = 0.25$ and $\alpha = 0.40$. (Source: Reprinted with the permission from Berthe et al. [45] copyright @ American Institute of Physics.)

processing and materials characterization under shock loading. Increase in the laser power density results in increase in the peak plasma pressure, which attains saturation at a particular power density; reduction in shock wave durations; and duration of the laser pulse reaching at the target surface after transmitting through the confined plasma becoming smaller than the original one. These results are same for 1.064, 0.532, and 0.355 µm lasers. However, the laser power density thresholds corresponding to the pressure saturation levels are lower for UV and greenlight, while the pressure durations decrease more strongly at 0.355 µm as compared to the 1.064 µm laser wavelength (Figure 6.1.1).

Peyer et al. [32] studied the influences of laser intensity and pulse duration on the plasma pressure generated in WCR. They observed that pressure increases linearly with the increase of laser irradiance but saturates at higher laser intensities due to the optical breakdown at water–air interface. At particular laser intensity, shorter pulse laser generates higher peak pressure in WCR as compared to longer laser pulses. Sakka et al. investigated thermodynamic parameters of liquid-confined graphite [46] and aluminum [47] plasmas using optical emission spectroscopy of corresponding plasma plumes. They measured plasma temperature using optical analysis of C_2 swan band and reported that plasma temperature is 4000–5000 K, the ablated species density is $10^{22}-10^{23}$ cm^{-3}, and the pressure is about 10 GPa when the second harmonic of a pulsed Nd:YAG laser operating at 10^{10} W cm^{-2} irradiance and 10 ns pulse duration irradiates graphite target in water.

Takada et al. [48] have studied the effect of additional external pressure on the optical emission intensity, volume, and pressure of liquid-confined plasma. They observed an enhancement in the optical emission intensity, compression in the

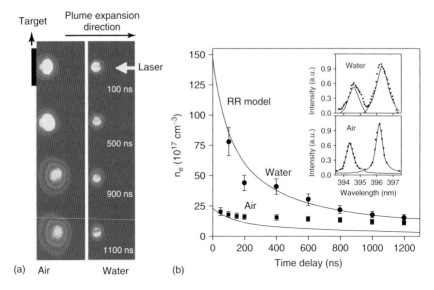

Figure 6.1.2 (a) Images of the laser-produced aluminum plasma plume in air and in water and (b) corresponding electron densities obtained by optical emission lines of aluminum plasma shown in the inset of (b). (Source: Reprinted with the permission from B. Kumar and R.K. Thareja [167], copyright @ American Institute of Physics.)

spatial distribution of optical emission intensity, and increase in the pressure inside the plasma plume with the application of additional external pressure.

Recently, Bhupesh and Thareja [167] have carried out a comparative investigation of plasma parameters and expansion dynamics for plasmas produced in air and water using optical emission spectroscopy and plume imaging, respectively. Figure 6.1.2 illustrates images of the aluminum plasma plume produced in air and water ambience captured at different instances; it shows higher confinement of plasma plume for PLA in liquids as compared to PLA in air. Electron densities were calculated using Stark broadening in the spectral lines using $n_e = (\Delta\lambda/2\,w) \times 10^{16}$; where $\Delta\lambda$ (angstrom) is the full width at half maximum (FWHM) and w is the electron impact parameter. It is observed that at a particular instant, electron density for plasma produced by PLA in water is much higher than that produced by PLA in air. Temporal profile of the electron density for plasma produced in water follows comparatively faster decay because of its higher collisional parameter as a consequence of confinement of electron and ion in smaller volume.

6.1.3.1.3 Analytical Models for Shock Laser Processing and Additional Pressure in the Plasma

Model of Fabbro and Coworkers Assumptions of the model of Fabbro and coworkers [39] : It is assumed that (i) the laser pulse is rectangular with intensity I_0 and pulse duration τ; (ii) laser energy E_{laser} is totally converted into internal energy, $E_{int} = E_{th}$ (thermal energy) $+ E_{ionization}$ (ionization), of the plasma; (iii) thermal and radiative losses are negligible; (iv) plasma is an ideal gas in thermodynamic

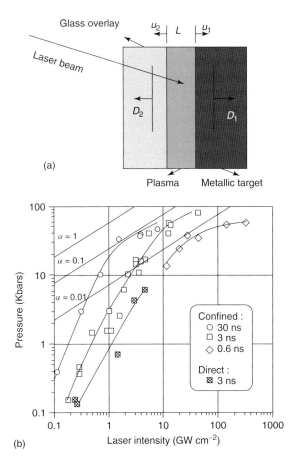

(a)

(b)

Figure 6.1.3 Variation of the peak pressure as a function of laser intensity for 30, 3, and 0.6 ns pulse durations in confined geometry. Results obtained in direct ablation and $\tau = 3$ ns have been reported. Comparison with Febro's model of Eq. (6.1.4) for different values of α is shown ($Z = 1.4 \times 10^6$ g cm^{-2} was used due to the glass/quartz interface). (Source: Reprinted with the permission from Fabbro et al. [39] Copyright @ American Institute of Physics.)

equilibrium so that the pressure is related to the thermal energy by $P = 2/3 E_{th} = 2/3\alpha E_{int}$; where $\alpha = E_{th}/E_{int}$; (v) target and confining media have constant shock impedances Z_1 and Z_2; and (vi) boundaries move with the shock wave particle velocities u_1 and u_2 and shock velocity in the target and confining media are D_1 and D_2, respectively, as shown in Figure 6.1.3a.

The model is based on two differential equations

$$\text{Energy conservation equation: } I(t) = P(t)\frac{dL}{dt} + \frac{d\left(E_{int}(t)L\right)}{dt} \tag{6.1.1}$$

$$\text{Expansion of the plasma openings: } \frac{dL(t)}{dt} = Z^* P(t) \tag{6.1.2}$$

where $Z^* = 2Z_1 Z_2/Z_1 + Z_2$ is the combined shock impedance of target material and confining media with $Z_1 = \rho_1 D_1$ and $Z_2 = \rho_2 D_2$ are shock impedances of target and confining medium, respectively, and ρ_1 and ρ_2 are densities of target material and confining medium, respectively. $P(t)$ is the transient pressure inside the plasma and L is the width of the plasma opening.

For constant laser intensity I_0 and pulse width τ, integration of above equations gives the maximum pressure P generated by the plasma

$$P = A \sqrt{Z I_0 \left(\frac{\alpha}{2\alpha + 3} \right)} \tag{6.1.3}$$

where A is a constant and I_0 is the absorbed laser energy.

Practically, Eq. (6.1.3) can be written as

$$P(GPa) = 0.01 \sqrt{Z \left(g\,cm^{-2} s \right) \times I_0 \left(GW\,cm^{-2} \right) \left(\frac{\alpha}{2\alpha + 3} \right)} \tag{6.1.4}$$

$$L_0 = L(\tau) = \frac{4\tau}{3} \sqrt{\frac{\alpha}{2\alpha + 4} \frac{I_o}{Z}} \tag{6.1.5}$$

Here, α can be obtained experimentally. Berthe *et al.* [45] and Wu and Shin [43] reported that values for α lie in the range of 0.1–0.5, and it is indicated in the work of Wu and Shin [43] that the low values of α cause negligible reflection of light at plasma–water interface for water-confined laser plasma.

For water-confined plasma, Eq. (6.1.4) reduces to $1.02 \sqrt{I_0 \left(GW cm^{-2} \right)}$ and size of the plasma plume $L(\tau)$ at the end of the laser pulse becomes $L(\tau)(\mu m) = 2 \times 10^3 \times P(GPa) \times \tau(ns)$. This model gives pressure, $P(\tau) = \sqrt{(m_1 I_0 \alpha)/2\tau(\alpha + 1)}$, at the end of the laser pulse and target velocity $V_{target}(\tau) = (2P(\tau) \times \tau)/m$, where m is the mass of the target. Figure 6.1.3b illustrates experimental data on the variation of the peak pressure as a function of laser irradiance for different pulse width lasers under confined and direct ablation. Experimental data is compared with the model given by Eq. (6.1.4) for different α values.

Shortcoming of Model of Fabbro and Coworkers   In their model, Fabbro and coworkers have assumed that ideal gas is present before the laser pulse without dealing with detailed mechanisms of plasma processing. However, threshold laser fluence is required before the plasma processing. It was considered that the mass of plasma remains constant throughout the processes, which is not true. There is certain dependency of shock impedance on the pressure, which is ignored in the model. Energy losses due to the radiation, conduction, and target modification are totally ignored. The current model is not suitable for shorter pulse lasers but works for thicker targets.

Model by Sollier et al.   Taking the shortcomings of the model by Fabbro and coworkers into account, Sollier *et al.* [40] numerically calculated transmittance of laser light through the water plasma, which incorporates losses due to the multi-photon ionization and diffusion and recombination of electrons. They considered plasma as a neutral gas from target material and accounted thermal losses in target–water system. They used the following mass balance equation to calculate plasma parameters:

$$\frac{dn(t)}{dt} = \frac{1}{L(t)} \left[n V_{abl}(t) - \frac{2}{Z} P(t) n(t) \right] \tag{6.1.6}$$

where $n(t)$ is the density of neutral at time t and V_{abl} is the ablation velocity from Heartz–Knudsen theory. They calculated plasma parameters using one-dimensional ACCIC (autoconsistent confined interaction code) from the experimentally determined laser absorption.

Model by Zhang et al. Zhang *et al.* [42] proposed a model for thermodynamic investigation of water-plasma-target system by considering that (i) plasma obeys the ideal gas law, (ii) target does not feel any thermal effect, that is, only the coating layer is vaporized, (iii) plasma expands only in axial direction and in the preliminary stage, plasma parameters such as density, internal energy, and pressure of the plasma are uniform throughout the plasma volume but vary with time, and (iv) shock pressure and velocities of coating layer and target are equal.

They divided water-plasma-target system into six strips (Figure 6.1.4a) namely (i) unshocked water, (ii) shocked water, (iii) plasma, (iv) coating layer, (v) shocked solid, and (vi) unshocked solid, where shocked and unshocked properties depend on the conservation of mass, energy, momentum, and shock speeds as follows:

$$\frac{U_w - U_{w_0}}{D_w - D_{w_0}} = 1 - \frac{\rho_{w_0}}{\rho_w} \tag{6.1.7}$$

$$P_{w_0} = \rho_{w_0} \left(D_w - D_{w_0} \right) \left(U_w - U_{w_0} \right) \tag{6.1.8}$$

$$\left(E_w + \frac{U_w^2}{2} \right) - \left(E_{w_0} + \frac{U_{w_0}^2}{2} \right) = \frac{1}{2} \left(P_w + P_{w_0} \right) \left(\frac{1}{\rho_{w_0}} - \frac{1}{\rho_w} \right) \tag{6.1.9}$$

$$D_w = D_{w_0} - S_w U_w \tag{6.1.10}$$

where E is the stored energy, D is the shock velocity, P is the pressure, and ρ is the density with subscripts w for unshocked water and w_0 for shocked water.

Mass and energy conservation equation were applied in the following form:

$$\rho_{P(t)} \int_0^t \left(U_{pW} + U_{pT} \right) dt = \int_0^t \left(MF_W + MF_C \right) dt \tag{6.1.11}$$

where U_{pW} and U_{pT} are particle velocities near water and target surfaces, respectively, while MF_W and MF_C are flow of masses from water and from coating into

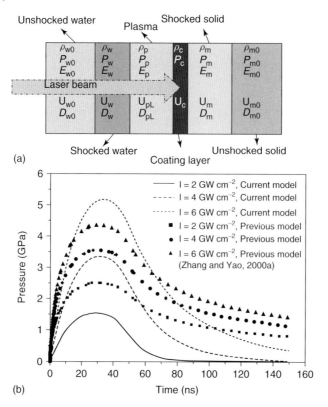

(a)

(b)

Figure 6.1.4 (a) Six regions used in the modeling and (b) temporal evolution of pressure in the plasma plume at different laser irradiances. (Source: Reprinted with the permission from Zhang *et al.* [42] copyright @ American Institute of Physics.)

the plasma, respectively.

$$\int_0^t A_P \times I(t)\mathrm{d}t = E_{Pt} + W_P - E_{MF} \tag{6.1.12}$$

where A_P, E_{Pt}, W_P, and E_{MF} denote fraction of energy absorbed by the plasma, total energy stored in the plasma, work done by the plasma, and energy exchanged through the mass flow, respectively. The set of equations were solved using mass and energy conservation relations at the interfaces. Figure 6.1.4b compares the one-dimensional shock pressure determined by using the current model (Eqs. (6.1.7–6.1.12)) and the model of Zhang and Yao [49]. The latter model assumes that a constant fraction $\alpha = 0.2$ of plasma energy is used to increase the shock pressure.

Model by Wu and Shin As shown in Figure 6.1.5a, Wu and Shin [43] considered (i) reflection of laser light at plasma-water and water-air interface, (ii) absorption of laser light in the plasma through the processes of inverse Bremsstrahlung

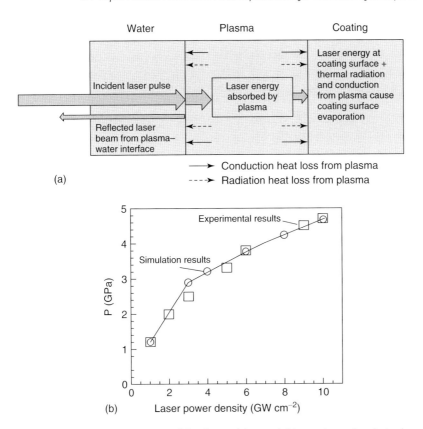

Figure 6.1.5 (a) Three regions used for the modeling and (b) experimental and simula-tion results of variation in the confined plasma pressure with laser power density. (Source: Reprinted with the permission from Wu and Shin [43] copyright @ American Institute of Physics.)

and photoionization, (iii) energy losses from the plasma through radiation and conduction processes, (iv) laser ablation of coating layer, water evaporation, and (v) plasma ionization and expansion, to propose a self closed thermal model for laser-shock-processing in water.

In their model, they assumed that (i) physical processes are one dimensional, (ii) plasma parameters such as density, temperature, and so on are uniform in space but vary with time, (iii) only electron-ion and electron-atom inverse Bremsstrahlung absorption and photoionization are responsible for the absorption of laser light by the plasma, (iv) all the electrons are assumed to have same temperature T_e and all the atoms and ions have same temperature of T_i (two temperature model), and (v) water molecules get dissociated into H and O atoms just after the evaporation. The pressure-caused receding velocities of the water and coating surfaces are calculated using shock wave relation as follows:

$$u_{w,pre} = \frac{P}{\rho_w D_w} = \frac{P}{\rho_w \left(D_{w_0} + s_w u_{w,pre} \right)} \qquad (6.1.13)$$

$$u_{c,pre} = \frac{P}{\rho_c D_c} = \frac{P}{\rho_w \left(D_{c0} + s_c u_{c,pre}\right)} \tag{6.1.14}$$

where ρ_w and ρ_c are densities and D_w and D_c are shock velocities of water and coating, respectively. D_{w0}, s_w, D_{c0}, and s_c are material property constants for shock velocity calculations.

The evaporation-caused surface receding velocity was calculated from the Hertz–Knudsen equation [50]

$$u_{cev} = \frac{1}{\rho_c} \beta p_0 \left(T_{cs}\right) \left(\frac{m_{c,a}}{2\pi k_B T_{cs}}\right)^{1/2} \tag{6.1.15}$$

where $m_{c,a}$ is the mass of coating atom or mass of target atom in the case of bare target, and the equilibrium vapor pressure is given in Ref. [50] as

$$p_0\left(T_{cs}\right) = p_\infty \exp\left[\frac{L_{ev,c}}{R_v T_{ev,c}} \left(1 - \frac{T_{ev,c}}{T_{cs}}\right)\right] \tag{6.1.16}$$

where R_v is the universal gas constant, p_∞ is the equilibrium vapor pressure at $T_{ev,c}$, and β is the evaporation coefficient, which is ~ 0.84.

Similarly, the water evaporation was calculated and given as follows:

$$u_{wev} = \frac{1}{\rho_w} \beta p_0 \left(T_{ws}\right) \left(\frac{m_{w,a}}{2\pi k_B T_{ws}}\right)^{1/2} \tag{6.1.17}$$

$$p_0\left(T_{ws}\right) = p_\infty \exp\left[\frac{L_{ev,w}}{R_v T_{ev,w}} \left(1 - \frac{T_{ev,w}}{T_{ws}}\right)\right] \tag{6.1.18}$$

where ρ_w, T_{ws}, $L_{ev,w}$, and $m_{w,a}$ are the density, temperature, latent heat of evaporation, and molecular mass of water, respectively. $p_0(T_{ws})$ is the water equilibrium vapor pressure at T_{ws} and p_∞ is the water equilibrium vapor pressure at $T_{ev,w}$. The total receding velocities of water and the coating surface were taken as a sum of the evaporation and plasma-pressure-caused boundary velocities. The calculation of the peak plasma pressure by this model was in good agreement with that from experiments in which laser power densities range from 1 to 10 GW cm^{-2} and for different combinations of pulse shape, wavelength, and duration: at power densities more than 25 GW cm^{-2} the measured pressure was lower than the calculated one. Figure 6.1.5b presents experimental and simulated results of effects of laser irradiance on peak plasma pressure.

Model by Etina Very recently, Etina [51] proposed a model for the generation of NPs by short and ultrashort PLA of solids under liquid environment by dividing the ablation process into three stages, namely, (i) early stage, in which absorption of laser energy, ejection of material, and plume formation occur, (ii) intermediate stage, which includes expansion of the plume under liquid confinement, and (iii) last stage, which consists of mixing of the plume with liquid, NP coalescence/aggregation process.

Assumptions: (i) The laser spot size is large enough so that one-dimensional approximation can be used for the early stage, (ii) collisional fragmentation and chemical reactions with surfactant molecules in the liquid are neglected, and (iii) top hat laser pulse.

1) **Early stage**: According to the model of Fabbro *et al.* [39] (Eqs. (6.1.1–6.1.5)) internal energy of the plasma $E_{int} = E_{th}$ (thermal energy) $+ E_{ionization}$ (ionization), where thermal energy $E_{th} = \alpha E_i$ is too small for ablations using short and ultrashort laser pulses. In addition, a fraction of ablated material ejects in the form of particles and clusters [52, 53]. For the early stage, Etina used two-shock-wave model of Fabbro *et al.* (Eqs. (6.1.1–6.1.5)) with the specific case of short laser pulses. Thus Eqs. (6.1.3) and (6.1.5) can be used for ultrashort pulses in liquid, but τ stands for electron–phonon/ion relaxation time in this case. Equations (6.1.3–6.1.5) state that the initial plume length, L, is smaller, whereas the initial pressure, P, is larger for ultrashort laser pulses than for longer ones.

2) **Intermediate stage**: In this stage, plasma plume expands adiabatically behind the shock wave and cavitation bubble; therefore the plume length and pressure follow the equations given below.

$$P(t) = P_0 \left[1 + \frac{\gamma + 1}{\tau} (t - \tau) \right]^{-\gamma/\gamma + 1} \tag{6.1.19}$$

$$L(t) = L_0 \left[1 + \frac{\gamma + 1}{\tau} (t - \tau) \right]^{1/\gamma + 1} \tag{6.1.20}$$

where $\gamma = 1 + 2\alpha/3$ is an adiabatic parameter. Here, we can see that the final pressure and length depend on their corresponding initial values and adiabatic parameter. Therefore, plume number density is higher at this stage, while plume temperature is lower for shorter laser pulses. These conditions favor the formation of smaller NPs, which is controlled by free energy as follows:

$$\Delta G(n, c) = -nk_B T \ln\left(\frac{c}{c_0}\right) + 4\pi a^2 n^{2/3}\sigma \tag{6.1.21}$$

where T is temperature in Kelvin, n is number of monomer/atom in the nucleus, a is effective radius of plume species, c and c_0 are concentration and equilibrium concentration, respectively, k_B is the Boltzmann constant, and σ is effective surface tension. The peak of nucleation barrier corresponds to the critical cluster size as follows:

$$n_c = \left[\frac{8\pi a^2 \sigma}{3k_B T \ln(S)} \right]^3, \quad S > 1 \text{ is supersaturation parameter} \tag{6.1.22}$$

The nuclei generated at this stage can grow due to the addition of both monomer and cluster and can evaporate due to the interaction with laser. Size of nuclei continues to increase inside the plasma plume until density of plasma plume drops and collision stops. The growth process dominates over the evaporation if temperature of cluster and plasma are not too high. Since in ultrafast (femtosecond and picosecond) lasers, the laser does not reheat the expanding plasma, cluster growth is a more favorable condition in ultrafast laser ablation. These nuclei, grown in the intermediate stage, act as seeds for the prolonged growth of particles in the later stage.

3) **Later stage**: At this stage species from the plasma plume, such as atoms and clusters, mix in the liquid to form a solution. The following processes of nucleation and growth in the solution depend on its initial conditions such as solute concentration, temperature, particle size, and density. For example, if solution is supersaturated, new nucleus can be formed in the solution, while if solution concentration is not too high, growth of particles produced in the intermediate stage would occur. The rate of formation of critical clusters at this stage is given by

$$\rho(t) = 4\pi \, an_c^{1/3} \, Dc^2 \exp\left[\frac{-\Delta G(n, c)}{k_B T}\right] \tag{6.1.23}$$

where D is the diffusion coefficient of target atoms in the liquid. If number of nuclei is sufficiently high for collisions among them, coalescence and aggregation occur, which causes increase in the particle size. Etina [51] described growth processes using the following three equations:

$$\frac{dN_1}{dt} = \rho(t) - \sum_{j=2}^{\infty} j \frac{dN_j}{dt} \quad (N_1 \text{ is the primary particle}) \tag{6.1.24}$$

$$\frac{dN_2}{dt} = fk_1 N_1^2 - k_2 N_1 N_2 \quad (N_2 \text{ are secondary particles with 2}$$
$$\text{primary particles}) \tag{6.1.25}$$

$$\frac{dN}{dt} = k_{s-1} N_1 N_{s-1} - K_s N_1 N_s \quad (N \text{ denotes particles with } s \geq 3$$
$$\text{primary particles}) \tag{6.1.26}$$

where $N_s(t)$ is the time-dependent number density of secondary particles having s number of primary particles; $K_s = 4\pi (R_1 + R_s)(D_1 + D_s)$ is attachment rate constant, $R_s = 1.2rs^{1/3}$, where r is the average radius of primary particle; and $D_s = D_1 s^{-1/3}$ is the diffusion constant of secondary particles having s number of primary particles. Here, the diffusion coefficient of primary particle is given by Stokes–Einstein relation, $D_1 = k_B T / 6\pi \eta\sigma$, which depends on the viscosity and temperature of liquid medium. Figure 6.1.6 illustrates the effects of solute concentration and water temperature on size distribution of Au NPs. It is shown theoretically that if solute concentration is small, increase in the water temperature causes increase in number of particles with constant hydrodynamic radius, while for higher solute concentration, particle radius increases with the increase of liquid medium temperature with constant number of particles.

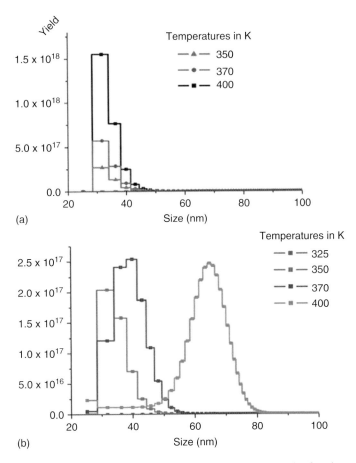

Figure 6.1.6 Size distribution of gold NPs obtained as a result of nucleation and aggregation in liquid (water) solutions at concentrations (a) 3.5×10^{21} m^{-3} and (b) 3.5×10^{22} m^{-3} at different solution temperatures. (Source: Reprinted with the permission Etina [51] copyright American Chemical Society.)

6.1.3.1.4 Effects of Laser and Liquid Media Parameters on the Kinetics of Laser Ablation at Solid–Liquid Interface: Rate of Ablation

Rate of material removal from the solid surfaces in the course of its laser irradiation depends on the wavelength, pulse width, laser irradiance of laser beam, and ablation environment. At particular laser parameters, higher absorption coefficient, that is, lower reflectance, $R = [(n_T - n_m)/(n_T + n_m)]^2$, where n_T and n_m are refractive indices of target and ablation medium respectively, of laser beam with target material causes higher rate of ablation. For a given target material, absorption coefficient for laser light at target–medium interface differs with the wavelength of laser and ablation media. Since liquids have higher refractive indices than gases, ablation of solid in liquids has higher absorption coefficient as compared to ablation in vacuum or gases. Figure 6.1.7a,b illustrates a comparison between

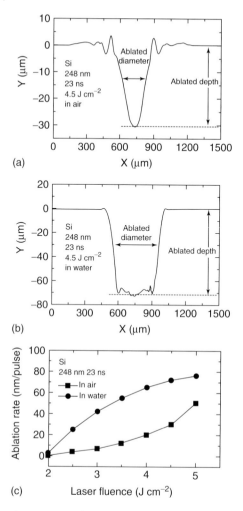

Figure 6.1.7 (a,b) Depth profiles of the ablated region of Si after 1000 pulse irradiation (a) in air and (b) in water, and (c) dependence of ablation rate on laser fluence in air and water ambient. (Source: Reprinted with the permission from Zhu *et al.* [54] Copyright @ American Institute of Physics.)

ablation craters formed on the silicon surface under air and water environment using a pulsed laser operating at 2.3 ns pulse width and 4.5 J cm^{-2} laser irradiance [54]. For the same laser parameters, a crater formed on the surface of silicon by PLA in water has almost twice the depth and diameter as compared to that obtained in air [54]. Figure 6.1.7c shows variation in the ablation rate of silicon with the laser irradiance under water and air environment [54]. Rate of ablation of Si in water is always higher as compared to the rate of ablation in air in the fluence range of 2–5 J cm^{-2}, but the difference in the ablation rate decreases at higher fluence because of the breakdown of water and water hammer effect [55]. Liquids have higher light absorption coefficients as compared to vacuum or gases; therefore PLA of target in liquids possesses significant absorption of laser light and hence reduced rate of ablation. Lower reflectivity of laser light at the solid–liquid interface favors higher rate of ablation, while absorption of laser light by water layers reduces ablation rate. Therefore, one should optimize the thickness of liquid layer in order to get the highest rate of ablation and hence highest rate of material processing. Zhu *et al.* [56] studied the effect of water layer thickness on the ablation rate. They observed increase in the rate of ablation with the increase in the liquid layer thickness in the range of 1.0–1.1 mm, with maximum ablation for a 1.1 mm thick water layer. For a 50 µm thin layer of water vapor deposit, the ablation rate was 19 nm/pulse, while it increased to 62 nm/pulse for 1.1 mm layer of liquid water. Ablation rate decreases from 62 to 0.5 nm/pulse with the increase in water layer thickness from 1.1 to 2.2 mm. They also observed that the ablation rate of Si can be greatly enhanced with a water layer 1–1.6 mm thickness. With a water layer thicker than 1.6 mm, the ablation rate drops below that in air. When laser irradiates silicon surface at water–silicon interface through the water layer, shock wave generated by laser–matter interaction emits into the water first, and an explosion into the water layer is produced by plasma plume. After this, the shock wave will decay into acoustic wave by air friction, which is termed as *"ablative piston effect"* [57]. This process enhances the ablation rate caused by high-temperature and high-pressure plasma plume. An optimal thickness of water layer provides a balance between ablative piston effect with plasma etching and laser absorption by water layer. If thickness of water layer is less than the optimal water thickness, ablative piston effect dominates, which causes high rate of ablation. On the other hand, when thickness of water layer is higher than the optimal thickness, laser absorption dominates over ablative piston effect, which decreases the rate of ablation.

For a given liquid medium and laser parameters, rate of material removal depends on the refractive index of target material as well as its physical properties such as specific heat, melting and boiling temperatures, and so on. Similarly, for a given target material and laser parameters, rate of ablation and particle generation depends on the transmission coefficient and refractive index of liquid medium at a given laser wavelength. Figure 6.1.8a shows variation in the ablation masses of Ag, Cu, Mg, and ZrO_2 materials with the pulse energy under the same experimental conditions in acetone [58]. It is observed from the reported curves that the ablation threshold in acetone is minimum for copper and maximum for ZrO_2. Ablation mass versus pulse energy curves for Ag and ZrO_2 targets in water and acetone liquid media are given in Figure 6.1.8b. From these curves, one can say that ablation threshold and rate of ablation highly depend on the liquid medium under which ablation is being performed. For example, ablation threshold for ZrO_2 lies at ~181 µJ for ablation in water, while it corresponds to ~222 µJ for PLA in acetone [58]. Slight decrease in the ablation mass with the increase in pulse energy is observed for PLA of Ag in water, while it increases for PLA of the same material in acetone.

The effective focal length of a convex lens decreases, if the distance between the lens and target is partially or fully filled by any liquid having refractive index

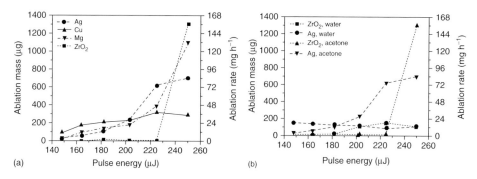

(a)

(b)

Figure 6.1.8 Dependence of NPs productivity on laser pulse energy for the ablation of (a) silver, copper, magnesium, and zirconia in acetone and (b) silver and zirconia in acetone and in water. Here 6 ps laser pulses at 1030 nm wavelength were focused though a 10 mm liquid layer using a focal length of 420 mm. (Source: Reprinted with the permission from Barsch *et al.* [58] Copyright @ Institute of Physics.)

higher than that of air. Therefore, for optimal rate of ablation, the distance of target from the lens should be lesser than its actual focal length. Owing to differences in the refractive indices of liquid media, the position of the target for optimal ablation also differs for different liquids. For example, acetone ($n\sim1.36$) has higher refractive index than water (~1.33); therefore, for the same focusing lens and laser wavelength, the target distance for optimal ablation should be smaller for PLA in acetone than PLA in water (Figure 6.1.9). In addition to this linear effect, at higher laser irradiances, *self-focusing*, a nonlinear phenomenon, takes place, which focuses laser beam at shorter distances than the actual focal length. Self-focusing of laser

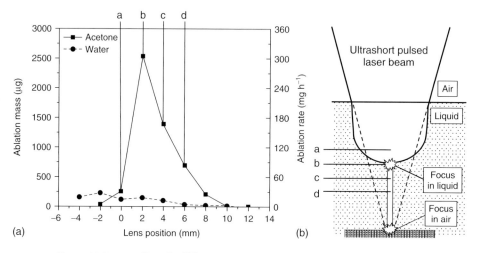

(a)

(b)

Figure 6.1.9 (a) Effect of self-focusing and hence position of the target on the rate of ablation of silver target into water and acetone and (b) sketch of focal displacement caused by nonlinear beam propagation. (Source: Reprinted with the permission from Ref. [58] Copyright @ Institute of Physics.)

light is the consequence of increase in the refractive index with laser field intensity as follows:

$$n = n_0 + \Delta n = n_0 + \gamma I = n_0 + n_2 \frac{E^2}{2} \tag{6.1.27}$$

where n_0 is the linear refractive index of the medium, Δn is refractive index change, n_2 is the nonlinear refractive index, I is the intensity of the light, and E is the amplitude of the electric field. The distance by which beam shrinkage due to self-focusing occurs is given by $d = (D/2) \sqrt{n_0/2n_2 I_0}$; where D is the maximum diameter of the laser beam and I_0 is the laser intensity [59].

For a given liquid medium, higher wavelength of the laser beam possesses lower refractive index ($n = A + B/\lambda^2$), therefore it needs lesser shrinkage in the position of target surface for optimal ablation as compared to the corresponding position in air. Tsuji et al. [60, 61] demonstrated influences of laser wavelength and focusing conditions on the ablation efficiency of Ag and Cu targets in water. For PLA of Ag, they observed that for unfocused beam, the rate of ablation was higher for shorter wavelength (355 > 532 > 1064), while in the case of PLA of Cu under unfocused beam, maximum ablation was registered for 532 nm wavelength (532 > 355 > 1064). Under the focused beam conditions, higher wavelength caused larger ablation of both the targets (1064 > 532 > 355 nm). Figure 6.1.10 illustrates that optimal ablation by 1064 nm laser light occurs at target position almost equal to the actual focal length of the lens, while for 532 and 355 nm laser beams, target positions are lesser than the actual focal length in air ($f_{lens,air} \approx f_{1064, water} > f_{532, water} > f_{355, water}$).

Pulse duration of the laser also has significant influence on the rate of ablation and hence the size, shape, and morphology of produced nanomaterials (Figure 6.1.11). Longer pulses have higher thermal effects and absorption in liquids, and hence lesser energy reaches at the target surface. Longer pulse length laser shows higher ablation rate for PLA of target at air–target interface, while shorter laser pulses exhibit faster materials removal for PLA at target–water interface. Sakka et al. [62] investigated the influences of pulse width on the ablation rate of copper target in water using optical emission spectroscopy, time-resolved plume imaging, shadowgraph of cavitation bubble, and measurement of pit volume at target surface. They observed more intense line emission, with self-reversed feature at the center of the line, but much smaller pit volume when longer laser pulse was used for the ablation of copper in water. More intense line emission and lesser pit volume are contradictory results for the longer laser pulses. They explained that longer pulse width lasers are absorbed by expanding plasma plume, therefore it heats plasma plume for longer times rather than removing material from surface. The small amount of material removed from the target surface by the front part of the longer (150 ns) laser pulse is reheated by the middle and trailing parts of laser pulse under strong confinement by water molecules, which causes excitation of more Cu atoms in the excited state and consequently, higher emission intensity. Summarizing, we can say that a higher proportion of longer laser pulse reheats plasma rather than removing material from the target surface, while maximum part of the shorter laser pulse is used in the ablation of target surface. They have also observed that a longer nanosecond pulse gives a larger and clearer hemispherical plume, while

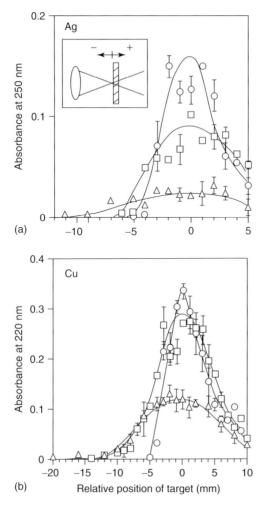

Figure 6.1.10 Relative position of the target surface from actual focal position of lens for PLA of (a) Ag and (b) Cu in water. (Source: Reprinted with the permission from Ref. [60] copyright @ Elsevier.)

a shorter pulse results in a smaller and rather flat emission image. In accordance with these, they explained that the highly excited plasma plume due to reheating by the longer laser pulse, results in a large cavitation bubble. While in the case of shorter nanosecond laser pulses, a large part of energy deposited into the surface of the target diffuses into the bulk of the target and is not used to heat up the bubble, which results in smaller-sized bubble.

Rate of material removal by PLA of solid target in liquid and hence nucleation/growth of NPs depends on the temperature of the liquid medium. Hong *et al.* [63] reported that colder liquids result in higher ablation rate. In their investigation, they used 20% aqueous solution of KOH or distilled water as the ablation medium

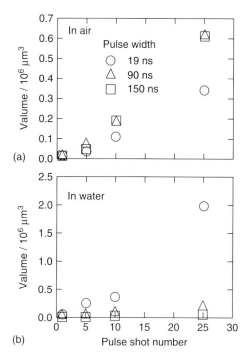

Figure 6.1.11 Volume of the ablation crater on the target as a function of pulse shot number (a) in air and (b) in water for different laser pulse widths. (Source: Reprinted with the permission from Sakka *et al.* [62] copyright @ Elsevier.)

for the ablation/etching of silicon in the temperature range of 4–30 °C. They observed ~37 nm/pulse rate of ablation for PLA in deionized (DI) water at the temperature of 10 °C (Figure 6.1.12). It changes slightly as the liquid temperature increases, with a variation of 5 nm/pulse. However, rate of ablation for PLA of Si in the KOH solution varied greatly with the variation in temperature. The ablation rate was ~75 nm/pulse at a liquid temperature of 6 °C, which reduces to ~22 nm/pulse at a liquid temperature of 30 °C (Figure 6.1.12). It demonstrated that the cooled KOH liquid environment enhances the laser ablation process greatly by removing more substrate materials.

6.1.3.1.5 Cavitations Bubble Formation and Related Effects

Cavitation is a dynamic phenomenon that occurs in the flowing liquid when local pressure is less than the saturated vapor pressure at ambient temperature. Cavitation bubbles have several destructive as well as constructive roles in the material processing. When a high-intensity laser is focused onto a solid target surface at the target–liquid interface through the transparent liquid, it causes generation of shock wave, cavitation bubble, and high-pressure plasma plume, as given in panel (A) of Figure 6.1.13. A cavitation bubble can be produced by high-pressure plasma and its radial motion is guided by pressure gradient between

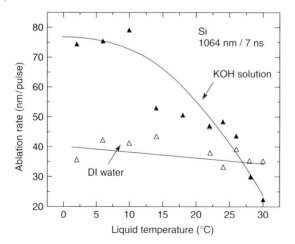

Figure 6.1.12 Variation in the rate of target ablation with the temperature of liquid medium. (Source: Reprinted with the permission from Hong *et al.* [63] copyright @ Springer.)

outside and inside the bubble. Rayleigh has theoretically developed a relationship between the maximum radius R_{max} of a spherical cavitation bubble in an infinite liquid and its collapse time T_c, which is given as follows:

$$R_{max} = \frac{T_c}{0.915\sqrt{\rho/(p_\infty - p_0)}} \tag{6.1.28}$$

where R_{max} is the maximum radius of the spherical bubble, T_c is the collapsing time of the bubble, ρ is the density of the fluid, p_∞ is the static pressure, and p_0 is the vapor pressure inside the bubble. There are two pathways of formation of cavitation bubble in the experiments of PLA of solids under liquid environment. First is the laser-induced photothermal effect that causes increase of target surface temperature above the boiling point of liquid, which causes generation of bubbles near the solid surface. The second phenomenon involves laser-induced optical breakdown of water near target surface to generate water plasma, which heats nearby water molecules above its boiling point and thus causes formation of vapor bubbles. Time-dependent shadow-graphs for the growth, decay, and collapse of cavitation bubble and shock wave during PLA of titanium metal in distilled water are shown in Figure 6.1.13B. Shadowgraph taken at 0.7 μs shows generation of shock wave and new small-sized cavitation bubbles at the target surface. Images at 20 and 120 μs show increase in size of cavitation bubbles, which start decreasing after 200 μs and finally collapse at ∼260 μs. Expansion of the plasma plume and nucleation of clusters occur inside these bubbles. Surfaces of long-lived cavitation bubbles support small NPs to fabricate large hollow spheres [64]. The change in the surface free energy for the combination of a nanocluster with the bubble interface is given by following equation:

$$\Delta G = -\pi r^2 \gamma (1 - \cos\theta)^2 \tag{6.1.29}$$

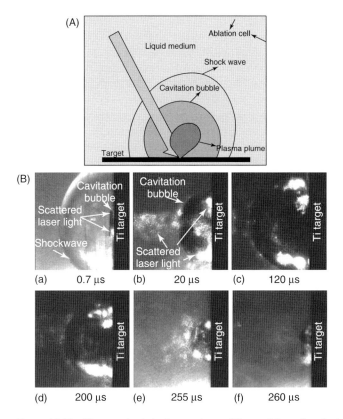

Figure 6.1.13 Time-resolved shadowgraphs at different delays after the laser pulse super-imposed with images of laser light scattering. (Source: Reprinted with the permission from Soliman *et al.* [64] Copyright @ The Japan Society of Applied Physics.)

where r is the radius of the NP, γ is the surface tension of bubble interface, and θ is the contact angle of particle with liquid. Reduction in the free energy provides driving force for the absorption of NPs, and once a particle is absorbed by the bubble interface, the energy barrier $|\Delta G|$ will obstruct its escape. Yan *et al.* [65] numerically calculated that for a 2 nm ZnO particles in water–ethanol mixture at 20 °C, $\gamma \approx 30 \times 10^3$ N m^{-1}, and that $\theta \approx 30°$ causes $|\Delta G| = 12.5 \times k_B T$; where k_B is the Boltzmann constant and $T = 293$ K, which is much larger than the thermal energy of the NP. This makes irreversible trapping of NPs by bubble. During shrinking of bubble, areal density of clusters increases, which could cause formation of network by NPs that are bonded by capillary attraction. This bonding generates a steric barrier that, by the interface motion, assembles NPs into hollow nanospheres (NSs).

Evolution, expansion, and collapse time of the cavitation bubble depends on the parameters of the laser (wavelength, irradiance, pulse width, and repletion rate) and liquid medium (surface tension, viscosity, vapor pressure, boiling point, etc.). Higher energy of the expanding plasma plume and its longer lifetime are directly associated with the size and lifetime of the cavitation bubble. Shorter pulse width laser delivers most of its energy to the target surface and very little to the expanding plasma plume, and hence little energy can be transferred to the surrounding liquid for the formation of bubble. In other words, PLA of solid target in liquids using shorter pulse width lasers produces cavitation bubbles of smaller sizes and shorter lifetimes. In contrast to this, longer pulse width laser transfers a large fraction of its energy for the reheating of the expanding plasma plume, and hence larger energy is transferred from the plasma to the surrounding liquid to form larger-sized and longer-lived cavitation bubble. Sakka et al. have used 19, 90, and 150 ns laser pulses for the PLA of copper in water to generate and diagnose cavity dynamics [62]. Study of bubble dynamics using time-resolved shadowgraphy showed that 150 ns pulse produced the most intense plasma emission, while 19 ns pulse produced the least. Bubbles produced in all the three cases achieved their maximum sizes at 100 μs, but size was maximum for 90 ns ($d_{90 ns} > d_{150 ns} > d_{15 ns}$) and minimum for 19 ns pulse durations. Cavitation bubbles formed by 19 ns laser pulse collapsed at 150 μs, while those generated with 90 and 150 μs pulses collapsed at 200 μs.

Yan et al. [65] demonstrated size reduction curves from the maximum sizes of three bubbles, produced by PLA of zinc target in water–ethanol mixture using an ArF laser operating at 5 J cm^{-2} energy and 10 Hz repetition rate (Figure 6.1.14a). Lifetimes for three bubbles of 10, 19, and 22 μm are 20, 66, and 99 s, respectively. For the trapping of NPs by the interface of cavitation bubble, the root-mean-square (rms) displacement of the particle, given by $d_{rms} = \sqrt{6DT_c}$ (here, $D = k_B T / 6\pi \mu r$ is the diffusion coefficient of the particle and μ is the dynamic viscosity of the liquid), due to the Brownian motion at collapse time, T_c, of the bubble should be larger than the size R_{max} of the bubble. For instance, a NP having 2 nm average size at 293 K temperature in water–ethanol mixture possesses $D = 4.2 \times 10^{-11}$ m^2 s^{-1} diffusion coefficient, and makes 71 μm rms displacement in 20 s, which is much larger than the size (10 μm) of cavitation bubble [65]. Therefore, enough number of clusters can approach at the bubble interface before its collapse in order to form an assembly of hollow sphere. The same sized particles in water at 293 K have diffusion coefficient of 1.07×10^{-10} m^2 s^{-1}, and they can approach only 24 nm in

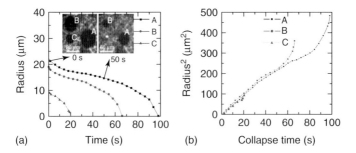

(a)   Time (s)   (b)   Collapse time (s)

Figure 6.1.14 (a) Time variation in the sizes of three different laser-induced cavitation bubbles in water–ethanol mixture with the insets shows optical images of bubbles and (b) square of the cavitation bubble radius versus time of their collapse. (Source: Reprinted with the permission from Yan et al. [65] copyright @ American Institute of Physics.)

the course of much smaller collapse time (925 ns) of cavitation bubble. Therefore, it is more difficult to diffuse enough NPs on the interface to fabricate hollow particles in water as compared to that in ethanol–water mixture. Figure 6.1.14b is a plot of the square of bubble radius against the collapse time, which is almost linear, with the exception of the early part of each collapse [65]. It may be due to the fact that in the early part the dynamic timescale of the bubble motion is larger than the diffusion timescale [66].

The ability of cavitation-bubble-assisted fabrication of hollow NPs from the assembly of small-sized solid NPs depends on the target–liquid system for a given experimental condition. Yan *et al.* fabricated hollow nanostructures of ZnO, Si, Ag, Fe–Ni alloy, and Al with the assistance of LP-PLA-induced cavitation bubbles [65]. Figure 6.1.15a,b shows hollow NSs of Ni–Fe alloy fabricated by PLA of bulk Ni–Fe alloy target material in ethanol–water mixture. Hollow NPs of Si were produced by PLA in pure water (Figure 6.1.15c,d), while it needs the addition of ethanol in water to prepare zinc oxide NSs under almost same experimental conditions. Yan *et al.* found it difficult to prepare hollow Ag NSs even in ethanol–water mixture, although it was not so difficult to prepare hollow NSs of ZnO, Si, and Fe–Ni alloy. In order to solve this issue, they added Triton X-100 (0.06% v/v) in the ethanol–water mixture and successfully prepared Ag NSs (Figure 6.1.15e,f). Addition of Triton X-100 probably further elongated the lifetime of the cavitation bubble by increasing the viscosity of the liquid medium. Yan *et al.* [65] used 0.05 M aqueous solution of SDS as ablation media for the generation of cavitation bubbles, which assisted as soft template for the fabrication of hollow NSs of Fe–Ni alloy. In another report, Yan *et al.* [67] reported the formation of Al_2O_3 hollow nano/microspheres, which were formed on the laser-induced bubbles from laser-vaporized/ionized liquid during ablation. PLA of Al target in water results in the formation of hollow Al_2O_3 nano/microparticles that are amorphous in nature. Proportion for the formation of hollow spheres in the product and crystallinity of the product increase with the addition of ethanol to water.

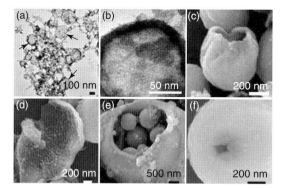

Figure 6.1.15 (a,b) TEM images of NPs produced fabricated by PLA of Fe-Ni alloy in ethanol–water mixture. SEM images of hollow nanospheres produced by PLA of (c,d) Si in water and (e,f) Ag in ethanol–water mixture with Triton X-100. (Source: Reprinted with the permission from Yan *et al.* [65] copyright @ American Institute of Physics.)

6.1.3.1.6 **Physical and Chemical Aspects of Laser Ablation in Liquid Media**

Laser ablation of solid surface at solid–liquid interface through the liquid and under liquid confinement photothermally heats and vaporizes the target surface. The nature and mechanism of surface vaporization/explosive ablation depends on the irradiance and pulse width of laser. Short and ultrafast laser pulses have comparatively higher irradiance and much lower pulse duration, which causes delivery of maximum part of the laser energy in a very short duration for explosive ablation of materials and thus plasma formation. Shorter laser pulses cannot initiate electron–phonon and phonon–phonon interactions in the target materials, and thus no/little thermal damage to target surface. Shorter laser pulses also exhibit little or no reheating of the expanding plasma plume, hence maximum part of the laser energy is transfered to the target surface. In contrast to the short/ultrashort lasers, interaction of nanosecond laser pulses with target surface creates electron–phonon and phonon–phonon interactions that generate lattice wave in the material. Thus the lattice waves generated can heat/vaporize/melt/damage the target surface beyond the focal spot. Middle and trailing parts of nanosecond laser pulses also heats, and get absorbed by, the expanding plasma plume, therefore, comparatively less amount of energy is transferred to the target surface for materials removal. As discussed in previous sections, in addition to laser-produced plasma, interaction of laser pulses with target materials under liquid confinement also creates shock waves in the liquid and target and cavitation bubble in the liquid. These shock waves and cavitation bubbles have several advantages in the materials processing such as surface cleaning, etching, and so on [68]. Laser-produced plasmas under liquid confinement have much higher temperature, pressure, and density as compared to plasmas produced by PLA in vacuum/gases. This HTHPHD plasma produces several metastable phases of nanostructure such as diamond, boron nitride, and so on, which is generally impossible for plasmas produced in normal conditions. Cavitation bubbles generated during PLA of solid in liquid can physically support as-synthesized smaller-sized particles in the solution on their interfaces to generate hollow nanostructures.

Strong visible–UV emission from laser-produced plasma under liquid confinement and/or heating of liquid molecules by expanding plasma plume at plasma–liquid interface ionizes/dissociates the neighboring liquid to generate plasma of liquid species, which may be termed *plasma-induced plasma* [69]. Here, laser-induced plasma has species from target material, while plasma-induced plasma contains species from liquid medium. Plasma-induced plasma is sandwiched by the expanding plasma plume and liquid layers; therefore it experiences strong pressure, which may causes intermixing of plasma-induced plasma (species from liquid medium) into the laser-produced plasma (species from the target material) and deep into the liquid. As proposed by Yang, four types of chemical reactions would occur between the species of target material and liquid media. First is the reaction among the plasma species inside the plasma plume that results in clusters of target species with same or different chemical compositions. The second type of reaction would occur between the species of target material and liquid medium inside the plasma plume, which might result in clusters having abundance of target species with traces of liquid medium. The third set of chemical reactions would take place between the species of target material and liquid medium at the plasma–liquid interface, which may result in target compound NPs with comparable chemical composition. The fourth and final type of chemical reaction might take place between the species of plasma plume, which are forcefully injected deep into liquid medium, and species from liquid inside the liquid. This type of chemical reaction would result in target compound NPs with abundance of species from the liquid medium.

The chemical composition of NPs produced by PLA of solid target under liquid confinement depends on the reactivity of target species and molecules from liquid medium. For example, PLA of noble metals such as Ag, Au, and Pt in liquids produces their elemental NPs [9–17, 60, 70–111], while PLA of active metals such as Zn, Ti, Al, and so on generates their compound nanostructures [18–20, 44, 65, 112–147]. Chemical composition and rate of chemical reactions between different species depends on the concentrations of species from target material and liquid medium. If laser irradiance is high enough to produce dense plasma plume of target material in short and ultrashort durations, then nucleation and growth of particles take place inside the plasma plume. In addition, if the rate of reaction among the species of target material is higher than that between species from target material and liquid medium then elemental metal core with metal compound shell NPs would occur.

6.1.3.1.7 Basics and Advancements in Liquid-Phase Laser Ablation of Solid Target Material

Some basic experimental arrangements for PLA/irradiation of a solid target immersed into the liquid (Figure 6.1.16a–d), liquid-suspended particles (Figure 6.1.16e,f), and solution of metal salts or inorganic/organic precursors of metals are shown. Vertically downward geometry of the laser beam and horizontal target/solution surface (Figure 6.1.16a,e) in open ablation vessel is mostly exploited for the ablation of solid target as well as liquid-suspended particles and solution of metal salts/precursors because of its simplest and cheapest experimental arrangement. Open surface of the liquid often causes splash and vaporization from its surface, which gets deposited on the focusing optics in the form of liquid droplets or condensed vapor. These deposits absorb/scatter laser light to weaken the rate of ablation and hence nanomaterials processing. Formation of splash and vaporization of liquid surface cause reduction in the thickness of water layer above the target surface, which can cause uncertainty in experimental findings if thickness of the liquid is not maintained during ablation with the addition of water. Maintaining the thickness of liquid layer becomes more problematic if volatile organic solvents are used as ablation media. Application of quartz lid on the ablation vessel is one of the simplest ways to prevent splash deposits on the surface of focusing lens and hence to maintain thickness of liquid layer. Formation of splash depends on the laser irradiance, focusing conditions, thickness of liquid layers, and so on; therefore, the process can be controlled by monitoring these parameters. In addition to these, formation of splash and hence

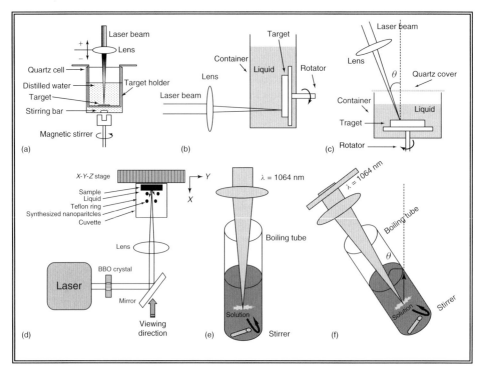

Figure 6.1.16 Some basic experimental arrangements for (a–d) PLA of solid target in liquids; (a) vertically downward beam [60], (b) horizontal beam [148], (c) beam inclined at $\theta°$ from the normal to the target surface, (d) vertically upward beam [149], and (e,f) PLA of liquid-suspended particles/powders (e) vertically downward geometry [151, 152], and (f) inclined at $\theta°$ from the normal [151].

the problem of maintaining the water thickness can be addressed by changing the direction of laser beam from vertical (Figure 6.1.16a) to horizontal (Figure 6.1.16b) configuration. There are some similarities and differences in the mechanism of plasma processing and its confinement among different arrangements of laser beam, target surface, and gravitational mass of water. For instance, in the cases of experimental arrangements shown in (Figure 6.1.16a,b), plasma expanding normal to the target surface is being irradiated/reheated by the trailing part of laser pulse. In spite of this, in vertically downward beam arrangement (Figure 6.1.16a), the gravitational mass of water works against the direction of expansion of plasma plume, while in horizontal beam arrangement, the gravitational mass acts normal to the direction of plasma expansion [60, 148]. For a set of given experimental parameters, interaction of gravitational mass of water with the expanding plasma plume for the arrangement shown in (Figure 6.1.16c) is similar to that presented in (Figure 6.1.16a), but here, the extent of interaction of laser beam with expanding plasma plume, that is, reheating of the plasma plume, is reduced because of the inclined geometry of the laser beam. The extent of reheating of the plasma can be further reduced by increasing the angle of laser beam from the normal to the of target surface. Experimental arrangement shown in (Figure 6.1.16d), where vertically upward laser beam and horizontal target surface is used [149], exhibits the same extent of laser–plasma interaction as the arrangements shown in (Figure 6.1.16a,b), but here, the direction of gravitational mass is in the direction of expansion of plasma plume. All the experimental arrangements shown in Figure 6.1.16a–d are also suitable for the laser ablation/irradiation of suspended NPs for their melting/vaporization/fragmentation and irradiation of metal salts/precursors for laser-induced chemical synthesis of nanomaterials. Figure 6.1.16e,f shows other arrangements for laser ablation/irradiation of suspended particles and liquid precursors [150], (and Figure 6.1.16f shows reduction of splash formation and deposition of liquid droplets and vapor on focusing optics [150].

There is some recent advancement in the conventional and simple experimental arrangement of PLA in liquids for the preparation of special type of nanostructures, to increase the yield, to functionalize the nanostructures and, in their advanced applications. Liu *et al.* [172, 173] have used electric-field-assisted laser ablation in liquids (EFLALs) for the shape transformation of GeO_2 nanostructures from spherical to cubical and spindlelike structure and generation of metastable phase of single crystalline Ge NPs. Barsikowaski and coworkers [132, 144] used flowing liquid and scanning optics techniques for the gram-scale synthesis of nanostructures and *in situ* [9, 10] and *ex situ* [10] surface functionalization of nanomaterials. Some groups have used two laser beams and controlled the preparation of nanomaterials by controlling delay between two pulses [78]. These advancements are discussed in the subsequent sections of this chapter.

6.1.3.1.8 Metal/Elemental and Metal/Compound Nanomaterials by LP-PLA of Solid Targets

Metal/Elemental NPs by Nonreactive LP-PLA of Solids   Here, species of target material in the plasma do not react with the surrounding liquid medium, thus the formed nanostructures have the same composition as the applied targets. This is the most basic configuration of an LP-PLA process and is used extensively. In principle, due to the very broad practicability of laser ablation in generation of plasma, this method is very versatile and can produce nanostructures of many kinds of materials as long as the corresponding targets can be obtained. Generally speaking, corresponding to the types of targets, two kinds of products can be obtained, noble metal and compound NPs. Noble metals such as silver, gold, and platinum are nonreactive with most of the liquids; therefore PLA of these metal targets in liquid media produces their corresponding NPs and assemblies. In other words, we can say that nonreactive laser ablation produces particles of single element or alloy materials depending on the stoichiometry of target. Nonreactive laser ablation illustrates a set of experimental parameters such as laser parameters, liquid media composition, and target parameters, which prevent any reaction between laser-produced plasmas and liquid molecules at plasma–liquid interface, inside the plasma, and inside the liquid. In a particular target material and liquid medium, a set of laser parameters may exhibit reactive conditions [18–20, 44, 65, 112–147], while others may show nonreactive conditions [9–11, 70–111, 153]. In most of the cases, PLA of Ag [11, 70–93], Au [9, 10, 14–16, 94–109], and Pt [17, 110, 111, 153] in most of the liquids produces their corresponding elemental metal NPs by nonreactive PLA, but in certain conditions, depending on laser and liquid media parameters, metal compounds are also produced on the surface of metal NPs through reactive PLA to synthesize core @ shell nanostructures. For example, Ag core @ Ag_2O shell NPs were produced by femtosecond PLA of Ag in water as well as in ethanol liquid media. Reactivity of the laser-produced metal plasma with the surrounding liquid molecules depends on its thermodynamics, which is monitored by laser parameters such as fluence. Thickness of the Ag_2O shell depends inversely on the laser fluence and attains a maximum thickness of 0.2 nm for a fluence of 60 J cm^{-2} and a minimum thickness of 0.04 nm corresponding to a laser fluence of 1000 J cm^{-2}. Active metals such as zinc, copper, and titanium can also exhibit nonreactive PLA under high surfactant concentrations and larger laser fluencies. Formation of pure zinc NPs at SDS concentrations higher than its critical micelle concentration (CMC is an example of nonreactive PLA for active metals. In the following subsections, we discuss nonreactive PLA for noble and active metals separately.

Preparation of Elemental NPs of Noble Metals   Silver, gold, and platinum have low reactivity with liquids and gases; therefore PLA of these metals in most of the liquids has a high probability for the synthesis of their corresponding elemental NPs. Table 6.1.1 shows PLA of noble metals for the synthesis of their elemental nanostructures.

Šišková *et al.* [70] have ablated Ag metal target in the aqueous media of some strongly adsorbing ions, which resulted in the formation of chemically modified Ag NPs. They used UV–visible absorption and SERS spectroscopy for the characterization of as-synthesized metal/organic hybrid NPs. Virgin coconut oil controls the particles size and prevents agglomeration of the ablated NPs [71]. Therefore the obtained NPs are stable for quite a long time. Smaller particles are produced for longer time of ablation, with the results of resizing of particles by melting/vaporization. Spherical Ag NPs of 4.84, 5.18, and 6.33 nm average sizes were obtained for PLA of Ag target in virgin coconut oil for 45, 30, and 15 min respectively [71]. Yang *et al.* [72] have fabricated Ag nanostructured films through electrophoric deposition (EPD) in the PLA-synthesized colloidal solution of Ag NPs. It is reported that the morphology of the produced film is tunable and controllable with the EPD parameters.

PLA of Ag target in water or aqueous media of electrolyte solutions (HCl, NaCl, NaOH, $AgNO_3$, and/or Na_2SO_3) using fundamental wavelength from Nd:YAG laser (20 ns, 10 Hz, 190 mJ/pulse, 20 min continuous ablation or 5 successive steps interrupted by 2 min breaks) produced spherical Ag NPs with different surface morphologies [73]. Presence of Cl$^-$ and/or OH$^-$ ions in the aqueous media prevents the formation of larger Ag NPs and aggregates. This effect was credited to an efficient adsorption of Cl$^-$ and/or OH$^-$ ions on surfaces of Ag NPs and an efficient buildup of the electric bilayer around particles. The presence of HCl, NaCl, and/or NaOH during PLA has led to the stabilization of the resulting Ag NPs, while the presence of $AgNO_3$ and $Na_2S_2O_3$ has shown a destabilizing effect. Amendola *et al.* [75] used different organic solvents such as acetonitrile (AN), *N,N*-dimethyl formamide (DMF), tetrahydrofuran (THF), or dimethyl sulfoxide (DMSO) as ablation media for the PLA of Ag target using 1064 nm wavelength from pulsed Nd:YAG laser operating at 9 ns pulse width and 10 Hz repetition rate. They have reported that NPs produced in AN have an average radius of 3.5 nm and a spheroid contribution of only 2%, while Ag NPs with 5 nm average size and 100% spheroidal contribution were produced in DMF. Core/shell NPs with 2.9 nm Ag core and 0.85 nm graphite shell were produced by PLA in THF. Similarly, core/shell NPs with 6 nm thickness of graphite layer and 3.9 nm average size for Ag core were synthesized by PLA in DMSO.

Mafuné *et al.* [12] studied effect of SDS concentrations, laser irradiance, and focusing conditions on the size, distribution, and abundance of Ag NPs produced by PLA of Ag target in aqueous media of SDS. They have reported that average size of NPs decreases, while dispersion first increases upto CMC and slightly decreases after that with the increase of SDS concentrations. Mean size of NPs increases with the increase of laser pulse energy, while dispersion remains almost constant. Abundance of particles increases when laser irradiance reaches a maximum and decreases after that. The value of laser pulse energy corresponding to maximum abundance depends on the laser spot size on the surface. Larger spot size results in relatively higher maximum abundance at higher pulse energy. At a particular laser irradiance, relative abundance increases quadratically with laser spot diameter. Relative abundance increases linearly with the increase in number of laser shots upto 50 000 shots, and deviates downward from linear beyond this. Figure 6.1.17 illustrates TEM images and corresponding size histograms for NPs produced by PLA of Ag target in aqueous media of different concentrations of SDS.

Petersen *et al.* [9] ablated Au target in the aqueous media of 1 or 5 μM of penetratin for *in situ* bioconjugation of Au NPs, which allowed rapid design of cell-penetrating nanomarkers for intracellular bioimaging. Significantly enhanced penetratin conjugation efficiencies could be observed at higher pH values, whereas the size and morphology of bioconjugates depends on the peptide concentration during ablation. Ana Menendez-Manjon *et al.* [95] studied the effect of water temperature on the hydrodynamic diameter of Au NPs produced by LAL. Temperature of liquid media affects its physical properties such as viscosity, surface tension, and compressibility, which drive volume, expansion velocity, and lifetime of cavitation bubbles. Sizes of the Au NPs increase with the increase of liquid medium temperature. Hydrodynamic diameter of Au NPs exponentially decays with compressibility of water, which depends on solvent temperature.

Saitow *et al.* [100] studied the effect of density of fluid on the size of Au NPs produced. They used supercritical CO_2 in high pressure (upto 20 MPa) cell as ablation medium for LAL and synthesized Au nanonecklaces and NSs. Morphology of produced Au nanomaterials depends on the density of fluids. Low density (ρ 0.2 g cm^{-3} and P 4.29 MPa) causes necklace-like morphology of 30 nm mean diameter, while higher density ($\rho = 1.7\,\mathrm{g\,cm^{-3}}$, $P = 14.5\,\mathrm{MPa}$) results in spheres of 500 nm average size. Sylvestre *et al.* [104] used water or aqueous media of various ionic solvents for LAL and studied surface chemistry of Au NPs produced by PLA of Au target in these solutions. PLA of Au targets in water produces Au NPs in the size range of 1–250 nm with partially oxidized surfaces. The hydroxylation of these Au−O compounds, followed by a proton loss to give surface Au−O⁻, resulted in the negative charging of NPs. The oxidized surface reacted efficiently with Cl⁻ and OH⁻ to augment its net surface charge, with reduced coalescence of NPs, due to electrostatic repulsion, and led to a significant reduction in NP size. This virtue resulted 7 ± 5 nm and 5.5 ± 4 nm sizes of Au NPs in 10 mM aqueous media of KCl, and NaCl, respectively and 8 ± 5 nm size for NaOH with pH 9. Use of *N*-propylamine controlled the size in the range of $5–8 \pm 4–7$ nm.

Solid and hollow spherical Pt NPs were produced by PLA of Pt target in water or aqueous solution of 0.05 M SDS using KrF excimer laser (248 nm, 30 ns, 10 Hz) at different laser irradiances [17]. Depending on the laser fluence, different kinds of hollow Pt structures were obtained that could be coalesced by micrograins, assembled by nanocrystals or with smooth shells. These hollow structures are considered to form on laser-produced bubbles in water, which provided thermodynamically preferred nucleation sites and diffusion sinks for Pt small clusters or particles. Mafuné *et al.* [153] have produced Pt NSs in the size range of 1–7 nm using PLA of Pt target in the aqueous media of different concentrations of SDS. The ablation rate and hence abundance of NPs changes with the concentration of SDS in solution. Size of the particles remains constant with increase of SDS concentration from 0 to 10^{-3} M, decreases sharply near CMC (10^{-2} M), and decreases slightly with SDS concentration above CMC.

Table 6.1.1 presents nanostructures produced by nonreactive LP-PLA of noble metal targets in the liquid media.

Preparation of Elemental NPs of Active Metals   Active metals such as Zn, Cd, Cu, Ti, Al, Mg, Pb, Fe, and so on usually react with the surrounding liquid molecules, vapors, and air and form their compound oxide, hydroxide, nitride, carbide, and other molecules on their surface. PLA of these solid targets immersed in the liquid generates plasma at solid–liquid interface, and species of the plasma react with the liquid molecules at plasma–liquid interface to generate their compound molecules and consequently their compound NPs. However, if the rate of generation of the plasma species is high and time for their clusterization into particles is small, then reaction between the plasma species inside the plasma dominates over the reaction between plasma species and liquid molecules at plasma–liquid interface. Under such conditions, particles of active metals [154] or active metal core/metal compound [112, 113] shell NPs are produced. PLA of active metals in the liquid with stabilizing/capping agents also limits reaction between plasma species and liquid molecules at plasma–liquid interface. There occurs competition between formation of clusters of plasma species with its surface capping by capping agents and reaction of plasma species with the liquid molecules. PLA of Ti in ethanol and dichloroethane produced cubic phase Ti nanocrystals of 35 nm average size [155].

Table 6.1.1 LP-PLA of noble metals in liquids for the preparation of their elemental nanomaterials.

Liquid parameters/target	Laser/ experimental parameters	Product obtained	Findings/ descriptions	References
PLA of Ag in liquid media				
Deionized water or bpy or 5, 10, 15, H2TMPyP (Silver plate)	Nd:YAG laser (1064 or 532 nm, 20 ns, 10 Hz), 5 J cm^{-2} for 1064 and 0.5 J cm^{-2} for 532 nm	Ag NPs, Ag NPs/bpy, and Ag NPs/H$_2$TMPyP hydrosol systems were produced	Under same experimental conditions, PLA in bpy produces comparatively smaller Ag NPs as than that produced in H$_2$TMPyP, while 1064 nm PLA in particular liquid medium such as water, bpy, or H$_2$TMPyP produces larger-sized particles with wider distribution than produced by 532 nm PLA. SERS sensitivity of Ag NPs in bpy produced by 1064 nm PLA was maximum for 5 min ablation, while it was maximum at 2 min for 532 nm PLA. For 1064 nm produced Ag particles, freshly prepared sol was more SERS active, while for particles produced with 532 nm, 30 min aged sample was more active compared to fresh ones	[70]
Virgin coconut oil	Pulsed Nd:YAG laser (1064 nm, 5 ns, 10 Hz), 162 J cm^{-2}, liquid height 4 ml, time 15, 30, and 45 min	Spherical silver NPs	Spherical Ag NPs having 4.84, 5.18, and 6.33 nm sizes were obtained for 45, 30, and 15 min ablation times, respectively. The virgin coconut oil controls the particles size and prevents agglomeration between the ablated NPs. Therefore, the obtained NPs are stable for quite a long time. Smaller particles produced for longer time ablation shows resizing of particles by melting/vaporization	[71]
DI water	Pulsed Nd:YAG laser (1064 nm, 10 ns, 10 Hz), 100 mJ/pulse, spot diameter 2 mm, time 30 min *Electrophoric deposition (EPD)* of colloidal solution: 1 cm × 1.5 cm Si electrodes separated by 35 mm, constant current 50 μA cm^{-2}, time 14 h	Ag NPs and their assembly into Ag nanoplates	As-synthesized colloidal solution contained spherical charged Ag NPs of 30 nm average diameter. After EPD, silver aggregates, composed of polygonal or regular Ag nanoplates of about 160 nm in edge length and 60 nm in thickness, were obtained. Formation of Ag nanoplates by EPD strongly depends on current density and nature of substrates. Increase in the current density leads to morphological change and size reduction of the particles on the substrate. Use of ITO substrate in place of Si under the same experimental conditions produces Ag nanoplates of different morphology	[72]

(continued overleaf)

Table 6.1.1 *(continued)*

Liquid parameters/target	Laser/experimental parameters	Product obtained	Findings/descriptions	References
Water or aqueous media of electrolyte solutions (HCl, NaCl, NaOH, AgNO$_3$, and/or Na$_2$SO$_3$)	Nd:YAG laser (1064 nm, 20 ns, 10 Hz), ~190 mJ/pulse, time 20 min continuous, or 5 successive steps interrupted by 2 min breaks	Spherical Ag NPs	Presence of Cl$^-$ and/or OH$^-$ ions in the aqueous media prevents the formation of larger Ag NPs and aggregates. This effect was credited to an efficient adsorption of Cl and/or OH anions on surfaces of Ag NPs and an efficient buildup of the electric bilayer around particles. The presence of HCl, NaCl, and/or NaOH during PLA has led to the stabilization of the resulting Ag NPs, while the presence of AgNO$_3$ and Na$_2$S$_2$O$_3$ has shown a destabilizing effect	[73]
DI water	Dual beam PLA: two Nd:YAG lasers (1064 nm, 5.5 ns, 10 Hz), liquid height = 10 mm, focal spot size 17 μm, ~11 kJ cm^{-2} and ~33 kJ cm^{-2} fluences for ablating laser at target surface. First laser is focused on target surface vertically downward, while second beam was focussed on the plasma plume at 2 mm above from the target surface and parallel to the target. Delay between two pulses was tuned 40 μs	Spherical Ag NPs	For PLA of Ag with fluence of 11 kJ cm^{-2} and single-beam approach, the mean particle size was about 29 nm, with majority of NPs in the range of 19–35 nm with some big ones as large as 50–60 nm. For double-beam approach at 11 kJ cm − 2, the particles were smaller, with the average size of about 18 nm, and the majority of the particles were in the 9–21 nm range with few big ones as large as 40 nm. For PLA with higher ~33 kJ cm^{-2} fluence and single-beam configuration, the particle sizes were smaller, with ~18 nm mean size, and the majority of the particles fell in the range of 10–22 nm, with some big ones as large as 40 nm. For double-beam approach, the mean particle size was larger (24.2 nm) and the majority of the particles were distributed from 14 to 35 nm, with some particles as big as 70 nm	[74]
Acetonitrile (AN) and N,N-dimethylform-amide (DMF), tetrahydrofuran (THF), dimethylsulfoxide (DMSO)	Nd:YAG laser (1064 nm, 9 ns, 10 Hz), time 10 min	Ag spherical NPs	NPs produced in AN have an average radius of 3.5 nm and a spheroid contribution of only 2% with σ_G = 1.5, while for DMF, 5 nm diameter, 100% spherical NPs were obtained. Ag NPs of 2.9 nm Ag core with 0.85 nm graphite shell have been produced by PLA in THF, while thickness of graphite layer more than 6 nm with the Ag core size of 3.9 nm is expected from PLA in DMSO. Functionalization of free Ag NPs by addition of organic molecules in AN solution has been presented	[75]

Medium	Laser parameters	Product	Description	Ref.
Water, aqueous media of 0.2 mM NaCl, KCl, and MgCl$_2$ salts	Nd:YAG laser (1064 nm, 10 ns, 10 Hz) 36 J cm^{-2}, time 10 min (*for synthesis*) Nd:YAG laser (532 and 355 nm, 6 ns, 10 Hz) 50 mJ cm^{-2} or fluorescent lamp (*for postirradiation*)	10–100 nm Ag NPs (as-synthesized)	As-synthesized colloidal solutions contain 10–100 nm spherical Ag NPs. Postirradiation of the colloid obtained by 355 and 532 nm laser light or by fluorescent lamp light in 0.2 mM NaCl solution yielded prismatic, rod-shaped Ag crystals; their growth was explained by photooxidation of Ag particles with twin planes by Cl$^-$ ions and following photoreduction of silver ions, although Ag crystals were found also on irradiation of colloids in pure water. Spectral change in absorption is more pronounced in NaCl-containing solution as compared to pure water and increases with laser irradiance	[76]
Acetone or distilled water	Nd:YAG laser (1064, 532 nm, 10 ns), irradiance 5×10^8 to 5×10^9 W cm^{-2} (*for synthesis*) Nd:YAG laser (532 and 266 nm), or Ti: Sapphire (800, 400 nm) (*Postirradiation*)	Ag NPs of 15–50 nm by 1064 nm Disclike particles $d > 50$ nm by 532	NPs produced in acetone were much more stable as compared to that obtained in water. Isolated Ag NPs and disclike agglomerates were observed. Postirradiation of obtained colloidal solutions by 266, 532, 400, and 800 nm laser light for 5 min at fluences 0.1, 0.5, and 0.6 J cm^{-2}, respectively, causes size reduction of single particles and transformation of nanodisc agglomerates to nanowire agglomerates is observed	[77]
Water or acetone	Nd:YAG, 532 and 1064 nm, 10 Hz), irradiance 0.5–5 GW cm^{-2}	Ag NPs	*DPLA*: NPs produced with DPLA has lower mean size, narrower size distribution, and higher colloidal stability as compared to single-beam PLA. It was concluded that more effective ablation can be achieved by DPLA as compared to SPLA. Selection of time delay between two laser pulses and their wavelengths provide size-controlled particle synthesis	[78]
Water	Synthesis: Nd:YAG laser (1064 nm 10 ns, 10 Hz), 38 J cm^{-2}, 10 min. Postirradiation: 355 nm, 4–12 mJ/pulse, 10 min)	Ag NPs	Spherical silver NPs with 30 nm average size have been produced. Postirradiation reduces particle size and causes fragmentation of Ag wires and sheets	[79]

(continued overleaf)

Table 6.1.1 (*continued*)

Liquid parameters/target	Laser/experimental parameters	Product obtained	Findings/descriptions	References
Aqueous media of 70 mM SDS	Nd:YAG laser (1064 nm) 350 mJ/pulse	Ag NPs	Spherical Ag NPs with SPR absorption at 405 nm were obtained. Dielectric constant of colloidal solution was 1.8167 and refractive index was 1.3478 as measured by Kretschmann-type SPR sensor. Dielectric constant decreases with increasing Ag NP concentration	[11]
Water	Nd:YAG laser (532 nm, 10 ns, 10 Hz), spot size; 0.6–1.3 mm, 340 mJ cm^{-2}	Ag spherical NPs	Ag NPs having average size 4.2 nm corresponding to 0.7 nm spot size were produced. Effect of laser spot size on the target surface on the properties of colloidal solution was studied. Smaller spot size produces colloid of higher stability	[80]
Ethylene glycol, water, or ethanol	Nd:YAG laser (532 nm, 10 ns, 10 Hz), 20 J cm^{-2}, time 30 min	Ag NPs	As-synthesized colloidal solution had NPs of different shapes, viz. rods, spheres, ellipsoidal, etc. After 15 d, only spherical particles remain in the solution. Z-scan measurement of colloidal solution using 397.5 nm light of 1.2 ps pulse width, resulted in third-order susceptibility of 5×10^{-8} esu, nonlinear refractive index 3×10^{-13} cm^2 W^{-1}, and nonlinear absorption coefficient of -1.5×10^{-9} cm W^{-1}	[81a,b]
Water	Ti:Sapphire laser (810 nm, 120 fs, 1 kHz), 500 µJ, spot size ∼100 µm	Ag NPs	Ag NPs of 5 nm average size with very narrow size distribution were obtained. Generation of much smaller Ag NPs as compared to Au NPs (12 nm) produced under same experimental conditions suggest Ag-induced generation of second harmonics (405 nm) and reduction of Ag NPs size with this 405 nm light because of the matching with SPR	[82]
Water	Ti:Sapphire (800 nm, 120 fs, 10 Hz); femtosecond beam Ti:Sapphire (800 nm, 8 ns, 10 Hz), pulse energy 4 mJ/pulse, unfocused beam spot size 4 mm. Continuous stirring	Ag NPs	Comparison of size, distribution stability, and efficiency of particles generated by PLA of nanosecond and femtosecond laser pulses was investigated. Synthesis efficiency by femtosecond pulses was significantly lower than that by nanosecond pulses, but particles produced with femtosecond PLA have smaller average size and narrower size distribution	[83]

Water	Nd:YAG laser (1064, 532, and 355 nm), 12 mJ/pulse, 36 J cm^{-2}, time 30 min Solution stirred during ablation	Ag spherical NPs	Ag NPs of average sizes 29, 26, and 12 nm with SD 13, 11, and 8 nm were produced by PLA in water using 1064, 532, and 355 nm laser pulses, respectively. Smaller-sized particles with narrower size distribution are the cause of photofragmentation of synthesized particles by self-absorption due to the of higher photon energy of shorter wavelength	[84]
Water or aqueous media of NaCl	Nd:YAG laser (1064 nm, 5 ns, 10 Hz), spot sized of focused beam 1 mm	Ag NPs	Nature of the SPR peak of Ag colloidal solution significantly depends on the concentration of NaCl in ablation media. Absorbance at 400 nm increases from 0.86 to 1.45 with the increase of NaCl concentration from 0 to 5 mM, while the FWHM decreases. This observation revealed that rate of particle synthesis as well as monodispersity increase with NaCl concentration. Absorbance as well as FWHM follows opposite trend for concentration >5 mM because of aggregation. Colloid prepared in pure water was stable for longer time as compared to those produced in NaCl solution	[85]
Water	Nd:YAG laser (355, 532 nm, 10 Hz) 130 mJ/pulse	Ag spherical particles	As-synthesized Ag particles were on an average 3.53 μm, but irregular and amorphous. Irradiation of suspension first lowers the size of particles, but size increases again because of the coagulation of smaller-sized NPs into larger ones. Critical lower size for Ag NPs to initiate coagulation was ≈50 nm	[86]
Water	Nd:YAG laser (355, 532, and 1064 nm 5–9 ns, 10 Hz) 0.1–1 J cm^{-2}	Ag spherical NPs	Ag NPs of 12 and 31 nm average sizes were produced by 532 and 1064 nm PLA, respectively, in water. Ablation efficiency depended significantly on the laser wavelength. The relation between the ablation efficiencies and the laser wavelength varied with the fluence of the laser light. It was suggested that the self-absorption of the shorter wavelength laser light by colloidal particles efficiently occurs at high laser fluence. It was proposed that the self-absorption occurred by intrapulse as well as interpulse processes	[60]

(continued overleaf)

Table 6.1.1 *(continued)*

Liquid parameters/target	Laser/ experimental parameters	Product obtained	Findings/ descriptions	References
Aqueous media of SDS 3 (below), 10 (near), and 50 mM (above) CMC	Pulsed Nd:YAG laser (532 nm, 10 ns, 10 Hz), 40–90 mJ/pulse spot size 1–3 mm Influences of SDS concentration, laser irradiance, and focusing conditions on the size, distribution, and abundance of Ag NPs were studied	Spherical Ag NPs	Average size of NPs decreases, while dispersion first increases upto CMC and slightly decreases after that, with the increase of SDS concentrations. Mean size of NPs increases with the increase of laser pulse energy, while dispersion remains almost constant. Abundance of particles increases with laser irradiation, reaches maximum, and decreases after that. The value of laser pulse energy corresponding to maximum abundance depends on the laser spot size on the surface. Larger spot size results in relatively higher maximum abundance at higher pulse energy. At a particular laser irradiance, relative abundance increases quadratically with laser spot diameter. Relative abundance increases linearly with increase in number of laser shots upto 50 000 shots and deviates downward from linear beyond this	[12]
Aqueous media of $C_nH_{2n+1}SO_4Na$ concentration 0–1 M	Pulsed Nd:YAG laser (532 nm, 10 ns, 10 Hz), 40–90 mJ/pulse spot size 1–3 mm	Spherical Ag NPs	Colloidal stability of Ag NPs with concentration and number of carbon atoms of surfactant was studied. Colloidal stability of NPs was found maximum for CMC of surfactant. Rate of precipitation decreases, attains a minimum value near CMC, and again increases after that with the increase of surfactant concentration. For all the surfactants, there are a concentration-independent portion $<10^{-6}$ M, and concentration-dependent region $>10^{-6}$ M NPs. Abundances of particles in solutions are lower at concentrations $>10^{-6}$ M for $n = 8$ and 10, while these are much higher for $n = 12$ and 16 than those of the concentration-independent portion	[13]
Water, methanol, and 2-propanol	Nd:YAG, (1064 or 532 nm), 247 and 397 mJ cm^{-2}	Spherical Ag NPs	Ag spherical NPs in the size range of 16.3–32.9 nm and 12.4–17.4 nm were produced by PLA in water and 2-propanol, respectively. Longer laser wavelength and smaller fluence resulted in smaller NPs. Colloid prepared in water was stable for several months, in methanol for 1 day, and in 2-propanol for at least 6 months	[87]

Medium	Laser conditions	Product	Description	Ref.
Water, aqueous media of 7×10^{-4} to 7×10^{-2} M NaCl, 10^{-5} M aqueous media of pht	Nd:YAG laser (1064 nm, 4 ps, 1 Hz), 40 mJ/pulse, 200 μm spot size	Ag NPs Mean size: 18, 13, 14, 10 nm	NPs with mean sizes 18, 13, 14, 10 nm were obtained by PLA of Ag in water + tppz, in NaCl + bpy, in NaCl (7×10^{-7} M), and in pht, respectively. SERS-active colloid/adsorbate systems prepared directly by laser ablation in an adsorbate (phthalazine) solution. Colloid-adsorbate (bpy or tppz) films on glass and Cu/C-grids fabricated. SPR peak lies at λ max = 406 nm (water), 410 nm (phthalazine added), 398 nm (7×10^{-4} M NaCl)	[88]
Water, water + NaCl 0.2–5 mM, and NaNO$_3$	Nd:YAG, 1064 nm 20 ns, 10 Hz, 10–30 mJ (liquids optionally stirred)	Ag NPs	Ag NPs with average sizes 12.3–14.8 nm have been produced. Smaller laser fluence produces smaller particles, and initially, larger particles get fragmented by laser irradiation. NaCl additive (but not NaNO$_3$) enhances the particles yield and provides smaller size, but causes their aggregation. Metalation of a free base porphyrin on laser-fabricated particles was faster and more stable in time (at least 10 mo) than that on chemically produced particles	[89]
He II (1.6 K)	Nd:YAG (532 nm, 10 and 20 Hz) 10–20 mJ	Ag atoms, clusters, particles; AgHe2-exciplexes	Ag clusters produced by Nd:YAG laser were further dissociated by XeCl laser operating at 308 nm, 10 Hz, 10 mJ/pulse. Linear He-Ag-He-exciplexes trapped in microcavities were found. Formation of AgHe$_2$-exciplexes was confirmed by *ab initio* calculations	[90]
Water	Nd:YAG laser (1064 nm, 10 Hz) 55 mJ/pulse, time 15 min	Ag NPs	Ag NPs with 20 nm average size was produced. Surface of colloids were modified by I and Br ions. Effect of this modification on plasma resonance frequency was small	[91]
He II (1.7 K)	Nd:YAG laser (532 and 355 nm, 10 and 20 Hz), 10–20 mJ	Atoms, clusters, and particles	Emission and absorption spectra and dynamics of neutral atoms, also residing at microscopic He bubbles were investigated	[92]
He II (1.7 K)	Nd:YAG laser (532 and 355 nm, 10 Hz), 10–20 mJ	Clusters and micron-sized particles	Larger particles were further dissociated by continuing laser irradiation; the dissociation was more effective at shorter wavelengths. Absorption and emission spectra of Ag$_2$ in UV–VIS region are presented	[93]

(continued overleaf)

Table 6.1.1 (*continued*)

Liquid parameters/target	Laser/ experimental parameters	Product obtained	Findings/ descriptions	References
LP-PLA of Au for the preparation of elemental Au NPs				
Water or 1 or 5 µM penetratin solutions at pH 9, 500 µl liquid volume	Ti:Sapphire (800 nm, 120 fs, 5 kHz), 100 µJ/pulse, target scan rate 1 mm s^{-1}, time 106 s	Bioconjugated Au NPs	*In situ* conjugation during laser ablation allows the rapid design of cell-penetrating nanomarkers for intracellular bioimaging. Significantly enhanced penetratin conjugation efficiencies could be observed at higher pH values, whereas the size and morphology of bioconjugates depends on the peptide concentration during ablation	[9]
Double distilled water	Nd:YAG laser (1064 nm, 9 ns, 10 Hz), 10 J cm^{-2}, time 30 min	Au NPs of 2 nm average diameter	The effect of addition of different concentrations of pyridine into colloidal solution of 2 nm Au NPs on their aggregation has been studied. BSA of 0.1 mg ml^{-1} was added to aggregated Au NPs for stabilization. Increasing the concentration of pyridine causes increase of aggregation. Plasmonic nanoaggregates having desired extinction spectra were selected for optical trapping. Larger aggregates cause more intense signal	[94]
Deionized water; temperature of liquid is varied 283–353 K	Ti:Sapphire laser (800 nm, 120 fs, 5 kHz), 0.17 J cm^{-2}, time 1 min, 76 mm min^{-1} scanning velocity	Au NPs (hydrodynamic diameter 50–100 nm)	Temperature of liquid media affects its physical properties such as viscosity, surface tension, and compressibility, which drive volume, expansion velocity, and lifetime of cavitation bubbles. Size of the Au NPs increases with the increase of media temperature. Hydrodynamic diameter of Au NPs exponentially decays with compressibility of water, which depends on solvent temperature	[95]
Deionized water	Ti:Sapphire laser (800 nm, 100 fs, 10 kHz), Nd:YAG laser (1064 and 532 nm, <7 ns, 10 Hz), rotating target 4.5° per s; For femtosecond irradiation 3.9–16 J cm^{-2}; for nanosecond irradiance 0.3–5.9 J cm^{-2}	Spherical Au NPs, NAGs, NFs, NRs of different aspect ratio	For femtosecond laser, spherical NPs of ~2.5 nm average diameter were produced with small percentage of large particles. At higher energies size of bigger NPs increases. With nanosecond beam of 532 nm wavelength, monodispersed colloidal solution was obtained. Increasing the laser irradiance causes increase of average diameter in the range of 3–10 nm. NPs, NRs, and nanofragments were produced when 1064 nm, nanosecond beam was used, showing comparatively more complex mechanism	[96]

Ethanol	Ti:Sapphire laser (800 nm), 200 fs, 1 kHz), two time-delayed pulses were generated by single using Michelson interferometer. time delay 0–10 ps, liquid height 1–2 mm	Spherical and polycrystalline Au NPs	Increasing time delay, t_d, from 0 to 10 ps between two pulses causes shorter wavelength, 538–533 nm, shift in SPR peak position, illustrating decrease in NP size. Increase of t_d in the range of 100–300 fs causes increase in NP size, while $t_d > 300$ produces NPs of smaller size. Size distribution of NPs also decreases with the increase in t_d between two pulses. Efficiency of NP generation increases with delay time in the range of 0 to ~2.5 ps and attains a constant value after ~2.5 ps. Ablation efficiency of DPLA with delay times <~2 ps was lower than that of SPLA, but higher for delay in the range of ~2–10 ps	[97]
Aqueous media of dextran or PEG or PNIPAM-SH	*For synthesis*: Ti:Sapphire (800 nm, 110 fs, 1 kHz), irradiance 0.4 J cm^{-2}, liquid height 1 cm, 0.2, and 1.10 J cm^{-2} *For growth*: same laser system, 2.6 cm path length, spot diameter 180 ± 20 μm, numerical aperture: 0.034, 0.047, and 0.12 continuous stirring, time 1 h	Functionalized Au NPs 2 to ~80 nm	Two step processes (i) laser synthesis of gold seeds and (ii) laser-induced growth were used for the synthesis of size-controlled, narrowly dispersed, and functional Au NPs in the size range of 2 to ~80 nm. Laser fluences of 0.2 and 1.10 J cm^{-2} during first step produces Au NPs with the rate of 0.01–0.14 mg min^{-1}. For dextran, mean size and dispersion of Au NPs decrease with the increase of dextran concentration. Spectral broadening of 800 nm at higher pulse energies and lower numerical aperture produced supercontinuum white light source, which may couple with the SPR of Au NPs for their resizing	[98]
Aqueous media of oligo-nucleotides (0–7.5 μM)	Ti:Sapphire (800 nm, 120 fs, 5 kHz), 400 μJ/pulse, 4 mm beam diameter, irradiance 0.35 J cm^{-2}, spot diameter 270 μm, scanning 1 mm s^{-1}	Bioconjugated Au NPs	*In situ* and *ex situ* bioconjugation of Au NPs were demonstrated with the presence and absence of oligoneucleotides in the ablation medium, respectively. Conjugation efficiency and surface coverage of NPs were highly influenced by concentrations of conjugative agents and NPs. The strength of the interaction of the oligonucleotide with the gold NP seems to be higher during *in situ* conjugation, and five times higher surface coverage values were obtained for laser bases *in situ* and *ex situ* techniques as compared to conventional chemical methods	[10]

(continued overleaf)

Table 6.1.1 (continued)

Liquid parameters/target	Laser/ experimental parameters	Product obtained	Findings/ descriptions	References
Aqueous media of PolyAminoAMines (PAMAM)	Nd:YAG laser (1064 or 532 or 355 nm), 15 mJ/pulse, 16 min	Dendrimer-capped Au NPs	Equal pulse energies (16 mJ/pulse) from all the three wavelengths were applied in PLA for 16 min. Colloidal solution produced by 532 nm wavelength has strongest SPR, while that produced with 1064 nm has least. Sizes for dendrimer-capped Au NPs are 5.2 ± 1.6 nm, 3.2 ± 1.0 nm, and 2.5 ± 1.5 nm for 1064, 532, and 355 nm PLA, respectively.	[99]
Supercritical CO_2 in high-pressure (upto 20 MPa) cell	Nd:YAG laser (532 nm, 20 Hz, 9 ns), 0.8 J cm^{-2}, time 5 min	Au nanonecklaces and nanospheres	Morphology of produced Au nanomaterials depends on the density of fluids. Low density ($\rho 0.2$ g cm^{-3} and P 4.29 MPa) resembles necklace-like morphology of 30 nm mean diameter, while higher density (1.7 g cm^{-3}, P 14.5 MPa) results in spheres of 500 nm average size	[100]
Water Post synthesis: Au NPs were mixed in aqueous media of SDS and CTAB	Nd:YAG laser (1064 nm, 10 Hz), 80 mJ/pulse energy, focal length 25 cm, 36 000 shots	Au NPs with oxidized surface atoms	Spherical Au NPs with average diameter of 11 ± 4 nm and negatively charged surface were produced. Surfactant-free NPs were stabilized in water because of surface charge. The ζ-potential and XPS measurements showed that a part of the surface gold atoms of the particle were oxidized, and the particle with the outer Stern layer is negatively charged. Addition of CTAB neutralized the surface charge and destabilized the gold solution	[101]
Water Target: prismatic Au particle array on quartz substrate	Ti:Sapphire (800 nm, 100 fs, 1 kHz); irradiation was done on the back of substrate; scan rate 1 mm s^{-1}, 4000–5000 pulse/particle	Photoejected Au nanoprism	Photothermal ejection of Au nanoprism in water was observed at a fraction of energy required for the ejection of NPs in air. Ejected particles in solution were observed to increase in thickness on irradiation instead of decreasing as ejected particles in air were previously observed to do. Nanoprism ejection was proposed to occur in two separate processes: (i) NP substrate dissociation and (ii) NP solvation and diffusion into solution	[102]

Dimethyl sulfoxide (DMSO), acetonitrile (AN), and tetrahydrofuran (THF)	*For synthesis:* Nd:YAG laser (1064 nm, 9 ns, 10 Hz), 10 J cm^{-2} *For resizing and reshaping:* Nd:YAG laser (532 nm, 1 Hz), 1–5 J cm^{-2}	Free and functionalized Au NPs	Colloidal solutions of Au NPs having SPR absorption maxima at 530, 528, and 522 nm; average sizes ± SD are 2.4 ± 0.9 nm, 4.1 ± 2.5 nm, and 1.8 ± 1.2 nm with 10, 34, and 14%; spherical contributions were obtained by PLA of Au targets in the aqueous media of DMSO, THF, and acetonitrile, respectively. Size and aggregation of NPs were controlled by postirradiation of colloidal solutions	[103]
Water or 0.1 N aqueous media of NaCl or KCl or NaNO$_3$ or HCl, or NaOH	Ti:Sapphire laser (800 nm, 120 fs, 1 kHz, maximum energy 1 mJ/pulse) height 12 mm, focal position 1.5 mm above the target surface, 0.2 mJ/pulse	Au NPs with partially oxidized surfaces	PLA of Au targets in water produces Au NPs in the size range of 1–250 nm with partially oxidized surfaces. The hydroxylation of these Au–O compounds, followed by a proton loss to give surface Au–O^{-}, resulted in the negative charging of NPs. Oxidized surface reacted efficiently with Cl^{-} and OH^{-} to augment its net surface charge, with reduced coalescence of NPs, due to electrostatic repulsion and led to a significant reduction in NP size. This virtue resulted in 7 ± 5 nm and 5.5 ± 4 nm sizes for 10 mM aqueous media of KCl and NaCl, respectively, and 8 ± 5 nm for NaOH, with pH 9. Use of N-propylamine controled size in the range of 5–8 ± 4–7 nm	[104]
10 mM aqueous media of β-CD with pH values 3–9	Ti:Sapphire laser (800 nm, 120 fs, 1 kHz) height 12 mm, fluence 600 J cm^{-2}	Au NPs	NPs produced in the aqueous media of 10 mM β-cyclodextrin with higher pH values have lower mean particle sizes, narrower dispersion, and hence higher stability. For example, for pH values 6 and 9, mean size of Au NPs was 2.5 ± 1.5 nm, while mean size of 20 nm and broader distributed NPs were produced for pH value 3. Surfaces of NPs were negatively charged for all pH values. Zeta potential decreases sharply with the increase of pH values in the range of 3–5 and became almost constant beyond this	[105]
n-Alkanes (n = 5–10) n-pentane to n-decane	Nd:YAG laser (532 nm, 5 ns, 10 Hz), height 35 mm, fluence 1–200 J cm^{-2}, spot size 0.28–1 × 10^{-4} cm^2, time 5 min	Au spherical and elongated NPs	At high laser fluence (200 J cm^{-2}), PLA of Au in all n-alkanes produces colloidal sols having two SPR components. First is close at ~530 in all samples, while the second shifts toward blue with increasing number of carbon atoms. Second component starts to appear just after 5 J cm^{-2} and proportionally increases in intensity with increasing laser fluence. The aspect ratio of the samples change from 4 to 6.5 going from n = 5–10. It should be noted that spherical NPs were produced in all the solutions with elongation at all laser fluencies	[106]

(continued overleaf)

Table 6.1.1 (*continued*)

Liquid parameters/target	Laser/ experimental parameters	Product obtained	Findings/ descriptions	References
Aqueous media of α, β, or γ CDs	Ti:Sapphire laser (800 nm, 120 fs, 1 kHz, maximum energy 1 mJ/pulse) height 12 mm	Spherical Au NPs	Size and distribution of particles produced in all the three types of CDs decreases with the increase of their concentrations. β-CD was most effective in reducing the size and dispersion of the gold NPs as well as their stabilization, followed by γ-CD and α-CD	[107]
For synthesis: aqueous media of 0.01 M SDS *For postirradiation*: Aqueous media of SDS with different concentrations	*For synthesis*: Nd:YAG laser (1064 nm), 800 mJ/pulse cm^2 *For postirradiation*: Nd:YAG laser (532 nm, 10 Hz), spot size 0.12 cm^2 on liquid surface, temperature 278 K	Spherical Au NPs	Colloidal solution of as-synthesized Au NPs with average size of ~8 nm was produced. Size reduction via postirradiation at given laser fluence of 532 nm at a particular time is more effective for higher (0.05 M) concentration of SDS. For example, Au NPs sizes were reduced upto 3.4 ± 0.6 and 1.7 ± 0.6 nm for 9×10^{-4} M and 0.05 M SDS solutions, respectively. For all the laser irradiances, SDS-concentration-dependent size reductions were observed for concentrations near its CMC, and these effects are more pronounced at higher laser irradiances	[14a,b]
For synthesis: deionized water Addition of SDS after synthesis for stabilization	*For synthesis*: Nd:YAG laser (532 nm, 10 Hz), time 5 min, within the ice bath or at room temperature *For postirradiation*: Nd:YAG laser (532 or 1064 nm, 10 Hz), power monitored	Networked Au NPs and twisted Au NRs	Au cross-linked nanonetworks and twisted NRs with the mean diameter of 6 ± 3 and 12 ± 3 nm were produced by PLA in DI water at ice bath and room temperature, respectively. Lower fluence of longer (1064 nm) laser irradiation of suspensions containing Au nanonetwork produces very large spherical NPs, while for twisted NRs it results in hole burning centered at 1064 nm, which may be the consequence of coincidence of longer SPR band with incident wavelength	[108]
Water	Cu vapor laser (510.6 nm, 20 ns, 15 kHz), height of liquid 5 mm, spot diameter 10 μm, laser fluence on target 10–30 J cm^{-2}, scan speed 1 mm s^{-1}, liquid flow 1 cm s^{-1}	Spherical and elongated Au NPs	Colloidal solution of Au NPs produced at higher (32 J cm^{-2}) laser fluence has two SPR maxima near 520 and 740 nm, while at lower fluence only one SPR peak is observed. At the same laser fluence under circulating liquid, there is no second absorption peak. Second peak for PLA in static liquid is due to the linear assembly of Au NPs, while liquid circulation may provide separation between particles and hence prevents self assembly	[109]

Aqueous media of SDS (10^{-9} to 1 M)	Nd:YAG laser (1064 or 532 nm, 10 Hz), spot size 1–3 mm	Spherical Au NPs	Size of the NPs decreases, while stability increases with the increase of SDS concentration. Relative abundance of NPs in the solution remains constant for $<10^{-5}$ and increase rapidly after that. Precipitation factor is higher for larger wavelength and larger laser irradiance, while for a given laser wavelength and laser fluence, it decreases with the increase of SDS concentration. For 1064 nm laser wavelength, relative abundance increases linearly with the increase of number of laser shots upto almost 3×10^{4} shots, while for 532 nm, it increases rapidly with laser shots upto some value and becomes constant after that	[15]
Aqueous media of SDS (0.01 M SDS)	Nd:YAG laser (1064 or 532 nm, 10 Hz), spot size 1–3 mm, 800 mJ cm^{-2} pulse	Spherical Au NPs	Au NPs with average size of 8 nm were produced by PLA of Au target in aqueous media of SDS. The average diameter of the NPs was found to decrease toward the smallest possible value as the laser shot increases, and the smallest possible diameter was found to decrease with the increase in laser fluence	[16]

LP-PLA of Pt NPs for the preparation of its elemental nanomaterials

Water or aqueous medium of 0.05 M SDS	KrF excimer laser (248 nm, 30 ns, 10 Hz); laser fluence 2.3, 3.6, and 6.8 J cm^{-2}; spot size 1.2 mm^{2}; liquid height 4 mm, time 10 and 60 min	Solid and hollow spherical Pt particles	Depending on the laser fluence, different kinds of hollow Pt structures were obtained that could be coalesced by micrograins and assembled by nanocrystals or with smooth shells. These hollow structures are considered to form on laser-produced bubbles in water, which provided thermodynamically preferred nucleation sites and diffusion sinks for Pt small clusters or particles	[17]
Water or acetonitrile or pentane	Nd:YAG laser (1064 nm)	Pt NPs	The stability of NPs in different liquids has been investigated in terms of their polarizability index, local electric molecular dipole moment, and viscosity	[110]

(continued overleaf)

Table 6.1.1 *(continued)*

Liquid parameters/target	Laser/experimental parameters	Product obtained	Findings/descriptions	References
Water	Nd:YAG laser (1064, 10 ns; 532 nm, 8 ns, and 355 nm, 7 ns, 10 Hz), fluence 1–16 J cm⁻², spot diameter 1 mm, time 15 min, 20 rpm rotation, 300 laser shots every point	Various nanostructures on target surface	It is a series of three papers. In the first paper, patterns and structures grown on the target surface are investigated. Ablation with 1064 nm resulted large crater (20 μm), while that with 355 nm wavelengths generates ripple surface pattern at low fluence, which changes to rough surfaces at higher fluencies. Ablation with 532 nm creates results of combination of 1064 and 355 nm. In the second paper, ablation rate and NPs size distribution are discussed. For all the laser wavelengths, fluence in the range of 10–70 J cm⁻² produce spherical platinum NPs in the size range of 1–30 nm. Size distributions are bimodal, attributing to thermal evaporation and explosive boiling mechanism. Peak of smaller particles always lie at 3 nm, which is because of the thermal evaporation. Peaks for larger particles lie in the range of 5–15 nm, which strongly depends on laser parameters and are attributed to the explosive boiling. The third paper of the series describes three different mechanism of materials removal from the surface at three different low (10 J cm⁻²), mid (10–70 J cm⁻²), and high fluence (>70 J cm⁻²) regimes	[111a,b,c]

APD, avalanche photodiode; XPS, X-ray photoelectron spectroscopy; bpy, 2,2′-bipyridine; H2TMPyP; 5, 10, 15, 20-tetrakis (1-methyl-4-pyridyl)-21H, 23H-porphine; SPLA, single-beam pulsed laser ablation; DPLA, dual beam pulsed laser ablation; CDs cyclodextrins; SD, standard deviation; Tppz, 2,3,5,6-tetrakis(2′-pyridyl)pyrazine; Bpy, 2,2′-bipyridine; Pht, phthalazine; BSA, bovine serum albumin; PEG, poly(ethylene) glycol; PANIPAM-SH; ω-dimercapto-poly(N-isopropylacrylamide; NAGs, nano-aggregates; CTAB, cetyltrimethylammonium bromide.

100 nm

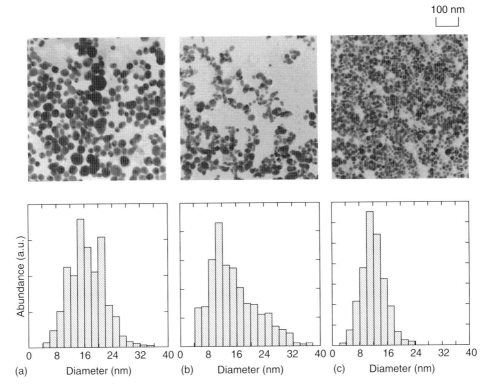

Figure 6.1.17 TEM images and correspond- of SDS. (Source: Reprinted with the permis-
ing size histograms of NPs produced by PLA sion from Mafuné *et al.* [12] © American
of Ag target in (a) 0.003 M, (b) 0.01 M, and Chemical Society.)
(c) 0.05 M concentrations of aqueous media

Synthesis of zinc core/ZnO shell NPs [112, 113] in the aqueous medium of SDS
near its CMC, while pure Zn NPs in the aqueous media of SDS with concentrations
higher than CMC [112]; spherical Al NPs by nanosecond PLA in ethanol, acetone,
and ethylene glycol [156]; Co NPs in 0.003 mM ethylene glycol [157]; Ni NPs in
water as well as 0.01 M aqueous medium of SDS [158]; and tungsten [159] NPs in
ethanol are some examples of nonreactive PLA of active metals.

Preparation of Target Compound NPs by Reactive LP-PLA of Solids Here, the species of
target material in the plasma plume react with the species of liquid medium plasma
at plasma–liquid interface or inside the plasma or in the liquid and produce target
compound nanomaterials. This technique has the ability to greatly expand the product composition such as metal oxide, hydroxide, carbide, nitrides, and composites
by choosing smart reaction between the selected target and liquid medium.

Preparation of Compound Nanostructures of Active Metals Laser-induced plasma creates plasma-induced plasma at plasma–liquid interface. Laser-induced plasma has

ionic species from target material while plasma-induced plasma contains ionic radical species from liquid media. In the case of PLA of active metals, active metal ions, usually cations of the plasma, rapidly react with the ions or radical, usually anions of liquid media, to synthesize metal compound nanostructures. For example; PLA of zinc in water or in alcohol produces zinc oxide [114] or $ZnO/Zn(OH)_2$ nanocomposites [115], or $ZnO/ZnOOH$ nanocomposite [116, 117] particles. Similarly, PLA of Ti in water produces TiO_x including TiO, TiO_2, and Ti_2O_3 NPs [118, 160], in liquid nitrogen TiN [161], in 2-propanol TiO and TiC [119, 162], and in *n*-hexane TiH [119] NPs, and so on. The ratio of target element to liquid media element in the NPs and the phase of NPs depend on the laser and liquid media parameters. PLA of Ti in water produces different nanostructures of titanium oxide at different laser irradiances [119, 120]. At lower laser irradiances (130 and 150 MW cm^{-2}), amorphous TiO, Ti_2O_3, and TiO_2 NPs with 5.2, 5.5 and 5.5 GPa internal strains, respectively, were produced, while at higher laser irradiance (110 GW cm^{-2}), multiwall nanotubes (MWNTs) of TiO, Ti_2O_3, and TiO_2 with 4.2 GPa internal stress and at 140 GW cm^{-2}, MWNTs of only TiO_2 with 4.5 GPa internal stress were produced. It is observed that content of elements from liquids is higher for smaller particles [120].

Golightly and Castleman [119] reported ablation of Ti in 2-propanol, water, ethanol, and *n*-hexane and produced TiC, TiO*x*, and TiH NPs. They observed that higher laser fluences produce larger-sized particles with wider size distribution in all liquid media. PLA in water gives a higher number of larger-sized (>70 nm) particles as compared to 2-propanol, while *n*-hexane produces particles <70 nm. In all the liquid media, smaller-sized particles have larger incorporation of solvent elements. For example, in 2-propanol, higher O and C contents were observed in particles <10 nm; particles in the size range of 10–30 nm have 1 : 1 Ti and C ratio, while above this size range, Ti : C > 1. Similarly, for PLA in water, particles larger than 50 nm have lower oxygen content, while particles smaller than 50 nm have Ti : O between 1.43 : 1 and 3 : 1. Particles having diameter <5 nm have average Ti : O ratio 1.04 : 1. Similar results for composition were observed in NPs produced by PLA in *n*-hexane.

Singh and Gopal [114] reported PLA of zinc in methanol using fundamentals of nanosecond Nd:YAG laser and produced drop-shaped zinc oxide quantum dots (QDs) that self-assembled into various dendritic nanostructures. The drop shape of these QDs electrostatically drives them to get assembled into various dendritic nanostructures. The dendritic assemblies are classified into three categories: (i) linear axis symmetrical branching, (ii) linear axis asymmetrical branching, and (iii) curvilinear axis asymmetrical branching. Figure 6.1.18 presents assemblies of drop-shaped zinc oxide into various dendritic nanostructures and schemes for their branching with axis. In another interesting work, Singh and Gopal reported LP-PLA of a cadmium rod placed at the bottom of a glass vessel containing aqueous media of SDS at different concentrations to produce a variety of cadmium hydroxide nanostructures from NPs to nanorods (NRs), nanotetrapods, nanoflower buds, and 2D and 3D nanoflowers [122]. It is suggested that initially produced spherical NPs get self-assembled into one-dimensional NRs, which themselves also get assembled into their successor, complex 3D nanoarchitectures. An aqueous medium of 20 mM SDS was found most suitable for the growth

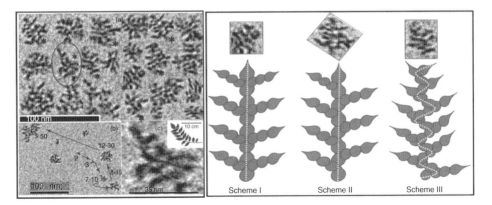

Figure 6.1.18 Assemblies of drop-shaped zinc oxide quantum dots into various dendritic nanostructures, and three possible branching schemes. (Source: Reprinted with the permission from Singh and Gopal [114] copyright @ Elsevier.)

of such nanostructures. They reported that increase in the concentration of surfactant induces the synthesis of higher-aspect-ratio one-dimensional NRs with a larger tendency of aggregation and agglomeration [122]. The rate of increase of agglomeration and aggregation with the surfactant concentration is so high that the nanomaterials produced in 100 mM surfactant lost their individual identity. Figure 6.1.19 illustrates various $Cd(OH)_2$ nanoarchitectures produced by LP-PLA of cadmium metal in the aqueous media of different concentrations of SDS.

Simultaneous flow of oxygen in the close vicinity of laser-produced plasma plume during PLA has the ability to increase oxygen content in the NPs, increase oxide/hydroxide ratio in nanocomposites, increase crystallinity of particles, and decrease their size and distribution [116]. PLA of active metals in water or alcohols first produces their hydroxide molecules. These hydroxide molecules either get dehydrated into oxide molecules by losing water through hydrothermal reaction process or clusterize to form hydroxide NPs. In the former case, oxide molecules clusterize to form oxide NPs, while in the latter case, hydroxide NPs are produced. In some of cases, oxide/hydroxide nanocomposites are produced [115–117]. PLA of zinc in water [115] or low concentration of surfactant [112, 116, 117] produces pure zinc oxide [112] or zinc oxide/hydroxide nanocomposite [116, 117], while PLA of Cd under almost same experimental conditions produces $Cd(OH)_2$ nanostructures [121, 122]. These $Cd(OH)_2$ nanostructures can be converted into CdO nanostructures through the postannealing process [121].

Concentration of surfactant in the liquid media controls the rate of reaction at the plasma–liquid interface before particle synthesis or at the particle–liquid interface after the production of NPs in solutions, as well as the colloidal stability of NPs. PLA in liquid without any surfactant or low concentration of surfactant (<CMC) allows reactions between plasma species and species from liquid media to dominate over reactions among plasma species, which results in the synthesis of pure metal compound NPs instead of metal NPs [112]. At the concentration of surfactant near their CMC, rate of reaction among the plasma species becomes

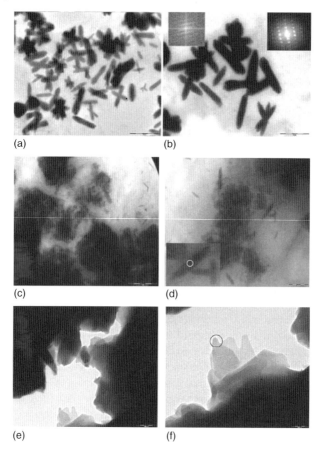

(a) (b)

(c) (d)

(e) (f)

Figure 6.1.19 TEM images of LP-PLA-produced Cd(OH)$_2$ nanoarchitectures in the aqueous media of (a,b) 20 mM, (c,d) 50 mM, and (e,f) 100 mM concentrations of SDS. Right Inset of Figure 6.1.19(b) shows Selective Area Electron Diffraction (SAED) pattern of the nanostructure shown in Figure 6.1.19, while left inset is Fourier Transform image of SAED pattern. Inset of Figure 6.1.19(d) shows High Resolution Image of nanostructures produced in the 50 mM aqueous medium of SDS. Circle in the Figure 6.1.19(f) shows sharp tip of the nanostructure produced in 100 mM aqueous medium of SDS. (Source: Reprinted with the permission from Singh and Gopal [122] copyright @ American Chemical Society.)

higher than the rate of reaction between the plasma species and species from liquid media, which results in the formation of metallic core with metal compound shell NPs. With aging of NPs in solution, thickness of shell layer increases. For PLA in the liquid media having concentrations of surfactant higher than the CMC, rate of reaction among the plasma species is much higher, which supports the formation of metal NPs with a strong protective layer of surfactant molecules. Capping with surfactant molecules prevents reaction between surface atoms of NPs with the species of liquid media at the particle–liquid interface, as well as disperses NPs in the solution to enhance their colloidal stability. Effects of surfactant concentration on the size and shape of NPs and stability of colloidal solution produced by LAL depend on the nature of charge on the NP's surface and nature of surfactant. For example, increase in the concentration of SDS decreases size and distribution and increases colloidal stability of particles with negative surface charge such as Zn or ZnO [112], while increases the size and rate of aggregation of particles with positive surface charge such as $Cd(OH)_2$ NPs [122].

Liang *et al.* reported reaction between laser-produced zinc plasma and water molecules for the synthesis of zinc hydroxide/DS inorganic/organic hybrid nanostructures. A layered zinc hydroxide/surfactant composite (ZnDS) was produced by PLA of Zn in the aqueous media having SDS at 1, 10, and 100 mM concentrations [163]. The ZnDS products were octagonal platelets with a single crystalline form in hexagonal symmetry. In the composite formation process, the charged zinc hydroxide molecules were produced through the strong reaction between zinc species and water molecules, which are assembled into layered structure step-by-step by surfactant molecules. Usui *et al.* [164, 165] reported that increase in the length of alkyl chain in sodium alkyl sulfate increases interplanner spacing and angle of tilt between two planes in layered $Zn(OH)_2$/DS nanocomposite. Yan *et al.* [166] controlled the morphology of zinc hydroxide/DS nanostructures by varying the composition of ethanol/water in liquid media. PLA of zinc in pure water produces ZnO, while in water/H_2O_2 mixture, it generates ZnO_2 NPs [134].

PLA of Al in aqueous media of NaOH produces spherical β-$NaAlO_2$ and γ-Al_2O_3 NPs. At higher concentrations of NaOH, amount of $NaAlO_2.4/2\,H_2O$ and β-$NaAlO_2$ NPs dominates over γ-Al_2O_3 NPs [123]. Ablation of Al in hydrogen-saturated liquid ethanol produces core shell NPs with multiple shell layers [124]. NPs have face-centered cubic (fcc) single crystalline Al as core, amorphous Al as middle layer, and Al_2O_3 as outer shell layer. The use of nanomaterials formed by reactive LAL of active metals to synthesize metal compound nanostructures is further described in detail in Chapter 6.2. Table 6.1.2 lists some recent important reports on the preparation of metal compound nanostructures by reactive LP-PLA of active metals.

Reactive PLA for Noble Metals　In most of the cases, PLA of noble metals such as Ag, Au, Pt in most of the liquids produces their elemental NPs. Here, some reports on the reactive PLA of noble metals are presented. PLA of Ag in the aqueous media of 10^{-5} to 0.1 M NaCl using KrF excimer laser (248 nm wavelength, 30 ns pulse width, 10 Hz, 7.5–15 J/cm^2 laser irradiance) produces AgCl nanocubes [168]. In this case, reaction between Ag^+ ions of the plasma and Cl^- ions from the liquid at the plasma–liquid interface depends on the time of ablation. AgCl nanocubes were produced for PLA in the aqueous media of 5×10^{-3} M NaCl for 20 min of ablation, while for only 5 min ablation, there was no AgCl particle. When NaCl concentration is larger than 0.05 M, soluble complex anions of the type $AgCl_{m+1}^{m-}$ may form, which reduces the formation of AgCl and Ag. Similarly, Ag_2O cubes, pyramids, triangular plates, pentagonal rods, and bars were obtained by PLA of bulk Ag in polysorbate 80 aqueous solutions at room temperature [169]. Polysorbate 80 played an important role in the formation of Ag_2O micro/nanostructures. Silver core/silver oxide shell

Table 6.1.2 LP-PLA of active metals for the preparation of metal compound nanostructures.

Liquid parameters/ experimental arrangements	Laser parameters	Product obtained	Findings	References
PLA of Ti in liquids				
2 M aqueous media of Ce(NO$_3$)$_3$.6H$_2$O	Nd:YAG laser (355 nm, 10 Hz, 7 ns) 25 mJ/pulse, 3 J cm^{-2}	CeO$_2$/TiO$_2$ NC	Spherical TiO$_2$ and nonspherical CeO$_2$ NPs with 6.35 nm size and 2.73 nm SD were produced	[125]
Aqueous media of NaAuCl$_4$ (0.4 mM) and AgNO$_3$ (0.45 mM)	Nd:YAG laser (1064 nm, 1.2 ms, 10 Hz), 1.277 × 10^7 W cm^{-2}, rotation of target 60 rpm	Au/TiO$_2$ and Ag/TiO$_2$ NCs	Au/Anatase TiO$_2$ NCs with mean sizes of 5.5 and 25 nm for Au and TiO$_2$, respectively. Au NPs were dispersed homogeneously on the surface of TiO$_2$ NPs. Au NPs grow epitaxially on anatase TiO$_2$ NPs. Real-time photographs were taken in order to understand the synthesis mechanism	[126]
Water/horizontal	Nd:YAG laser (532 nm, 15 Hz, 4 mJ). Energy is varied by moving f = 5 cm lens in the beam direction	Spherical TiO$_2$ and Ti$_2$O$_3$ NPs	High fluence under focused condition results in rutile phase (4–150 nm), while lower fluence under defocused condition produces anatase phase (2–45 nm) of TiO$_2$ NPs. Under all the conditions, Ti$_2$O$_3$ are coproduced because of oxygen vacancies	[127]
Water/vertical	Nd:YAG (1064 nm, 10 Hz) 13 × 10^7, 15 × 10^7, 11 × 10^{10}, and 14 × 10^{10} W cm^{-2} intensity	TiO$_x$ NCA and multiwall nano tubes (MWNT)	At lower irradiances of 130 and 150 MW cm^{-2}, amorphous TiO and Ti$_2$O$_3$,TiO$_2$ NCAs with 5.2 and 5.5 GPa internal strains, respectively, were produced. At higher irradiance of 110 GW cm^{-2} MWNTs of TiO and Ti$_2$O$_3$,TiO$_2$ with 4.2 GPa internal stress, while at 140 GW cm^{-2} MWNT of only TiO$_2$ with 4.5 GPa internal were produced	[120]
Water, vertical	Nd:YAG laser (1064 nm, 10 ns, 10 Hz), 35 mJ/pulse, spot size = 0.5 mm	Anatase TiO$_2$ NPs	Spherical, anatase TiO$_{2-x}$ NPs with 3–5 nm sizes were produced. Particle size increases with the aging in the solution	[118]
Water, thickness 2–3 mm/vertical	Nd:YAG laser (532 nm, 15 ns and 15 Hz) Ip = 4.1–12.4 J cm^{-2}	Spherical TiO$_x$ NPs	Particles produced at larger fluencies have larger size and broad size distribution but are crystalline, while at lower fluencies, smaller sizes but amorphous	[128]

			Power	Number of shots/ scanning speed	Liquid medium	Diameter ± SD (nm)	
Water and aqueous media of 0.01 M SDS	CW fiber laser (1070 nm) 200 and 250 W were focused. Two types of experiments. (i) 1 s continuous ablation and stop, repeating the process for 5 and 20 times. (ii) Continuous ablation with 1 mm s⁻¹ and 5 mm s⁻¹ scanning speed	TiO₂ Sspherical NPs. Particles are more crystalline with higher rutile phase in the medium of 0.01 M SDS as compared to DI water	250 W	5 shots of 1 s	DI water 0.01 M SDS	23 ± 13 28 ± 12	[18]
			250 W	20 shots of 1 s	DI water 0.01 M SDS	15 ± 6 17 ± 7	
			250 W 250 W	1 mm s⁻¹ 5 mm s⁻¹	0.01 M SDS 0.01 M SDS	18 ± 11 12 ± 5	
0.05 M aqueous solution of PVP (poly(vinylpyrrolidone))	Nd:YAG (1064 nm, 10 Hz, 10 ns), 80 mJ/pulse, spot size on the target = 2 mm	Rutile TiO₂ spherical NPs	Rutile titanium dioxide NPs having medium crystallinity and good photocatalytic activity were obtained				[129]
Liquid N₂	Nd:YAG (1064 nm, 10 Hz, 6 ns), 1–30 J cm⁻² laser irradiance	—	Spherical TiN NPs having sizes in the range of several microns to several tenths of nanometers with 0.4 : 1 N/Ti ratio at the center of particle and moderate crystallinity were obtained				[130]

(continued overleaf)

Table 6.1.2 (continued)

Liquid parameters/ experimental arrangements	Laser parameters	Product obtained	Findings	References
2-Propanol, water, ethanol, n-hexane	Nd:YAG (532 nm, 10 Hz), 20–100 mJ/pulse, laser spot on target surface = 1 mm	TiC with TiO and TiH in 2-propanol; TiO$_2$ in water, and Ti with amorphous carbon deposits in hexane	Higher fluence produces larger-sized particles with higher distribution in all the three liquid media. PLA in water gives higher number of larger (>70 nm) particles as compared to 2-propanol, while n-hexane produces particles <70 nm. In all the media, smaller particles have larger incorporation of solvent elements. For example, in 2-propanol, higher O and C contents in the particles of <10 nm, particles in the range of 10–30 nm have 1:1 Ti and C ratio, while above this, Ti:C >1. Similarly, for water, particles larger than 50 nm have lower oxygen content, while particles smaller than 50 nm have Ti:O ratio between 1.43:1 and 3:1. Particles below 5 nm diameter have average Ti:O ratio of 1.04:1. Similar results for PLA in n-hexane	[119]
Water and aqueous media of, 0.001, 0.01, 0.1 M SDS	Nd:YAG laser (355 nm, 10 Hz), 150 mJ/pulse, focal spot size = 1 mm	Anatase TiO$_2$ spherical NPs	NPs produced in aqueous media of 0.01 M SDS had highest crystallinity, was 3 nm in size, and had maximum yield	[19]

PLA of Zn and Cd in liquids

Methanol	Nd:YAG laser (1064 nm, 40 mJ/pulse, 10 ns, 10 Hz), 0.6 mm spot diameter, 5 cm height of liquid above target surface, time 60 min	Drop-shaped zinc oxide quantum dots and their self-assembly	Drop-shaped zinc oxide quantum dots (QDs) having 6 ± 2.4 nm average length, 3.5 ± 1.4 nm average width, and 1.69 ± 0.4 nm average aspect ratio were obtained. Drop shape of these QDs electrostatically drives them to get self-assembled into various dendritic nanostructures. The dendritic assemblies are classified into three categories: (i) linear axis symmetrical branching, (ii) linear axis asymmetrical branching, and (iii) curvilinear axis asymmetrical branching	[114]
Water with or without continuous flow of oxygen	Nd:YAG laser (532 nm, 40 mJ/pulse, 10 ns, 10 Hz); oxygen was injected close to the plasma plume through jet. Time of ablation 60 min	$ZnO/Zn(OH)_2$ nanocomposites	Effect of injection of oxygen in the close vicinity of plasma plume on the size, shape, morphology, composition, and crystallinity of product was investigated. Injection of oxygen causes decrease in particle size, distribution, and ratio of $Zn(OH)_2/ZnO$, while increase in the crystallinity of product	[115]
DI water/vertical, 5 mm height/vertical	Nd:YAG laser (1064, 16 ns, 10 Hz, and 532 nm) 1.7×10^{11} W cm^{-2} peak power (2.6×10^4 W cm^{-2} average power) for 1064 and 8.3×10^{10} W cm^{-2} peak power (1.3×10^4 average power) for 532 nm	ZnO and ε-$Zn(OH)_2$ nanocomposite	Nanocomposite materials produced with 1.7×10^{11} W cm^{-2} (1064 nm excitation) and 8.3×10^{10} W cm^{-2} (532 nm excitation) were platy and equiaxed in shape, respectively. W-ZnO nanoplates showed well-developed {0001} terraces for mutual coalescence and {1010} and {1120} edges for lateral growth. The produced composites have internal compressive stress upto 0.7 GPa and band gap of 3.1 eV	[131]
Ethanol–water mixture (0.24 : 1)/vertical	KrF excimer laser (248 nm, 10 Hz) fluence 8.4 J cm^{-2}	ZnOx; $1 \geq x \geq 0$	ZnO_x nanoclusters produced by PLA in ethanol–water mixture got assembled into hollow NPs, while no assembly was observed in the solution produced by PLA in distilled water. Absorption of ZnO_x clusters at the interface of laser-induced bubbles produced hollow NPs	[65]

(continued overleaf)

Table 6.1.2 (continued)

Liquid parameters/ experimental arrangements	Laser parameters	Product obtained	Findings	References
Tetrahydrofuran (THF) with 0.5% w TPU stabilizer// horizontal	Picosecond laser system (515 nm, 7 ps, variable energy upto 125 µJ, variable frequency upto 200 kHz). Beam was coupled with scanner optics that can scan an area of 16×16 mm^2 with maximum lateral scan speed of 3.75 m s^{-1} (Exp. Setup fig. 6.1.23).	Zn/ZnO NPs	Variation in repetition rate (temporal delay) and interpulse distance (spatial distance between consecutive pulses as controlled by scan speed) strongly affect ablation efficiency. The ablation in pure THF caused fast aggregation of generated NPs, while after adding TPU in THF produced stable NPs of average size 6.5 and 4.5 nm FWHM. After 200 µs, the cavitation bubble collapsed, causing the minimal time spacing between two subsequent pulses to avoid shielding by the cavitation bubble	[132]
0.04 M solution of SDS in water and 0.04 M SDS in 1:4 mixture of water and ethanol	KrF excimer laser (248 nm, 10 Hz, 30 ns) 4.3 J cm^{-2} fluence, time = 20 s	3D Zn(OH)$_2$/SDS nanostructures	Three-dimensional zinc hydroxide/dodecylsulfate (Zn(OH)$_2$/DS) nanostructures self-assembled by nanolayers have been fabricated by PLA of bulk Zn in ethanol–water solution of sodium DS. The presence of ethanol causes defects in the hydroxyl group sites of Zn(OH)$_2$ that partially form ZnO structures, which facilitate heterogeneous nucleation of secondary nanolayers, resulting in 3D nanostructures	[20]
(i) Ethanol/water (10:1 v/v), (ii) mercaptoethanol, (iii) ethanol/water (10:1 v/v) with ferrocene, and (iv) dodecyl mercaptane; depth = 5 mm	Nd:YAF millisecond laser system (1064 nm, 0.6 ms, 1 Hz); fluence = 10^6 W cm^{-2}: time = 5 min. In all solutions except in (iii) 1 Hz repetition rate was used. For media (iii) 20 Hz repetition rate was used	Zn/ZnO CSNPs in solution (i), ZnO HNPs in solution (ii), ZnO hollow NPs in solution (iii), and ZnS nanocrystals in solution (iv)	Mechanism of particle synthesis using millisecond laser system is slightly different than nanosecond and femtosecond systems. With the increase of laser repetition rate particle size decreases, while increase of liquid reactivity produces larger and hollow nanosystems. Synthesized particles showed room temperature ferromagnetism	[44a,b]
Distilled water/ZnO target, beam vertically upward	Ti:Sapphire femtosecond laser system (775 nm, 180 fs, 1 kHz), 11 J cm^{-2} fluence	ZnO spherical NPs	Zinc oxide spherical NPs with average diameter of 5.4 nm and SD of 0.5 nm is obtained. Produced sample was used for the modification of electronic property of PEDOT:PSS	[133]

10 mM aqueous media of SDS	Nd:YAG laser (1064 nm, 10 ns, 10 Hz), 35 mJ/pulse, time = 2 h, spot size = 0.5 mm	Zn/ZnO core shell NPs	Spherical as well as nonspherical NPs in the size range of 20–200 nm were produced. Obtained colloidal solution emits intense green-yellow (520–600 nm) and violet (420–478) bands with the excitation by 514 nm. Upconversion is observed	[113]
Water and 3% H_2O_2 solution in water	Nd:YAG laser (355 nm, 8 ns, 10 Hz), energy was varied upto 300 mJ/pulse spot size = 0.08 mm, time = 40 min	ZnO and ZnO_2 NPs were produced in water and water/H_2O_2 media, respectively	Particles produced in H_2O_2/water mixture were much smaller than those produced in pure water. Air annealing of ZnO_2 particles at 200° for 8 h converts them into ZnO particles. Annealing at higher temperature produced larger ZnO particles with larger lattice parameters	[134]
DI water: Al was ablated in the suspension obtained by PLA of Zn in water	Nd:YAG laser (1064 nm, 5.5 ns, 10 Hz), fluence = 0.265 J cm^{-2}	Zn-Al layered double hydroxide nanostructures were produced	Zn_4Al_2-LDH nanostructures having hexagonal lattice with $a_0 = 3.07 A°$ and $c_0 = 15.12 A°$. The average diameter of these structures was about 500 nm and the thickness of a single layer was ~6.0 nm	[135]
DI water	Nd:YAG laser (1064 nm, 10 ns, 10 Hz), 20, 70, and 120 mJ/pulse energy	ZnO nanostructures	Crystallinity and particle size of the products increases with the increase of laser power. At low power thin nano-flakes with some spherical NPs, at medium power irregular polycrystalline powder with sizes <20 nm, while at higher power nanorods with small aspect ratio were produced	[136]
Aqueous media of SDS 0.05 mM. Continuous flow of pure oxygen close to the plume	Nd:YAG laser (532 nm/8 ns and 355 nm/5 ns, 10 Hz), 60–100 mJ/pulse for 532 nm and 25–70 mJ/pulse for 355 nm. Spot size = 0.5 mm	ZnO/ZnOOH nanocomposite	The ratio of ZnO ZnOOH, average particle size, and distribution decrease with the increase of laser irradiance for both the wavelengths. The rate of decrease is faster for 355 nm as compared to 532 nm wavelength. Injection of oxygen induces new mechanism in particle synthesis	[116, 117]
DI water, methanol, ethanol; target as ZnO pellet	Nd:YAG laser (355 nm, 9 ns, 10 Hz), spot size = 2 mm, energy = 25, 35, and 45 mJ/pulse, time = 1 h. For water PLA was done at 45 mJ/pulse for 1, 2, and 3 h	ZnO spherical fluorescent NPs	Average particle size and distribution increases with the increase of laser fluence. Comparatively larger-sized particles with narrower size distribution were produced with increase of ablation time. Produced ZnO NPs have oxygen vacancies, which was reduced by flow of oxygen during ablation. NPs produced without and with flow of oxygen emit yellow and violet lights, respectively	[137]

(continued overleaf)

Table 6.1.2 (continued)

Liquid parameters/ experimental arrangements	Laser parameters	Product obtained	Findings	References
Aqueous media of SDS 0.1–100 mM	Nd:YAG, (1064 nm, 10 ns, 10 Hz), 70, 50, and 35 mJ/pulse. Spot = 2 mm, time = 30 min	ZnO and Zn/ZnO core/shell spherical NPs	Below CMC (10 mM), pure zinc oxide NPs, while above CMC, Zn core/ZnO shell NPs were produced. Size and distribution of particles; nature of aggregation, ZnO shell thickness of Zn core/ZnO shell particles, as well as PL emission intensity decreases with the increase of SDS concentration [112]. Spherical Zn/ZnO core/shell NPs produced in 0.05 M SDS aggregated into ZnO-tree-like nanostructures with enhanced PL emissions [138]. At particular SDS (0.05 M) concentration, shell thickness increases with the increase of laser irradiance [139]	[112, 138, 139]
DI water at 40, 60, and 80 °C temperatures and aqueous media of LDA and CTAB at 80 °C	Nd:YAG laser (355 nm, 7 ns, 10 Hz), 3.2 J cm^{-2}, spot size = 1 mm, time = 40 min	Spherical ZnO NPs and ZnO nanorods with different aspect ratios	Aggregate of ZnO spherical NPs, at 40 °C, while small polyhedron and rodlike nanostructures were observed by PLA in DI water at 60 and 80 °C. Columnar nanostructures with 500–600 nm in length and 200 nm diameter are produced. Ablation at RT and aging at 80 °C for 40 min after ablation also produced hexagonal rodlike structures. Rod-shaped crystals, smaller than those obtained from DI water, were also produced in both the surfactant solutions at 80 °C temperature	[140]
Aqueous media of SDS (0–0.1 M) at different concentrations	Nd:YAG laser (355 nm, 7 ns, 10 Hz), sopt size = 2.25 mm^2, 100 mJ/pulse	ZnO spherical NPs and layered ZnDS nanocomposite	A layered zinc hydroxide/surfactant composite (ZnDS) was produced by PLA of Zn in the aqueous media containing SDS at 1, 10, and 100 mM concentrations, while ZnO particles in DI water. The ZnDS products were octagonal platelets with a single crystalline form in hexagonal symmetry	[141]

DI water and aqueous media of CTAB, SDS, LDA, and OGM near their critical micelle concentrations	Nd:YAG laser (355 nm, 5–7 ns, 10 Hz), spot size = 1.5 mm², 6.7 J cm⁻²	ZnO spherical NPs and Zn(OH)₂/SDS layered nanocomposite	ZnO spherical NPs were produced in all the liquid media except in the aqueous media of SDS. Average size and distribution of ZnO produced in DI water was 38 and 16 nm, respectively. Mean size, distribution, as well as excitonic and defect-related emission intensity of particles are highly dependent on the nature and concentration of surfactants	[142]
Double distilled water/*target* Cd rod	Nd:YAG laser (1064 nm, 10 ns, 10 Hz), 35 mJ/pulse, spot size = 0.6 mm, time = 60 min	Cd(OH)₂/CdO spherical NPs CdO spherical NPs and rods	As-synthesized product was Cd(OH)₂ spherical NPs with 18.6 nm average, which converted into CdO spherical NPs and nanorods after heat treatment at 350 °C for 9 h	[121]
Aqueous media of SDS (20, 50, and 100 mM)/*target Cd rod*	Nd:YAG laser (1064 nm, 10 ns, 10 Hz), 55 mJ/pulse, spot diameter 0.6 mm, time 60 min	γ-Cd(OH)₂ nanoarchitec-tures	γ-Cd(OH)₂ nanostructures including spherical NPs, nanorods (aspect ratio 2.4–4.0), 2D and 3D nanoflowers, nanoflower buds, nanotetrapods, and nanosheets were observed in 20 mM SDS solution without any aggregate. The aspect ratio, yield of the product, and rate of aggregation/agglomeration-synthesized nanomaterials increased with the increase of surfactant concentrations	[122]

PLA of Al in liquids for preparation of its compound nanostructures

Aqueous media of 0.05–1 M NaOH	Nd:YAG laser (1064 nm, 16 ns, 10 Hz), 850 mJ/pulse, average power density 2.8 × 10⁴ W cm⁻², and peak power density 1.8 × 10¹¹ W cm⁻²	β-NaAlO₂ and γ-Al₂O₃ spherical NPs	At higher NaOH concentrations, produced nanocomposite has dominance of NaAlO₂.4/2 H₂O and β-NaAlO₂ NPs as compared to γ-Al₂O₃ NPs. Upon settling on the silica substrate at room temperature, the NaAlO₂ 4/5H₂O tended to develop as micrometer-sized plates for the sample fabricated under a relatively low (i.e., 0.05 M) NaOH concentration	[123]
Liquid ethanol saturated with H₂ at atmospheric pressure	Nd:YAG laser (1064 nm, 30 ps, 10 Hz), 4 J cm⁻²	Core shell NPs with several layers	Cavities are observed more frequently in NPs produced by PLA in H₂ saturated ethanol as compared to pure ethanol. NPs produced in the former case have fcc Al single crystalline structure as core, amorphous Al as middle layer, and Al₂O₃ as outer shell layer. Hydrogen dissolves in Al plasma at higher temperature (>660°) and is released after solidification, leaving a cavity behind	[124a,b]

(continued overleaf)

Table 6.1.2 (continued)

Liquid parameters/ experimental arrangements	Laser parameters	Product obtained	Findings	References
Water	Nd:YAG laser (1064 nm, 8 ns, 10 Hz), 10–55 J cm^{-2}	Al and Al$_2$O$_3$ spherical NPs	The NPs are spherical with broad size distribution centered at ~25 nm. The electron density of the plasma plume in liquid is higher compared to the air phase	[167]
Distilled water and ethanol–water mixture (water:ethanol = 3 : 1)	KrF excimer laser (248 nm, 30 ns, 10 Hz) fluence = 2.3 J cm^{-2}	Solid and hollow Al$_2$O$_3$ amorphous micro/nano particles	Ratio of hollow to solid particles increases with the increase of size illustrating that bigger particles are hollower. Abundance of the hollow particles increases by adding ethanol in water. Trapping of laser-synthesized small particles on the surface of laser-induced cavitations bubbles is responsible for creation of hollow particles	[67]
Air and distilled water/flowed on the target surface with the speed of 190 ml min^{-1}, Al$_2$O$_3$ pellet	Nd:YAG laser (1064 nm, 40–55 ns, 0.5–20 kHz), laser beam scanning speed = 10–1400 mm s^{-1}	Al$_2$O$_3$ spherical NPs	Laser ablation rate in water is much higher as compared to that in air. Ablation rate in air decreases with beam scan rate (interpulse distance). while it increases with the increase of scan rate upto 500 mm s^{-1} (interpulse distance of 175 µm) and decreases after that. Ablation rate (nanogram/pulse) in air is almost independent of the repetition rate, while it decreases with the decrease of repetition rate	[143]
Distilled water, Al$_2$O$_3$ pellet, horizontal geometry, liquid flow, liquid layer thickness = 2–10 mm	Nd:YLF laser (1047 nm, 20–60 ns, 4.6 mJ), rep. rate 4–15 kHz	Al$_2$O$_3$ spherical NPs	At a particular liquid thickness and scan rate, NP productivity increases linearly upto 50 J cm^{-2}, saturated upto 71 J cm^{-2} and further increases quadratic dependence. NP's productivity (milligram/hour) increases linearly with the decrease of liquid layer thickness in the range of 8 to 2 mm. Almost 20% of the productivity rate has been increased due to the 190 ml min^{-1} liquid flow as compared to stationary liquid. The interpulse delay and repetition rate dependence on the NP's productivity are same as described above	[144]

DI water, Al target	Nd:YAG laser (1064 nm, TEM_{00}, 16 ns, 10 Hz), 850 mJ/pulse, average, and peak power densities were 2.8×10^4 and 1.8×10^{11}, respectively	H^+ and Al^{++} codoped spherical Al_2O_3 NPs	The as-formed γ- and θ-Al_2O_3 NPs are mainly 10–100 nm in size and have a significant internal compressive stress (10 GPa) because of a significant shock loading effect in water. The γ-Al_2O_3 NPs are nearly spherical in shape but become cuboctahedra when grown to ~100 nm to exhibit more facets as a result of martensitic $\gamma \rightarrow \theta$ transformation following the crystallographic relationship $(311)\theta//(022)\gamma$; $(024)\theta//(311)\gamma$	[145]
Distilled water, α-Al_2O_3 single crystal	Pulsed N_2 laser (337 nm, TEM_{10}, 10 ns) 7 mJ/pulse, spot size = 100 μm	α-Al_2O_3 NPs	Most of the produced material is irregularly shaped, while some small particles are spherical	[146]
DI water/Mg, Al, and Zn target separately, liquid height = 10 mm	Nd:YAG (1064 nm, 5.5 ns, 10 Hz) laser, 0.265 J cm^{-2}	$Al(OH)_2$, $Mg(OH)_2$ NPs, Mg-Al LDH and Zn-Al LDH	PLA of Al in DI water produces bayerite and gibbsite NPs with triangular, rectangular, and spherical shapes. Size distribution of spherical NPs was from 1 to 10 nm, while distribution for other shapes was 5–300 nm. PLA of Al target in the PLA-synthesized Mg-DI and Zn-DI water suspensions produces gel-like well-crystallized Mg-Al LDH and hexagonal Zn_4Al_2 LDH, respectively	[135]
Distilled water/Al target	Nd:YAG laser (532 nm, 5 ns, 10 Hz), unfocused beam was used for ablation of Al in 5 ml water, 80 mJ/pulse, t = 10 and 20 min	Simultaneous yield of bayerite, gibbsite, and boehmite spherical, triangular, and rod-shaped NPs	Through the course of aging, bayerite, gibbsite, and boehmite particles were formed simultaneously. These crystals exhibited three different morphologies: triangular (bayerite), rectangular (gibbsite), and fibrous (boehmite). The boehmite products can be generated solely by increasing the solution temperature	[147]

CW, continuous wave; NC, nanocomposite; NCA, nanochain aggregates; TPU; thermoplastic polyurethane; -LDH, layered dihydroxide; CSNPs, Core/shell NPs; HNPs, Hollow Nanoparticles; PEDOT:PSS, (poly(3,4-ethylenedioxythiophene):poly(4-styrenesulfonate)); LDA, lauryl dimethylaminoacetic acid betaine; OGM, octaethylene glycol monododecyl ether.

NPs were produced by femtosecond PLA of Ag in water and ethanol [170]. It was found that the dominant Ag_2O effective thickness is inversely proportional to the fluence, reaching a maximum of 0.2 nm for a fluence of 60 J cm^{-2} and a minimum of 0.04 nm for a fluence of 1000 J cm^{-2}. This shows that lower laser fluence is more suitable for reaction between plasma and surrounding liquid.

PLA of Ag in the aqueous media of KBr (0.1 M) as well as in the mixture of aqueous solution of KBr (0.1 M) and aqueous solution of SDS (0.01 M) produced AgBr NPs with irregular spherical shapes [171]. Layered AgBr-based inorganic/organic nanocomposites were prepared by PLA of Ag target in the aqueous media such as CTAB, tetradecyltrimethylammonium bromide ($C_{14}H_{29}N$ + (CH_3)$_3$ Br-) (TTAB), and stearyltrimethylammonium bromide ($C_{18}H_{37}N$ + (CH_3)$_3$ Br-) (STAB). The obtained AgBr nanostructures are 2D, and the size of nanosheets can be changed with surfactant concentration. Nanocomposite was composed of alternating organic and inorganic lattice in which cationic surfactant molecules were adsorbed between the anionic AgBr interlayers with an interdigitated bilayer arrangement. Basal spacing increases linearly with the increase in number of carbon atoms in surfactant.

6.1.3.1.9 Some Recent Advancements in Liquid-Phase Pulsed Laser Ablation of Solid Targets

Electric-Field-Assisted LP-PLA of Solid Target Laser-produced plasma has electrons and ions of target materials, which are expanding normal to the target surface. For PLA in liquid media, this hot plasma could ionize liquid molecules to produce molecular ions at the plasma–liquid interface, which may be termed as *plasma-induced plasma*. Application of external electric and magnetic fields parallel or perpendicular to the direction of expansion of plasma plume can drive and monitor motion of charged particles of the laser-produced plasma, which can control (i) kinetics of reaction at plasma–liquid interface for the synthesis of product molecules, (ii) clustering of these product molecules into molecular assembly of controlled shape, and (iii) assembling of these NPs into larger nanoarchitectures of desired shapes and sizes. Since surface of NPs are charged, application of electric field during the ablation and/or after the synthesis of colloidal solution of NPs can affect size, shape, and assembly of NPs. Employment of external electric field also assists in the fabrication of high-pressure nanophase, that is, metastable phase of nanostructures that cannot be synthesized under normal experimental conditions. Figure 6.1.20 illustrates an experimental arrangement for the EFLAL [172].

Liu *et al.* [172] used the experimental arrangement shown in Figure 6.1.20, which includes a quartz chamber with Ge target at the bottom immersed into deionized water, a DC electric field brought by two quadrate parallel electrodes separated by 1.6 cm, and a second harmonic 532 nm from pulsed Nd:YAG laser. The electric field, perpendicular to the direction of expansion of the plasma plume, can be varied in the range of $9.06 \times 10^2 - 2 \times 10^3$ V m^{-1}. Magnitude of electric field drives the size and shape of the produced nanostructures. For example, application of 14.5 V DC voltage generated nano- and microcubes of size range 200–400 nm, while that of 32 V DC voltage produced spindlelike GeO_2 nanostructures. Figure 6.1.21 illustrates the SEM images of GeO_2 nanostructures obtained with conventional LAL, and

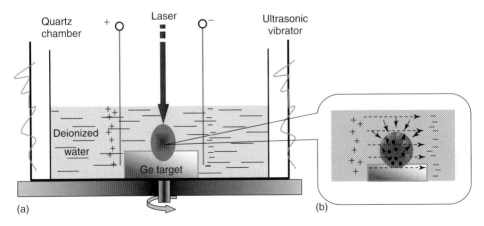

Figure 6.1.20 Experimental arrangement for electric-field-assisted pulsed laser ablation in liquid (EFLAL). (Source: Reprinted with the permission from Liu *et al.* [172] copyright @ American Chemical Society.)

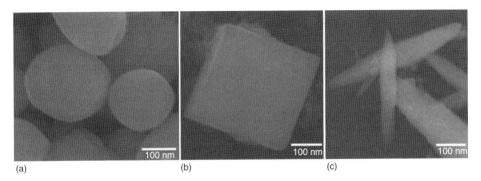

Figure 6.1.21 SEM images of nanostructures produced by (a) conventional LAL, and electric-field-assisted LAL with (b) 14.5 V, and (c) 32 V DC voltages. (Source: Reprinted with the permission from Liu *et al.* [172] Copyright @ American Chemical Society.)

under EFLAL. In the absence of external electric field almost all the particles are spherical in shape with the sizes lying in the range 300–400 nm. Generation of nano/microcubes and spindlelike structures using external electric field shows that electric field plays an important role in the fabrication of high-index facets of nano/microcubes and spindlelike particles. Spherical particles produced in the absence of external electric field are of pure germanium, while those obtained with the use of electric field are of germanium oxide (GeO_2). Change in the composition of particles and valance of Ge with the electric field concludes that electric field helps in the laser-induced breakdown/decomposition of water molecules and generation of oxygen. Therefore, use of electric field is a crucial parameter that significantly affects formation and shape evolution of GeO_2 nano/microcubes and spindles.

The same group of authors had used a slightly different experimental geometry, where a single crystalline Ge target biased with a DC voltage of −47 V was held at 5 mm above a silicon substrate, which was biased with a DC voltage of 47 V. The whole system was immersed into liquid toluene and maintained at room temperature [173]. Second harmonics of pulsed Nd:YAG laser operating at 10 ns pulse width, 1 Hz repetition rate, and 10^{10} W cm^{-2} laser irradiance was used for the ablation of Ge target for 60 h. PLA of single crystalline cubic Ge in toluene under the assistance of applied DC electric field generated single crystalline spherical Ge NPs of metastable tetragonal phase along with some inherent cubic Ge residue NPs [173].

Liquid-Phase Laser Ablation of Solid with Fast Scanning Optics   Low rate of productivity is one of the main problems associated with PLA in liquids and require optimization of laser parameters for scaled synthesis. Increase in time of ablation, laser irradiance, and repletion rate are some of the laser parameters, while thickness of liquid above target surface, liquid density, surface tension, and viscosity are a few liquid media parameters to be optimized to get high productivity. With the passage of time of ablation, the concentration of suspended NPs in the solution increases. These NPs absorb and scatter a portion of laser radiation, which results in lesser number of photons at the target surface and hence, significantly reduced rate of ablation and particle synthesis. Rate of NP production usually decreases with the increase of time of ablation. Use of circulating liquid above the target surface overcomes this problem and allows higher productivity upto gram scale. Circulating liquid eliminates not only NPs but also air bubbles, which also scatter laser radiation, from the ablation zone. Formation of cavitation bubble, its long (>100 μs) lifetime, and large physical size (50–200 μm) centered at ablation zone prevents the next pulse and/or the trailing portion of the current pulse from reaching at the target surface. Increase of the laser's repletion rate can increase the production rate of NPs, but it is limited by the formation of cavitation bubble. There is no interaction between laser pulse and cavitation bubble for low repetition rate such as 10 Hz (0.1 s interpulse time separation $>>$ cavitation bubble lifetime) laser systems. However, as we increase the repletion rate, the trailing part of current and/or front part or whole of next pulse might be prevented by cavitation bubble. For example, for a laser system with 10 kHz repletion rate (100 μs interpulse distance), there might be interaction between laser pulse and cavitation bubble, which limits the rate of synthesis of NPs. Use of scanning optics that can help laser beam to cross spatial (50–200 μm) and temporal dimension (100 μs) of cavitation bubble allows gram-scale synthesis of NPs by PLA in liquids. In the particular case of 100 μm size of cavitation bubble with 5 kHz (200 μs interpulse delay) repetition rate laser system, 500 mm/s scanning speed is required so that the current pulse can cross the cavitation bubble produced by the previous pulse.

Sajti *et al.* [144] optimized laser fluence, thickness of liquid layer, flow of liquid, pulse overlap, and laser repetition rate to maximize productivity of NPs by PLA in liquid. Mechanism of ablation and productivity of NPs strongly depends on the laser energy delivered into the target material. For 6 mm liquid layer and 80 μm spot diameter, NPs' productivity increases almost linearly with the pulse energy in the range of 0.7–3.0 mJ/pulse (14–50 J cm^{-2} laser fluence). When pulse energy is above 3.55 mJ/pulse, an interesting quadratic dependence of productivity on laser pulse energy is observed. They have obtained 445 mg h^{-1} productivity of NPs using 3.9 mJ/pulse energy with 5 kHz repetition rate, 6 mm liquid layer thickness, and 120 mm s^{-1} scan speed of focusing lens. Liquid layer thickness is another parameter that affects NPs productivity. At given experimental parameters of 3.8 mJ/pulse energy, 5 kHz repetition rate, 120 mm s^{-1} scan speed, 80 μm focal spot diameter, and 116 ml min^{-1} liquid flow, they have reported decrease of productivity of NPs with the increase of liquid layer thickness above the target in the range of 2.5–8.0 mm. They reported an enhancement of 20% in the particles' productivity with the use of 190 ml/min liquid flow over stationary liquid. In the stationary liquid, the generated NPs disperse into the liquid through very slow diffusion and Brownian motion; therefore they interact with the laser pulse. In contrast to this, liquid flow causes rapid removal of NPs and air bubbles from the ablation zone as well as quick dispersion of particles into whole liquid volume. Increase of liquid flow rate at when other experimental parameters are constant caused almost no further enhancement in ablation efficiency.

Scan speed of the focusing optics determines the distance between the current and next laser pulse. Heat and shock waves generated by current laser pulse usually heats larger area of target surface than that covered by cavitation bubble. Preprocessing of target surface by heat and shock waves generated from the current laser pulse increases next laser pulse to target interaction. Interaction of next laser pulse beyond the cavitation bubble generated by current pulse, but within its heat and shock-wave-affected zone, causes higher ablation rate than the pulse either falling on the cavitation bubble or beyond the heat and shock-wave-affected zone. At 4 kHz repetition rate, NP productivity increases with the increase of spatial interpulse distance, reaches maximum at 125 μm (500 mm s^{-1} scan speed), and starts decreasing beyond this (Figure 6.1.22). They have reported maximum NP productivity of 1.265 g h^{-1} in this optimized condition. NP productivity increases constantly in the scan speed range of 50–500 mm s^{-1} from 0.402 to 1.265 g h^{-1}. At a given interpulse distance, laser fluence, and liquid layer thickness, NPs productivity is significantly affected by variation in the laser repetition rate. Increase in pulse repletion rate from 4 to 9 kHz exhibits particle productivity of 7.6 and 1 mg min^{-1}, respectively.

Wagener et al. [132] have used the same technique of bypassing of cavitation bubble with circulating liquid during PLA of zinc in THF as shown in Figure 6.1.23. They have reported that ablation efficiency is maximum when the distance between focal spot and target is almost zero. They have also used variation in scan speed of f-θ lens and repetition rate of laser source to vary spatial and temporal separations between two subsequent laser pulses. Ablation efficiency verses repetition rate has two distinguishable regions: at lower repetition rate (<5 kHz) the ablation efficiency forms a plateau and shows almost constant value with the repetition rate, while it drops exponentially at higher repetition rates. At a given spatial interpulse separation, ablation efficiency increases quickly with the increase of interpulse temporal delay upto ∼200 μs, and attains a constant value beyond this. For 10 kHz repetition rate, ablation efficiency increases with the increase in spatial separation

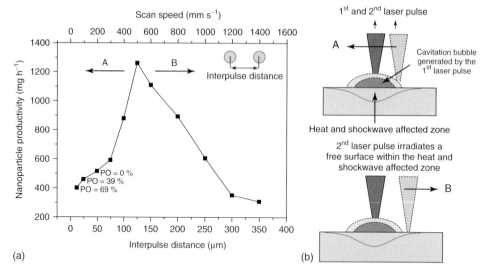

(a)   (b)

Figure 6.1.22 (a) Variation in the productivity of NPs with interpulse distance and scan speed using laser pulse energy of 4.6 mJ/pulse, 4 kHz repetition rate, and 4 mm thickness of liquid layer. PO shows pulse overlap in percentage. (b) Schemes of ablation, laser pulse, cavitation bubble, and heat- and shock-wave-affected zones. (Source: Reprinted with the permission from Sajti *et al.* [144] copyright @ American Chemical Society.)

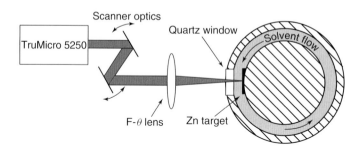

Figure 6.1.23 Experimental arrangement for LAL with circulating liquid and scanning laser beam. (Source: Reprinted with the permission from Wagener *et al.* [132] copyright @ American Chemical Society.)

between two subsequent laser pulses, attains a maximum value at ~180 μm, and starts decreasing beyond this. Low repetition, say 1 kHz, laser source under similar experimental conditions has comparatively higher ablation efficiency at a given spatial interpulse separation, and attains maximum value at comparatively much lower interpulse distance. Figure 6.1.24 exhibits variation in the ablation efficiency with spatial interpulse distance for laser sources having 1 and 10 kHz repetition rates. Hydrodynamic diameter of NPs decreases exponentially with the increase of spatial interpulse distance between two subsequent laser pulses.

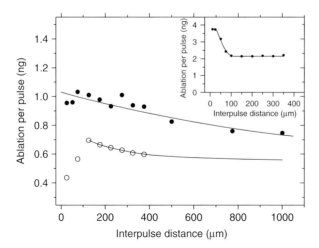

Figure 6.1.24 Effect of interpulse distance on per pulse ablation amount at 1 kHz (filled circles), and 10 kHz open circles (125 µJ/pulse). Inset shows effect of interpulse distance on per pulse ablation in ambient air at 10 kHz repetition rate and 125 µJ/pulse energy. (Source: Reprinted with the permission from Wagener et al. [132] © American Chemical Society.)

Liquid-Phase Laser Ablation of Solid with Circulating Liquid   Laser ablation of solids in liquid phase generates NP suspension whose concentration increases with the increase of ablation duration. These suspended NPs absorb and scatter incident laser light to reduce the rate of ablation and hence, the rate of NP generation. Low rate of productivity of NPs is one of the major problems of PLA in liquid for scaled synthesis of particles. Circulation of the liquid under controlled flow has the ability to enhance particle generation upto some extent. Liquid flow also facilitates (i) dissipation of thermal energy into the liquid, which is advantageous for getting better colloids and (ii) minimizes the influence of thermal lensing [174]. Barcikowski *et al.* [132, 144] have developed liquid flow systems for the enhancement of particle yield and for the fast *ex situ* biofunctionalization [175] of the generated NPs. They have compared results of PLA in stationary and flowing liquids and optimized velocity of liquid flow along the target surface in order to induce sample cooling and fast removal of particles and bubbles, that is, scatterers and absorbers, from the ablation zone. The NP generation mass rate by femtosecond ablation in flowing liquid attained $0.83 \pm 0.1 \, \mu g \, s^{-1}$, while only $0.22 \pm 0.1 \, \mu g \, s^{-1}$ was achieved in stationary liquid under the same experimental conditions. It is also observed that the picosecond laser attains higher ablation rate than femtosecond laser.

Sajti *et al.* [175] used a novel liquid flow system for the fast *ex situ* functionalization of LP-PLA in synthesized gold NPs, while Wagener *et al.* [132] used closed circuit circular liquid flow system for the PLA of zinc in circulating THF. Chen *et al.* [176] have used PLA of zinc and Eu_2O_3 targets under liquid flow for the preparation of functionalized zinc oxide and Eu_2O_3 NPs, respectively. For the PLA of zinc they used ethanol solution of acac, or cyclohexane solution of 0.5% polystyrene or ethyl butyrate solution of 0.5% polymethyl methacrylate (PMMA) as flowing liquid. In the case of PLA of Eu_2O_3 target thenoyltrifluoroacetone (TTA) or 1,10-phenanthroline (phen) was used as a flowing liquid for the preparation of functionalized particles.

Liquid-Phase Laser Ablation of Solid with Multiple Laser Beams Multiple beams obtained by furcation of a single laser beam or multiple beams where each beam is from a separate laser are recently being used to improve the rate of particle synthesis and for the preparation of particles of comparatively smaller average sizes with narrower size distribution. Rate of NP generation and characteristics of produced particles depend on the delay time and angle between two pulses falling on the solid surface. Depending on the time of delay, plasma, cavitation bubble, shock wave, or NPs produced by first laser beam can be excited or reheated by the second laser pulse.

Phuoc *et al.* [177] used two perpendicular nanosecond Nd:YAG laser beams operating on the same laser parameters (1064 nm wavelength, 5.5 ns pulse width, and 10 Hz repetition rate), where first beam ablates the Ag target surface and the second beam reheats the produced plasma at 2 mm distance from the target surface and after 40 μs from the first pulse. It was reported that at lower irradiance, double pulse PLA results in smaller average size than single pulse PLA, but at higher fluence, the results was reversed. Figure 6.1.25 illustrates experimental arrangement for dual beam PLA in liquids. Histograms of Ag NPs obtained by single and dual pulse PLA at 0.09 and 0.265 J cm^{-2} laser irradiances are given in Figure 6.1.26. In another work, Burakov *et al.* [78] reported that dual pulse PLA can produce lower mean size, narrower size distribution, and higher colloidal stability than single-beam PLA. Very recently, Giacomo *et al.* [178] used single and double-beam PLAs of titanium and graphite targets in distilled water at high pressure (1–146 atm) and studied the effects of external pressure on the temperature and pressure of laser of Ti plasma produced and on the morphology and phase of carbon nanostructures. They reported that at a given time after ablation, double-pulse-produced plasma has much higher degree of ionization that

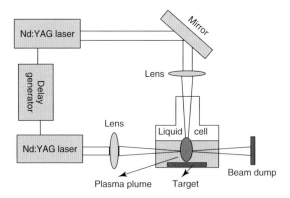

Figure 6.1.25 Experimental arrangement for dual beam PLA in liquids.

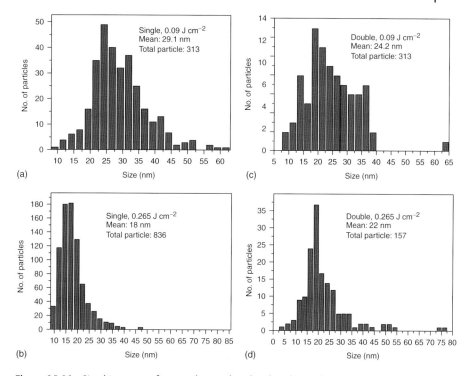

Figure 6.1.26 Size histograms for particles produced with (a,b) single and (c,d) dual beam PLA of Ag in deionized water at (a,c) 0.09 J cm^{-2} and (b,d) 0.265 J cm^{-2}. (Source: Reprinted with the permission from Phuoc *et al.* [177] copyright @ Elsevier.)

produced by single pulse PLA. As a consequence of the confinement effect of liquid on laser-produced plasma, they reported that plasma produced by single laser pulse has higher number density of atoms and that fast cooling occurs because of the efficient energy exchange between the plasma species and the surrounding water. In contrast to this, plasma produced with double pulse expands inside the cavitation bubble, thus allowing the plasma to preserve its energy longer. This results in a longer plasma lifetime and a thermodynamic course of particles aggregation.

6.1.3.2 Laser-Induced Melting and Fragmentation of Liquid-Suspended Particles

Small-sized particles with narrow size distribution are required in several scientific and technological applications such as in sensing and catalysis because of their larger surface area and in data storage because of their higher density on the surface. Most of the physical and chemical processes result in large-sized particles with wide dispersion, and therefore the need for resizing and size narrowing. Photoinduced resizing and/or reshaping of NPs has been a subject of recent intense interest, with a range of results favoring thermal ablation [152, 158, 179–193] or a nonthermal pathway involving charging of the NPs [9–11, 18–150, 153–178, 194–202] or creation of strong fields [203–205]. If colloidal solution of nano/micro particles, or powders/flakes dispersed in the solution are used as a target in place of the usual solid target placed at the bottom of liquid container, some very interesting phenomena would occur. Nano/microparticles or powder/flakes could undergo melting/vaporization or fragmentation to synthesize smaller-sized particles of same or different shapes, phase, and composition. As a remarkable function, this method can be used to reproduce or modify the size, shape, composition, and phase of NPs. Through careful selection of colloidal solution of NPs produced by LAL or chemical methods and laser parameters, two different effects can be obtained. First, the laser irradiation of mixture colloid composed of two kinds of noble metal NPs could lead to alloy or core/shell particle formation. Second, laser irradiation of the single kind of colloid could lead to fragmentation of primal nanoparticles and result in smaller NPs. Laser-induced melting/vaporization or fragmentation of NPs produced by preceding laser pulses in LP-PLA of solid target may also be caused by succeeding laser pulses. Three mechanisms are proposed for the resizing and/or reshaping of the nanostructures and/or their surface modification using laser irradiation. First is the laser-induced melting and/or vaporization of larger size particles and resolidification/rearrangements of molten globule and/or atomic species into smaller or larger-sized nanostructures with same or different shapes depending on the experimental conditions. The second and third routes are based on the nonthermal excitation of the particles using ultrashort laser pulses and take place below the melting point of particle. Second mechanism is laser-induced Coulombic explosion, which involves ejection of photo or thermal electrons from the surface of particles that leaves behind positive charges. These induced surface charges drive electrostatic repulsion between different parts of the surfaces and consequently, fragmentation of a single larger particle into several smaller ones with shape different from their mother particles, which may finally become spherical to minimize their surface energy. Kamat *et al.* [194] adopted this model in order to interpret 355 nm picosecond laser-induced fragmentation of silver NPs on the basis of observed hydrated electrons in the transient absorption measurements. Later, Mafuné and coworkers [195, 196] observed hydrated electrons in the nanosecond laser irradiation of SDS-suspended gold NPs and reported higher size reduction efficiencies for 355 nm irradiation as compared to the surface plasmon resonance (SPR) band excitation of gold NPs. The higher rate of size reduction was explained on the basis of interband electronic excitation-relaxation that provides higher absorption cross section for incident photons than the SPR absorption cross section. Moreover, they observed gold ions in the mass spectroscopy and suggested that the thermionic emission of electrons, which is a consequence of interband excitation, is responsible for the faster rate of fragmentation of gold NPs [195]. In the mean time, Koda and coworkers [206] proposed photothermal evaporation concept for the size reduction of chemically prepared gold NPs. Inasawa and coworkers performed schematic and detailed investigation in support of Koda's work, and proposed that size reduction in picoseconds irradiation of water-suspended gold NPs takes place by a layer-by-layer mechanism on the basis of observed bimodal distribution of particles [207]. The third mechanism of size reduction is quite different from photothermal ablation and Coulombic explosion and is based on the field enhancement near curved surfaces [203]. Experimental conditions to follow these mechanisms are quite different. The first method requires a laser irradiance that can make the temperature of the mother target higher than its boiling point to vaporize the surface atoms/molecules, the second needs either laser photon energy higher than the sum of work function of target material and ionization potential or needs a much higher laser irradiance for multiphoton ionization for the ejection of photoelectrons, and the third mechanism takes place with ultrashort laser pulses. Laser-induced fragmentation may also be achieved with laser wavelength having photon energy much less than the work function, through multiphoton ionization, which is a nonlinear process, and requires very high, $\sim 10^{17}-10^{18}$ W cm^{-2}, laser intensity. Such high intensity can be achieved from shorter pulses such as pulses from picosecond or femtosecond laser systems. Recent theoretical investigations by Werner and Hashimoto [179] suggested that the fragmentation is the dominating process for femtosecond irradiation, while purely thermal evaporation occurs for nanosecond irradiation, even at 355 nm wavelength excitation. They proposed an improved working model for the size reduction of Au NPs through the processes of heat dissipation, bubble formation, and electron phonon dynamics. Pyatenko *et al.* [208] also supported similar phenomenon. They demonstrated that electron ejection from the surface of NPs can be practically neglected for all possible parameters of a nanosecond laser used under unfocused condition. Different strategies of laser treatment of liquid-suspended particles can not only reduce average size of particles but also increase their dimension larger than the mother particles. For example, ultrafast laser irradiation of ethanol-suspended copper flakes converted them into copper nanowires (NWs) and copper NSs [197]. The diameter of NWs increases at the rate of 10.2–54.2 nm min^{-1} with the increase in the time of laser irradiation. After 10 min of irradiation, all NWs were converted into copper NSs. Photothermal and nonthermal processes of size/shape modification of liquid-suspended particles are described in detail in the subsequent sections:

6.1.3.2.1 Laser-Induced Melting/Vaporization
In this section, some theoretical models in support of experimental work and recent important reports on laser-induced melting/vaporization of liquid-suspended particles are presented.

Model by Takami et al. Takami *et al.* [206] proposed an analytical model for the estimation of temperature and maximum diameter of particles as a result of laser irradiation of water-suspended gold NPs. A laser beam operating at the irradiance of E_{in} (J s^{-1}), irradiates volume, V, of the solution of suspended particles having concentration C (g cm^{-3}). If E_{abs} (J s^{-1}) gets absorbed ($E_{abs} = E_{in} - E_{Tr}$; where E_{Tr} is the transmitted energy) in the solution of volume V, then energy, Q (J g^{-1} pulse^{-1}), absorbed by unit mass of the particles per pulse is given by $Q = E_{abs.}/RCV$; where R is the repetition rate of the laser source. For example, if 5 mJ/pulse energy from 7 ns pulse width and 10 Hz laser is absorbed in the 5 cm^3 solution of NPs having 2×10^{-2} g cm^{-3} concentration, then the estimated value of Q will be $\sim 10^{-5}$ J g^{-1} pulse^{-1}. Laser energy delivered to the suspended particles is lost because of the two main heat transfer processes. First is the conductive/convective process, and second is radiative heat transfer to the surrounding liquid. In the case of conductive/convective heat loss process, if temperature of suspended particle rises much higher than the temperature of surrounding liquid, then boiling heat transfer is considered. The boiling heat flux q from the common material to the water is almost $10^6 \, \mathrm{J \, m^{-2} \, s^{-1}}$ if temperature difference between water and materials is of the order of 10^3 degrees [209]. Radiative heat transfer from particles suspended to the liquid follows blackbody radiation whose flux is δT^4, where δ is the Stefan–Boltzmann constant with the assumption for unit value of emissivity. Conductive/convective loss from 45 nm gold particle suspended in water at its boiling point is estimated to be $4.5 \times 10^{-17} \, \mathrm{J}$ per particle per 7 ns, which is the pulse duration of laser [206]. The radiative loss from the same size of the gold particle at boiling temperature is estimated to be $2.3 \times 10^{-16} \, \mathrm{J}$ per particle per 7 ns. Sum of the energy loss from the particle to the surrounding liquid due to the radiative and conductive/convective thermal processes is negligible as compared to the per pulse energy delivered to unit mass of particle. Therefore, almost all the energy absorbed by large (>10 nm) particles is used in raising its temperature. That temperature T of the particles increased after irradiation with a single laser pulse can be estimated by following equations:

$$T = \frac{(Q - \Delta H_{\mathrm{melt}} - \Delta H_{\mathrm{evp}})}{C_{\mathrm{P}}} + 293 \qquad (6.1.30)$$

$$T = \frac{(Q - \Delta H_{\mathrm{melt}})}{C_{\mathrm{P}}} + 293 \qquad (6.1.31)$$

where, ΔH_{melt} is heat of fusion, ΔH_{vap} is the heat of vaporization, and C_{p} is the specific heat of the material of particle at constant pressure. The first equation is valid if temperature of particles increases above the vaporization temperature of the material, while the second equation holds if the temperature lies between the melting and vaporization points. If thermal diffusion length, $l_{\mathrm{th}} = (\alpha \tau)^{1/2}$, of the particle (where α is defined as $\alpha = k/\rho c_{\mathrm{p}}$ with thermal conductivity being k, density ρ, specific heat capacity of material of particles c_{p}, and the pulse width of the laser τ) is larger than the size of particle and temperature is above the vaporization point or in between melting and vaporization points, then whole of the volume of irradiated particle gets vaporized into atoms or melted, respectively, under irradiation by a single laser pulse. However, if size of the particle is larger than the thermal diffusion length, then partial evaporation or melting of surface of particles occurs in the time duration of single laser pulse. Evaporated atoms will undergo rearrangement in the same or different geometry depending on their energies to form same or different phase of NPs relative to the mother particles. Figure 6.1.27a,b illustrates variations in the estimated temperature, and maximum size, respectively of gold NPs with the absorbed laser energy [206].

According to Sambles [210], with the support from Reiss and Wilson [211] and Curzon [212], the melting temperature of particles follows the equation

$$\frac{H_{\mathrm{m}}}{2M} \frac{\rho_{\mathrm{s}}}{T_0} (T_0 - T_r) = \frac{\gamma_{\mathrm{sl}}}{r - p} + \frac{\gamma_{\mathrm{l}}}{r} \left(1 - \frac{\rho_{\mathrm{s}}}{\rho_{\mathrm{l}}}\right) \qquad (6.1.32)$$

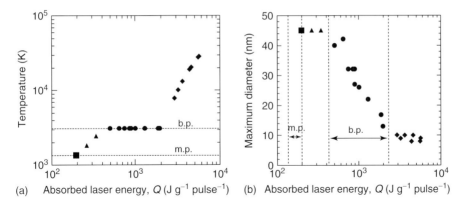

(a) Absorbed laser energy, Q (J g^{-1} pulse^{-1}) (b) Absorbed laser energy, Q (J g^{-1} pulse^{-1})

Figure 6.1.27 Dependence of (a) estimated temperature of target gold and (b) maximum diameter of product gold particles on absorbed laser energy. (Source: Reprinted with the permission from Takami *et al.* [206] copyright @ American Chemical Society.)

where notations H_m, T_0, and T_r stand for latent heat of fusion, bulk melting point, and melting point at radius r, respectively and ρ_s, ρ_l, γ_l, and γ_{sl} stand for density of solid, density of liquid, surface energy of liquid, and mean solid–liquid interface energy. M is the molecular mass and p is the relevant skin depth. Equation (6.1.32) illustrates that melting temperature of particles decreases rapidly with the decrease in size.

Takami *et al.* [206] reported that if the temperature of 45 nm gold particles is less than the melting point, change in neither the size nor the shape of particles is observed, while if the temperature is in between the melting and boiling points, only change in shape occurs [206] (Figure 6.1.27b). However, when the temperature reaches boiling point, both size and shape of particles change and the maximum diameter depends on the absorbed laser energy. Last, when the temperature of particle is higher than the boiling point, that is, the absorbed laser energy is high enough to completely vaporize the whole volume of particle into atoms, the size of particles attains a constant value [206] (Figure 6.1.27b).

Model by Insawa et al. In their work, Takami *et al.* [206] assumed that there is almost negligible heat loss from the gold particle surface to the surrounding water in the nanosecond duration. However, Link *et al.* [213] and Hodak *et al.* [214] reported that the heat loss from the particle surface to the surrounding medium occurs within 100–200 ps. These fast heat loss processes should be considered for the estimation of temperature and particle size in the nanosecond laser-induced melting/vaporization of liquid-suspended particles. Inasawa *et al.* [207, 215] modified the model of Takami *et al.* [206], considering the heat dissipation from particle surface during laser pulses in order to study the dependence of maximum size of particles on the laser irradiance. They have accounted three processes, namely, (i) photon energy absorbed by the particle, (ii) rise in the temperature of the particle, and (iii) heat dissipation from particle surface to the

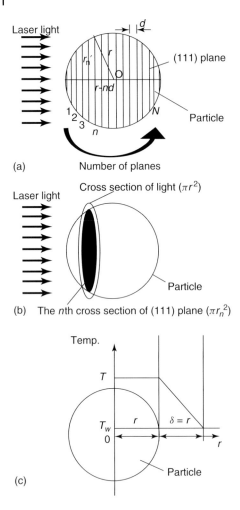

Figure 6.1.28 Schematic illustration of laser plane wave falling on the (a) (111) plane of the Au NPs; (b) cross section of *n*th (111) plane; (c) and variation of temperature inside and outside the particle. (Source: Reprinted with the permission from Inasawa *et al.* [207] copyright @ American Chemical Society.)

surrounding liquid, in order to heat balance of particles. Insawa *et al.* assumed that (i) a gold particle is a spherical single crystal with its (111) planes perpendicular to the direction of laser light (Figure 6.1.28a), (ii) thermodynamic properties of Au NPs are the same as those of its corresponding bulk material, (iii) laser light is cylindrical with uniform intensity throughout the cross section, (iv) temperature is uniform inside the particle, (v) the thickness of temperature boundary layer is equal to the particle radius, and (vi) temperature of the surrounding is uniform outside the boundary layer. Considering that the laser light is being absorbed by each (111) plane of the particle, they used the Lambert–Beer law to estimate the photon energy, J_1, absorbed by a single particle in infinitesimal time, Δt, as follows:

$$J_1 = \pi r^2 \frac{F}{\tau} \left\{ 1 - \prod_{n=1}^{N} \left(1 - \frac{r_n'^2}{r^2} \eta \varepsilon' \right) \right\} \Delta t \qquad (6.1.33)$$

where F is laser irradiance, τ is pulse width, r is radius of gold particle, N is the total number of (111) planes in the particle, $r_n' = \sqrt{r^2 - (r - nd)^2}$ is the radius of the nth (111) plane (Figure 6.1.28), η is the fraction of the area of the plane occupied by gold atoms, and ε' is the absorption coefficient of a gold atom. Heat flux J_2 from the surface of particle to the surrounding liquid is calculated using boundary layer model and given as follows:

$$J_2 = 4\pi r^2 \frac{k}{r} (T - T_w) \Delta t \qquad (6.1.34)$$

where T, T_w, and k are temperature of particle, temperature of surrounding water, and thermal conductivity of surrounding.

Using Eqs. (6.1.33) and (6.1.34), considering ΔT as a temperature change in the particle and principle of energy conservation, they obtained differential equation (Eq. (6.1.36)) as follows:

$$\frac{4}{3}\pi r^3 \frac{\rho}{M} C_P \Delta T = \pi r^2 \frac{F}{\tau} \left\{ 1 - \prod_{n=1}^{N} \left(1 - \frac{r_n'^2}{r^2} \eta \varepsilon' \right) \right\} \Delta t - 4\pi rk (T - T_w) \Delta t \quad (6.1.35)$$

where symbols M, ρ, and C_P stand for atomic weight, density, and specific heat of gold, respectively. Equation (6.1.35) gives the rate of change of temperature of particle with time as follows:

$$\frac{\partial T}{\partial t} = \frac{3FM}{4\tau \rho C_P r} \left\{ 1 - \prod_{n=1}^{N} \left(1 - \frac{r_n'^2}{r^2} \eta \varepsilon' \right) \right\} - \frac{3Mk}{\rho C_P r^2} (T - T_w) \qquad (6.1.36)$$

Using initial condition $T = T_w$ at $t = 0$, Eq. (6.1.36) gives temperature of the particles as a function of time as follows:

$$T = T_w + \frac{Fr}{4\tau k} \left\{ 1 - \prod_{n=1}^{N} \left(1 - \frac{r_n'^2}{r^2} \eta \varepsilon' \right) \right\} \times \left(1 - \exp\left(\frac{3MK}{\rho C_P r^2} t \right) \right) \qquad (6.1.37)$$

Using Eq. (6.1.37), one can calculate times t_m and t_b required to raise the temperature of particle to the melting temperature T_m and boiling temperature T_b, respectively, from the surrounding temperature T_w. The temperature of particle never increases until $t = t_m'$, the time when the particle absorbs ΔH_{melt}, the enthalpy of melting.

$$t_m' = t_m + \frac{\frac{4\pi r^2 \rho}{2M} \Delta H_{melt}}{\pi r^2 \frac{F}{\tau} \left\{ 1 - \prod\limits_{n=1}^{N} 1 - \frac{r_n'^2}{r^2} \eta \varepsilon' \right\} - 4\pi rk (T_m - T_w)} \qquad (6.1.38)$$

After the time $t = t_m'$, temperature T of the particle increases until it reaches T_b, the boiling temperature, at time $t = t_b$, which is given as follows:

$$t_b = t_m' - \frac{1}{B} \ln \frac{A - B(T_b - T_w)}{A - B(T_m - T_w)} \qquad (6.1.39)$$

where $A = \frac{2FM}{4\tau\rho Cpr}\left\{1 - \prod_{n=1}^{N}\left(1 - \frac{r_n'^2}{r^2}\eta\varepsilon'\right)\right\}$ and $B = \frac{3M\lambda}{\rho Cpr^2}$.

Figure 6.1.24 illustrates the temporal evolution of particle temperature for different sizes of gold particles having packing fraction $\eta = 0.91$, interplanar spacing $d = 2.36 \times 10^{-8}$ cm, $M = 197$, density $= 19.3$ g cm^{-3}, and $Cp = 25.3$ J mol^{-1} K^{-1} irradiated with laser light of 532 nm wavelength and 7 ns pulse duration in the liquid with temperature $T_w = 293$ K, thermal conductivity 7.99×10^{-4} W cm^{-1} K^{-1}. It is observed from the Figure 6.1.29 that particles having sizes <15 nm attain boiling temperature within the pulse duration of laser and are therefore subjected to size change. Particles of 14 nm diameter could not reach boiling temperature within the pulse duration, therefore their size remained unchanged. Therefore, for a given laser irradiance, there exist a critical size, and particles having sizes equal or less than the critical diameter will remain unchanged in the system. For small particles, heat dissipation from particles to the surrounding environment dominates; therefore one needs higher laser irradiance to heat the particle to the boiling temperature to reduce its size. This mechanism is also supported by experimental work of Singh *et al.* [152] on laser-induced melting/vaporization of selenium NPs. The minimum laser fluence that can attain the boiling threshold, $F_{threshold}$, and starts size reduction is given by following relation:

$$F_{threshold} = \frac{(T_b - T_w)\,C_p + \Delta H_{melt}}{\pi r^2 \left\{1 - \prod_{n=1}^{N}\left(1 - \frac{r_n'^2}{r^2}\eta\varepsilon'\right)\right\}}\frac{4\pi r^3}{3M} \tag{6.1.40}$$

Insawa *et al.* irradiated Au NPs with different irradiances of 532 nm laser and observed that experimental variation in the maximum diameter of Au NP with the laser irradiance is in good agreement with the simulated data obtained from their model, but calculated $F_{threshold} = 3 \times 10^{-2}$ J cm^{-2} pulse^{-1} for 45 nm gold particle is lower than the experimental observations for 532 and 308 nm lasers. They have also calculated variation in the maximum diameter with laser fluence for

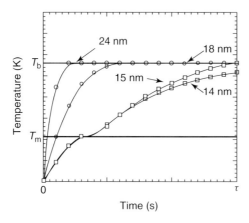

Figure 6.1.29 Temporal evolution of temperature for different sizes of gold NPs. (Source: Reprinted with the permission from Inasawa *et al.* [207] copyright @ American Chemical Society.)

different temperatures of surrounding liquid and reported that for a given laser irradiance, one can get smaller value of maximum diameter for lower surrounding temperature.

Two Step Size Reduction Model (Singh et al.) Singh et al. [152] proposed that rate of particle size reduction strongly depends on the size of particles and have different values above and below the critical size (R_c) (size at which strong quantum confinement occurs) of the particle. For the selenium particles

$$\frac{dR}{dt} = aR \qquad \text{for } R > R_c \tag{6.1.41}$$

$$\frac{dR}{dt} = \beta R \qquad \text{for } R < R_c \tag{6.1.42}$$

From Eqs. (6.1.42) and (6.1.43) we have

$$R(t) = R_0 e^{-\alpha t} \qquad \text{for } R > R_c \tag{6.1.43}$$

$$R(t) = R_0 e^{-\beta t} \qquad \text{for } R < R_c \tag{6.1.44}$$

where R_0 is radius of particle before laser irradiation. They irradiated 69.0 nm size of Se particles dispersed in double distilled water using fundamental wavelength of pulsed Nd:YAG laser and reported 0.44, 0.12, and 3.75 nm values for α, β, and R_c, respectively. They obtained size reduction constants ($\tau = 1/\alpha$ or $1/\beta$) of 2.27 and 8.33 s and $T_{1/2}$ (time at which size becomes half of its initial value) of 1.57 and 5.57 min for particles above and below the critical size, respectively.

Two Temperature Model (Hodak et al.) Takami et al. [206] ignored the heat loss from particles to the surrounding liquid environment, which holds good for the excitation of particles with picosecond and femtosecond lasers. Therefore, temperature calculated using the Eqs. (6.1.30) and (6.1.31) falls close to the initial temperatures expected for the picoseconds laser excitations. However, in nanosecond experiments, the maximum temperature obtained will be lower than that predicted since significant heat dissipation takes place in the course of excitation. Hodak et al. [214] proposed a model in which Newton's law of cooling for the energy exchange between conducting electrons and lattice, and between lattice and surrounding liquid, is used. This model is based on the following coupled differential equations:

$$C_e(T_e) \frac{\partial T_e}{\partial t} = -g(T_e - T_l) + F(t) \tag{6.1.45}$$

$$C_l(T_l) \frac{\partial T_l}{\partial t} = -g(T_e - T_l) - \frac{(T_l - T_m)}{\tau} \tag{6.1.46}$$

where T_e, T_l, and T_m (298 K) are electronic, lattice, and media temperatures, respectively; g is the electron–phonon coupling constant; $C_l(T_l)$ and $C_e(T_e)$ are

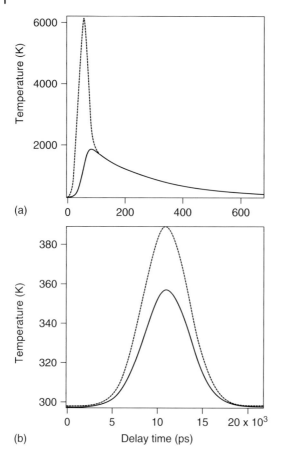

(a)

(b)

Delay time (ps)

Figure 6.1.30 Temporal evolution of electron temperature T_e (dotted line) and lattice temperature T_l (full line) calculated for 20 nm Au particles excited with (a) 30 ps and (b) 5 ns laser pulses with $E_{abs} = 2.05$ mJ/pulse. (Source: Reprinted with the permission from Hodak *et al.* [214] copyright @ American Chemical Society.)

the lattice and electron heat capacities, respectively; $F(t) = \left[\sigma_{\text{Mie}} \big/ \left(\tfrac{4}{3}\right) \pi R_{np}^3\right] I_L(t)$ is the energy exchange with laser pulse, where σ_{Mie} is the Mie absorption cross section and $I_L(t)$ is the laser intensity; and τ is a phenomenological time constant that accounts for heat dissipation to the solvent. Electronic heat capacity is given by $C_e(T_e) = \gamma T_e$; where $\gamma = 66$ J K^{-2} m^{-3} for Au. In the proposed model, laser pulse transfers heat to the electrons, electrons equilibrate with the lattice in the time scale given by C_e/g, and the hot electron/phonon system equilibrates with the surrounding liquid. Here it is assumed that the electron and ion temperatures are uniform inside the particle and that electrons do not directly couple with the surrounding liquid environment. Figure 6.1.30 presents electron and lattice temperature results of Eqs. (6.1.45) and (6.1.46) for 20 nm Au particles excited with 30 ps and 5 ns laser pulses with $E_{abs} = 2.05$ mJ/pulse.

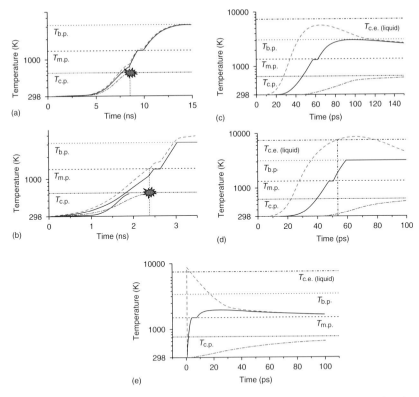

Figure 6.1.31 Temporal evolution of electron temperature (T_e) (dashed curve), lattice temperature (T_l) (solid curve), and surrounding liquid temperature (T_m) (dashed–dotted curve) at particle–water interface for 55 nm Au particle interacting with nanosecond laser pulses operating at 355 nm wavelength, 5 ns pulse width, and (a) 28 mJ cm^{-2}, and (b) 52 J cm^{-2} laser irradiances; picosecond laser pulses interacting with 355 nm wavelength, 30 ps pulse width, and (c) 18 mJ cm^{-2}, and (b) 52 J cm^{-2} laser irradiances; and (f) femtosecond laser pulses operating at 300 fs pulse width, 400 nm wavelength, and 10 mJ cm^{-2} laser irradiance. (Source: Reprinted with the permission from Werner and Hashimoto [179] copyright @ American Chemical Society.)

Modified Two Temperature Model by Giammanco et al. F. Giammanco and coworkers [216] used stretched exponential decay function, $Ae^{-(t/\tau)^\beta}$, where β is a factor that depends on the size of particle, proposed by Hu and Hartland [217] for dissipation of heat from particle to the surrounding media and modified the second term of Eq. (6.1.46) with $0.3\,(T_l - T_m)/\tau$ for $t < \tau$ and $0.3(T_l - T_m)/(t^{0.3}\tau^{0.7})$ for $t > \tau$. After that they added the third equation for the temperature of surrounding medium as follows:

$$
\begin{aligned}
\frac{\partial T_m}{\partial t} &= \frac{0.3\,(T_l - T_m)}{\tau} \quad \text{for } t < \tau \\
&= \frac{0.3\,(T_l - T_m)}{t^{0.3}\tau^{0.7}} \quad \text{for } t > \tau
\end{aligned}
\tag{6.1.47}
$$

They used values of electron capacity C_e, electron–phonon coupling, g, and chemical potential, μ, from the very recent report of Lin and Zhigilei [218], for the calculation of electron and lattice temperatures using Eqs. (6.1.45–6.1.47).

Improved Two Temperature Model of Werner and Hashimoto   Similar to the Giammanco *et al.*, Werner and Hashimoto [179] also used stretched exponential decay function and added the following equation for the energy dissipation from the particle surface to the surrounding medium:

$$\frac{\partial T_m(r, t)}{\partial t} = \frac{k_m}{C_m} \frac{1}{r^2} \frac{\partial}{\partial x} \left(r^2 \frac{\partial T_m(r, t)}{\partial r} \right) + F \tag{6.1.48}$$

where F is the heat loss term that has been experimentally determined by Plech and coworkers [219] and Juve and coworkers [220]. Plech and coworkers introduced thermal conductance, h, as a fitting parameter for the cooling process of gold NPs in the aqueous solution and obtained a value of $105 \times 10^6 \, \text{W m}^{-2} \, \text{K}^{-1}$. Cooling time also showed parabolic dependence on the particle size. The heat loss term, F, is given by following equation:

$$F = \frac{3h}{R(T_l)} [T_l - T_m(R)] \tag{6.1.49}$$

where T_m describes maximum medium temperature at particle–medium interface. The NP–liquid interface builds a layer of a few nanometers surrounding the NP surface with an average temperature defined by $T_m(R)$. Depending on the distance, r, from the NP–liquid interface, the surrounding liquid medium gives a temperature distribution described by Eq. (6.1.48). Werner and Hashimoto calculated temporal evolution of temperatures of electrons, lattice, and surrounding liquid medium using Eqs. (6.1.45–6.1.48) for a 55 nm diameter gold particle interacting with nanosecond (5 ns pulse width, 355 nm wavelength with 28 mJ cm^{-2} or 52 J cm^{-2} laser irradiances), picosecond (30 ps, 355 nm, 18 mJ cm^{-2} or 56 J cm^{-2}), or femtosecond (300 fs, 400 nm, 10 mJ cm^{-2}) laser pulses. Figure 6.1.31 illustrates the temporal evolution of electron, lattice, and surrounding liquid medium temperature for 55 nm gold particle irradiated with nanosecond, picosecond, and femtosecond laser pulses.

6.1.3.2.2   Some Reports on Laser-Induced Melting/Vaporization of Nanoparticles

Link *et al.* [213] have reported shape transformation of Au NRs in micellar solution through laser photothermal melting and fragmentation of liquid-suspended Au NRs and studied their dependence on laser fluence (microjoules to millijoules) and pulse width (911 fs and 7 ns). They observed that at a particular laser wavelength ($\lambda = 800$ nm) and fluence (40 µJ/pulse), femtosecond laser (100 fs, 400 MW) converted all the NRs into NSs, while nanosecond (7 ns, 6 kW) laser converted them into NRs of shorter aspect ratios, whose length decreases with the increase of irradiation time. In another set of experiments using 20 mJ/pulse (3 MW) energy from 7 ns pulse width laser system and 200 nJ/pulse (2 MW) energy for femtosecond system, they had observed that nanosecond laser converted all the NRs into NSs, while femtosecond laser converted NRs of longer aspect ratio into shorter

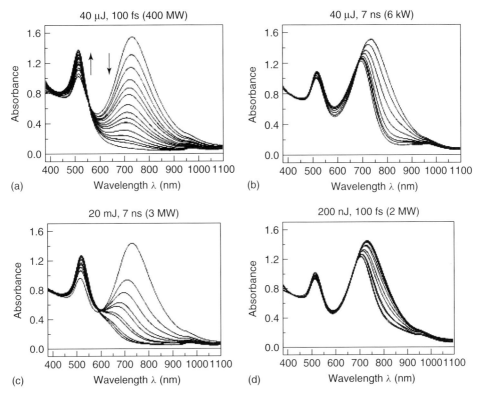

Figure 6.1.32 UV–visible absorption spectra of Au NRs for different time of irradiation using (a) 40 μJ, 100 fs (400 MW), (b) 40 μJ, 7 ns (6 kW), (c) 20 mJ, 7 ns (3 MW), and (d) 200 nJ, 100 fs (2 MW) laser pulses. (Source: Reprinted with the permission from Link *et al.* [213] Copyright @ American Chemical Society.)

ones. Figure 6.1.32 illustrates UV–visible absorption spectra of colloidal solution of Au nanomaterials produced after laser irradiation of Au NRs under different experimental conditions for different times of irradiations. The high power (400 MW) of femtosecond laser could lead to multiphoton absorption, surface ionization, and Coulombic explosion that convert NRs into NSs of comparable volumes. On the other hand, lower irradiance from nanosecond (6 kW and 3 MW) and femtosecond (2 MW) lasers heats NRs above the melting/boiling point and converts them into NRs of shorter aspect ratios (Figure 6.1.32b,d) and NSs. Takami *et al.* [206] irradiated water-suspended chemically prepared Au NPs and investigated the effects of irradiation time and laser fluence on the resulting size and temperature of Au NPs. Hodak *et al.* [214] used nanosecond and picosecond laser irradiation of water-suspended Au core/Ag shell or Ag core/Au shell NPs to produce Au:Ag alloy NPs. Similarly, Chen and Yeh [221] irradiated a mixture of colloidal solution of gold and silver NPs, while Zhang *et al.* [180] irradiated a mixture of gold and silver powders for the synthesis of alloy particles.

In an interesting report that appeared in *Science* in 2001, Jin *et al.* [222] reported photoinduced conversion of silver NSs into nanoprisms through irradiation of silver NSs with 40 W conventional fluorescent light source. Mafuné *et al.* [223] observed formation of Au nanonetworks and small Au NPs through 532 nm laser irradiation of LAL-synthesized water or aqueous medium of SDS-suspended ∼20 nm Au NPs. Irradiation of Au NPs suspended in water or low concentration of SDS solution suspended caused network formation, where the size of network increases with the irradiation time. However, irradiation of particles suspended in higher concentration of SDS solutions resulted in Au NPs of smaller sizes. Kawasaki and Masuda [181] irradiated water-suspended Au flakes of >10 μm sizes using 1 J cm^{-2} energy of nanosecond 532 nm laser light to obtain fine Au NPs through the intermediate stage of submicron-sized spherical Au particles. Gelesky *et al.* [224] irradiated Pd or Rh NPs suspended in 1-*n*-butyl-3-methylimidazolium hexafluorophosphate (BMI.PF6) ionic liquid using nanosecond laser having 532 nm wavelength, 8 ns pulse width, and 200 mJ/pulse energy to produce corresponding fine NPs. Usui *et al.* [182] irradiated water-suspended ITO (Indium doped SnO$_2$) NPs using different irradiances from nanosecond laser working at 355 nm wavelength and 7 ns pulse width and studied the optical properties of colloidal solution obtained.

Singh *et al.* [152] synthesized α-Se QDs by laser irradiation of water-suspended β-Se NPs of 69 nm average diameter. They reported that the size of the produced selenium QDs follow second-order exponential decay function of irradiation time, while rate of size reduction, da/dt, is directly dependent on the diameter, a, of the instantaneous QDs, very similar to radioactive decay model. They achieved minimum 2.74 ± 2.32 nm diameter for selenium QDs in 15 min of laser irradiation and reported that ∼3.75 ± 0.15 nm size may be the quantum confinement limit for Se QDs. Surface defect density of the selenium QDs increases, while defect/electron trap level energy decreases with the time of laser irradiation. They proposed that size reduction follows the trend of $a(t) = a(0)e^{-t/\tau}$, where a_0 and τ are the initial diameter of the raw selenium particles and the size reduction constant, respectively. Size reduction constants are 2.27 ± 0.12 and 8.33 ± 0.2, with corresponding $T_{1/2} = (\tau \ln 2)$ values (time require to get half the diameter of its initial value) being 1.57 ± 0.12 and 5.75 ± 0.2 min for sizes above and below the quantum confined size (≈3.75 ± 0.15 nm), respectively. Almost 6.5 ± 1.5 min of irradiation was required to attain quantum confined size, after which size reduction rate reduced too low and attained a final size of 2.55 ± 1.1 nm at this particular experimental condition. At this size value, rate of reduction, da/dt, becomes zero and size reduction process stops. Figure 6.1.33 presents photographs and variation in the size and band gap of Se QDs with time of irradiation. Figure 6.1.34 illustrates the rate of decrease, da/dt, in the size of particles with size, when the particle is above (Figure 6.1.34a) and below (Figure 6.1.34b) its quantum confined size (≈3.75 ± 0.15 nm). The filled square shown in Figure 6.1.34a corresponds to the region below the quantum confined size. Rate of decrease in the size of Se particles is 0.44 and 0.12 nm min^{-1} when the size is above and below, respectively, the quantum confined size. The first step of faster size reduction starts from 69 nm size of Se (raw size) and stops at quantum confined size, while the second stage

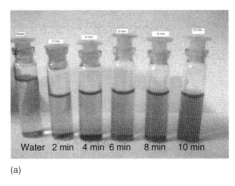

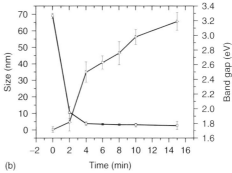

(a)

(b)

Figure 6.1.33 (a) Photographs and (b) sizes and corresponding band gap energies of colloidal solution of Se QDs obtained after different times of laser irradiation. (Source: Reprinted with the permission from Singh et al. [152] copyright @ American Chemical Society.)

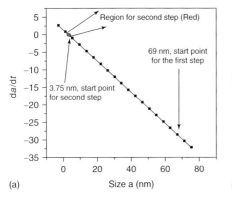

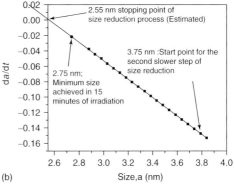

(a)

(b)

Figure 6.1.34 Variation in the rate of size reduction with the size of Selenium QDs (a) above and (b) below the quantum confinement ($\approx$3.75 $\pm$ 0.15 nm) size of Se particle.

of slower size reduction starts at quantum confined size, attains 2.75 nm at 15 min of total ablation, and should cease at a critical size of 2.55 nm (estimated size) at given experimental conditions.

The mechanism behind the attainment of a critical size at a given experimental conditions at which size reduction process ceases should be made clear. With the decrease of size of particles, extinction coefficient at the laser wavelength decreases, while heat loss through radiation and conduction processes from the surface of particles increases. Decrease in the absorption coefficient for laser radiation and increase in heat loss factor would not increase the temperature of particles above melting/vaporization point, which stops the size reduction process. In order to restart the size reduction process, one has to increase laser irradiance.

Zeng et al. [183] produced zinc oxide QDs of 1–8 nm size through laser irradiation of colloidal solution of zinc oxide hollow spheres. They reported that the size of NPs decreases rapidly with the increase of irradiation time in the range of 0–10 min and starts increasing in the time range of 15–30 min. Three steps including fragmentation in the irradiation time range of 0–10 min, growth in the range of 10–20 min, and ripening in the duration of 20–30 min were proposed. The residual thermal effect facilitated Ostwald ripening through incorporation of small nanocrystals into large ones.

Nanosecond laser irradiation of CuO powder into acetone and 2-propanol reduces their size and is a simple and effective pathway for the synthesis of colloidal solution of copper and Cu_2O NPs [184]. Kawasaki [184] irradiated acetone-suspended CuO powder using 1064 nm wavelength from nanosecond laser and reported the generation of ~10 nm Cu NPs with the production rate of ~1 mg min^{-1}. The produced copper NPs undergo rapid aerobic oxidation and convert copper particles into Cu_2O particles. Very similarly, Yeh *et al.* [225] irradiated 2-propanol-suspended CuO powder by using either 1064 nm or 532 nm wavelengths laser beams from nanosecond Nd:YAG laser.

6.1.3.2.3 Laser-Induced Coulombic Explosion of Nanoparticles

Coulombic explosion of particles occur when Coulombic repulsion among the surface charges on the particle exceeds attractive interatomic cohesive forces of the particle. Higher value of surface charge on the particle causes larger instabilities of the particles toward fragmentation. Ejection of electrons from the particle surface via any means leaves positive charges on the particle surface behind them. Laser irradiation of NPs can eject electrons from the particles' surface through three different physical mechanisms: (i) multiphoton ionization, (ii) photoelectric effect, and (iii) thermionic emission. At a given experimental condition, electron ejection can take place via more than one of the three processes.

Electron Ejection through Multiphoton Ionization Multiphoton ionization of atoms/molecules or clusters is a nonlinear optical process and requires comparatively higher laser irradiance. Threshold value of laser intensity, which is required for the ejection of electrons from particle's surface through the process of multiphoton ionization, depends on the multiphoton parameter (X) [226], and can be derived from Fermi energy (E_F) of free electron gas in a metal particle as follows:

$$E_F = \frac{\hbar^2}{2m} \left(\frac{3\pi^2 N_e}{V_P} \right)^{2/3} \tag{6.1.50}$$

where N_e is number of free electrons in volume V_P of the particle. The multiphoton parameter is given by

$$X = \frac{e E_0 k_F}{m_e} \omega^2 \tag{6.1.51}$$

where e and m_e are electronic charge and mass, E_0 is the amplitude of the electric field in laser plane wave, and $k_F = (2m_e E_F)^{1/2}/\hbar$ is the radius of Fermi sphere. Multiphoton ionization will be possible when value of X is larger than unity.

Intensity, I_0, of the laser can be described in terms of magnitudes of its electric and magnetic field vectors as $I_0 = E_0 H_0 = E_0^2 \sqrt{\varepsilon_0/\mu_0}$ or $E_0 = I_0^{1/2}(\mu_0/\varepsilon_0)^{1/4}$, where

μ_0 and ε_0 are permeability and permittivity of free space, respectively. Using E_0 and Eq. (6.1.47), the critical intensity, $(I_0)_{cr}$, can be given in terms of critical multiphoton parameter $(X)_{cr}$ as follows:

$$(I_0)_{cr} = \frac{X_{cr}^2 \omega^4 (\varepsilon_0/\mu_0)^{1/2} m_e^2}{e^2 k_F^2} \tag{6.1.52}$$

Fore silver and gold particles, each atom has single conduction electron; therefore, the total number of conducting electrons in Ag and Au NPs is equal to the number of atoms in the particle. Silver and gold particles have fcc structure with similar lattice constants of $a_{Ag} \approx a_{Au} = 0.408$ nm [227]. Hence the number of atoms per spherical particle of volume $V_p = \pi d_p^3/6$; where d_p is the particle diameter, will be

$$N_a = N_e = \frac{2\pi}{3} \left(\frac{d_p}{a}\right)^3 \tag{6.1.53}$$

Fermi energy and Fermi radius of silver and gold particles will be $E_F = 24.1^2/(2m_e a^2) = 8.87 \times 10^{-19}$J and $k_F = 4.91/a = 1.2 \times 10^8$cm^{-1}. Putting the value of k_F in Eq. (6.1.52) gives the value of critical laser intensity as follows:

$$(I_0)_{cr} = \frac{X_{cr}^2 (\varepsilon_0/\mu_0)^{1/2} m_e^2}{24.1e^2} a^2 \omega^4 \tag{6.1.54}$$

Equation (6.1.54) illustrates that the value of critical intensity depends on the laser wavelength, $\lambda = 2\pi c/\omega$. For the minimum value of $X_{cr} = 1$ for multiphoton ionization to occur, Pyatenko *et al.* [208] estimated 6×10^{15} W m^{-2}, 9×10^{16} Wm^{-2}, and 5×10^{17} W m^{-2} values for the critical irradiances corresponding to 1064, 532, and 355 nm wavelengths of Nd:YAG laser, respectively, and concluded that ejection of electrons through multiphoton ionization is not possible with any wavelength of the nanosecond Nd:YAG laser beam under unfocused condition. They also checked the possibility of multiphoton ionization by comparing photon density, $n_{ph} = I_0/\hbar\omega c$, in the laser beam with the number of free electrons, $n_e = N_e/v_p = 4/a^3$, and defined a new parameter $Y = n_{ph}/n_e = (I_0 a^3)/4\hbar\omega c$, which should have at least unit value for one-photon ionization. Critical value of laser intensity for photoionization should be $(I_0)_{cr} = 4\hbar\omega c/a^3$, which will be $3 \times 10^{17}, 7 \times 10^{17}$, and 1×10^{18} for 1064, 532, and 355 nm, respectively, wavelengths of Nd:YAG laser. All the three values are much higher than the maximum corresponding possible values of Nd:YAG laser intensities under unfocused condition; therefore the possibility of multiphoton ionization can be ruled out for the nanosecond laser irradiation of particles under unfocused condition.

Electron Ejection through Photoelectric Effect Photoelectronic emission of hot electrons from the surface of particle occurs when particle absorbs photons of energy $\hbar\omega$ and thermal energy, ε_T, exceeds $E_F + W - \hbar\omega$, where W is work function of the material of particle. Photoelectronic emission of electron occurs through the process of interband and intraband (Drude) absorption. According to the Mie theory [228, 229], the extinction and scattering cross section of a sphere with radius r and complex refractive index m relative to the external medium, water in this case, are given by

$$\sigma_{\text{EXT}} = \pi r^2 \left(\frac{2}{q^2}\right) \sum_{j=1}^{\infty} (2j+1) \left\{ \text{Re}\left(a_j + b_j\right) \right\} \tag{6.1.55}$$

$$\sigma_{\text{EXT}} = \pi r^2 \left(\frac{2}{q^2}\right) \sum_{j=1}^{\infty} (2j+1) \left\{ \left|a_j^2\right| + \left|b_j^2\right| \right\} \tag{6.1.56}$$

where q is the dimensionless size parameter defined by $2\pi r/\lambda$ and expansion coefficients, a_j and b_j, are given by complex functions of q and mq and represent oscillating electric and magnetic multipoles, respectively. For the values of $q \ll 1$, the expansion series converges very rapidly and results in almost same values for extinction and absorption coefficients. For the 69 nm average size of selenium particles with refractive index ~1 and 1064 nm laser light, values of q, m, and mq become less than unity. With the decrease of the particle size, scattering of 1064 nm laser light may dominate over its absorption, which may be one of the causes of attaining critical size limit in the process of laser-induced size reduction. When size of NPs are much smaller than the wavelength of laser light and the extinction and absorption coefficients have the same values, the absorption cross section in terms of the angular frequency is estimated by the following expression:

$$\sigma(\omega) = 9\varepsilon_m^{3/2} V\omega \frac{\varepsilon_2(\omega)}{c\left[\varepsilon_1(\omega) + 2\varepsilon_1\right]^2 + \varepsilon_2^2(\omega)} \tag{6.1.57}$$

where V is the particle volume, c is the velocity of light, ε_m is the frequency-independent medium dielectric constant, and $\varepsilon_1(\omega)$ and $\varepsilon_2(\omega)$ are the real and imaginary parts, respectively, of the dielectric constant of the particle and have interband and free electronic contributions. Contribution of interband electronic transitions in absorption cross section of nanomaterials is almost the same as that for bulk materials; therefore size dependence of $\sigma(\omega)$ in NPs mainly incorporates free electronic Drude term contribution, which itself depends on the contribution from inelastic scattering of free electrons [230] by the particle surface and free electron plasma frequency, defined by $(Ne^2/\varepsilon_0 m_{\text{eff}})^{1/2}$, where N is the free electron density, e is the electronic charge, ε_0 is the permittivity of vacuum media, and m_{eff} is the effective mass of the electron in the NP system. The inelastic scattering can be expressed by increase of scattering frequency from ω_0 for the bulk to ~v_F/a for the NPs, where v_F is the Fermi velocity and a is the particle diameter. Incorporation of Drude term and free electron plasma frequency contributions in the basic expression (Eq. (6.1.55)) for the absorption cross section are termed as *mean free path* and free electron density corrections in the position and shape of absorption spectrum of NPs. With the decrease of particle diameter, a, both contributory factors of the Drude term, that is, free electron plasma frequency (free electron density, N, increases due to decrease of the particle volume) as well as scattering frequency (~v_F/a) increases, which causes increase of the real and imaginary parts of the dielectric constant and hence decrease of the absorption cross section, $\sigma(\omega)$, and extinction coefficient with decrease of the particle size. Grua *et al.* [231] irradiated 5.2 nm gold NPs using 351 nm light and observed that the relative number of photoelectrons is smaller than the number of thermoelectrons due to the relatively low concentration of highly excited electrons that can be ejected through the photoelectric process.

Electron Ejection through Thermionic Emission Laser irradiation of NPs causes increase in its electron and lattice temperature as described by two temperature model (Eqs. (6.1.45) and (6.1.46)). Efficiency of thermionic emission of electrons depends on its temperature. Equations (6.1.45) and (6.1.46) can be written as follows:

$$\frac{d\Delta T}{dt} = -a\Delta T + b \tag{6.1.58}$$

where $a = g\left(\frac{1}{C_e(T_e)} + \frac{1}{C_l(T_l)}\right)$ and $b = \frac{F(t)}{C_e(T_e)} + \frac{T_l - T_m}{C_l(T_l)\tau}$

For the time period shorter than the laser pulse duration, heat loss from the particle surface to the liquid medium is too small as compared to the laser energy absorbed by it. Electron heat capacity C_e is much smaller than the lattice heat capacity C_l, therefore coefficient a reduces to $a = g/C_e = g/\gamma T_e$ and $b = F(t)/\gamma T_e$. Since electron temperature rises very fast as compared to the lattice temperature, initially, $T_l \ll T_e$, which makes $T_e \approx \Delta T$. Thus $a = g/\gamma \Delta T = a^*/\Delta T$ and $b = F(t)/\gamma \Delta T = b^*/\Delta T$. Now Eq. (6.1.58) can be transformed as follows:

$$\frac{d\Delta T}{dt} = -a^* + \frac{b^*}{\Delta T} \tag{6.1.59}$$

The analytical solution $\ln\left(1 - \left(a^*/b^*\right)\Delta T\right) + \left(a^*/b^*\right)\Delta T = -\left(a^{*2}/b^*\right)$ of Eq. (6.1.59) shows exponential growth of ΔT upto its maximum value $\Delta T_{max} = b^*/a^* = F(t)/g$ and time t_0 needed to obtain ΔT_{max} is $(\gamma/g)\Delta T_{max}$. Here, $F(t) = \left[\sigma_{Mie}/\left(\frac{4}{3}\right)\pi R_{np}^3\right]I_L(t)$ and electron–phonon coupling constant g can be obtained from the literature to calculate maximum temperature of the particle after the irradiation with the laser pulse. After the completion of this fast electronic heating process, comparatively slow lattice heating process becomes dominant, while electron temperature rises slowly to maintain temperature difference ΔT constant. Pyatenko *et al.* [208] assumed that once the particle melts, the heat transfer between electrons and melted particle remains the same as that of electron–phonon coupling before melting to calculate maximum temperature of electron gas in different sizes of silver and gold particles, irradiated with different wavelengths of Nd:YAG laser (Figure 6.1.35).

Thermionic emission of electron takes place when thermal energy, ε_T, exceeds $E_F + W$. The fraction of electron crossing the threshold energy $E_F + W$ can be estimated for any temperature using the following equation:

$$\frac{n_{E_F+W}}{N_E} = \frac{\int_{E_F+W}^{\infty} \sqrt{E} f_E dE}{\int_0^{\infty} \sqrt{E} f_E dE} \tag{6.1.60}$$

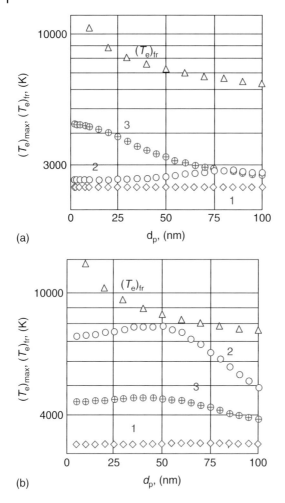

(a)

(b)

Figure 6.1.35 Maximum temperatures of electron gas in (a) silver and (b) gold NPs of different sizes irradiated with (1) 1064 nm, (2) 532 nm, and (3) 355 nm wavelengths of an Nd:YAG laser The fragmentation temperatures, $(T_e)_{fr}$, are depicted for comparison. (Source: Reprinted with the permission from Pyatenko *et al.* [208] copyright @ American Chemical Society.)

where $f_E = 1/(1 + \exp[E - \mu]/kT)$ is the Fermi–Dirac distribution function and $\sqrt{E}$ shows the density of states around energy E. If temperature of the particle system is much lower than the Fermi temperature $T_F = E_F/k$, one can use Fermi energy E_F in place of chemical potential (μ). Critical fraction of electrons required for the fragmentation of particles can be estimated by fissility parameter X using liquid drop model. According to the liquid drop model, a highly charged particle becomes unstable when the repulsive Coulombic forces exceed the attractive cohesive force and undergo Coulombic explosion [232]. Quantitative expression from Coulombic explosion is defined by the criteria of fissility as $X = E_c/2E_s$, where E_c and E_s are the Coulomb energy and surface energy of particle, respectively [232]. NPs having multiple charged surfaces are supposed to rapidly disintegrate into smaller particles through Coulombic explosion if $X \geq 1$, while the processes of thermal evaporation as well as Coulomb explosion occurs competitively when fissility parameter lies in the range $0.3 < X < 1$. Only the process of thermal evaporation takes place when $X < 0.3$. The fissility parameter is given by the expression $X = (q^2/n)/(16\pi r_{WS}^3 \sigma_S/e^2)$; where q, n, r_{WS}, and σ_S are the charge state, the cluster size, the Wigner–Seits radius (1.65 Å), and the surface tension of the particle, respectively [232]. Pyatenko et al. estimated the minimum laser intensity that would be needed for the fragmentation of silver and gold particles of different diameters as shown in Figure 6.1.36.

Kamat et al. [194] irradiated water-suspended silver NPs using picoseconds pulsed laser operating at 355 nm wavelength and 18 ps pulse width and studied size reduction mechanism through time-resolved transient absorption experiments in a picosecond laser flash photolysis apparatus in pump-probe geometry. They observed that biphotonic absorption process was responsible for the photoejection of electrons and Coulombic explosion. Fujiwara et al. [233] reported picosecond (18 ps, 532 nm) laser irradiation of water-suspended thionicotinamide (TNA)-capped gold NPs. They observed formation of larger-sized NPs during short-term laser irradiation though the melting of aggregates. At higher laser irradiance and/or long-term irradiation they observed fragmentation of particles. Giusti et al. [234] irradiated Au NPs using intense picosecond laser pulses of 532 nm wavelength and reported that two-photon absorption process is responsible for the ejection of photoelectrons from the PAMAM G5 (fifth-generation ethylenediamine-core poly(amidoamine))-capped Au NPs, and consequently, their fragmentation into smaller-sized NPs. Yamada et al. [195] and Muto et al. [101] reported Coulombic explosion of Au NPs for size reduction through the nanosecond laser irradiation of liquid-suspended Au particles on the basis of thermionic emission of electrons. The Au NPs are thermally excited by the nanosecond laser pulse, and then they are multiply ionized by thermionic emission. It was observed that the charge on the Au NPs increases with the increase of laser irradiance.

6.1.3.2.4 Field Enhancement near Curved Surfaces

When particle is irradiated with light, electrons in the particle oscillate with the oscillating field of the laser wave and interact with the lattice through electron–phonon coupling. Electron in the particles experiences electrostatic force ($F = eE_{laser}$), which is different for particles of different sizes and shapes. For the constant field of the laser $E_{0, laser}$ particles of smaller radius, or high curvature corners of the polygonal particles, experience enhanced field by several orders of magnitude than the incident field. This results in higher electron–phonon coupling and hence faster rise in the temperature of the smaller-sized particles or corners/edges of the polygonal particles. This causes melting of the smaller-sized particles and corners of the polygons, which converts polygon-shaped particles into spherical ones. Melting of

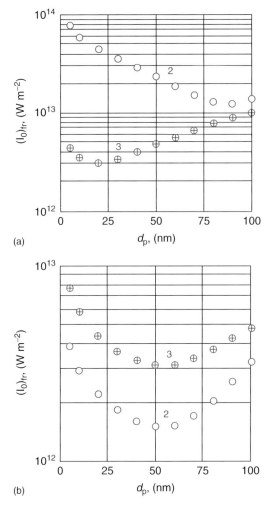

(a)

(b)

Figure 6.1.36 Minimum laser intensities needed to heat the electron gas in (a) silver and (b) gold NPs of different diameters that can raise particle temperature up to fragmentation temperature, $(T_e)_{fr}$. Calculations were made for (1) 1064 nm, (2) 532 nm, and (3) 355 nm wavelengths of wavelengths of a Nd:YAG laser. (Source: Reprinted with the permission from Pyatenko *et al.* [208] copyright @ American Chemical Society.)

the particles surface with the laser pulses having pulse width shorter than the times for electron–phonon and phonon–phonon coupling and enhancement of the field near curved surface is nonthermal in nature like Coulombic explosion. Intensity of the electric field near the surface of the spherical particles is different at different points and depends on the polarization of laser field. Hubert *et al.* [204] irradiated an array of polymer-covered 75 nm Ag NPs with a periodicity of 500 nm using linearly polarized laser light of 532 nm wavelength and observed highest field intensity and

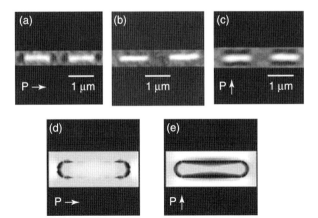

Figure 6.1.37 Topographic images and theoretical calculations of optical near fields around ellipsoidal gold particles. Atomic force microscopy (AFM images recorded (b) before and (a,c) after the irradiation of ellipsoidal gold particles covered with DR1MA/MMA. The light polarization direction is (a) parallel and (c) perpendicular to the long axis of the ellipsoids. Theoretical negative images of the field intensity for (d) parallel and (e) perpendicular polarization. (Source: Reprinted with the permission from Hubert *et al.* [205] copyright @ American Chemical Society.)

melting near the particle surfaces that were parallel to the laser polarization. They irradiated polymer-covered ellipsoidal gold particles using 514 nm light, with the polarization vector parallel or perpendicular to the larger axis of the particle. In the case of polarization parallel to the long axis of the ellipsoid, maximum intensity and hence melting was at the center of the particle, while for perpendicular polarization these are located at both ends of the particle (Figure 6.1.37). In a similar way, they irradiated an array of polymer-covered Ag NPs of 100 nm diameter with circularly polarized light of 514 nm and observed an inner as well as outer lobe of maximum intensity around the particle. They proposed that the outer lobe is the consequence of far-field interference of diffraction orders (Figure 6.1.38). Leiderer *et al.* [205] irradiated an array of triangular Ag NPs with 150 fs, 800 nm laser light and observed that the maximum field intensity and melting occurs at the corners of the particle. Plech *et al.* [203] presented observation of 150 fs and 800 nm laser ablation from 38 nm gold particles suspended in water using time-resolved X-ray scattering. They also observed highest field enhancement at the both surfaces of spherical particle parallel to the direction of polarization of laser field vector.

6.1.3.2.5 Applications of Laser-Induced Melting/Vaporization/Fragmentation in the Preparation of Alloy and Core/Shell NPs

Laser-induced photothermal melting/vaporization of metal NPs suspended in liquid media provides an efficient pathway for the synthesis of alloy NPs. Irradiation of a colloidal solution of two or more elemental NPs can melt/vaporize them to form their alloy. Chen and Yeh [221] have prepared Ag–Au alloy NPs of different

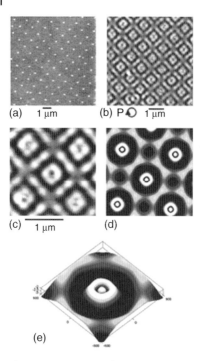

Figure 6.1.38 Atomic force microscopy (AFM) images recorded (a) before and (b) after irradiation of silver particles covered with DR1MA/MMA using 514 nm laser wavelength, 100 mW cm^{-2} irradiance and 30 min irradiation time. (c) Blow-up of (b). (d) Theoretical negative field image corresponding to (c), and (e) is a surface plot around one of the particles. (Source: Reprinted with the permission from Hubert *et al.* [205] copyright @ American Chemical Society.)

compositions using laser irradiation of mixture of their colloidal solution in different proportions. The SPR absorption of alloy NPs lies in between the SPR peaks of Au and Ag NPs with its position λ_{max} depending on the composition of Au and Ag in the alloy. Zhang *et al.* [180] produced Ag–Au alloy NPs using 532 nm laser irradiation of heterogeneous suspension of Ag (0.5–1 μm) and Au (0.8–1.5 μm) powders in the aqueous media of 0.05 M SDS. The position of SPR absorption band λ_{max} shifted linearly toward the longer wavelength side but in between the SPR peaks of Au and Ag NPs with the increase of Au compositions. Color of colloidal solution of Au NPs and corresponding UV–visible absorption are presented in Figure 6.1.39. Zhang *et al.* [235] synthesized Fe core @ Au shell NPs using second harmonic of nanosecond Nd:YAG laser for the irradiation of mixture of Au and Fe NPs in the solution of 25 ml water and 15 ml octane. Besner and Meunier [199] irradiated a mixture of colloidal solutions of PLA-produced Ag and Au NPs using femtosecond laser beam from Ti:Sapphire laser (800 nm, 110 fs, 1 kHz) with 0.3 J cm^{-2} irradiance. Molar ratio of Au/Ag NPs in the mixture was varied from zero to unity to produce Au–Ag alloy NPs of different composition.

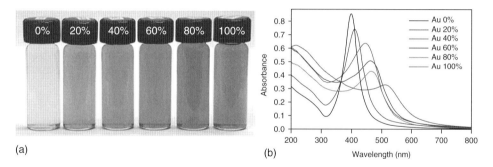

Figure 6.1.39 Photographs of colloid solution of Ag:Au alloy NPs of different compositions with corresponding UV–visible SPR absorption spectra. (Source: Reprinted with the permission from Zhang *et al.* [180] copyright @ American Chemical Society.)

Boyer *et al.* [200] reported Co core @ Au shell NPs by femtosecond laser irradiation of mixture of colloidal solutions of PLA-produced Co and Au NPs. They observed that Au atoms replace oxidized cobalt atoms on the surface of Co NPs and increase degree of ferromagnetism.

Table 6.1.3 presents laser-induced melting/vaporization and fragmentation of NPs.

6.1.3.3 Laser Irradiation of Metal Salts or Liquid Precursors

Laser irradiation of aqueous media of metal salts or liquid precursors is a laser-induced bottom-up approach that generates colloidal solution of NPs [237–243]. Here, metal slats are photochemically reduced to the neutral atoms M^0, which undergo clustering to form NPs. The following two methods are used for photochemical decomposition of metal salts or liquid precursors.

6.1.3.3.1 Photochemical Synthesis

Photochemical synthesis [237–241] involves generation of neutral metal atoms through photodissociation of metal salts in the solutions without any supporting substrate. This process can be done either by direct photoreduction [237–239] of metal salt/complex or by reduction of metal ions using the photochemically generated intermediates, such as excited molecules and radicals [240, 241]. The second method is termed as *photosensitization* in the synthesis of NPs and utilizes a photosensitive reagent that generates intermediate by photoreduction. Photosensitization is faster and more efficient as compared to direct photoreduction. The produced intermediate product reduces the metal ions to generate M^0. Photochemical synthesis is also a clean, quick, and simple process that enables (i) controlled synthesis of NPs and (ii) synthesis of NPs in any medium such as sol, emulsion, glass, surfactant micelles, polymers, and so on. Figure 6.1.40 illustrates schemes for direct photoreduction and photosensitized reduction for photochemical synthesis of NPs.

Table 6.1.3 Laser-induced melting/vaporization and fragmentation of liquid-suspended particles.

Liquid	Powder	Laser parameters	Product	Description	References
Water	60 ± 8 nm Au NPs	Ti:Sapphire (400 nm, 150 fs, 1 kHz/100 Hz) 150 W Xe lamp for *in situ* absorption. Pump-probe spectroscopy	Smaller-sized Au NPs	Electron temperature T_e, lattice temperature T_l, and media temperature T_m at the particle liquid interface for 60 nm diameter Au NPs irradiated by 400 nm wavelength, 150 fs laser pulses are simulated. Reshaping and fragmentation of NPs were observed. Coulombic explosion mechanism through thermionic emission is responsible for NP fragmentation	[185]
Water	55 nm Au NPs	Theoretical modeling and simulation (400 nm, 300 fs, 10 mJ cm^{-2})	Smaller-sized NPs	An improved working model for the size reduction of Au NPs through the processes of heat dissipation, bubble formation, and electron phonon dynamics is proposed. This model suggests that fragmentation is dominant process for femtosecond irradiation, while purely thermal evaporation occurs for nanosecond irradiation even at 355 nm wavelength excitation	[179]
Acetone/35 mg powder in 3 ml acetone	Dendritic shaped CuO powder	Nd:YAG laser (1064 nm, 5 ns, 10 Hz), 1 J cm^{-2}, unfocused beam, horizontal beam irradiation; continuous N$_2$ flow	Cu and Cu$_2$O NPs	Laser-induced decomposition of CuO powder into Cu NPs, followed by aerobic oxidation of these NPs into Cu$_2$O NPs with aging are reported. Effects of different ketonic solvents on the dispersivity and productivity of NPs were also investigated. Cu and Cu$_2$O showed comparatively higher stability in acetone as compared to other ketones	[184]

Distilled water	ZnO hollow spheres 30 nm size	Nd:YAG laser (1064 nm, 10 ns, 10 Hz), 50 mJ/pulse	ZnO solid NPs	Amorphous ZnO hollow spheres of 30 nm diameter converted into crystalline ultrafine ZnO QDs of 1–8 nm diameter with increased Zn/O ratio. Sizes of the QDs were 2.3 ± 0.5 nm and 3.1 ± 1.2 nm for the laser irradiation of 10 and 20 min, respectively	[183]
Distilled water	PbTe powder (20–100 μm)	Nd:YAG laser (1064 nm, 532 nm, 266 nm, 15–60 J cm^{-2} at focal point, time 1–5 min	PbTe NPs	Spherical PbTe NPs of 4–6 nm are produced by irradiation of micrometer-sized PbTe particles. Effects of wavelength, time duration, and laser irradiance and the properties of colloidal solution are studied	[186]
Water	Hollow Au NPs	CW Ti:Sapphire laser 780 nm; FS Ti:Sapphire (810 ns, 752 Hz), 2 mm beam for both; ND filters of 0.1, 0.3, and 0.5 were used for energy	Smaller-sized hollow Au NPs and solid Au NPs	Two processes were proposed for spectral hole burning (i) NIR-absorbing hollow Au NPs converts into smaller blue-absorbing hollow Au NPs and (ii) breakdown of hollow Au NPs into solid Au NPs. The branching ratio between these two processes is found to be dependent on the peak power of the femtosecond laser pulses with lower peak power favoring the first process, while higher peak power supporting the second process	[198]
Aqueous media of 1 g/l dextrin	Mixture of PLAL produced Ag and Au NPs	Ti:Sapphire laser (800 nm, 110 fs, 1 kHz), 0.3 J cm^{-2}; height of liquid 5 mm	Ag-Au alloy NPs	PLA-produced colloidal solution of Au and Ag NPs having 3.3 ± 1.5 and 4.4 ± 2.7 nm average sizes, respectively, were mixed together, with Au:Ag molar ratio varying between 0 and 1, and resulting solution was subjected to further laser irradiation. SOB and LOB were used for alloying. Oxidation resistance of alloy NPs increases with the increase of Au composition	[199]

(continued overleaf)

Table 6.1.3 (*continued*)

Liquid	Powder	Laser parameters	Product	Description	References
Acetone	Mixture of PLAL produced Au and Co colloidal NPs	Ti:Sapphire laser (800 nm, 120 fs, 1 kHz), 0.3 J cm^{-2}; height of liquid 5 mm, flow of N$_2$ during irradiation	Co-Au alloy or Co core/Au shell NPs	PLA-produced colloidal solution of Co and Au NPs were mixed and irradiated with femtosecond laser pulses with continuous flow of N$_2$ to reduce the oxidation. For Co composition >93%, Au clusters are distributed randomly in the NPs without any affect on the average Co magnetization with M_s much less than that of pure Co NPs. For Co fraction in the range of 65–93%, magnetization increases with the decrease of Co concentration, which is possible due to the formation of Au shell layer	[200]
Water	β-Se NPs with 69 nm average diameter	Nd:YAG laser (1064 nm, 10 ns, 10 Hz), 35 mJ/pulse, spot size = 0.5 mm	α-Se QDs	α-Se QDs with average size 10–2.7 nm were produced for pulsed laser irradiation of water-suspended β-Se NPs. The size of the produced Se QDs follows a second-order exponential decay function of irradiation time, while the rate of size reduction, da/dt, is directly dependent on the diameter, a, of the instantaneous QDs, very similar to the radioactive decay model. Surface defect density of the selenium QDs increases, while defect/electron trap level energy decreases with the time of laser irradiation	[152]

Water or H_2O_2 or N_2H_4	Si, SiO, SiOC powders (0.2 g powder/3 ml water)	Nd:YAG laser (1064 nm, 1.2 ms, 20 Hz), 12.3×10^6 W cm^{-2}; spot size 0.2 mm	Si and SiC NPs	Laser irradiation of SiO powder suspended in water produces Si spherical and irregular-shaped NPs. NPs produced with stirring during ablation produces more homogeneous and smaller particles. Irradiation of Si powder in 30% H_2O_2 also produces Si NPs with irregular shapes. Irradiation of SiO in N_2H_4 and SiOC produces Si NPs with amorphous phase and SiC NPs in the crystalline phase, respectively. Solid-phase separation method for target materials is proposed	[187]
Water	Triangular Ag NPs, 110 nm edge, 14 nm thick	Nd:YAG laser (355 nm, 6 ns, 10 Hz), 3 mJ/pulse, spot size 5 mm	Hexagonal and spherical Ag NPs	SERS activity of product NPs decreases with laser irradiance showing dependency of irradiance on shape transformation. Electrophoretic mobility decreases with irradiation time, indicating that the surface charge density of Ag NPs has been modified by laser pulses	[188]
Ethanol	h-GaN powder	Nd:YAG laser (532 nm, 10 ns, 10 Hz), 10^{10} W cm^{-2}	c-GaN NPs	Irradiation of GaN powder suspended in the ethanol induced transformation of phase from hexagonal to cubic. GaN nanocrystals are also produced during this hexagonal to cubic phase transition. Cathodoluminescence spectroscopy was used for product characterization	[189]

(continued overleaf)

Table 6.1.3 (*continued*)

Liquid	Powder	Laser parameters	Product	Description	References
Distilled water (1.5 mg/3 ml)	Au or Ag flakes thick 0.1–0.2 μm	Nd:YAG laser (1064 nm, 532 nm, 10 ns, 10 Hz), spot size 3–3.2 mm; 40–100 mJ/pulse; 1.3 J/cm² for 100 mJ]	Ag and Au NPs	Systematic investigations of generation of NPs from suspended flakes, and fragmentation of thus produced particles into smaller ones have been investigated with *in situ* extinction spectroscopy. There are two main findings: (i) significant spectral changes in SPR peak characteristic during laser irradiation with 532 and 1064 nm laser pulses owing to the synthesis of much smaller-sized NPs and (ii) laser-induced agglomeration of NPs leading to immediate flocculation	[190]
Water or DMSO solution	PLA-synthesized Au NPs	Nd:YAG laser (532 nm, 9 ns, 10 Hz), fluence 10–500 mJ cm⁻²; time 15 min	Au NPs of smaller sizes	Laser irradiation of PLA-synthesized Au NPs colloidal solutions were used for reduction in NPs sizes, while KCl and THF solutions are added to induce particle aggregation to get larger NPs. No change in size was observed for laser fluence in the range 0g 10–25 mJ/pulse, while size of NPs decreases quickly in the range of 10–100 mJ/pulse and comparatively slowly after that. Combination of addition of KCl and THF with the laser irradiation can provide desired NPs size	[191]

Ethanol	Cu flakes 5 μm size, 100 nm thick	Ti:Sapphire laser (780 nm, 215 fs, 1 kHz), beam waist ~4 μm, fluence 3.5×10^3 J cm^{-2}	Copper NWs and copper NSs	Copper flakes were converted into copper NWs after femtosecond laser irradiation for times in between 1 and 5 min. The diameter of NWs increase with the increase of laser irradiation. Diameter of NWs increases with the time from 10.2 to 54.2 nm min^{-1}. After 10 min of irradiation, all NWs were converted into copper NSs. Although a portion of surface of NWs was oxidized into Cu$_2$O, NSs were pure copper	[197]
Aqueous media of dextran	Colloidal solution of Au NPs	Ti:Sapphire laser (800 nm, 140 fs, 1 kHz), continuous stirring	Smaller-sized Au NSs	Supercontinuum radiation was produced through nonlinear interaction of femtosecond radiation with liquid media. This supercontinuum fragments large colloids and reduces agglomeration	[201]
Water (0.8 mg ITO in 40 ml water)	ITO NPs 30 nm average size; Sn:In = 1:9	Nd:YAG laser (355 nm, 7 ns), 20–150 mJ/pulse, stirring	Smaller-sized ITO NSs	Size of the NSs decrease from 30.6 to 6.0 nm with increase of fluence from 0 to 150 mJ/pulse. Zeta potential values increase from negative to positive with the increase of laser irradiance. Optical properties of colloidal solutions were investigated	[182]
25 ml water + 15 ml octane	Mixture of Au and Fe NPs	Nd:YAG laser (532 nm, 20 Hz), 65 mJ/pulse, beam diameter 3.5 mm, unfocused	Fe core/Au shell NPs	Chemically synthesized 0.067 g of Fe NPs and 0.30 g Au powder was added into a mixture of water and octane. CTAB (0.02 M) surfactant and 1-butanol (10 ml) as cosurfactant were used as for stabilization. Irradiation of this mixture using 532 nm melted Au NPs and deposited on the surface of Fe to produce core shell NSs	[235]

(continued overleaf)

Table 6.1.3 *(continued)*

Liquid	Powder	Laser parameters	Product	Description	References
Aqueous media of SDS (2×10^{-4} to 5×10^{-2} M)	Colloidal solution of 0.2 mM PLA-synthesized Au NPs	Nd:YAG laser (355 nm, 10 ns, 10 Hz), fluence 94.3 MW cm^{-2}, stirring; Xe lamp for transient absorption	Smaller-sized Au NPs	Pump-probe transient spectroscopy technique was used to study ablation dynamics in a solution of different concentration of SDS. Absorbance at the ablation point decreases with delay time, revealing that NPs diffuse from ablation point. Higher concentration of SDS generates larger surface charge, hence more coulombic explosion and fission, and hence smaller NPs. Liquid drop model is used for explaining the results	[195]
Deionized water (liter concern)	Zeolite μ-particles 0.06 g/l concentration	Nd:YAG laser (1064, 532, and 355 nm, 8 ns, 10 Hz), 1–50 mJ/pulse	Zeolite NPs	Zeolite LTA NPs were produced by laser irradiation of water-suspended Zeolite LTA μ-particles. Larger (>200 nm) NPs are crystalline and irregular in shape, while smaller (<50 nm) NPs are spherical in shape and comparatively amorphous, unlike original crystal structure. Shorter wavelengths have higher fragmentation efficiency because of larger absorption for zeolite crystals	[192]
Water	Au NPs of 38 nm diameter	Ti:Sapphire (400 nm, 100 fs, 986.2 Hz), 0–0.25 J cm^{-2}	Smaller-sized Au NPs	A new laser-induced resizing/reshaping mechanism different from photothermal melting/vaporization and Coulombic explosion is explored. The proposed mechanism is based on the enhancement of laser electric field near curved surfaces of particles. This approach of material modification can take place below the particle's melting point	[203]

Liquid	Target	Laser conditions	Product	Remarks	Ref.
Water	Au flakes (2–4 mg/6 ml)	Nd:YAG laser (532 nm, 5 ns, 10 Hz), 35 mJ/pulse, ~1 J/cm²	<10 nm Au NPs	Pulsed laser irradiation in combination with time-dependent extinction spectra for the transformation of Au flakes to subnanometer-sized particles via micrometer-sized intermediate stage and diagnosis of mechanism. Surface of Au NPs are negatively charged, which makes them stable without any surfactant	[181]
Ionic solvent BMI PF6	Pd and Rh aggregates	Nd:YAG laser (532 nm, 8 ns, 20 Hz), 200 mJ cm⁻², time 120 min	Spherical Pd and Rh NPs	Pd (average 12 nm) and Rh (average 15 nm) NPs produced by hydrogen reduction in BMI PF6 ionic liquid were large aggregates before laser irradiation. However, Pd (4.2 ± 0.8 nm) and Rh (7.2 ± 1.3 nm) NPs of regular shapes were produced after laser irradiation	[193]
Water	Ag nanoprisms	Ti:Sapphire (800 nm, 120 fs, 1 kHz, 400 mW), time 0–10 min	Ag NPs ~10 nm (for 10 min irradiation)	After laser irradiation even for 1 min the SPR peak at 720 nm corresponding to the main resonance of Ag nanoprism is strongly damped with progressive increase of characteristic SPR of spherical Ag NPs at 405 nm. Size of spherical Ag NPs and their abundance increases with increase of time	[202]
0.05 M aqueous solution of SDS	Ag (0.5–1 μm) and Au (0.8–1.5 μm) powder	Nd:YAG laser (532 nm, 30 Hz), 590 mJ cm⁻², t: 30 min, unfocused	Ag-Au alloy NPs	Plasmon peak shifts toward red with the increase of Au fraction from 0 to 100% for pure Ag, Au-Ag alloy, and pure Au NPs. Mean size and SD for Au-Ag alloy for 1:1 composition was 5.4 and 2.1 nm, respectively.	[180]

(continued overleaf)

Table 6.1.3 *(continued)*

Liquid	Powder	Laser parameters	Product	Description	References
Aqueous medium of SDS or SDBS, or SOS	Au NPs of ~20 nm size by PLA in water	Nd:YAG laser (532 nm, 10 Hz), spot size 0.023 cm^2	Au nanonetwork	SPR peaks of Au colloidal solution shifts toward blue with the evolution of new IR peak in the range of 700–1000 nm, which shows decrease in the diameter of NPs and formation of linear structure. Shift of IR peaks toward longer wavelength side with number of pulses illustrates increase of network length. Au NPs to Au nanonetwork transformation is found more suitable for lower SDS concentrations and higher laser fluence	[223]
Water	TiO$_2$ powder 5 μm diameter TiO$_2$ NPs ~2 nm particles	Excimer laser (308 nm, 15 ns, 5 Hz), spot size: 27 mm^2 for tight focused	TiO$_2$ NPs with 10 nm average size and narrow distribution	There are two types of raw particles: (i) commercially available TiO$_2$ powder in the size range of 170–1050 nm and (ii) chemically synthesized NPs of 2 nm average size. UV (308 nm) laser irradiation of both the samples turns to the TiO$_2$ NPs of 10 nm diameters with narrow size distribution	[236]
Aqueous media	13.7 nm Au, and 16.8 nm Ag NPs solution	Nd:YAG laser (532 nm), 2.45 × 10^{12} mJ cm^{-2}	Au-Ag alloy	Colloidal solution of chemically synthesized Au and Ag NPs were mixed in the molar ratio of 1 : 2, 1 : 1, and 2 : 1. Laser irradiation of these mixtures causes increase of SPR in between the SPRs of Au and Ag NPs with time of irradiation and disappearance of SPR peaks of constituents. Irradiation of mixture with molar ratios of 1 : 1 produces network of Au/Ag alloy NRs, while ablation for 25 min fragments these networks into very small and monodispersed alloy NPs	[221]

CW, continuous wave; SOB, strong optical breakdown; LOB, low optical breakdown.

Examples for direct photoreduction of $AgClO_4$ in aqueous and alcoholic liquid media for the synthesis of Ag NPs are given as follows:

$$Ag^+ + H_2O \xrightarrow{h\nu} Ag^0 + H^+ + OH^\bullet \qquad (1)$$

$$nAg0 \longrightarrow (Ag^0)_n \qquad (2)$$

Aqueous medium

$$Ag^+ + RCH_2OH \xrightarrow{h\nu} Ag^0 + H^+ + RC^\bullet HOH \qquad (1)$$

$$Ag^+ + RC^\bullet HOH \xrightarrow{h\nu} Ag^0 + H^+ + RCHO \qquad (2)$$

Alcoholic medium

$$nAg^0 \longrightarrow (Ag^0)_n \qquad (3)$$

$$(6.1.61)$$

In the following example, dimethyl ketone acts as a photosensitizer for the reduction of M^+ ions into M^0 nonvalent atoms for the synthesis of $(M^0)_n$ NPs. Here, an intermediate radical $(CH3)_2 C^\bullet OH$ reduces metal ions into neutral atom.

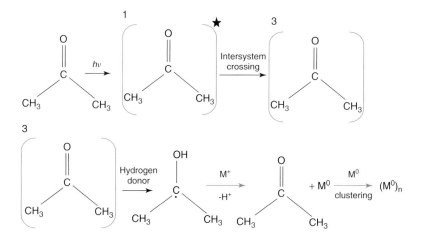

6.1.3.3.2 Photocatalytic Deposition

Reduction of metal ions M^+ adsorbed on the surface of semiconductor substrate has the capability of synthesis of metal semiconductor composite or semiconductor core/metal shell NPs [245–247]. Electron and hole pairs are created via excitation of semiconductor substrate with the photons having energy higher than the band gap energy of substrate. The electrons thus produced reduce surface-adsorbed metal ions at the semiconductor–liquid interface. A hole scavenger is added into the solution, which captures holes from it. Titanium dioxide is one of the best substrate for photocatalytic decomposition of noble metal ions on its surface to synthesize TiO_2 core/Au shell, TiO_2 core/Ag shell, or TiO_2 core/Cu shell NPs [242, 243, 248, 249]. A similar approach can be used for the synthesis of ZnO core/metal shell NPs systems [250].

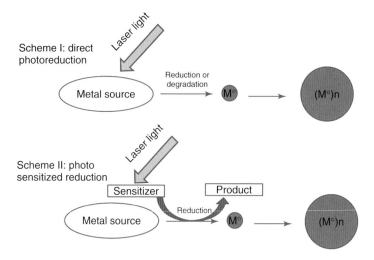

Figure 6.1.40 Schemes for direct photoreduction and photosensitized reduction of solution of metal salts or liquid precursors of metal source. (Source: Reprinted with the permission from Sakamoto *et al.* [244] copyright @ Elsevier.)

6.1.4
Applications of Nanomaterials Produced by Liquid–Phase Pulsed Laser Ablation/Irradiation

Owing to the inherent properties of LP-PLA in the generation of chemically pure and highly stable colloidal sols of any organic/inorganic targets in the wide variety of liquids with the combination of *in situ* functionalization of surface, particles produced by this method have applications as catalyst, as SERS-active substrates, as drug delivery agent, as fluorescent markers, in the photoablation therapy (PAT) of cancerous cells, as nanofertilizers to stimulate germination of seeds and growth of plants, and so on. Laser-induced melting/vaporization and fragmentation of liquid-suspended NPs has the capability of resizing, reshaping, phase transformation, and generation of more stable colloids.

6.1.4.1 Applications in PV Solar Cells

Svrcek *et al.* [256] carried out nanosecond laser irradiation of water- or methanol-suspended electrochemically synthesized silicon NPs for surface modification and applied these processed particles in the fabrication of silicon photovoltaic (PV)solar cells. They fabricated three types of laser-produced silicon-NP-based solar cells. In the first case, they used anatase TiO_2 nanotubes as host template for laser-generated Si-nc for the fabrication of inorganic sensitized solar cell architectures. In this type of architectures, Si-nc produced by the irradiation of Si particles in ethanol showed external quantum efficiency (EQE) several orders higher than that of those produced in water. Difference in the EQE is more significant in the shorter (>3.0 eV) wavelength of solar spectrum.

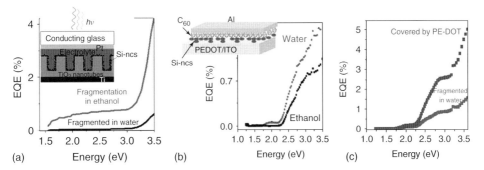

Figure 6.1.41 External quantum efficiency (EQE) of PLA-produced silicon nanocrystal. (a) Anatase TiO₂-nanotube-based inorganic Si-nanocrystal- sensitized silicon solar, (b) fullerenes-interface-based inorganic/organic hybrid solar cell, and (c) functionalizing PLA-produced Si nanocrystals of the cell (b) with PEDOT. (Source: Reprinted with the permission from Svrcek *et al.* [244] copyright @ American Chemical Society.)

In the second solar cell architecture, they made Si-nc – fullerenes photoactive interface for the fabrication of hybrid inorganic/organic solar cells. In the third architecture, they improved the efficiency of hybrid inorganic/organic solar cell by functionalizing Si-nc surface with water-soluble poly (3,4-ethylenedioxythiophene), that is, PEDOT. In the second type of PV architectures, that is, Si-nc/fullerenes hybrid solar cells, unlike TiO₂-Si-nc-based cells, Si-nc produced by laser irradiation in water shows higher EQE as compared to those produced in ethanol and this difference increases in the shorter wavelength of solar spectrum. The EQE of Si-nc/fullerenes cell further increased when Si-nc produced by irradiation of Si in water was functionalized with PEDOT. Figure 6.1.41 presents the EQE of the three types of PV cells fabricated from laser-ablation-synthesized Si nanocrystals.

6.1.4.2 *In situ* Functionalization for Biological Applications

LP-PLA represents a powerful tool for the generation of pure NPs, avoiding chemical precursors, reducing agents, and stabilizing ligands. The bare surface of the charged NPs makes them highly available for functionalization and as a result, especially interesting for biomedical applications. It possesses the capability of *in situ* surface functionalization for the generation of bioconjugated NPs of noble metals and semi-conductors for various medical and biological applications. Addition of surfactant molecules, polymers, or any other biomolecules in the ablation medium during PLA increases the colloidal stability of NPs with their in situ surface functionalization. Salmaso *et al.* [22] decorated the surface of LP-PLA-produced Au NPs with thermore-sponsive thiol terminated poly-N-isopropylacrylamide-co-acrylamide co-polymer to fabricate temperature-tunable colloidal system switchable for enhanced cellular uptake. In the cell culture studies, they reported that the polymer-decorated Au NPs located in human breast adenocarcinoma MCF7 cells treated at 40 °C (12 000 Au NP/cell) have more than 80-fold greater uptake as compared to cells treated at 34 °C (140 Au NPs/cell). Figure 6.1.42 illustrates that about 6400 and 6100 naked Au NPs were taken up per cell after incubation at 34 °C and 40 °C, respectively.

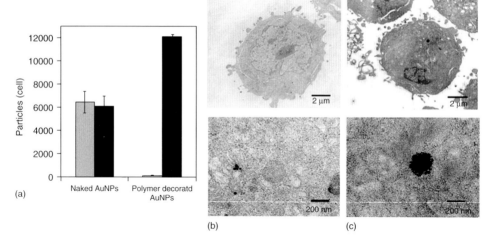

(a)

(b)   (c)

Figure 6.1.42 Cell uptake of Au and polymer-conjugated Au NP. The uptake of particles by MCF7 cell lines is shown in (a) with light gray and black bars expressing numbers of Au NP per cell on samples incubated at 34 and 40 °C respectively. TEM images depict MCF7 cells following incubation with naked Au NP (b) and polymer-decorated Au NP (c) at 40 °C. (Source: Reprinted with the permission from Salmaso *et al.* [22], *J. Mater. Chem.*, 2009, **19**, 1608–1615 copyright @ Royal Society of Chemistry.)

However, in the case of functionalized NPs, cell uptake was negligible at 34 °C (140 NPs per cell), while the cell uptake was very high (12 000 particles per cell) when incubation was done at 40 °C.

Hahn *et al.* [23] synthesized silver and copper nanomaterials using LP-PLA of corresponding targets in polymer-doped organic liquids to produce customized drug release systems. A strategy was reported in order to determine the therapeutic window for cells relevant for cochlear implant electrodes, defined by the viability of L929 fibroblasts, PC12 neuronal cells, and spiral ganglion cells on different concentrations of silver and copper ions. They found that hexane doped with 1% silicone resin is a suitable liquid medium for the synthesis of a nanocomposite with a constant ion release rate. Copper ions with 100 µmol/l or silver ions with 10 µmol/l cause inhibition of tissue growth without affecting growth of neural cell. Petersen *et al.* [9, 10] ablated Au foil in the aqueous media of penetratin for the *in situ* surface functionalization/bioconjugation of gold NPs. Net charge on the penetratin and zeta potential of Au NPs were found to be dependent on the pH of aqueous media. Conjugation efficiency increases with the increase in the pH value of the solution. With the addition of penetratin, size and distribution of primary Au NPs decreases, while that of secondary particles/aggregates of larger size increases. They studied penetration efficiency of bare Au NPs and Au-penetrtin bioconjugate NPs which were produced by *in situ* functionalization during PLA in 5 µM penetratin solution on bovine endothelial cells. They observed the presence of Au NPs in up to 100% of coincubated cells if the particles were penetratin-conjugated, whereas ligand-free Au NPs were found in approximately half of the cells. Stelzig *et al.* [21]

reported production of silver and copper particles by laser ablation in an organic solvent and their *in situ* functionalization with amphiphilic copolymers bearing fluorinated side chains. Beside the stabilization of NPs, the fluorinated side chains make the modified particles compatible with a perfluorinated matrix, which results in a homogeneous distribution of the particles in the matrix.

6.1.4.3 Semiconductor NPs as Fluorescent Markers

High temperature, high pressure, and high density of the LP-PLA-produced plasma under strong liquid confinement causes fast quenching of the ablated species into particles. Reactions between plasma species and liquid at plasma–liquid interface and among the plasma species inside the plasma plume take place under highly nonequilibrium conditions, which creates high density of point defects in the particles. These point defects in the wide band gap semiconductor NPs such as ZnO, TiO_2, and so on fascinate them with the ability of visible luminescence. Therefore LP-PLA-produced semiconductor NPs exhibit unique luminescent properties that are not observable for NPs produced by conventional chemical methods [113, 114, 118, 122, 152]. These defect levels can act as an intermediate energy state for the two-photon absorption, that is, upconversion process for the generation of higher energy photons [113]. The highly luminescent semiconductor NPs that are so produced can be used as fluorescent markers for cell staining and its detection in fluorescence microscopy. Figure 6.1.43 illustrates photoluminescence spectra of PLA-synthesized $Cd(OH)_2$ nanoarchitectures [122] in the aqueous media containing 20 and 50 mM SDS and laser-induced melting/vaporization-generated selenium QDs [152] in distilled water. Highly luminescent NPs of semiconductor and rare-earth-doped oxides are synthesized by LP-PLA of their corresponding target materials or irradiation of liquid-suspended powders. Gong *et al.* [251] produced CdS NPs under different focusing conditions using femtosecond irradiation of CdS target in water. Amans *et al.* [252] produced luminescent NPs of doped oxides such as $Eu^{+3} : Y_2O_3$, $Eu^{+3}:Gd_2O_3$, and $Ce^{+3}:Y_3Al_5O_{12}$ using PLA of corresponding target into an aqueous solution of 2-[2-(2-methoxyethoxy)ethoxy]acetic acid (MEEAA).

6.1.4.4 Surface-Enhanced Raman Scattering (SERS) Active Substrates

Detection of trace and biohazards is of potential interest for the pollution monitoring, waste water treatment, homeland security, and so on. Raman spectroscopy has the potential to detect variety of organic and biomolecules. Electromagnetic enhancement of Raman signal known as *Surface-Enhanced Raman Scattering* of an analyte adsorbed on the metal surface has the potential for detection up to the single molecule level. Since noble metal NPs produced with PLA in liquids has surfaces free from any undesired chemical contaminants, they have higher efficiency for capturing/adsorbing analyte molecules as compared to their chemically produced counterparts [153].

Muniz-Miranda *et al.* [253] used PLA in water produced colloidal solution of copper NPs as a SERS substrate for the observation of phen and bipy analytes. Šišková *et al.* [70] reported PLA of Ag target in the aqueous media of some adsorbing

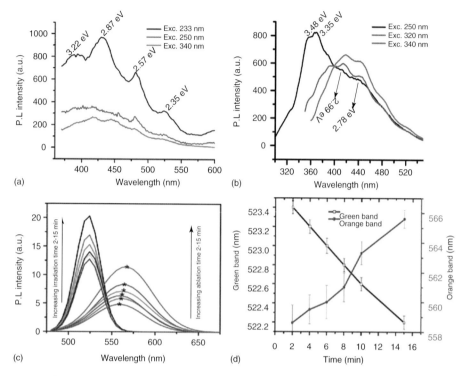

Figure 6.1.43 Photoluminescence spectra from PLA-synthesized Cd(OH)$_2$ Nanoarchitectures in aqueous media of (a) 20 mM and (b) 50 mM SDS. (c) Photoluminescence spectra from Se QDs produced by laser-induced melting/vaporization of water-suspended Se NPs for different times of irradiation and (d) variation in the position of green and orange bands of (c) with time. (Source: (a,b) Reprinted with the permission from Singh and Gopal [122] copyright @ American Chemical Society. (c,d) Reprinted with the permission from Singh *et al.* [152] copyright @ American Chemical Society.)

ions, which resulted in the formation of chemically modified Ag NPs. SERS activity of the produced Ag NPs depends on the time of ablation and laser wavelength used for ablation. The SERS activity of NPs produced by PLA using 1064 nm laser beam was higher than that of those produced by 532 nm beam. SERS activity decreases with the decrease in the size of particle as a result of intermittent ablation (Figure 6.1.44). SERS activity of 30 min aged Ag NPs was higher than that of fresh particles. These observations show that larger particles are more SERS sensitive.

6.1.4.5 Nanofertilizer for Seed Germination and Growth Stimulation

Use of nanomaterials in the area of plant science is relatively new, but highly emerging with the expectation that these small particles might increase productivity of crops to fulfill the future demands of the unexpectedly growing population of Asian countries and might have the capability to provide fresher foods for European countries. Nanoscale materials also have the ability to increase oxygen

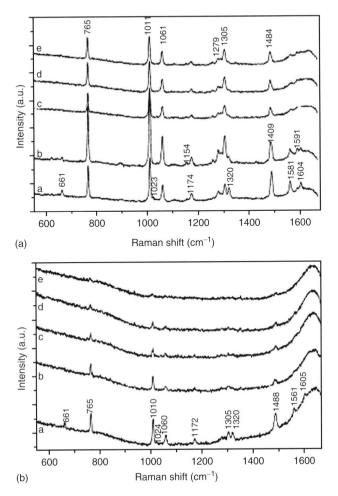

Figure 6.1.44 SERS spectra of Ag NP/bpy systems throughout the stages ($a = 2$ min, $b = 2 + 3$ min, $c = 2 + 3 + 3$ min, $d = 2 + 3 + 3 + 5$ min, $e = 2 + 3 + 3 + 5 + 7$ min) of intermittent LA/NF (laser ablation/ nanoparticle fragmentation) performed at (a) 1064 nm and (b) 532 nm.

evolution capacity of plants to buildup better biosphere. Singh *et al.* [254] used PLA-synthesized anatase TiO_2 NPs as a nanofertilizer for the stimulation of seed germination and plant growth in *Brassica oleracea var. capitata*. It is observed that certain concentration of colloidal solution of anatase TiO_2 NPs stimulates seed germination and growth of *B. oleracea* (Figure 6.1.45). Radical length in the germinated seeds, physical parameters (number of leaves, leaf area, height, etc.) of plants, and the amount of chlorophyll a, chlorophyll b, and carotenoid increased with the increase of anatase TiO_2 NPs concentrations; attained their maximum values near 1.5 mM, and started decreasing at higher (>2.0 mM) concentrations. Amount of protein is higher for lower and higher concentrations (maximum at 0.5

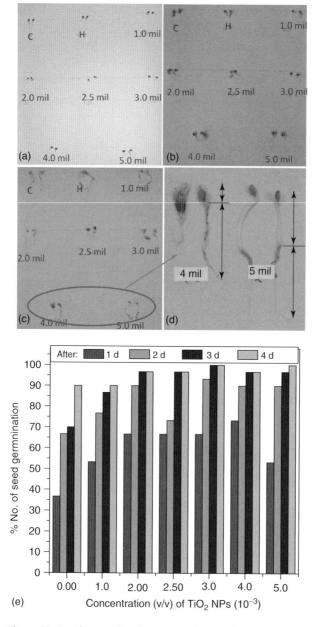

Figure 6.1.45 Photographs of TiO$_2$-treated seed germination at (a) 3 days, (b) 4 days, and (c) 5 days of sowing, and (d) magnified image of 4 mM (mil), and 5 mM TiO$_2$-treated *Brassica oleracea* var. *capitata* seeds after 5 days of sowing. (e) Histogram of percentage seed germinated after 1-4 days of sowing for different concentrations of TiO$_2$ NPs. Here number of mM (mil) is the volume (ml) of colloidal solution of TiO$_2$ in 1 l of Hogland solution.

and 3 mM), while minimum at intermediate concentration. Superoxide dismutase (SOD) activity increases with concentration upto 2.5 mM, and decreases after this, while peroxidase (POX) activity attains maximum at 0.5 mM. Catalase (CAT) activity is higher for higher (>2.5 mM) concentrations. Summarizing these, lower amount (<2.5 mM) is beneficial for the growth, germination, and biochemical parameters and overall health of *Brassica oleracea* var. *capitata*. High colloidal stability of laser-synthesized anatase TiO_2 NPs facilitated their dispersity and availability in the plant growth media for longer time.

6.1.4.6 Other Applications

Capability of producing highly stable colloidal solution of NPs is one of the best advantages of LP-PLA, and it has applications in chemical industry as catalyst, for photodegradation of organic molecules for waste water treatment, as nanofluid or ferrofluid, for injection in the cells either directly or impulsively using laser pulses. One of the problems usually faced in catalytic applications of transition metal NPs is their instability toward aggregation and formation of larger particles, which changes the catalytic properties or can even lead to inactive materials. Higher colloidal stability of PLA-produced transition metals solved issues due to the aggregation of NPs. Any ionic liquid, such as BMI.PF6, can be used as a liquid medium for the generation of stable catalytic NPs using LP-PLA. Gelesky *et al.* [224] performed laser-induced fragmentation of Rh and Pd particles for the generation of stable colloidal solution of smaller NPs. Since bulk powder of any material suitable for fluorescent, photoconductive, or energy harvesting batteries [151] or for bone implant [255] can be easily converted into their NPs or QDs using laser ablation of their pellets or laser-induced melting/fragmentation of liquid-suspended powder, this technique has the potential to generate NPs for a wide range of applications.

6.1.5
Conclusion and Future Prospects

A detailed review of the experimental and theoretical investigation of PLA of solid targets in liquids and laser-induced melting/vaporization or fragmentation of liquid-suspended particles for the generation of NPs has been presented. Laser-induced photochemical reduction of metal salts/a liquid precursor has been also discussed in brief. Since LP-PLA can generate NPs from elemental, binary, ternary, as well as complex multielemental targets or powders in desired chemical compositions for their applications in various field of science and technology, this method of nanomaterial processing can be used as an ecofriendly, simple, and quick process of nanomaterial sampling in the future.

Acknowledgments

Dr. Singh is thankful to the Irish Research Centre for Science, Engineering and Technology (IRCSET) for providing EMPOWER postdoctoral fellowship for financial assistance during compilation of this chapter.

References

1. Maiman, T.H. (1960) *Nature*, **187**, 493.
2. Harilal, S.S., Bindhu, C.V., Tillack, M.S., Najmabadi, F., and Gaeris, A.C. (2003) *J. Appl. Phys.*, **93**, 2380–2388.
3. Perriere, J., Boulmer-Leborgne, C., Benzerga, R., and Tricot, S. (2007) *J. Phys. D: Appl. Phys.*, **40**, 7069–7076.
4. Leitner, A., Rogers, C.T., Price, J.C., Rudman, D.A., and Herman, D.R. (1998) *Appl. Phys. Lett.*, **72**, 3065–3067.
5. Huang, C.N., Chen, S.Y., and Shen, P. (2007) *J. Phys. Chem. C*, **111**, 3322–3327.
6. Zbroniec, L., Sasaki, T., and Koshizaki, N. (2004) *Appl. Phys. A*, **79**, 1783–1787.
7. Pa'szti, Z., Peto, G., Horvath, Z.E., and Karacs, A. (1997) *J. Phys. Chem. B*, **101**, 2109–2115.
8. Yudasaka, M., Kokai, F., Takahashi, K., Yamada, R., Sensui, N., Ichihashi, T., and Iijima, S. (1999) *J. Phys. Chem. B*, **103**, 3576–3581.
9. Petersen, S., Barchanski, A., Taylor, U., Klein, S., Rath, D., and Barcikowski, S. (2011) *J. Phys. Chem. C*, **115**, 5152–5159.
10. Petersen, S. and Barcikowski, S. (2009) *J. Phys. Chem. C*, **113**, 19830–19835.
11. Chen, T.-C., Su, W.-K., and Lin, Y.-L. (2004) *Jpn. J. Appl. Phys.*, **43**, L119–L122.
12. Mafuné, F., Kohno, J., Takeda, Y., Kondow, T., and Sawabe, H. (2000) *J. Phys. Chem. B*, **104**, 8333–8337.
13. Mafuné, F., Kohno, J., Takeda, Y., Kondow, T., and Sawabe, H. (2000) *J. Phys. Chem. B*, **104**, 9111–9117.
14. (a) Mafuné, F., Kohno, J., Takeda, Y., and Kondow, T. (2002) *J. Phys. Chem. B*, **106**, 7575–7577 (b) Mafuné, F., Kohno, J., Takeda, Y., and Kondow, T. (2002) *J. Phys. Chem. B*, **106**, 8555–8561.
15. Mafuné, F., Kohno, J., Takeda, Y., Kondow, T., and Sawabe, H. (2001) *J. Phys. Chem. B*, **105**, 5114–5120.
16. Mafuné, F., Kohno, J., Takeda, Y., and Kondow, T. (2001) *J. Phys. Chem. B*, **105**, 9050–9056.
17. Yan, Z., Bao, R., and Chrisey, D.B. (2010) *Nanotechnology*, **21**, 145609.
18. Liu, Z., Yuan, Y., Khan, S., Abdolvand, A., Whitehead, D., Schmidt, M., and Li, L. (2009) *J. Micromech. Microeng.*, **19**, 054008.
19. Liang, C.H., Shimizu, Y., Sasaki, T., and Koshizaki, N. (2005) *Appl. Phys. A*, **80**, 819–822.
20. Yan, Z., Bao, R., and Chrisey, D.B. (2010) *Chem. Phys. Lett.*, **497**, 205–207.
21. Stelzig, S.H., Menneking, C., Hoffmann, M.S., Eisele, K., Barcikowski, S., Klapper, M., and Müllen, K. (2011) *Eur. Polym. J*, **47**, 662–667.
22. Salmaso, S., Caliceti, P., Amendola, V., Meneghetti, M., Magnusson, J.P., Pasparakis, G., and Alexander, C. (2009) *J. Mater. Chem.*, **19**, 1608–1615.
23. Hahn, A., Stover, T., Paasche, G., Lobler, M., Sternberg, K., Rohm, H., and Barcikowski, S. (2010) *Adv. Eng. Mater.*, **12**, B156–B162.
24. Muniz-Miranda, M., Gellini, C., and Giorgetti, E. (2011) *J. Phys. Chem. C*, **115** (12), 5021–5027.
25. Messina, E., Cavallaro, E., Cacciola, A., Saija, R., Borghese, F., Denti, P., Fazio, B., D'Andrea, C., Gucciardi, P.G., Iati, M.A., Meneghetti, M., Compagnini, G., Amendola, V., and Marago, O.M. (2011) *J. Phys. Chem.*, **115**, 5115–5122.
26. Patil, P.P., Phase, D.M., Kulkarni, S.A., Ghaisas, S.V., Kulkarni, S.K., Kanrtkar, S.M. *et al.* (1987) *Phys. Rev. Lett.*, **58**, 238.
27. Ogale, S.B. (1988) *Thin Solid Films*, **163**, 215.
28. Ogale, S.B., Malshe, A.P., Kanetkar, S.M., and Kshirsagar, S.T. (1992) *Solid State Commun.*, **84**, 371.
29. Devaux, D., Fabbro, R., Tollier, L., and Bartnicki, E. (1993) *J. Appl. Phys.*, **74**, 2268.
30. Peyre, P. and Fabbro, R. (1995) *Opt. Quantum Electron.*, **27**, 1213.
31. Berthe, L., Fabbro, R., Peyer, P., Tollier, L., and Bartnicki, E. (1997) *J. Appl. Phys.*, **82**, 2826.

32. Peyer, P., Scherpereel, X., and Fabbro, R. (1998) *J. Mater. Sci.*, **33**, 1421.

33. Berthe, L., Fabbro, R., Peyer, P., and Bartnicki, E. (1999) *J. Appl. Phys.*, **85**, 7552.

34. Berthe, L., Sollier, A., Fabbro, R., Peyer, P., and Bartnicki, E. (2000) *J. Phys. D: Appl. Phys.*, **33**, 2142.

35. Peyer, P., Berthe, L., Fabbro, R., and Sollier, A. (2000) *J. Phys. D: Appl. Phys.*, **30**, 498.

36. Sollier, A., Berthe, L., and Fabbro, R. (2001) *Eur. Phys. J. Appl. Phys.*, **16**, 131.

37. Fairand, B.P. and Clauer, A.H. (1979) *J. Appl. Phys.*, **50**, 14971502.

38. Griffin, R.D., Justus, B.L., Campillo, A.J., and Goldberg, L.S. (1986) *J. Appl. Phys.*, **59**(6), 1968–1971.

39. Fabbro, R., Fournier, J., Ballard, P., Devaux, D., and Virmont, J. (1990) *J. Appl. Phys.*, **68**, 775–784.

40. Sollier, A., Berthe, L., and Fabbro, R. (2001) *Eur. Phys. J. Appl. Phys.*, **16**, 131–139.

41. Colvin, J.D., Ault, E.R., King, W.E., and Zimmerman, I.H. (2003) *Phys. Plasmas*, **10**, 2940–2947.

42. Zhang, W., Yao, Y.L., and Noyan, I.C. (2004) *J. Manuf. Sci. Eng.*, **126**, 10–17.

43. Wu, B. and Shin, Y.C. (2005) *J. Appl. Phys.*, **97**, 113517.

44. (a) Niu, K.Y., Yang, J., Kulinich, S.A., Sun, J., Li, H., and Du, X.W. (2010) *J. Am. Chem. Soc.*, **132**, 9814–9819 (b) Niu, K.Y., Yang, J., Kulinich, S.A., Sun, J., and Du, X.W. (2010) *Langmuir*, **26**, 16652–16657.

45. Berthe, L., Fabbro, R., Peyre, P., and Bartnicki, E. (1999) *J. Appl. Phys.*, **85**, 7552–7555.

46. Sakka, T., Iwanaga, S., and Ogata, Y.H. (2000) *J. Chem. Phys.*, **112**, 8645–8653.

47. Sakka, T., Takatani, K., Ogata, Y.H., and Mabuchi, M. (2002) *J. Phys. D: Appl. Phys.*, **35**, 65–73.

48. Takada, N., Nakano, T., and Sasaki, K. (2009) *Appl. Surf. Sci.*, **255**, 9572.

49. Zhang, W. and Yao, Y.L. (2000) *Proc. ICALEO'00, Laser Mater. Proces.*, **89**, E183–E192.

50. Lu, Q., Jeong, S.H., Greif, R., and Russo, R.E. (2002) *Appl. Phys. Lett.*, **80**, 3072.

51. Etina, T.E. (2011) *J. Phys. Chem. C*, **115**, 5044–5048.

52. Leveugle, E. and Zhigilei, L.V. (2004) *Appl. Phys. A*, **79**, 753.

53. Povarnitsyn, M.E., Itina, T.E., Sentis, M., Khishenko, K.V., and Levashov, P.R. (2007) *Phys. Rev. B*, **75**, 235414.

54. Zhu, S., Lu, Y.F., Hong, M.H., and Chen, X.Y. (2001) *J. Appl. Phys.*, **89**, 2400–2403.

55. Isselin, J.C., Alloncle, A.P., and Autric, M. (1998) *J. Appl. Phys.*, **84**, 5766.

56. Zhu, S., Lu, Y.F., and Hong, M.H. (2001) *Appl. Phys. Lett.*, **79**, 1396–1398.

57. Leung, W.P. and Tam, A.C. (1992) *Appl. Phys. Lett.*, **60**, 23.

58. Barsch, N., Jakobi, J., Weiler, S., and Barcikowski, S. (2009) *Nanotechnology*, **20**, 445603–445611.

59. Lallemand, P. and Bloembergen, N. (1965) *Phys. Rev. Lett.*, **15**, 1010–1012.

60. Tsuji, T., Iryo, K., Nishimura, Y., and Tsuji, M. (2001) *J. Photochem. Photobiol. A: Chem.*, **145**, 201–207.

61. Tsuji, T., Iryo, K., Ohta, H., and Nishimura, Y. (2000) *Jpn. J. Appl. Phys.*, **39**, L981–L983.

62. Sakka, T., Masai, S., and Fukami, K. (2009) *Spectrochim. Acta B*, **64**, 981–985.

63. Ogata, Y.H., Hong, M.H., Ng, K.Y., Xie, Q., Shi, L.P., and Chong, T.C. (2008) *Appl. Phys. A*, **93**, 153.

64. Soliman, W., Takada, N., and Sasaki, K. (2010) *Appl. Phys. Express*, **3**, 035201–035203.

65. Yan, Z., Bao, R., Wright, R.N., and Chrisey, D.B. (2010) *Appl. Phys. Lett.*, **97**, 124106–124108.

66. Brenner, M.P., Hilgenfeldt, S., and Lohse, D. (2002) *Rev. Mod. Phys.*, **74**, 425.

67. Yan, Z., Bao, R., Huang, Y., and Chrisey, D.B. (2010) *J. Phys. Chem. C*, **114**, 11370–11374.

68. Kruusing, A. (2008) *Handbook of Liquid-Assisted Laser Processing*, Elsevier, ISBN-13:978-0-08-044498-7.

69. Yang, G. (2007) *Prog. Mater. Sci.*, **52**, 648–698.

70. Šišková, K., Vlckova, B., Turpin, P.-Y., Thorel, A., and Prochazka, M. (2011) *J. Phys. Chem. C*, **115**, 5404.

71. Zamiri, R., Azmi, B.Z., Sadrolhosseini, A.R., Ahangar, H.A., Zaidan, A.W., and Mahdi, M.A. (2011) *Int. J. Nanomed.*, **6**, 71–75.

72. Yang, S., Cai, W., Liu, G., and Zeng, H. (2009) *J. Phys. Chem. C*, **113**, 7692–7696.

73. Siskova, K., Vlckova, B., Turpin, P.-Y., and Fayet, C. (2008) *J. Phys. Chem. C*, **112**, 4435–4443.

74. Phuoc, T.X., Soong, Y., and Chyu, M.K. (2007) *Opt. Lasers Eng.*, **45**, 1099–1106.

75. Amendola, V., Polizzi, S., and Meneghetti, M. (2007) *Langmuir*, **23**, 6766–6770.

76. Tsuji, T., Okazaki, Y., Higuchi, T., and Tsuji, M. (2006) *J. Photochem. Photobiol. A*, **183**, 297–303.

77. Tarasenko, N.V., Butsen, A.V., and Nevar, E.A. (2005) *Appl. Surf. Sci.*, **247**, 418–422.

78. Burakov, V.S., Tarasenko, N.V., Butsen, A.V., Rozantsev, V.A., and Nedel'ko, M.I. (2005) *Eur. Phys. J. Appl. Phys.*, **30**, 107–112.

79. Takeshi, T. and Masaharu, T. (2005) *Rev. Laser Eng.*, **33**, 36–40.

80. Pyatenko, A., Shimokawa, K., Yamaguchi, M., Nishimura, O., and Suzuki, M. (2004) *Appl. Phys. A*, **79**, 803–806.

81. (a) Ganeev, R.A., Baba, M., Ryasnyansky, A.I., Suzuki, M., and Kuroda, H. (2004) *Opt. Commun.*, **240**, 437–448 (b) Ganeev, R.A. (2005) *J. Opt. A*, **7**, 717–733.

82. Shafeev, G.A., Freysz, E., and Verduraz, F.B. (2004) *Appl. Phys. A*, **78**, 307–309.

83. Tsuji, T., Kakita, T., and Tsuji, M. (2003) *Appl. Surf. Sci.*, **206**, 314–320.

84. Tsuji, T., Iryo, K., Watanabe, N., and Tsuji, M. (2002) *Appl. Surf. Sci.*, **202**, 80–85.

85. Bae, C.H., Nam, S.H., and Park, S.M. (2002) *Appl. Surf. Sci.*, **197198**, 628–634.

86. Brause, R., Möltgen, H., and Kleinermanns, K. (2002) *Appl. Phys. B*, **75**, 711–716.

87. Jeon, J.-S. and Yeh, C.-S. (1998) *J. Chinese Chem. Soc.*, **45**, 721–726.

88. Srnová, I., Procházka, M., Vlcková, B., Štìpánek, J., and Malý, P. (1998) *Langmuir*, **14**, 4666–4670.

89. Procházka, M., Mojzeš, P., Štìpánek, J., Vlcková, B., and Turpin, P.--Y. (1997) *Anal. Chem.*, **69**, 5103–5108.

90. Persson, J.L., Hui, Q., Jakubek, Z.J., Nakamura, M., and Takami, M. (1996) *Phys. Rev. Lett.*, **76**, 1501–1504.

91. Sibbald, M.S., Chumanov, G., and Cotton, T.M. (1996) *J. Phys. Chem.*, **100**, 4672–4678.

92. Hui, Q., Persson, J.L., Beijersbergen, J.H.M., and Takami, M. (1995) *Z. Phys. B*, **98**, 353–357.

93. Persson, J.L., Hui, Q., Nakamura, M., and Takami, M. (1995) *Phys. Rev. A*, **52**, 2011–2015.

94. Messina, E., Cavallaro, E., Cacciola, A., Iati, M.A., Gucciardi, P.G., Borghese, F., Denti, P., Saija, R., Compagnini, G., Meneghetti, M., Amendola, V., and Marago, O.M. (2011) *ACS Nano.*, **5**, 905–913.

95. Menendez-Manjon, A., Chichkov, B.N., and Barcikowski, S. (2010) *J. Phys. Chem. C*, **114**, 2499–2504.

96. Kalyva, M., Bertoni, G., Milionis, A., Cingolani, R., and Athanassiou, A. (2010) *Microsc. Res. Tech.*, **73**, 937–943.

97. Axente, E., Barberoglou, M., Kuzmin, P.G., Magoulakis, E., Loukakos, P.A., Stratakis, E., Shafeev, G.A., and Fotakis, C. (2010) Size distribution of Au NPs generated by laser ablation of a gold target in liquid with time-delayed femtosecond pulses *http://arxiv.org/abs/1008.0374* (accessed February 2010).

98. Besner, S., Kabashin, A.V., Winnik, F.M., and Meunier, M. (2009) *J. Phys. Chem. C*, **113**, 9526–9531.

99. Giorgetti, E., Giusti, A., Giammanco, F., Marsili, P., and Laza, S. (2009) *Molecules*, **14**, 3731–3753.

100. Saitow, K., Yamamura, T., and Minami, T. (2008) *J. Phys. Chem. C*, **112**, 18340–18349.

101. Muto, H., Yamada, K., Miyajima, K., and Mafuné, F. (2007) *J. Phys. Chem. C*, **111**, 17221–17226.

102. Tabor, C., Qian, W., and El-Sayed, M.A. (2007) *J. Phys. Chem. C*, **111**, 8934–8941.

103. Amendola, V., Polizzi, S., and Meneghetti, M. (2006) *J. Phys. Chem. B*, **110**, 7232–7237.

104. Sylvestre, J.-P., Poulin, S., Kabashin, A.V., Sacher, E., Meunier, M., and Luong, J.H.T. (2004) *J. Phys. Chem. B*, **108**, 16864–16869.

105. Sylvestre, J.-P., Kabashin, A.V., Sacher, E., Meunier, M., and Luong, J.H.T. (2004) *J. Am. Chem. Soc.*, **126**, 7176–7177.

106. Compagnini, G., Scalisi, A.A., and Puglisi, O. (2003) *J. Appl. Phys.*, **94**, 7874.

107. Kabashin, A.V., Meunier, M., Kingston, C., and Luong, J.H.T. (2003) *J. Phys. Chem. B*, **107**, 4527–4531.

108. Chen, C.-D., Yeh, Y.-T., and Wang, C.R.C. (2001) *J. Phys. Chem. Solids*, **62**, 1587–1597.

109. Simakin, A.V., Voronov, V.V., Shafeev, G.A., Brayner, R., and Bozon-Verduraz, F. (2001) *Chem. Phys. Lett.*, **348**, 182–186.

110. Maaza, M., Chambalo, H., Ekambaram, S., Nemraoui, O., Ngom, B.D., and Manyala, N. (2008) *Int. J. Nanopart.*, **1**, 212–223.

111. (a) Nichols, W.T., Sasaki, T., and Koshizaki, N. (2006) *J. Appl. Phys.*, **100**, 114911 (b) Nichols, W.T., Sasaki, T., and Koshizaki, N. (2006) *J. Appl. Phys.*, **100**, 114912 (c) Nichols, W.T., Sasaki, T., and Koshizaki, N. (2006) *J. Appl. Phys.*, **100**, 114913.

112. Zeng, H., Cai, W., Li, Y., Hu, J., and Liu, P. (2005) *J. Phys. Chem. B*, **109**, 18260.

113. Singh, S.C., Swarnkar, R.K., and Gopal, R. (2010) *Bull. Mater. Sci.*, **33**, 21–26.

114. Singh, S.C. and Gopal, R. (2010) *Appl. Surf. Sci.*, doi: 10.1016/j.apsusc.2011.05.018

115. Singh, S.C. (2011) *J. Nanopart. Res.*, doi: 10.1007/s11051-011-0359-2

116. Singh, S.C. and Gopal, R. (2008) *J. Phys. Chem. C*, **112**, 2812.

117. Singh, S.C. and Gopal, R. (2008) *Phys. E*, **40**, 724.

118. Singh, S.C., Swarnkar, R.K., and Gopal, R. (2009) *J. Nanosci. Nanotechnol.*, **9**, 5367.

119. Golightly, J.S. and Castleman, A.W. (2006) *J. Phys. Chem. B*, **110**, 19979.

120. Huang, C.-N., Bow, J.-S., Zheng, Y., Chen, S.-Y., Ho, N.J., and Shen, P. (2010) *Nanoscale Res. Lett.*, **5**, 972.

121. Singh, S.C., Swarnkar, R.K., and Gopal, R. (2009) *J. Nanopart. Res.*, **11**, 1831.

122. Singh, S.C. and Gopal, R. (2010) *J. Phys. Chem. C*, **114**, 9277.

123. Liu, I.L., Lin, B.C., Chen, S.Y., and Shen, P. (2011) *J. Phys. Chem. C*, **115**, 4994–5002.

124. (a) Viau, G., Collière, V., Lacroix, L.-M., and Shafeev, G.A. (2011) *Chem. Phys. Lett.*, **501**, 419–422 (b) Stratakis, E., Barberoglou, M., Fotakis, C., Viau, G., Garcia, C., and Shafeev, G.A. (2009) *Opt. Express*, **17**, 12650–12659.

125. Lee, B.-H., Nakayama, T., Tokoi, Y., Suzuki, T., and Niihara, K. (2011) *J. Alloys Compd.*, **509**, 1231–1235.

126. Lin, F., Yang, J., Lu, S.-H., Niu, K.-Y., Liu, Y., Sun, J., and Du, X.-W. (2010) *J. Mater. Chem.*, **20**, 1103–1106.

127. Nath, A., Laha, S.S., and Khare, A. (2010) *Integr. Ferroelectr.*, **121**, 58–64.

128. Nikolov, A.S., Atanasov, P.A., Milev, D.R., Stoyanchov, T.R., Deleva, A.D., and Peshev, Z.Y. (2009) *Appl. Surf. Sci.*, **255**, 5351–5354.

129. Liu, P., Cai, W., Fang, M., Li, Z., Zeng, H., Hu, J., Luo, X., and Jing, W. (2009) *Nanotechnology*, **20**, 285707.

130. Takada, N., Sasaki, T., and Sasaki, K. (2008) *Appl. Phys. A*, **93**, 833–836.

131. Lin, B.C., Shen, P., and Chen, S.Y. (2011) *J. Phys. Chem. C*, **115**, 5003–5010.

132. Wagener, P., Schwenke, A., Chichkov, B.N., and Barcikowski, S. (2010) *J. Phys. Chem. C*, **114**, 7618–7625.

133. Semaltianos, N.G., Logothetidis, S., Hastas, N., Perrie, W., Romani, S., Potter, R.J., Dearden, G., Watkins, K.G., French, P., and Sharp, M. (2010) *Chem. Phys. Lett.*, **484**, 283–289.

134. Gondal, M.A., Drmosh, Q.A., Yamani, Z.H., and Saleh, T.A. (2009) *Appl. Surf. Sci.*, **256**, 298–304.

135. Hur, T.-B., Phuoc, T.X., and Chyu, M.K. (2009) *Opt. Lasers Eng.*, **47**, 695–700.

136. Zhang, X., Zeng, H., and Cai, W. (2009) *Mater. Lett.*, **63**, 191–193.

137. Ajimsha, R.S., Anoop, G., Aravind, A., and Jayaraj, M.K. (2008) *Electrochem. Solid State Lett.*, **11**, K14–K17.

138. Zeng, H., Liu, P., Cai, W., Cao, X., and Yang, S. (2007) *Cryst. Growth Des.*, **7**, 1092–1097.

139. Zeng, H., Li, Z., Cai, W., Cao, B., Liu, P., and Yang, S. (2007) *J. Phys. Chem. B*, **111**, 14311–14317.

140. Ishikawa, Y., Shimizu, Y., Sasaki, T., and Koshizaki, N. (2006) *J. Colloid Interface Sci.*, **300**, 612–615.

141. Liang, C., Shimizu, Y., Masuda, M., Sasaki, T., and Koshizaki, N. (2004) *Chem. Mater.*, **16**, 963–965.

142. Usui, H., Shimizu, Y., Sasaki, T., and Koshizaki, N. (2005) *J. Phys. Chem. B*, **109**, 120–124.

143. Sajti, C.L., Sattari, R., Chichkov, B., and Barcikowski, S. (2010) *Appl. Phys. A*, **100**, 203–206.

144. Sajti, C.L., Sattari, R., Chichkov, B.N., and Barcikowski, S. (2010) *J. Phys. Chem. C*, **114**, 2421–2427.

145. Liu, I.L., Shen, P., and Chen, S.Y. (2010) *J. Phys. Chem. C*, **114**, 7751–7757.

146. Musaev, O.R., Midgley, A.E., Wrobel, J.M., and Kruger, M.B. (2010) *Chem. Phys. Lett.*, **487**, 81–83.

147. Lee, Y.-P., Liu, Y.-H., and Yeh, C.-S. (1999) *Phys. Chem. Chem. Phys.*, **1**, 4681–4686.

148. Fojtik, A. and Henglein, A. (1993) *Ber. Bunsen-Ges. Phys. Chem. Chem. Phys.*, **97**, 252–254.

149. Semaltianos, N.G., Logothetidis, S., Perrie, W., Romani, S., Potter, R.J., Edwardson, S.P., French, P., Sharp, M., Dearden, G., and Watkins, K.G. (2010) *J. Nanopart. Res.*, **12**, 573–580.

150. Tsuji, T., Tatsuyama, Y., Tsuji, M., Ishida, K., Okada, S., and Yamaki, J. (2007) *Mater. Lett.*, **61**, 2062–2065.

151. Tsuji, T., Tatsuyama, Y., Tsuji, M., Ishida, K., Okada, S., and Yamaki, J. (2007) *Mater. Lett.*, **61**, 2062–2065.

152. Singh, S.C., Mishra, S.K., Srivastava, R.K., and Gopal, R. (2010) *J. Phys. Chem. C*, **114**, 17374–17384.

153. Mafuné, F., Kohno, J., Takeda, Y., and Kondow, T. (2003) *J. Phys. Chem. B*, **107**, 4218.

154. Singh, S.C. and Gopal, R. (2007) *Bull. Mater. Sci.*, **30**, 291.

155. Dolgaev, S.I. (2002) *Appl. Surf. Sci.*, **186**, 546.

156. Baladi, A. and Mamoory, R.S. (2010) *Appl. Surf. Sci.*, **256**, 7559.

157. Jhang, J. and Lan, C.Q. (2008) *Mater. Lett.*, **62**, 1521.

158. Kima, S., Yooa, B.K., Chuna, K., Kanga, W., Choob, J., Gongc, M.-S., and Jooa, S.-W. (2005) *J. Mol. Catal. A: Chem.*, **226**, 231.

159. Lima, M.S.F., Ladário, F.P., and Riva, R. (2006) *Appl. Surf. Sci.*, **252**, 4420.

160. Nikolov, A.S., Atanasov, P.A., Milev, D.R., Stoyanchov, T.R., Deleva, A.D., and Peshev, Z.Y. (2009) *Appl. Surf. Sci.*, **255**, 5351.

161. Takada, N., Sasaki, T., and Sasaki, K. (2008) *Appl. Phys. A: Mater. Sci. Process.*, **93**, 833.

162. Dolgaev, S.I., Simakin, A.V., Voronov, V.V., Safeev, G.A., and Bozon-Verduraz, F. (2002) *Appl. Surf. Sci.*, **186**, 546.

163. Liang, C., Shimizu, Y., Masuda, M., Sasaki, T., and Koshizaki, N. (2004) *Chem. Mater.*, **16**, 963.

164. Usui, H., Sasaki, T., and Koshizaki, N. (2005) *Chem. Lett.*, **34**, 700.

165. Usui, H., Sasaki, T., and Koshizaki, N. (2006) *Chem. Lett.*, **35**, 752.

166. Yan, Z., Bao, R., and Chrisey, D.B. (2010) *Chem. Phys. Lett.*, **497**, 205.

167. Kumar, B. and Thareja, R.K. (2010) *J. Appl. Phys.*, **108**, 064906–064911.

168. Yan, Z., Compagnini, G., and Chrisey, D.B. (2011) *J. Phys. Chem. C*, doi: 10.1021/jp109240s

169. Yan, Z., Bao, R., and Chrisey, D.B. (2011) *Langmuir*, **27**, 851.

170. Schinca, D.C., Scaffardi, L.B., Videla, F.A., Torchia, G.A., Moreno, P., and Roso, L. (2009) *J. Phys. D: Appl. Phys.*, **42**, 215102.

171. He, C., Sasaki, T., Zhou, Y., Shimizu, Y., Masuda, M., and Koshizaki, N. (2007) *Adv. Funct. Mater.*, **17**, 3554.

172. Liu, P., Wang, C.X., Chen, X.Y., and Yang, G.W. (2008) *J. Phys. Chem. C*, **112**, 13450.

173. Liu, P., Cao, Y.L., Chen, X.Y., and Yang, G.W. (2009) *Cryst. Growth Des.*, **9**, 1390.

174. Barcikowski, S., Menéndez-Manjón, A., and Chichkov, B. (2007) *Appl. Phys. Lett.*, **91**, 083113.

175. Sajti, C.L., Barchanski, A., Wagener, P., Klein, S., and Barcikowski, S. (2011) *J. Phys. Chem. C*, **115**, 5094–5101.

176. Chen, Q., Shi, S., and Zhang, W. (2009) *Mater. Chem. Phys.*, **114**, 58–62.

177. Phuoc, T.X., Soonga, Y., and Chyu, M.K. (1997) *Opt. Lasers Eng.*, **45**, 1099–1106.

178. Giacomo, A.D., Bonis, A.D., Aglio, M.D., Pascale, O.D., Gaudiuso, R., Orlando, S., Santagata, A., Senesi, G.S., Taccogna, F., and Teghil, R. (2011) *J. Phys. Chem. C*, **115**, 5123–5130.

179. Werner, D. and Hashimoto, S. (2011) *J. Phys. Chem. C*, **115**, 5063–5072.

180. Zhang, J., Worley, J., Denommee, S., Kingston, C., Jakubek, Z.J., Deslandes, Y., Post, M., and Simard, B. (2003) *J. Phys. Chem. B*, **107**, 6920.

181. Kawasaki, M. and Masuda, K. (2005) *J. Phys. Chem. B*, **109**, 9379.

182. Usui, H., Sasaki, T., and Koshizaki, N. (2006) *J. Phys. Chem. B*, **110**, 12890–12895.

183. Zeng, H., Yang, S., and Cai, W. (2011) *J. Phys. Chem. C*, **115**, 5038–5043.

184. Kawasaki, M. (2011) *J. Phys. Chem. C*, **115**, 5165–5173.

185. Werner, D., Furube, A., Okamoto, T., and Hashimoto, S. (2011) *J. Phys. Chem. C*, **115**, 8503–8512.

186. Chubilleau, C., Lenoir, B., Migot, S., and Dauscher, A. (2011) *J. Colloid Interface Sci.*, **357**, 13–17.

187. Qin, W.-J., Kulinich, S.A., Yang, X.-B., Sun, J., and Du, X.-W. (2009) *J. Appl. Phys.*, **106**, 114318.

188. Kim, J.-Y., Kim, S.J., and Jang, D.-J. (2009) *J. Sep. Sci.*, **32**, 4161–4166.

189. Liu, P., Cao, Y.L., Cui, H., Chen, X.Y., and Yang, G.W. (2008) *Cryst. Growth Des.*, **8**, 559–563.

190. Werner, D., Hashimoto, S., Tomita, T., Matsuo, S., and Makita, Y. (2008) *J. Phys. Chem. C*, **112**, 16801–16808.

191. Amendola, V. and Meneghetti, M. (2007) *J. Mater. Chem.*, **17**, 4705–4710.

192. Nichols, W.T., Kodaira, T., Sasaki, Y., Shimizu, Y., Sasaki, T., and Koshizaki, N. (2006) *J. Phys. Chem. B*, **110**, 83–89.

193. Gelesky, M.A., Umpierre, A.P., Machado, G., Correia, R.R.B., Magno, W.C., Morais, J., Ebeling, G., and Dupont, J. (2005) *J. Am. Chem. Soc.*, **127**, 4588–4589.

194. Kamat, P.V., Flumiani, M., and Hartland, G.V. (1998) *J. Phys. Chem. B*, **102**, 3123–3128.

195. Yamada, K., Tokumoto, Y., Nagata, T., and Mafuné, F. (2006) *J. Phys. Chem. B*, **110**, 11751–11756.

196. Yamada, K., Miyajima, K., and Mafuné, F. (2007) *J. Phys. Chem. C*, **111**, 11246–11251.

197. Shimotsuma, Y., Yuasa, T., Homma, H., Sakakura, M., Nakao, A., Miura, K., Hirao, K., Kawasaki, M., Qiu, J., and Kazansky, P.G. (2007) *Chem. Mater.*, **19**, 1206.

198. Wheeler, D.A., Newhouse, R.J., Wang, H., Zou, S., and Zhang, J.Z. (2010) *J. Phys. Chem. C*, **114**, 18126–18133.

199. Besner, S. and Meunier, M. (2010) *J. Phys. Chem. C*, **114**, 10403–10409.

200. Boyer, P., Menard, D., and Meunier, M. (2010) *J. Phys. Chem. C*, **114**, 13497–13500.

201. Besner, S., Kabashin, A.V., and Meunier, M. (2006) *Appl. Phys. Lett.*, **89**, 233122.

202. Zhao, Q., Hou, L., Zhao, C., Gu, S., Huang, R., and Ren, S. (2004) *Laser Phys. Lett.*, **1**, 115–117.

203. Plech, A., Kotaidis, V., Lorenc, M., and Boneberg, J. (2006) *Nat. Phys.*, **2**, 44–47.

204. Hubert, C., Rumyantseva, A., Lerondel, G., Grand, J., Kostcheev, S., Billot, L., Vial, A., Bachelot, R., and Royer, P. (2005) *Nano Lett.*, **5**, 615–619.

205. Leiderer, P., Bartels, C., König-Birk, J., Mosbacher, M., and Boneberg, J. (2004) *Appl. Phys. Lett.*, **85**, 5370–5372.

206. Takami, A., Kurita, H., and Koda, S. (1999) *J. Phys. Chem. B*, **103**, 1226–1232.

207. Inasawa, S., Sugiyama, M., and Yamaguchi, Y. (2005) *J. Phys. Chem. B*, **109**, 9404–9410.

208. Pyatenko, A., Yamaguchi, M., and Suzuki, M. (2009) *J. Phys. Chem. C*, **113**, 9078–9085.

209. Incropera, F.P. and Dewitt, D.P. (1996) *Fundamentals of Heat and Mass Transfer*, 4th edn, John Wiley & Sons, Inc., New York.

210. Sambles, J.R. (1971) *Proc. R. Soc. Lond. A*, **324**, 339–351.

211. Reiss, H. and Wilson, I.B. (1948) *J. Coll. Sci.*, **3**, 551–561.

212. Curzon, A.E. (1959–1960) The use of electron diffraction in the study of (1) Melting and supercooling of thin films; and (2) Magnetic crystals. PhD Thesis. University of London (Imperial College of Science and Technology).

213. Link, S., Burda, C., Nikoobakht, B., and El-Sayed, M.A. (2000) *J. Phys. Chem. B*, **104**, 6152–6163.

214. Hodak, J.H., Henglein, A., Giersig, M., and Hartland, G.V. (2000) *J. Phys. Chem. B*, **104**, 11708–11718.

215. Inasawa, S., Sugiyama, M., Noda, S., and Yamaguchi, Y. (2006) *J. Phys. Chem. B*, **110**, 3114–3119.

216. Giammanco, F., Giorgetti, E., Marsili, P., and Giusti, A. (2010) *J. Phys. Chem. C*, **114**, 3354–3363.

217. Hu, M. and Hartland, G.V. (2002) *J. Phys. Chem. B*, **106**, 7029–7033.

218. Lin, Z. and Zhigilei, L.V. (2008) *Phys. Rev. B*, **77**, 075133, *http://www.faculty.virginia.edu/CompMat/electron-phonon-coupling/*.

219. Plech, A., Kotaidis, V., Gresillon, S., Dahmen, C., and Von Plessen, G. (2004) *Phys. Rev. B*, **70**, 195423.

220. Juve, V., Scardamaglia, M., Maioli, P., Crut, A., Merabia, S., Joly, L., Del Fatti, N., and Vallee, F. (2009) *Phys. Rev. B*, **80**, 195406.

221. Chen, Y.-H. and Yeh, C.-S. (2001) *Chem. Commun.*, 371.

222. Jin, R., Cao, Y.W., Mirkin, C.A., Kelly, K.L., Schatz, G.C., and Zheng, J.G. (2001) *Science*, **294**, 1901.

223. Mafuné, F., Kohno, J., Takeda, Y., and Kondow, T. (2003) *J. Phys. Chem. B*, **107**, 12589.

224. Gelesky, M.A., Umpierre, A.P., Machado, G., Correia, R.R.B., Magno, W.C., Morais, J., Ebeling, G., and Dupont, J. (2005) *J. Am. Chem. Soc.*, **127**, 4588.

225. Yeh, M.-S., Yang, Y.-S., Lee, Y.-P., Lee, H.-F., Yeh, Y.-H., and Yeh, C.-S. (1999) *J. Phys. Chem. B*, **103**, 6851.

226. Lugovskoy, A. and Bray, I. (1999) *Phys. Rev. B*, **60**, 3279.

227. Davey, W.P. (1925) *Phys. Rev.*, **25**, 753.

228. Born, M. and Wolf, M. (1975) *Principles of Optics*, Chapter 13, 5th edn, Pergamon Press, Oxford.

229. Kerker, M. (1985) *J. Colloid Interface Sci.*, **105**, 297.

230. (a) Doyle, W.T. (1958) *Phys. Rev. B*, **111**, 1067 (b) Fragstein, C.V. and Kreibig, U.Z. (1969) *Z. Phys.*, **224**, 306 (c) Alvarez, M.M., Khoury, J.T., Schaaff, T.G., Shafigullin, M.N., Vezmar, I., and Whetten, R.L.J. (1997) *Phys. Chem. B*, **101**, 3706.

231. Grua, P., Moreeuw, J., Bercegol, H., Jonusauskas, G., and Vallee, F. (2003) *Phys. Rev. B*, **68**, 035424.

232. (a) Naher, U. Bjornholm, S., Frauendorf, S., Garcias, F., and Guet, C. (1997) *Phys. Rep.*, **285**, 245 (b) Saunders, W.A. (1992) *Phys. Rev. A*, **11**, 46.

233. Fujiwara, H., Yanagida, S., and Kamat, P.V. (1999) *J. Phys. Chem. B*, **103**, 2589–2591.

234. Giusti, A., Giorgetti, E., Laza, S., Marsili, P., and Giammanco, F. (2007) *J. Phys. Chem. C*, **111**, 14984.

235. Zhang, J., Post, M., Veres, T., Jakubek, Z.J., Guan, J., Wang, D., Normandin, F., Deslandes, Y., and Simard, B. (2006) *J. Phys. Chem. B*, **110**, 7122.

236. Sugiyama, M., Okazaki, H., and Koda, S. (2002) *Jpn. J. Appl. Phys.*, **41**, 4666–4674.

237. Hada, H., Yonezawa, Y., Yoshida, A., and Kurakake, A. (1976) *J. Phys. Chem.*, **80**, 2728.

238. Kurihara, K., Kizling, J., Stenius, P., and Fendler, J.H. (1983) *J. Am. Chem. Soc.*, **105**, 2574.

239. Eustis, S., Hsu, H.Y., and El-Sayed, M.A. (2005) *J. Phys. Chem. B*, **109**, 4811.

240. Marciniak, B. and Buono-Core, G.E. (1990) *J. Photochem. Photobiol. A*, **52**, 1.

241. Sakamoto, M., Tachikawa, T., Fujitsuka, M., and Majima, T. (2006) *Langmuir*, **22**, 6361.

242. Tauster, S.J., Fung, S.C., and Garten, R.L. (1978) *J. Am. Chem. Soc.*, **100**, 170.

243. Fleischauer, P.D., Kan, H.K.A., and Shepard, J.R. (1972) *J. Am. Chem. Soc.*, **94**, 283.

244. Sakamoto, M., Fujistuka, M., and Majima, T. (2009) *J. Photochem. Photobiol. C: Photochem. Rev.*, **10**, 33.

245. Kraeutler, B. and Bard, A.J. (1978) *J. Am. Chem. Soc.*, **100**, 4317.

246. Bard, A.J. and Fox, M.A. (1995) *Acc. Chem. Res.*, **28**, 141.

247. Kamat, P.V. (2002) *J. Phys. Chem. B*, **106**, 7729.

248. Hada, H., Yonezawa, Y., and Saikawa, M. (1982) *J. Chem. Soc. Faraday. Trans.*, **26**, 2677.

249. Hada, H., Yonezawa, Y., and Saikawa, M. (1982) *Bull. Chem. Soc. Jpn.*, **55**, 2010.

250. Wood, A., Giersig, M., and Mulvaney, P. (2001) *J. Phys. Chem. B*, **105**, 8810.

251. Gong, W., Zheng, Z., Zheng, J., Gao, W., Hu, X., and Ren, X. (2008) *J. Phys. Chem. C*, **112**, 9983–9987.

252. Amans, D., Malaterre, C., Diouf, M., Mancini, C., Chaput, F., Ledoux, G., Breton, G., Guillin, Y., Dujardin, C., Masenelli-Varlot, K., and Perriat, P. (2011) *J. Phys. Chem. C*, **115**, 5131–5139.

253. Muniz-Miranda, M., Gellini, C., and Giorgetti, E. (2011) *J. Phys. Chem. C*, **115**, 5021–5027.

254. Singh, D., Kumar, S., Singh, S.C., Lal, B., and Singh, N.B. (2012) *Sci. Adv. Mater.*, **4**, 522–531.

255. Musaev, O.R., Dusevich, V., Wieliczka, D.M., Wrobel, J.M., and Kruger, M.B. (2008) *J. Appl. Phys.*, **104**, 084316–084320.

256. Svrcek, V., Mariotti, D., Nagai, T., Shibata, Y., Turkevych, I., and Kondo, M. (2011) *J. Phys. Chem. C*, **115**, 5084–5093.

6.2
Synthesis of Metal Compound Nanoparticles by Laser Ablation in Liquid

Haibo Zeng, Shikuan Yang, and Weiping Cai

6.2.1
Introduction

As a type of nanomaterials with quasi-zero dimension, nanoparticles (NPs) have become one of the most important bases of nanoscience and nanotechnology. On the one hand, due to a series of physical and chemical effects related to reduced dimension, NPs have found a vast range of applications. On the other hand, NPs have been used to assemble various high-level nanostructures and nanodevices, a typical example for highly ordered superlattice and NP-based light-emitting diodes (LEDs). Among the various types of NPs, semiconductor NPs have occupied the central position since their emergence and have become more and more significant because of their roles in nano-optics, nanoelectronics, and nano-optoelectronics [1–4].

Up to now, a large number of methods have been developed to fabricate NPs, such as the sol–gel method [5], chemical deposition [6, 7], vapor-phase transport [8], and metal-organic chemical vapor deposition (CVD) [9]. However, both low-temperature solution and high-temperature vapor methods are not free of disadvantages. For example, colloidal ZnO nanocrystals, formed by chemical deposition, tend to aggregate because of their high surface energy. On the other hand, the serious requirements of matching substrate greatly restrict the adaptability of the high-temperature epitaxial method. Therefore, it is still a big challenge to directly fabricate semiconductor NPs with controlled parameters for demands of optics, optoelectronics, and spintronics.

As a conventional method, laser ablation was successfully developed as an effective and highly qualified technique for growth of thin films in an atmosphere or vacuum environment, as shown in Figure 6.2.1a. The obtained thin films usually have accurate stoichiometric proportion controlled by used targets, less impurities, and high crystallinity. During the initial stage of nanoscience, laser ablation was combined with CVD, as shown in Figure 6.2.1b, to fabricate nanowires and carbon nanotubes. It is in recent years that laser ablation in liquid (LAL) was implemented to fabricate colloidal solution with NPs, as shown in Figure 6.2.1c. In 1993, Henglein and Cotton [10–12] applied a pulsed laser to ablate pure metal targets in various solvents to form colloidal solutions containing metal NPs. Since then, LAL has been developed into an important method for metal, semiconductor, and even polymer NPs.

Laser ablation of solids in a liquid environment is a simple and versatile method of synthesizing NPs. The corresponding studies were mainly focused on two subjects: (i) preparation of NPs via laser ablation of (mostly, noble) metal targets [13–20] and (ii) laser-induced modification of the size and shape of NPs [21–26]. However, there have been very few reports on the synthesis of NPs using active main group

Nanomaterials: Processing and Characterization with Lasers, First Edition.
Edited by Subhash Chandra Singh, Haibo Zeng, Chunlei Guo, and Weiping Cai.
© 2012 Wiley-VCH Verlag GmbH & Co. KGaA. Published 2012 by Wiley-VCH Verlag GmbH & Co. KGaA.

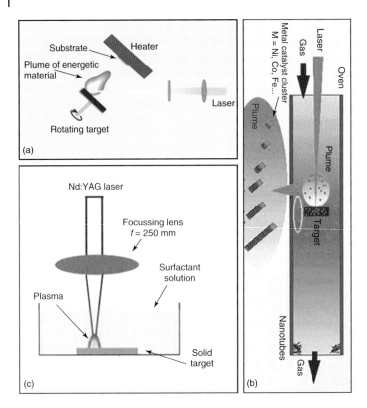

Figure 6.2.1 Schematic representations of laser ablation in different atmospheres and for different purposes: (a) conventional laser ablation for thin-film growth, (b) combination of laser ablation and CVD for nanowires and nanotubes growth, and (c) laser ablation in liquid for colloidal growth.

or transition metal targets. Recently, Koshizaki and coworkers [27–29]successfully synthesized several active metal oxides and hydroxides. In this case, strong reactions take place in the aqueous solution, such as the reaction between water molecules with the ablated species, since the ablated active species are electronically excited and hence highly reactive. On the other hand, the effect of such aqueous oxidation should be controllable through surface modification by surfactant coverage and manipulation of laser parameters. Hence, it is possible to obtain the metal oxide–metal composite NPs and a series of NPs with special microstructure or metastable phases by rapid reactive quenching with surfactant aqueous solution. Finally, through adjusting the liquid medium, the LAL generation of NPs can be expanded into carbide and nitride NPs although the reports are very few.

Recently, we successfully prepared ZnO–Zn NPs with controllable composition and size by laser ablation of pure zinc metal in pure water or in aqueous solution of ionic surfactant [30–33]. The formation mechanism was investigated in this system. After that, the LAL fabrication of NPs was extended to other semiconductor materials, such as Si [34, 35], β-SiC [36], FeO [37], and TiO_2 [38]. These NPs in solutions were found to be charged and have unique redox behaviors [39]. According to these surface features, the NPs can be assembled into complex nanostructures and nanoarrays [40–42]. Corresponding to the special microstructure, these NPs exhibited some interesting properties and improved performances, such as tunable blue luminescence and photocatalysis efficiency. These results could advance our understanding of the characteristics of NPs by LAL and extend the applications of LAL in obtaining functionalized NPs.

In this chapter, we review the LAL generation of compound semiconductor NPs, including oxide, carbide, and nitride. We first describe NP formation and control through LAL using ZnO NPs as a typical example and then introduce the expansion to carbide and nitride NPs. The content in this chapter demonstrates that LAL is a very effective method to controllably fabricate semiconductor NPs.

6.2.2
Synthesis of Nanoparticles by LAL

6.2.2.1 Oxide Nanoparticles

In this section, we describe the LAL generation of ZnO NPs and the control on their size and composition as a typical example of the LAL generation of oxide semiconductor NPs.

As an important wide band gap semiconductor, wurtzite ZnO, with a band gap energy of 3.37 eV at room temperature and exciton binding energy of 60 meV, possesses many important applications in electronic and optical devices [43, 44], especially in optoelectronic applications such as the UV/blue lasing media [45]. Therefore, the controlled synthesis of various ZnO nanostructures, such as nanocrystals, nanowires, nanobelts, and other complex nanoarchitectures [46–49], has been extensively reported. How to control these structures has become the most important issue in ZnO nanostructures. Besides the usually considered size and morphology, due to the important influence of the microstructure, such as core/shell structure, lattice ordering, intrinsic defects, and doping, on the properties and hence applications [50–53], microstructural control is of great significance, but seldom has investigation been reported from this consideration in the mass literatures. Herein, the ZnO NPs fabricated by LAL are of special microstructure and provide a kind of model materials for the understanding of these scientific problems.

The laser ablation of a metal target was performed in an aqueous solution with sodium dodecyl sulfate (SDS), as shown in Figure 6.2.1c. Briefly, a zinc plate (99.99%) was fixed on a bracket in a glass vessel filled with 10 ml 0.05 M SDS (99.5%) aqueous solution, which was continuously stirred. The plate was located at 4 mm from the solution surface in the solution and then was ablated for 30 min by the first harmonic of a Nd:YAG pulsed laser (wavelength 1064 nm, frequency 10 Hz, and pulse duration 10 ns) with the powers from 35 to 100 mJ/pulse and the spot size about 2 mm in diameter. After ablation, all the colloidal suspensions were centrifuged at 14 000 rpm. Then, the obtained products or powders were rinsed

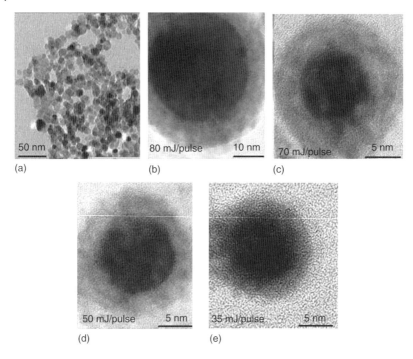

Figure 6.2.2 (a) Typical TEM image of Zn/ZnO NPs by LAL with 0.05 M SDS and 70 mJ/pulse laser and (b–e) single-particle HRTEM images, showing ZnO shell thickness evolution with the applied laser power. (Source: Reproduced from Refs. [30, 33] with permission from the American Chemical Society (ACS).)

with ethanol several times to remove the covered surfactant and dried at room temperature.

Figure 6.2.2a shows the transmission electron microscopic (TEM) image of typical product by 0.05 M and 70 mJ/pulse. They are well dispersed NPs with 20 nm average diameter. In fact, the diameter of NPs can be controlled within 50 nm and the composition can be adjusted from pure ZnO to Zn–ZnO composite through adjusting the surfactant concentration. In order to control the microstructure of NPs by LAL, the surfactant concentration was set as a constant of 0.05 M, and then the applied laser power was adjusted from high power to near threshold. The laser power dependence of the shell thickness is very obvious from the high-solution transmission electron microscopy (HRTEM) examinations in Figure 6.2.2b–e. It can be found that the thickness of the ZnO shell sharply reduced from 12 to 2.5 nm with the decrease of the laser power from 100 to 35 mJ/pulse.

More importantly, the lattice microstructure of the ZnO nanoshell also exhibits strong dependence on the applied laser power with the change in shell thickness. Typically, the ZnO shells were composed of many ultrafine nanocrystals, as well as a great deal of disorderly arranged areas and boundaries surrounding these nanocrystals. With the laser power decreased (as well as for the shell thickness),

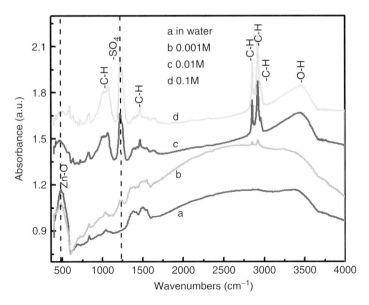

Figure 6.2.3 FT-IR spectra of Zn/ZnO NPs by LAL with different SDS concentrations. (Source: Reproduced from Ref. [30] with permission from ACS.)

the nanocrystals in the shells became fewer in quantity and smaller in size. On the other hand, the disorder area enlarged, especially in nanoshells, by the very low power. Obviously, such disorder degree variance could be induced by the applied laser power in the LAL process and would affect a lot of properties of the nanomaterials.

Figure 6.2.3 shows the Fourier-transformed-infrared (FT-IR) spectra collected from the ZnO NPs by LAL. There is a strong band at 460 cm^{-1} for all samples, which is associated with the characteristic vibration mode of Zn–O bonds [54]. The broad absorption peak centered at 3445 cm^{-1} corresponds to the –OH groups and water molecules [55]. The bands at 1061, 1468, 2849, 2920, and 2952 cm^{-1} are attributed to –C–H stretching and bending modes [56]. The band at 1228 cm^{-1} is related to –SO^4 of SDS molecules [56]. The absorption peak belonging to Zn–O bonds decreases with the increase in SDS concentration, whereas the absorption peak corresponding to surfactant molecules increases, which indirectly confirms that the ZnO relative amount reduces with the increase in SDS. The FT-IR measurements coincide well with the XRD results. These results demonstrate that the formation and control of NPs by LAL is a combinational result of plasma dynamics and surfactant function.

Although the fundamental mechanism regarding nanostructure formation by LAL is still not fully understood, according to previous studies of laser ablation at the liquid–solid interface, interaction between pulsed laser light and the target can create local high-temperature and high-pressure plasma plumes above the target surface. Various chemical reactions and physical processes will thus take place among the ablated metal species, solvent molecules, and surfactant molecules,

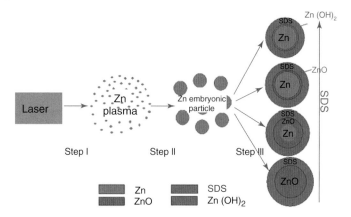

Figure 6.2.4 Illustration of the formation process for ZnO/Zn composite NPs by LAL. (Source: Reproduced from Ref. [30] with permission from ACS.)

which induce the formation of NPs in solution. The structure, morphology, size, and, hence, properties of NPs will differ for different media including solvents and surfactants.

Here, the formation of ZnO–Zn composite NPs could be described in three steps, as illustrated in Figure 6.2.4. (i) The high-temperature and high-pressure zinc plasma is produced in the solid–liquid interface immediately after the interaction between pulsed laser and the metal target. (ii) The subsequent ultrasonic and adiabatic expansion of the zinc plasma, due to high temperature and high pressure, will induce the formation and growth of zinc clusters or particles in the interface region between the solution and the plume because of cooling. (iii) With the gradual extinguishment of the plasma, the formed zinc particles encounter the solvent and surfactant molecules, which induces some chemical reactions and capping effects. The final structure and morphology of the particles are dependent on the SDS concentration in the solution or the competition between aqueous oxidation of zinc particles and SDS protection. The chemical reactions in pure water can be expressed as follows:

$$Zn_{(target)} \xrightarrow{\text{laser}} Zn_{(plasma)}{}^+ + 2e^- \tag{6.2.1}$$

$$Zn_{plasma}{}^+ + 2e^- \longrightarrow Zn_{(cluster)} \tag{6.2.2}$$

$$Zn_{(clusters)} + 2H_2O \longrightarrow Zn(OH)_2 + H_2 \uparrow \longrightarrow ZnO + 2H_2O \tag{6.2.3}$$

Because of the local high-temperature environment, the aqueous oxidation reaction is very intense for the highly active Zn clusters, leading to the formation of the initial oxidation product $Zn(OH)_2$ that can easily be decomposed to ZnO due to the thermal effect. The existence of a small amount of metal zinc in the final product in pure water is due to the incomplete oxidation for some larger Zn particles formed in the initial stage, leading to Zn/ZnO core–shell structured composite NPs.

As described above, the formation of NPs by LAL was the result of competition between aqueous oxidation and surfactant protection of Zn clusters produced in the high-temperature and high-pressure zinc plasma. The SDS concentration determined the content and the surface adsorption rate of DS$^-$ ions and hence the extent of the protection effect on the zinc clusters. However, the applied laser power directly determined the intensity, density, and lifetime of the high-temperature and high-pressure zinc plasma, which not only affected the formation of the primal zinc clusters but also the subsequent oxidation effect due to the correlative thermal effect.

As to the change from pure ZnO to Zn/ZnO core/shell NPs with SDS, when SDS concentration (C) was very low (C < critical micelle concentration (CMC)), the aqueous oxidation effect would be dominant (Eq. (6.2.3)), leading to pure ZnO NPs. However, when C > CMC, the SDS molecules would form a bilayer and even surface micelles on the particle surface, and the protection of SDS became dominant, which would significantly decrease the oxidation rate and result in the formation of Zn/ZnO core/shell NPs. If the concentration of SDS was fixed, the ability of the surfactant protection could be taken as invariable. With decreased laser power, the density and intensity of the hot plasma would be diminished, which would induce the formation of less zinc clusters and particles (Eqs. (6.2.1) and (6.2.2)) and greatly depress the laser-induced thermal effect. These would lead to the reduction of the particle diameter and oxide shell thickness with the decrease of the applied laser power.

In fact, in the LAL process, the highly nonequilibrium feature of the ultrarapid reactive quenching of hot plasma plays an important role, especially for the microstructure, which includes the high-temperature and high-pressure extreme conditions and ultrarapid reactive quenching within a single laser pulse. Such extreme conditions in a confined space enable the formation of various defects, greatly deviate from the equilibrium process, make the species at a highly excited state, and lead to the defect formation energy being no longer an obstacle. Such nonequilibrium processes and conditions result in the emergence of abundant disorderly arranged atoms in the interfaces when the ultrafine nanocrystals are formed in the shells. Finally, when the applied laser power decreases, the disorder degree of shells will become more intensive because of the more rapidly quenching and more incomplete oxidization reaction.

These results demonstrate that, through parameter adjusting, LAL can be used to obtain not only pure metal oxide NPs but also a series of metal/oxide core/shell NPs with controlled shell thickness and disorder degree. Such a microstructure control would be expected to greatly affect some physical and chemical properties of nanomaterials, even obtaining nanostructures with novel functions.

The LAL generation and control of oxide NPs could be readily extended to other metal oxides, such as FeO and TiO$_2$, when we changed the metal targets in our aimed materials.

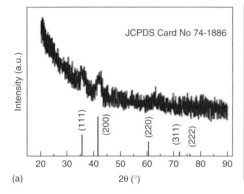
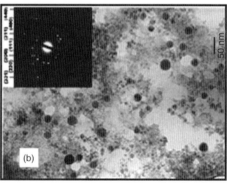

Figure 6.2.5 (a,b) XRD pattern and TEM images of FeO NPs by LAL. (Source: Reproduced from Ref. [37] with permission from ACS.)

FeO NPs can be obtained by LAL in the following manner. An iron plate (99.99%, 1.5 cm × 1 cm) was first fixed on a bracket in a quartz glass vessel filled with 10 ml PVP (polyvinylpyrrolidone) ((K-30, Mw) 40 000, Aldrich) aqueous solution, which was continuously stirred. The PVP concentration varies from 0.01 to 0.2 M. The plate was located 4 mm below the solution surface and then irradiated for 60 min by the first harmonic of a Nd:YAG pulsed laser (wavelength 1064 nm, frequency 10 Hz, and pulse duration 10 ns) with power about 80 mJ/pulse and spot size about 2 mm in diameter on the target. After irradiation, the solutions were centrifuged at 14 000 rpm. The obtained powder products were ultrasonically rinsed with ethanol six times, which was enough to remove the surfactant molecules on the particles, and then draught dried at room temperature.

The typical characterizations of FeO NPs by LAL are presented in Figure 6.2.5. It was found that the higher the concentration of the PVP solution, the more stable the prepared colloidal solutions. The colloidal solutions are stable without aggregation for several days, more than one week, and more than one month for 0.001, 0.02, and 0.1 M PVP solution, respectively.

As for the PVP concentration dependence of the particle size, it can be attributed to the capping effect of PVP on the particles. The commercially available PVP molecules are terminated in the hydroxyl group because of the involvement of water and hydrogen peroxide in polymerization [57, 58]. PVP in solution adsorbs on the FeO particles. Such surface capping will prevent not only growth of the particles but also the coalescence among particles due to the repulsive interaction, leading to a stable colloidal solution and small-sized particles. Obviously, the higher PVP concentration will lead to more PVP molecules in solution attaching on the surface of the FeO NPs, inducing a more stable colloidal solution and smaller particle size.

Furthermore, LAL is also effective in producing TiO_2 NPs using metal Ti foils as ablated target, which can be used as photocatalysis materials. Figure 6.2.6 illustrates the XRD pattern of the products prepared in 0.05 M PVP solution. Although the diffraction intensity is relatively weak, the pattern is well in agreement with the

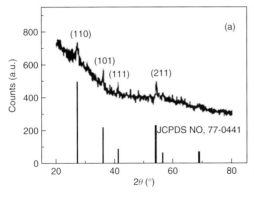
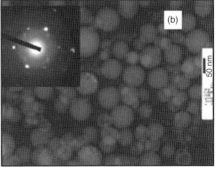

Figure 6.2.6 (a,b) XRD pattern and TEM images of TiO$_2$ NPs by LAL. (Source: Reproduced from Ref. [38] with permission from the Institute of Physics (IOP).)

standard diffraction pattern of rutile TiO$_2$ powders (JCPDS file no. 77-0441). The morphology and the structure were examined. Low-magnification TEM observation revealed that the products consist of nearly spherical particles about 55 nm in mean size and 17 nm in standard deviation, as shown in Figure 6.2.10b. No obvious aggregations were observed in the as-prepared sample. The corresponding selected-area electron diffraction (SAED) pattern shows the rutile crystal structure. Further, high-resolution TEM examination showed the crystalline particles with clear lattice fringes as well as an amount of amorphous material. From the local magnified image, we can thus determine the fringe separations of 0.32 nm, which corresponds to the {110} rutile TiO$_2$ crystal planes.

As an important photoelectrochemical semiconductor, the performance of TiO$_2$ is greatly dependent on the phase selection. Crystalline TiO$_2$ has three phases: rutile (tetragonal, c/a < 1), anatase (tetragonal c/a > 1), and brookite (orthorhombic) [59–62], of which rutile is the most thermodynamically stable phase (generally in the range of 600–1855 °C), whereas anatase and brookite are metastable and readily transformed to rutile when heated. These works demonstrate that LAL is suitable for the phase-selected growth of nanocrystalline TiO$_2$.

6.2.2.2 Carbide Nanoparticles

For the LAL experimental configuration, there are three key components: laser source, ablated target, and liquid medium. Besides changing the target materials, adjusting the liquid medium can also be applied to fabricate novel NPs as demonstrated below with products changing from pure Si NPs to SiC NPs.

Si NPs can be obtained by LAL in the following manner. The cleaned (111) silicon wafer was used as a target and immersed in 0.03 M SDS aqueous solution and irradiated by Nd:YAG laser operated at 1064 nm with pulse duration of 10 ns and frequency of 10 Hz, as previously described. The laser fluence was about 150 mJ/pulse. The XRD pattern of products by LAL Si target in aqueous solution

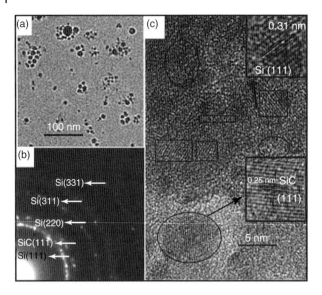

Figure 6.2.7 (a) TEM image, (b) SAED pattern, and (c) HRTEM image of Si and β-SiC NPs by LAL of Si target in ethanol. (Source: Reproduced from [36] with permission from the Royal Society of Chemistry (RSC).)

can be indexed as a silicon phase. The TEM image shows that the Si powders consist of about 10–20 nm NPs.

Interestingly, if we change the liquid medium from aqueous solution to organic agents such as ethanol and toluene and prolong the LAL time, β-SiC NPs can be obtained by LAL. Figure 6.2.7 shows the products synthesized by LAL Si target in ethanol for 2 h. It is obvious that the products contain both Si and β-SiC NPs. Through etching products with HF (4 wt%) +H_2O_2 (6 wt%) for 24 h, pure β-SiC quantum dots were obtained as a result of the selective etching of the Si component. Many methods have been developed to produce the nanostructured β-SiC, such as CVD, physical vapor transport, and the reduction–carburization route. All these methods need high temperatures and sometimes toxic species such as SiH_4, CH_4, and $SiCl_4$. Furthermore, such acquired β-SiC particles are relatively large in size and show insignificant quantum confinement effect. Herein, with the noncontaminant LAL method, we successfully fabricated β-SiC quantum dots with size around 3 nm.

The formation of the β-SiC quantum dots could be understood as follows. When the laser beam irradiates the silicon target, the high-temperature and high-pressure Si plasma will be formed quickly on the solid–liquid interface. Subsequently, such Si plasma will ultrasonically and adiabatically expand, leading to the cooling of the plume region and the formation of Si clusters. At the same time, C–O and C–C ligands in ethanol molecules, at the interface between the silicon plasma plume and ethanol media, will be broken to form carbon atoms. Subsequently, the carbon atoms formed will react with Si clusters at the local high-temperature region and result in the formation of β-SiC quantum dots.

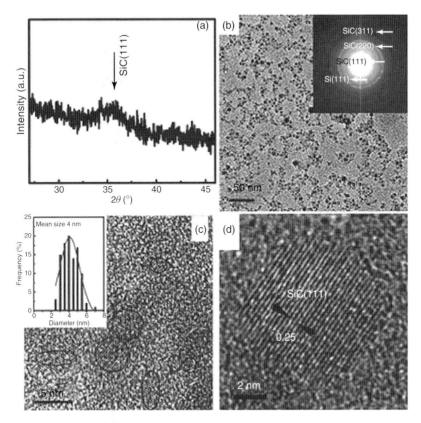

Figure 6.2.8 NPs by LAL of Si target in ethanol and toluene with volume ratio of 7 : 1. (a) XRD. (b) TEM. Inset: SAED. (c) HRTEM. Inset: size distribution. (d) Typical SiC quantum dots with 0.25 nm plane distance. (Source: Reproduced from Ref. [36] with permission from RSC.)

On the basis of the understanding of this mechanism, we found that if laser ablation was conducted in the mixture solution of ethanol and toluene with a volume ratio of 7 : 1, β-SiC quantum dots will become the dominant component, as shown in Figure 6.2.8. The production rate of β-SiC quantum dots can be obviously improved. The mean size of the quantum dots is 4 nm with 0.9 nm standard deviation.

In the case of laser ablation of iron in ethanol [63], NPs with a well-defined spherical shape and an average size of 19 ± 11 nm can be fabricated, as shown in Figure 6.2.9.

High-resolution TEM analysis shows that NPs are prevalently monocrystalline with an interplanar distance of 0.21 nm. Such an interplanar distance observed by HRTEM is compatible with both metal iron and iron carbide nanocrystals. Powder XRD analysis on the same sample clearly indicates the presence of the Fe_3C phase. These results are in agreement with the absorption band at 330 nm

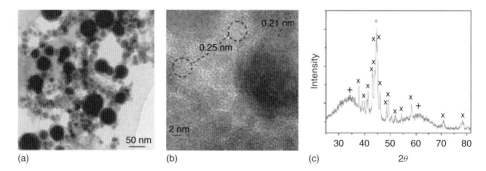

(a) (b) (c) 2θ

Figure 6.2.9 (a–c) TEM, HRTEM images, and XRD pattern of Fe$_3$C NPs by LAL Fe target in EtOH. Powder XRD analysis showing the reflections corresponding to the Fe3C (x), magnetite (+), and cubic Fe (*) compositions. (Source: Reproduced from Ref. [63] with permission from ACS.)

observed by UV–vis spectroscopy characteristics of nanometric iron carbides. Two broad bands centered at 35° and 60° are also present in the XRD pattern. These peaks are compatible with the presence of very small crystals of magnetite (Fe$_3$O$_4$) or maghemite (γ-Fe$_2$O$_3$). Another less intense peak at 45° can be ascribed to traces of cubic Fe. The average crystal size estimated by the Debye–Scherrer formula is on the order of 1 nm for the oxide phase, in close agreement with what is observed by HRTEM.

Therefore, LAL of iron in ethanol yields two different populations of NPs, the first one composed of Fe$_3$C with a size of tens of nanometers and the second composed of magnetite with sizes on the order of 1 nm. The iron carbide composition is very interesting because the bulk saturation magnetization of Fe$_3$C is 140 emu g^{-1}, higher than bulk magnetite or maghemite (<90 emu g^{-1}) although lower than metal iron (<220 emu g^{-1}) [64, 65]. Contrary to iron oxides and metal iron, however, iron carbide is stable in air up to 200 °C, and it is highly resistant to acidic dissolution; therefore, it can be easily purified from the other compounds by acidic treatment. The good magnetic response of NPs obtained by LAL in EtOH is qualitatively confirmed by their migration in a few minutes toward an NdFeB magnet placed in the proximity of the solution.

Laser ablation produces a plasma plume at high temperature and pressure containing highly ionized and excited species from both the target and the solvent. When solvents react with the ablated material, NPs with composition different from the target are formed during the nucleation and growth processes. For instance, LAL of gold in CH$_2$Cl$_2$ yielded gold chloride NPs [66]. Even when the solvent does not react with the ablated material, it has an important role in the determination of the average size of NPs or can form a matrix embedding NPs, as shown for gold and silver NPs obtained by LAL in various organic solvents [67–69]. Both circumstances take place during the LAL of iron in the six organic solvents. The formation of Fe$_3$C NPs in EtOH is an example of a solvent that is chemically active with the target atoms in the plasma plume. The reactivity of ethanol with iron atoms at high temperature to form carbides was also reported for electric plasma discharge in liquids and is a consequence of the chemical species formed during the thermal decomposition of ethanol molecules [70, 71].

6.2.2.3 Nitride Nanoparticles

Besides oxide and carbide NPs, nitride NPs can be also fabricated by LAL, such as BN and TiN NPs. It is worth noting that LAL forms some nitride NPs with novel phases, which are hard to obtain under usual conditions. This should be related to the special high-temperature and high-pressure conditions of hot plasma formed by LAL.

For the LAL formation of c-BN nanocrystals [72], the second harmonic was produced by a Q-switched Nd:YAG laser with wavelength 532 nm, pulse width 10 ns, repetition frequency 5 Hz, and power density 10^{10} W cm^{-2}. The solid target was an h-BN wafer (96%), and high purity (99.5%) acetone was used as the reactive liquid.

According to the XRD pattern of products shown in Figure 6.2.10, it can be seen that there are four characteristic peaks at $43.16°$, $74.16°$, $90.08°$, and $136.1°$, which correspond to the (1 1 1), (2 2 0), (3 1 1), and (3 3 1) crystalline planes of c-BN, respectively. The characteristic peaks of h-BN, which were stronger than that of c-BN, were also observed, which demonstrated that the obtained sample was not pure. Therefore, these analyses showed that the c-BN crystals can be synthesized as one component of composite product by LAL.

The TEM morphologies of the prepared c-BN crystals revealed that most of the resultant c-BN crystals were quasi-spherical and their diameters varied from 30 to 80 nm. The quasi-spherical shape of the resultant c-BN nanocrystals was similar to the kind of polycrystalline morphology usually found in nanocrystals of diamond and related materials.

The formation of c-BN nanocrystals at an h-BN solid–acetone liquid can be understood as follows. The laser-induced plasma is first generated at the liquid–solid interface; when pulsed laser ablates the solid target, it contains some species, for example, B, N, B–N with sp^2 bonding, and their ions, from the laser-ablated solid.

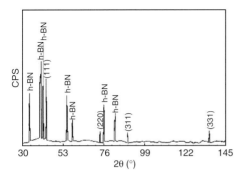 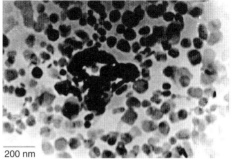

Figure 6.2.10 (a,b) XRD pattern and TEM image of c-BN NPs by LAL h-BN target in acetone. (Source: Reproduced from Ref. [72] with permission from RSC.)

Subsequently, owing to the laser-induced high pressure, the plasma is driven into a higher temperature, higher density, and higher pressure state; moreover, the chemical reactions between those species can occur in the laser-induced plasma. Since the c-BN phase with sp^3 bonding is a stable phase under conditions of high temperature and high pressure in the T–P phase diagram and h-BN phase with sp^2 bonding is a metastable phase under the same conditions, c-BN nuclei generate on the chemical reactions that have taken place in the plasma. Meanwhile, the laser-induced plasma, which is generated at the plasma–liquid interface and contains the species, for example, OH, H, and their ions, from the confined liquid, rapidly dissolves into the laser-induced plasma. Then, these species can be involved in those reactions, resulting in c-BN nuclei being formed and may enhance c-BN nuclei forming and growing. For example, these OH- and H^+ ions can promote the transformation of $sp^2 \rightarrow sp^3$ by suppressing sp^2 bonding on pulsed laser ablation at the gas–liquid interface [73]. As a result, the c-BN nuclei grow into c-BN crystals with the plasma rapidly being quenched in the confined liquid. Because the growth times (plasma quenching time) of the nuclei are very short, the diameter of the grown crystals is in the nanometer scale. It is noted that the synthesis reaction on LAL has one distinct feature, that is, the finally synthesized products generally have a metastable structure under conditions of normal temperature and pressure [74, 75]. Therefore, LAL obviously provides an efficient way to synthesize nanocrystalline boron nitride as a cubic phase.

Besides using nitride targets, changing the liquid medium from the usually used water or aqueous solution to liquid nitrogen in the LAL process is also effective to fabricate nitride NPs as reported by Takada *et al.* [76].

In the experimental apparatus, a stainless steel chamber with a doubled structure was used to realize laser ablation of titanium target in liquid nitrogen. The inside of the inner chamber of the doubled structure was filled with liquid nitrogen, and the space sandwiched by the inner and outer chambers was evacuated using a turbomolecular pump to avoid the condensation of water vapor on the optical windows. In addition, in order to prevent the dissolution of oxygen into liquid nitrogen, the space above the liquid nitrogen was purged by pure nitrogen gas. YAG laser pulses at the fundamental wavelength of 1.06 μm irradiated titanium and silicon targets immersed in liquid nitrogen.

The NPs synthesized from a titanium target were firstly checked by SEM and EDS. The sizes of the particles were distributed from several micrometers to several tens of nanometers. The shapes of all particles were spherical, indicating that the particles were produced from a melted target. According to EDS analysis, the atomic ratio of N/Ti was ~0.4 and the fractional abundance of oxygen was much lower than that of nitrogen. These results indicate the production of TiN particles and negligible post-oxidation after exposure to air.

The crystal structure of the synthesized particles was further examined by XRD and TEM, as shown in Figure 6.2.11. Figure 6.2.11a shows an XRD spectrum of particles produced from a titanium target, together with an XRD spectrum of a virgin target for the sake of comparison. The XRD spectrum of the virgin target was mainly occupied by the peaks of α-Ti. In contrast, the XRD spectrum

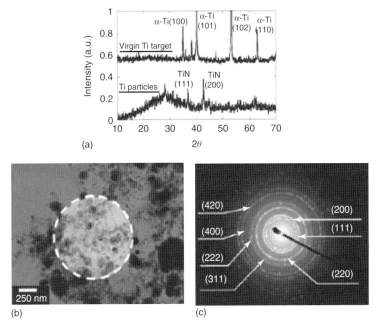

(a)

(b)

(c)

Figure 6.2.11 (a–c) The used setup, SEM image, and EDS line profile of TiN NPs by LAL of Ti in nitrogen medium. (Source: Reproduced from Ref. [76] with permission from RSC.)

of the particles had no α-Ti peaks. Instead of the α-Ti peaks, the peaks of TiN corresponding to (111) and (200) can be identified in the XRD spectrum of particles. This result indicates that no metallic titanium particles are included in the sample. However, the broad peak in the XRD spectrum of particles suggests the existence of amorphous titanium and TiN particles in the sample. The crystal structure of the particles is understood more clearly from the electron diffraction pattern in the TEM analysis. Figure 6.2.11b,c shows a TEM image and the corresponding electron diffraction pattern (inside the dashed circle in Figure 6.2.11b) of the particles. From Figure 6.2.11c, it is clear that the particles have the polycrystalline structure of TiN, including (220), (311), (222), and so on. It has been reported that laser ablation of a Ti target in water is an effective method for synthesizing polycrystalline TiO2 particles [77]. In contrast, the present experimental results reveal that LAL nitrogen is useful for synthesizing polycrystalline TiN particles from a titanium target. Although the synthesis mechanism is not understood yet, it is speculated that the nitride layer formed on the titanium target may contribute to the production of TiN particles. On the other hand, the ablation depth of the target by one laser pulse in this experimental condition was ∼10 nm, which was shallower than the nitride layer on the target. Hence, it is expected that both Ti and N are ejected from the target. Atomic nitrogen is provided from both the ambient liquid nitrogen and the nitride target under this experimental condition.

6.2.3
Conclusions

This chapter gives a comprehensive review of the progress of laser ablation generation of oxide, carbide, and nitride NPs in liquid medium. These results demonstrate that LAL is very facile and versatile for the fabrication of semiconductor NPs. The oxide NPs are the most spontaneous products of LAL of metal targets in water and in aqueous solutions according to the formation of plasma and oxidation reaction. However, it is very interesting that, through adjusting the ablated targets into aiming compounds or changing the liquid medium into other organic agents, carbide and nitride NPs can be effectively fabricated by LAL. The formation mechanism involves the variation of plasma composition from targets and liquid medium.

Furthermore, besides the fabrication, the morphology, composition, and size of formed NPs can be easily controlled through adjusting the various LAL parameters, including laser power, surfactant, liquid medium type, and so on. Most importantly, special microstructure and metastable phases are usually found in the formed NPs by LAL and can be attributed to the extreme conditions (high temperature and high pressure) and high unequilibrium process (ultrafast chemical reaction and thermal quenching), which are induced by the instantaneous hot plasma of LAL. Such extreme conditions and unequilibrium process result in the special microstructure of the obtained nanomaterials, especially the partial disorder atom arrangement and abundant defect states, and hence in defect-related properties, such as blue luminescence.

Considering the versatility of LAL in the formation of various NPs and the special microstructure and unique phases of formed NPs, LAL will receive more and more attention and will play a more important role in the fabrication of functional NPs for various applications and in the fundamental studies of some materials physics problems in the near future.

Acknowledgments

This work is financially supported by the Natural Science Foundation of China (Grant Nos. **10604055** and **50831005**) and the National Basic Research Program of China (Grant No **2007CB936604**).

References

1. Subramanian, V., Wolf, E., and Kamat, P.V. (2001) *J. Phys. Chem. B*, **105**, 11439.
2. Cozzoli, P.D., Comparelli, R., Fanizza, E., Curri, M.L., Agostiano, A., and Laub, D. (2004) *J. Am. Chem. Soc.*, **126**, 3868.
3. Dawson, A. and Kamat, P.V. (2001) *J. Phys. Chem. B*, **105**, 960.
4. Hsu, Y.J. and Lu, S.Y. (2004) *Langmuir*, **20**, 23.
5. Spanhel, L. and Anderson, M.A. (1991) *J. Am. Chem. Soc.*, **113**, 2826.
6. Fu, Y.S., Du, X.W., Kulinich, S.A., Qiu, J.S., Qin, W.J., Li, R., Sun, J., and Liu, J. (2007) *J. Am. Chem. Soc.*, **129**, 16029.

7. Norberg, N.S., Kittilstved, K.R., Amonette, J.E.R., Kukkadapu, K., Schwartz, D.A., and Gamelin, D.R. (2004) *J. Am. Chem. Soc.*, **126**, 9387.

8. Lu, J.G., Ye, Z.Z., Zhang, Y.Z., Liang, Q.L., Fujita, S., and Wang, Z.L. (2006) *Appl. Phys. Lett.*, **89**, 023122.

9. Kim, S.W., Ueda, M., Funato, M., Fujita, S., and Fujita, S. (2005) *J. Appl. Phys.*, **97**, 104316.

10. Fojtik, A. and Henglein, A. (1993) *Ber. Bunsen Ges. Phys. Chem.*, **97**, 252.

11. Neddersen, J., Chumanov, G., and Cotton, T.M. (1993) *Appl. Spectrosc.*, **47**, 1959.

12. Sibbald, M.S., Chumanov, G., and Cotton, T.M. (1996) *J. Phys. Chem.*, **100**, 4672.

13. Pyatenko, A., Shimokawa, K., Yamaguchi, M., Nishimura, O., and Suzuki, M. (2004) *Appl. Phys. A*, **79**, 803.

14. Mafune, F., Kohno, J.Y., Takeda, Y., and Kondow, T. (2000) *J. Phys. Chem. B*, **35**, 8335.

15. Mafune, F., Kohno, J.Y., Takeda, Y., and Kondow, T. (2000) *J. Phys. Chem. B*, **104**, 9111.

16. Mafune, F., Kohno, J.Y., Takeda, Y., and Kondow, T. (2001) *J. Phys. Chem. B*, **105**, 5114.

17. Mafune, F., Kohno, J.Y., Takeda, Y., and Kondow, T. (2002) *J. Phys. Chem. B*, **31**, 7577.

18. Mafune, F., Kohno, J.Y., Takeda, Y., and Kondow, T. (2003) *J. Phys. Chem. B*, **107**, 4218.

19. Takeuchi, Y., Ida, T., and Kimura, K. (1997) *J. Phys. Chem. B*, **101**, 1322.

20. Sylvestre, J.P., Kabashin, A.V., Sacher, E., Meunier, M., and Luong, J.H. (2004) *J. Am. Chem. Soc.*, **126**, 7176.

21. Mafune, F., Kohno, J.Y., Takeda, Y., and Kondow, T. (2001) *J. Phys. Chem. B*, **105**, 9050.

22. Mafune, F., Kohno, J.Y., Takeda, Y., and Kondow, T. (2003) *J. Am. Chem. Soc.*, **125**, 1686.

23. Satoh, N., Hasegawa, H., Tsujii, K., and Kimura, K. (1994) *J. Phys. Chem. B*, **98**, 2143.

24. Aguirre, C.M., Moran, C.E., Young, J.F., and Halas, N.J. (2004) *J. Phys. Chem. B*, **108**, 7040.

25. Link, S., Burda, C., Nikoobakht, B., and Sayed, M.A.E. (2000) *J. Phys. Chem. B*, **104**, 6152.

26. Bosbach, J., Martin, D., Stietz, F., Wenzel, T., and Trager, F. (1999) *Appl. Phys. Lett.*, **74**, 2605.

27. Liang, C.H., Shimizu, Y., Sasaki, T., and Koshizaki, N. (2003) *J. Phys. Chem. B*, **107**, 9220.

28. Liang, C.H., Shimizu, Y., Masuda, M., Sasaki, T., and Koshizaki, N. (2004) *Chem. Mater.*, **16**, 963.

29. Usui, H., Shimizu, Y., Sasaki, T., and Koshizaki, N. (2005) *J. Phys. Chem. B*, **109** (1), 120.

30. Zeng, H.B., Cai, W.P., Li, Y., Hu, J.L., and Liu, P.S. (2005) *J. Phys. Chem. B*, **109**, 18260.

31. Zeng, H.B., Cai, W.P., Hu, J.L., Duan, G.T., Liu, P.S., and Li, Y. (2006) *Appl. Phys. Lett.*, **88**, 171910.

32. Zeng, H.B., Cai, W.P., Cao, B.Q., Hu, J.L., Li, Y., and Liu, P.S. (2006) *Appl. Phys. Lett.*, **88**, 181905.

33. Zeng, H.B., Li, Z.G., Cai, W.P., Cao, B.Q., Liu, P.S., and Yang, S.K. (2007) *J. Phys. Chem. B*, **111**, 14311.

34. Yang, S.K., Cai, W.P., Zeng, H.B., and Li, Z.G. (2008) *J. Appl. Phys.*, **104**, 023516–023520.

35. Yang, S.K., Cai, W.P., Zhang, H.W., Xu, X.X., and Zeng, H.B. (2009) *J. Phys. Chem. C*, **113**, 19091.

36. Yang, S.K., Cai, W.P., Zeng, H.B., and Xu, X.X. (2009) *J. Mater. Chem.*, **19**, 7119.

37. Liu, P.S., Cai, W.P., and Zeng, H.B. (2008) *J. Phys. Chem. C*, **112**, 3261.

38. Liu, P.S., Cai, W.P., Fang, M., Li, Z.G., Zeng, H.B., Hu, J.L., Luo, X.D., and Jing, W.P. (2009) *Nanotechnology*, **20**, 285707.

39. Yang, S.K., Cai, W.P., Liu, G.Q., Zeng, H.B., and Liu, P.S. (2009) *J. Phys. Chem. C*, **113**, 6480.

40. Yang, S.K., Cai, W.P., Yang, J.L., and Zeng, H.B. (2009) *Langmuir*, **25**, 8287.

41. Zeng, H.B., Cai, W.P., Liu, P.S., Xu, X.X., Zhou, H.J., Klingshirn, C., and Kalt, H. (2008) *ACS Nano*, **2**, 1661.

42. Zeng, H.B., Liu, P.S., Cai, W.P., Yang, S.K., and Xu, X.X. (2008) *J. Phys. Chem. C*, **112**, 19620.

43. Dai, Z.R., Pan, Z.W., and Wang, Z.L. (2003) *Adv. Funct. Mater.*, **13**, 9.

44. Wong, E.M. and Searson, P.C. (1997) *Appl. Phys. Lett.*, **4**, 2939.

45. Lau, S.P., Yang, H.Y., Yu, S.F., Yuen, C., Leong, E.S.P., Li, H.D., and Hng H.H. (2005) *Small*, **10**, 956.

46. Cao, B., Teng, X., Heo, S.H., Li, Y., Cho, S.O., Li, G., and Cai, W. (2007) *J. Phys. Chem. C*, **111**, 2470.

47. Li, L., Pan, S., Dou, X., Zhu, Y., Huang, X., Yang, Y., Li, G., and Zhang, L. (2007) *J. Phys. Chem. C*, **111**, 7288–7291.

48. Shen, G.Z., Bando, Y., and Lee, C.J. (2005) *J. Phys. Chem. B*, **109**, 10779.

49. Mo, M.S., Yu, J.C., Zhang, L.Z., and Li, S.A. (2005) *Adv. Mater.*, **17**, 756.

50. Yu, K., Zaman, B., Romanova, S., Wang, D.S., and Ripmeester, J.A. (2005) *Small*, **3**, 332.

51. Ding, Y., Gao, P.X., and Wang, Z.L. (2004) *J. Am. Chem. Soc.*, **126**, 2066.

52. Mahan, G.D. (1983) *J. Appl. Phys.*, **54** (7), 3825.

53. Radovanovic, P.V., Norberg, N.S., McNally, K.E., and Gamelin, D.R. (2002) *J. Am. Chem. Soc.*, **124**, 15192.

54. Cozzoli, P.D., Curri, M.L., Agostiano, A., Leo, G., and Lomascolo, M. (2003) *J. Phys. Chem. B*, **107**, 4756.

55. Kooli, F., Chsem, I.C., and Vucelic, W. (1996) *Chem. Mater.*, **8**, 1969.

56. Clearfield, A., Kieke, M., Kwan, J., Colon, J.L., and Wang, R.C.J. (1991) *Inclusion Phenom. Mol. Recognit. Chem.*, **11**, 361.

57. Washio, I., Xiong, Y.J., Yin, Y.D., and Xia, Y.N. (2006) *Adv. Mater.*, **18**, 1745.

58. Xiong, Y.J., Washio, I., Chen, J.Y., Cai, H.G., Li, Z.Y., and Xia, Y.N. (2006) *Langmuir*, **22**, 8563.

59. Wang, C., Deng, Z.X., and Li, Y. (2001) *Inorg. Chem.*, **40**, 5210.

60. Kumar, P.M., Badrinarayanan, S., and Sastry, M. (2000) *Thin Solid Films*, **358**, 122.

61. Sen, S., Mahanty, S., Roy, S., Heintz, O., Bourgeois, S., and Chaumont, D. (2005) *Thin Solid Films*, **474**, 245.

62. Wu, J.J. and Yu, C.C. (2004) *J. Phys. Chem. B*, **108**, 3377.

63. Amendola, V., Riello, P., and Meneghetti, M. (2011) *J. Phys. Chem. C*, **115**, 5140.

64. Giordano, C., Kraupner, A., Wimbush, S.C., and Antonietti, M. (2010) *Small*, **6**, 1859.

65. Herrmann, I.K., Grass, R.N., Mazunin, D., and Stark, W.J. (2009) *Chem. Mater.*, **21**, 3275.

66. Compagnini, G., Scalisi, A.A., and Puglisi, O. (2002) *Phys. Chem. Chem. Phys.*, **4**, 2787.

67. Amendola, V., Polizzi, S., and Meneghetti, M. (2006) *J. Phys. Chem. B*, **110**, 7232.

68. Amendola, V., Polizzi, S., and Meneghetti, M. (2007) *Langmuir*, **23**, 6766.

69. Amendola, V., Rizzi, G.A., Polizzi, S., and Meneghetti, M. (2005) *J. Phys. Chem. B*, **109**, 23125.

70. Sergiienko, R., Shibata, E., Akase, Z., Suwa, H., Nakamura, T., and Shindo, D. (2006) *Mater. Chem. Phys.*, **98**, 34.

71. Park, J., Xu, Z.F., and Lin, M.C. (2003) *J. Chem. Phys.*, **118**, 9990.

72. Wang, J.B., Yang, G.W., Zhang, C.Y., Zhong, X.L., and Ren, Z.A. (2003) *Chem. Phys. Lett.*, **367**, 10.

73. Xiao, R.F. (1995) *Appl. Phys. Lett.*, **67**, 3117.

74. Wang, J.B. and Yang, G.W. (1999) *J. Phys. Condens. Matter*, **11**, 7089.

75. Yang, G.W. and Wang, J.B. (2001) *Appl. Phys. A*, **A72**, 475.

76. Takada, N., Sasaki, T., and Sasaki, K. (2008) *Appl. Phys. A*, **93**, 833.

77. Dolgaev, S.I., Simakin, A.V., Voronov, V.V., Shafeev, G.A., and Bozon-Verduraz, F. (2002) *Appl. Surf. Sci.*, **186**, 546.

6.3
Synthesis of Fourth Group (C, Si, and Ge) Nanoparticles by Laser Ablation in Liquids

Minghui Hong, Guoxin Chen, and Tow Chong Chong

6.3.1
Laser Ablation in Liquid (LAL)

6.3.1.1 Introduction
Since the rediscovery of carbon nanotubes in the early 1990s, a lot of fabrication techniques were developed for preparing nanoparticles with controlled phases, sizes, shapes, and properties. But conventional methods, such as gas-phase processes, often produce agglomerated particles with sizes up to several micrometers, while chemical methods often induce impurities to the nanoparticles by catalysts and by-products. Fabrication of nanomaterials in liquids has been widely investigated in the past two decades [1]. Laser ablation in liquid (LAL) was proved to be a convenient, effective, clean, and safe method to control synthesis of nanomaterials. It is also a promising nanoparticle preparation method for constructing large arrays of complicated hierarchical and ordered superstructures using the bottom-up strategy. The fabrication methods and processes play significant roles in the properties and functionalities of nanomaterials. To control the unique physical and chemical properties of nanoparticles, it is essential to investigate and develop novel nanofabrication techniques and the associated fabrication processes. Thus, the LAL method was studied extensively with a variety of materials, including metals, semiconductors, and carbon.

Previously, LAL was studied in two research directions: improving fatigue performance of metallic components with enhanced shock wave effect [2] and medical applications with reduced shock wave effect [3]. The interaction of laser materials in a liquid environment was also applied on laser-induced surface change for the chemical etching of semiconductors [4]. Interesting metastable nanomaterial synthesis employing LAL was developed in 1987 [5, 6]. Nitridation of titanium in ammonia and liquid nitrogen and oxidation of iron in water were studied with pulsed laser treatment. A mechanism of high-power pulsed-laser-induced reactive quenching at the liquid–solid interface was applied to describe the laser processing. The investigation covered mainly laser-induced bulk material change.

The synthesis of particulates using LAL was firstly reported by Ogale *et al.* [7]. Diamond microparticles with sizes ranging from 5 to 20 μm were produced by ablating a graphite target in benzene using a ruby laser. Fojtik and Henglein [8] used a pulsed ruby laser to ablate metal films, silicon films, carbon films, and graphite particles in different solvents. Carbon nanoparticles about 1–3 nm in diameter were prepared in water, 2-propanol, and c-hexane, respectively. Ultrasmall carbon clusters of fullerenes (C_{60}, C_{70}, and C_{80}) were detected from the colloidal solutions. Colloidal silicon with a remarkable absorption band around 230 nm was fabricated by the laser ablation of a silicon film in c-hexane. After that, a lot of research was

Nanomaterials: Processing and Characterization with Lasers, First Edition.
Edited by Subhash Chandra Singh, Haibo Zeng, Chunlei Guo, and Weiping Cai.
© 2012 Wiley-VCH Verlag GmbH & Co. KGaA. Published 2012 by Wiley-VCH Verlag GmbH & Co. KGaA.

intensively carried out to fabricate nanoparticles of a variety of materials in different solvents. The preparation work of some special nanomaterials, such as noble metal nanoparticles, has been summarized in review articles [9, 10] and a comprehensive LAL handbook [11]. A recent review of LAL on fundamental concepts, physical, and chemical aspects as well as its applications in the synthesis of nanomaterials has been summarized in a systematic review article [12]. There is a timely review paper devoted to the recent developments of the LAL methods that are assisted with inorganic salts and electrical fields [13]. LAL has developed to be a promising technique of nanomaterial synthesis and functional nanostructure fabrication. This chapter covers the recent progress of LAL for the fourth group elements in aspects of fabrication methodology.

Generally speaking, LAL for nanomaterial synthesis may be conducted in several ways by using different types of source materials: solids, liquids, and suspensions. The final products may be produced in different forms: film, micro-/nanoparticle colloids, and gas. Much of the research in this area has been done using laser irradiation of a solid target in liquids to fabricate nanoparticle colloids. For the process of nanoparticle fabrication, usually an immersed solid target is applied (Figure 6.3.1). Laser ablation occurs at the liquid–solid interface with a strong confinement by the surrounding liquid.

Pulsed laser as a pure energy source with strong peak power is extensively used in diverse and broad material processing applications. The mechanism of laser ablation is determined by laser characteristics and the physical and chemical properties of target materials. The target materials absorb laser light and convert the electromagnetic energy into electronic excitation. Then, the excited electrons decay by releasing their energy into the lattice/molecules and transform to thermal, chemical, and mechanical energy and eventually cause ablation if the input laser energy is above the threshold of material ablation. Laser ablation is a

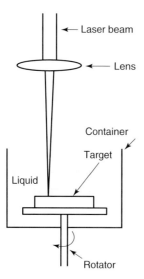

Figure 6.3.1 Schematic diagram of a LAL setup.

complex process involving plasma plume formation and expansion. The processing performance is always affected by the working environment. There are some liquid environments and experimental configurations with which the workpiece is in contact with liquids. Laser ablation usually takes a few minutes to several hours, depending on the source materials, laser parameters, and expected final products. After laser irradiation, the color or transparency of the liquid changes that indicates the formation of nanoparticle colloids.

6.3.1.2 Dynamic Process

LAL is conducted at the solid–liquid interface by a pulsed laser. Nanosecond Nd:YAG laser is often chosen as the energy source because of its availability. Other laser sources, such as pulsed ruby laser, excimer laser, Ti:Sapphire femtosecond laser, and copper vapor laser were also reported as producing nanoparticles in liquids. Recently, Liu and coauthors [14, 15] fabricated metal oxide nanoparticles using a continuous wave (CW) fiber laser. The process of laser ablation varies with pulse duration (or laser interaction time with target materials for CW lasers): vaporization for CW and microsecond lasers, plasma formation for lasers in the nanosecond scale and beyond. Nanosecond laser ablation is a unique process to produce a strong plasma by an inverse bremsstrahlung absorption. When the laser intensity is high enough (e.g., $10^7 - 10^{12}$ W cm^{-2}), the later part energy of a laser pulse can be absorbed by the plasma plume formed in the earlier part of the laser pulse. The strong plasma can support the growth of special crystalline nanoparticles, such as diamond nanoparticles.

LAL is a complex and explosive process for the synthesis of nanoparticles. It includes the formation of plasma plume, shock wave in the liquid generated by plume expansion, shock wave in the target generated by plume recoil, cavitation bubbles, and the dynamic growth of nanoparticles as illustrated in Figure 6.3.2. Investigation on the processes of materials' ejection, nucleation, and aggregation is significant for controlled fabrication of nanoparticles. Fabbro, Berthe, and coworkers [16–19] studied the formation of plasma, shock waves, and laser-induced pressure in water. Through the measurement of peak pressures using high-resolution piezoelectric

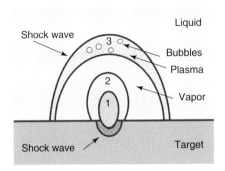

Figure 6.3.2 Sketch of the laser plasma plume formation induced by LAL at different stages: (1) initial, (2) expansion, and (3) saturation.

X-cut quartz crystal gauges and a fast oscilloscope, the relation between laser parameters and peak pressure as well as laser shock waves was established.

Laser power intensity (or laser fluence with a specific pulse duration) is the main parameter responsible for materials and liquid breakdown and plasma formation. When the power intensity reaches a threshold of $3-4$ GW cm^{-2}, laser ablation can create a plasma to induce an ionization avalanche process [17]. This threshold is sensitive to pulse duration and increases strongly for short pulse durations. For a 3 ns pulse, the threshold value was detected to have a 10-fold increase in comparison to a 25 ns Gaussian pulse [17]. Higher input laser power intensity tends to induce a stronger confinement effect, which leads to high-pressure and high-temperature plasma, and further gives rise to a higher ablation rate. Figure 6.3.3 shows the ablation rates of silicon and glassy carbon in air and water. Compared to the ablation rate in air, there is a slight enhancement of the ablation rate in water starting at 0.2 and 2.0 J cm^{-2} for glassy carbon and silicon, respectively. The ablation rate increases nearly 50% in water for glassy carbon at a laser fluence of 0.8 J cm^{-2}, while it increases above 500% in water for silicon at a laser fluence of 3.1 J cm^{-2}.

Laser wavelength can influence the plasma-induced pressure [19]. Peak pressure was detected as a function of laser power intensity obtained in water at 355,

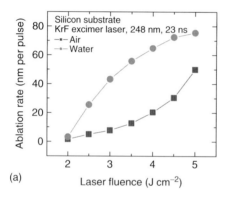

(a)

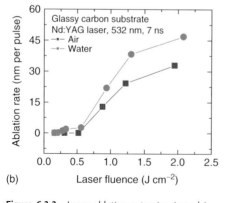

(b)

Figure 6.3.3 Laser ablation rates in air and in water for (a) silicon and (b) glassy carbon.

532, and 1064 nm laser wavelengths. The peak pressure measurement indicates that the generation of laser plasma in a confining regime is promoted by shorter wavelengths in ultraviolet (UV) and visible light bands with higher interaction efficiency. Below a threshold power intensity about 10 GW cm^{-2}, the experimental curves coincide well with the simulation results. The high interaction efficiency at deep UV wavelengths was investigated by 248 nm excimer laser ablation of a silicon wafer in water [20, 21]. By considering that laser-generated plasma in water induces shock waves, and the plasma shares similar state equations as the perfect gas, the laser-ablation-induced pressure can be estimated and compared to that in air [21]. The enhancement ratio of plasma pressures in water is 6.5 times at a laser fluence of 0.8 J cm^{-2} for glassy carbon under a Nd:YAG laser ablation and 5.8 times at a laser fluence of 4.5 J cm^{-2} for silicon under a KrF excimer laser ablation.

Laser wavelength used for LAL is varied from UV to near-infrared (NIR) bands. It is a critical parameter for optical breakdown of the liquid as different liquids have different absorption bands. Water has an absorption peak at the NIR band around 1000 nm, while organic solutions such as toluene have strong absorption in the UV band. For the case of water mixed with nanoparticles, the breakdown threshold at nanosecond laser irradiation is around 10 GW cm^{-2} at 532 nm, while it is 10 times lower for 1064 nm at around 1 GW cm^{-2} [22]. As the breakdown of liquid may involve nanoparticle growth, proper selection of laser sources can help control the size distribution and shape uniformity of the nanoparticles.

Liquid layer thickness plays a significant role in the LAL process. It was demonstrated that the ablation rate is strongly dependent on the liquid thickness [23]. The ablation rate increases with the liquid thickness before going over a threshold thickness value. After that the ablation rate is either saturated or reduced with the further increase in liquid thickness. With a 248 nm KrF excimer laser irradiation, the ablation rate of silicon is greatly increased in a water-confinement geometry with a peak enhancement at an optimal water layer thickness of 1.1 mm and at a laser fluence of 3.1 J cm^{-2} as shown in Figure 6.3.4 [20]. Usually for short and ultrashort laser pulses at nanosecond scale and beyond, the created plasma plume at the leading edge of a laser pulse does not affect laser irradiation in the remaining

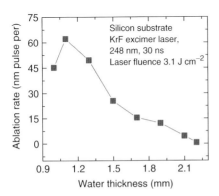

Figure 6.3.4 Ablation rates of silicon at different water layer thicknesses.

edge. The decrease in the ablation rate in liquids that are thicker than the optimal liquid thickness may be attributed to laser energy loss due to light scattering as well as nonlinear optical limiting effect induced by nanoparticles. The local density of nanoparticles explosively rises under intense laser light at the focusing spot. In a liquid environment, the transmission of laser light follows the rule of the Beer–Lambert law and decreases quickly with the increase in nanoparticle density. When the laser fluence is high enough, the presence of nanoparticles can bring about a nonlinear optical limiting effect to induce a further decrease in the laser energy absorbed by the target [24].

6.3.1.3 Growth Mechanism of Nanoparticles by LAL

Particle formation during pulsed laser ablation of a target in a gas environment was elucidated in detail as mechanisms of removal due to laser-ablation-induced shock waves, forcible ejection of surface matter, splashing of molten surface layer, and condensation from plasma species [25]. Only the particles formed by a condensing process from plasma species have sizes likely in the nanometer range. Liquid can provide much stronger confinement of the plasma plume than gas. The nanoparticles produced in liquid confinement experience a much more intense thermodynamic and kinetic process of the generation, expansion, and condensation of the plasma plume. The thermal and kinetic process of LAL can be described as follows:

- Laser ablation of a solid target in liquid
  - plasma formation and expansion
  - shock wave generation
  - plasma-induced pressure
  - plasma temperature increase.

LAL induces high particle density (including plasma species, nuclei, and nanoparticles), high pressure, and high temperature. At the end of a laser pulse, the plasma reaches a thermodynamic state under the strong confinement of liquid. In the plasma created by laser ablation of an isotropic graphite target in water, the induced pressure can easily go up to 10 GPa with a density of plasma species around $10^{22} - 10^{23}$ cm^{-3} and a temperature of 4000–5000 K [12]. The cavitation bubbles created by laser plasma also contribute to the temperature rise of the plasma plume. Shadowgraphy results of the ablation reaction of a metal target in water revealed the fast expansion and collapse of the cavitation bubbles at a speed up to 400 m s^{-1} [26]. It is interesting to find that a snapping shrimp can eject high-speed cavitation bubbles with a velocity around 25 m s^{-1} [27]. The maximum size ($\sim$3.5 mm) and lifetime ($\sim$300 μs) of bubbles are comparable to that produced by LAL ($\sim$3 mm and 300 μs, respectively). At the point of collapse, the cavitation bubbles created by snapping shrimps have an extremely high temperature above 5000 K [28]. So, it can be deduced that bubbles produced by much stronger LAL processes should have a much higher temperature above 5000 K.

Bubble formation has been monitored with a time-resolved transmission method during the LAL process [29]. As shown in Figure 6.3.5, the transmission is constant

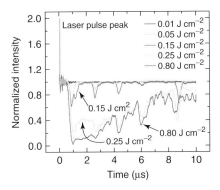

Figure 6.3.5 Time-resolved evolution of transmission during laser ablation at different laser fluences.

at a low laser fluence of $0.01\,\mathrm{J\,cm^{-2}}$. The formation of bubbles was observed at a laser fluence of $0.05\,\mathrm{J\,cm^{-2}}$. The bubbles collapse to induce the disturbance shown as valleys in the transmission spectra. When the photolytic removal of carbon species begins at $0.15\,\mathrm{J\,cm^{-2}}$, there is a sharp drop in transmission. The generation of carbon species explosively promotes bubble formation.

At the extreme condition provided by LAL, chemical reactions that cannot be realized in an ambient environment take place, with the formation of the metastable phase of the ablated species, reactions between the species from the laser-ablated target materials and liquid molecules or their plasma species. By choosing proper target–liquid systems, nanoparticles of target materials or allotropes and new nanoparticles by the combination of the elements of the target and the liquid can be fabricated [12]. For example, amorphous carbon and diamond nanoparticles were prepared by laser ablation of a graphite target in water [7, 30–33], while carbon nitride nanocrystals were synthesized from a pure graphite target in an ammonia solution [34].

The mechanism of nanoparticle growth is not yet well elucidated as nanoparticle generation is a complex process involving the generation of plasma, rapid expansion of plasma plume, shock wave, and strong interaction of plasma species that are instantaneously quenched in a highly confined liquid environment. The plasma plume has high pressure, high temperature, and high density of species, consisting of a mixture of energetic species including atoms, molecules, ions, clusters, and ultrafine particulates. The strong hyperthermal reaction between the ablated energetic species and the confining background liquid molecules promotes the nucleation and aggregation. This is the so-called reactive quenching process [7]. The nanoparticle size is decided by the availability of energetic species in the close vicinity of the formed nucleus. Mafuné *et al.* [35] proposed a dynamic formation mechanism to explain the growth phenomenon. With strong laser ablation, the rapid ejection of "embryonic plasma species", as well as a consecutive nanoparticle growth, competes with growth termination factors, such as protection from solvent molecules and depletion of species. The nanoparticles continue to explosively grow until the dense species in their close vicinity are completely depleted.

They may have a second-phase growth when species are supplied through diffusion, but this second-phase growth is relatively slow and can be interrupted by solvent molecules. It was reported that the solvent plays a significant role in controlling the final size and shape of nanoparticles [24, 36]. For high-repetition-rate laser ablation with sufficiently high input energy, the nanoparticles may be fragmentized [37–39]. Then, the growth process cycles and reaches a new equilibrium status. The size of nanoparticles can be finely tuned by properly selecting laser parameters. In addition, bubbles created by plasma and fast vaporization of surrounding liquid also contribute to nanoparticle growth due to hydrodynamic instabilities at the interface [26]. It means that LAL is a process far from thermodynamic equilibrium.

6.3.1.4 **LAL Process**

As mentioned above, the ablation rate of LAL is greatly enhanced by liquid confinement. However, for brittle materials, such as silicon and glassy carbon, the target is easily broken when the laser fluence is much higher than, for example, $1\,J\,cm^{-2}$ for glassy carbon plates and $5.0\,J\,cm^{-2}$ for silicon wafers. The target is often broken into small pieces in water after several hundred pulses of laser irradiation due to a "water hammer" effect [40]. The water hammer, with a pressure up to a few hundred mega pascal, damages the rear surface of the substrate. The laser-induced shock wave breaks the target into micrometer-sized particulates, while the water hammer fractures the rear surface into small pieces by repetitive laser ablation. In order to reduce microparticles and avoid breaking the target, the lowest applicable laser fluence is often used to make carbon nanoparticles with small size distribution, for instance, $0.8\,J\,cm^{-2}$ for laser ablation of a glassy carbon plate in water [41]. However, for the production of diamond nanoparticles, very high laser fluence up to $1000\,J\,cm^{-2}$ is needed to provide the extreme reaction condition [30].

The target material is usually placed in an open container for LAL processing as illustrated in Figure 6.3.1. It has a simple design that allows the laser plume to be monitored easily by detecting optical signals emitted. However, under intense laser irradiation, splashing of liquid due to laser-ablation-induced shock wave is often observed. The splashed liquid with nanoparticles and/or organic components can contaminate the optics. The contaminant diffuses and attenuates the incident laser light and leads to energy loss and low productivity. As the splashed liquid droplets contain nanoparticles, there is more or less a loss of nanomaterials. The liquid also needs to be constantly added to keep a constant depth. To prevent liquid splashing, a sealed container is often used [8, 42].

Liquid splashing is caused by several factors. Laser fluence is a crucial factor, which is usually very high for producing special nanoparticles, such as diamond nanoparticles. The thickness of the liquid layer on top of the target material plays a significant role in splashing. For the case of laser ablation of a silicon target in water, a water thickness of 1.1 mm generates the strongest acoustic wave and results in severe water splashing at a laser fluence of $3.1\,J\,cm^{-2}$ [20]. But when the water depth is over 2.2 mm, the splashing is reduced greatly. Laser spot size is another important factor that affects liquid splashing. At fixed laser fluences, a larger spot needs higher input laser energy to induce splashing. Besides tuning

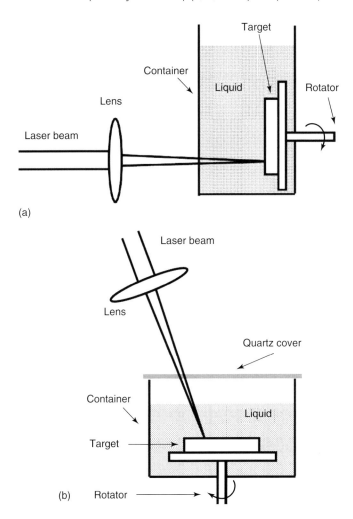

Figure 6.3.6 Experimental setups to reduce liquid splashing: (a) side ablation (Source: Adapted from Ref. [8]) and (b) inclined ablation of a target.

laser parameters, the configuration of the irradiation angle and direction of laser beam can be changed to reduce splashing [43, 44], as sketched in Figure 6.3.6a. In addition, a quartz window helps splashing control. The experimental configuration is shown in Figure 6.3.6b.

LAL often generates audible acoustic waves. Material removal takes place during laser ablation and leads to plasma generation and shock wave formation. In the beginning, the shock wave transports into the water layer, while the plasma causes water explosion. In the following stage after penetrating from water into air, the shock wave decays and turns into acoustic waves as a result of air friction. This process is described as an "ablative piston" effect. The audible acoustic wave generated during the laser ablation can be detected by a wideband microphone.

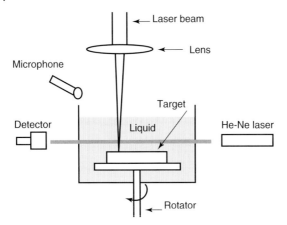

Figure 6.3.7 Experimental setup for *in situ* monitoring of the LAL process.

The first peak-to-peak amplitude of the acoustic waves generated in a liquid is 25% higher than that generated in air. Using acoustic wave measurement, liquid thickness (from liquid surface to target surface) and ablation rate can be checked in real time by proper calibration of the laser ablation rate and the first peak-to-peak amplitude [20, 21].

LAL needs *in situ* monitoring of the process, especially for long time processing up to several hours. The liquid loss due to vaporization and splashing can be verified by acoustic wave measurement. The concentration of nanoparticles generated is examined using a He–Ne laser that passes through the liquid sample during laser ablation. The transparence of the sample decreases with the increase in nanoparticle concentration. The final concentration is determined by UV–vis spectrum of liquid solution. The experimental setup for *in situ* monitoring of the LAL process is presented in Figure 6.3.7.

After the laser ablation, the obtained colloidal solution was stirred to make a uniform distribution of nanoparticles. Then, the liquid solution was collected from the container with the rest stored tightly in a Pyrex vial. To ensure the uniformity of particle dispersion and avoid possible aggregations, the sample was vibrated with an ultrasonic wave before preparing samples for measurement. Condensing and dilution may be applied to change the concentration of nanoparticles.

6.3.1.5 Nanoparticle Control
LALs can be used to produce nanoparticles for a wide range of applications. The size and shape as well as the crystal phase of nanoparticles are significant factors for determining their unique features of property, quality, and function. How to grow nanoparticles by LAL with controlled features is still under extensive research. For example, size and shape control of metal nanoparticles have been reported by Kondow *et al.* [35]. The nanoparticle size increases almost linearly with the laser power. The produced nanoparticles can be dissociated into smaller ones by multipulse repetitive ablation at proper laser fluences. The nanoparticle

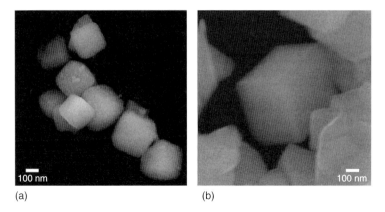

(a)   (b)

Figure 6.3.8 Nanocubes synthesized by inorganic-salt-assisted LAL: SEM images of (a) carbon nanocubes [13] and (b) and silicon nanocubes [47].

shape can be tuned by melting at a relatively low laser fluence [45] and an appropriated wavelength [46]. Surfactant molecules can protect and stabilize the formed nanoparticles from explosive growth. A liquid with a higher concentration of surfactants favors the production of smaller nanoparticles [35]. These results provide good references for the controlled synthesis of nanoparticles of silicon, carbon, and germanium by tuning laser parameters and changing the liquid environment using surfactant and inorganic salts. For example, carbon nanocubes with C_8-like structures and silicon nanocubes with zinc blende structures were constructed using inorganic-salt-assisted LAL as shown in Figure 6.3.8. The ions of the inorganic salt preferably adsorb onto a certain crystallographic plane to initialize the oriented nanocube growth.

An electric field can be a significant factor affecting nanoparticle growth during the nucleation process [48]. In order to manipulate the nanoparticles, an external electrical field can be applied during the LAL process [13]. As demonstrated in Figure 6.3.9, the size and shape as well as chemical components of nanoproducts

Figure 6.3.9 Ge and GeO_2 nanoparticles synthesized by electrical-field-assisted LAL: (a) Ge nanoparticles synthesized without applying an electric field, (b) GeO_2 nanocubes synthesized at a voltage of 14.5 V, and (c) GeO_2 nanospindles synthesized at a voltage of 32.0 V [13].

are flexibly controlled by the variation of the voltage of the external electrical field. The nanoparticles have interesting shape-dependent luminescence [13]. With the existence of an external electrical field, the surface electrostatic potential of the forming crystallite affects the nuclei–nuclei interactions to provoke an orientated formation of the crystalline planes. The nanoparticles grow in an anisotropic way from a primary nucleus. The electrical field can electrolyze water to supply the active oxygen element, which interacts with the laser plasma and involves in the process of nanoparticles growth to form germanium oxide.

6.3.1.6 Safety Matters

Almost all organic solutions are flammable. Some of them even cause health risks to users. The usage of organic solutions should follow the descriptions on the Material Safety Data Sheet (MSDS) and the laboratory rules for handling hazard chemicals. To avoid fire and explosion, it is strongly recommended that an inert gas flow is applied so as to protect the processing container. For toxic solutions (such as benzene) and highly volatile solutions (such as acetone), it is safer to conduct the LAL process in a fume cupboard or a cabinet connected to the exhaust system.

6.3.2
Carbon Nanoparticles

Extensive research has been carried out on carbon nanoparticles in recent decades since the successful fabrication of diamonds by Ogale *et al.* [7]. Different types of carbon nanoparticles, including amorphous carbon, diamonds, short carbon chains, fullerenes, carbon nanotubes, and carbide were prepared by LAL. The experimental parameters used in the literature for producing carbon nanoparticles are summarized in Table 6.3.1.

6.3.2.1 Diamond Nanoparticles

Diamond is one of the naturally highly stable carbon allotropes. The synthesis of diamond nanoparticles has significant impacts as diamond has unique physical and chemical properties that cannot be found in other materials. The transition from other carbon allotropes to diamond requires high pressure and high temperature due to the high energetic barrier. For example, the transition between graphite and diamond requires a pressure around 10 GPa and a temperature around 4000 K as shown in the phase diagram [61]. Laser ablation can provide such a transition condition at a high pressure up to 15 GPa and a high temperature up to 5000 K [62–64]. It was well established that the LAL opens a new route toward dynamic synthesis of diamond nanoparticles, although productivity can be increased further.

The reported LAL parameters for growing diamond nanoparticles are summarized in Table 6.3.1. Experimentally, there are three critical parameters in fabrication of diamond nanoparticles: source materials, laser parameters, and liquids.

1) **Source material effect on nanoparticle growth:** The conversion from graphite to diamond takes place via a minimal displacement of atoms for the transformation, which means conversion paths are from a hexagonal graphite lattice to

Table 6.3.1 Carbon nanoparticles produced by laser ablation in different liquids.

Liquid parameters/ Experimental arrangement	Target material	Laser parameters	Product obtained	Findings	References
Benzene 3–4 mm thick	Polished pyrolytic graphite	Ruby (694 nm, 30 ns, single pulse) 20 J cm^{-2}	Diamond	Diamond particulates with sizes of 5–20 μm were produced on the target surface	[7]
Water 5 mm thick	HOPG, polymorphous graphite, arc-discharging deposit, glassy carbon, and imitation diamond	Nd:YAG (532 nm, 7 ns) 5 × 10^8 W cm^{-2}	Diamond and other carbon allotropes	Spherical carbon nanoparticles and diamond nanospherulite were synthesized. The formation of the carbon nanospheres becomes easier with increasing the degree of chaos and the amount of sp^3 bonding in target materials. Carbon nanotubes were observed	[32]
Twice-distilled water 1–2 mm thick	Spectroscopically pure polycrystalline graphite	Nd:YAG (532 nm, 10 ns, 5 Hz, 30 min) 10^{10} W cm^{-2}	Diamond	Diamond grains with sizes of 300–400 nm were synthesized. The nanograins have both hexagonal and cubic diamond structures	[33]
Twice-distilled water 1–2 mm thick	Spectroscopically pure polycrystalline graphite	Nd:YAG (532 nm, 10 ns, 5 Hz, 30 min) 10^{10} W cm^{-2}, 200–350 mJ/pulse	Diamond	The metastable rhombohedral graphite phase was observed as the intermediate phase in the conversion from hexagonal graphite to cubic diamond	[49]
Absolute acetone 1–2 mm thick. The target was rotated at 20 rpm	Spectroscopically pure polycrystalline graphite	Nd:YAG (532 nm, 10 ns, 5 Hz, 45 min) 10^{10} W cm^{-2}, 200–350 mJ/pulse	Diamond and graphite	Diamond nanocrystals were prepared in a mixture of cubic and hexagonal phases. The ratios of nanodiamonds and graphite to the whole prepared powders were about 5 and 95%, respectively. A new Raman line at 926 cm^{-1} and a Raman line at 1100 cm^{-1} for nanodiamond were observed	[50]

(continued overleaf)

Table 6.3.1 (Continued)

Liquid parameters/ Experimental arrangement	Target material	Laser parameters	Product obtained	Findings	References
Deionized water 2–3 mm thick/vertically downward	700 nm amorphous carbon film deposited by filtered cathode vacuum arc method on a 5 × 5 mm² single-crystalline silicon	Nd:YAG (532 nm, 10 ns) ~5.7 × 10^8 W cm⁻², diameter of focus spot about 1.8–2.2 μm, two to three laser pulses at each spot	Diamond	Diamond phase was created in the amorphous carbon films with an increase of the sp³ content localized in the treated sample. The nanocrystalline diamond-embedded amorphous carbon films show improvement in field emission performance	[51]
Water or cyclohexane (with Ar gas flow) 1 cm thick/vertically downward	Polished pure graphite disk	Nd:YAG (532 nm, 10 ns, 10 Hz, 15 min) 130 mJ/pulse, 66 J cm⁻², 25-cm-focal-length lens	Diamond	The majority (~95%) of the product is graphite. Diamond nanocrystals were observed in small patches at the edge of the larger pieces of graphite. Diamond can be produced in non-oxygen-containing liquid	[31]
5–10 ml of deionized water with a thickness of 5 mm/vertically downward	Graphite (99.99%)	Nd:YAG (532 nm, 15 ns, 30–60 min) 125 mJ/pulse, ~28 J cm⁻², laser spot about 0.5 mm, and 50 mm focal length of a biconvex lens	Diamond	The product is a mixture of amorphous carbon and graphite with a small amount of hexagonal crystalline diamond nanostructures (~5%). Strong emission peaks from atomic H were detected	[42]

Water	Pyrolytic graphite	Nd:YAG (355 nm, 5 ns, 10 Hz) 80 mJ], 40–1000 J cm^{-2}, 200 mm focal length, spot size on the target surface 0.1–0.5 mm	Diamond	Diamond–graphite core–shell nanoparticles were fabricated with core sizes of 4–15 nm and shell thickness of 3–4 nm	[30]
Water, 2-propanol, c-hexane/ horizontal	125 μm carbon film without substrate	Ruby (694 nm, 30 ns, single pulse, 10–100 shots) 27 J cm^{-2}, laser spot area of 1.5–18 mm^2	Amorphous carbon	Carbon nanoparticles having graphite structure with sizes of 1–3 nm were obtained	[8]
Ethanol, water	Hot-pressed boron carbide (B$_{4,3}$C)	Nd:YAG (1064 nm, 10–15 ns, 1 Hz, 120 min) 0.5–1.0 J/pulse, 100 mm focal length	Amorphous carbon	Encapsulated carbon spherules with sizes from 70 nm to 2.8 μm were produced. The graphite core is covered with a boron shell. Crystalline carbon nanofibers with length of ~1 μm and diameter of 20–60 nm were also observed	[52]
Water, ethanol, tetrahydrofuran	Glassy carbon, pyrolytic graphite	Nd:YAG (532 nm, 7 ns, 10 Hz, 5 min) 0.8–1 J cm^{-2}, 250 mm focal length	Amorphous carbon	Uniform carbon nanoparticles below 30 nm were prepared. Tetrahydrofuran can help form stable carbon nanoparticle colloids	[24, 41]
Isopropyl alcohol ((CH$_3$)$_2$CHOH)/ horizontal	Graphite	Nd:YAG (1064 nm, 3.5 ns, 30 Hz, 30 min) 10 mJ/pulse, 10^8 W cm^{-2}, laser spot size of ~100 μm	Amorphous carbon	Crystalline carbon nanoparticles assemble to form rose-shaped carbon particles with sizes of about 2 μm, which further assemble to cracknel-shaped carbon microparticles with a size of about 20 μm	[43]

(continued overleaf)

Table 6.3.1 *(Continued)*

Liquid parameters/ Experimental arrangement	Target material	Laser parameters	Product obtained	Findings	References
Twice-distilled water 2–3 mm thick	100–150 nm amorphous carbon films on single-crystalline silicon substrates	Nd:YAG (532 nm, 10 ns, >20 min) 10^9 W cm^{-2}, target rotating at 15 rpm	Carbon nanocubes	Carbon nanocubes were produced in a gray product mixed with amorphous carbon. TEM examination showed that the carbon nanocubes are single crystals with a C_8 crystalline (bcc) structure	[53]
Mixture solution of twice-distilled water, ethanol, acetone with inorganic salts (≤4 mM) such as KCl (99.5%) and NaCl (99.5%) 2–3 mm	150–300 nm amorphous carbon films on silicon substrates	Nd:YAG (532 nm, 10 ns, ≤10 Hz, 15 min) 10^{10} W cm^{-2}	Carbon micro-/ nanocubes	Carbon micro-/nanocubes with slightly truncated corners were generated with an average edge length of about 1 μm. The size can be controlled by processing in a vibrator with oscillation frequency of 60 kHz, which reduces the size distribution	[36]
Twice-distilled water	HOPG	Nd:YAG (532 nm, 7 ns, 10 Hz, 60 min) 5×10^8 W cm^{-2}, laser spot area of 1 mm^2	Carbon nanotubes	Carbon nanotubes were prepared without the addition of any catalyst. Perfect graphite layer structure favors the formation of nanotubes	[54]

Target	Product	Laser conditions	Liquid	Results	Ref.
Graphite	Carbon nanotubes	Nd:YAG (1024 nm, 10 Hz) 4.5−6.0 × 10^9 W cm^−2	Fe or Ni nano-sol prepared by laser ablation of metal in ethanol. The typical flow rate is 1.2 ml min^−1, which keeps liquid thickness about 1 mm	Carbon nanotubes were prepared with the presence of Fe or Ni nano-sol flows as catalysts. Nanotube formation only occurred within the fluence range of 4.5−6.0 × 10^9 W cm^−2. The nanotubes are 15−45 nm in diameter and a few micrometers in length. Without the addition of catalyst, the final product is carbon fiber	[55]
Polycrystalline graphite	Carbon nitride	Nd:YAG (532 nm, 10 ns, 5 Hz, 30 min) 10^10 W cm^−2	25% ammonia solution 1−2 mm thick	Carbon nitride nanocrystals with a cubic-C_3N_4 phase were observed	[34]
Graphite (99.99%), sealed in a stainless steel cell with rotation speed 700 rpm	Carbon nitride	Nd:YAG (532 nm, 15 ns, 10 Hz, 300 min) 25−125 mJ/pulse, a 25 mm focal length lens and a spot diameter ~0.5 mm	5 ml of 35% ammonia solution with a thickness of ~5 mm	Carbon nitride nanoparticles composed of crystalline α- or β-C_3N_4. The morphology of the crystalline material can be manipulated by changing laser ablation time and laser fluence. Ordered leaflike structures (30−50 nm by 200 nm), interconnected networks, large mesoscale clusters, and multifold-symmetry flowerlike structures, can also be fabricated via a self-assembly ordered scheme using nanoparticles/nanorods (10 nm by 200 nm) as building blocks	[56]
Graphite (99.99%) with diameter 16 mm by 4 mm height, sealed in a stainless steel cell with rotation speed 700 rpm	Carbon nitride	Nd:YAG (532 nm, 15 ns, 10 Hz, 60−720 min) 100 mJ/pulse, a 25 mm focal length lens and a spot diameter ~0.5 mm	5 ml of 35% ammonia solution with a thickness of ~5 mm	Carbon nitride nanoparticles composed of crystalline α- or β-C_3N_4 were produced. The nanoparticles change shapes with ablation time from spherical shape (a few minutes), nanorods (~60 min), "leaflike" nanostructures (60−420 min), disordered micrometer-scale complex crystallites in an open network (>420 min), to "flowerlike" structures (>720 min). Longer ablation times increase the length of the nanoleaf and nitrogen component. The band gap can be tuned in a range of 3.90−4.05 eV by varying the ablation time	[57]

(continued overleaf)

Table 6.3.1 (*Continued*)

Liquid parameters/ Experimental arrangement	Target material	Laser parameters	Product obtained	Findings	References
3–5 ml of 25–35% ammonia solution (about 5–10 mm liquid layer)	Graphite (99.99%) with diameter 16 mm by 4 mm height, sealed in a stainless steel cell	Nd:YAG (532 nm, 15 ns, 10 Hz, 30–600 min) 50–125 mJ/pulse, a 25 mm focal length lens and a spot diameter ~0.5 mm	Carbon nitride	Self-assembled 3D nanostructures and microstructures of crystalline carbon nitride were generated. The ammonia concentration, reaction time, and evaporation rate of the liquid affect the morphology	[58]
5 ml of 25–35% ammonia solution ~5 mm liquid layer	Graphite (99.99%) with diameter 16 mm by 4 mm height, sealed in a stainless steel cell	Nd:YAG (532 nm, 15 ns, 10 Hz, 30–720 min) 25–125 mJ/pulse, a 25 mm focal length lens and a spot diameter ~0.5 mm	Carbon nitride	The nanostructure morphology is tunable by varying the ablation time, laser energy, and ammonia concentration. Different types of symmetric organizations can be obtained by controlling the drying time of the carbon nitride colloid deposited onto substrates	[59]
5 ml solution of 35% ammonia ~5 mm liquid layer	Graphite (99.99%)	Nd:YAG (532 nm, 15 ns, 10 Hz, 120–300 min) 15 J cm^{-2}, a 25 mm focal length lens and a spot diameter ~0.5 mm	Carbon nitride	Carbon nitride nanoparticles in α-C_3N_4 phase were prepared with a size distribution of 8–12 nm. Longer ablation time leads to single-crystalline nanorod-like structures	[60]

HOPG, highly oriented pyrolytic graphite.

a hexagonal diamond lattice and from a rhombohedral graphite lattice to a cubic diamond lattice [33]. The metastable hexagonal graphite may be converted
into cubic diamond via an intermediate phase of rhombohedral graphite [49].
The presence of microcrystalline graphite can favor the transformation from
graphite to diamond.

2) **Laser parameter effect on nanoparticle formation:** The laser parameters were
   tuned to create the dynamic conditions of diamond growth as shown in
   Figure 6.3.10. Laser power intensity in the range of $5 \times 10^8 - 2 \times 10^{11} \ W \ cm^{-2}$
   is normally used to fabricate diamond nanoparticles. With a short laser pulse
   at nanosecond scale, the condition of diamond growth falls in the region
   proposed by Yang *et al.* [49, 65] as illustrated in Figure 6.3.10. It indicates that
   diamond nanoparticles form at a high tolerance of temperature. However, for
   a long laser pulse at microsecond scale, the generation of diamond for the laser
   ablation of suspensions occurs in a narrow phase window around 4000 K, as
   suggested by Tian *et al.* [61]. The diamond productivity rises with the increase
   in laser power intensity [49].

3) **Liquid effect on nanoparticle preparation:** For aqueous solutions, the intensive
   laser ablation as well as heat exchange from the target to the liquid can cause
   liquid breakdown to form high-temperature and high-pressure vapor with
   dense ions OH^- and H^+. Emission from atomic C and C^+ was observed in
   wavelength-resolved emission spectra of the laser plume induced by 532 nm
   laser excitation [42]. The reaction among water and these energetic ablated
   C atoms, ions, or clusters can produce excited hydrogen atoms or ions. It
   is proposed that hydrogen is a required element for graphite-to-diamond
   conversion. In addition, direct laser breakdown of the water and the collapse of
   cavitation bubbles are other sources of ions OH^- and H^+. These high-density
   ions can crucially suppress the generation of graphitic sp^2 bondings and
   promote the diamond growth [49]. Theoretically, diamond nanoparticles at a
   radius of 2–5 nm would be easily synthesized via a process of dynamic growth
   in water [62]. The occurrence of diamond nanoparticles from 40 to 200 nm

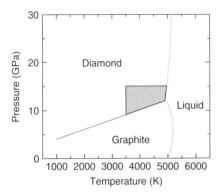

Figure 6.3.10 Pressure–temperature phase diagram of carbon. (Source: Adapted from
Bundy *et al.* [66].)

was also predicted and produced [63, 65]. Micrometer-sized diamond particles were successfully made by laser ablation of carbon targets in benzene [7, 32]. In organic solutions, the existence of cyclic and aromatic compounds can boost the diamond nuclei to grow. As the arrangement of carbon atoms in these molecules is similar to that of diamond, the reconstruction of carbon atoms is relatively easy [67]. Cyclohexane [31] and benzene [7] were successfully used as quenching liquids for the synthesis of diamond nanoparticles.

6.3.2.2 Amorphous Carbon Nanoparticles

Amorphous carbon nanoparticles have been found in various systems as a key component ranging from interstellar dust, arc-generated soot to vacuum-deposited thin films. Recent studies discovered that amorphous carbon nanoparticles have unique nonlinear optical properties [24, 41]. Particularly, the carbon nanoparticles prepared in water possess a strong nonlinear optical limiting property that is comparable to excellent optical limiting materials, such as carbon nanotubes

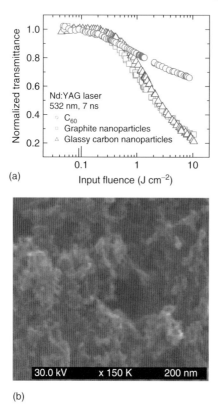

(a)

(b)

Figure 6.3.11 (a) Optical limiting performance of carbon nanoparticles in water with references of C_{60} toward 532 nm laser beam and (b) SEM image of carbon nanoparticles obtained by laser ablation in water.

and fullerenes (C_{60}), as displayed in Figure 6.3.11a. The amorphous carbon nanoparticles are spherical in shape as shown in Figure 6.3.11b.

Compared to the LAL preparation of diamond nanoparticles, which needs a high laser power intensity around $5 \times 10^8 - 2 \times 10^{11}$ W cm^{-2}, amorphous carbon nanoparticles can usually be grown at a laser power intensity as low as around 1×10^8 W cm^{-2} [29]. The graphite target as well as its microcrystals were converted into small nanoparticles without phase transition happening in the LAL process [43]. Amorphous carbon nanoparticles can also be made at high-power intensities. The main product of the LAL for growing diamond nanoparticles is amorphous carbon nanoparticles. These amorphous carbon nanoparticles are unavoidably formed during the LAL even with special inorganic salt for manipulating the nanoparticle shape [13].

Source materials, laser power intensities, and liquids are key experimental parameters for the formation of amorphous carbon nanoparticles:

1) **Target material effect on bonding structures of amorphous carbon nanoparticles:** The target materials used include carbon film, glassy (vitreous) carbon plate, pyrolytic graphite, polymorphous graphite, highly oriented pyrolytic graphite (HOPG), arc-discharging deposit, and imitation diamond. The pyrolytic graphite is easily cracked under laser irradiation because of the strong water hammer effect. The size of micrographite crystals in glassy carbon is close to that of pyrolytic graphite. Compared to pyrolytic graphite, glassy carbon has a higher density bonding structure similar to diamond bondings. This strong diamondlike bondings (sp^3 bonds) and unique microstructure of glassy carbon provide high strength and hardness to avoid the cracks during the laser ablation. Pyrolytic graphite has relatively weak carbon–carbon bondings (sp^2 graphitic bonds) and is likely to form micrometer-sized particulates. In addition, the sp^3 bonding in the carbon targets, such as arc-discharging deposit, glassy carbon, and diamond, tends to assist the formation of spherical carbon nanoparticles. However, a few of spherical carbon nanoparticles were observed for sp^2-rich graphitic targets, such as HOPG and polymorphous graphite [32]. Micrometer-thick carbon films were used to produce carbon colloids in water, 2-propanol, and c-hexane [8]. Carbon nanoparticles obtained are in graphite phase with sizes ranging from 1 to 3 nm in diameter.

2) **Laser power intensity effect on nanoparticle size:** Amorphous carbon nanoparticles with sizes ranging from 40 to 60 nm were fabricated at a laser power intensity around 5×10^8 W cm^{-2} [32]. However, for a low laser power intensity around 1×10^8 W cm^{-2}, most of the amorphous carbon nanoparticles have a size below 30 nm [29]. Figure 6.3.12 shows Raman spectra of a glassy carbon target and the carbon nanoparticles fabricated. The D band of the nanoparticles is shifted with a broader full-width half maximum (FWHM). The shifting and broadening of the D band indicate an increased level of disorder and a decrease in the graphitic domain size. The Raman spectra do not have a characteristic signal of diamond at 1332 cm^{-1}. The results indicate the formation of nanoparticles in amorphous phase.

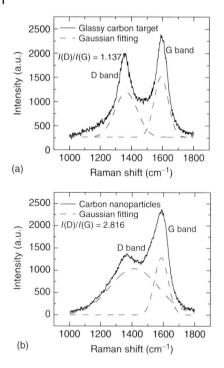

Figure 6.3.12 Raman spectra of (a) glassy carbon target and (b) carbon nanoparticles. Both Raman spectra are fitted with Gaussian distributions [29].

3) **Liquid effect on nanoparticle size:** The size of carbon nanoparticles prepared in deionized water ranges from 10 to 20 nm. Selected area diffraction (SAD) confirms that the obtained nanoparticles are in amorphous phase. However, for carbon nanoparticles produced in ethanol, the size of carbon nanoparticles distributes in a range from 5 to 30 nm [29]. Liu and coauthors [52] reported a wider size range from ~70 nm to 2.8 μm for laser ablation of a boron carbide target in ethanol. Carbon microparticles with sizes ranging from 1 to 10 μm were made from graphite target by laser ablation in isopropanol [43]. The difference in size distribution may come from the difference in thermodynamic properties among water, ethanol, and isopropanol. Compared to ethanol and isopropanol, water has the highest dielectric constant, boiling point, surface tension, density, and thermal conductivity, as shown in Table 6.3.2. High specific heat and high thermal conductivity induce fast creation of plasma and cavitation bubbles. High surface tension generates strong confinement conditions, thus leading to a narrow size distribution.

For some special solvents, such as tetrahydrofuran (THF), laser-plasma-induced high temperature and high pressure can initialize a polymerization process of THF molecules with the possible initiators of carbon clusters, such as C_1 and C_2. The polymerization of THF on carbon nanoparticle surfaces can stabilize the nanoparticles at its early growth stage. The typical carbon nanoparticles produced

Table 6.3.2 Thermodynamic parameters of water, ethanol, and isopropanol [68].

	Water	Ethanol	Isopropanol
Dielectric constant [69]	80.37	25.30	20.18
Boiling point ($^\circ$C) [69]	100.0	78.5	82.4
Density (g cm^{-3}) [70]	1.000	0.789	0.786
Refractive index (n) [70]	1.333	1.361	1.378
Specific heat (J g^{-1} K^{-1}) [71]	4.18	2.44	2.60
Surface tension (dyn cm^{-1}) [69]	73.05	22.75	23.78
Thermal conductivity (W m^{-1} K^{-1}) [71]	0.598	0.169	0.135

in THF have an average diameter around 6.5 nm. The samples obtained behave with strong nonlinear optical limiting properties showing a double exponential decay of transmission with the increase in input laser fluence [24]. It is interesting to note that after two months of the synthesis, carbon nanoparticles made in ethanol formed large aggregates with an average size around 500 μm. However, there is no obvious change for carbon nanoparticles fabricated in THF, as shown in Figure 6.3.13.

6.3.2.3 Carbon Nanocrystals

LAL provides an extreme reaction condition to form new carbon forms. As the sp^2 carbon bonding is metastable under this condition with high plasma density, high temperature, and high pressure, it tends to convert into stable sp^3 carbon bonding [53]. Figure 6.3.14a,b shows the bright-field transmission electron microscopic (TEM) images of the synthesized carbon nanocrystals in twice-distilled water. The measured interplane spacing of the nanocrystal is consistent with the calculated value of the (020) plane for the C$_8$ structures.

(a)   (b)

Figure 6.3.13 Photos of carbon nanoparticles synthesized by laser ablation in (a) ethanol and in (b) THF.

Figure 6.3.14 Carbon nanocrystals prepared by LAL: (a) TEM bright-field images of the synthesized body-centered cubic carbon nanocrystals and (b) corresponding high-resolution transmission electron microscopy (HRTEM) image with fast Fourier transform analysis [53]; (c) bright-field TEM and (d) HRTEM images of a carbon nanocube [36].

LAL is a simple technique to control the size of nanoparticles during the fabrication process. As discussed above, carbon nanoparticles produced under violent laser ablation are usually spherical. Controlled synthesis of nonspherical nanoparticles has been studied extensively for metallic particles [72]. After the LAL process, the as-prepared nanoparticles were irradiated by nonfocusing laser light. Laser-induced morphology conversion produces crystal-shaped nanoparticles in nonspherical shapes including nanoprisms and nanorods. The shape change was improved with the addition of inorganic salt (NaCl) into a nanoparticle solution. Intense laser light induces an oxidative etching – photoreduction reaction with the participation of chloride ions. The shape of carbon nanoparticles can also be manipulated by adding ions to the host liquid. Figure 6.3.14 shows carbon nanocrystals prepared by LAL with the addition of very low concentration inorganic salt solutions. The host solvent is a mixture of twice-distilled water, ethanol, and acetone with the addition of very low concentration inorganic salt solutions ($\leq$4 mM), such as KCl (99.5%) and NaCl (99.5%). As shown in Figure 6.3.14c,d, C_8-like carbon crystalline phase was successfully fabricated in 3D structures using salt-assisted LAL [36]. The extreme condition provided by LAL can induce nucleation and the phase transitions for the rapidly quenched plasma species. The salt ions participate in the nucleation and growth of nanocrystals in two ways: acting as a template to guide the carbon nuclei to grow into a nanocube and strongly adsorbing into a certain crystallographic plane and obstructing the crystal growth perpendicularly to this plane. The template effect and preferential adsorption lead to an oriental crystal growth and a final cubic structure in single-crystalline phase.

6.3.2.4 Synthesis of Other Carbon Nanomaterials by LAL

As mentioned in the previous sections, LAL provides an extreme condition with high temperature, high pressure, and high density of plasma species. The final product is normally a mixture of different types of carbon nanoparticles in different phases, shapes, or with different chemical components.

1) **Carbon nitride nanocrystals:** Strong chemical reactions could occur within the laser-induced plasma plume and at the interface between the liquid and the laser-induced plasma. Nanoparticles of metastable phase compounds can be prepared by the proper selection of the reaction systems. Yang and Wang [34] prepared column-polyhedron-shape nanocrystals with an average size of about 50 nm. The phase of nanocrystals is cubic – C_3N_4 structures. The productivity of carbon nitride nanocrystals is low compared to amorphous carbon nanoparticles in graphite phase. Yang *et al.* [60] prepared ultrafine carbon nitride nanoparticles with an average particle size of 8–12 nm at a laser processing time of 2 h. The nanoparticles transfer into anisotropy shapes as nanorods with extended processing time up to 5 h. It was confirmed that the nanorods are formed either by the aggregation of nanoparticles or by slow growth. The experimental parameters for preparing carbon nitride nanocrystals and their assembly structures are listed in Table 6.3.3.

2) **Carbon nanotubes:** Carbon nanotubes were fabricated by the laser ablation of HOPG in pure water [54]. The nearly perfect layer structures of HOPG favor the formation of carbon nanotubes. It was noticed that nanotubes are generated only when laser interacts perpendicularly with HOPG (0001) plane. Carbon nanotubes were also observed in the samples, which were prepared by the LAL for fabricating diamond nanoparticles [32]. Chen and coworkers [55] reported the formation of carbon nanotubes by a two-step LAL method. The first step is to prepare sol by laser ablation of a metal target in ethanol. In the second step, the metal sol is used as a catalyst flowing above the graphite target during the LAL process for preparing carbon nanotubes, and most of the obtained carbon nanotubes have a diameter of 15–45 nm and a length of a few micrometers. Typical carbon nanotubes using Fe nano-sol and Ni nano-sol catalysts are shown in Figure 6.3.15.

Table 6.3.3 Silicon and germanium nanoparticles produced by laser ablation in liquids.

Liquid parameters/ Experimental arrangement	Target material	Laser parameters	Product obtained	Findings	References
c-Hexane	Thin silicon film deposited on a thin glass plate (5 × 20 × 0.5 mm)	Ruby (694 nm, 30 ns, single pulse, 10–100 shots)	Silicon	Absorption of colloidal silicon increases with decreasing wavelength at visible and UV bands with an absorption peak at 230 nm	[8]
Ethanol, water, dichloroethane	Silicon crystal	Cu-vapor (510.5 nm, 20 ns, 15 Hz) 1–2 J cm^{-2}, with a spot diameter of 50 μm	Silicon	The size of silicon nanoparticles is in the range of 200–1000 nm with nanocrystals in the range of 60–80 nm, which was not affected by the liquid and the surface active agents. Silicon suboxide was observed	[73]
Supercritical liquid CO$_2$ (99.99%)/vertically downwards	Bulk silicon crystal (99.9999%)	Nd:YAG (532 nm, 20 Hz, 5 min) 30 mJ/pulse	Silicon	The hydrodynamic radius of generated clusters was estimated as <3 nm using dynamic light-scattering method. Absorption spectra of silicon clusters can be tuned by varying the pressure of CO$_2$	[74]
10 ml of deionized water	p-type crystalline silicon wafer (100), resistivity of 0.1 Ω cm, and thickness of 0.525 mm	Nd:YAG (355 nm, 8 ns, 30 Hz, 30 min) 0.07–6 mJ/pulse, laser spot of 1 mm	Silicon	Spherical silicon nanoparticles with a size distribution ranging from 2 to 100 nm and 60 nm in average were prepared at low fluences. Nonspherical aggregates were obtained at increased fluences	[75]

Liquid nitrogen	Crystalline silicon wafer with (100) orientation	Nd:YAG (1060 nm, 6 ns, 10 Hz, 30 min) 1–30 J cm^{-2}, focal length of 60 mm	Silicon	The synthesized silicon nanoparticles have sizes ranging from several tens to hundreds of nanometers with a polycrystalline structure and a pure cubic silicon phase	[76]
Water, glycerol (at 50 °C), and liquid nitrogen	(100) p-type crystalline silicon wafer, resistivity of 10 Ω cm, and thickness of 0.5 mm, surface size ~15 × 15 mm^2	Nd:YAG (1064 nm, 30 ps), 10 Hz (water and glycerol) 2 Hz (liquid nitrogen) 30 min (water) 60 min (glycerol and liquid nitrogen) focal lengths of 38 mm (water and glycerol) and 100 mm (liquid nitrogen) focus spot size of 0.3 mm (water and liquid nitrogen) and 0.1 mm (glycerol) 5 mJ or 0.7 J cm^{-2} (water) 10 mJ or 6 J cm^{-2} (glycerol) 10 mJ or 0.7 J cm^{-2} (liquid nitrogen)	Silicon	The maximum of the size distribution and the mean size of silicon nanoparticles obtained in water, glycerol, and liquid nitrogen are 20 and 23 nm, 12 and 24 nm, 7 and 9 nm, respectively	[77]

(continued overleaf)

Table 6.3.3 (*Continued*)

Liquid parameters/ Experimental arrangement	Target material	Laser parameters	Product obtained	Findings	References
8 ml of deionized water, ethanol, or a mixture of the two liquids	Silicon wafer with (111) plane	Nd:YAG (1064 nm, 10 ns, 10 Hz, 30 min) 50–200 mJ/pulse, 1–30 J cm^{-2}, spot size ~2 mm in diameter, focal length of 150 mm	Silicon	Stable colloid solution can be stored without any precipitation for two months. Nanoparticles prepared in ethanol have a smaller mean size and are more stable with a less blueshifted absorption edge than that in water. The mean size reduced gradually in ethanol and first decreased and then increased in water with the increase of laser fluence. Nanoparticles prepared in ethanol are single crystalline while that prepared in water are multicrystalline	[78]
Ethanol (95%) layer 2–3 mm thick with a volume of 10 ml	Single-crystal silicon	Ti:sapphire (800 nm, 35–900 fs, 1000 Hz, 10 min), single-pulse energy 2.5 mJ, constant at 4 J cm^{-2}, peak power of the laser radiation on the target varied from $4.3 \times 10^9 - 1.1 \times 10^{11}$ W cm^{-2}	Silicon	Spherical silicon nanoparticles obtained with sizes from a few nanometers to about 200 nm with a majority of 30–100 nm for long pulse durations. Silicon oxide and carbide were observed. The proportion of oxidized silicon is weak and increases when the particle size decreases. The solvent strongly affects the optical properties of the nanoparticles	[79]
5 ml of deionized water (18.2 MΩ cm), D$_2$O	n-type (100) silicon wafer	Ti:sapphire (800 nm, 120 fs, 1000 Hz, 30 min) pulse energy 0.1–1.0 mJ, a focal length of 75 mm and a spot diameter of ~1 mm	Silicon	Silicon nanoparticles have a mean size around 2.4 nm, which increase up to a few tens of nanometers with laser fluence. The nanoparticles obtained can generate singlet oxygen (^{1}O$_2$) under at various wavelengths and present no photobleaching	[80]

Toluene (99.5%) A DC electrical voltage of −47 V is applied to the Ge target, while +47 V applied to a Si substrate. Distance between the Ge target and the substrate surface is about 5 mm	Single-crystalline cubic germanium target (99.99%)	Nd:YAG (532 nm, 10 ns, 1 Hz, 3600 min) 10^{10} W cm^{-2}, laser spot <0.5 mm	Germanium	Single-crystal germanium nanoparticles were obtained with a metastable tetragonal structure. The nanoparticle diameters are in the range of 10–40 nm with a maximal probability about 20 nm in the size distribution	[81]
Deionized water with two quadrate parallel electrodes at a distance of 1.6 cm applied above the target (electric field intensity of 9.06×10^2–2.0×10^3 V m^{-1})	Single-crystalline germanium target (99.99%)	Nd:YAG (532 nm, 10 ns, 5 Hz, 180 min) 150 mJ/pulse	Germanium oxide	Germanium oxide (GeO$_2$) nanocubes have smooth surfaces in a perfect 3D geometry. The lengths of nanocube edge range from 200 to 500 nm. Micro-/nanospindles in the range of 200–400 nm were observed. The nanostructures have an α-phase hexagonal structure. Redshift of emission wavelength in the cathodoluminescence spectra was observed with the shape evolution from cube to spindle	[82]

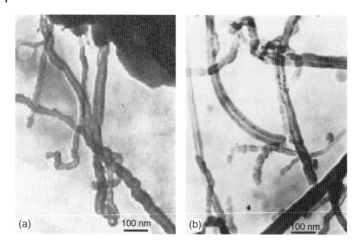

Figure 6.3.15 TEM images of carbon nanotubes prepared by pulsed laser ablation at the solid–liquid interface: (a) sample using Fe as a catalyst and (b) sample using Ni as a catalyst [55].

6.3.3
Silicon Nanoparticles

Silicon nanoparticles have attracted much research interest because of their promising applications in microelectronics, optoelectronics, and biophotonics [83]. LAL preparation of silicon nanoparticles was conducted for the first time by Fojitik and Henglein [8]. The target of a silicon thin film deposited on a thin glass plate (5 × 20 × 0.5 mm) was ablated by a ruby laser in c-hexane. The research on LAL synthesis of silicon nanoparticles was also conducted by a few other groups. It shows that sizes, shapes, and properties of silicon nanoparticles are influenced more or less by the nature of the liquid and laser parameters. The experimental parameters used in the literature for fabricating silicon nanoparticles are summarized in Table 6.3.3.

1) **Liquid effect on the preparation of silicon nanoparticles:** Bozon-Verduraz *et al.* [73] prepared polycrystalline silicon nanoparticles using copper vapor laser ablation of a silicon substrate in water, dichloroethane, and ethanol. As shown in Table 6.3.4, the microcrystal size of silicon nanoparticles is not influenced much by the nature of the liquid the surface active agent, and the laser fluence. The influence of different liquids and laser parameters on nanoparticle growth was not studied. An atomic force microscopy (AFM) image of silicon nanoparticles prepared in ethanol is presented in Figure 6.3.16. The average nanoparticle size is 500 nm with a size distribution from 200 to 1000 nm. Besides silicon, silicon oxide was detected from the diffractogram of the solid phase obtained after the by drying the aqueous sol.

A recent study indicates that the liquid medium plays a significant role in the size and microstructure of the silicon nanoparticles [78]. As displayed

Table 6.3.4 Influence of experimental conditions on the microcrystal size of silicon nanoparticles [73].

Liquid	Laser fluence (J cm^{-2})	Microcrystal size (nm)
Water	1.2	84 (no SAS[a]), 80 (LSb[b]), 74 (PVP[c])
	0.7	74 (no SAS[a])
Dichloroethane	1.5	66
	0.7	60
Ethanol	1.2	78
	0.7	76

[a] Surface active substance.
[b] Na lauryl sulfate.
[c] Polyvinylpyrrolidone, $M = 10^4$.

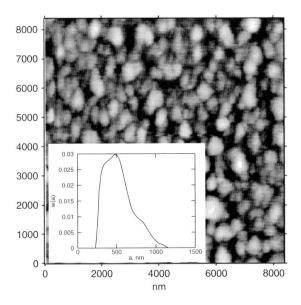

Figure 6.3.16 AFM image of nanoparticles obtained by the laser ablation of a crystalline silicon substrate in ethanol (the scale unit is nanometer). The inset shows the size distribution centered around 500 nm [73].

in Figure 6.3.17a, the mean size of the silicon nanoparticles fabricated in water is larger than that prepared in ethanol. TEM examination shows that silicon nanoparticles prepared in ethanol are superior to those acquired in water. Ethanol provides ultrafine and well-dispersed nanoparticles. Further investigation of the laser-synthesized silicon nanoparticles in the mixture of water and ethanol at different volume ratios (or mole fraction) shows that the mean size of the nanoparticles increases when the volume of water increases

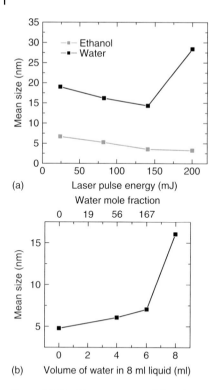

(a)

(b)

Figure 6.3.17 The mean size of the silicon nanoparticles synthesized in water and ethanol as a function of (a) laser fluence and (b) the volume of water in the binary mixture with ethanol [78].

as shown in Figure 6.3.17b. In addition, the size of the microcrystals of nanoparticles (while not nanoparticles themselves) grown in water is smaller than that obtained in ethanol.

Saitow [74] reported a modified method to generate silicon nanoparticles by using supercritical fluid CO_2 as a host liquid. The electronic structures of silicon nanoparticles can be tuned by only changing the fluid pressure and/or temperature during the laser ablation. Sasaki *et al.* used liquid nitrogen as a reaction medium to produce silicon nanoparticles. The synthesized nanoparticles have a size distribution ranging from tens to hundreds of nanometers. It is interesting to find that the purity of silicon nanoparticles is high without the formation of nitridation and oxidation, as the laser ablation of silicon in active liquids, such as water, produces silicon oxides [73, 75], while the laser ablation of silicon in organic solvents creates silicon carbide [79] and other materials, such as diamond particulates [67]. The nanoparticles can be synthesized in a controlled way at required sizes by laser ablation in properly selected liquid media [77].

2) **Effect of laser parameters on the growth of silicon nanoparticles:** Laser pulse duration is a significant parameter for the formation of spherical silicon

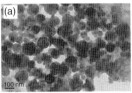 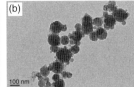 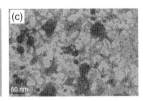

Figure 6.3.18 TEM images of silicon nanoparticles prepared by laser ablation of silicon target at different pulse durations of (a) 900, (b) 100, and (c) 35 fs [79].

nanoparticles during the laser ablation of a single-crystalline silicon wafer in liquid, especially when the pulse duration (from 35 to 900 fs) is short in comparison to electron–phonon relaxation typically on the order of picoseconds. The width of the size distribution of obtained nanoparticles decreases with the decrease in the laser pulse duration [79]. Typical TEM images of silicon nanoparticles prepared at different pulse durations are displayed in Figure 6.3.18.

Laser power intensity is another significant parameter that affects the size and shape of final products. At a fixed experimental configuration, the laser power intensity can be changed by varying laser fluence or pulse energy. Generally, the increase of laser fluence leads to an increase of nanoparticle size and size dispersion. As shown in Figure 6.3.17a, the mean size of silicon nanoparticles decreases with the increase of laser fluence. But when laser fluence increases above a critical value, seriously agglomeration may happen for nanoparticles produced in water. To produce ultrafine nanoparticles with a narrow size distribution, laser ablation should be conducted at low or near-threshold (0.05 J cm^{-2}) fluences, which gives rise to the formation of nanoparticles with a mean size of 2.4 nm [80].

6.3.4
Germanium Nanoparticles

Germanium nanoparticles have attracted much research attention recently because of their interesting properties and unique applications on visible photoluminescence (PL) and photovoltaic devices. Yang and coauthors [13, 81] successfully prepared spherical germanium nanoparticles with trapped metastable tetragonal nanophase. A novel electrical-field-assisted LAL method was applied to grow the germanium nanoparticles in toluene: a bias of -47 V was applied on the germanium target, while a $+47$ V bias was applied on the single-crystalline Si substrate, which was placed 5 mm opposite to the target. As demonstrated in Figure 6.3.19, the size distribution of germanium nanocrystals has a peak at 17–18 nm. These nanocrystals are mainly composed of Ge (93%) with an exterior oxide layer.

Further study indicates that changing the electrode configuration and voltage of the external electrical field can manipulate the sizes, shapes, and chemical components of germanium nanoparticles [82]. Spherical germanium nanoparticles transfer to germanium oxide (GeO_2) nanocubes, germanium oxide nanospindles

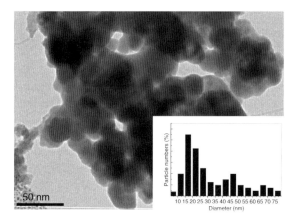

Figure 6.3.19 Bright-field TEM image of the synthesized products; a size distribution analysis based on TEM is shown in the inset [81].

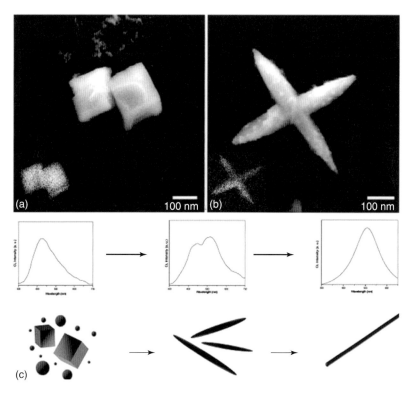

Figure 6.3.20 SEM images of GeO₂ nanostructures: (a) two single cubes and (b) two spindles. The insets are the corresponding CL images. (c) Sketches of luminescence shift with various nanomaterial shapes [82].

with increasing electrical fields. The high-index facet GeO_2 nanocubes have sharp edges with the lengths in the range of 200–500 nm and nanospindle in the range of 200–400 nm. The nanoparticles show interesting redshift of emission wavelengths when the nanoparticle shape evolutes from nanocube to nanospindle. Figure 6.3.20a,b shows images of nanocubes, nanospindles, and their corresponding cathodoluminescence (CL) spectroscopy images displayed as insets. The illustration of shape-induced luminescence shift of GeO_2 nanostructures is shown in Figure 6.3.20c.

The experimental parameters used in the literature for growing germanium nanoparticles are summarized in Table 6.3.3.

6.3.5
Conclusions

LAL opens a new route for the synthesis of novel functional nanoparticles. Although the physical and chemical principles of nanoparticle growth during the LAL process need to be elucidated in details, this innovative nanofabrication technique has been successfully used for synthesizing unique nanoparticles of carbon, silicon, germanium, and their compounds. The extreme condition of high density of plasma species in high temperature and high pressure induced by strong liquid confinement provides the possibility of thermodynamic nucleation and kinetic growth, phase transition, and chemical reaction of the nanomaterials that are normally not achievable in an ambient environment. LAL has been proved to be an essentially simple and clean synthesis method with simple experimental setups and varieties of source materials and can be flexibly chosen to design more target–liquid systems to synthesize desired nanoparticles.

Acknowledgments

The authors acknowledge funding provided by the Temasek Defence Systems Institute research grants (Project No: TDSI/11-013/1A).

References

1. Barcikowski, S., Devesa, F., and Moldenhauer, K. (2009) *J. Nanopart. Res.*, **11**, 1883–1893.
2. Ding, K. and Ye, L. (2006) *Laser Shock Peening Performance and Process Simulation*, CRC Press, Boca Raton, Boston, New York, Washington, DC.
3. Bäuerle, D. (2000) *Laser Processing and Chemistry*, Springer, Berlin and New York.
4. Loper, G.L. and Tabat, M.D. (1985) *J. Appl. Phys.*, **58**, 3649.
5. Patil, P., Phase, D., Kulkarni, S., Ghaisas, S., Kulkarni, S., Kanetkar, S., Ogale, S., and Bhide, V. (1987) *Phys. Rev. Lett.*, **58**, 238–241.
6. Ogale, S.B., Polman, A., Quentin, F.O.P., Roorda, S., and Saris, F.W. (1987) *Appl. Phys. Lett.*, **50**, 138.
7. Ogale, S., Malshe, A., Kanetkar, S., and Kshirsagar, S. (1992) *Solid State Commun.*, **84**, 371–373.

8. Fojtik, A. and Henglein, A. (1993) *Ber. Bunsen-Ges. Phys. Chem. Chem. Phys.*, **97**, 252–254.

9. Simakin, A., Voronov, V., Kirichenko, N., and Shafeev, G. (2004) *Appl. Phys. A*, **79**, 1127–1132.

10. Georgiou, S. and Koubenakis, A. (2003) *Chem. Rev.*, **103**, 349–394.

11. Kruusing, A. (2007) *Handbook of Liquids-Assisted Laser Processing*, Elsevier Science, Amsterdam and Boston.

12. Yang, G. (2007) *Prog. Mater. Sci.*, **52**, 648–698.

13. Liu, P., Cui, H., Wang, C.X., and Yang, G.W. (2010) *Phys. Chem. Chem. Phys.*, **12**, 3942.

14. Khan, S.Z., Yuan, Y., Abdolvand, A., Schmidt, M., Crouse, P., Li, L., Liu, Z., Sharp, M., and Watkins, K.G. (2009) *J. Nanopart. Res.*, **11**, 1421–1427.

15. Liu, Z., Yuan, Y., Khan, S., Abdolvand, A., Whitehead, D., Schmidt, M., and Li, L. (2009) *J. Micromech. Microeng.*, **19**, 054008.

16. Devaux, D., Fabbro, R., Tollier, L., and Bartnicki, E. (1993) *J. Appl. Phys.*, **74**, 2268.

17. Peyre, P. and Fabbro, R. (1995) *Opt. Quantum Electron.*, **27**, 1213–1229.

18. Berthe, L., Fabbro, R., Peyre, P., Tollier, L., and Bartnicki, E. (1997) *J. Appl. Phys.*, **82**, 2826–2832.

19. Berthe, L., Fabbro, R., Peyre, P., and Bartnicki, E. (1999) *J. Appl. Phys.*, **85**, 7552–7555.

20. Zhu, S., Lu, Y.F., and Hong, M.H. (2001) *Appl. Phys. Lett.*, **79**, 1396.

21. Zhu, S., Lu, Y.F., Hong, M.H., and Chen, X.Y. (2001) *J. Appl. Phys.*, **89**, 2400.

22. Kennedy, P.K., Hammer, D.X., and Rockwell, B.A. (1997) *Prog. Quantum Electron.*, **21**, 155–248.

23. Kim, D., Oh, B., and Lee, H. (2004) *Appl. Surf. Sci.*, **222**, 138–147.

24. Chen, G.X., Hong, M.H., Ong, T.S., Lam, M., Chen, W.Z., Elim, H.I., Ji, W., and Chong, T.C. (2004) *Carbon*, **42**, 2735–2737.

25. Chrisey, D.B. and Hubler, G.K. (1994) *Pulsed Laser Deposition of Thin Films*, Wiley-Interscience, New York.

26. Tsuji, T., Tsuboi, Y., Kitamura, N., and Tsuji, M. (2004) *Appl. Surf. Sci.*, **229**, 365–371.

27. Versluis, M., Schmitz, B., von der Heydt, A., and Lohse, D. (2000) *Science*, **289**, 2114–2117.

28. Lohse, D., Schmitz, B., and Versluis, M. (2001) *Nature*, **413**, 477–478.

29. Chen, G.X. (2005) Laser-synthesized carbon nanoparticles for nonlinear optical limiting effect. Thesis. National University of Singapore.

30. Amans, D., Chenus, A., Ledoux, G., Dujardin, C., Reynaud, C., Sublemontier, O., Masenelli-Varlot, K., and Guillois, O. (2009) *Diamond Relat. Mater.*, **18**, 177–180.

31. Pearce, S.R.J., Henley, S.J., Claeyssens, F., May, P.W., Hallam, K.R., Smith, J.A., and Rosser, K.N. (2004) *Diamond Relat. Mater.*, **13**, 661–665.

32. Wang, Y.H., Huang, Q.J., Chen, Z., Huang, R.B., and Zheng, L.S. (1997) *Sci. China Ser. B*, **40**, 608–615.

33. Yang, G.W., Wang, J.B., and Liu, Q.X. (1998) *J. Phys.: Condens. Matter*, **10**, 7923–7927.

34. Yang, G.W. and Wang, J.B. (2000) *Appl. Phys. A*, **71**, 343–344.

35. Mafuné, F., Kohno, J., Takeda, Y., Kondow, T., and Sawabe, H. (2000) *J. Phys. Chem. B*, **104**, 9111–9117.

36. Liu, P., Cao, Y.L., Wang, C.X., Chen, X.Y., and Yang, G.W. (2008) *Nano Lett.*, **8**, 2570–2575.

37. Mafuné, F., Kohno, J., Takeda, Y., and Kondow, T. (2002) *J. Phys. Chem. B*, **106**, 8555–8561.

38. Kwong, H.Y., Wong, M.H., Leung, C.W., Wong, Y.W., and Wong, K.H. (2010) *J. Appl. Phys.*, **108**, 034304.

39. Tsuji, M., Kuboyama, S., Matsuzaki, T., and Tsuji, T. (2003) *Carbon*, **41**, 2141–2148.

40. Isselin, J., Alloncle, A., and Autric, M. (1998) *J. Appl. Phys.*, **84**, 5766.

41. Chen, G.X., Hong, M.H., Chong, T.C., Elim, H.I., Ma, G.H., and Ji, W. (2004) *J. Appl. Phys.*, **95**, 1455–1459.

42. Yang, L., May, P.W., Yin, L., Smith, J.A., and Rosser, K.N. (2007) *Diamond Relat. Mater.*, **16**, 725–729.

43. Kitazawa, S., Abe, H., and Yamamoto, S. (2005) *J. Phys. Chem. Solids*, **66**, 555–559.

44. Tsuji, T., Tatsuyama, Y., Tsuji, M., Ishida, K., Okada, S., and Yamaki, J. (2007) *Mater. Lett.*, **61**, 2062–2065.

45. Link, S., Burda, C., Nikoobakht, B., and El-Sayed, M.A. (2000) *J. Phys. Chem. B*, **104**, 6152–6163.

46. Takami, A., Kurita, H., and Koda, S. (1999) *J. Phys. Chem. B*, **103**, 1226–1232.

47. Liu, P., Cao, Y.L., Cui, H., Chen, X.Y., and Yang, G.W. (2008) *Chem. Mater.*, **20**, 494–502.

48. Kashchiev, D. (2000) *Nucleation: Basic Theory with Applications*, Butterworth-Heinemann, Oxford.

49. Yang, G. and Wang, J. (2001) *Appl. Phys. A*, **72**, 475–479.

50. Wang, J.B., Zhang, C.Y., Zhong, X.L., and Yang, G.W. (2002) *Chem. Phys. Lett.*, **361**, 86–90.

51. Liu, P., Wang, C., Chen, J., Xu, N., and Yang, G. (2009) *J. Phys. Chem. C*, **113**, 12154–12161.

52. Liu, C.H., Peng, W., and Sheng, L.M. (2001) *Carbon*, **39**, 144–147.

53. Liu, P., Cui, H., and Yang, G.W. (2008) *Cryst. Growth Des.*, **8**, 581–586.

54. Wang, Y., Zhang, Q., Liu, Z., Huang, R., and Zheng, L. (1996) *Acta Phys. Chim. Sin.*, **12**, 905–909.

55. Chen, C., Chen, W., and Zhang, Y. (2005) *Phys. E*, **28**, 121–127.

56. Yang, L., May, P.W., Yin, L., Brown, R., and Scott, T.B. (2006) *Chem. Mater.*, **18**, 5058–5064.

57. Yang, L., May, P.W., Yin, L., Scott, T., Smith, J.A., and Rosser, K.N. (2006) *Nanotechnology*, **17**, 5798–5804.

58. Yang, L., May, P.W., Huang, Y., and Yin, L. (2007) *J. Mater. Chem.*, **17**, 1255–1257.

59. Yang, L., May, P.W., Yin, L., Huang, Y., Smith, J.A., and Scott, T.B. (2007) *Nanotechnology*, **18**, 335605.

60. Yang, L., May, P.W., Yin, L., Smith, J.A., and Rosser, K.N. (2007) *J. Nanopart. Res.*, **9**, 1181–1185.

61. Tian, F., Sun, J., Hu, S.L., and Du, X.W. (2008) *J. Appl. Phys.*, **104**, 096102.

62. Wang, C.X., Yang, Y.H., and Yang, G.W. (2005) *J. Appl. Phys.*, **97**, 066104.

63. Wang, J.B. and Yang, G.W. (1999) *J. Phys.: Condens. Matter*, **11**, 7089–7094.

64. Wang, C.X., Yang, Y.H., Liu, Q.X., Yang, G.W., Mao, Y.L., and Yan, X.H. (2004) *Appl. Phys. Lett.*, **84**, 1471–1473.

65. Wang, C.X., Liu, P., Cui, H., and Yang, G.W. (2005) *Appl. Phys. Lett.*, **87**, 201913.

66. Bundy, F.P., Bassett, W.A., Weathers, M.S., Hemley, R.J., Mao, H.K., and Goncharov, A.F. (1996) *Carbon*, **34**, 141–153.

67. Lu, Y.F., Huang, S.M., Wang, X.B., and Shen, Z.X. (1998) *Appl. Phys. A*, **66**, 543–547.

68. Tiwari, S.K., Joshi, M.P., Laghate, M., and Mehendale, S.C. (2002) *Opt. Laser Technol.*, **34**, 487–491.

69. Lide, D.R. (1995) *CRC Handbook of Chemistry and Physics*, 76th edn, CRC Press, Boca Raton.

70. (1999) Aldrich Chemical Company Catalog.

71. Bialkowski, S.E. (1996) *Photothermal Spectroscopy Methods for Chemical Analysis*, John Wiley & Sons, Inc., New York.

72. Tsuji, T., Okazaki, Y., Higuchi, T., and Tsuji, M. (2006) *J. Photochem. Photobiol. A*, **183**, 297–303.

73. Dolgaev, S.I., Simakin, A.V., Voronov, V.V., Shafeev, G.A., and Bozon-Verduraz, F. (2002) *Appl. Surf. Sci.*, **186**, 546–551.

74. Saitow, K. (2005) *J. Phys. Chem. B*, **109**, 3731–3733.

75. Svrcek, V., Sasaki, T., Shimizu, Y., and Koshizaki, N. (2006) *Appl. Phys. Lett.*, **89**, 213113–213115.

76. Takada, N., Sasaki, T., and Sasaki, K. (2008) *Appl. Phys. A*, **93**, 833–836.

77. Perminov, P.A., Dzhun, I.O., Ezhov, A.A., Zabotnov, S.V., Golovan, L.A., Panov, V.I., and Kashkarov, P.K. (2010) *Bull. Russ. Acad. Sci.: Phys.*, **74**, 93–95.

78. Yang, S., Cai, W., Zhang, H., Xu, X., and Zeng, H. (2009) *J. Phys. Chem. C*, **113**, 19091–19095.

79. Kuzmin, P.G., Shafeev, G.A., Bukin, V.V., Garnov, S.V., Farcau, C., Carles, R., Warot-Fonrose, B., Guieu, V., and Viau, G. (2010) *J. Phys. Chem. C*, **114**, 15266–15273.

80. Rioux, D., Laferriere, M., Douplik, A., Shah, D., Lilge, L., Kabashin, A.V., and

Meunier, M.M. (2009) *J. Biomed. Opt.*, **14**, 21010.

81. Liu, P., Cao, Y.L., Chen, X.Y., and Yang, G.W. (2009) *Cryst. Growth Des.*, **9**, 1390–1393.

82. Liu, P., Wang, C.X., Chen, X.Y., and Yang, G.W. (2008) *J. Phys. Chem. C*, **112**, 13450–13456.

83. Kumar, V. (2007) *Nanosilicon*, Elsevier Science, Amsterdam and Boston.

Part II
Nanomaterials: Laser-Based Characterization Techniques

Nanomaterials: Processing and Characterization with Lasers, First Edition.
Edited by Subhash Chandra Singh, Haibo Zeng, Chunlei Guo, and Weiping Cai.
© 2012 Wiley-VCH Verlag GmbH & Co. KGaA. Published 2012 by Wiley-VCH Verlag GmbH & Co. KGaA.

7
Raman Spectroscopy: Basics and Applications

7.1
Raman Spectroscopy and its Application in the Characterization of Semiconductor Devices

Patrick J. McNally

7.1.1
Introduction

Complementary metal oxide semiconductor (CMOS) microprocessor physical gate lengths are expected to decrease in size from 27 nm in 2010 to 13 nm in 2018. Major gaps in front-end processing (FEP) metrologies for these nanoscale dimensions have been identified and are highlighted in the recent 2009 International Technology Roadmap for Semiconductors [1, 2]. These also include metrologies for III–V and SiGe channels and for silicon on insulator (SOI) materials systems. These include the need for local stress and strain metrology, particularly along a line running from the source along the length of the active channel toward the drain region. As we shall discuss later, this turns out to be very challenging, even for the most advanced Raman spectroscopic techniques and is closely related to the doping regimes in a real CMOS device. Across what is termed as the *micro-area level*, the expectation for confocal Raman spectroscopy is a stress sensitivity of 20 MPa, equivalent to a 0.02% strain sensitivity in Si, across a measurement zone ~150 × 150 nm [2] (Figure 7.1.1).

The strain from standard CMOS processes, such as the shallow trench isolation (STI) formation, oxidation, silicide formation, and contact etch stop layer (CESL) fabrication, all affect transistor performance [3]. The tailored application of strain to the active channel region of the metal oxide semiconductor field-effect transistor (MOSFET) has recently become a necessary practice. This has largely been in reaction to short-channel effects and the architectural and processing changes they have necessitated. The sense of strain, be it tensile or compressive, is dictated by the local band structure modification required. For example, tensile stress is typically used to enhance n-channel MOSFET performance, while compressive stress is used to enhance p-channel devices.

The motivation for the use of strained Si structures is highlighted in the equation that describes the linear region for $I_{DS} - V_{DS}$ operation of a typical MOSFET:

Nanomaterials: Processing and Characterization with Lasers, First Edition.
Edited by Subhash Chandra Singh, Haibo Zeng, Chunlei Guo, and Weiping Cai.
© 2012 Wiley-VCH Verlag GmbH & Co. KGaA. Published 2012 by Wiley-VCH Verlag GmbH & Co. KGaA.

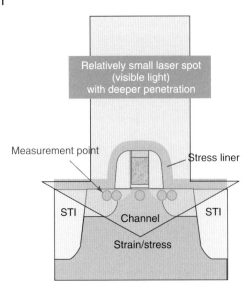

Figure 7.1.1 Conception of the use of micro-Raman spectroscopy for nondestructive channel stress/strain measurements [2].

$$I_{DS} = \frac{W}{L_G} \mu_{app} C_G (V_{GS} - V_T) V_{DS} \tag{7.1.1}$$

where I_{DS} is the drain–source current, V_{DS} the applied drain–source voltage, V_{GS} the gate–source voltage, V_T the threshold voltage, C_G the gate capacitance, W the device gate width, and L_G the device gate length. The apparent channel mobility, μ_{app}, is the combination of the ballistic mobility μ_B and the bulk mobility μ_0 as expressed by Matthiessen's rule [4]. In the first order, the crucial term in the equation is μ_{app}. For a given overdrive voltage ($V_{GS} - V_T$), the switching speed of the device is determined by the maximum available drive current (I_{DS}). Notwithstanding the device geometry (via W and L_G) and the constraints on allowable voltages, the greater value of μ_{app} for a particular carrier, the faster will be the ultimate device switching speed. This is a major driver in the quest for faster CMOS devices and materials. Silicon and SiGe and strained silicon on insulator (SSOI) materials systems all employ strategies to strain (uniaxially and biaxially) the channel regions to enhance both electron and hole mobilities. For future materials, the focus will most likely be on the replacement of silicon with high-mobility channel materials such as Ge and III–V compound semiconductors and on the fabrication of three-dimensional structures, for example, FINFETs.

It is against this backdrop that strained Si technology has advanced, as it provides a means of improving device performance independent of scaling. Initially, this performance enhancement manifested itself through increased conduction electron mobilities, but subsequently, improved dopant activation and diffusion properties have also become evident. In 1985, Abstreiter *et al.* [5] identified the two-dimensional electron gas (2DEG) nature of conduction in strained Si pseudomorphically grown on $Si_{1-x}Ge_x$ virtual substrates. They also identified the lifting of the threefold degeneracy of the conduction band. In 1991, Fitzgerald *et al.* [6] introduced the concept of compositional grading in $Si_{1-x}Ge_x$ virtual substrates. This technique reduced the dislocation density in the active strained Si (ε-Si) region by two orders of magnitude to 10^6 cm^{-2} and Welser *et al.* [7] reported the first MOSFET based on this technology. In 1998, Rim *et al.* [8] demonstrated the advantages of strained Si in device structures by revealing a 45% enhancement in transconductance in short-channel devices with channel lengths of 0.1 μm. More recent research into n-channel strained Si/SiGe structures has focused on industrial integration and scalability. To this end, interest turned to the effects of industrial processing on strained Si and the effect of strain on critical dopant properties. In 1996, Kringhoj *et al.* [9] experimentally demonstrated the retardation of Sb diffusion in ε-Si, an effect previously predicted in theoretical work by Cowern *et al.* [10].

7.1.2
Raman Scattering in Semiconductors

Micro-Raman spectroscopy (uRS) has been used to characterize semiconductor materials since their electronic and vibrational properties are reproducible. The characterization can be performed without contacting or damaging the sample, which is very useful for microelectronic devices. The Raman technique can be used to examine crystalline orientation, doping level, stress, and so on in the semiconductor device. Table 7.1.1 is an outline of the electronic and vibrational properties of cubic semiconductors at room temperature.

It was discovered in the late 1970s [11] that uRS was particularly useful for local stress and/or strain measurements in semiconductor devices. It is well known that stress generated in the crystal can change the equilibrium position of the atoms. The phonon frequencies associated with the distorted crystal can be calculated from the force constants of the crystal between the perfect and distorted parts. For a complete analysis of the theory underpinning stress measurement in semiconductors and structures fabricated thereupon, the reader is referred to

Table 7.1.1 The structural, electronic, and vibrational properties of some cubic semiconductors at room temperature.

Material	Lattice parameter (Å)	Indirect gap (eV)	Direct gap (eV)	ω_{LO} (cm^{-1})	ω_{TO} (cm^{-1})
Diamond	3.56683	5.4	6.5	1332	
Si	3.43086	1.12	4.25	521	
Ge	5.65	0.66	0.81	303	
GaAs	5.6534	–	1.429	292	269

Frequencies of longitudinal optical (ω_{LO}) and transverse optical (ω_{TO}) phonons given at $q \cong 0$.

a series of excellent papers by De Wolf *et al.* [12–15]. Briefly, the vibrations of a crystal are described in terms of quantized wavelike collective atomic motions known as *phonons*. These quantized phonon vibrational normal modes can give rise to a variation in the electrical susceptibility tensor (χ) of the crystal. The interaction of the incoming photonic electric fields with this spatially and temporally varying χ can give rise to "Raman scattering." If one invokes quantum mechanics, the Raman scattering involves the destruction of an incident probe photon of frequency w_i and the creation of a new photon with the frequency w_o. The difference $w_i - w_o$ is traditionally called the *Raman shift* and is measured in spectroscopic wave number notation as relative wave numbers (R cm^{-1}). This is traditionally abbreviated to cm^{-1} where it is understood that this refers to the relative wave number shift of the outgoing photon(s) with respect to the incident photon(s) wave number. Since the values of these Raman shifts is determined by local atomic arrangements, their measurement can be used to determine the local stress/strain conditions as they have a direct impact on local atomic arrangements.

Mechanical strain or stress may affect the frequencies of the Raman modes and lift their degeneracy. One of the first papers addressing theoretically the effect of stress on the Raman modes was that by Ganesan *et al.* [16]. They showed that the frequencies of the three optical modes in the presence of the strain, to terms linear in the strain, can be obtained by solving the following secular equation [16, 17]

$$\begin{vmatrix} p\varepsilon_{11} + q(\varepsilon_{22} + \varepsilon_{33}) - \lambda & 2r\varepsilon_{12} & 2r\varepsilon_{13} \\ 2r\varepsilon_{12} & p\varepsilon_{22} + q(\varepsilon_{33} + \varepsilon_{11}) - \lambda & 2r\varepsilon_{23} \\ 2r\varepsilon_{13} & 2r\varepsilon_{23} & p\varepsilon_{33} + q(\varepsilon_{11} + \varepsilon_{22}) - \lambda \end{vmatrix} = 0$$

(7.1.2)

Here, the deformation potentials are indicated by p, q, and r and ε_{ij} are the strain tensor components. The eigenvalues λ_j in Eq. (7.1.2) can be related to the difference between the Raman frequency of each mode in the presence of stress (ω_j) ($j = 1, 2, 3$) and the absence of stress (ω_{j0})

$$\text{That is, } \lambda_j = \omega_j^2 - \omega_{j0}^2 \quad \text{or} \quad \Delta w_j = \omega_j - \omega_{j0} \approx \frac{\lambda_j}{2\omega_{j0}}$$

(7.1.3)

For example, in the case of uniaxial stress along the [100] direction, the application of Hooke's law gives $e_{11} = S_{11}\sigma$, $e_{22} = S_{12}\sigma$, and $e_{33} = S_{12}\sigma$, where the S_{ij} are the elastic compliance tensor elements of silicon. For backscattering from a (001) surface, only the third Raman mode is observed and the relationship between the shift of this mode and the stress is given by solving Eqs. (7.1.2) and (7.1.3)

$$\Delta\omega_3 = \frac{1}{2\omega_0}[pS_{12} + q(S_{11} + S_{12})]\sigma$$

(7.1.4)

For the materials data for Si given in [15], one finds that the Raman mode shift is given by the simple expression

$$\Delta\omega_3 \, (\text{cm}^{-1}) = -2 \times 10^{-9}\sigma \, (\text{Pa})$$

(7.1.5)

In the case of biaxial stress in the $x-y$ plane, with stress components σ_{xx} and σ_{yy}, this becomes

$$\Delta\omega_3 \, (\text{cm}^{-1}) = -4 \times 10^{-9} \left(\frac{\sigma_{xx} + \sigma_{yy}}{2} \right) (\text{Pa}) \qquad (7.1.6)$$

Thus, it can be seen that a compressive uniaxial or biaxial stress will result in an increase in the Raman frequency, whereas a tensile stress will cause a decrease.

7.1.3
Micro-Raman Spectroscopy: Microscale Applications

For the past 20 years or so, uRS has been used successfully for the analysis of Si CMOS device and integrated circuit technology. Until recently, studies have been carried out mainly in the micrometer or submicrometer spatial scale, albeit very successfully. A few typical applications are outlined below.

The stress distribution in silicon with LOCOS (local oxidation of silicon) has been evaluated using uRS by Kobayashi *et al.* and De Wolf *et al.* [18, 19] who measured the shift of the Si Raman band for various widths of the active area with an accuracy of ~ 0.05 cm^{-1}. Figure 7.1.2 shows the Raman shift as a function of the distance

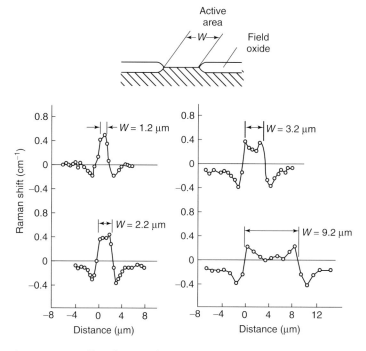

Figure 7.1.2 Profiles of Raman frequencies relative to that of stress-free Si for LOCOS structures with different active area widths (*W*). Positive Raman shifts (blueshifts) are indicative of compressive stresses, while negative Raman shifts (redshifts) indicate tensile stress. (Source: From [18].)

[18]. The Raman band of the silicon surface in the oxidized region shows a redshift, but it shows a blueshift in the active area. The sign of the stress is reversed at the LOCOS edges. This figure also shows that the stress in the center of the active area decreases with increasing width in the active area.

De Wolf *et al.* [19] also found that compressive stress was present in the Si substrates underlying the Si_3N_4 LOCOS mask lines and that the magnitude of this stress was directly dependent on the width of the lines. In addition, large tensile stresses tended to be concentrated under the LOCOS "bird's beak" regions and these stresses increased with a decrease in the bird's beak length and with an increase in the Si_3N_4 thickness. Residual stresses in metal interconnect lines in ICs are of great importance since these can lead to interfacial delamination and cracking effects, possible ingress of moisture, and ultimately, device failure. Many authors have used X-ray diffraction (XRD) techniques to evaluate these metal line stresses. However, the X-ray penetration depths are often much larger than the line thicknesses (this is also the case for multilayer metallization and low-*k* dielectric systems), and the X-ray probe areas can be up to several square millimeters in the spatial extent. Thus any measurement only represents an average, and local spatial stress measurements are impossible. However, uRS laser probe beams are typically as small as 1 μm in diameter leading to much better X – Y spatial resolutions. De Wolf *et al.* [20] applied this to the evaluation of local mechanical stresses near tungsten metallization on Si substrates. Kanatharana *et al.* [21] used uRS to study the strain fields imposed on underlying silicon substrates due to lead–tin solder bumps in ball grid array packaging. Pre- and post-reflowed samples were examined. For the pre-reflowed samples, uniaxial compressive stresses of the order of 200 MPa were observed near the edge of the under bump metallization (UBM). However, a tensile stress on the order of 300 MPa was found for post-reflowed samples. The magnitudes and spatial distribution of the stresses were found to be in agreement with finite element elasticity modeling.

7.1.4
Raman Spectroscopy Approaches the Nanoscale

In recent years, however, there have been significant advances toward developing uRS for nanoscale applications. These efforts have taken place on two main fronts: (i) using UV laser probes to ensure nanometer-scale probe depths in, for example, Si, Ge, SiGe, and III–V materials and (ii) the use of surface plasmon effects to enhance surface strain sensitivities. Examples of both these approaches are outlined.

It is useful to reiterate that strained Si, grown on relaxed SiGe virtual substrates [22] or strained by device capping layers [3], is a very important technology for ultrahigh speed Si CMOS IC technology. The strained Si epitaxial layers or active channel layers are typically very thin, on the order of 10 nm or so. One technological approach using uRS to measure these ultrathin layers is to use a "conventional" visible light laser probe combined with signal processing techniques to enhance the data from the uppermost 10 nm or so within the substrate. One such approach was used by Dobrosz *et al.* [23] to determine if any strain relaxation had occurred during the processing of strained Si on SiGe virtual substrates. The authors reported the use of uRS to study epitaxial $Si_{1-x}Ge_x$ alloys ($x = 0.1–0.3$) covered with a ~10 nm thick tensile Si cap layer. A 514.5 nm wavelength probe laser was used (spot size ~1 μm), which has an optical penetration depth into the substrate of many hundreds of nanometers. This implied that the contribution of the relatively small volume of the strained Si cap would be masked by the much larger signal from the total integrated volume signal. However, the signal from the Si cap was enhanced using two methods. The first method was to obtain a difference spectrum before and after selectively etching the Si cap and the second method involved software-based peak fitting and deconvolution. The data from both methods agreed closely. The peak-fitting technique is, however, much faster and is by definition a nondestructive technique. Perova *et al.* confirmed the efficacy of this approach [24], but they caution that very careful spectral fitting is required.

The key problem with the aforementioned approaches is the relatively deep penetration depths of the visible wavelength probe laser beams. One, albeit rather expensive, alternative is to use ultraviolet (UV) wavelength lasers with an appropriate optical train to ensure that the laser spot size approaches its theoretical smallest diffraction-limited dimension, *D*. This is given by De Wolf *et al.* [25]

$$D \approx \frac{0.88\lambda}{\text{NA}} \tag{7.1.7}$$

where λ is the wavelength of the laser light and NA is the numerical aperture of the objective lens. Modern UV Raman systems consist of objective lenses with $NA \approx 0.9$, for example, at 364 nm wavelengths. This, combined with the use of a confocal aperture at the intermediate image plane between the microscope and the spectrometer, thus rejecting light from undesired focal planes, can ensure an X–Y spatial resolution on the order of $D \approx 360$ nm or so.

Dombrowski *et al.* [26] used UV uRS to study local mechanical stress enhancements in STI trenches in Si substrates. These STI trenches were backfilled with an oxide filler deposited using a complex series of spin-on glass steps, annealing and densification steps in either a wet steam (850 °C) or a dry (1050 °C) ambience. The authors used a 364 nm UV probe laser that penetrates only 15 nm into the Si substrate. They found that very large local stresses up to 0.8 GPa built up at the STI edges during densification in the steam ambience.

Himcinschi *et al.* [27] used UV Raman spectroscopy (He-Cd laser with excitation wavelength of 325 nm producing a penetration depth in Si of 9 nm) to investigate the near-surface relaxation of strained Si, which had been subjected to a series of etch steps to form high-aspect ratio pillars resembling gate stacks in a CMOS fabrication process. Tensile-strained Si layers were epitaxially deposited on fully relaxed $Si_{0.78}Ge_{0.22}$ virtual substrate epitaxial layers on silicon substrates. Periodic arrays of pillars, 100 nm high, with X–Y spatial dimensions of 150 nm × 150 nm and 150 nm × 750 nm, respectively, were fabricated into the strained Si layers by electron beam lithography and reactive ion etching. Their Raman measurements showed that the nanopatterning yields a relaxation of strain of ~33% in the large pillars and ∼53% in the small pillars with respect to the initial ∼0.95% strain in the unpatterned strained Si layer. This could have an impact on channel strain engineering since such nanopatterning will become increasingly relevant in advanced CMOS fabrication and processes whereby any initiated stress relief could have deleterious knock-on consequences for carrier mobility engineering.

One of the limitations of Raman spectroscopy, when compared to, for example, Fourier transform infrared (FTIR) or UV–Visible spectroscopy, is the fact that signal-to-noise ratios are very small. This is primarily due to the very small Raman scattering cross section for most atoms and molecules [28]. The situation is even worse in the case where one is attempting to sieve out the signal from a very thin overlayer from that of the remaining bulk substrate, which is a very ubiquitous situation in nano-CMOS applications. However, the discovery of surface-enhanced Raman scattering (SERS) nearly 30 years ago opened new possibilities for true surface metrology [29]. In SERS, one observes an enormous enhancement of the intensity of the Raman signals from a thin layer of the material under test (in the original experiments, this was actually an adsorbed analyte) if it is adjacent to a metal surface, which is locally rough on the nanoscale. It is now thought that this magnification of the Raman signals is electromagnetic in nature and it is related to the excitation of collective oscillations of the conduction electrons, known as *surface plasmons* [30], in the underlying metal nanoparticles. These collective oscillations result in the enhancement of the local field experienced by atoms/molecules adsorbed on the surface of the nanoparticles.

Recent work by Hayazawa *et al.* [31] highlighted the successful use of SERS for strain metrology on thin layers of strained Si on SiGe virtual substrates for CMOS applications. The strained Si layer was covered with a 10 nm thick evaporated silver island film on a 30 nm thick strained silicon epilayer. Even though they used a 488 nm laser, whose optical penetration depth into the sample was of the order of 400 nm, the Raman signal from the strained silicon was detected via SERS and was easily observed compared to the overwhelming background signal from the underlying SiGe/Si layers. The authors also speculated on a potentially nondestructive embodiment of this technique using a tip-enhanced Raman microscope with a sharpened metallic probe tip.

7.1.5
Confocal Raman Spectroscopy – Applications to Future Sub-22 nm Node CMOS Technology

As outlined earlier, confocal UV Raman spectroscopy is currently considered to be a possible nondestructive metrology tool for the assessment of damage, strain, and, perhaps, doping for future CMOS devices. Since Anastassakis' groundbreaking work in 1970 [17], Raman scattering has been widely used as a strain characterization technique in the semiconductor industry. As device scaling continues according to the ITRS Roadmap, finite element modeling of stress and the resulting electrical properties is a key aspect of process development and metrology, and an accurate stress metrology can help calibrate these simulations. We highlight some recent results in the quest to use UV confocal uRS to accurately measure stress in highly doped n- and p-type Si. In this recent work, UV uRS and high-resolution X-ray diffraction (HRXRD) were used as complementary, independent stress characterization tools for a range of strained Si materials. These included low-energy (2 keV) Sb ion implanted strained Si for low straggle, highly activated n^+-type Si fabrication and B implants (500 eV) with Ge preamorphization for p^+-Si fabrication. The full details of the fabrication regimes can be found in [32, 33]. Following dopant implantation, good agreement is found between the magnitudes of strain measured by the two techniques. However, following dopant activation by annealing, strain relaxation is detected by HRXRD but not by micro-Raman. This discrepancy mainly arises from an anomalous redshift in the Si Raman peak position originating from the high levels of doping achieved in the samples. This redshift is observed for both Sb and B dopants and is greater by nearly an order of magnitude for p-type doping. Clearly, the unusual Raman redshifting is not strain related as the direction of Raman shift for both n- and p-type doping is the same!

A more complete explanation of this phenomenon can be elucidated if one considers that a number of factors may contribute to this observed Si Raman peak shift, including stress, phonon confinement effects [17], and, crucially, dopant carrier concentrations [34]. Observed asymmetries in the Si Raman peak and the redshift of the peak position in the as-implanted samples is indicative of confinement effects resulting from damage to the top Si layer during ion implantation. In a perfect crystalline semiconductor, the long-range periodicity of the lattice makes the correlation length of the normal-mode vibrations infinite, giving rise to the momentum selection rule, which limits Raman scattering to zone-center ($q = 0$) optical phonon modes. The bulk, plane-wave-like phonon wave function cannot exist within a small crystallite or when the crystal is damaged. The spatial confinement, via the Heisenberg uncertainty principle, results in light scattering from a nominally zone-center phonon, whose wave vector has an uncertainty Δq and whose energy thus has an uncertainty $\Delta \omega$. Thus, the phonon's spatial confinement results in a broadening of the Raman scattering features reflecting the uncertainty in its energy and a redshift of the Raman peak position occurs with increasing spatial confinement, as lower-energy bulklike phonons are incorporated into the wave function describing the nominally zone-center confined phonon. This confinement model, first developed by Richter *et al.* [35] and Campbell and Fauchet [36], is adapted from the models of Huang *et al.* [37] and Macia *et al.* [38] for these highly doped samples. Assuming a constant correlation length L in the scattering volume, the intensity of the first-order Raman band of silicon is given by

$$I(\omega) \propto \int_0^{2\pi/a_0} \frac{d^3 q \exp\left(-L^2 q^2 / 16\pi^2\right)}{\left(\omega - \omega_0 - \omega\left(\vec{q}\right)\right)^2 + \left(\Gamma_0/2\right)^2} \tag{7.1.8}$$

where a_0 is the lattice constant of Si and Γ_0 is the Raman intrinsic line width of crystalline-Si (c-Si). $\omega\left(\vec{q}\right)$ is the phonon dispersion relation, taken as

$$\omega\left(\vec{q}\right) = \omega_p - 120 \left(\frac{q}{q_0}\right)^2 \tag{7.1.9}$$

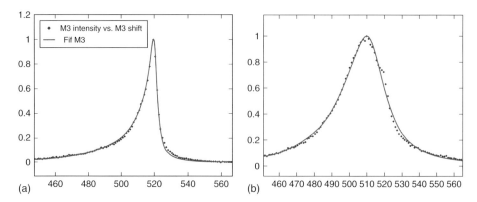

(a) (b)

Figure 7.1.3 Experimental Raman data fitted with the confinement model (a) c-Si as implanted fitted with Eq. (7.1.8) and (b) c-Si fRTP (700/1300 °C) annealed sample fitted with Eq. (7.1.10).

where ω_p is the wave number of the first-order Raman band in the absence of disorder effects (520 cm^{-1}) and $q_0 = 2\pi/a_0$. The parameter L assumes that in the presence of damage the phonon is restricted to a sphere of diameter L [35]. As L decreases, there is an increase in the peak asymmetry and a slight shift of the peak to lower wave numbers. The parameter ω_0 is introduced to represent the other contributions to the Raman line shift, namely stress and carrier concentration effects. The experimental data were fitted by varying the parameters ω_0 and L, so that the asymmetrical broadening and line shift due to residual damage (contained in the parameter L) is decoupled from other contributions to the line shift contained in the parameter ω_0. Fits were performed using the Matlab curve fitting toolbox. In all cases, the R^2 correlation coefficient is >0.98. By way of example, the success of the fitting procedure is shown in Figure 7.1.3a for a c-Si sample after boron implantation.

Following the activation anneal, we have seen a broadening of the Raman Si peak due to the dopant and a slight asymmetry in the peak shape suggesting some remaining damage. The asymmetry due to the damage is included in the model in the fitted parameter L. By introducing a *broadening* term, B, to Eq. (7.1.8) accounting for symmetrical broadening of the Raman peak due to doping, we can fit the experimental data of the annealed samples according to Eq. (7.1.10)

$$I(\omega) \propto \int_0^{2\pi/a_0} \frac{d^3q \exp\left(-L^2 q^2/16\pi^2\right)}{\left(\omega - \omega_0 - \omega(\vec{q})\right)^2 + \left((\Gamma_0 + B)/2\right)^2} \tag{7.1.10}$$

A typical fit for the annealed samples is shown in Figure 7.1.3b for a c-Si sample that has undergone a flash-assisted rapid thermal process (fRTP) anneal at 700/1300 °C. By applying this model, we find the Raman peak shift due to confinement and the fitted ω_0 parameter is the total Raman peak shift due to stress and carrier concentration effects.

It appears that the net Raman shift excluding strain and confinement effects results from the high doping levels in the samples under test. These were in

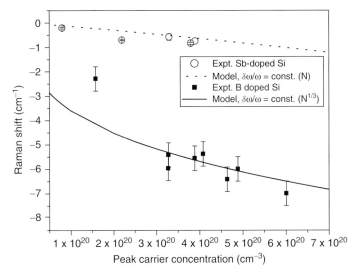

Figure 7.1.4 Experimental Raman peak shift as a function of peak carrier concentration for p-type B-doped bulk Si and n-type Sb-doped bulk Si. The model fitted is based on the theory in [34].

fact highly doped n- and p-type samples that were being studied for source–drain extension CMOS technology. In p-type Si, a strong dependence of the Raman frequency on carrier concentration resulting from the hole redistribution between the light and heavy hole bands has been reported [34]. The Raman redshift $\delta\omega$ and carrier concentration N are related according to $\delta\omega/\omega \propto N$ for n-type doping and $\delta\omega/\omega \propto N^{1/3}$ for p-type doping [34, 39]. The continuous anodic oxidation technique (CAOT) [40] was used to measure carrier concentration profiles from the measured resistivities using the carrier mobility values obtained experimentally. The peak carrier concentration values are plotted in Figure 7.1.4 as a function of the measured Raman peak shifts for the doped Si. In agreement with the theory [34], the Raman Si peak shift due to high dopant concentration observed is nearly an order of magnitude greater for the p-type dopants than for the n-type dopants.

The spatial resolution of standard confocal uRS is limited by the spot size of the laser, typically >300 nm. These spatial resolutions could be enhanced greatly by using, for example, tip-enhanced Raman spectroscopy (TERS) for transistor-level stress measurements with a sensitivity as low as 50 MPa [31]. TERS utilizes the coupling of a sharp metal AFM tip with the Raman microscope. In this case, the spatial resolution is determined by the size of the tip apex. As the physical gate length of an MOS transistor is already in the 22 nm range, which is comparable to the area probed by a TERS tip, we caution that it will be difficult to selectively probe the strain in the channel region without also probing the highly doped source–drain extension, or indeed the actual source–drain, regions. As the channel dimensions continue to shrink, the relative contribution from the source and drain regions to the overall Raman signal will increase. In highly doped p-type bulk Si materials, we expect the very large Si Raman peak shift that we have observed due to doping to obscure the much smaller peak shift due to strains. The scenario is further complicated when probing intentionally strained channel regions. Therefore, it is crucial to understand and model the impact of high doping levels on the Raman spectrum of Si in order to reliably extract the required strain data.

7.1.6
Conclusion

Confocal uRS has proved to be a very useful metrology tool in CMOS technology for the past two decades or so. While its X–Y spatial resolution has been in the micrometer range, we highlight recent advances which show that uRS can continue to play a role in advanced nanoscale metrology as CMOS gate lengths and junction depths are aggressively pushed into sub-22 nodes. The use of shorter wavelength UV probe lasers coupled with surface enhancement techniques may pave the way for uRS to be a prime metrology tool, as envisaged in the 2009 IRTS Roadmap [1]. However, we caution that data interpretation is very important in analyzing, for example, strain data for device channel regions, especially as device dimensions shrink and local dopant levels remain very high.

Acknowledgments

PMN acknowledges Science Foundation Ireland funding under the Investigator Programme Grant # **05/IN/I656** and the Strategic Research Cluster Programme ("Precision" 08/SRC/I1411). This work was conducted under the framework of the INSPIRE programme, funded by the Irish Government's Programme for Research in Third Level Institutions, Cycle 4, National Development Plan 2007–2013.

References

1. (2009) International technology roadmap for semiconductors (ITRS), *http://www.itrs.net* (accessed 17 June 2010).

2. International Technology Roadmap for Semiconductors (2010) Winter Conference Proceedings, HsinChu, Taiwan, 16 December, 2010, *http://www.itrs.net/Links/2009Winter/Presentations.html* (accessed 29 November 2011).

3. Chidambaram, P.R., Bowen, C., Chakravarthi, S., Machala, C., and Wise, R. (2006) Fundamentals of silicon material properties for successful exploitation of strain engineering in modern CMOS manufacturing. *IEEE Trans. Electron Dev.*, **53** (5), 944–963.

4. Liu, Y., Pal, H.S., Lundstrom, M.S., Kim, D.H., del Alamo, J.A., and Antoniadis, D.A. (2010) in *Fundamentals of III-V Semiconductor MOSFETs*, Chapter 3 (eds S. Oktyabrsky and P.D. Ye), Springer, ISBN-978-1-4419-1546-7, doi: 10.1007/978-1-4419-1547-4

5. Abstreiter, G., Brugger, H., Wolf, T., Jorke, H., and Herzog, H.J. (1985) Strain-Induced Two-Dimensional Electron Gas in Selectively Doped Si/Si_xGe_{1-x} Superlattices. *Phys. Rev. Lett.*, **54**, 2441.

6. Fitzgerald, E.A., Xie, Y.-H., Green, M.L., Brasen, D., Kortan, A.R., Michel, J.,

Mii, Y.-J. and Weir, B.E. (1991) Totally relaxed Ge$_x$Si$_{1-x}$ layers with low threading dislocation densities grown on Si substrates. *Appl. Phys. Lett.*, **59**, 811–813.

7. Welser, J., Hoyt, J.L., and Gibbons, J.F. (1992), International Electron Devices Meeting, San Francisco, CA, pp. 1000–1002.

8. Rim, K., Hoyt, J.L., and Gibbons, J.F. (1998) Technical Digest-International Electron Devices Meeting, San Francisco, CA, pp. 707–710.

9. Kringhoj, P., Larsen, A.N., and Shirayev, S.Y. (1996) Diffusion of Sb in Strained and Relaxed Si and SiGe. *Phys. Rev. Lett.*, **76**, 3372.

10. Cowern, N.E.B., Zalm, P.C., van der Sluis, P., Gravesteijn, D.J., and de Boer, W.B. (1994) Diffusion in strained Si(Ge). *Phys. Rev. Lett.*, **72**, 2585.

11. Anastassakis, E. (1980) in *Dynamical Properties of Solids* (eds G.K. Horton and A.A. Maradudin), North-Holland, p. 157, *et seq.*

12. De Wolf, I., Maes, H.E., and Jones, S.K. (1996) Stress measurements in silicon devices through Raman spectroscopy: bridging the gap between theory and experiment. *J. Appl. Phys.*, **79** (9), 7148–7156.

13. De Wolf, I. and Anastassakis, E. (1999) Addendum: stress measurements in silicon devices through Raman spectroscopy: bridging the gap between theory and experiment. *J. Appl. Phys.*, **85** (10), 7484–7485.

14. De Wolf, I., Norström, H., and Maes, H.E. (1993) Process-induced mechanical stress in isolation structures studied by micro-Raman spectroscopy. *J. Appl. Phys.*, **74** (7), 4490–4500.

15. De Wolf, I. (1996) Micro-Raman spectroscopy to study local mechanical stress in silicon integrated circuits. *Semicond. Sci. Technol.*, **11**, 139–154.

16. Ganesan, S., Maradudin, A.A., and Oitmaa, J. (1970) A lattice theory of morphic effects in crystals of the diamond structure. *Ann. Phys.*, **56**, 556–594.

17. Anastassakis, E., Pinczuk, A., Burstein, E., Pollak, F.H., and Cardona, M. (1970) Effect of static uniaxial stress on the Raman spectrum of silicon. *Solid State Commun.*, **8**, 133–138.

18. Kobayashi, K., Inoue, Y., Nishimura, T., Nishioka, T., Arima, H., Hirayama, M., and Matsukawa, T. (1987) Stress measurement of LOCOS structure using microscopic Raman spectroscopy. Extended Abstracts of the 19th Conference on Solid State Devicesand Materials, Tokyo, pp. 323–326.

19. De Wolf, I., Vanhellemont, J., Romano-Rodríguez, A., Norström, H., and Maes, H.E. (1992) Micro-Raman study of stress distribution in local isolation structures and correlation with transmission electron microscopy. *J. Appl. Phys.*, **71** (2), 898–906.

20. De Wolf, I., Ignat, M., Pozza, G., Maniguet, L., and Maes, H.E. (1999) Analysis of local mechanical stresses in an near tungsten lines on silicon substrate. *J. Appl. Phys.*, **85** (9), 6477–6485.

21. Kanatharana, J., Pérez-Camacho, J.J., Buckley, T., McNally, P.J., Tuomi, T., Riikonen, J., Danilewsky, A.N., O'Hare, M., Lowney, D., Chen, W., Rantamäki, R., and Knuuttila, L. (2002) Examination of mechanical stresses in silicon substrates due to lead-tin solder bumps via micro-Raman spectroscopy and finite element modeling. *Semicond. Sci. Technol.*, **17**, 1255–1260.

22. Lee, M.L., Fitzgerald, E.A., Bulsara, M.T., Currie, M.T., and Lochtefeld, A. (2005) Strained Si, SiGe, and Ge channels for high-mobility metal-oxide-semiconductor field-effect transistors. *J. Appl. Phys.*, **97**, 011101.

23. Dobrosz, P., Bull, S.J., Olsen, S.H., and O'Neill, A.G. (2005) The use of Raman spectroscopy to identify strain and strain relaxation in strained Si/SiGe structures. *Surf. Coat. Technol.*, **200**, 1755–1760.

24. Perova, T.S., Lyutovich, K., Kasper, E., Waldron, A., Oehme, M., and Moore, R.A. (2006) Stress determination in strained-Si grown on ultra-thin SiGe virtual substrates. *Mater. Sci. Eng. B*, **135**, 192–194.

25. De Wolf, I., Chen, J., Rasras, M., Merlijn van Spengen, W., and Simons, V. (1999) High-resolution stress and temperature measurements in semiconductor devices using micro-Raman

spectroscopy. *Proc. SPIE Conf. Adv. Photonic Sens. Appl.*, **3897**, 239–252.

26. Dombrowski, K.F., Dietrich, B., De Wolf, I., Rooyackers, R., and Badenes, G. (2001) *Microelectron. Reliab.*, **41**, 511–515.

27. Himcinschi, C., Radu, I., Singh, R., Erfurth, W., Milenin, A.P., Reiche, M., Christiansen, S.H., and Gösele, U. (2006) Relaxation of strain in patterned strained silicon investigated by UV Raman spectroscopy. *Mater. Sci. Eng. B*, **135**, 184–187.

28. Ko, H., Singamaneni, S., and Tsukruk, V.V. (2008) Nanostructured surfaces and assemblies as SERS media. *Small*, **4** (10), 1576–1599.

29. Jeanmaire, D.L. and Van Duyne, R.P. (1977) Surface Raman spectroelectrochemistry. Part 1. Heterocyclic, aromatic, and aliphatic amines adsorbed on the anodized silver electrode. *J. Electroanal. Chem.*, **84**, 1–20.

30. Knoll, A. (1998) Interfaces and thin films as seen by bound electromagnetic waves. *Ann. Rev. Phys. Chem.*, **49**, 569–638.

31. Hayazawa, N., Motohashi, M., Saito, Y., and Kawata, S. (2005) Highly sensitive strain detection in strained silicon by surface-enhanced Raman spectroscopy. *Appl. Phys. Lett.*, **86**, 263114.

32. O'Reilly, L., Horan, K., McNally, P.J., Timans, P.J., Reyes, J., Prussin, S., Bennett, N.S., Cowern, N.E.B., Gelpey, J., McCoy, S., Lerch, W., Paul, S., and Bolze, D. (2009) Raman metrology for Advanced CMOS devices – advantages and challenges. Proceedings of the International Workshop on INSIGHT in Semiconductor Device Fabrication, Metrology, and Modeling (INSIGHT-2009), Napa, CA, April 26–29.

33. O'Reilly, L., Horan, K., McNally, P.J., Bennett, N.S., Cowern, N.E.B., Lankinen, A., Sealy, B.J., Gwilliam, R.M., Noakes, T.C.Q., and Bailey, P. (2008) Constraints on micro-Raman strain metrology for highly doped strained Si materials. *Appl. Phys. Lett.*, **92**, 233506.

34. Cerdeira, F. and Cardona, M. (1972) Effect of carrier concentration on the Raman frequencies of Si and Ge. *Phys. Rev. B*, **5**, 1440–1454.

35. Richter, H., Wang, Z.P., and Ley, L. (1981) The one phonon Raman spectrum in microcrystalline silicon. *Solid State Commun.*, **39**, 625.

36. Campbell, I.H. and Fauchet, P.M. (1986) The effects of microcrystal size and shape on the one phonon Raman spectra of crystalline semiconductors. *Solid State Commun.*, **58**, 739.

37. Huang, X., Ninio, F., Brown, L.J., and Prawer, S. (1995) Raman scattering studies of surface modification in 1.5 MeV Si-implanted silicon. *J. Appl. Phys.*, **77**, 5910.

38. Macía, J., Martín, E., Pérez-Rodríguez, A., Jiménez, J., Morante, J.R., Aspar, B., and Margail, J. (1997) Raman microstructural analysis of silicon-on-insulator formed by high dose oxygen ion implantation: As-implanted structures. *J. Appl. Phys.*, **82**, 3730.

39. Wei, W. (2007) One- and two-phonon Raman scattering from hydrogenated nanocrystalline silicon films. *Vacuum*, **81**, 857.

40. Prussin, S. and Reyes, J. (2009) The evaluation of state-of-the-art front end structures by differential Hall effect continuous anodic oxidation technique. *ECS Trans.*, **19** (1), 135–146.

7.2
Effect of Particle Size Reduction on Raman Spectra

Vasant G. Sathe

7.2.1
Introduction

Raman spectra of reduced-dimension materials show changes when compared with the bulk material. Raman spectra reflect changes in local structure of the material and is hence a sensitive tool for the study of nanostructure. Infact, Raman spectra have become a very popular technique with the advent of nanoparticle research. Raman spectrum depends on the symmetry of the molecule and crystal structure, hence judicial use of polarized Raman spectroscopy can give information about phonon as well as electronic symmetry properties of the material. The use of resonance phenomena in resonant Raman spectroscopy provides additional information on the modification of electronic band structure with reduction in dimensionality. As the particle size reduces, the system enters a regime where quantum mechanical effect becomes very important. This is reflected in the form of confinement of phonons that is very effectively traced in Raman scattering studies. The determination of particle size using Raman scattering has been attempted by many researchers with very limited success because of inherent theoretical limitations that are discussed further in this chapter. In this chapter, we give several examples of confinement and laser heating effects in oxide nanoparticles (rods, particles, and thin films) and resonant Raman spectroscopy of semiconducting nanoparticles and films. As such, Raman spectroscopy is extremely popular and provides important information about carbon nanotubes and other carbon materials. However, it is not included in this chapter, and readers may consult a number reviews available [1–4].

Figure 7.2.1 shows Raman spectra of bulk silicon single crystal along with those of silicon nanoparticles greater than 10 nm and less than 10 nm. The Raman spectra for the bulk single crystal show a sharp well-defined first-order optical phonon peak at 520.6 cm^{-1} characteristic of silicon structure along with broad second-order acoustic mode around 302 cm^{-1} and extremely broad second-order optical mode at 975 cm^{-1}. The Raman spectra of the nanoparticles also show all these three peaks but with a marked difference. In the spectra of nanoparticles, the peaks are shifted considerably to the lower wave number side (redshifted) when compared to the peak position for the bulk material. The first-order peak for the nanoparticles showed considerable broadening, asymmetry, and reduced intensity. These spectra are collected in a few minutes and immediately give qualitative information about the presence of nanocrystals. One can easily conclude that as the particle size decreases the redshift and broadening increases. Thus, Raman spectroscopy is a very fast and easy technique to obtain quality information about the characterization of nanomaterials. It does not require any sample preparation, which makes it a very popular handy technique. If

Nanomaterials: Processing and Characterization with Lasers, First Edition.
Edited by Subhash Chandra Singh, Haibo Zeng, Chunlei Guo, and Weiping Cai.
© 2012 Wiley-VCH Verlag GmbH & Co. KGaA. Published 2012 by Wiley-VCH Verlag GmbH & Co. KGaA.

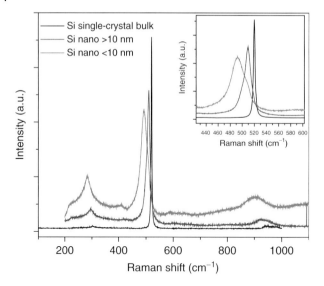

Figure 7.2.1 The Raman spectra of Si single crystal and nanoparticles of size ∼10 and ∼3–8 nm obtained with a micro-Raman instrument equipped with excitation laser of wavelength 488 nm. The well-studied second-order acoustic phonon mode, first-order optical phonon mode, and second-order optical phonon mode are observed for all the samples. The inset shows the region around first-order mode.

one observes the inset of Figure 7.2.1, the Raman band for samples with less than 10 nm particle size shows a shoulder on the right-hand side, indicating the presence of two bunches of particle sizes. Thus, Raman spectra are very sensitive to the reduction in particle size, and one can obtain qualitative information by first-hand observation of the spectra. The reasons for the observed changes and methods for quantitative estimation are given in the subsequent sections.

7.2.2
Nanoparticles and Phonon Confinement

In principle, a *nanoparticle material* can be defined as any particle that has a size less than 1000 nm. However, the quantum effects or effects where surface contributions are comparable with the bulk are observed in much smaller particle sizes, typically less than 50 nm. The modification in properties due to confinement effects is observed in particles of sizes less than 10 nm. In general, the surface effects take over the bulk properties when the particle size is reduced below a particular size. In order to observe the confinement effects, the particle size should be ideally smaller than the DeBroglie wavelength. Many times, it is observed that even when the particle size is higher than the critical size defined by the above criteria; it still shows quantum mechanical effects because of the presence of defects. Defects also localize the phonon wave function concomitant with the confinement effects and are dealt in more details later.

Generally, inelastic neutron scattering experiments are used to get complete dispersion relation of the phonons in a crystal. This is because in neutron scattering experiments the entire Brillouin zone can participate following momentum conservation rules. A thermal neutron typically has wavelength of $\sim 2-3$ Å giving scattering vector of $4\pi \sin(\theta)/\lambda$ that is comparable to the size of the Brillouin zone $\sim \pi/a$; hence during the inelastic process, the entire Brillouin zone participates giving full information of the dispersion of phonons. On the other hand, in a Raman scattering experiment, usually visible light of around 5000 Å is used for the scattering process, giving a wave vector of $\mathbf{k}_o \sim 2\pi/5000$ Å^{-1}. The maximum wave vector transfer in a backscattering geometry is $2\mathbf{k}_o$ that is, thus, 1000 times smaller than the size of the Brillouin zone. Hence a very small portion near the center of the Brillouin zone ($q = 0$) participates in the Raman process. This is the famous $q = 0$ selection rule applicable for the Raman scattering for crystalline solids. In an amorphous material, this selection rule is not applicable as the Brillouin zone is not well defined. The wave vector in such a Brillouin zone is not defined; hence, momentum conservation can be achieved even when the wave vector of the incident photon interacts with any part of the defective Brillouin zone. In general, the dispersion relation of the optical phonons in the Brillouin zone is negative (Figure 7.2.2); this means that phonons of smaller energy dominate in the scattering process when the entire Brillouin zone participates in the scattering process. This is reflected in the downshift of Raman frequency and broad phonon mode for amorphous solids where the participation of entire Brillouin zone is expected with relaxation of $q = 0$ selection rule compared to corresponding crystalline materials. For example, single-crystal Si gives well-defined Raman peak corresponding to triply degenerate optical phonon mode at ~ 520 cm^{-1} and full width at half maximum (FWHM) of

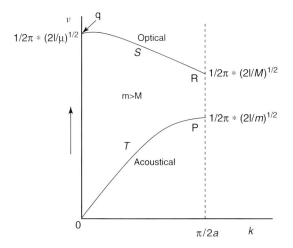

Figure 7.2.2 Dispersion curves for acoustic and optical phonon modes in a Brillouin zone.

$\sim$2.5 cm^{-1}, while amorphous Si gives an extremely broad Raman band with trailing asymmetric nature toward lower frequency around 480 cm^{-1}. The degeneracy of the optical modes gets lifted at higher wave vectors in the Brillouin zone, which is further responsible for the broadness of the observed Raman band in amorphous Si.

The critical size of the nanoparticle below which the properties start to differ from the bulk depends on many factors. It can be broadly categorized into (i) where surfaces become important and (ii) where the confinement effects become observable. The latter factor relaxes the $q = 0$ selection rule, and hence optic phonons other than $q = 0$ and with lower frequencies also start participating in the scattering process. This results in redshifting of the Raman band and asymmetric broadening. Naturally, the amount of redshift and broadening will increase as the particle size decreases as more and more region of Brillouin zone participates in the process. We consider a few illustrative examples showing this effect in the later part of this chapter.

When the surface starts playing dominating role compared to the bulk, the Raman spectra show extra surface-induced modes of acoustic frequencies. They fall very close to Reyleigh line and are known as *low-frequency phonons* ($<$100 cm^{-1}). In most of the cases, they are extremely close to Rayleigh frequency $<$10 cm^{-1} and hence need special instrumentation (interferometers) to observe. They are also of weak intensity when compared to optical phonon modes. However, low-frequency mode is important to understand electrical and thermal properties of materials. These acoustic modes possess positive dispersion and a nearly linear dispersion curve near the Brillouin zone center (Figure 7.2.2). When the size of the nanoparticle is less than the phonon mean free path, the acoustic phonon spectrum undergoes a strong modification. The frequency of these acoustic modes increases on decreasing the particle size. Here, modeling the increase in frequency with decrease in particle size is easy as the dispersion curve is nearly linear (Figure 7.2.2). Such a model calculation has been attempted for ZnO prismatic nanoparticles by Chessaing *et al.* [5]. The experimental frequencies for various particle sizes are compared with those obtained in the framework of linear elastic theory. In a homogeneous elastic sphere, the low-frequency eigenmodes can be categorized into (i) spheroidal and (ii) torsional modes. Following Duval [6], these modes are further classified by two integers n and l, denoting the harmonic and angular momentum numbers, respectively. Only breathing ($l = 0$) and quadrupolar ($l = 2$) modes are Raman active, and these two modes obey different polarization selection rules facilitating assignment of peak symmetry. The prismatic shape of nanoparticles leads to anisotropic elastic tensor, and it is shown in Ref. [5] that if it is ignored it leads to an error of $\sim$2 cm^{-1} in the value of breathing and extensional frequencies. The frequency versus inverse of particle size leads to linear relationship. There are reports of size estimation of nanoparticles from the frequency of confined acoustic modes [7], and it is claimed that this method gives the estimation of particle size with reasonable accuracy. However, this is a branch of Brillouin scattering experiments that needs special measurement techniques. Here, we limit our discussion to optical phonon modes that can be

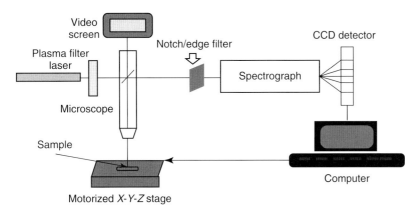

Figure 7.2.3 Schematics of a dispersive Raman spectrometer.

measured easily without any sample preparation and normal Raman spectrometers (Figure 7.2.3).

The confinement of phonon can be in one dimension, as in the case of a very thin film where the phonons are confined along the thickness of the film; in two dimension, as in the case of a nanorods and tubes where phonons are restricted in two directions and are free to move only in one direction, that is, along the length of the rod/tube; or three dimensional (movement in zero dimension) as in the case of a very small nanoparticle or quantum dots.

7.2.3
Theoretical Considerations of Optical Phonon Confinement

Analysis of Raman band position and peak shape can in principle give an estimation of nanoparticle size. At present, a precise *ab initio* determination of the vibration dynamics in nanostructures is limited to crystals below 1 nm because of computational power availability. Therefore, several semiempirical and phenomenological models have been developed in order to relate the observed changes in the Raman spectra to the crystal size. The most popular are elastic sphere model (ESM) [6, 8] and phonon confinement model (PCM) [9, 10]. According to Gouadec and Colomban [11], the ESM model is best suited for particle sizes around 5 and less than 10 nm, while PCM explains the Raman spectrum of nanocrystallites of 10 nm and higher. We present below summary of PCM that is widely used and describe attempts of improvement in the model.

The basic model of optical phonon confinement was described by Richter *et al.* [9] that was then developed further by Campbell *et al.* and many other workers. The main ingredient of this model is a relaxation of the conservation of crystal momentum ($q = 0$ selection rule) in the creation and annealation of phonons in nanoparticles. The wave function of phonon that extends to infinity in a normal crystal lattice is now restricted in a nanocrystal of size L. This means that the wave function should decay close to the boundary of the nanoparticle, that is, in a distance *L*, which is only tens of lattice parameters. Richter *et al.* [9] proposed a simple modification of the normal periodic wave function by multiplying a coefficient function $\psi'(\boldsymbol{q_0}, \boldsymbol{r})$ to the infinite periodic function $\phi(\boldsymbol{q_0}, \boldsymbol{r})$ such that the new wave function ψ has a magnitude of

$$\left|\psi(\boldsymbol{q_0}, \boldsymbol{r})\right|^2 = A^2 e^{-r^2 / \left(\frac{L}{2}\right)^2} \tag{7.2.1}$$

The new wave function ψ is thus localized to $|r| \leq L$ in the form of a Gaussian distribution. The coefficient function $\psi'(\boldsymbol{q_0}, \boldsymbol{r})$ represents confinement function and essentially defines the range of *q* vectors participating in the process. In other words, the infinitely extending periodic wave function is weighted by a Gaussian function $W(r, L) = e^{-2r^2 / L^2}$ such that the new wave function decays rapidly at the boundary of the nanoparticle. This function is thus composed of Fourier coefficients $C(\boldsymbol{q_0}, \boldsymbol{q})$ that can be evaluated by simple Fourier algebra as

$$C(\boldsymbol{q_0}, \boldsymbol{q}) = \frac{AL}{(2\pi)^{3/2}} e^{-\frac{1}{2}\left(\frac{L}{2}\right)^2 (q - q_0)^2} \tag{7.2.2}$$

In this way, the confined phonon wave function is no longer an eigenfunctions of phonon wave vector $\boldsymbol{q_0}(q = 0)$ but a superposition of eigenfunctions with wave vectors in the interval $|q - q_0| \leq 1/2\,L$ centered at $\boldsymbol{q_0}$. The transition probability reflected in the form of intensity contribution in the spectral response in Raman spectroscopy due to $\boldsymbol{q} \neq \boldsymbol{q_0}$ is given by

$$\left|<\boldsymbol{q_0}\left|\mathbf{O}\right|\boldsymbol{q}>\right|^2 = \left|<\boldsymbol{q_0}\left|\mathbf{O}\right|\boldsymbol{q_0}>\right|^2 \times C\left(\boldsymbol{q}, \boldsymbol{q_0}\right) \tag{7.2.3}$$

where **O** is the phonon–photon interaction operator. The magnitude of $C(\boldsymbol{q}, \boldsymbol{q_0})$ depends on the value of *L* and diminishes rapidly as *L* becomes larger and becomes zero for *L* > 100 nm. In addition, the magnitude of $C(\boldsymbol{q_0}, \boldsymbol{q})$ diminishes as *q* goes away from $\boldsymbol{q_0}$, this results in the asymmetry of the phonon band at lower wave numbers. Away from $q = 0$, the degeneracy of the optical phonon is lifted into the transverse optical (TO) and the longitudinal optical (LO) phonons, and the dispersion of both the branches is negative, resulting in decrease in Raman frequency as contributions away from $\boldsymbol{q_0}$ become prominent.

From relation (Eq. (7.2.2)), it is clear that the amplitude of phonon wave reduces to 1/*e* at the boundary of the nanoparticle. Campbell and Fauchet [10] argued that there is no physical reason for this form of confinement or its particular value at the boundary of nanoparticles. For spherical nanoparticles and Gaussian weighting function (Eqs. (7.2.1) and (7.2.2)), the first-order Raman spectrum, $I(\omega)$, is

$$I(\omega, L) \cong \int \frac{d^3 q |C(o, q)|^2}{\left(\omega - \omega\left(q\right)\right)^2 + (\Gamma_0/2)^2} \tag{7.2.4}$$

where $\omega(q)$ is the phonon dispersion curve and Γ_0 is the natural line width. Here, in order to simplify the calculations, spherical Brillouin zone and isotropic dispersion curve are assumed. These assumptions are justified because only a small region of the Brillouin zone, centered at Γ point, contributes to the scattering [10].

Table 7.2.1 The weighting functions used in different models with the infinitely extending periodic wave function to force the confinement of wave function, corresponding Fourier coefficient and their values at the boundary.

References	Weighting function	Fourier coefficient	Value at the boundary	
[9]	$W(r, L) = e^{-2r^2/L^2}$	$|c(o, q)|^2 \cong e^{\left(-\frac{q^2 L^2}{4}\right)}$	$1/e$	(i)
[10]	$W(r, L) = \dfrac{\sin\left(\frac{2\pi r}{L}\right)}{2\pi r/L}$	$|c(o, q)|^2 \cong \dfrac{\sin^2\left(\frac{qL}{2}\right)}{\left(4\pi^2 - q^2 L^2\right)^2}$	0	(ii)
[12][a]	$W(r, L) = \dfrac{\sin\left(\frac{2\pi r}{L}\right)}{2\pi r/L}$	$|c(o, q)|^2 \cong \dfrac{\sin^2\left(\frac{qL}{2}\right)}{q^2 \left(4\pi^2 - q^2 L^2\right)^2}$	0	
[10]	$W(r, L) = e^{-4\pi^2 r/L}$	$|c(o, q)|^2 \cong \dfrac{1}{\left(16\pi^2 - q^2 L^2\right)^2}$	$e^{-4\pi^2}$	(iii)
[10]	$W(r, L) = e^{-g\pi^2 r^2/L^2}$	$|c(o, q)|^2 \cong e^{-q^2 L^2/16\pi^2}$	$e^{-4\pi^2}$	(iv)

[a]Authors of Ref. [12] claim that the Fourier coefficient given by Campbell et al. is incorrect for (ii) and coefficients according to Ref. [12] are given in the next line.

In order to include the effects due to nonspherical column and thin films and to further improve the Richter's PCM, Campbell and Fauchet [10] used different weighting functions and evaluated the corresponding Fourier coefficients. The different weighting functions, Fourier coefficients, and their values at the boundary are given in Table 8.2.1.

The fitting of the Si nanocrystals for shift and width of the Raman line as a function of shape and size of the nanocrystals improves significantly when the Gaussian function (Eq. (7.2.4)) is used. This confinement condition gives a good agreement for Raman peak position and shape data from Iqbal and Veprek [13] for Si nanocrystals and by Tiong *et al.* [14] on the results of ion implantation in GaAs. This choice of weighting function gives more rigid confinement with Gaussian confinement function.

Tiong *et al.* [14] developed a spatial correlation model for disordered material where the confinement of the phonons is due to defects produced by ion implantation, and the model follows spatial correlation model originally developed for amorphous materials by Shuker and Gammon [15]. The confinement model by shuker and Gammon successfully showed that short coherence length in amorphous materials results in first-order Raman scattering intensity in terms of the density of states function and known frequency-dependent amplitudes.

It is observed that the shape and dimensionality of the nanocrystallites has appreciable effect on the confinement. In thin films where the confinement is only in one direction, the frequency shift and broadening of the Raman bands is the smallest. The shift and broadening is found to be relatively more for a rod- or wirelike structure where the confinement is in two directions and is maximum in very small particles or dots where confinement is in all the three directions.

7.2.3.1 Effect of Particle Size Distribution

In all the models discussed so far, the distribution in crystallite size is not accounted. However, nanoparticles prepared by any method essentially have size distribution, and only for very ideal narrow distribution, its effects can be neglected. In most of the cases, particle size distribution and disorder are inherently present [16, 17] and the size distribution produces a dispersion of the lattice parameter and appearance of inhomogeneous strain. The combined effects of inhomogeneous strain and phonon confinement can produce an even larger shift and asymmetric broadening of the Raman modes. Several attempts are made to include effect of size distribution on the Raman line shapes. Islam *et al.* [18, 19] considered Gaussian distribution of crystallite sizes with a deviation of σ across mean particle size of L_0 in the form of a Gauss function given simply as

$$\Phi(L) = \frac{1}{\sqrt{2\pi\sigma^2}} e^{\left[-\frac{(L-L_0)^2}{2\sigma^2}\right]}$$

(7.2.5)

Here, crystals are assumed to be spherical. The total Raman intensity profile for the ensemble of nanocrystallites can be written using Eqs. (7.2.4) and (7.2.5) as

$$I(\omega, L_0, \sigma) = \int \Phi(L) I(\omega, L) dL$$

(7.2.6)

By replacing $\Phi(L)$ from Eq. (7.2.4) and after rearranging, Eq. (7.2.6) becomes

$$I(\omega, L_0, \sigma) \propto \frac{f(q) q^2 \exp\left[-\frac{q^2 L_0^2 f^2(q)}{2\alpha}\right]}{[\omega - \omega(q)]^2 + (\Gamma_0/2)^2} \times \left\{1 + erf\left[\frac{L_0 f(q)}{\sigma\sqrt{2}}\right]\right\}$$

$$\text{where } f(q) = \frac{1}{\sqrt{\left(1 + \frac{q^2\sigma^2}{\alpha}\right)}}$$

(7.2.7)

The erf (error function) can be neglected for $L_0 > 3\sigma$ [19]. Then, the resultant expression is similar to Raman intensity expression with an additional parameter $f(q)$, which incorporates the deviation of particle size parameter σ into the Raman intensity profile.

Initially, Bottani *et al.* [16] calculated deviation in the particle size from the bar chart obtained by transmission electron microscopy (TEM) and included in the intensity profile of the Raman scattering. This attempt followed treatment by Gonzalez-Hernandez *et al.* [20] where the effective particle size determined by TEM and Raman scattering was compared. In both these studies on Si and Ge quantum dots, the Gaussian weighting function of Richter *et al.* [9] was used to consider the effect of confinement as described before.

7.2.3.2 Estimation of Dispersion Curve

In all these models, it is assumed that the particle size reduction relaxes Raman selection rules but does not significantly modify the phonon dispersion curves of the reference three-dimensional system. It is also necessary to know or assume a function for dispersion curve of the optical phonon. In general, an isotropic and trigonometrical function (sine or cosine function) for the dispersion curve of the optical phonon is assumed, and further, to simplify the calculations a spherical Brillouin zone is assumed. For example, in gallium alloys [21], the dispersion curve is simulated by taking analytical model relationship based on a one-dimensional linear chain model [22] as

$$\omega^2(q) = A + \{A^2 - B[1 - \cos(\pi q)]\}^{1/2} \tag{7.2.8}$$

where the coefficients A and B are calculated by fitting the dispersion curve of the corresponding three-dimensional compound.

7.2.3.3 Limitations of Phonon Confinement Model

As such Gaussian weighting function is the most popular and is widely used by many researchers [21, 23–31] in simulating Raman scattering results of nanostructures. Line softening and broadening observed in Raman spectra of diamond [24, 32], graphite [33], Ge [34], and CdSe [35] nanoparticles was also explained by considering this model. The model partially succeeded in explaining the Raman signal, and some part of the peak shape and position got unaccounted by the PCM. Richter *et al.* [9] and then following him many researchers [36–38] attributed the unaccounted Raman signal to the contribution from amorphous phase. However, according to Gouadec and Colomban [39], the PCM has the following inherent weaknesses that give poor fitting of the experimental results depending on particle size and shape involved.

1) PCM assumes isotropic dispersion in Brillouin zone. This hypothesis is valid only closed to BZ center and in materials with homogeneous structures in all directions of space (spherical model).
2) PCM assumes uniform size and shape that is instrumental in consideration of Gaussian weighting function. In practice, the size and shape in nanoparticles has wide distribution depending on the method of preparation [16, 17].
3) PCM does not account for surface phonons, which is important when particle size is reduced below 10 nm [40–47].
4) The PCM is arbitrary in nature and possesses no physical meaning.

In order to test the validity of above Gaussian PCM, a bond polarizability model is used to calculate the Raman spectra of Si nanocrystals [48, 49–51] from the eigenvalues and eigenfunctions given by the microscopic calculations based on the partial density of state. In this model, spherical particles are considered and the Raman spectra are obtained from the contributions of each bond in the system considered, based on the calculated eigenvalues and eigenvectors. The Raman intensity in the XY polarization for the backscattering geometry is given by Zi *et al.* [12]

$$I_{XY}(\omega) \propto [n(\omega) + 1] \Sigma_j \delta\left(\omega - \omega_j(q)\right) |\Delta\alpha_{XY}(jq)|^2 \tag{7.2.9}$$

Here, $\Delta\alpha_{XY}(jq) = \Sigma_i^{\text{bonds}} \Delta\alpha_{XY}(i|jq)$ represents the variation of XY component of polarizability tensor (α) due to a phonon mode j with wave vector $q \sim 0$ that is equal to the sum of the contributions from each bond in the whole spherical particle.

7.2.4
Experimental Setup for Confocal Micro-Raman Spectroscopy

Unlike the past century, nowadays, compact spectrometers, attached with an optical microscope, are becoming more and more popular because of many advantages over conventional double or triple monochromator spectrometers. The advantages are easy operation, fast data collection, easy alignment, and most importantly, ability to record Raman images. Here, we describe, in brief, the general principle and schematics of a micro-Raman spectrometer (Figure 7.2.3).

The fundamental requirement of any Raman spectrometer is an excitation laser source, a monochromator, and a detector. In a micro-Raman instrument, the laser photons guided by several optical elements and filtered for laser plasma lines by interference filter are focused on a spot of $1–10\,\mu m$ using an optical microscope on a sample. The back-reflected photons are captured by the same objective lens of the microscope and passed to a notch/edge filter to filter the Rayleigh lines. The Raman signal is then passed through a confocal hole and spectrograph consisting of a grating and a charge-coupled detector (CCD). The laser spot size on the sample can be varied by choosing appropriate objective lens: $10\times, 50\times$, or $100\times$; a $100\times$ objective lens tight focuses the laser spot to around 1 μm. Limitations induced due to diffraction of light, the smallest dimension over which a laser spot is focused which dictates the lateral resolution can be calculated using Rayleigh criterion [52]: $R = 0.61 \times \lambda/NA$; where NA stands for numerical aperture of the objective lens. Even when the very best objective lens and the smallest possible wavelength are chosen for measurement, the lateral and axial resolutions achieved are around $0.5\,\mu m$. Thus, even using a confocal hole, an axial differentiation of typically micrometers can be made with the micro-Raman spectrometers.

7.2.5
Case Studies of Raman Spectroscopy of Nanomaterials

7.2.5.1 Resonant Raman Spectroscopy of CdS and CdSe Nanoparticles
The optical properties of semiconductor nanoparticles have been extensively studied because of a wide variety of applications as a photosensitive material. Applications include phosphors, light emitting diodes, and solar cell applications. Understanding its optical and vibrational properties can greatly enhance the device performance. It also carries fundamental importance in understanding of properties of strongly confined phonons and electrons. Raman spectroscopy is extensively used for studying its optical and vibrational properties as it is a very efficient tool and gives added information about electron–phonon interaction, which is important for a better understanding of the electrical and optical properties of confined nanostructure semiconductors. Raman studies of CdS nanoparticles [53–55] revealed a redshift and broadening of the single-Raman-active LO mode around 300 cm^{-1} characteristic of phonon confinement. The redshift and broadening increases as the particle size decreases.

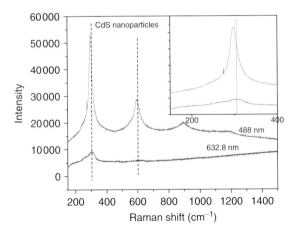

Figure 7.2.4 The Raman spectra of CdS nanoparticles recorded using two excitation wavelengths, 488 and 632.8 nm. The inset facilitates the comparison of first-order Raman band position for the two excitation wavelengths.

Figure 7.2.4 shows the Raman spectra collected using 632.8 and 488 nm excitation wavelengths in backscattering geometry. Two features can be noticed prominently from the figure. The first one is the spectra recorded with 488 nm wavelength is extremely intense when compared with the spectra collected using 632.8 nm wavelength. This indicates resonance behavior of the CdS nanoparticles for 488 nm wavelength. Absorption curves reported for CdS in literatures show a bandgap of around 2.2–3 eV in CdS nanoparticles and vary with varying particle size [55–57]. The second feature to notice is the redshift of the Raman band position for 488 nm line compared with 632.8 nm line and is clearly illustrated in the zoomed first-order spectra shown in the inset of the figure. Thus, the Raman peak position of these CdS nanoparticles showed particle size dependence of the redshift. Such a dependence is reported for CdSe nanoparticles [58] and is attributed to size-selective resonance Raman scattering. Here, the dispersion of the nanocrystallites size is taken into consideration and the dependence of the LO phonon frequency on the nanoparticle diameter is considered due to the phonon confinement effect. As mentioned previously, due to phonon confinement nanoparticles of different sizes produce different amount of shift. Such a size-selective resonant Raman scattering is also reported for other semiconductor nanoparticles in literature [59–62].

Thus, in the present case, it can be concluded that the 488 wavelength produces size-selective resonant Raman spectra. The spectra collected with 488 showed 1LO, 2LO, 3LO and 4LO peaks. It is interesting to note that the spectra collected using 632.8 nm excitation wavelength is broad and asymmetric in nature, indicating broad distribution in the particle size. According to the absorption spectra, 633 nm falls below the absorption threshold corresponding to bandgap absorption and hence the resultant spectra are nonresonant in nature. Thus, the spectrum collected using 632.8 nm reflects average spectrum from nanoparticles of all the sizes present in the system and the redshift observed in these spectra represents the average particle size of the ensemble. The asymmetry in the peak shape reflects the asymmetric distribution of particle sizes. Depending on the distribution of the particle size, a distribution of phonon peak frequency is present in this spectrum and the first-order LO peak is a convolution of contribution from various particle sizes. On the other hand, when the spectrum is collected using 488 nm wavelength, only very few nanoparticles of size having bandgap of ~2.52 eV participate in the resonant Raman spectrum. As the intensity of resonant Raman spectroscopy is orders of magnitude higher than the nonresonant component due to other particle sizes, it dominates the resultant spectrum. Therefore, the resonant Raman spectrum is not a convolution of contributions from all the particles but rather a signature contribution from very small ensembles of particles with a very narrow distribution. This results in a relatively narrow band width of the Raman spectra as seen in the Figure 7.2.4. The redshift observed in the case of 488 nm is thus due to phonon confinement corresponding to this narrow size distribution of particles participating in the resonant Raman spectrum. Thus, this method provides an indirect way of studying the size-dependent vibrational properties of nanoparticles within a single ensemble by tuning the λ_{exc} over its absorption band.

Dzhagan *et al.* [58] studied the resonant Raman scattering of small CdSe nanoparticles in polymer films. They considered the size selectivity of the Raman scattering under resonant conditions as one of the reasons for the scatter of the experimental data in the literature for the LO phonon frequency dependence on the nanoparticle size. As mentioned earlier, the redshift observed for a similar ensemble of nanoparticles can differ by few cm^{-1} depending on the excitation wavelength used. This can lead to different conclusions about particle size deduced by using PCMs described earlier and their size distribution. Our present study concludes that the frequency shift and width observed in a ensemble of nanoparticles strongly depend on the excitation wavelength used. It is particularly important near resonance conditions and resonant Raman scattering can be used for size-selective studies of nanoparticles.

7.2.5.2 CeO$_2$ Nanostructures

Reduction of particle size can lead to several other associated effects that can again contribute to the peak shift and broadening of the Raman bands apart from the contributions due to phonon confinement. As the particle size reduces, the lattice parameter also shows changes, which in turn can affect the peak position. Other factors include strain, size distribution, defects, and life time effects due to phonon relaxation process in reduced dimension and particle size.

All these effects were considered in the Raman study of CeO$_2$ nanoparticles. Graham *et al.* [63] and Weber *et al.* [64] observed redshift and large broadening of the first-order Raman peak; the combined results of the two studies showed a linear variation of the half-width of the Raman line as a function of the inverse crystallite size. They attempted to explain the results using the PCM (Eq. (7.2.4)) over the three equally weighted branches using theoretically calculated dispersion relation for CeO_2 crystal, but the calculated widths were found to be very small compared to the experimental results. The simulated results could only explain qualitatively the trend observed experimentally. They attributed the failure of this model to address very large broadening of the line with decreasing particle size to the phonon life time reduction induced by surface and boundary scattering effects as predicted by Nemanich *et al.* in nanoparticles of boron nitride (BN) [65]. Spanier *et al.* [66] carried out systematic size-dependent Raman scattering studies of CeO_{2-y} nanoparticles as a function of temperature. They could explain the large widths and redshift observed for the triply degenerate first-order optical phonon mode F_{2g} at 464 cm^{-1} in the Raman studies. It was clearly shown that simple PCM fails to simulate the experimentally observed redshift and line broadening and line shapes. They developed a comprehensive model that includes dispersion in particle size, dispersion in lattice parameter, and the presence of inhomogeneous strain.

The effect of increasing lattice parameter producing uniform strain inside the particle with decreasing size is included in the PCM by introducing a additional term $\Delta\omega_i(q, L)$. The Raman scattering intensity is then given by

$$I(\omega, L) \propto \Sigma \int \frac{\exp\left(\frac{-q^2 L^2}{8\beta}\right)}{\{\omega - [\omega_i(q) + \Delta\omega_i(q, L)]\}^2 + (\Gamma/2)^2} d^3q \qquad (7.2.10)$$

The change in line position of the Raman mode centered at ω_i due to increasing lattice parameter can be written as $\Delta\omega_i(q, L) = -3\gamma_i(q)\omega_i(q) \times [\Delta a/a_0]$, where γ_i denotes the mode Gruneisen parameter and the measured variation of lattice parameter "a" on particle size L is fitted by $a = a_0 + k/L^2$; a_0 is the lattice constant of the bulk. Using this shifted frequency gives the effect of average frequency alone, and Eq. (7.2.10) gives the combined effect of average strain and confinement effects.

The dispersion in particle size again leads to dispersion in lattice constants and, therefore, inhomogeneous strain. This distribution for L is approximated by taking a Gaussian distribution (as described in Eqs. (7.2.5) and (7.2.6)), with a width of $\Delta L = 0.44 L_0$ by Spanier *et al.* [66] and $\Delta L = 0.52 L_0$ by Popovic *et al.* [67]. It is concluded that the increasing lattice constant with decreasing particle size explains the redshift while linewidth change is fairly well explained by the inhomogeneous strain broadening associated with dispersion in particle size and by phonon confinement. It is shown that the phonon coupling is not responsible for the redshift and broadening of nanoparticles. Extra source of broadening appears due to vacancies present in the oxygen site. Popovic *et al.* [67] have used a higher degree of phonon confinement $W(r,L) \sim L/8\sqrt{5}$ and could achieve a very good agreement between the experimental and theoretical curves concluding that the peak position and line width of the CeO_2 can be very well explained by combined effect of phonon confinement with higher degree of confinement and inhomogeneous strain.

7.2.5.3 ZnO Nanostructures

As mentioned in Section 7.2.2 as the particle size decreases, a larger portion of Brillouin zone participates in the scattering process. In such case, the extent of peak shift and line broadening is expected to depend on the shape of the dispersion curve. For phonon branches with large dispersion, the peak shift and broadening is expected to be large, while for phonon branches with small dispersion, it should be small. This was shown in ZnO nanoparticles for the first time by studying phonons of different symmetries [68]. ZnO crystallizes in wurtzite structure (C_{6v}) and three Raman active modes $A_1 = 393$ cm^{-1}, $E_1 = 591$ cm^{-1}, and $E_2 = 465$ cm^{-1}. Owing to characteristic of this structure, the frequencies of both LO and TO phonons split into two frequencies with symmetries A_1 and E_1. In ZnO, in addition to LO and TO phonon modes, there are two nonpolar Raman active phonon modes with symmetry E_2. The E_1 (LO) mode shows a much larger redshift (7 cm^{-1}) and broadening (60 cm^{-1}) when compared to E_2 mode redshift (1 cm^{-1}) and broadening (12 cm^{-1}) for nanoparticles. The widely different broadening of E_1 and E_2 arises because of widely different widths of the corresponding dispersion curves. Therefore, broader the dispersion curve, broader will be the corresponding Raman band for a given size of nanoparticle [69]. It may also be noted that these effects are prominent when the particle size is smaller than ~5 nm. As mentioned earlier, it is shown theoretically [70] that the PCM is applicable for small covalent nanocrystals but it cannot be applied to ionic ZnO nanoparticles, quantum dots with size greater than ~4 nm. This is because the polar optical phonons in ZnO are almost nondispersive in the region close to center of Brillouin zone.

Apart from redshift and broadening, from the size dependence of the integrated intensity ratio of the second-order and first-order A1 LO Raman scattering, one can evaluate the coupling strength of the electron–phonon coupling within the Franck-Condon approximation [71, 72]. The electronic oscillation strength distribution over the nth phonon mode is defined as $I \sim s^n e^{-S}/n!$, in which S is the Huang-Rhys parameter, and also can be used to express the strength of the coupling between the electron and the LO phonon.

7.2.6
Effect of Laser Heating in Nanoparticles

7.2.6.1 ZnO Nanostructures

The heat generated and in consequence the rise in temperature of the sample in a micro-Raman experiment is a very important factor when analyzing Raman data. The enhancement of temperature is observed to be very high in nanoparticle samples compared to bulk. In fact, in many Raman studies, the observed redshift and broadening due to particle size reduction differs enormously from the theoretically predicted shifts and broadening by PCM effects because of extra contribution due to local temperature enhancement.

In ZnO, in nonresonant condition, the E_2 (high) peak in the spectrum of ZnO quantum dots (20 nm) is redshifted by 3 cm^{-1} from its position in the bulk ZnO spectrum [73]. As mentioned earlier, the quantum confinement effect is expected in nanoparticles of size less than 20 nm. Therefore, the observed redshift of 3 cm^{-1} cannot be attributed to the phonon confinement alone. The local temperature is measured by recording the stokes and antistokes Raman specta and then using standard relation of intensity ratio of Stokes and anti-Stokes bands; $I_S/I_{AS} \approx \exp\left[\frac{\omega}{k_\beta T}\right]$. In this study [73], the resulting temperature was found to be less than 50 °C under visible excitation nonresonant sources on a powdered sample. It is concluded that heating in nonresonant condition cannot be responsible for the observed frequency shift, and it is attributed to the presence of intrinsic defects such as impurities and oxygen vacancies. As such, E_2 (high) originates from the oxygen vibration and thus known to be sensitive to the oxygen stoichiometry of ZnO [74, 75]. On the other hand when excitation wavelength in UV region (325 nm) satisfying resonant condition is used in a micro-Raman mode, a large redshift is observed that increases with increasing laser power. This amount of shift cannot be attributed to only lattice defects. The experiment was carried out in two configurations, focusing the laser to a spot of 11 and 1.6 μm^2 by choosing different objective lenses. For the first case, the redshift was limited to 7 cm^{-1} for 20 mW incident laser power that reaches about 14 cm^{-1} for only 10 mW of incident laser power for the later case. The estimated temperature rise for the 14 cm^{-1} redshift was found to be around 700 °C. The enormous rise in temperature is attributed to the lower density (8% compared to ZnO crystal) and resulting large amount of air gaps between the quantum dots giving very small thermal conductivity of the illuminated spot. However, such a dramatic wavelength dependence of the laser heating is puzzling.

7.2.6.2 Effect of Laser Heating and Quantum Confinement in NiFe$_2$O$_4$ Nanostructures

The effect of laser heating and phonon confinement is studied in details in NiFe$_2$O$_4$ nanoparticles prepared by wet chemical method. The samples were in all the three forms, bulk powder, bulk pellet, thin film, and nanoparticles of different sizes. In most of the earlier reports the effect of power density was attributed to temperature rise [76–78]. However, in our recent study, it is shown that apart from thermal factors electronic and other excitations also significantly affect Raman spectra [79]. Shebanova and Lazor had observed oxidation of magnetite as a function of increasing laser power in single crystal as well as powder samples [80]. It was shown in this report that the oxidation is completed in two steps, first metastable γ-Fe$_2$O$_3$ is formed leading to second and final product of hematite.

Like magnetite, NiFe$_2$O$_4$ (NFO) compound crystallizes in inverse spinel structure [81]. In this structure, the tetrahedral site is fully occupied by Fe^{3+}, while the octahedral site is occupied by Ni^{2+} and Fe^{3+} ions. It is reported that the site inversion takes place as the particle size is reduced [82, 83]. It shows ferrimagnetism originating from antiparallel spins between Fe^{3+} at tetrahedral site and Ni^{2+} at the octahedral site.

The samples in the form of nanoparticles were prepared by sol–gel auto-combustion method. The as-prepared particles were then calcined at different temperatures to get nanoparticles of different sizes. Finally, when the nanoparticles were sintered at 1200 °C single-phase pure bulk Ni ferrite resulted. The samples were characterized by X-ray diffraction. The bulk NFO powder was compressed in the form of pellet and sintered at 1200 °C for 4 h. The oriented thin film of NFO was grown on the single-crystal $LaAlO_3$ (LAO)(001) substrate using KrF excimer laser ($\lambda = 248$ nm). Raman measurements were carried out in backscattering geometry using a Ar^+ excitation source having wavelength 488 nm coupled with a Labram-HR800 micro-Raman spectrometer equipped with a ×50 objective, appropriate edge filter, and pelter cooled CCD detector. The spectral resolution of the spectrometer when collected using an 1800 g mm^{-1} grating monochromator is ~1 cm^{-1}.

Figure 7.2.5 shows the X-ray diffraction patterns collected in $\theta - 2\theta$ geometry for nanoparticles, bulk sample, and thin-film sample (from bottom to top). The bulk sample showed sharp well-defined peaks that can be indexed by taking spinel structure. The nanoparticle sample showed broad peaks when compared to bulk sample. The particle size for this sample was calculated using standard Debye Sherrrer

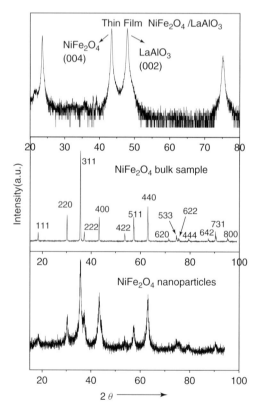

Figure 7.2.5 X-ray diffraction pattern of the $NiFe_2O_4$ nanoparticles, bulk, and the thin film. The nanoparticles showed broad peaks compared to the bulk compound. The thin films showed only peaks of (004) apart from peaks from the substrate, confirming highly oriented growth.

Table 7.2.2 Various parameters for $NiFe_2O_4$ sample calculated from the XRD and Raman spectra.

X-ray diffraction parameters			Raman spectroscopy parameters		
Sample	FWHM of the 400 peak	Lattice constant	Laser power (mW)	Peak position	FWHM
Nano particles	1.4506 ± 0.0609	8.3530 ± 0.0037	0.1	698.023 ± 0.125	44.153 ± 0.203
			40	665.421 ± 0.355	58.545 ± 0.804
Bulk	0.2612 ± 0.0015	8.3394 ± 0.0002	0.1	703.11 ± 0.080	30.212 ± 0.312
			40	673.155 ± 0.169	85.923 ± 0.852
Thin film	0.2237 ± 0.0031	8.3201 ± 0.0012	0.5	701.014 ± 0.012	701.014 ± 0.012
			40	692.5 ± 0.0011	40.893 ± 0.531

formula [84] from the FWHM of the diffraction peaks after subtracting the instrumental contribution and found to be ~9 nm. The film showed only one peak (004) besides the contribution from substrate in the diffraction pattern plotted on a logarithmic scale. The particle size calculated from the peak width is more than 100 nm.

The FWHM of the (004) peak and lattice constants are given in Table 8.2.2. Figure 7.2.6c shows Raman spectra of bulk compound between 150 and 750 cm^{-1} recorded using a 488 nm excitation source. The spectra were collected at various incident laser powers from 0.1 to 40 mW. The power mentioned in this chapter is at the laser head and is expected to reduce significantly when it falls on the sample. The laser was focused to a spot size of ~2μm on the sample. The spectra showed five Raman bands typical of an inverse spinel structure of $NiFe_2O_4$. The Raman spectrum is of good quality and matches with that reported earlier [85, 86]. $NiFe_2O_4$ compound crystallizes in a spinel structure of space group Fd-3m, and group theoretical calculations result in five Raman-active bands, namely $A_{1g} + E_g + 3T_{2g}$ [87]. The spectra match well with those for Fe_3O_4 exhibiting similar structure [88]. The Raman bands for Fe_3O_4 are sharper and well defined, while Raman bands of $NiFe_2O_4$ (Figure 7.2.6c) showed a shoulderlike feature at the left side of all the Raman active bands. In other words, all the Raman bands showed double-peak-like features and is well reported in literature [89]. When the structures of Fe_3O_4 and $NiFe_2O_4$ are compared, it is seen that both are identical crystallographically except that in Fe_3O_4 all the tetrahedral and octahedral sites are occupied by Fe ions and that in $NiFe_2O_4$ the octahedral sites are occupied by Ni or Fe ions. This results in a distortion-free structure and uniform Fe–O bond distances in Fe_3O_4 compound. On the other hand, in $NiFe_2O_4$, due to differences in ionic radii of Ni and Fe ions, the Fe/Ni–O bond distance shows a considerable distribution. This means that the local structure in two cases is different, and Raman spectroscopy being a local-structure-sensitive tool detects these changes very effectively. This distribution

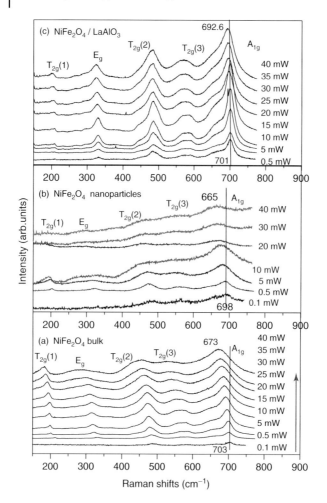

Figure 7.2.6 Raman spectra of $NiFe_2O_4$ bulk sample (a), nanocrystalline sample (b), and $NiFe_2O_4$ thin film on $LaAlO_3$ substrate (c) in the spectral range $100-750$ cm^{-1} ($\lambda = 488$ nm). The spectra are collected at various laser powers from 0.1 to 40 mW.

in bond distances probably results in double-peak-like structure in $NiFe_2O_4$, one peak representing the unit cell with all Fe ions and the other one, the unit cell with mixed Fe and Ni ions.

As the incident laser power increases, the disorder in the structure increases that results in increase in Raman band width and redshift of the bands as shown in Figure 7.2.7. After 25 mW of incident laser power, the thermal disorder takes over the site disorder for bulk sample and the double-peak-like shape merges in a single broad Raman band (Figure 7.2.6a). The Raman band that occurs at 703 cm^{-1} for 0.1 mW of incident laser power occurs at 673 cm^{-1} for incident laser power of 40 mW. This means that the bands shift by ($\sim$30 cm^{-1}) with an increase of $\sim$40 mW of laser power.

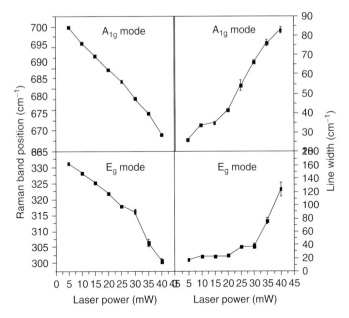

Figure 7.2.7 The change in Raman band position and line width of A_{1g} and E_g modes as a function of incident laser power.

Figure 7.2.6b shows Raman spectra at increasing laser power for nanocrystalline sample. Here again with increasing laser power, the Raman bands show similar trend as that for the bulk compound. The position of the A_{1g} Raman band at the lowest laser power of 0.1 mW is found to be at 698 cm^{-1}, which is 5 cm^{-1} lower compared to the bulk value. This redshift is attributed to homogeneous strain due to increased lattice parameter of the nanoparticles compared to bulk compound and confinement effects. The changes in Raman line position due to changes in the lattice parameter because of particle size can be estimated as $\Delta\omega_i = -3\gamma_i(q)\omega_i(q)(\Delta a/a_0)$, where the mode Gruneisen parameter is $\gamma_i = -\frac{d\ln\omega_i}{d\ln V} = \left(\frac{B}{\omega_i}\right)\left(\frac{d\omega_i}{dP}\right)$ [66] as mentioned in Section 7.2.5.2. The value of γ_i for the 703 cm^{-1} mode of 0.42 is taken based on the value for CoFe$_2$O$_4$ obtained from high-pressure Raman studies [90]. The two compounds match well in structural and other properties, and hence to the first approximation, we assume the Gruneisen parameter for the two compounds to have similar value. Here, Δa represents difference in lattice parameter of the nanoparticles compared to bulk lattice parameter (a_0) and ω_i denote Raman mode position. Using this formula, the value of $\Delta\omega_i$ is found to be $\sim$1.5 cm^{-1}. Therefore, the remaining redshift (3.5 cm^{-1}) is attributed to phonon confinement and inhomogeneous strain due to distribution in particle size. As such, a phonon confinement is expected to produce asymmetric line shape [9]. However, in the present case as mentioned earlier, all the Raman bands are double-peak-like structures. Hence, the asymmetry in line shape due to confinement is difficult to discern and may be buried in the broad band. The Raman bands further shift by 33 cm^{-1} as the incident laser power was raised to 40 mW. This shift can be attributed to the thermal effects in the form of rise in temperature of nanoparticles because of very poor thermal conductivity of nanoparticles compared to bulk. Taking the intrinsic A$_{1g}$ mode wave number at room temperature to be 703 cm^{-1}, the nanoparticles exhibit the largest peak shift of 38 cm^{-1}. One can then also argue that the peak position even at 0.1 mW of the incident laser power is affected and redshifted due to thermal effects. However, a shift of 5 cm^{-1} due to 0.1 mW of laser power and further, just a shift of 33 cm^{-1} due to 400 times more power are unlikely. This confirms that the redshift observed for nanoparticle when compared to bulk at 0.1 mW of incident laser power is due to strain and confinement effects.

On the contrary, the effect of incident laser power on thin-film samples is feeble. Figure 7.2.6c shows the Raman spectra for NiFe$_2$O$_4$ films grown on LaAlO$_3$ single-crystal substrate. Upto 25 mW of incident laser power, the Raman band position and band width remains nearly unchanged. When the incident laser power is increased to 40 mW, it showed a redshift of 8 cm^{-1}. In order to observe the effect of laser heating in powder and pellet form, a highly compacted pellet of bulk compound was prepared and sintered at 1200 °C for 4 h. The Raman spectra collected at various laser powers from 0.5 to 10 mW are shown in Figure 7.2.8. The position of the Raman band at 0.5 mW is same as that for the powder sample and the pellet sample showed similar behavior with increasing laser power. The peak positions and width were extracted by fitting the observed spectra by multiple lorentzian peak shapes, and the results are given in Table 8.2.2.

In short Ni ferrite showed a redshift of different magnitudes and increase in band width as the incident laser power is increased. The thin-film samples are very

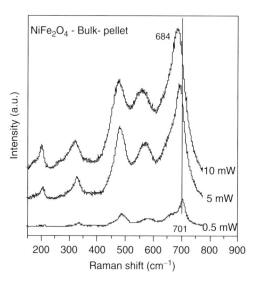

Figure 7.2.8 Raman spectra of NiFe$_2$O$_4$ bulk compound pressed in the form of pellet at various incident laser powers.

robust and showed feeble effect of laser power variation. The other common feature in all these measurements is the absence of new feature as the incident laser power is increased up to 40 mW. In one of our earlier studies on $La_{1-x}Sr_xMnO_3$ series of single crystals, we observed new features that were interpreted in the form of structural transformation [79]. Shebanova and Lazor observed transformation of Fe_3O_4 phase to hematite phase due to oxidation as a result of increase in incident laser power. They identified the newly formed hematite phase by observation of new features in Raman band as the laser power is increased [80].

In the present experiments on all types of samples, we do not observe any such phase transformation. In order to further confirm the stability of compounds, we carried out the experiments with increasing and decreasing laser power, while the laser was focused on the same spot of the sample that showed reversible behavior. This again confirms the lack of oxidation of Ni ferrite for these incident laser powers. We argue that if Ni ferrites get oxidized, say at the highest laser power of 40 mW, and form a stable second phase there is no reason that it will get reduced immediately as the laser power is decreased. Therefore, we conclude that the crystallographic structure as well as chemical composition of this compound remains unaffected due to incident laser power.

7.2.7
Summary and Future Directions

In summary, the effect of particle size reduction on Raman spectra is discussed here. Reduction in particle size below a critical size typically less than ~20 nm is reflected in the laser Raman spectrum in the form of redshift and asymmetric broadening. This makes laser Raman spectroscopy a very promising tool for the characterization of nanostructures as in this technique sample preparation is not needed. The shift and broadening can be understood in the simple framework of confinement of phonons in the nanosize particles. This can be achieved by simply invoking Gaussian confinement model. In this chapter, attempt is made to provide these phenomenological confinement models and various modifications in the form of inclusions such as effects due to particle size distribution, distribution in lattice constants, inhomogeneous strain, and thermal effects. In this way, the Raman band position and line shape can be simulated for a particular ensemble of nanoparticles. These calculations of particle size by considering Gaussian crystal size distribution and confinement models along with the effect of laser heating and strain provide a very easy and nondistructive method for characterization. However, this model is oversimplification due to the assumption that nanocrystals of size less than 20 nm, having extensive surface reconstruction, lattice defects, and dangling bonds are assumed to have the phonon density of state for the bulk material and the model only relaxes $q = 0$ selection rule. This may lead to significant errors in the determination of particle size, and further understanding is required to measure strain, changes in lattice constant due to variation in particle size, and its effects on the Raman spectra of nanoparticles.

It is important that the thermal effects in the form of temperature rise and inhomogeneous strain are also considered when characterizing nanoparticles by laser Raman spectroscopy. It is more so when a micro-Raman instrument is used. In this chapter, it is shown that even a very small incident laser power of 1 mW is sufficient to raise the temperature of the sample significantly, which will contribute to redshift and broadening.

At present, micro-Raman instruments provide spatial resolution of 1 μm, which is insufficient for nanoparticle research. The micro-Raman scattering technique is promising, but it lacks the spatial resolution at the nanoscale due to the diffraction limit (Section 7.2.4) from the probing light. Recently, tip-enhanced Raman scattering is shown to provide very high spatial resolution far beyond the diffraction limits of the probing light. This technique utilizes the known principles of surface-enhanced Raman scattering with a very sharp metal quoted tip touching the surface of the sample. Raman signal only from the vicinity of this tip is surface enhanced and thus providing spatial resolution less than 100 nm. Efforts are on to reduce it further to few nanometers so that optical properties of individual nanoparticles can be studied.

Acknowledgments

The author would like to express his sincere thanks to Dr. P. Chaddah for his support and encouragements. Ms. Anju Ahlawat and Mr. Dileep Mishra is acknowledged for carrying out experiments.

References

1. Dresselhaus, M.S., Dresselhaus, G., Saito, R., and Jario, A. (2005) *Phys. Rep.*, **409** (2), 47.
2. Jario, A., Pimenta, M.A., Souza Filho, A.G., Saito, R., Dresselhaus, G., and Dresselhaus, M.S. (2003) *New J. Phys.*, **5**, 139.
3. Dresselhaus, M.S., Jario, A., Hofmann, M., Dresselhaus, G., and Saito, R. (2010) *Nano Lett.*, **10**, 751.
4. Malard, L.M., Pimenta, M.A., Dresselhaus, G., and Dresselhaus, M.S. (2009) *Phys. Rep.*, **473**, 51.
5. Chassaing, P.M., Demangeot, F., Combe, N., Saint-Macary, L., Kahn, M.L., and Chaudret, B. (2009) *Phys. Rev. B*, **79**, 155314.
6. Duval, E. (1992) *Phys. Rev. B*, **46**, 5795.
7. Yadav, H.K., Gupta, V., Sreenivas, K., Singh, S.P., Sundarakannan, B., and Katiyar, R.S. (2006) *Phys. Rev. Lett.*, **97**, 085502.
8. Verma, P., Cordts, W., Irmer, G., and Monecke, J. (1999) *Phys. Rev. B*, **60**, 5778.
9. Richter, H., Wang, Z.P., and Ley, L. (1981) *Solid State Commun.*, **39**, 625.
10. Campbell, I.H. and Fauchet, P.M. (1986) *Solid State commun.*, **58**, 739.
11. Gouadec, G. and Colomban, P. (2007) *J. Raman Spectrosc.*, **38**, 598.
12. Zi, J., Zhang, K., and Xie, X. (1997) *Phys. Rev. B*, **55**, 9263.
13. Iqbal, Z. and Veprek, S. (1982) *J. Phys. C: Solid State Phys.*, **15**, 377.
14. Tiong, K.K., Amirharaj, P.M., Pollak, F.H., and Aspnes, D.E. (1984) *Appl. Phys. Lett.*, **44**, 122.
15. Shuker, R. and Gammon, R.W. (1970) *Phys. Rev. Lett.*, **25**, 222.
16. Bottani, C.E., Mantini, C., Milani, P., Manfredini, M., Stella, A., Tognini, P., Cheyssac, P., and Kofman, R. (1986) *Appl. Phys. Lett.*, **69**, 2409.

17. Binder, M., Edelmann, T., Metzger, T.H., Mauckner, G., Goerigk, G., and Peisl, J. (1996) *Thin Solid Films*, **276**, 65.

18. Islam, M.N. and Kumar, S. (2001) *Appl. Phys. Lett.*, **78**, 715.

19. Islam, M.N., Pradhan, A., and Kumar, S. (2005) *J. Appl. Phys.*, **98**, 024309.

20. Gonzalez-Hernandez, J., Azarbayejani, G.H., Tsu, R., and Pollak, F.H. (1985) *Appl. Phys. Lett.*, **47**, 1350.

21. Parayanthal, P. and Pollak, F.H. (1984) *Phys. Rev. Lett.*, **52**, 1822.

22. Kittel, C. (1967) *Introduction to Solid State Physics*, 3rd edn, Chapter 5, John Wiley & Sons, Inc., New York.

23. Roca, E., Giner, C.T., and Cardona, M. (1994) *Phys. Rev. B*, **49**, 13704.

24. Ager, J.W., Viers, D.K., and Rosenblatt, G.M. (1991) *Phys. Rev. B*, **43**, 6491.

25. Boppart, H., van Straaten, J., and Silvera, J.F. (1985) *Phys. Rev. B*, **32**, 1423.

26. Grimsditch, M.H., Anastassakis, E., and Cardona, M. (1978) *Phys. Rev. B*, **18**, 901.

27. Fischer, A., Anthony, L., and Compaan, A.D. (1998) *Appl. Phys. Lett.*, **72**, 2559.

28. Carles, R., Mlayah, A., Amjoud, M.B., Reynes, A., and Morancho, R. (1992) *Jap. J. Appl. Phys.*, **31**, 3511.

29. Rohmfeld, S., Hundhausen, M., and Ley, L. (1984) *Phys. Status Solidi B*, **215**, 115.

30. Zhang, W.F., He, Y.L., Zhang, M.S., Yin, Z., and Chen, Q. (2000) *J. Phys. D*, **33**, 912.

31. Fujii, M., Hayashi, S., and Yamamoto, K. (1991) *Jap. J. Appl. Phys.*, **30**, 687.

32. Oswald, S., Mochalin, V.N., Havel, M., Yushin, G., and Gogotsi, Y. (2009) *Phys. Rev. B*, **80**, 075419.

33. Nakamura, K., Fujitsuka, M., and Kitajima, M. (1990) *Phys. Rev. B*, **41**, 12260.

34. Fujii, M., Hayashi, S., and Yamamoto, K. (1990) *Appl. Phys. Lett.*, **57**, 2692.

35. Tanaka, A., Onari, S., and Arai, T. (1992) *Phys. Rev. B*, **45**, 6587.

36. Havel, M., Baron, D., and Colomban, P. (2004) *J. Mater. Sci.*, **39**, 6183.

37. Havel, M. and Colomban, P. (2005) *J. Raman Spectrosc.*, **34**, 786.

38. Li, B., Yu, D., and Zhang, S.L. (1999) *Phys. Rev. B*, **59**, 1645.

39. Gouadec, G. and Colomban, P. (2007) *Prog. Cryst. Growth Charact. Mater.*, **53**, 1.

40. Gomonnai, A.V., Azhniuk, Y.M., Yukhymchuk, V.O., Kranjeec, M., and Lopushansky, V.V. (2003) *Phys. Status Solidi B*, **239**, 490.

41. Yan, Y., Zhang, S.L., Fan, S., Han, W., Meng, G., and Zhang, L. (2003) *Solid State Commun.*, **126**, 649.

42. Zhang, L., Xie, H.J., and Chen, C.Y. (2002) *Eur. Phys. J. B*, **27**, 577.

43. Xiong, Q., Wang, J., Reese, O., Lew-Yan-Voon, L.C., and Eklund, P.C. (2004) *Nano Lett.*, **4**, 1991.

44. Roy, A. and Sood, A.K. (1996) *Phys. Rev. B*, **53**, 12127.

45. Klein, M.C., Hache, F., Ricard, D., and Flytzanis, C. (1990) *Phys. Rev. B*, **42**, 11123.

46. Hayashi, S. and Kanamori, H. (1982) *Phys. Rev. B*, **26**, 7079.

47. Fonoberov, V.A. and Balandin, A.A. (2005) *J. Phys. Condens. Matter*, **17**, 1085.

48. Zi, J., Zhang, K., and Xie, X. (1997) *Phys. Rev. B*, **55** (15), 9263.

49. Jusserand, B. and Cardona, M. (1989) in *Light Scattering in Solids V* (eds M. Cardona and G. Guntherodt), Springer, Berlin, p. 49.

50. Maradudin, A.A. and Burstein, E. (1967) *Phys. Rev.*, **164**, 1081.

51. Go, S., Bilz, H., and Cardona, M. (1975) *Phys. Rev. Lett.*, **34**, 580.

52. Rayleigh, L. (1879) *Philos. Mag.*, **8**, 261.

53. Lee, J. and Tsakalakos, T. (1997) *Nanostruct. Mater.*, **8**, 381.

54. Rolo, A.G., Vieira, L.G., Gomes, M.J.M., and Ribeiro, J.L. (1998) *Thin Solid Films*, **312**, 348.

55. Balandin, A., Wang, K.L., Koulin, N., and Bandyopadhyay, S. (2000) *Appl. Phys. Lett.*, **76**, 137.

56. Abdi, A., Titova, L.V., Smith, L.M., Jackson, H.E., Yarrison-Rice, J.M., Lensch, J.L., and Lauhon, L.J. (2006) *Appl. Phys. Lett.*, **88**, 043118.

57. Zeiri, L., Patla, I., Acharya, S., Golan, Y., and Efrima, S. (2007) *J. Phys. Chem. C*, **111**, 11843.

58. Dzhagan, V.M., Valakh, M.Y., Raevskaya, A.E., Stroyuk, A.L., Kuchmiy, S.Y., and Zahn, D.R.T. (2008) *Nanotechnology*, **19**, 305707.

59. Alivisatos, A.P., Harris, T.D., Carrol, P.J., Staigerwald, M.L., and Brus, L.E. (1989) *J. Chem. Phys.*, **90**, 3463.

60. Trallero-Giner, C., Debernardi, A., Cardona, M., Menendez-Proupin, E., and Ekimov, A.I. (1997) *Phys. Rev. B*, **57**, 4664.

61. Milekhin, A.G., Nikiforov, A.I., Pchelyakov, O.P., Schulze, S., and Zahn, D.R.T. (2002) *Nanotechnology*, **13**, 55.

62. de Paula, A.M., Barbosa, L.C., Cruz, C.H.B., Alves, O.L., Sanjurjo, J.A., and Cesar, C.L. (1996) *Appl. Phys. Lett.*, **69**, 357.

63. Graham, G.W., Weber, W.H., Peters, C.R., and Usmen, R. (1991) *J. Catal.*, **130**, 310.

64. Weber, W.H., Hass, K.C., and McBride, J.R. (1997) *Phys. Rev. B*, **48**, 178.

65. Nemanich, R.J., Solin, S.A., and Martin, R.M. (1981) *Phys. Rev. B*, **23**, 6348.

66. Spanier, J.E., Robinson, R.D., Zhang, F., Chan, S.W., and Herman, I.P. (2001) *Phys. Rev. B*, **64**, 245407.

67. Popovic, Z.V., Dohcevic-Mitrovic, Z., Konstantinovic, M.J., and Scepanovic, M. (2007) *J. Raman Spectrosc.*, **38**, 750.

68. Rajalakshmi, M., Arora, A.K., Bendre, B.S., and Mahamuni, S. (2000) *J. Appl. Phys.*, **87**, 2445.

69. Arora, A.K., Rajalakshmi, M., Ravindran, T.R., and Sivasubramanian, V. (2007) *J. Raman Spectrosc.*, **38**, 604.

70. Fonoberov, V.A. and Baladin, A.A. (2004) *Phys. Rev. B*, **70**, 233205.

71. Huang, K. and Rhys, A. (1950) *Proc. R. Soc. London, Ser. A*, **204**, 406.

72. Thangavel, R., Moirangthem, R.S., Lee, W.S., Chang, Y.C., Wei, P.K., and Kumar, J. (2010) *J. Raman Spectrosc.*, doi: 10.1002/jrs.2599

73. Alim, K.A., Fonoberov, V.A., and Balandin, A.A. (2005) *Appl. Phys. Lett.*, **86**, 053103.

74. Serrano, J., Manjon, F.J., Romero, A.H., Widulle, F., Lauck, R., and Cardona, M. (2003) *Phys. Rev. Lett.*, **90**, 055510.

75. Scepanovic, M., Grujic-Brojcin, M., Vojisavljevic, K., Bernik, S., and Sreckovic, T. (2010) *J. Raman Spectrosc.*, **41**, 914, doi: 10.1002/jrs.2546

76. Tinnemans, S.J., Kox, M.H.F., Sletering, M.W., Nijhuis, T.A., Visser, T., and Weckhuysen, B.M. (2006) *Phys. Chem. Chem. Phys.*, **8**, 2413.

77. Laikhtman, A. and Hoffman, A. (1997) *J. Appl. Phys.*, **82**, 243.

78. Kagi, H., Tsuchida, I., Wakatsuki, M., Takahashi, K., Kamimura, N., Iuchi, K., and Wada, H. (1994) *Geochim. Cosmochim. Acta*, **58**, 3527.

79. Sathe, V.G., Dubey, A., Rawat, R., Naralikar, A.V., and Prabhakaran, D. (2009) *J. Phys. Condens. Matter*, **21**, 075603.

80. Shebanova, O.N. and Lazor, P. (2003) *J. Raman Spectrosc.*, **34**, 845.

81. Kavas, H., Baykal, A., Toprak, M.S., Köseoglu, Y., Sertkol, M., and Aktas, B. (2009) *J. Alloys Compd.*, **479**, 49.

82. Ponpandian, N., Balaya, P., and Narayanasamy, A. (2002) *J. Phys. Condens. Matter*, **14**, 3221.

83. Chinnasamy, C.N., Narayanasamy, A., Ponpandian, N., Chattopadhayay, K., Shinoda, K., Jeyadevan, B., Thhji, K., Nakatsuka, K., Furubayashi, T., and Nakatani, I. (2001) *Phys. Rev. B*, **63**, 184108.

84. Cullity, B.D. and Stock, S.R. (2001) *Elements of X-Ray Diffraction*, 3rd edn, Prentice Hall Publishers, New York, p. 664.

85. Shia, Y., Dinga, J., Shenb, Z.X., Sunb, W.X., and Wangb, L. (2000) *Solid State Commun.*, **115**, 237.

86. Graves, P.R., Johnston, C., and Campaniello, J.J. (1988) *Mater. Res. Bull.*, **23**, 1651.

87. Kreisel, J., Lucazeau, G., and Vincent, H. (1998) *J. Solid State Chem.*, **137**, 127.

88. Sousa, M.H., Tourinho, F.A., and Rubim, J.C. (2000) *J. Raman Spectrosc.*, **31**, 185.

89. Zhou, Z.H., Xue, J.M., and Wang, J. (2002) *J. Appl. Phys.*, **91**, 6015.

90. Wang, Z., Downs, R.T., Pischedda, V., Shetty, R., Saxena, S.K., Zha, C.S., Zhao, Y.S., Schiferl, D., and Waskowska, A. (2003) *Phys. Rev. B*, **68**, 094101.

8
Size Determination of Nanoparticles by Dynamic Light Scattering

Haruhisa Kato

8.1
Introduction

In the past decade, a large number of experimental and theoretical studies on nanoscale materials have been widely performed. Nanoscale materials are currently a subject of interest because of their attractive physical, photochemical, and catalytic properties. As the potential usefulness of nanomaterials can increase with decreases in their size, the number of studies on the reduction of the size of nanoscale materials is growing [1–10]. For example, a good photocatalyst requires a large catalytic surface area, since the size of the primary catalyst nanoparticle defines the surface area available for adsorption and decomposition of the target molecules. Commonly, commercial nanoparticles are provided as dry powder and the size of the primary nanoparticles can be determined by the Brunauer, Emmett, and Teller (BET) method or by microscopy. However, the primary nanoparticles tend to aggregate when a nanoparticle dispersion (e.g., a cosmetic liquid) is prepared, since the high ionic strength of the solution, along with electrostatic and van der Waals particle interactions, results in the production of larger, aggregated secondary particles. In this case, the secondary particle size is important because the van der Waals interaction is roughly directly proportional to the particle size, attributing to the stability of such commercial nanoparticle dispersions.

Although nanoscale materials possess many desirable, novel, and superior properties, they may pose an entirely new risk to human health [11–20]. Unexpected adverse effects of nanomaterials on human health are a serious issue that has attracted much attention. Many international organizations and researchers have therefore carried out toxicity assessments of various nanoparticles, such as metal, metal oxide, fullerenes, and carbon nanotubes [12, 15, 16, 18, 21]. These studies have shown that it is possible for nanoscale particles to penetrate the human body and induce toxicity [22]. In contrast, other researchers have reported that they have observed little toxicity due to nanoparticles [23]. Poor characterization of the secondary particle sizes in suspension could be a reason for such discrepancies in the toxicity studies; therefore, the determination of the size of secondary nanoparticles in dispersion is significant.

Nanomaterials: Processing and Characterization with Lasers, First Edition.
Edited by Subhash Chandra Singh, Haibo Zeng, Chunlei Guo, and Weiping Cai.
© 2012 Wiley-VCH Verlag GmbH & Co. KGaA. Published 2012 by Wiley-VCH Verlag GmbH & Co. KGaA.

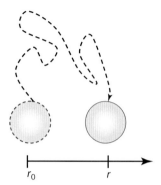

Figure 8.1 Illustration of the Brownian motion of a nanoparticle in two dimensions. The root mean square displacement is described in Eq. (8.1).

r_0 r

Dynamic light scattering (DLS) is widely used to determine the size of Brownian nanoparticles in colloidal suspensions in the nano- and submicrometer ranges [24–28]. When the particles are dispersed in a liquid, they are in constant random motion, that is, Brownian motion, in which a given particle undergoes random position changes in time. Assuming the distribution for the displacement of a particle undergoing Brownian motion in aqueous solution is a Gaussian function, the average root mean square displacement in two dimensions can be calculated using the concentration of the particles by

$$\langle (r - r_0)^2 \rangle = 2Dt \tag{8.1}$$

where t is the diffusion time (Figure 8.1). The diffusion of spherical particles can be described by the Stokes–Einstein equation

$$D = \frac{k_B T}{3\pi \eta d} \tag{8.2}$$

where k_B is the Boltzmann constant, T is the absolute temperature, η is the viscosity of the solvent, D is the diffusion coefficient of the particles, and d is the diameter of the particles. According to the Stokes–Einstein assumption, small particles move quickly by diffusion (Figure 8.2).

Using DLS, these diffusion phenomena and the Stokes–Einstein assumption are utilized in the determination of secondary nanoparticle size in solution. In this method, the diffusion coefficient of the nanoparticles is determined first and then the size of the particles is calculated from the diffusion coefficient using the Stokes–Einstein assumption (Eq. (8.2)). The practical upper limit of the particle size determined using the DLS method is around 1–3 µm because of the dependence on Brownian motion and relaxation time. Usually, the upper limit can be probed with the available laser power. The observed size range corresponds well with a large number of commercial nanoparticle dispersion applications. In addition, an advantage of using the DLS method compared with, for example, microscopy is the possibility of *in situ* size determination of *secondary* nanoparticles in the suspension. The DLS method is therefore one of the most powerful and useful tools to determine colloidal secondary particle size in suspension.

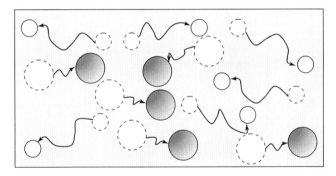

Figure 8.2 Diffusion (e.g., self-diffusion, interdiffusion, and hydrodynamic diffusion) is the most crucial process by which thermodynamics manifests itself in nature and technology. Small particles move more quickly by diffusion than bigger ones.

8.2
General Principles of DLS (Photon Correlation Spectroscopy)

In DLS, coherent light from an incident laser beam illuminates a liquid medium in which small particles are suspended. When the particle size is much smaller (e.g., <10%) than the wavelength of light, the scattered radiation from the particles will be in phase. If the incident light is unpolarized with irradiance I_0, the scattered irradiance I_s at an angle θ to the incident light source can be described by the Rayleigh light scattering theory [29–31]. The Rayleigh light scattering intensity I_s observed for a uniform size distribution of nanoparticles is given by

$$I_s = I_0 \frac{c}{2r^2} \left(\frac{2\pi}{\lambda}\right)^4 \left(\frac{d}{2}\right)^6 \left|\frac{m^2 - 1}{m^2 + 2}\right|^2 (1 + \cos^2 \theta) \tag{8.3}$$

where c is the number of nanoparticles, d the diameter of the spherical nanoparticles, m the refractive index of the particle relative to that of the medium, r the distance to the particles, and λ the wavelength of the polarized laser. The scattered light intensity follows an inverse square law $(1/r^2)$ and is proportional to $1/\lambda^4$. The scattered light is partially polarized perpendicular relative to the incident beam even if the incident light is unpolarized, with the last term $(1 + \cos^2 \theta)$, in Eq. (8.3), representing the vertical (1) and horizontal $(\cos^2 \theta)$ components of the scattering relative to the incident light source (Figure 8.3). The vertically polarized light has equal intensity at all scattering angles, while the intensity of the horizontal component depends on the scattering angle and is zero at 90°. When sufficiently small spherical particles are illuminated by an unpolarized light source, the scattered light is completely polarized at a scattering angle of 90°.

Using DLS, information about the dynamics of a particle in suspension is extracted from this Rayleigh light scattering spectrum. Consequently, measurement of the variation in intensity and frequency of the scattered light from particles can be correlated to the Brownian motion and dynamics of the particles. DLS is ideally suited for particles smaller than 1 µm and requires a dilute particle dispersion

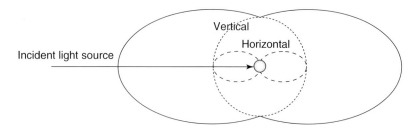

Figure 8.3 Angular distribution of Rayleigh light scattering by a spherical nanoparticle. The solid line represents the total light scattering, the dotted line represents vertically polarized light, and the dashed line represents horizontally polarized light.

because of the dependence on random Brownian motion and relaxation time. DLS uses the phenomenon by which light scattered from a moving particle will have a slightly different frequency from the incident light source, known as the *Doppler effect*. When the particle suspension is illuminated by a laser source, random Brownian motion of the suspended particles induces the scattered light to vary randomly in frequency, resulting in random fluctuations in the scattered light intensity, as shown in Figure 8.4a.

In order to detect the scattered light, optical mixing techniques (i.e., homodyning and heterodyning) are commonly used. In these methods, no filter is inserted between the scattering medium and the photomultiplier; therefore, the scattered light profile impinges directly onto the photomultiplier. The scattered light intensity is then analyzed by autocorrelation, measured via pulses from a photomultiplier tube, or an avalanche photodiode detector, which has higher sensitivity. In the homodyne (or self-beat) detection method, only scattered light reaches the photocathode. In the heterodyne method, a local oscillator (i.e., a portion of the unscattered laser source) is mixed with the scattered light at the photocathode surface. These two detection methods are illustrated in Figure 8.4b,c, respectively. The homodyne method is the more convenient and widely used detection scheme for DLS apparatus because of the difficulty of stabilizing two light sources in the heterodyne method.

In light scattering experiments, the incident light source field is sufficiently weak that the system can be assumed to respond linearly to it. The change of the light field caused by the scattering of light by particles, and the response of the equilibrium system to this weak incident field, can be described accurately. The resultant scattering pattern comprises scattering by individual particles and scattering by the wider many-particle systems, which changes in time because of the Brownian motion of particles. Time-dependent correlation functions are therefore used in DLS measurements to provide a concise method for expressing the degree to which the two dynamical properties are correlated over a period of time. Illustrations of the intensities of scattered light from nanoparticle dispersions as a function of time are shown in Figure 8.5a,b. The measured scattered intensities

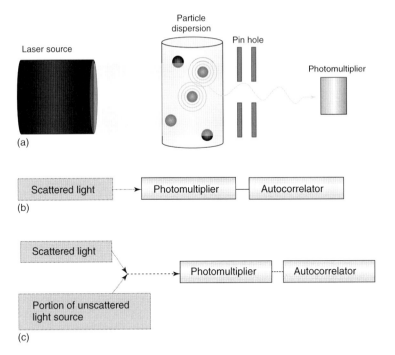

Figure 8.4 Schematic of a dynamic light scattering (DLS) system. (a) Scattered intensity in typical DLS setup, (b) illustration of the homodyne (or self-beat) detection technique, and (c) illustration of the heterodyne detection technique.

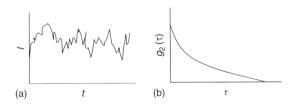

Figure 8.5 (a) Illustration of the intensity change in light scattered from a dispersion of different sized nanoparticles as a function of time. (b) Illustration of the time-averaged autocorrelation function of scattered light as described by Eq. (8.6).

of the light field can be simply time averaged as shown in Eq. (8.4).

$$\bar{I}(t_0, \tau) = \frac{1}{\tau} \int_{t_0}^{t_0+\tau} dt I(t) \tag{8.4}$$

Here, t_0 is the time at which the measurement was initiated and τ is the time over which it is averaged. The average becomes meaningful only if τ is sufficiently large compared to the period of fluctuation, indicating that the ideal experimental time tends toward infinity and that this infinite time average of intensities is independent of t_0. Because of its independence of t_0, the noise signal of light scattered intensities, $I(t)$, can be expressed as

$$\langle I \rangle = \lim_{\tau \to \infty} \frac{1}{\tau} \int_0^\tau dt I(t) \tag{8.5}$$

In light scattering experiments, $I(t)$ and $I(t + \tau)$ should have quite different values. However, when τ is sufficiently small compared to the observed time, $I(t)$ and $I(t + \tau)$ can take fairly similar values. The correlation between the two values can be defined by

$$g_2(\tau) = \frac{\langle I(t)\, I(t + \tau) \rangle_t}{\langle I(t) \rangle_t^2} \tag{8.6}$$

where $g_2(\tau)$ is known as the *light scattering intensity autocorrelation function*. In homodyne detection, $g_2(\tau)$ is related to the electric field autocorrelation function, using the assumption that its instantaneous current output is proportional to the square of the incident electric field, and is defined by

$$g_2(\tau) = 1 + \beta \left| g_1(\tau) \right|^2 \tag{8.7}$$

where β is a fundamental constant having a maximum value of 1 [32]. $g_1(\tau)$ is known as the *homodyne correlation function*.

On the other hand, in the heterodyne method, the autocorrelation function of the scattered light can be defined by

$$g_2(\tau) = 1 + \frac{2\beta \langle I_s \rangle g_1(\tau)}{I_{LO}} \tag{8.8}$$

where $\langle I_s \rangle$ is the average scattered light intensity, I_{LO} is the scattered light intensity of a constant local oscillator, and $g_1(\tau)$ is the heterodyne correlation function. In this case, it is assumed that fluctuations of the local oscillator field are negligible and that the local oscillator field and the scattered field are statistically independent. When the local oscillator intensity lies between the extremes $I_{LO} = 0$ in the homodyne case and $I_{LO} \gg \langle I_s \rangle$ in the heterodyne case, $g_2(\tau)$ will be a weighted sum of $g_1(\tau)$ and $\left| g_1(\tau) \right|^2$.

Assuming a *monodisperse* dilute particle suspension (i.e., all the particles are the same size and shape) and that there is no interaction between the particles, $g_1(\tau)$ can be represented by

$$\ln(g_1(\tau)) = -\Gamma \tau \tag{8.9}$$

$$\Gamma = DQ^2 \tag{8.10}$$

and

$$Q = \frac{4\pi n_0}{\lambda_0 \sin\left(\frac{\theta}{2}\right)} \tag{8.11}$$

where θ is the observed angle, λ_0 is the wavelength of the laser source in vacuum, n_0 is the refractive index of the medium solvent, and D is the diffusion coefficient of particles in the medium. The delay rate, Γ, is related to the translational diffusion coefficient of the particles. Q is the modulus of the scattering vector and is given by

Eq. (8.11). According to the Stokes–Einstein relation (Eq. (8.2)), the hydrodynamic diameter, d, of the particles can then be calculated from D via DLS, assuming that there is no size distribution.

When the particle dispersion is *polydisperse*, that is, the particles have some finite size distributions, the correlation function is represented as

$$g_1(\tau) = \int_0^\infty G(\Gamma) \exp(-\Gamma\tau) d\Gamma \tag{8.12}$$

and

$$\int_0^\infty g_1(\Gamma) d\Gamma = 1 \tag{8.13}$$

For DLS analysis of polydisperse particles, the cumulant method is readily used if the particle distribution is narrow and monodisperse. In this method, the light scattering intensity correlation function is expanded around the averaged value, $\overline{\Gamma}$, yielding a polynomial series, which is normally truncated after the quadratic term in the series for the sake of simplicity, yielding

$$g_2(\tau) \approx 1 + \beta \exp\left(-2\overline{\Gamma}\tau + \mu_2\tau^2\right) \tag{8.14}$$

In the cumulant method [33, 34], the electric field autocorrelation function is described by

$$\ln|g_1(\tau)| = \sum_{m=1}^\infty K_m \frac{(-\tau)^m}{m!} \tag{8.15}$$

$$K_1 = \overline{\Gamma} \tag{8.16}$$

$$K_2 = \overline{(\Gamma - \overline{\Gamma})^2} \tag{8.17}$$

where K_1 is the slope of the plot of $\ln|g_2(\tau)|$ versus τ and K_1/K_2 is related to the diameter of the particles in the dispersion. The polydispersity index [28], PI, is defined as

$$PI = \frac{2K_2}{\overline{\Gamma}^2} \tag{8.18}$$

This value represents the size distribution of particles measured by DLS analysis for a *monomodal* disperse particle solution. However, the interpretation of the PI is not important; therefore, this value is used only to monitor the qualitative change of the size distribution of particles.

For a *multimodal*-size-distributed particle dispersion, a Laplace inversion analytical method is generally used to determine the size distribution by DLS [35–37]. The outline of this method is only briefly described here because unfortunately there is no standardized algorithm, and the methods are strongly dependent on the specific DLS equipment used. However, the outline of the method is simple. Equation (8.11) is fixed by a particle size range and divided into a discrete set of respective fixed particle sizes, after which the light-scattering intensity-weighted fraction of particles can be determined. In this method, the size distribution will therefore be calculated as a discrete set of particle diameters, with corresponding light scattering intensity-weighted fractions. Such a stepped analysis is mathematically unstable, producing incoherent analytical results in some cases.

In principle, the concentration of the particle suspension in DLS measurements must be sufficiently low so that a multiple scattering effect is avoided. Unless the concentration is small, any scattered photon has a nonzero probability of being rescattered on its continued path through the suspension. The autocorrelation function of such multiple light scattering decays faster, resulting in the estimated particle size being smaller than the true particle size. Because of this, in theory, the concentration of the particles should be low, and the optical path must be sufficiently short to avoid multiple scattering. However, sufficiently small concentrations of particles induce unstable scattering, and the DLS measurement result can easily be affected by small amounts of dust. One of the solutions to this problem is the extrapolation of apparent diffusion coefficients (observed in sufficient numbers for different concentrations) to infinite dilution, yielding more reliable values than those obtained for just one concentration [38].

8.3
Particle Size Standards Applied to DLS

Recently, numerous institutions, organizations, and companies have produced commercial particle standards of various sizes and different materials, as summarized in Table 8.1. These have been produced because there are industrial manufacturing standards covering the general principles and procedures for instrument makers pertaining to accurate size determination using DLS instruments. In addition, the particle standards are utilized by general users on DLS systems for validation and performance evaluation of their instruments and to examine the validity of their procedures. The particle standards have different size distributions, distinguished as "narrow" and "wide" in Table 8.1. Also available are particle standards produced by national institutes. These "secondary" particle standards are all calibrated against the primary standards developed by international standards and accreditation agencies.

As shown in Table 8.1, polystyrene latex aqueous particle dispersions are widely produced in various sizes and with varying size distributions, since the particles are stable in aqueous dispersion without being affected by gravitational sedimentation. In addition, glass beads of varying sizes and size distribution are also widely produced in larger size ranges. The standard particles produced using these two materials form an ideal spherical shape and are used not only for DLS but also for transmission electron microscopy (TEM) and in particle counter systems. The National Institute of Standards and Technology (NIST) has produced a gold nanoparticle reference material that has a diameter of ~10 nm, which is the smallest particle standard available worldwide. The size of this gold particle standard, as determined by DLS, is ~14 nm. Even smaller particle size standards

Table 8.1 Summary of relevant standard particles for dynamic light scattering.

Institution	Model number	Size range	Material	Size distribution	Certified method	Traceability
NIST (USA)	1690	895 nm	Polystyrene latex	Narrow	LD	–
	1691	269 nm	Polystyrene latex	Narrow	TEM	–
	1692, 1961, 1965	2.982–29.64 μm	Polystyrene latex	Narrow	CDF	–
	1963	100.7 nm	Polystyrene latex	Narrow	DMA	–
	1964	63.9 nm	Polystyrene latex	Narrow	DMA	–
	1021, 1003	2–12, 20–50 μm	Glass bead	Wide	LD, ESZ	–
	1004b–1019b	53–125, 850–200 μm	Glass bead	Wide	SEM	–
	659	0.2–10 μm	Silicon nitride	Wide	Gravity-induced sedimentation	–
	1978, 1982	0.2–10, 10–150 μm	Zirconium oxide	Wide	SEM	–
	1984, 1985	9–30, 18–55 μm	Tungsten carbide	Wide	SEM	–
	8011	10 nm	Gold	Narrow	SEN	–
Duke Scientifics (USA)	3020A–3040A	22–41 nm	Polystyrene latex	Narrow	DLS	NIST
	3050A–3900A	50–903 nm	Polystyrene latex	Narrow	TEM	NIST
Polysciences (USA)	64004–64120	40 nm to 9 μm	Polystyrene latex	Narrow	Photocentrifuge method	NIST
	64130–64235	10–175 μm	Polystyrene latex	Narrow	Single-particle optical sensing	NIST

(continued overleaf)

Table 8.1 *(Continued)*

Institution	Model number	Size range	Material	Size distribution	Certified method	Traceability
Bangs Laboratories (USA)	PN02N–PN00N	21 nm to 7 μm	Polystyrene latex	Narrow	TEM, SEM, DLS	NIST
	BB01N–BB05N	1–10, 100–999 μm	PMMA bead	Wide	TEM, SEM, DLS	
	SS02N–SA05N	0.16–0.54 μm	Silica	Narrow	TEM, SEM, DLS	
	NT02N–NT40N	40 nm to 175 μm	Polystyrene latex	Narrow	Single-particle optical sensing	NIST
Interfacial Dynamics (USA)	S37200–D37472	20 nm to 15 μm		Narrow		NIST
Seradyn (Daw Chemical) (USA)	OptBind	83–855 nm	Polystyrene latex	Narrow	DLS	
EC JRC IRMM (BEL)	BCR-165–BCR-167	2–9.6 μm	Polystyrene latex	Narrow		–
	BCR-066–BCR-132	350–3500 nm, 1400–5000 μm	Glass bead	Wide		–
Magshere (USA)	–	1–26 μm	Polystyrene latex	Narrow	DLS	–
APPIE (JP)	MBP1-10, MBP10–100	1–10, 10–100 μm	Glass bead	Wide	Sedimentation balance method	–
JFCC (JP)	RP-1–RP-5	1–20 μm	Silicon nitride, aluminum oxide, silicon carbide, silicon nitride, barium titanate	–	LD	–

JST (JP)	STADEX	29–143 nm	Polystyrene latex	Narrow	DMA	NMIJ, NIST
	STADEX	144 nm to 2.005 μm	Polystyrene latex	Narrow	TEM	NMIJ, NIST
	DYNOSPHERES	3.21–7.123 μm	Polystyrene latex	Narrow	TEM	NMIJ, NIST
	DYNOSPHERES	10.04–95.6 μm	Polystyrene latex	Narrow	OPM	NMIJ, NIST
Whitehouse Scientific (UK)	MS0009–MS0589	9.18–589 μm	Glass bead	Narrow	Photocentrifuge method, SEM	NIST, NPL
	PS180–PS240	0.1–1, 500–2000 μm	Polystyrene latex	Wide	Photocentrifuge method, SEM	NIST, NPL, BCR
Micromod (GER)	Micromer	15 nm to 12 μm	Polystyrene latex	Narrow	DLS	–
	Sicastar	50 nm to 1.5 μm	Silica	Narrow	DLS	–
Microparticle GmbH (GER)	MF-ST-2.0–MF-ST-10.5	2.04–10.44 μm	Polystyrene latex	Narrow	TEM, SEM, DLS, gravity-induced sedimentation	NIST, BCR
	PS-ST-0.1–PS-ST-250.0	105 nm to 244 μm	Polystyrene latex	Narrow	TEM, SEM, DLS, gravity-induced sedimentation	NIST, BCR

Abbreviation: DMA, differential mobility analyzer; PMMA, polymethyl methacrylate; LD, laser diffraction; CDF, center distance finding; OPM, optical microscope.

are desired by DLS users. Such particle standards could be significant in allowing DLS users to achieve some degree of agreement on size determination using DLS methods.

8.4
Unique DLS Instruments

Herein, two unique, but significant DLS systems are briefly introduced. Because of the convenience of DLS measurements in the determination of particle size in the liquid phase, DLS analytical systems are continually being improved.

8.4.1
Single-Mode Fiber-Optic Dynamic Light Scattering

This technique is not novel and is rooted in DLS experiments from more than 20 years ago [39–46]. However, this technique still plays a significant role in the improvement of DLS systems. Single-mode fibers are optical waveguides with small core diameters that are comparable with the wavelength of light. Using a single-mode waveguide, only one mode of the electric field is propagated without multiple scattering effects, that is, the transverse structure of the field is determined by the properties of the waveguide. Because of the truly selective nature of fiber-optic single-mode detection of the scattering field, it is possible to achieve the ideal mixing efficiency to obtain the autocorrelation function without having to sacrifice signal strength. Indeed, using the single-mode fiber-optic DLS method, the systems have become more compact, more robust, and less expensive. The single-mode fiber-optic DLS system should enable observation of the dynamics of highly concentrated particle dispersions. However, subtle differences in the optical apparatus, setup, and procedure will have significant effects on the obtained results. Although this should be taken into account when using single-mode fiber-optic DLS detection, well-optimized systems are available on the market.

This method affords a convenient way to make DLS measurements, for example, as an online detector for size determination of fractions separated by size-exclusion chromatography or by flow field-flow fractionation. For such an online size detector, there still remains the problem of shear flow, since this induces decorrelation terms at small timescales compared to the Brownian motion of the particle, although this effect will be dependent on the size of particles. In addition, because the diffusion coefficient is defined by the ratio of the force due to the chemical potential gradient to the frictional coefficient between the particle and the solvent (i.e., nanoscale friction under liquid motion), the situation differs from that under steady state. Thus, it was proposed that online DLS measurements could be difficult in principle, but despite this, the online DLS detector has proved to be quantitatively useful.

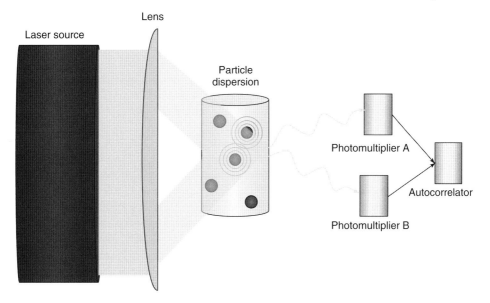

Figure 8.6 Schematic of the photon cross-correlation spectroscopy (PCCS) dynamic light scattering (DLS) system.

8.4.2
Photon Cross-Correlation Spectroscopy (PCCS)

In the photon cross-correlation spectroscopy (PCCS) method, two different lasers are focused on the same sample scattering volume, resulting in two separate, respective scattering patterns [47–52]. The experimental layout of the PCCS system is illustrated in Figure 8.6. Significantly, this cross-correlation method can, in principle, completely eliminate any contribution from multiple light scattering from highly concentrated particle dispersions. In theory, the two separated light scattering signals in this system should be the same. However, in practice, they differ to varying degrees, because of the different noise levels and the degree of multiple scattering. Therefore, under ideal and well-controlled conditions, the cross correlation of the two scattering signals works to remove the influence of noise and multiple scattering, making it possible to extract information about the particles even in highly concentrated particle dispersions. This measurement system has been greatly improved for the analysis of highly concentrated particle dispersions using the DLS method.

8.5
Sample Characterization Using DLS Measurements of Nanoparticles

It is interesting to compare the DLS results determined using different commercial DLS instruments, but the results are not easy to interpret because of the instability of the apparent results. A characterization of various standard particle samples measured by three different DLS systems follows. This proposed method could be a protocol that would allow researchers to achieve some degree of agreement on the size of nanoparticles using different DLS systems [53, 54].

8.5.1
DLS Instruments

Three different commercial DLS instruments were used in this study. The DLS7000 (Otsuka Electronics Co., LTD.) has a goniometer system with a 10 mW HeNe laser at a wavelength of 632.8 nm. Multiple tau digital correlation was measured at a minimum sampling time of 0.1 μs. The measurements were performed at a scattering angle of 90°. A sample quartz cell was set in a silicon oil bath where the refractive index of the oil is almost the same as the cell. The measurement temperature was regulated at $25.0 \pm 0.1\,°C$. The FPAR1000, Fiber-Optics Particle Analyzer (Otsuka Electronics Co., Ltd.), was used with a 5 mW semiconductor laser at a wavelength of 658 nm. Measurements were performed at a scattering angle of 90°, and a sample cell consisting of a polyethylene–polypropylene alloy was used. The measurement temperature was regulated at $25.0 \pm 0.1\,°C$ using the Peltier method. The Nanotrack UPA 150 (Microtrack Co.) particle analyzer was used with a 3 mW semiconductor laser at a wavelength of 780 nm. The temperature of the sample cell was regulated using the thermostatic bath. The measurements were performed at a scattering angle of almost 180°. All instruments were used in a clean booth kept at constant room temperature ($23.0 \pm 0.3\,°C$) and at a humidity of $40 \pm 3\%$. The DLS7000 and FPAR1000 systems use the homodyne method, and the Nanotrack system uses the heterodyne configuration.

DLS measurements were carried out on a TiO_2 nanoparticle suspension in a culture medium. Three separate measurements were carried out, and a simple average value was recorded. Viscosity measurements of the dispersion medium were carried out using an Uberode viscometer, since viscosity is a significant factor in determining the deviation of the size of particles using DLS. The temperature was regulated at $25.0 \pm 0.1\,°C$.

8.5.2
Size Determination of Particles in Suspension

Using the three DLS instruments, we determined particle size using two different definitions: (i) a light scattering, intensity-averaged diameter (d_l) and (ii) a number-averaged diameter (d_n). The particle size determined using DLS is the intensity-averaged diameter, d_l. The number-averaged diameter is calculated from the intensity-averaged result. If a simple assumption is made (e.g., size distribution and sphere structure), the two particle size definitions are useful for the following reasons. Firstly, the number-averaged diameter can be compared with the diameter obtained from electron microscopy. The sizes of the particle standards are often determined by the number-averaged diameter; therefore, this value is useful in validation of DLS instruments, although the particle standards should have a narrow size distribution somewhere in between the number-averaged and the intensity-averaged diameters. Secondly, taking the ratio between d_l and d_n makes it possible to estimate a qualitative size distribution of the particles, because a large difference between the two diameters implies a wide size distribution, similar to the case of the polydispersity index. For example, assuming a logarithmic normal particle size distribution with a peak diameter (d_n) of 150 nm and a standard deviation of 40 nm, the particles would have a d_l of 227 nm. In contrast, a particle that has a d_n of 150 nm and a standard deviation of 20 nm would have a d_l of 167 nm. These calculations show that a large difference between the two different averaged diameters indicates a wide size distribution of particles.

Herein, two different types of analytical procedures for a constrained regularization program are used. One is the "cumulant" or a monomodal analytical method, as described in the ISO 13321 standard. The other is the Laplace inversion analytical method or a multimodal analytical method. If the values determined using these two different analytical techniques agree, it strongly suggests that the size distribution of secondary nanoparticles is an ideal monomodal distribution, like a Gaussian distribution. The abbreviations for the observed diameters with different DLS instruments and analytical procedures used in this study are as follows. In the parenthesis, D, F, and N represent measurements from the DLS7000, FPAR1000, and Nanotrack systems, respectively. C represents the cumulant or monomodal analytical procedure, and M stands for the "Marquardt/Contin" or multimodal analytical procedure. For example, $d_n(D,C)$ represents a number-averaged diameter determined using the DLS7000 and a cumulant or monomodal analytical procedure.

The two different averaged sizes of the secondary nanoparticles in this study are calculated by

$$d_n = \frac{\sum_{i=1}^{N} d_n\,(D,\,C)_i}{N} \tag{8.19}$$

and

$$d_l = \frac{\sum_{i=1}^{N} d_l\,(D,\,C)_i}{N} \tag{8.20}$$

where N is the number of DLS measurements. In this DLS measurement, $d_n(D,C)$ and $d_l(D,C)$ were used because the TiO$_2$ particles in the dispersion exhibit good agreement between the raw and calculated data of the photon correlation function.

8.5.3
Concept of Identifying and Analyzing Uncertainty in the Size of the Secondary Nanoparticles

At first, a standard of the estimation of the uncertainty, often called a *GUM* and published by standards bodies such as the Bureau International des Poids et Mesures (BIPM) and the International Organization for Standardization (ISO), was employed. The uncertainty of d_n and d_l for the nanoparticles in the cell culture medium was determined as described in the following sections on the basis of the values of $d_n(D,C)$ and $d_l(D,C)$.

8.5.3.1 Change in Size of the Secondary Nanoparticles during a Time Period

The standard uncertainties of the size of the secondary nanoparticles, $u_{time}(d_l)$ and $u_{time}(d_n)$, arising from repeated DLS measurements, coupled with a change in the size of the particles during the *in vitro* toxicity assessment are given by

$$u_{time}(d_n) = \sqrt{\frac{\sum_{i-1}^{N}(d_n(D,C)_i - d_n)^2}{N-1}}$$

(8.21)

and

$$u_{time}(d_l) = \sqrt{\frac{\sum_{i-1}^{N}(d_l(D,C)_i - d_l)^2}{N-1}}$$

(8.22)

where N is the number of measurements. Although these uncertainties include the repeatability of the measurements, the main effect is on the stability of the suspension of the secondary nanoparticles. For example, when agglomeration of particles occurs in the experimental period, the uncertainty can be large.

8.5.3.2 Difference in Size Determined by Different DLS Instruments

The standard uncertainties caused by the difference in secondary nanoparticle size due to the differences in DLS instruments are given by

$$u_{app}(d_n) = \frac{\sqrt{(d_n(D,C) - d_n)^2 + (d_n(F,C) - d_n)^2 + (d_n(N,C) - d_n)^2}}{\sqrt{N_a - 1}}$$

(8.23)

and

$$u_{app}(d_l) = \frac{\sqrt{(d_n(D,C) - d_n)^2 + (d_n(F,C) - d_n)^2 + (d_n(N,C) - d_n)^2}}{\sqrt{N_a - 1}}$$

(8.24)

where N_a is the number of DLS instruments. This is the case because the uncertainty equation is an indicator of the difference of observed sizes relative to the commercial instrument used. Furthermore, this definition of the uncertainty should be useful for comparisons of size determinations between various organizations using various DLS instruments.

8.5.3.3 Difference in Size Determined by Different DLS Instruments

The standard uncertainties caused by a difference in the size of secondary nanoparticles determined using different DLS analytical procedures are given by

$$u_{\text{method}}\left(d_n\right) = \frac{\left|d_n\left(D, C\right) - d_n\left(F, C\right)\right|}{\sqrt{3}} \tag{8.25}$$

and

$$u_{\text{method}}\left(d_l\right) = \frac{\left|d_l\left(D, C\right) - d_l\left(F, C\right)\right|}{\sqrt{3}} \tag{8.26}$$

In this study, a small number of analytical procedures were employed. The standard deviation was calculated assuming a rectangular distribution for the variation, which applies to a type B uncertainty evaluation in the GUM (guide to the expression of uncertainty in measurement). As described earlier, a small uncertainty indicates a monomodal size distribution of the secondary nanoparticles.

8.5.4
Calculation of Combined Uncertainty

The combined standard uncertainties of the size of the secondary nanoparticles are calculated using

$$u\left(d_n\right) = \sqrt{u_{\text{time}}^2\left(d_n\right) + u_{\text{app}}^2\left(d_n\right) + u_{\text{method}}^2\left(d_n\right)} \tag{8.27}$$

and

$$u\left(d_l\right) = \sqrt{u_{\text{time}}^2\left(d_l\right) + u_{\text{app}}^2\left(d_l\right) + u_{\text{method}}^2\left(d_l\right)} \tag{8.28}$$

8.6
Result of DLS Characterization

The results of the size of the TiO_2 particles in suspension using these characterization techniques are summarized in Table 8.2. For small values of u_{time}, as shown in Table 8.2, agglomeration need not be considered in these dispersions. In addition, when the raw and calculated data of the photon correlation function obtained by DLS agreed sufficiently, like in the monomodal size distribution case, interestingly, $u_{\text{app}}(d_l)$ and $u_{\text{method}}(d_l)$ were sufficiently small. In contrast, $u_{\text{app}}(d_n)$ and $u_{\text{method}}(d_n)$ were larger, indicating that the different calculation algorithms of the different DLS instruments affect the transformation from d_l to d_n. Herein, clearly, d_l is the preferable size parameter when different workers determine particle sizes using different DLS instruments, since the values of d_l give similar values, independent of the DLS instruments employed. This analysis confirms that d_l is an important parameter for size determination by DLS methods for the comparison of particle size between different researchers and instruments.

Table 8.2 Summary of TiO$_2$ particle suspension properties obtained by DLS.

Sample	Size (nm)		u (nm)	u_{time} (nm)	u_{app} (nm)	u_{method} (nm)
A	d_l	159.0	7.2	5.2	3.0	4.0
	d_n	103.3	18.8	5.1	18.0	2.0
B	d_l	147.7	8.3	7.2	3.3	2.4
	d_n	103.9	17.7	13.7	8.4	7.5
C	d_l	190.6	6.5	4.0	4.2	3.0
	d_n	122.9	15.4	11.7	7.1	6.9
D	d_l	224.5	10.1	8.0	3.4	5.2
	d_n	148.5	18.1	12.3	7.8	10.8
E	d_l	261.0	7.4	3.7	2.4	5.9
	d_n	162.3	16.9	8.3	9.4	11.3

8.7
Conclusion

There has been an unprecedented increase in the number of studies related to nanosized materials. Accurate determination of sizes of nanomaterials is a key to the development of bio- and nanotechnologies since the sizes determine many of the physical and chemical properties of the functional materials. The DLS method is widely used as an effective technique to determine average size of the Brownian nanoparticles in colloidal suspensions. The principle of DLS based on the photon correlation spectroscopic method is presented first, and the recent development and the practical usage of DLS are contained in the latter. While more convenient DLS apparatuses have been desired in the nanotechnological field, the precise size determination of nanomaterials using DLS or other metrological techniques has become necessary for considering the recent nanotoxicity problem. The DLS metrology has a great possibility in the innovation in this issue.

References

1. Kasuy, A., Milczarek, G., Dmitruk, I., Barnakov, Y., Czajka, R., Perales, O., Liu, X., Tohji, K., Jeyadevan, B., Shinoda, K., Ogawa, T., Arai, T., Hihara, T., and Sumiyama, K. (2002) *Colloids Surf. A*, **202**, 291.

2. Tran, N.T., Campbell, C.G., and Shi, F.G. (2006) *Appl. Opt.*, **45**, 7557.

3. Maira, A.J., Yeung, K.L., Lee, C.Y., Yue, P.L., and Chan, C.K. (2000) *J. Catal.*, **192**, 185.

4. Kambe, S., Nakade, S., Wada, Y., Kitamura, T., and Yanagida, S. (2002) *J. Mater. Chem.*, **12**, 723.

5. Yano, H., Inukai, J., Uchida, H., Watanabe, M., Babu, P.K., Kobayashi, T., Chung, J.H., Oldfield, E., and Wieckowski, A. (2006) *Phys. Chem. Chem. Phys.*, **8**, 4932.

6. Chen, B., Penwell, D., Benedetti, L.R., Jeanloz, R., and Kruger, M.B. (2002) *Phys. Rev. B*, **66**, 144101-1.

7. Doyle, A.M., Shaikhutdinov, S.K., and Freund, H. (2005) *Angew. Chem. Int. Ed.*, **44**, 629.

8. Gribb, A.A. and Banfield, J.F. (1997) *Am. Mineral.*, **82**, 717.

9. Duran, J., Rajchenbach, J., and Clément, E. (1993) *Phys. Rev. Lett.*, **70**, 2431.

10. Sarkar, J., Pal, P., and Talapatra, G.B. (2005) *Chem. Phys. Lett.*, **401**, 400.

11. Warheit, D.B. (2006) *Carbon*, **44**, 1064.

12. Oberdörster, G., Oberdörster, E., and Oberdörster, J. (2005) *Nanotoxicology*, **113**, 823.

13. Stern, S.T. and McNeil, S.E. (2008) *Toxicol. Sci.*, **101**, 4.

14. Donaldson, K., Tran, L., Jimenez, L.A., Duffin, R., Newby, D.E., Mills, N., NacNee, W., and Stone, V. (2005) *Part. Fibre Toxicol.*, **2**, 10.

15. Li, N., Xia, T., and Nel, A.E. (2008) *Free Radical Biol. Med.*, **44**, 1689.

16. Davoren, M., Herzog, E., Casey, A., Cottineau, B., Chamber, G., Byrne, H.J., and Lyng, F.M. (2007) *Toxicol. In Vitro*, **21**, 438.

17. Oberdörster, G., Maynard, A., Donaldson, K., Castranova, V., Fitzpartrick, J., Ausman, K., Carter, J., Karn, B., Kreyling, W., Lai, D., Olin, S., Monteiro-Riviere, N., Warheit, D.B., and Yang, H. (2005) *Part. Fibre Toxicol.*, **2**, 8.

18. Sayes, C.M., Marchione, A., Reed, K.L., and Warheit, D.B. (2007) *Nano Lett.*, **7**, 2399.

19. Fujita, K., Horie, M., Kato, H., Endoh, S., Nakamura, A., Miyauchi, A., Yamamoto, K., Kinugasa, S., Nishio, K., Yoshida, Y., Iwahashi, H., and Nakanishi, J. (2009) *Toxicol. Lett.*, **191**, 109.

20. Horie, M., Nishio, K., Fujita, K., Kato, H., Endoh, S., Suzuki, M., Nakamura, A., Miyauchi, A., Kinugasa, S., Yamamoto, K., Iwahashi, H., Murayama, H., and Yoshida, Y. (2010) *Toxicol. In vitro*, **24**, 1629.

21. Buford, M.C., Hamilton, R.F. Jr., and Holian, A. (2007) *Part. Fibre Toxicol.*, **4**, 6.

22. Sayes, C.M., Wahi, R., Kurian, P.A., Lie, Y., West, J.L., Ausman, K.D., Warheit, D.B., and Colvin, V.L. (2006) *Toxicol. Sci.*, **92**, 174.

23. Gurr, J.R., Wang, A.S., Chen, C.H., and Jan, K.Y. (2005) *Toxicology*, **213**, 66.

24. Berne, B.J. and Pecora, R. (2000) *Dynamic Light Scattering: With Applications to Chemistry, Biology, and Physics*, Dover Publications, New York.

25. Chu, B. (2007) *Laser Light Scattering: Basic Principles and Practice*, 2nd edn, Dover Pubications, New York.

26. Pecora, R. (1985) *Dynamic Light Scattering: Applications of Photon Correlation Spectroscopy*, Plenum, New York.

27. ISO ISO 13321:1996. *Particle Size Analysis-Photon Correlation Spectroscopy (diluted suspensions)*, ISO TC24/SC4.

28. ISO ISO 22412:2008. *Particle Size Analysis – Dynamic Light Scattering*, ISO TC24/SC4.

29. Rayleigh, L. (1881) *Philos. Mag.*, **12**, 81.

30. Rayleigh, L. (1914) *Proc. Roy. Soc.*, **A90**, 219.

31. Rayleigh, L. (1918) *Proc. Roy. Soc.*, **A94**, 296.

32. Glauber, R.J. (1963) *Phys. Rev.*, **131**, 2766.

33. Kopperl, D.E. (1972) *J. Chem. Phys.*, **57**, 4814.

34. Brown, J.C., Pusey, P.N., and Dietz, R.J. (1975) *J. Chem. Phys.*, **62**, 1136.

35. Provencher, S.W. (1982) *Comput. Phys. Commun.*, **27**, 213.

36. Provencher, S.W., Hendrix, J., Maeyer, L.D., and Paulussen, N. (1978) *Chem. Phys.*, **69**, 4273.

37. Provencher, S.W. (1979) *Macromol. Chem.*, **180**, 201.

38. Takahashi, K., Kato, H., Saito, T., Matsuyama, S., and Kinugasa, S. (2008) *Part. Part. Syst. Charact.*, **25**, 31.

39. Auweter, H. and Horn, D.F. (1985) *J. Colloid Interface Sci.*, **105**, 399.

40. Brown, R.G.W. (1987) *Appl. Opt.*, **26**, 4846.

41. Wiese, H. and Horn, D.F. (1991) *J. Chem. Phys.*, **94**, 6429.

42. Bremer, L., Deriemaeker, L., Finsy, R., Gelade, E., and Joosten, J.G.H. (1993) *Langmuir*, **9**, 2008.

43. Rièka, J. (1993) *Appl. Opt.*, **32**, 2860.

44. Thomas, J.C. (1989) *Langmuir*, **5**, 1350.

45. Dhadwal, H.S., Ansari, R.R., and Meyer, W.V. (1991) *Rev. Sci. Instrum.*, **62**, 2963.

46. Destremaunt, F., Salmon, J., Qi, L., and Chapel, J. (2009) *Lab Chip*, **9**, 3289.

47. Schätzel, K.J. (1991) *Mod. Opt.*, **38**, 1849.

48. Aberle, L.B., Staude, W., and Hennemann, O.-D. (1999) *Phys. Chem. Chem. Phys.*, **1**, 3917.

49. Overbeck, E. and Sinn, C. (1999) *J. Mod. Opt.*, **46**, 303.

50. Aberle, L.B., Hülstede, P., Wiegand, S., Schröer, W., and Staude, W. (1998) *Appl. Opt.*, **37**, 6511.

51. Aberle, L.B., Kleemeier, M., Hülstede, P., Wiegand, S., Schröer, W., and Staude, W. (1999) *J. Phys. D. Appl. Phys.*, **32**, 22.

52. Pusey, P.N. (1999) *Curr. Opin. Colloid Interface Sci.*, **4**, 177.

53. Kato, H., Suzuki, M., Fujita, K., Horie, M., Endoh, S., Yoshida, Y., Iwahashi, H., Takahashi, K., Nakamura, A., and Kinugasa, S. (2009) *Toxicol. In vitro*, **23**, 927.

54. Kato, H., Fujita, K., Horie, M., Suzuki, M., Nakamura, A., Endoh, S., Yoshida, Y., Iwahashi, H., Takahashi, K., and Kinugasa, S. (2010) *Toxicol. In vitro*, **24**, 1009.

9
Photolumniscence/Fluorescence Spectroscopic Technique for Nanomaterials Characterizations

9.1
Application of Photoluminescence Spectroscopy in the Characterizations of Nanomaterials

Bingqiang Cao, Haibo Gong, Haibo Zeng, and Weiping Cai

9.1.1
Introduction

Light emission through any process other than blackbody radiation is called *luminescence* and requires external excitation as it is a nonequilibrium process [1]. It is a mechanism through which the excited samples relax to the equilibrium state. In comparison with optical reflection and transmission spectroscopes, luminescence process concentrates on the relaxation of the sample and often provides us more information. For example, impurities of low concentrations in semiconductors are impossible to detect through reflection spectroscopy and more difficult to detect by absorption spectroscopy. However, through luminescence spectroscopy measurements such information could be obtained. Therefore, luminescence spectroscopy is an important and powerful tool to analyze the optical behavior of materials, especially for phosphors and semiconductors. Luminescence processes originate from different excitations that can produce free electrons/holes and their pairs (e−h). Such carriers may recombine across the bandgap or through defect- and impurity-related intermediate levels and emit photons. The excitations can be (i) an intense monochromatic light from a halogen lamp or laser source, (ii) an incident electron beam, (iii) electrical injection of electrons and/or holes through contacts, and (iv) thermal excitation. These emission procedures are known as photoluminescence (PL), cathodoluminescence (CL), electroluminescence (EL), and thermoluminescence (TL), respectively [2]. Electroluminescence is the most difficult to obtain because of the complexity of producing appropriate contacts. However, in terms of the application, EL is the most important since a light emitter has to be able to produce light in an efficient manner under electrical excitation. TL is a technique used with insulators and wide-gap materials and is not widely used. PL and CL are discussed in the following parts in detail with emphases on their applications to characterize the semiconductor nanomaterials.

Nanomaterials: Processing and Characterization with Lasers, First Edition.
Edited by Subhash Chandra Singh, Haibo Zeng, Chunlei Guo, and Weiping Cai.
© 2012 Wiley-VCH Verlag GmbH & Co. KGaA. Published 2012 by Wiley-VCH Verlag GmbH & Co. KGaA.

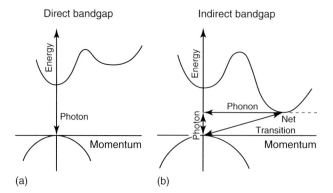

Figure 9.1.1 (a,b) An emission schematic of the differences between a direct bandgap and an indirect bandgap semiconductor with an $E(k)$ diagram.

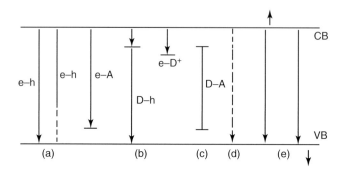

Figure 9.1.2 A schematic of the possible emission process in semiconductors with direct bandgap. (a) Bandgap emission, (b) CB/VB defect recombination, (c) donor–acceptor pair emission, (d) multiphonon emission process, and (e) Auger recombination.

PL is a result of incident photon absorption that generates electron–hole pairs and emits a photon of a different wavelength. The incident photons, when absorbed, excite electrons usually from the valence band into the conduction band through momentum-conserving processes because the photon momentum is negligible. The electrons and holes thermalize to the lowest energy state of their respective bands via phonon emission before recombining across the fundamental bandgap or the defect levels within the bandgap and emitting photons of the corresponding energies. The emitted photon has energy equal to the difference in the energies of states occupied by electrons and holes before recombination, as shown in Figure 9.1.1a. For semiconductors with indirect bandgap as shown in Figure 9.1.1b, the energy relaxation predominantly takes place through phonons, which makes this family of semiconductors inefficient light emitters. For semiconductors with direct bandgap, all possible recombination processes are schematically illustrated in Figure 9.1.2 [3].

PL spectroscopy is an extremely powerful nondestructive technique for assessing material quality [4] including intrinsic electronic transitions between energy bands and extrinsic electronic transitions at impurities and defects of organic molecules, semiconductors, and insulators. It is quite possibly the most widely used technique when developing a new material system, especially for semiconductors. It is nondestructive and requires virtually no sample preparation or no complex device structures. Moreover, variation of different parameters, for example, temperature, time, or pump power, can be used to obtain band offsets, identify various transitions, and even explore the structural quality of the material. The time evolution of the PL signal can be used to accurately determine the Auger recombination coefficient in other material systems [5].

The study of luminescence from condensed matter is not only of scientific but also of technological interest because it forms the basis of solid-state lasers and it is important for display panels in electronic equipment, lighting, and paints. Moreover, PL frequently provides a nondestructive technique for materials characterization or research in materials.

As a comprehensive survey of PL spectroscopy used on all material systems is beyond the scope of this chapter, only a limited number of topics on PL applications in nanomaterials, in particular, semiconductor nanomaterials are addressed. We first introduce the PL experimental techniques in Section 9.1.2. Then, in Section 9.1.3, we discuss the general PL spectroscopy applications in characterizations of nanomaterial ensembles, for example, room-temperature photoluminescence (RT-PL) and photoluminescence excitation (PLE) spectroscopes, temperature-dependent PL spectroscopy, time-resolved photoluminescence (TRPL) spectroscopy, and excitation-dependent PL spectroscopy. In Section 9.1.4, a special PL spectroscopy, microPL, and its applications in characterizing single nanomaterials are discussed, together with CL, which is also a powerful tool to study the optical properties of single nanomaterials.

9.1.2
Experimental Techniques

A typical and standard PL setup is illustrated in Figure 9.1.3, which is composed of excitation sources, optical elements to focus excitation on the sample and collect the luminescence light, a spectrometer (or filter) to analyze the luminescence, and a detector followed by a data acquisition system. The excitation system usually involves a broad-spectrum source, either a combined tungsten filament for the visible spectrum and deuterium lamp for the UV or a halogen lamp [6]. The lamp emission is passed through a grating monochromator and therefore selectively excites the luminescence. Band-pass or band-edge filters are generally required to eliminate the unwanted second- and higher-order diffraction maxima from the grating. In other cases, to minimize the effect of scattered light, different lasers with their intrinsically narrow linewidths and high intensities, such as He-Cd laser, YAG laser, and semiconductor diode lasers, are chosen as excitation source but suffer from the disadvantage of a fixed wavelength. The sample under investigation is usually fixed in a dark box in a strain-free manner, and sometimes, a cryostat is integrated to provide the desired low temperature near liquid helium or nitrogen.

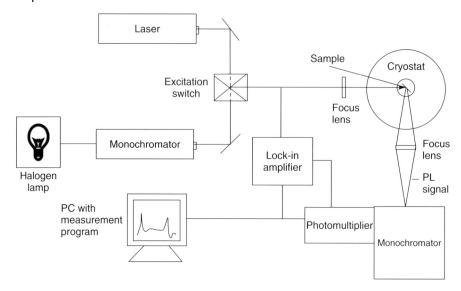

Figure 9.1.3 Schematic view of a generally applied experimental setup for PL spectroscopy measurements with a cryostat station.

Widely used cryostats are closed-cycle refrigerators or cold fingers of helium or nitrogen pipe. As thermally activated nonradiative recombination processes were minimized at low temperature, more spectroscopic information can be obtained from the low-temperature PL spectrum. The luminescence signal is efficiently gathered and focused by optical lens into a second grating monochromator to monitor and analyze the luminescence. The final detector may be a photomultiplier (PMT) or a nitrogen-cooled charge-coupled device (CCD) or photodiode array for improved data collection efficiency at multiple wavelengths. Finally, both the excitation and emission signals were processed with a laptop using some software and a PL spectrum was obtained.

With the similar experimental setup, if the photon energy of the exciting light beam is scanned and, at the same time, some characteristic PL peak intensity is monitored, we can measure the PLE spectrum. PLE may give us some information on the absorption spectrum or on the excited states. In the PLE spectrum, some peaks at positions of stronger absorption can be detected because more electron–hole pairs are created, which means these energies are usually more efficient to excite the monitored PL emission.

TRPL spectroscopy is one of the most powerful tools to investigate the dynamical progress of excited semiconductor samples [6]. To obtain time resolution, one needs a pulsed or temporally modulated excitation source and a time-resolving detection system. Possible excitation sources are modulated continuous-wave lasers, flashlamps, and Q-switched or mode-locked lasers. Till now, two main time-resolving detectors have been developed. One is fluorimeter that uses a fast modulator and phase-sensitive detection to measure the phase shift between luminescence and excitation signals. A time resolution of 50 ps can be measured in this way. Another is the time-correlated single-photon-counting technique (TCSPT), which can measure decay constants in the picosecond to nanosecond range [7]. TCSPT is based on the detection of single photons of a periodical light signal, the measurement of the detection times of the individual photons and the reconstruction of the waveform from the individual time measurements. In this method, the excitation comes from a fast laser pulse, and the light level reaching the PMT or microchannel plate detector is reduced to such a low level that less than one photon per excitation pulse is detected. The time delay between the photon detection and the time of the pulse is measured, and a histogram of the number of detected photons versus arrival time taken over a large number of excitation pulses is plotted.

9.1.3
Applications of General PL Spectroscopy on Nanomaterial Ensembles

9.1.3.1   Room-Temperature PL and PLE Spectroscopy

The main parameters of interest for RT-PL measurements are the wavelength of emission peak, peak intensity, linewidth, and integrated intensity. The wavelength of the emission peak is of obvious interest for the measured semiconductor sample as it is quite close to the bandgap if the defect-related emissions could be excluded. The other three figures of merit allow for a relative analysis of material quality between samples. Peak intensity is a direct indicator of the semiconductor optical quality. The lower the defect density, the stronger is the peak PL signal if the sample qualities and PL experimental conditions are similar. From the PL linewidth usually measured at half of the maximum, some information about defect density, surface/interface quality, and alloy disorder can be obtained. If the PL peak is typically wide and shows obviously asymmetry, the peaks must be carefully fitted with Gaussian or Lorentzian line shapes [8]. The integrated intensity is somewhat intermediate between the peak intensity and linewidth cases as it depends on both the height and width of the peak. In high-quality samples, the three figures of merit generally track one another.

For PL spectrum measurements of nanomaterials ensembles, it is almost similar to that of bulk or thin-film samples. Figure 9.1.4a shows some ZnO nanoplatelets that are typically tens of nanometers in thickness and several micrometers in dimension on a silicon substrate. Figure 9.1.4b shows the RT-PL spectra of nanosheets that are measured at excitation of the 325 nm He-Cd laser. The full laser exciting power intensity (I_0) is about 2 kW cm^{-2}. A sharp PL peak appears in the ultraviolet region centered at 380 nm with the full width at half maximum (FWHM) of 17 nm, which is near the bandgap of ZnO (3.37 eV at room temperature (RT)). Generally, it is attributed to the bandgap emission or free exciton emission. For direct bandgap semiconductor, the energy $\hbar\varpi$ of excitonic emission can be expressed by $\hbar\varpi = E_g - R^*$, where E_g is the bandgap and R^* is the effective Rydber constant of excitons. For ZnO, the excitonic emission peak is expected at 3.31 eV, which is 60 meV (R^*) lower than the ZnO bandgap of 3.37 eV. However, the emission peak

(a)

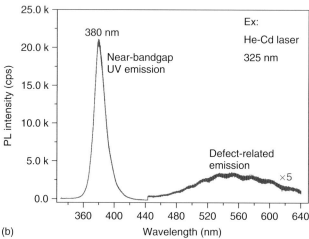

(b)

Figure 9.1.4 (a) Scanning electron microscopy image of ZnO nanoplate ensembles on a silicon substrate and (b) the room-temperature PL spectrum of the ZnO nanoplates of (a) excited with a He-Cd laser.

usually shows a slight redshift (40 meV) compared with the theoretical position, which was caused by the heating effect of laser excitation light [9].

At the same time, a rather broad visible emission centered at about 540 nm is also detected. Commonly, this visible emission of ZnO at RT is attributed to the deep-level defects in ZnO crystals, such as oxygen vacancies and/or zinc interstitials. For such ZnO nanoplates, the green emission is much weaker than the UV intrinsic emission. The strong and sharp UV emission in contrast to the weak green emission indicates that the measured ZnO samples are of good crystal quality.

Another powerful way to obtain information about the PL process is the study of the dependence on excitation wavelength of the luminescence spectrum. This is

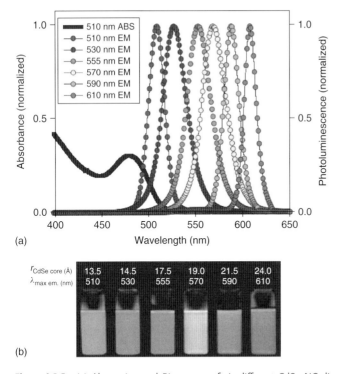

Figure 9.1.5 (a) Absorption and PL spectra of six different CdSe NC dispersions. The black line is the absorption NC of the 510 nm, and others are the PL spectra with different sizes. (b) Photo images the size-tunable luminescence properties and spectral range of the six NC dispersions as correspondingly shown in (a). (Source: From Ref. [10].)

usually performed by recording the PLE spectrum , the excitation wavelength being scanned at fixed luminescence wavelength. Another kind of nanomaterial ensemble typically studied with PL and PLE is semiconductor nanocrystals (NCs) or quantum dots (QDs) [11]. The optical properties of semiconductor NCs are determined by their sizes, shapes, surfaces and bulk defects, impurities, and crystallinity. The luminescence dependence on the NC sizes mainly arises from the quantum size confinement effect, which modifies the density of states near the band edges. When the size of the semiconductor NCs becomes smaller than their Bohr radius, the spatial size confinement of the motion of electrons leads to an increase of their bandgaps. In addition, the actual energy of bandgap absorption becomes sensitive to the size and shape of the NCs. Experimentally, the PL emissions of NCs of different sizes exhibit a series of blueshift toward higher energy from the band edge, as compared to the typical value of the corresponding bulk semiconductors. A typical example of CdSe NCs with different sizes is shown in Figure 9.1.5.

9.1.3.2 Temperature-Dependent PL Spectroscopy

In general, sample temperature is an important parameter that influences the PL properties of semiconductors since line narrowing can be expected when the measured temperature decreases. Moreover, as some nonradiative recombinations can be effectively suppressed when the sample temperature decreases, new ground or excited states can appear at low temperature and a PL spectrum with fine peak structures can be observed. On the other hand, by analyzing the temperature-dependent PL, information about the photodynamics and PL physical origin can be obtained. At present, optical cryostats allow easy operation from room temperature down to a few Kelvins using commercial closed-cycle refrigerators or liquid helium.

Figure 9.1.6 compares the RT-PL and low-temperature (10 K) PL of ZnO nanobelts grown on the silicon substrate with chemical vapor deposition. At room

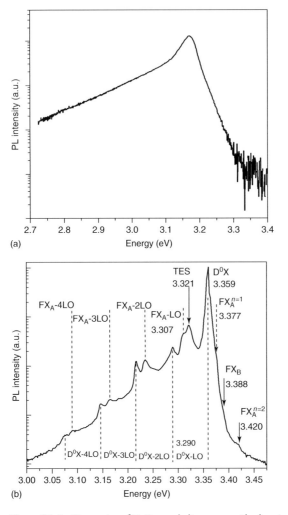

Figure 9.1.6 PL spectra of ZnO nanobelts grown with chemical vapor deposition on silicon substrates (a) measured at room temperature excited with a 355 nm YAG laser and (b) measured at 10 K excited with a He-Cd laser.

temperature, the PL only shows a featureless wide emission peak in the bandgap region. However, the PL spectrum at $T = 10$ K shows more complicated emission peaks. Figure 9.1.6b is dominated by the emission peak at 3.359 eV because of typical donor-bound excitons (D^0X), which is possibly caused by unintentional aluminum dopant. On the high-energy side of this peak, three small shoulders are observed, which can be assigned to the free exciton recombinations as indexed with $FX_{A,B}(n = 1,2)$. Two groups of phonon replica assigned as D^0X-nLO and FX-nLO ($n = 1, 2, 3, 4$) are observed with the energy separation of 72 meV (LO-phonon energy) below the D^0X and FX peaks, respectively. The peak at 3.33 eV is most probably due to a two-electron transition (TET).

The temperature-dependent PL spectra from 10 to 150 K are shown in Figure 9.1.7. It is found that the intensities of the two series of peaks associated with FX and D^0X exhibit an opposite dependence on temperature. The intensities of FX and its related phonon replicas increase with the increasing temperature, while the intensities of D^0X and its replicas decrease and are not detectable at temperature over 70 K. The fast intensity reduction of D^0X is a result of the rapid thermal ionization of bound excitons with increasing temperature. Therefore, more free excitons occupy the ground states. So, from the low-temperature and temperature-dependent PL spectra, the detailed optical characteristics including all kinds of possible recombination process and their dynamics can be well identified.

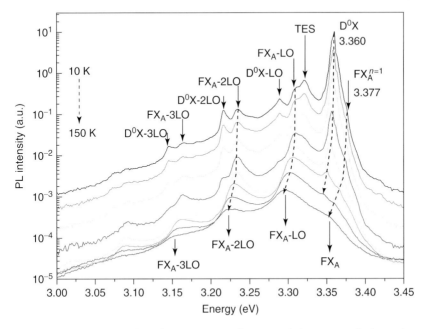

Figure 9.1.7 Temperature-dependent PL spectra of ZnO nanobelts grown with chemical vapor deposition on silicon substrates from 10 to 150 K excited with a He-Cd laser.

9.1.3.3 Time-Resolved PL Spectroscopy

TRPL is another kind of important PL spectroscopy. It is very effective to explore the properties of trapping states. On one hand, the luminescence decay can be detected to compare the lifetime of different trappers to study their emission dynamics. On the other hand, the PL curves can also be collected at a different time after excitation; this helps to directly observe the changes of PL bands. Taking a typical example, Foreman *et al.* [12] measured TRPL of different ZnO samples. Continuous-wave PL spectroscopy was performed by photoexciting the samples with a He-Cd laser operating at 325 nm. The PL was collected and refocused onto an all-silica optical fiber using a complementary pair of aluminum off-axis parabolic mirrors. The collected PL was dispersed by a 30 cm imaging spectrometer and measured using a liquid-nitrogen-cooled CCD camera. However, TRPL spectroscopy was performed by photoexciting the samples with ∼100 fs pulses from a 1 kHz, wavelength-tunable optical parametric amplifier tuned to 3.81 eV and by routing the collected PL to a Hamamatsu streak camera with a resolution of 30 ps.

The time-resolved, spectrally integrated band-edge PL of ZnO micropowder, doped micropowder, and nanostructure are presented in Figure 9.1.8. Obviously, the total lifetimes have very large difference. According to the following equations, the decay lines can be fitted by two lifetime components, as listed in Table 9.1.1.

$$I(t) = A_1 e^{-t/\tau_1} + A_2 e^{-t/\tau_2}, \quad A_1 + A_2 = 1$$

The undoped micropowder exhibits the strongest band-edge emission and the weakest defect emission. Although it has the slowest biexponential decay lifetimes, the low quantum efficiency indicates that nonradiative recombination

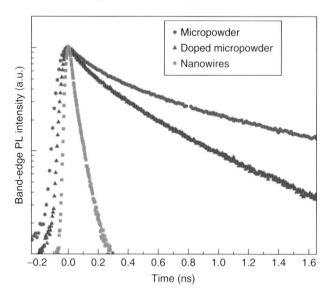

Figure 9.1.8 Time-resolved, spectrally integrated band-edge photoluminescence of ZnO micropowders, sulfur-doped micropowders, and nanostructure. (Source: From Ref. [12].)

Table 9.1.1 TRPL biexponential decay characteristics of band-edge emission for undoped and S-doped ZnO micropowders and nanowires.

	I_D/I_B	UV decay			QE (%)
		τ_1 (ns)	τ_2 (ns)	A_1	
Undoped powder	0.08	0.198	1.08	0.48	7
ZnO:S powder	320	0.112	0.481	0.37	65
ZnO:S wires	1600	<0.040	–	–	30

Source: From Ref. [12].

dominates band-edge relaxation. Doping the micropowder with sulfur significantly enhances the energy transfer from the band edge to the defect states responsible for visible emission, resulting in reduced band-edge emission, faster band-edge decay, much brighter visible emission, and dramatically increased quantum efficiency. Clearly, the defect-mediated decay channel responsible for visible emission favorably competes with the deleterious nonradiative decay channels. However, when the doped ZnO is formed into nanowires, the band-edge emission decay accelerates beyond the temporal resolution of the instrument and the spectrally integrated quantum efficiency drops. Clearly, nanostructuring increases nonradiative carrier relaxation, thus undermining the channel favorable for visible emission. By TRPL study, it is concluded that as the nanostructure surface-to-volume ratio increases, the nonradiative pathways increasingly competes with the sulfur-induced defects responsible for the bright visible emission.

9.1.3.4 Excitation-Dependent PL Spectroscopy

For the PLE source, excitation energy determined by the wavelength and pumping density determined by the power of excitation source are two important parameters. By changing these two parameters, PL can be measured under different excitation conditions, which are both useful extensions for general PL spectroscopy. For wavelength-dependent PL measurement, a series of wavelengths with insufficient and excessive energies are applied to excite the sample in the normal PL measurement mode. Such PL results can be used to study the correlation between the excited states and emission energy and hence help to explore the luminescence transition mechanisms and can also be very effective to explore the characters of defect levels located in the bandgap.

The following is an example of our recent results about defect-related PL of ZnO nanoparticles (NPs) by wavelength-dependent PL measurement. The selected ZnO NPs have an average diameter of 20 nm and were synthesized by laser ablation in liquid [14]. Strong blue emissions can be obtained from these ZnO NPs, but the mechanism of this kind of defect-related PL is still unclear. Figure 9.1.9 presents the excitation-dependent PL spectra of such ZnO NPs. The blue PL intensities exhibit

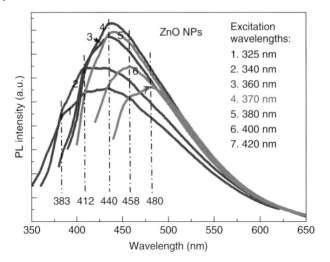

Figure 9.1.9 Time-resolved excitation wavelength-dependent photoluminescence of ZnO nanoparticles. (Source: From Ref. [13].)

nonlinear increase–decrease dependence on excitation – first they increase, then reach saturation under bandgap energy (E_g) excitation, and then decrease, but still effectively emit with excitation energy (E_{ex}) smaller than E_g. Among these universal excitation-dependent features, the most important is that $E_{ex} = E_g$ energy is the optimal excitation and $E_{ex} < E_g$ energy can still effectively excite blue emissions.

The fact that blue emissions can be effectively excited by $E_{ex} < E_g$ energy demonstrates that the excited states and initial states of corresponding transitions could be located below the conduction band edge, not at or above the conduction band edge. Under $E_{ex} = E_g$ excitation, the electrons can be first be excited up to conduction band, then nonradiatively transit into above initial states, and then radiatively transit and emit blue emissions. The fixed emitting wavelengths indicate several different energy gaps from initial states to end states, such as 3.0 eV (412 nm), 2.8 eV (440 nm), and 2.7 eV (458 nm). Combining such excitation-dependent PL spectra and other analyses such as reported defect levels and formation energy, the observed initial state was suggested to be correlated with Zn interstitials, and there could be several derivative levels with lower energy involved in possible localization or coupling with other defects.

With the development of high-power laser, the excitation density can be tuned to as high as gigawatts per square centimeters. Under high excitation, the optical properties of materials will show some different characters, which are called *nonlinear optics*. This is a totally different research topic in optics and is not discussed in this chapter. However, it should be noted that some groups have found that under excitation density of few kilowatts per square centimeters or megawatts per square centimeters, nanomaterials show some nonlinear optic properties, such as random lasers [15] and nanowire lasers [16].

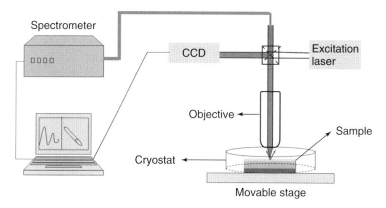

Figure 9.1.10 Schematic view of a typical microPL experimental setup.

9.1.4
Applications of MicroPL Spectroscopy on Single Nanomaterial

9.1.4.1 MicroPL Spectroscopy and Its Applications on Single Nanomaterial
The main difficulty of the PL measurements on single nanomaterials lies in focusing excitation laser on single nanomaterials at the nanoscale. At the same time, the emission signal is rather weak. To overcome these difficulties, microPL is usually built with an optical microscope to focus and detect the excitation and emission light, respectively, as shown in Figure 9.1.10. Moreover, a CCD is also integrated for sample imaging. For typical microPL, its spatial resolution is about 1 μm. Therefore, microPL is very useful to characterize single nanomaterials, such as single semiconductor nanowires [17], carbon nanotubes [18], and graphene [19]. The applications of microPL in single nanomaterials including PL, PLE, TRPL, and temperature-dependent PL are similar to those of general PL spectroscopy.

Figure 9.1.11a shows a far-field image of a single GaN nanowire, which was back illuminated with a lamp. The localization of bright emission at the ends of the wire suggests strong wave-guiding behavior of the Fabry–Perot nanowire laser [20]. Figure 9.1.11b shows the microPL spectrum of the single GaN nanowire. The broad and weak spectrum was excited with a 1 mW continuous-wave excitation (He-Cd laser).However, when the sample was excited with a 1 μJ cm^{-2} pulsed excitation using Ti: sapphire laser and optical parametric amplifier, several sharp (<1.0 nm) features appeared in the spectrum, indicating the onset of stimulated emission, which is consistent with observed laser actions in Figure 9.1.11a.

9.1.4.2 CL Spectroscopy
CL refers to luminescence from a sample excited by an electron beam, which is usually measured by means of the system based on a scanning electron microscope (SEM) [21]. The electron beam is emitted from an electron gun of the SEM, collected by electron lenses, and then focused on a sample. The luminescence

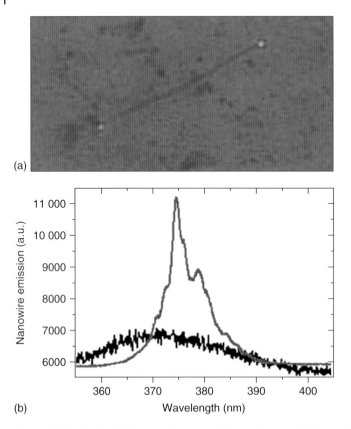

Figure 9.1.11 (a) Far-field image of a single GaN nanolaser and (b) microPL spectrum of the single GaN nanowire. (Source: From Ref. [20].)

from the excited sample is collected by an ellipsoidal mirror and then sent to a spectrometer equipped with a CCD camera, as schematically shown in Figure 9.1.12. Using SEM for sample location, CL usually has a lateral resolution better than 10 nm as the de Broglie wavelength of electrons is much shorter than light wavelengths. On the other hand, the SEM electron beam can be focused on a spot size of tens of nanometers and scanned over the area of the sample. By CL intensity mapping, one can obtain not only spectroscopic information but also spatial details. Therefore, CL is very appropriate for characterizations of single nanostructure because of its high resolution and intensity scanning beyond the optical diffraction limit.

9.1.4.3 Applications of CL in Single Nanomaterials

Figure 9.1.13a,b shows the proposed structure of ZnO/ZnMgO quantum well (QW) nanowire grown with a two-step pulsed laser deposition (PLD) and its corresponding SEM image. Clearly, it is almost impossible to measure such single

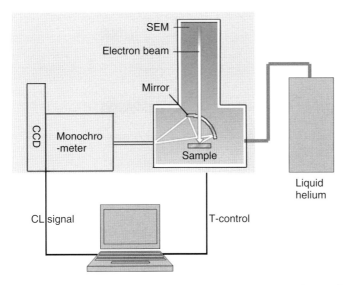

Figure 9.1.12 Schematic view of a generally applied experimental setup for the CL spectroscopy measurements with a cryostat station equipped on a scanning electrony microscope.

nanowires with traditional PL. For microPL, it is also difficult as, with optical microscopy, the single nanowire cannot be well focused. In this case, CL would be a very powerful tool to access the quality and uniformity of such core–shell nanowire heterostructure. To focus the CL measurement on a single nanowire, the sample is $60°$ tiled in the SEM chamber. So, by focusing the excitation electron beam on a single nanowire, CL spectrum can be detected from the different parts of a single nanowire, as shown in Figure 9.1.13c. In addition to the emissions from the ZnO core at 3.36 eV and ZnMgO barrier at 3.52 eV, another new peak at 3.45 eV is detected from the different parts of the wire. These emissions are caused by the exciton recombination confined in the radial nanowire QW. At the same time, a film QW with bigger thickness is also detected from the substrate.

Spatially resolved CL intensity mapping is another useful method to study the carrier distribution in the QW nanowire. Figure 9.1.14a,b shows the SEM image and a typical CL spectrum of the scanned QW nanowire. Figure 9.1.14c is the CL intensity map monitored at different peak energies of ZnO, QW, and ZnMgO barrier. The intensity map of the ZnO core is very homogeneous, which reflects the high-quality crystal of the core nanowire and homogeneous distribution of excitons. The intensity map of the QW distributes along the whole nanowire from the tip to the bottom. Moreover, the QW intensity of the tip part is a little stronger. It means that both the radial QW around the side face and axial QW on the tip of the nanowire are grown. However, the ZnMgO barrier CL intensity of the wire stem is much weaker than that of the tip part. The radial ZnMgO does not show up because the carriers are quickly captured into the radial QW and recombine radiatively. By high-resolution special and spectral CL, the relation

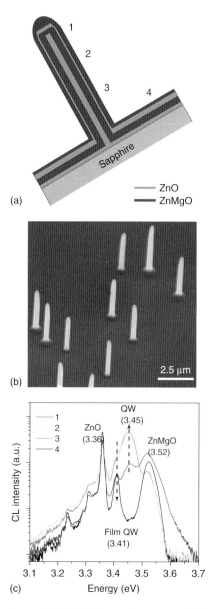

Figure 9.1.13 (a) Schematic structure of the proposed ZnO/ZnMgO quantum well nanowire, (b) SEM image of the low-density ZnO quantum well nanowire, and (c) CL spectrum measured from a single quantum well nanowire at $T = 10$ K. (Source: From Ref. [22].)

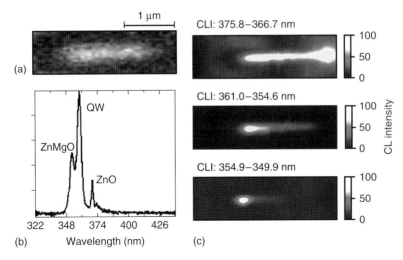

Figure 9.1.14 (a) An SEM image and (b) a typical CL spectrum measured from a single ZnO QW (15 pulses) nanowire; the ablated laser pulses for QW growth are 15; and (c) CL intensity maps of a single ZnO/ZnMgO QW nanowire. From top to bottom, the intensity maps correspond to the CL of ZnO, QW, and ZnMgO, respectively.

between the detailed optical properties and its structure characters can be clearly demonstrated.

9.1.5
Conclusions

As a traditional optical characterization technique, PL spectroscopy continues to be a powerful tool for the investigation of optical properties of nanomaterials. In this review chapter, not only experimental equipments but also information that properties or material parameters can be obtained with respective techniques are summarized. The foregoing examples demonstrate well the breadth of novel experiments that can be performed with different kinds of PL methods. We still believe that, because of their scientific and technological interests, the study of luminescence properties of materials, including nanomaterials, will show new developments and findings in the future. More additional reference can be find in Ref. [23].

Acknowledgments

This work is supported by NSFC (51002065, 11174112) and Shandong provincial science foundation (BS2010CL003). BC thanks the Taishan Scholar Professorship (TSHW20091007) tenured at the University of Jinan. The Program for New Century Excellent Talents in University and the Research Foundation for Returned Overseas Students, Ministry of Education, China, are also acknowledged.

References

1. Morkoç, H. and Özgür, Ü. (2009) *Zinc Oxide: Fundamental, Materials and Device Technologies*, John Wiley & Sons, Inc.
2. Czichos, H., Saito, T., and Smith, L. (eds) (2006) *Handbook of Materials Measurement and Methods*, Springer.
3. Klingshirn, C. (2001) *Semiconductor Optics*, Springer.
4. Willardson, R.K. and Beer, A.C. (eds) (1972) *Semiconductors and Semimetals*, Academic Press.
5. Henimi, M. (2005) *Dilute Nitride Semiconductor*, Elsevier.
6. Kasap, S. and Capper, P. (eds) (2006) *Handbook of Electronic and Photonic Materials*, Springer.
7. Becker, W. (2005) *Advanced Time-Correlated Single Photon Counting Techniques*, Springer.
8. Cao, B.Q., Cai, W.P., and Zeng, H.B. (2006) *Appl. Phys. Lett.*, **88**, 161101.
9. Cao, B.Q., Cai, W.P., Li, Y., Sun, F.Q., and Zhang, L.D. (2005) *Nanotechnology*, **16**, 1734.
10. Medintz, I.L., Uyeda, H.T., Goldman, E.R., and Mattoussi, H. (2005) *Nat. Mater.*, **4**, 435.
11. Wiederrecht, G.P. (ed.) (2010) *Handbook of Nanoscale Optics and Electronics*, Elsevier.
12. Foreman, J.V., Li, J., Peng, H.Y., Choi, S., Everitt, H.O., and Liu, J. (2006) *Nano. Lett.*, **6**, 1126.
13. Zeng, H.B., Yang, S.K., Xu, X.X., and Cai, W.P. (2009) *Appl. Phys. Lett.*, **95**, 191904.
14. Zeng, H.B., Cai, W.P., Li, Y., Hu, J.L., and Liu, P.S. (2005) *J. Phys. Chem. B*, **109**, 18260.
15. Cao, H., Xu, J.Y., Zhang, D.Z., Chang, S.H., Ho, S.T., Seelig, E.W., Liu, X., and Chang, R.P.H. (2000) *Phys. Rev. Lett.*, **84**, 5584.
16. Gargas, D.J., Toimil-Molares, M.E., and Yang, P.D. (2009) *J. Am. Chem. Soc.*, **131**, 2125.
17. Czekalla, C., Sturm, C., Schmidt-Grund, R., Cao, B.Q., Lorenz, M., and Grundmann, M. (2008) *Appl. Phys. Lett.*, **92**, 241102.
18. Lefebvre, J., Fraser, J.M., Finnie, P., and Homma, Y. (2004) *Phys. Rev. B*, **69**, 075403.
19. Gokus, T., Nair, R.R., Bonetti, A., Böhmler, M., Lombardo, A., Novoselov, K.S., Geim, A.K., Ferrari, A.C., and Hartschuh, A. (2009) *ACS Nano*, **3**, 3963.
20. Johnson, J.C., Choi, H.J., Knutsen, K.P., Schaller, R.D., Yang, P.D., and Saykally, R.J. (2005) *Nat. Mater.*, **1**, 106.
21. Czichos, H., Saito, T., and Smith, L. (eds) (2006) *Springer Handbook of Materials Measurement Methods*, Springer.
22. Cao, B.Q., Zúñiga-Pérez, J., Boukos, N., Czekalla, C., Hilmer, H., Lenzner, J., Travlos, A., Lorenz, M., and Grundmann, M. (2009) *Nanotechnology*, **20**, 305701.
23. Additional reference (a) Singh, J. (ed.) (2006) *Optical Properties of Condensed Matter and Applications*, John Wiley & Sons, Inc.; (b) Toyozawa, Y. (2003) *Optical Process in Solids*, Cambridge University Press; (c) Fox, M. (2001) *Optical Properties of Solids*, Oxford University Press; (d) Demtröder, W. (2002) *Laser Spectroscopy*, Springer.

9.2
Fluorescence Correlation Spectroscopy of Nanomaterials

Kaushal Kumar, Luigi Sanguigno, Filippo Causa, and Paolo Antonio Netti

9.2.1
Introduction

Nanotechnology refers to the research and technology developments of atomic and molecular entities at nanometer length scale. From the past two decades, the research on nanomaterials has grown explosively, and nowadays, it has become a matter of much debate because of its huge potential in future technological advancements [1, 2]. Nanotechnology may be able to create many new materials and devices with a vast range of applications, such as in medicine, electronics, biomaterials, energy production, and in other fields where materials are being used. A decade ago, nanoparticles were studied because of their size-dependent physical and chemical properties, but now they have entered into the commercial exploration age for use in our medicines, cosmetic products, biomedical diagnosis, and so on. Nanoscale structures and materials (e.g., nanoparticles, nanowires, nanofibers, and nanotubes) have been vastly explored in biological applications such as biosensing, biological separation, molecular imaging, and anticancer therapy because of their novel properties and new functions that differ from their bulk counterparts [3–5]. Particularly, their high volume/surface ratio, surface tailorability, improved solubility, and multifunctionality open many new possibilities for biomedicine [3]. Moreover, nanomaterials offer remarkable opportunities to study and regulate the complex biological processes for biomedical applications in an unprecedented manner, and intense research is going on in this direction. The typical cell size of living organisms is $\sim 10\,\mu m$, and it contains submicrometer-sized components. These submicrometer-sized components can easily be probed by the use of the nanoparticles and allow us to detect cellular machinery without introducing too much interference [6]. Understanding biological processes at the nanoscale level is a strong driving force behind the development of nanotechnology in biological areas [6].

The popularity of word "nano" is not too old, although there are some evidences of nanomaterials in the past [7]. Earlier studies on nanosized materials concentrated on the synthesis nanoparticles in order to produce size- and shape-controlled particles of various materials. But now, we have wise synthesis routes, and only efforts are needed to apply nanoparticles into practical devices. As mentioned above, several products based on nanomaterials are now available commercially, but the research does not end here. As society has begun using consumer products based on nanomaterials in greater quantities, the interest in the broader implications of these products has also grown. A number of serious concerns have been raised about what effects these will have on our society and what action, if any, is appropriate to mitigate these risks. The central question is whether the unknown risks of engineered nanoparticles can spoil the established benefits. At present, it

Nanomaterials: Processing and Characterization with Lasers, First Edition.
Edited by Subhash Chandra Singh, Haibo Zeng, Chunlei Guo, and Weiping Cai.
© 2012 Wiley-VCH Verlag GmbH & Co. KGaA. Published 2012 by Wiley-VCH Verlag GmbH & Co. KGaA.

is proved that nanomaterials can be a health hazard for humans [8, 9], and it opens a new field of research, *nanotoxicology* [10, 11]. Recent study shows that inhaled ultrafine particles exert respiratory effects; they translocate from the lung into the circulatory system [8, 12] and result in cardiovascular problems. In future, this issue will be more prominent and major funding would be needed to research on health and environment friendly products. Because of its vast capability in improving device efficiency, research will go ahead on both positive and negative impacts of nanotechnology. At present, there are numerous applications of various types of nanomaterials inside cells ranging from cell imaging and cell tracking to cancer treatment. Here we mention some examples of biomedical applications where nanoparticles are in use.

- Fluorescent biological labels [13]
- Drug and gene delivery [14]
- Biodetection of pathogens [15]
- Detection of proteins [16]
- Probing of DNA structure [17]
- Tissue engineering [18]
- Tumor destruction via heating (hyperthermia) [19]
- MRI contrast enhancement [20].

Detailed description of the above said applications can be found in respective references. The possibility of using nanoparticles as superior labels and sensors in biological studies has sparked widespread interest. In the above-mentioned examples, the nanoparticles come closer to the cell and interaction between the cell components and nanoparticles gives rise to specific signals, which are detected by external devices. In order to apply nanoparticles in these areas, it is necessary to understand diffusion and localization of nanoparticles in the cells. The uptake and cellular location of nanoparticles is a field of major interest in the context of drug delivery [21]. Metal nanoparticles have also found increasing application in live cell imaging [22].

In order to probe the biomolecular interactions, scientists rely on a group of techniques. Since the biological process occurs in a dilute medium, it is highly essential to investigate these processes at the single-molecule level where single/few nanoparticles interact with target. Conventional ensemble techniques used for investigating biomolecular interactions such as mass spectrometry, X-ray crystallography, NMR, and so on provide information on bulk samples and in nonphysiological conditions. These techniques do not have the capability to detect the molecules at single-molecule level. But the advancement of science and technology in recent decades has led to the emergence of biophysical techniques that are capable of probing single molecules in very dilute solutions in real time. By focusing on an individual molecule in space and time, such analyses provide quantitative information about force properties, conformational dynamics, molecular interactions, and temporal changes with its microenvironment that could otherwise be hidden in ensemble experiments. Molecular dynamics can be studied without having to bring the ensemble population into a nonequilibrium state. Furthermore, because of the use of smaller sample volume, the high spatial resolution of single-molecule techniques provides the opportunity to look at rare molecular events that exist only in highly localized regions of the cell.

One approach to single-molecule detection is the optical method based on fluorescence detection. Fluorescence techniques are noninvasive and nondestructive to biological samples. They can be performed in real time at ambient or physiological temperatures. Their versatility with the molecular environment implies that they may be applied *in vitro* or *in vivo*. Since in very dilute systems, fluorescence emission intensity becomes very low, systems based on the measurement of emitted intensity become appropriate. In subnanomolar concentrations, another group of fluorescence methods that monitor the fluorescence intensity fluctuations of single molecules moving in and out of a confined illuminated volume have been developed [23, 24]. This method is known as fluorescence fluctuation spectroscopy (FFS), which provides information that lies hidden in the fluctuating signal such as dynamic processes, chemical kinetics, or molecular interactions. One of the first FFS method was introduced by Magde, Elson, and Webb in 1972 [25] and is known as fluorescence correlation spectroscopy (FCS). It uses intensity fluctuations of fluorescent particles diffusing through a focused laser beam, to characterize translational diffusion coefficients and chemical rate constants. The improvement of this technique for single-molecule sensitivity was achieved by using a confocal microscope with a high numerical aperture (NA) objective and single-photon counting avalanche photodiodes (APDs) as detectors. Since then, it has become an increasingly popular technique for the study of dynamics in physiological environment at thermodynamic equilibrium. The technique was originally developed to study the bimolecular interactions at physiological regime, but now this technique is also used to study the dynamics of inorganic fluorescent nanomaterials and has vast potential to characterize nanomaterials.

Besides this, another variant of this technique emerged recently and is based on the detection of the dark part of the emitted energy from excited molecule. The FCS technique needs that particles/molecules under probe have good fluorescence quantum yield. Many particles and molecules do not show good fluorescence emission (bright part) property; rather a major part of the excitation goes through nonradiative way (dark part). Thus, nonfluorescent particles/molecules need to be tagged with other fluorescent molecules to make FCS measurement possible. This is the limitation of FCS for nonideal fluorophores. The new variant known as photothermal correlation spectroscopy (PhCS) is suitable for nonfluorescent molecules and eliminates the necessity of molecular tagging [26, 27]. Particularly, it is suitable for absorbing semiconducting and metallic nanoparticles [27]. Analogous to FCS, it measures the time correlation function of the particle's photothermal signal, which is directly proportional to the light absorption by the particle and is thus named PhCS. Since absorption by metal nanoparticles neither saturates nor photobleaches at reasonable excitation intensities or has complicated photophysics, PhCS is free from the limitation encountered by FCS.

In this chapter, the authors intend to make familiar readers with this technique who want to study the dynamical behavior of nanoparticles in solution and nanoparticle–cell interaction. The chapter would also provide guidelines to the readers in establishing own homebuilt FCS setup.

9.2.1.1 What FCS Can Do for Nanoparticles?

FCS can perform two tasks: first, testing the stability of single or multicomponent nanoparticles in a liquid medium, and second, finding out interaction mechanisms of nanoparticles with others such as cellular molecules. The first criterion for the applicability of nanoparticles is that they should be stable for a specific time period. For cellular and many other applications, the formation of stable solutions of nanoparticles in buffer media is important, and not all the nanoparticles have been found to be stable as hydrosols in solutions. The stability of hydrosols strongly depends on the surface potential of the particles, the zeta potential, pH, and salt concentration of the solvent. FCS can reveal the stability of nanoparticles in an aqueous medium and has now become a common tool for characterizing the properties of nanoparticles in solution [28]. This technique is particularly useful in the characterization of fluorescent nanoparticles because FCS is based on fluorescence emission. FCS can measure the concentration, brightness, hydrodynamic radius, and monodispersity of the fluorescent probe in a single measurement, allowing for rapid characterization of probes in solution [29]. Cellular interaction with nanoparticles depends on various physical properties of nanoparticles such as size, shape, surface charge density, surface chemistry, and degree of aggregation, and understanding how these properties influence the interaction mechanism will lead to a predictive model for nanoparticle–cell interaction. Any phenomenon that alters the motion of nanoparticles could be probed using FCS technique.

9.2.1.2 Fluorescence Is a Tool for FCS

A typical fluorescent molecule emits 10^5 photons before photobleaching in water and at rates up to 10^9 s^{-1} (at least during microsecond bursts before ground state depletion by intersystem crossing to excited triplet states). Modern photon detectors, laser excitation, and high NA microscopy optics allow collection of 3% of the emitted fluorescence photons. Sometimes, 100 photons can be detected when a single molecule in solution is diffusing (in less than a millisecond) through the focus of a laser beam tuned to excite the target fluorophore. So recognition and identification of the individual target molecule above the background fluorescence of the buffer can be easily done in appropriate liquid solvents. The photon detection rate from individual molecules is used by FCS for the determination of fundamental dynamical parameters. Measurement scales from microseconds to seconds are easily accessible and cover transport coefficients, chemical kinetics, and recognition of aggregation in samples.

9.2.1.3 How Does FCS Work?

In an FCS experiment, the fluorescence emitted from a small, optically well-defined open-volume element of a solution in equilibrium is monitored as a function of time (Figure 9.2.1a). The recorded fluorescence emission signal is proportional to the number of fluorescent molecules in the probe volume. This number fluctuates

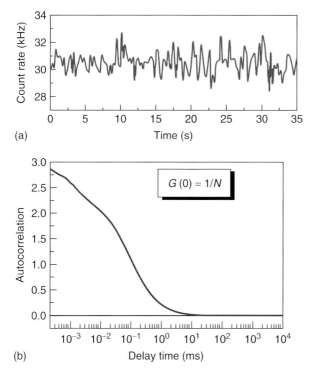

Figure 9.2.1 (a) The fluorescence signal (raw data) recorded from the volume fluctuates in time. (b) A temporal autocorrelation of this fluctuating signal exhibits characteristic decay times corresponding to the relaxation times for chemical kinetics and diffusion.

about its equilibrium value as molecules diffuse in and out of the volume. The temporal autocorrelation of the fluorescence signal fluctuation (Figure 9.2.1b) yields the timescale of such dynamics, and its variance yields the average number of independent fluorophores (N) in the probe volume. The correlation functions contain information about chemical reaction kinetics, coefficients of diffusion, and equilibrium chemical concentrations. Mathematically, the normalized autocorrelation function $G(T)$ is calculated as the time average of the product of the fluctuations in detected fluorescence at every time t and the fluctuations at delayed times $t + dT$, normalized by the squared time average of the fluorescence emission.

9.2.1.4 Basic Theory of FCS

Ideally, FCS works in very dilute samples, in average, one particle per excitation volume, and measures fluorescence intensity fluctuation. In a typical FCS setup, a laser beam is focused on a solution under thermal equilibrium. The focused laser beam, wavelength, and detection optics define the probe volume of the system. The pinhole spatially filters the emitted fluorescent light to ensure that only light from the focus is detected. The concept of confocal arrangement can

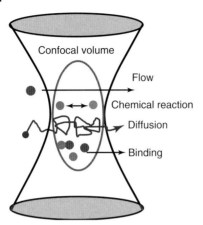

Figure 9.2.2 Fluorescent solute molecules (circles) fluctuate in a small elliptical volume about an average value as molecules diffuse in and out of the volume. Figure shows various dynamical processes that FCS can measure.

be found in Ref. [28]. A typical FCS setup measures various dynamical processes as depicted in Figure 9.2.2. As fluorescent particles move in and out of this minute region, the total fluorescence $F(t)$ detected will fluctuate in time due to the change in the total number of molecules, $N(t)$, present in the probe volume. The time-dependent autocorrelation function of $F(t)$ contains information about the average duration and amplitude of the fluctuations. The decay rate and the shape of the autocorrelation function, $G(\tau)$, give the mechanism of the process and the residence time of the particles inside the probe volume. The magnitude of $G(\tau)$ provides information about the number density of the fluorescent molecules or particles in the probe volume. The total detected fluorescence $F(t)$ is proportional to the number of particles in the focal volume, $N(t)$. The normalized autocorrelation function is defined as

$$G(\tau) = \frac{\langle \delta F(t + \tau)\delta F(t)\rangle}{\langle F(t)\rangle^2} \tag{9.2.1}$$

where $\delta F(t) = F(t) - \langle F(t)\rangle$ is the fluctuation of the fluorescence signal $F(t)$ at about its average value and $\langle \ \rangle$ denotes an ensemble average. The fluctuation correlation function requires the calculation of the intensity time average before calculating the correlation functions. δF can be defined as

$$\delta F(t) = KQ \int_{\text{all space}} \delta C(r, t)I(r)\phi(r)\mathrm{d}r \tag{9.2.2}$$

where $\delta C(r, t)$ is the concentration fluctuation, $I(r)$ is the excitation intensity profile, $\phi(r)$ is the collection efficiency function that characterizes the spatial filtering effect of the pinhole on the point spread function (PSF), K is the detection efficiency of the instrument, and Q is the product of the absorption coefficient and the molecular quantum yield of the particle. The PSF of the optical design describes the intensity distribution of the image of a point emitter [30, 31].

The product of $I(r)$ and $\phi(r)$ gives the molecule detection function (MDF) (r) that determines the spatial distribution of the effective sample volume. The MDF depends on the intensity distribution of the focused laser illumination and the efficiency of photons detected from a fluorescent molecule. The factors K and Q can be combined to fluorescence yield parameter, η that is determined by the photon counts per molecule per second.

In diffraction-limited illumination and detection scheme, the PSF is described by Bessel function, while for an underfilled back aperture of objective illumination, it is approximated as a Gaussian–Gaussian–Lorentzian (x, y, z) intensity profile. The PSF of a microscope objective is then convoluted with the circular pinhole function to give MDF (r) and then approximated as 3D Gaussian illumination intensity profile

$$\text{MDF}(r) = I_0 e^{-2r^2/r_0^2} e^{-z^2/z_0^2} \tag{9.2.3}$$

where $I_0 = 2P/(\pi r_0^2)$ is the excitation intensity at the center of the laser beam waist with laser power, P; z is the distance along the axial direction from the focal plane, and z_0 is the intensity dropped to $1/e^2$; r_0 is the diffraction-limited beam waist radius at $z = 0$ and is given by

$$r_0 = \frac{0.61\lambda}{\text{NA}} \tag{9.2.4}$$

where NA is the numerical aperture of the microscope objective and λ is the excitation wavelength. Equation (9.2.3) along with Eq. (9.2.4) gives effective excitation volume as

$$V_{\text{eff}} = \pi^{3/2} r_0^2 z_0 \tag{9.2.5}$$

Using Eqs. (9.2.1–9.2.5) and at $\tau = 0$, the correlation function can be expressed as

$$G(0) = \frac{\langle \delta F(t) \rangle^2}{\langle F(t) \rangle^2} = \frac{1}{V_{\text{eff}} \langle C \rangle} = \frac{1}{\langle N \rangle} \tag{9.2.6}$$

Thus, inverse of $G(0)$ gives the absolute number of particles in the probe volume. In Eq. (9.2.6), it is taken that the number of photons fluctuates according to Poisson statistics, $\langle \delta N^2 \rangle = \langle N \rangle$.

The concentration fluctuation, $\delta C(r, t)$, in Eq. (9.2.2) is governed by Fick's law

$$\nabla^2 \delta C(r, t) = \frac{1}{D} \frac{\partial \delta C(r, t)}{\partial t} \tag{9.2.7}$$

where D is the translational diffusion constant and from the Stokes–Einstein relation can be expressed as

$$D = \frac{k_B T}{6\pi R \eta} \tag{9.2.8}$$

where k_B is the Boltzmann constant, T is the absolute temperature, η is the viscosity of the medium, and R is the hydrodynamic radius of the diffusing particle. By measuring the diffusion constant, D, one can obtain the size of the particle. Equation (9.2.8) is valid for spherical particles, but with some approximation, it applies to the particles that deviate from spherical symmetry. In many cases, the relative change in D, and not the absolute value of R itself, is the parameter of interest. From Eqs. (9.2.1), (9.2.7), and (9.2.8), the autocorrelation function for translational diffusion can be calculated as

$$G(\tau) = \frac{1}{N} \left[1 + \frac{\tau}{\tau_D}\right]^{-1} \left[1 + \frac{\tau}{\tau_D}\left(\frac{r_0}{z_0}\right)^2\right]^{-1/2} \tag{9.2.9}$$

where N is the mean number of particles in the effective observation volume and can be related to $\langle N \rangle$ by Eq. (9.2.6). The first term in Eq. (9.2.9) defines translational diffusion (xy-direction), and the second term, in z-direction. If diffusion occurs in two dimensions, the second term would be zero.

The translational diffusion, τ_D, is given by:

$$\tau_D = \frac{r_0^2}{4D} \tag{9.2.10}$$

Actually, the experimental value of D (Eq. (9.2.8)) can be obtained by measuring the translational diffusion constant, τ_D, of a molecule diffusing through axial radius, r_0, of excitation volume. Using a reference fluorophore sample for calibration with a known diffusion coefficient such as Rhodamine 6G ($D_{Rho} = 4.15 \times 10^{-6}$ cm^2 s^{-1}), the D_s of the molecule under investigation can be obtained as follows

$$D_s = \frac{\tau_{Rho}}{\tau_s} D_{Rho} \tag{9.2.11}$$

FCS measures τ_D by analyzing the fluorescence fluctuations obtained from excitation volume in an equilibrium solution. A least square fit of the model function, Eq. (9.2.9), to the experimental autocorrelation curve yields the free parameters, N, τ_D, and aspect ratio, $\omega = r_0/z_0$. If there are two different fluorescent particles A and B diffusing in the excitation volume, it is possible to calculate a cross-correlation function G_{AB} according to Eq. (9.2.1)

$$G_{AB}(\tau) = \frac{\langle \delta F_A(t+\tau)\delta F_B(t)\rangle}{\langle F_A(t)\rangle\langle F_B(t)\rangle} \tag{9.2.12}$$

If A and B are independently diffusing species, then G_{AB} is always zero. However, if they are associated with each other, G_{AB} is analogous to an autocorrelation function for this complex. Thus, cross correlation has the remarkable power of isolating signal from the complex AB even in a solution containing considerable amounts of free A and B.

9.2.2
Instrumentation

The FCS instrumentation depends on the type of experiment to be performed. So one should look for a multipurpose setup that can be easily modified without much effort and expertise. Hence most people want to fabricate a homemade FCS setup. FCS measurement basically requires good focusing of the laser beam, next, efficient collection, and then, detection of collected fluorescence from the specimen with fast computation of the autocorrelation of the detected signal. There are two kinds of FCS instruments: one uses continuous wave excitation and the other uses pulsed excitation. FCS using pulsed excitation [32] has some advantages over the continuous wave FCS but it costs more due to the use of pulsed lasers. Picoquant [33] provides the best solution for time-gated time-resolved FCS microscopy. The company also provides components to upgrade various laser scanning microscopes, for example, Olympus, Nikon, Leica, and so on. Owing to the simplicity in fabrication, FCS using continuous wave laser has been described in this chapter. An FCS instrument should meet the following requirements: (i) best excitation of molecules in a small region of the sample, (ii) efficient collection of the fluorescence from that region, (iii) rejection of stray light and any background fluorescence, (iv) focusing the fluorescence into a pinhole, (v) detection of the fluorescence with a high quantum efficiency detector, and (vi) fast computation of the autocorrelation function of the obtained signal. The authors are describing here how to effectively implement these steps to bring the instrument in the working condition. An FCS instrument without microscope can be easily built in the laboratory with less effort. For nanoparticles' dynamics studies, a microscope is not required. Some companies are providing cheap FCS spectrometers with/without microscopes [33–36]. The required components with specifications can be found in well described method in Ref. [37]. If one wants measure in cells, a conventional fluorescence microscope would be necessary. A commercial confocal or multiphoton laser scanning microscope can also be modified for FCS measurements in a laboratory using the same components. The motivation for building an FCS instrument in-house, aside from considerable cost saving, comes from the extraordinary flexibility it offers the user in tailoring the instrument to particular experimental needs. Changing between different excitation lasers and different modes of excitation (single or multiphoton) requires interchange of different components and easy realignment of the instrument. Only homebuilt instrument allows a large flexibility to tailor it according to the requirement of the experimenter.

9.2.2.1 Components of the Setup
Before describing the optical design, the main components of the system are discussed first.

9.2.2.1.1 Excitation
The efficient detection of fluorescence from small illumination volume is the most crucial part of the FCS experiment. To achieve this, it is necessary to select appropriate excitation wavelength in order to get good emission from the sample. The tight focusing and exact shape of the focusing volume should be known in order to calculate the diffusion constant. For these reasons, lasers are the perfect light sources as they emit light with high degree of directionality and monochromaticity.

A Gaussian beam profile of a laser beam allows the estimation of focal volume. For single-photon experiments, low-power (<50 mW) continuous-wave lasers are sufficient. A multiline argon ion laser (blue at 488 nm and green at 514 nm) may extend the excitation choice. Lasers emitting in TEM_{00} mode are the ideal choice. Nowadays, cheap diode lasers are available in the market. The TEM_{00} mode from diode lasers can be achieved by the use of single-mode optical fiber.

For focusing the laser beam at the sample, a microscope objective lens is necessary. When working with samples in water medium and using silica glass as sample holder, water immersion objective with refractive index 1.2 is the best choice. Smaller spot size at the focus requires a larger beam diameter at the back aperture of the objective lens, implying that an expansion of the laser beam is usually necessary. The smallest focal volume can be achieved by overfilling the back aperture of the objective lens. The focus thus achieved is nearly diffraction limited, but the focal plane intensity distribution is not described by a simple analytical function. A Gaussian intensity distribution at the focal plane can be obtained by underfilling the back aperture of the microscope objective, but this yields a slightly larger observation area for the experiment.

9.2.2.1.2 Collection

The fluorescence emitted from the sample needs to be collected using the same microscope objective lens that is used for focusing the excitation beam. Higher NA of the lens ensures efficient collection of the fluorescence. In addition, this epifluorescence geometry automatically ensures that the collected fluorescence is decoupled from the forward-moving excitation light.

9.2.2.1.3 Filters

The fluorescence collected by a microscope objective needs to be separated from the excitation light path. A suitable dichroic, typically a multilayered coated dielectric thin film that transmits one range of wavelengths and reflects another range, optics is used for this purpose. A suitably selected dichroic reflects the excitation beam toward the microscope objective and transmits the fluorescence coming from the objective. After the dichroic, fluorescence filters are required. These are another specially coated optic designed to transmit light only within a small wavelength window. These two filters need to be chosen carefully to effectively cut off the excitation light and reduce the nonspecific background while maximizing the transmission of the fluorophore emission to the detector.

9.2.2.1.4 Focusing to a Pinhole

The next step is to focus the filtered fluorescence at the pinhole, and it can be done with a high-quality achromatic lens. Light originating from the focus passes through the pinhole aperture, while light from other regions is preferentially blocked. The probe volume in FCS experiments is a convolution of this detection profile with the illumination profile at the focal spot. Introducing a pinhole in the beam path is very critical, and an easy way is to use a multimode fiber, where the fiber face acts as the pinhole. Changing the fiber allows easy alteration of the pinhole aperture size without the need for substantial realignment of the instrument.

9.2.2.1.5 Detection of the Fluorescence

The FCS measurements need very sensitive detectors, and the preferred detector is single-photon-counting APDs (avalanche photo-diodes). Silicon APDs have high quantum efficiency over a wide range in visible spectrum. The peak quantum efficiency of a Si-APD from PerkinElmer is >80%. The PicoQuant also provides APDs with high detection efficiency. Using a fiber-coupled APD makes the alignment and the light shielding of the instrument simpler. Data processing requires that the output of the detector be in the form of TTL (transistor-transistor logic) pulses. Photon-counting APDs and PMTs (photo-multiplier tubes) with built-in high-voltage (HV) power supply, amplifier, discriminator, and TTL logic output are easily available.

9.2.2.1.6 Analysis of the Detector Signal

The detector signal in form of either DC voltage or TTL pulses (in the case of pulsed excitation) is directly fed to a specialized digital signal processing card (autocorrelator card) housed in a personal computer that can perform quasi-real-time autocorrelation of the incoming signal. The commercially available autocorrelator cards are typically supplied with convenient driver software, which contain a few standard models for fitting limited types of FCS data. However, the data can be fitted with user-defined fitting models. In place of hardware correlators, the software correlators (data acquisition, DAQ cards) can also perform fast computing and contain the photon statistics. So these days, systems using DAQ cards are becoming more and more popular.

9.2.2.2 Construction of the Instrument

A ray diagram for the FCS instrument in the inverted geometry incorporating the components mentioned above is given in Figure 9.2.3, below.

The beam from laser aperture is first expanded by a combination of lenses L1 and L2 and then passed through aperture I2. An aperture is introduced between the combinations of lens to cut the diffracted light. Choosing the focal length depends on the laser beam diameter and back aperture diameter of the objective lens. The distance between the lenses adds up to the sum of their focal lengths to ensure proper beam collimation. The beam diameter is chosen to exactly fill the back aperture plane of the objective to achieve a minimal spot size with Gaussian profile in the sample. A filter wheel containing a series of neutral-density (ND) filters is placed in the excitation beam path so that the intensity of the excitation light can be easily adjusted. After ND filter, the expanded beam falls on the correctly chosen dichroic mirror (DM) and gets reflected vertically at right angle. Finally, the beam passes through high-NA water immersion objective. The objective focuses the light on the sample, and the redshifted fluorescence emanating from the sample is collected by the same objective, transmitted through the DM, and the falls on the mirror (M). The fluorescence is now filtered by emission filter (EM),

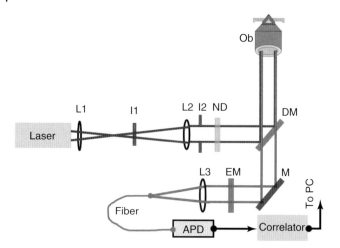

Figure 9.2.3 Block diagram of the proposed single-focus FCS.

focused by lens L3 (150 mm), and then coupled to the fiber through fiber coupler. The front face of the fiber will work as the pinhole or aperture. The fiber holder should allow the fiber to move, a few millimeters of translation along the radial (*x* and *y*) directions and 10 mm translation along the axial (*z*) direction. The other end of the fiber is connected with single-photon-counting detector (APD). The APD feeds signal to the data correlator card, which is connected to the computer. The excitation wavelength, the dichroic, and the EM filter must be chosen according to the excitation and emission spectra of the fluorescent probe used. The inverted geometry (Figure 9.2.3) is most suitable for liquid samples. The whole assembly needs to be mounted on a vibration-free optical bench. For a perpendicular deflection of the beam, cubic optical mounts housing rotatable dichroic holders are required. A fiber holder mount with three-axis movement is required for alignment.

For the cross-correlation experiments (investigating association of two kinds of particles having different emission wavelengths), an extra dichroic at an angle of 45° to the vertical axis is required. It will separate the two emission wavelength ranges. A second APD detector will be needed to couple the second emission to the second input channel of the autocorrelator card. The card can then either cross-correlate these two signals or autocorrelate them separately.

A spring-loaded sample mounting stage with micrometer-resolution vertical axis translation would be required to change the distance between the sample and the objective. The sample in the form of a droplet may be placed on a thin coverslip or on a coverslip-bottom petri dish on the sample stage. It requires very small amount of sample. The evaporation of the liquid may be prevented by covering the liquid droplet with a small dark cap and sealing it airtight with petroleum jelly or grease.

9.2.2.2.1 FCS Setup with Microscope

When FCS requires measurement inside the cell then an imaging setup is required (Figure 9.2.4). Any fluorescence microscope can incorporate FCS. Through

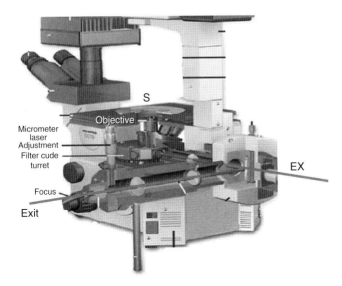

Figure 9.2.4 Optical path in the microscope (Olympus IX7).

fluorescence microscopy, an enlarged image of the sample can be obtained in transmission mode, and the laser spot defines the position of FCS measurement. Fluorescence microscopes, however, have lower resolution, and the FCS points cannot be taken with great accuracy. In order to achieve greater accuracy, a laser confocal scanning system could be used.

Instructions for modifying the Nikon TE300 inverted microscope can be found in Ref. [37]. A similar procedure can be adopted to modify other microscopes also. Here, in Figure 9.2.4, the optical path for Olympus IX7 inverted microscope is shown. The halogen lamp connected to the back port has been removed and is used for allowing entry of the laser light for FCS measurements. Outer diachroic mirror can be used to separate out the fluorescence. The side port is also available for the same, and either one works efficiently. Cells are plated on coverslip-bottom petri dishes mounted on the microscope stage. The focus can be moved along all three axes with precision, which allows us to choose the region of interest in the cells. The cells can be visualized using upper illumination and viewing through the eyepiece. While viewing the cells, utmost care should be taken to avoid scattered laser light from entering the eyes.

9.2.2.2.2 Prism-Based Fluorescence Cross-Correlation Spectrometer (FCCS)

Single-wavelength fluorescence cross-correlation spectrometers (SW-FCCSs) use conventional instrumentation including dichroic mirrors and emission filter sets to select the desired emission wavelengths or separate them into different detection channels as mentioned in the above setup. For multiple wavelength detection, multiple dichroics and emission filters will have to be used. This not only complicates the setup but also amplifies the intensity losses because of nonideal transmission, principally surface reflections through each optical component. Commercially available emission filters and dichroics usually have broad spectral bandwidths and rise/fall bandwidths, respectively. Except each filter is designed to pass a limited wavelength region of the optical spectrum. The difficulty of balancing between optimizing signal detection and reducing spectral cross talk will increase with each additional detection channel. To overcome these problems, a dispersive element such as a diffraction grating could be used to spectrally separate the emission light.

For cross-correlation studies, prism-based spectrometer could be designed, constructed, and combined with an FCS system with single-laser excitation, and it is easy. Dispersion by the prism spectrometer causes a wavelength-dependent deflection angle such that the fluorescence signal can be focused on well-separated spots for the spectral ranges of interest. An optical fiber scanned through these foci selects different spectral ranges for detection and autocorrelation of standard and tandem dyes. The single fiber is then replaced with an optic fiber array to detect signals from two channels for cross correlation. This is an important step for detecting different kinds of nanoparticles and their interaction with each other in same media.

9.2.2.2.3 Prism Spectrometer

A schematic diagram of the prism-based fluorescence correlation spectrometer is shown in Figure 9.2.5. An argon-ion laser emitting at an excitation wavelength of 488 nm can be used for the excitation of several fluorophores. The laser beam diameter is expanded with two planar convex lenses L1 and L2 and coupled to the back aperture of the objective mounted onto a home-fabricated assembly or commercial setups. Fluorescence emission from the sample is collected by the objective and separated from the backscattered excitation light with a dichroic mirror (DM). The fluorescence is focused by the microscope tube lens L3 into the pinhole. An achromat L4 collimates the emission light, which then passes a $30°$ isosceles prism dispersing the fluorescence light. The focusing lens L5 brings the dispersed wavelengths into focus at different positions in the focal plane. A magnified image of the pinhole is formed for each wavelength, distributed in the image plane. The desired wavelength range is defined by the core diameter and the position of the optic fiber at the image plane.

There are several factors influencing the desired wavelength range to be detected: (i) The core diameter of the optic fiber acting as a slit for the spectrometer. The core diameter determines the spectral bandwidth, while the distance between the fiber cores determines the size of spectral channel separation. (ii) The focal length of the focusing lens. Longer focal length increases the linear deflection of the wavelengths at the image plane. (iii) The size or angle of the prism. A larger prism will have higher dispersion than a smaller prism. (iv) The spot size in the image plane. To achieve good spectral filtering, it is important to keep the ratio of the core diameter to spot diameter high while keeping the focusing NA below the acceptance angle of the fiber. For more details readers can go through Ref. [38].

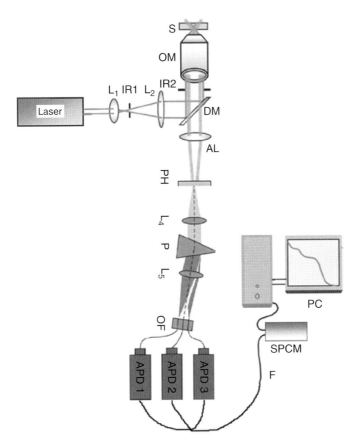

Figure 9.2.5 Optical setup of prism-based FCCS spectrometer.

9.2.3
Instrument Optimization and Performing FCS Experiments

9.2.3.1 Aligning and Optimizing the Setup
The laser beam used for excitation is made to pass through the center of all the optics and roughly through the center of two alignment irises IR1 and IR2 (Figure 9.2.3). IR1 is always present in the setup, while IR2 is placed in front of the objective holder during the initial alignment procedure. The expanded excitation beam is then collimated by adjusting the distance between the telescope lenses. The microscope objective is placed in this collimated beam path (after removing IR2). The beam axis needs to be aligned to the optical axis of the microscope objective lens. The excitation spot without the microscope objective lens in place is marked on a piece of paper placed in front. After the objective lens is placed, the divergent beam should be centered at this marked spot. This alignment is achieved by using the adjustment screws on the objective holder piece. The alignment on the detection

side is the most crucial. As stated earlier, the preferred way of performing FCS is with a fiber-coupled APD detector. The fiber face needs to be properly aligned with the focus of the fluorescence. Initially, the fiber is placed roughly at the distance where the focus of the achromat lens is estimated to be. Then, a concentrated solution (10^{-5} M) of a standard fluorescent dye (e.g., rhodamine B) can be used as the sample, and the fiber face is placed at the fluorescent spot, which is now easily visible. Then, the fiber holder's $x-y$ positioning is adjusted until the fluorescence couples to the fiber and can be seen at the other end of the fiber with the naked eye. Then, a very dilute (a few nanomolar) solution of the dye, which emits enough fluorescence to be easily detectable by the detector without damaging it, is used for the finer alignment. With this solution, the fluorescence and autocorrelation are measured as a function of the z-position (the fluorescence signal is maximized in x and y at each z-point). Finally, the fiber is placed at a position on the z-axis where fluorescence per particle (per-particle brightness) shows the maximum value. The sample placed on the microscope stage is then translated along the z-direction to find the appropriate distance between the sample and the objective lens. If there is a mismatch between the refractive indices of the immersion medium and the sample, the probe volume becomes progressively larger as the focus moves deeper into the sample. This lowers the autocorrelation magnitude. However, FCS experiments with solution specimens are conveniently carried out deep inside the sample, where the autocorrelation value does not change sharply with changes in the sample z-position. More details of the instrument parametric studies can be found in the Ref. [38]. Several other important requirements need to be fulfilled during the establishment of the instrument. These are discussed in the following sections.

9.2.3.1.1 Light Isolation

In FCS, the level of the detected signal can be much less than that emitted by a single molecule. So, even an apparently "small" amount of stray light can overwhelm the instrument and can damage the ultrasensitive detector. Black beam tubes are used to cover the space between the optics holders. The instrument skeleton is then wrapped with black electrical insulating tape to make it light tight. Special care is taken so that no light can leak around the band-pass-emission filter placed in front of the detector. In addition, the whole instrument with the detector is put inside a black cover box to allow experiments to continue under normal room light. Alignment of the fiber can be controlled with knobs outside the box that are connected to the micrometer adjustments of the fiber holder through steel cables. It is inconvenient to completely cover a microscope; hence, the experiments incorporating the microscope are carried out in a dark room.

9.2.3.1.2 Vibration Isolation

In FCS, we measure the fluctuation in the fluorescence caused by diffusive movement of molecules in a microscopic volume. This demands vibrationally isolated systems that can produce a stationary excitation spot. The instrument is mounted on a vibrationally isolated laser table after construction. If such a table is not available, a heavy table with its feet placed in sandboxes can be used. A heavy optical breadboard (thick metal sheet with tapped holes in a square grid pattern) placed on a thick layer of foam on top of this table is well insulated from most of the vibrations.

9.2.3.1.3 Electrical Isolation

This is important if the laboratory's electrical supply is noisy. The detectors used for FCS measurements are very sensitive, prone to damage from light and electrical surges, and costly. A DC-to-DC converter, connected to a car battery, can supply power to the detectors, thus completely isolating the detectors from main fluctuations.

9.2.3.2 Preparing the Sample for FCS

The water to be used for making the sample solutions should be distilled twice before the experiment. Fluorescent dye solution for calibration purpose can be made by sonicating them in the required solvent using a tabletop sonicator. Sonication ensures that no aggregates are left in the sample.

9.2.4
Some FCS Studies on Nanomaterial Characterizations

In the recent years, several works have been done on the FCS studies of nanoparticles. These studies are becoming important because of the reasons mentioned in the Section 9.2.1. In the following paragraphs, some cases of the recent studies have been discussed.

Akcakir *et al.* [39] used FCS to measure the number density, brightness, and size of electrochemically etched Si into a colloid of ultrasmall blue luminescent nanoparticles. The results showed particle size of 1 nm, in close agreement with that obtained by direct imaging using transmission electron microscopy. They used mode-locked, femtosecond, titanium sapphire near-infrared laser, 150 fs duration at a repetition rate of 80 MHz, with average power around 20 mW at the target. The beam was focused on a spot of 0.7 μm diameter using a lens with NA 1.3, giving an average intensity of 5×10^6 W cm^{-2}. Fluctuations in the fluorescence signal were detected in photon-counting mode by either a photomultiplier or an APD. They measured fluorescence at the wavelength of second harmonic of excitation wavelength.

Neugart *et al.* [40] have studied the diffusion of diamond nanoparticles inside the cell. This is an area of increasing importance in modern life science. These types of studies can help to design new drug molecules for better efficiency. Various surface-coated metal nanoparticles have been used in the cell, but their cytotoxicity is a major concern. As an alternative, carbon or, more specifically, diamond nanoparticles, would be an option. The authors have used luminescent properties of defects in 40 nm sized diamond nanoparticles for their study. For cellular applications, the nanoparticle solution should be stable, and the authors have used different media to study the nanoparticle stability. Figure 9.2.6 shows the

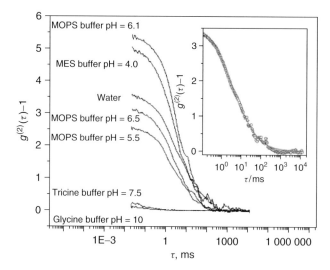

Figure 9.2.6 FCS curves of fluorescent nanodiamonds in different buffer solutions of different pH. Inset shows FCS curve in water with a numerical two-component fit.

correlation curves for different buffer solutions at various pH levels. It is apparent from the figure that the amplitude of the FCS curve depends on the pH value of the buffer solution. When pH becomes larger than 7.5, no correlation has been seen. The authors concluded that disappearance of the FCS curve at larger pH is because of nanocrystal aggregation. In the inset of Figure 9.2.6, the dual-component FCS fitting has been done for a sample in water buffer. Two decay times were obtained: one fast component of 2.5 ms and another slow component of 30 ms. It was found that the stability of the nanoparticles depends on the zeta potential. When the zeta potential becomes low, particle aggregation occurs. Further, to improve the stability of hydrosols, diamonds have been treated with sodium dodecyl sulfate (SDS) as a surfactant. Figure 9.2.7 shows an example of nanodiamonds in phosphate-buffered saline (PBS) buffer with and without SDS treatment. The authors found that nanodiamonds without surfactants precipitate in PBS buffer solutions, while those with SDS form a stable hydrosol.

The conclusion is that nanodiamonds with SDS form stable hydrosols in PBS at pH ranging between 5.7 and 7.8.

First, the authors used stable colloidal solution of these 40 nm sized nanoparticles inside HeLa cells to study the distribution and dynamics and found that nanoparticles got immobilized just after uptake by the cells. Next, they used 4 nm biotin-coated diamond nanoparticles along with streptavidin and found good correlation curve as shown in Figure 9.2.8.

Rochira *et al.* [41] studied CdSe/ZnS quantum dots through FCS and compared the photophysical properties with Alexa488 dye. They found that quantum dots showed numerous transitions between bright and dark states, especially at high

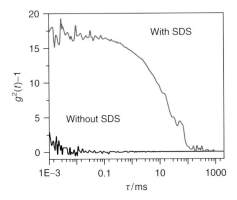

Figure 9.2.7 Diffusion of nanodiamonds with and without surfactant in phosphate-buffered saline (PBS) at pH = 7.2.

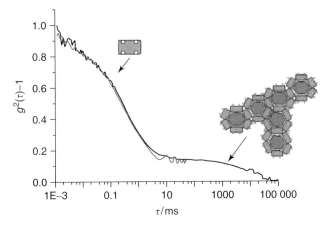

Figure 9.2.8 Correlation curve of biotinylated nanodiamonds in cells. Gray curve is Alexa488-labeled streptavidin alone. The first fast decaying part in the mixture is due to un-bound streptavidin, and the slower component is due to the nanodiamonds bound with streptavidin.

illumination intensities and that the results suggested possibilities for significant improvement of quantum dots for biological applications by adjustments of manufacturing techniques and environmental conditions.

Ow *et al.* [42] studied the photophysical properties of core–shell silica particles through multiphoton FCS. They performed FCS measurements first on a series of parent fluorophores, cores, and core–shell particles made from a single synthesis and then on tetramethylrhodamine isothiocyanate (TRITC) dye and dye-based silica particles. They found that although containing multiple fluorophores, the per-particle brightness of the cores is less than that of core–shell particles and the free dye molecules (Figure 9.2.9). They also demonstrated that the addition of

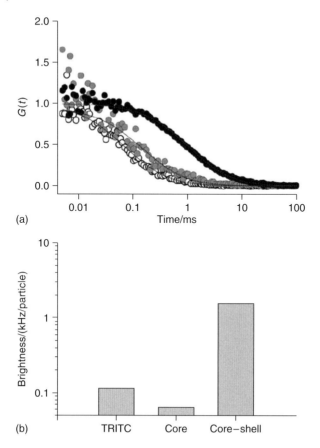

(a)

(b)

Figure 9.2.9 (a) FCS autocorrelation curves for free TRITC dye (white), TRITC-based silica core particles (gray), and 30 nm core–shell TRITC-based silica nanoparticles (black) and (b) a comparison of per-particle brightness values for TRITC dye, core particles, and core–shell nanoparticles.

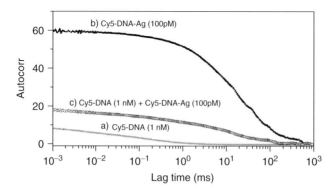

Figure 9.2.10 Autocorrelation of Cy5-DNA, bound to a silver particle and the mixture.

the silica shell onto the core significantly enhances the brightness of the core–shell particles.

Ray *et al.* [43] have used fluorescence lifetime correlation spectroscopy (FLCS) to separate the FCS contributions from fluorophores and metal-conjugated fluorophores. They suggest that FLCS is a powerful method for investigating metal–fluorophore interaction at the single molecule level and for separating two different species from a mixture solution emitting at the same wavelength. Figure 9.2.10 shows the measured correlation functions for free Cy5-DNA (1 nM), Cy5-DNA-Ag-particle (100 pM), and a mixture of both in solution. The correlation function for the bound Cy5 can be seen to be strongly shifted to longer times because of the slower diffusion of the Cy5-DNA-Ag-particle as compared to the free Cy5-DNA.

9.2.5
Conclusions and Future Prospects

FCS is a laser spectroscopic tool whose versatility and ease of implementation has opened up new possibilities to understand the behavior of molecules at the single-molecule level. Nanoparticles can be studied well with this technique. It is minimally invasive due to use of low-power light radiation, and is thus extremely useful for investigating soft biological systems. As this method is concerned with fluctuations around the thermodynamic equilibrium, no external stress has to be applied to determine the relaxation parameters. The confocal setup promises high spatial resolution, which is combined with its inherently high temporal resolution to render it complementary to most other fluorescence techniques. A large number of parameters can be determined by FCS, among them the mobility constants and concentrations, fast internal dynamics, and photophysical processes.

This chapter discusses FCS introduction, experimental technique, and its potential in nanomaterials characterization. However, the possibilities are vast and in future, there will be new testing of its instrumentation and research on understanding of nanoparticle photophysics. After Medge *et al.* (1974), many improvements in the technique have been made, but there is still room for more improvement. Very recently, dual-focus FCS technique has been introduced, which provides us absolute diffusion constant of diffusing molecules and has also made it possible to measure unknown focal volume [45, 46]. Also, the necessary condition that nanoparticle/molecule must be fluorescent during FCS measurement has been eliminated by the introduction of new PhCS. FCS studies on nanomaterials demand much attention since nowadays medical science utilizes more and more nanoparticles. A clear understanding is necessary to eliminate after-use health risks, and FCS is the most suitable tool for this.

Acknowledgments

Kaushal Kumar gratefully acknowledges the Italian Institute of Technology (IIT), Genova, Italy, for financial support.

References

1. Hodge, G.A., Bowman, D., and Ludlow, K. (eds) (2007) *New Global Frontiers in Regulation: the Age of Nanotechnology*, Edward Elgar Publishing.

2. Roco, M.C. (1999) Nanoparticles and nanotechnology research. *J. Nanopart. Res.*, **1**, 1.

3. Roco, M.C. (2003) Nanotechnology: convergence with modern biology and medicine. *Curr. Opin. Biotechnol.*, **14**, 337.

4. Kawasaki, E.S. and Player, A. (2005) Nanotechnology, nanomedicine, and the development of new, effective therapies for cancer. *Nanomed. Nanotechnol. Biol. Med.*, **1**, 101.

5. Sahoo, S.K., Parveen, A.S., and Panda, J.J. (2007) The present and future of nanotechnology in human health care. *Nanomed. Nanotechnol. Biol. Med.*, **3**, 20.

6. Gaoa, J. and Xu, B. (2009) Nano, applications of nanomaterials inside cells. *Nano Today*, **4**, 37.

7. Rodriguez, J.A. and Garcia, M.F. (2007) *Synthesis, Properties, and Applications of Oxide Nanomaterials*, Wiley-Interscience, Hoboken, NJ.

8. Colvin, V.L. (2003) The potential environmental impact of engineered nanomaterials. *Nat. Biotechnol.*, **21**, 1166.

9. Hoet, P.H.M., Brüske-Hohlfeld, I., and Salata, O.V. (2004) Nanoparticles-known and unknown health risks. *J. Nanobiotechnol.*, **2**, 12.

10. Oberdorster, G., Stone, V., and Donaldson, K. (2007) Toxicology of nanoparticles: a historical perspective. *Nanotoxicology*, **1**, 2–25.

11. Oberdörster, G., Oberdörster, E., and Oberdörster, J. (2005) Nanotoxicology: an emerging discipline evolving from studies of ultrafine particles. *Environ. Health Perspect.*, **113**, 823.

12. Oberdörster, G. (2001) Pulmonary effects of inhaled ultrafine particles. *Int. Arch. Occup. Environ. Health*, **74**, 1.

13. Bruchez, M., Moronne, M., Gin, P., Weiss, S., and Alivisatos, A.P. (1998) Semiconductor nanocrystals as fluorescent biological labels. *Science*, **281**, 2013.

14. Mah, C., Zolotukhin, I., Fraites, T.J., Dobson, J., Batich, C., and Byrne, B.J. (2000) Microsphere- mediated delivery of recombinant AAV vectors in vitro and in vivo. *Mol. Ther.*, **1**, S239.

15. Edelstein, R.L., Tamanaha, C.R., Sheehan, P.E., Miller, M.M., Baselt, D.R., Whitman, L.J., and Colton, R.J. (2000) The BARC biosensor applied to the detection of biological warfare agents. *Biosens. Bioelectron.*, **14**, 805.

16. Nam, J.M., Thaxton, C.C., and Mirkin, C.A. (2003) Nanoparticles-based bio-bar codes for the ultrasensitive detection of proteins. *Science*, **301**, 1884.

17. Mahtab, R., Rogers, J.P., and Murphy, C.J. (1995) Protein-sized quantum dot luminescence can distinguish between "straight", "bent", and "kinked" oligonucleotides. *J. Am. Chem. Soc.*, **117**, 9099.

18. Ma, J., Wong, H., Kong, L.B., and Peng, K.W. (2003) Biomimetic processing of nanocrystallite bioactive apatite coating on titanium. *Nanotechnology*, **14**, 619.

19. Yoshida, J. and Kobayashi, T. (1999) Intracellular hyperthermia for cancer using magnetite cationic liposomes. *J. Magn. Magn. Mater.*, **194**, 176.

20. Weissleder, R., Elizondo, G., Wittenburg, J., Rabito, C.A., Bengele, H.H., and Josephson, L. (1990) Ultrasmallsuperparamagnetic iron oxide: characterization of a new class of contrast agents for MRimaging. *Radiology*, **175**, 489.

21. Luo, D. and Saltzman, W.M. (2000) Synthetic DNA delivery systems. *Nat. Biotechnol.*, **18**, 33.

22. Berciaud, S., Lasne, D., Blab, G.A., Cognet, L., and Lounis, B. (2006) Photothermal heterodyne imaging of individual metallic nanoparticles: theory versus experiment. *Phys. Rev.*, **B73**, 045424.

23. Plakhotnik, T., Donley, E.A., and Wild, U.P. (1997) Single molecule spectroscopy. *Annu. Rev. Phys. Chem.*, **48**, 181.

24. Moerner, W.E. (2002) A dozen years of single-molecule spectroscopy in physics, chemistry, and biophysics. *J. Phys. Chem. B*, **106**, 910.

25. Magde, D., Elson, E., and Webb, W.W. (1972) Thermodynamic fluctuations in a reacting system--measurement by fluorescence correlation spectroscopy. *Phys. Rev. Lett.*, **29**, 705.

26. Octeau, V., Cognet, L., Duchesne, L., Lasne, D., Schaeffer, N., Fernig, D.G., and Lounis, B. (2009) Photothermal absorption correlation spectroscopy. *ACS Nano*, **3**, 345.

27. Paulo, P.M.R., Gaiduk, A., Kulzer, F., Gabby Krens, S.F., Spaink, H.P., Schmidt, T., and Orrit, M. (2009) Photothermal correlation spectroscopy of gold nanoparticles in solution. *J. Phys. Chem. C*, **113**, 11451.

28. Thompson, N.L. (1991) *Topics in Fluorescence Spectroscopy: Techniques*, Plenum Press, New York, pp. 337–378.

29. Larson, D.R., Zipfel, W.R., Williams, R.M., Clark, S.W., Bruchez, M.P., Wise, F.W., and Webb, W.W. (2003) *Science*, **300**, 1434.

30. Qian, H. and Elson, E. (1991) Analysis of confocal laser-microscope optics for 3-d fluorescence correlation spectroscopy. *Appl. Opt.*, **30**, 1185.

31. Rigler, R., Mets, U., Widengren, J., and Kask, P. (1993) Fluorescence correlation spectroscopy with high count rate and low background-analysis of translational diffusion. *Eur. Biophys. J.*, **22**, 69.

32. Wahl, M.B.M., Rahn, H.J., Erdmann, R., and Enderlein, J. (2002) Time-resolved fluorescence correlation spectroscopy. *Chem. Phys. Lett.*, **353**, 439.

33. *http://www.picoquant.com/*.

34. *http://www.iss.com/microscopy/ instruments/albaFCS.html*.

35. *http://www.fcsxpert.com/products/ fluorescence-correlation-spectrometer.html*.

36. *http://sales.hamamatsu.com/en/products/ system division/spectroscopy/fluorescence-correlation-spectroscopy.php*.

37. Sengupta, P., Balaji, J., and Maiti, S. (2002) Measuring diffusion in cell membranes by fluorescence correlation spectroscopy. *Methods*, **27**, 374.

38. Chin, H.L. (2006) *Development of a fluorescence correlation spectroscopy method for the study of biomolecular interactions*, Ph. D. Thesis, Department of Chemistry, National University of Singapore.

39. Akcakir, O., Therrien, J., Belomoin, G., Barry, N., Muller, J.D., Gratton, E., and Nayfeh, M. (2000) Detection of luminescent single ultrasmall silicon nanoparticles using fluctuation correlation spectroscopy. *Appl. Phys. Lett.*, **76**, 1857.

40. Neugart, F., Zappe, A., Jelezko, F., Tietz, C., Paul Boudou, J., and Wrachtrup, A.K.J. (2007) Dynamics of diamond nanoparticles in solution and cells. *Nano Lett.*, **7**, 3588.

41. Jennifer, A., Rochira, A., Gudheti, M.V., Gould, T.J., Laughlin, R.R., Nadeau, J.L., and Hess, S.T. (2007) Fluorescence intermittency limits brightness in CdSe/ZnS nanoparticles quantified by fluorescence correlation spectroscopy. *J. Phys. Chem. C*, **111**, 1695.

42. Ow, H., Larson, D., Srivastava, M., Baird, B., Webb, W., and Wiesner, U. (2005) Bright and stable core-shell fluorescent silica nanoparticles. *Nano Lett.*, **5**, 113.

43. Ray, K., Zhang, J., and Lakowicz, J.R. (2009) Fluorophore conjugated silver nanoparticles: a time-resolved fluorescence correlation spectroscopic study. *Proc. Soc. Photo. Opt. Instrum. Eng.*, **7185**, 71850C, doi: 10.1117/12.808958

44. Magde, D., Elson, E.L., and Webb, W.W. (1974) Fluorescence correlation spectroscopy-II. An experimental realization. *Biopolymers*, **13**, 29.

45. Korlann, Y., Dertinger, T., Michalet, X., Weiss, S., and Enderlein, J. (2008) Measuring diffusion with polarization-modulation dual-focus fluorescence correlation spectroscopy. *Opt. Express*, **16**, 14609.

46. Didier, P., Godet, J., and Mély, Y. (2009) Two-photon two-focus fluorescence correlation spectroscopy with a tunable distance between the excitation volumes. *J. Fluoresc.*, **19**, 561.

9.3
Time-Resolved Photoluminescence Spectroscopy of Nanomaterials

Yashashchandra Dwivedi

9.3.1
Introduction

Spectroscopic characterizations of materials basically deal with the problems that explore the molecular structure and relaxation dynamics of the excited species. To determine the molecular structure, eigenstates of total Hamiltonian and position of defect level, typical optical processes such as absorption, emission, and scattering were analyzed using several techniques such as UV–Vis–NIR absorption, FT-IR, photoluminescence, thermoluminescence, cathodoluminescence, X-ray spectroscopy, Raman scattering, and so on. However, the investigation of the relaxation process is multifaceted because of the possible interaction of excited molecules/ions (via intramolecular or intermolecular), which leads to redistribution of excited species and significantly affects the radiative relaxation process and consequently the variation observed in photoluminescence intensity. Hence it is highly desirable to understand the relaxation dynamics of excited species, which is possible by using the time-resolved photoluminescence (TRPL) technique. TRPL is a nondestructive, unique way of studying materials because the temporal information combined with spectral data can elucidate the dynamics of carriers involved in optical transitions.

In photoluminescence spectroscopy, the energy needed to change the electron distribution in a molecule is of the order of several electron volts; consequently, the photons used to excite the sample should have higher energies (or at least of this order), so that these can impart excess energy into the material in a process called *photoexcitation*. One way by which this excess energy can be dissipated by the sample is through the emission of light or photoluminescence. Photoluminescence includes radiative and nonradiative processes. A radiative process involves the molecule losing its excitation energy as a photon, while in the nonradiative process the excess energy is transformed into heat that can be measured in a variety of ways, for example, thermal lensing or photoacoustic techniques. As shown in Figure 9.3.1, the fluorescent state of a molecule can directly decay to the ground state (resonance fluorescence) or a lower state through a process called *internal conversion*, before radiatively dropping to the ground state (fluorescence). It can also decay nonradiatively through a process termed *intersystem crossing* to a triplet state in a radiationless transition. This naturally causes a longer duration between excitation and emission, and because of longer duration of the emission, a delayed fluorescence is sometimes observed even after several minutes.

Although analysis of relaxation processes in an organic molecule and in nanostructured material should not be treated in the same way, various optical processes involved in the loss of excitation energy can be understood on the basis of a Jabolonski-like diagram (Figure 9.3.1). Typically, decay of photoexcited valence

Nanomaterials: Processing and Characterization with Lasers, First Edition.
Edited by Subhash Chandra Singh, Haibo Zeng, Chunlei Guo, and Weiping Cai.
© 2012 Wiley-VCH Verlag GmbH & Co. KGaA. Published 2012 by Wiley-VCH Verlag GmbH & Co. KGaA.

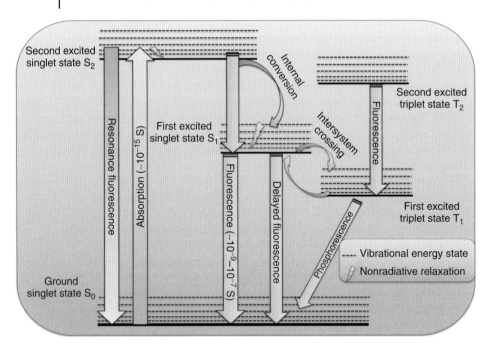

Figure 9.3.1 Jabolonski-like diagram and the different radiative and nonradiative relaxation processes in an organic molecule.

electrons takes place within $10^{-3}-10^{-9}$ s, while high-energy atomic states populated by inner shell excitation decay in a much shorter time (10^{-15} s). The transition between states of same parity (symmetry) may persist for approximately milliseconds. Relaxation of excited electrons in metals or semiconductor materials occurs within a timescale of $10^{-13}-10^{-15}$ s, and excited states of atomic nuclei can even decay in times shorter than 10^{-18} s. In case of intermolecular interaction, energy of one excited species is transferred to another neutral or excited species in the timescale of $\sim10^{-9}-10^{-15}$ s, although the time required is typically a function of concentration, distance between the interacting species, temperature and so on.

Owing to nonradiative processes, intensity of radiative emission gets quenched. Excited species can lose their excitation by nonradiative process either through the aid of coupled vibrational and electronic energy states or through collisions with other atoms or molecules. The former process is known as *static quenching*, while the latter is termed as *dynamic quenching*.

When a material is illuminated by a laser or lamp, a small population of ground state atoms or molecules is excited. Let us assume that on excitation with laser, a temporary concentration of excited-state molecules $[X^*]$ at some moment in time is generated. Let us further assume that X^* is strongly fluorescent and that we can follow the intensity of this fluorescence later using a sensitive light detector such as a photomultiplier tube or semiconductor diodes. If there are no quenching agents present in the system (i.e., in the absence of species that can quench the fluorescence through collisions) then X^* can return to the ground state through a normal fluorescence (Eq. (9.3.1)) emitting a photon of energy $h\nu$ or by a nonradiative decay process (Eq. (9.3.2)) through coupling of vibrational and electronic levels.

$$X^* \xrightarrow{K_1} X + h\nu \tag{9.3.1}$$

$$X^* \xrightarrow{K_2} X \tag{9.3.2}$$

where k_1 and k_2 are the rate constants for radiative and nonradiative processes. If only these two paths to the ground state are available, the rate equation for $[X^*]$ can be written as

$$\begin{aligned} \frac{d[X^*]}{dt} &= -k_1[X^*] - k_2[X^*] \\ \frac{d[X^*]}{dt} &= -(k_1 + k_2)[X^*] \end{aligned} \tag{9.3.3}$$

On integrating Eq. (9.3.3) with respect to presume conditions $t = 0$ and $[X^*] = [X^*]_0$ results in

$$[X^*] = [X^*]_0\, e^{-(k_1+k_2)t} \tag{9.3.4}$$

It is evident that the rate constants k_1 and k_2 quantify the relaxation rate. Hence, fluorescence lifetime in the absence of a quencher can be expressed as

$$\tau_0 = \frac{1}{(k_1 + k_2)} \tag{9.3.5}$$

Summing the expressions (9.3.4) and (9.3.5),

$$[X^*] = [X^*]_0\, e^{-t/\tau_0} \tag{9.3.6}$$

Through this way, fluorescence lifetime (τ_0) can be defined as the amount of time that it takes for the fluorescence intensity to decay to $1/e$ of its initial value. τ_0 is the mean spontaneous lifetime of the state, which is related to the total transition probability A by $A = 1/\tau_0$. Typical values of the lifetime of excited valence electrons lie in the nanosecond range. In the presence of a quenching species (Q), a third process will come into play, which enforces an excited species X^* to relax to the ground state through nonradiative process

$$X^* + Q \xrightarrow{K_q} X \tag{9.3.7}$$

Including Eq. (9.3.7), the rate equation for $[X^*]$ (Eq. (9.3.3)) becomes

$$\frac{d[X^*]}{dt} = -(k_1 + k_2 + k_q[Q])[X^*] \tag{9.3.8}$$

where k_q is the quenching constant. Assuming $[Q] \gg [X^*]$ and integrating Eq. (9.3.8), we have

$$[X^*] = [X^*]_0\, e^{-(k_1+k_2+k_q[Q])t} \tag{9.3.9}$$

Hence, the fluorescence lifetime in the presence of the quencher will be

$$\tau = \frac{1}{(k_1 + k_2 + k_q[Q])} \tag{9.3.10}$$

Equation (9.3.6) represents the ideal case where the excitation pulse is infinitely narrow. If the pulse has a finite width, the exponential fluorescence decay (Eq. (9.3.6)) will be convoluted with the pulse profile function $B(t)$, resulting in the following more general expression

$$[X^*] = [X^*]_0 \int_0^t B(t - t')e^{-t'/\tau_0} dt' \tag{9.3.11}$$

Considering the case of nanomaterials, nonradiative transition may be influenced in the presence of defect sites, size, and shape of the nanostructure and the ambient conditions. As the size approaches the exciton Bohr radius, optical properties critical to device applications, such as bandgap and photoluminescence lifetime, are affected greatly because of the quantum confinement effect [1–3]. In response of confinement, excitons in geometrically confined systems exhibit different properties as compared to three-dimensional excitons [4, 5].

Excitation of a nonequilibrium density of electrons and holes, excitons, or any other number of quasi-particles will ultimately lead to the recombination and simultaneous emission of photons, and this is called *radiative recombination*. If the recombination proceeds with the emission of phonons instead of photons, then the recombination is termed as *nonradiative recombination*. Detailed information about the different recombination processes can be accessed from Refs. [6, 7]. Both radiative and nonradiative rate of recombination affect the observed luminescent intensity. Recalling Eq. (9.3.5), a photoluminescence decay rate is a sum of the radiative and nonradiative decay rates

$$\frac{1}{\tau_{\text{Lifetime}}} = \frac{1}{\tau_{\text{Radiative}}} + \frac{1}{\tau_{\text{Nonradiative}}} \tag{9.3.12}$$

Radiative and nonradiative processes are competitive processes among which the process that dominates can be estimated by TRPL using suitable reference material. Since nonradiative processes are thought to be more likely at the surface, the effective lifetime is often written as a combination of surface and bulk lifetimes [8]

$$\frac{1}{\tau_{\text{eff}}} = \frac{1}{\tau_{\text{bulk}}} + \frac{1}{\tau_{\text{surf}}} \tag{9.3.13}$$

The transformation of the bulk material into nanostructures can enhance nonradiative surface-mediated trapping through defect states and collisional effects. Such effects, including Auger recombination [9], can also occur on a short timescale, competing with radiative decay [10]. In addition, nonradiative processes in the system invoked by changing the defect concentration of the sample in such a way that the concentration increases linearly over the growth time, and surface states cause nonradiative decay. Stimulated emission can also influence the short time dynamics, often obscuring nonradiative decay mechanisms [11].

Size-dependent radiative decay can be differentiated into three different regions [12]:

1) If a nanoparticle is smaller than the Bohr diameter of the exciton, quantum confinement effect may lead to size-dependent oscillator strength.
2) If a nanoparticle is bigger than the exciton Bohr radius but smaller than the wavelength of the light, then scattering with the surface provides a path for an exciton to decay radiatively.
3) If the size of the nanoparticle approaches the dimension of the wavelengths, exciton polariton effects come into play.

Theoretical analysis showed that the rate of radiative recombination decreases as the size increases in regions (1) and (3), whereas it behaves oppositely in region (2) [13]. According to the computational result for a quantum dot including the exciton–polariton effect, the size-dependent recombination rate is

$$\Gamma_0 = \frac{\sqrt{2\pi}}{12} \omega_{LT} \left(\frac{2\pi}{\lambda_0} \right)^3 \langle r \rangle^3 \exp \left(-8 \, \epsilon_b \, \frac{\pi^2 \langle r \rangle^2}{\lambda_0^2} \right) \tag{9.3.14}$$

Here, ε_b is the dielectric constant, ω_{LT} the longitudinal–transverse splitting frequency, and $<r>$ the size ($1/e$) of the exciton wave function. For a nanosphere radius of 17 nm, ω_{LT} is 1 meV and λ_0 is 350 nm and there is a minimum radiative lifetime; 260 ps is obtained for a nanosphere radius of 17 nm, and it increases monotonically as the size increases [12]. In addition, the perturbation of the electron and hole states by quantum confinement produces unique dynamics that are strongly dependent on the nanoparticle size [14].

The strain in crystal could be clearly evidenced by several techniques, namely, shifts in the photoluminescence, photoreflectance, and line positions [15, 16]. As the energy of the free exciton associated with the top valence band varies linearly with the in-plane and axial components of the strain tensor, lifetime could be a good tool to predict the expected strain in the crystal lattice [16]. TRPL measurements could be an additional measure of crystal quality to correlate sample thickness with incorporated strain. TRPL decay measurements on the samples showed that radiative lifetime increased as the sample thickness increased, verifying that the crystal quality improves with reduced strain [17].

9.3.1.1 Example

In case of high-quality GaN crystals grown on sapphire by hydride vapor-phase epitaxy, the thermal expansion mismatch between sapphire and GaN produces strain in the GaN crystal as it is cooled from the growth temperature to room temperature. The variation in lifetimes for transition energies of both the *A* free exciton energies and a neutral-donor-bound exciton $(D + X)$ as a function of strain is given in Figure 9.3.2 [18].

The decay data display single exponential decays for all of the transitions over several times the radiative recombination lifetime. The D^0, X lifetime is greater than the free exciton lifetime as is usually observed [19]. The lifetime increases with sample thickness, as the surface strain decreases. This suggests that the factors

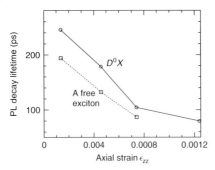

Figure 9.3.2 Decay lifetimes of the A-band free exciton and the donor-bound exciton as a function of the axial strain ϵ_{zz}. PL, photoluminescence [18].

contributing to strain also introduce recombination paths as well as nonradiative decay processes. Slower exciton decay was also observed when the layer thickness was increased in MOCVD-grown GaN [20].

9.3.2
Experimental Methods of TRPL

To measure the lifetime of the emitting state, two techniques are known, that is, time-domain techniques and frequency-domain techniques. The time-domain techniques are direct techniques. It measure fluorescence decay curves (i.e., fluorescence intensity as a function of time) directly and estimate lifetime by proper exponential fitting function of the obtained decay curve. The phase technique is totally different from that of the former in terms of measurement and the method of detection. It utilizes a sinusoidally modulated light for the excitation of the fluorescence. The emitted light also becomes modulated at the same frequency, but it will show some shift (delay) relative to the excitation light. This shift is called a *phase shift*, and it contains the information about the lifetime. Depending on the lifetime, the fluorescence will show a decreased depth of the sinusoidal modulation relative to the excitation light.

In practice, it is difficult to specify the lower limit of the lifetime that can be measured by an instrument. Factors such as quantum yield, fluorophore concentration, and decay kinetics can affect the measurement. In order to demonstrate performance, one must use a well-characterized standard or propose a special convincing protocol. TRPL is measured by exciting luminescence from a sample with a pulsed light source, and then measuring the subsequent decay in photoluminescence as a function of time. The time resolution achieved is determined by the quality of excitation source and the detector. A wide variety of experimental configurations can accomplish this. Most experiments excite the sample with a pulsed laser source and detect the photoluminescence with a photodiode, streak camera, or photomultiplier tube set up for upconversion or single-photon counting. The system response time, wavelength range, sensitivity, operational difficulty, and cost vary widely for each configuration. In order to obtain precise fluorescence life-time, the profile of instrument response function (IRF) (excitation pulse) has to be measured in addition to the fluorescence decay. This is because the lamp (or laser) pulse has a finite-temporal width, which distorts the intrinsic fluorescence decay response from the sample. This effect is called *convolution*. In a typical experiment, two curves are measured: the IRF using a scatterer solution and the decay curve. Analysis is then performed by convoluting the IRF with a model function (e.g., a single exponential decay or a double exponential decay or some other function) and then comparing the result with the experimental decay. This is done by an iterative numerical procedure until the best agreement with the experimental decay curve is achieved.

In case of nanosecond lifetime measurements, digital storage oscilloscopes with sample rates higher than 1 Gs s^{-1}, which follow the electrical signal with a time resolution of nanoseconds, are used. It can be used for single-shot and repetitive events. The time resolution of only a few nanoseconds can be achieved with this system. A boxcar integrator stores the signal during a time window. This apparatus is mainly used for the gating of the temporal signal with a low repetition rate up to 1 KHz, for example, pump-probe experiments using a regenerative amplifier. It is used for repetitive events, and subnanosecond time resolution can be measured.

Time-resolved measurements in picosecond resolution are performed by making use of time resolution of detectors. In recent times, pump-probe photoinduced absorption, single-photon counting, photoluminescence upconversion, and photoluminescence imaging techniques are mostly used for picosecond lifetime measurements. These techniques have been tremendously exploited to probe exciton dynamics, energy transfer, and radiative decay channels in organic and nanomaterials. For these experiments, a laser with a broad spectral gain profile is used as a light source, and the wavelength is tuned across the gain profile by wavelength-selecting elements (such as prisms, gratings, or interferometers) inside the laser cavity. For this, Ti:sapphire lasers of femtosecond pulse width, which pump by continuous argon ion laser or green laser source, are used. An optical frequency tripler further extends this range to include 233–327 nm. Additional excitation wavelengths are obtained using a harmonic generator for frequency doubling and tripling or an optical parameter oscillator. The pulse energy of Ti:sapphire can be improved by the conventional amplification method in which population of gain medium is inverted while pumped by a powerful laser or by parametric amplification. In case of optical parametric amplification, nonlinear optical crystals are used, which split pump photon into two photons, called the *signal and idler*; both the signal and the idler can be tuned over a wide range in the visible or infrared region, by seeding the system with a wavelength-tunable laser and/or by using an optical parametric oscillator and changing the phase-matching conditions of the crystal. The laser repetition rate can be lowered by a pulse picker . An acousto-optic pulse picker is used to lower the laser pulse rate for longer lifetime fluorophores. A brief introduction of the different techniques used for TRPL spectroscopy is discussed in the following sections.

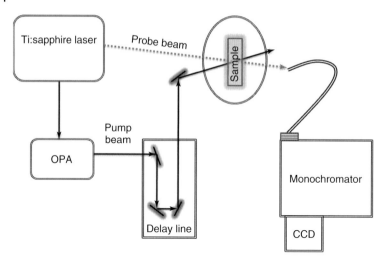

Figure 9.3.3 A schematic diagram of the pump-probe method.

9.3.2.1 Pump-Probe Technique

The pump-probe technique is the most commonly used technique to study transient phenomena. This technique needs simple optical pathways and detection with the advantage to probe transients up to the laser pulse width. The technique uses two femtosecond beams: a pump to initiate absorption in the sample and a probe beam, which is a fraction of the pump at a different wavelength, entering the sample at a later time delay and monitors an optical property. By sweeping the time delay between the pump and probe pulses using a variable optical delay line placed on the pump beam optical pathway, it is possible to assemble measurements as a function of time. The wavelength of the excitation pulse (pump) is chosen by adjusting the nonlinear crystal of an optical parametric amplifier (OPA). The probe pulse is typically a UV, visible, or infrared pulse in which a snap-shot spectrum is taken as a function of the delay time. Often, the probe pulse is generated from a portion of the excitation beam, but it can also be an independently generated electromagnetic pulse (Figure 9.3.3).

9.3.2.2 Single-Photon Counting Technique

Time-correlated single-photon counting (TCSPC) is based on the detection of single photons of a periodical light signal, the measurement of the detection times of the individual photons, and the reconstruction of the waveform from the individual time measurements [21, 22]. TCSPC method generates a histogram that represents the fluorescence lifetime. This is an efficient process because it counts photons and records arrival time in picoseconds, which directly represents the fluorescent decay. Time to amplifier converter (TAC) is used with a multichannel analyzer (MCA) to perform TCSPC. The TAC produces an electronic pulse with a height that is proportional to the time difference between the start pulse triggered by the excitation pulse and the stop pulse from the detector. The MCA receives the electronic pulse from the TAC and converts the voltage into a channel address number of a storage memory. A histogram, which corresponds to the decay curves, is built up in the MCA with increasing numbers of events. The counting rate should be smaller than ~1%, so that the probability of a simultaneous arrival of two photons is negligible (<0.01%). Thus, a light source with a high repetition rate, for example, a mode-locked laser is required. The time resolution of this technique is determined by the jitter of the electric circuit, not by the width of the electric signal from the photosensors. Therefore, time resolution of tens of picoseconds can be achieved with a specially designed PMT. TAC is used for repetitive events and for single-photon counting. The time resolution with this system is ~10 ps.

The principle of the TCPSC technique is quite complicated; however, it is frequently used as it has several remarkable benefits

- The time resolution of TCSPC is limited by the transit time spread, not by the width of the output pulse of the detector. With fast MCP PMTs, an instrument response width of less than 30 ps is achieved.
- TCSPC has a near-perfect counting efficiency and therefore achieves optimum signal-to-noise ratio for a given number of detected photons.
- TCSPC is able to record the signals from several detectors simultaneously.
- TCSPC can be combined with a fast scanning technique and therefore be used as a high-resolution high-efficiency lifetime imaging (FLIM) technique in confocal and two-photon laser scanning microscopes.
- TCSPC is able to acquire fluorescence lifetime and fluorescence correlation data simultaneously [23].
- State-of-the-art TCSPC devices achieve count rates in the MHz range and acquisition times down to a few milliseconds.

9.3.2.3 TRPL Imaging Technique

The imaging of ultrafast photoluminescence or slower processes (fluorescence, phosphorescence) can readily be carried out with special optoelectronic equipment. The technique allows one to record simultaneously the intensity and the spectral and the temporal responses of fluorescence. A streak camera is widely used in time-resolved spectroscopy, because it enables us to obtain temporal and spectral information simultaneously. A schematic diagram of the streak camera is given in Figure 9.3.4. It should be pointed out that no other device devoted to the direct detection of light has better temporal resolution than a streak camera. The device can be considered as a time–space converter.

A spectrometer is usually installed before the streak camera in order to disperse the incoming light horizontally. The spectrally dispersed light impinges on a photocathode from which photoelectrons are emitted. The electrons are accelerated and temporally dispersed by the deflection electrodes subjected to a rapidly changing sweep voltage in the vertical direction. Then, the spectrally and temporally dispersed electrons hit a multichannel plate, which multiplies electrons while keeping their spatial distribution. The multiplied electrons irradiate a phosphor screen on which a so-called streak image appears. The image is recorded with a CCD camera.

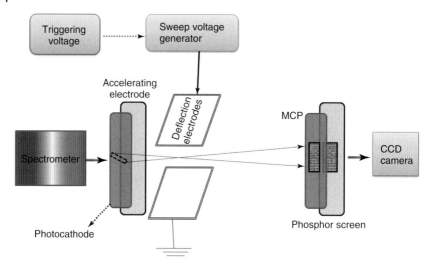

Figure 9.3.4 Schematic diagram of a streak camera.

Synchroscan streak cameras used for the mode-locked lasers with high repetition rates of ∼100 MHz achieve picosecond time resolution.

9.3.2.4 Nonlinear Optical Techniques

9.3.2.4.1 Kerr Gate Technique

An extension of the optical gating technique is to use a "Kerr gate," which allows the optical signal to be collected before the (slower) fluorescence signal overwhelms it. Optical Kerr shutters utilize the optical Kerr effect in which a Kerr active medium is used as an optical gate (Figure 9.3.5).

In this method, a strong laser pulse induces birefringence in the Kerr medium, so that the plane of the polarization of the incident light determined by a polarizer P1 is rotated. Thus the incident light, normally blocked by a crossed polarizer P2, can pass through the P2. This configuration is inserted between the collection lenses and the spectrometer. The time-resolved spectrum is then obtained by changing

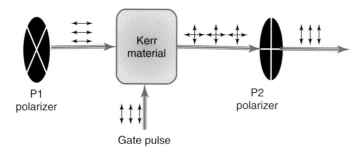

Figure 9.3.5 Principle of Kerr shutter.

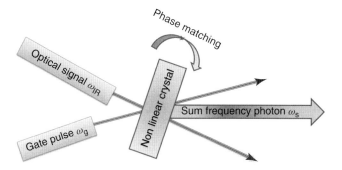

Figure 9.3.6 Schematic diagram of upconversion time-resolved spectroscopy.

the delay between the gate and incident pulses. Time resolution of approximately subpicosecond is possible.

9.3.2.4.2 Upconversion Technique

Upconversion (or optical mixing) is one of the most widely utilized methods for femtosecond time-resolved measurements [24]. With the advent of improved versions of the Ti:sapphire laser, the reliability of such experiments increases. Upconversion spectroscopy is based on sum frequency generation (SFG), which is used to generate higher frequency from the lower frequency. The upconverted photon ω_s from IR optical signal ω_{IR} is emitted when the gating laser pulse ω_g irradiates the nonlinear crystal, shown in Figure 9.3.6. The spontaneous emission is mixed with another femtosecond pulse (gate) in a nonlinear crystal to generate the sum frequency of the gate and optical signal. The intensity of the generated signal is proportional to the intensity of the optical signal that is temporally overlapped with the gate pulse. Therefore, by measuring the upconverted signal while changing the delay time between the pump and the gate pulses, we can obtain a replica of the time-resolved emission. A combination of the pump pulse and the nonlinear crystal acts as an optical gate similar to the boxcar integrator. By sweeping the delay time of the pump pulse, a temporal profile of the luminescence is obtained. A time-resolved spectrum is obtained by scanning the crystal angle (and monochromator) for the phase-matching condition. The time resolution of ~100 fs is possible.

9.3.3
Case Study of ZnO

Despite the commercial success of the III–V gallium nitride (GaN) material system, interest in the II–VI semiconductor zinc oxide (ZnO) was renewed in the late 1990s when room temperature, optically pumped lasing was demonstrated for ZnO thin films [25–29]. Zinc oxide is recognized as a promising material for advanced photonics because of its wide bandgap (3.37 eV) and high exciton binding energy of 60 meV. The bulk exciton's binding energy of ZnO (60 meV) is much higher than that of GaN (24 meV) and ZnSe (20 meV) that are close to kT at room temperature (26 meV), which is extensively used for the development of room temperature optoelectronic devices in the short wavelength range (green, blue, and UV), information storage and sensors [30–33]. The following are applications of optoelectronics: as blue-color-light-emitting phosphors [30], as nanorod UV light emitters [31], as fluorescence labels in medicine and biology, in controlling units as UV photodetectors and as high-flame detectors [32], as nanosensors of various gases, and also in the cosmetic industry, as a component of sun screens, are envisioned [33]. The high exciton binding energy of ZnO would allow excitonic transitions even at room temperature, which could mean high radiative recombination efficiency for spontaneous emission as well as a lower threshold voltage for laser emission. Studies have been carried out to fine-tune the properties of ZnO and to adopt it for different applications; for example, the bandgap of ZnO is modified to use as UV detectors and emitters. Recently, long lasting afterglow has been reported in ZnO nanocrystals, which is caused by the spin-dependent tunneling recombination and can last for a very long time (several hours) after switching off the ultraviolet (UV) light excitation [34]. Considering the vast commercial application in the field of display devices, biomedical, and others, it is necessary to standardize the ZnO samples using the precise TRPL technique for the quality control of the final product. In the following section, a brief about ZnO photoluminescence and a study using this technique are mentioned.

9.3.3.1 Origin of ZnO Photoluminescence

Nanometer-sized particles have very different physical and chemical properties from bulk materials; consequently, the optical behavior of ZnO is entirely different compared to its bulk counterparts [35]. The observed variation is not only due to increased surface area but also due to the changes of surface properties such as surface defects, adsorbed impurities etc. When the crystallite dimension of a semiconductor particle falls below a critical radius of approximately 50 nm, the charge carriers appear to behave quantum mechanically like simple particles in a box. This confinement produces a quantization of discrete electronic states and increases the effective bandgap of the semiconductor [36]. As a result, the band edges shift to yield larger redox potentials. The solvent reorganization free energy for charge transfer to a substrate, however, remains unchanged. The increased driving force resulting from the increased redox potential and the unchanged solvent reorganization free energy in size-quantized systems are expected to increase the rate constant of charge transfer and may result increased photoefficiencies of systems in which the rate-limiting step is charge transfer.

The optical and electronic properties of semiconductors can be further tuned by varying the size of the particles in the range below 10 nm. There is a great variety of the ZnO nanostructure morphologies [37] that has already been demonstrated, such as nanowires [38–40], nanoribbons [41], nanorods [42–44], tetrapod nanowires [45], microtubes [46], nanoneedles [47], nanohelixes, nanorings [48], and dendritic nanowire arrays [49].

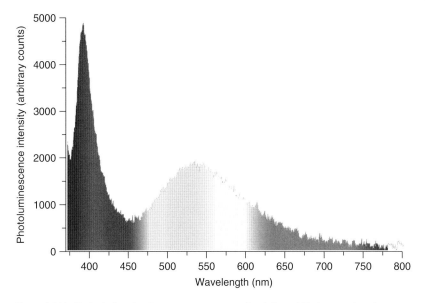

Figure 9.3.7 Typical photoluminescence spectrum of rod-shaped ZnO nanophosphor on excitation with 355 nm laser radiation.

Despite the myriad of potential applications of ZnO nanomaterials, several perplexing problems are yet to be solved, which evoked research among the scientists. One of the most discussed problems, still a matter of debate, is the chemical and structural origins of the visible luminescence from pure ZnO crystal. In general, photoluminescence spectra of ZnO crystals contain a sharp UV emission band (3.3 eV at 295 K) and a broad band in visible region (centered at ~2.5 eV) due to defects and/or impurities (Figure 9.3.7). In bulk ZnO and in thin films of good quality, the intensity of visible emission is of several orders of magnitude weaker than that of the band edge emission [50–53]. For ZnO nanostructures, however, the intensity of the defect emission can be much stronger. The most accepted explanation [54] of the defect emission is the recombination of electrons trapped at singly ionized oxygen vacancies with valence band holes. One other possible explanation [55] in literature is the recombining of electrons in the conduction band and/or shallow donor states with holes, which have been trapped at oxygen vacancies. As a detailed description about the origin of defect is beyond the scope of the chapter, hence only a brief introduction is given here: the crystal structure of ZnO is a relatively open structure, with a hexagonal close-packed lattice where Zn atoms occupy half of the tetrahedral sites, while all the octahedral sites are empty. This facilitates plenty of sites for ZnO to accommodate intrinsic (namely Zn interstitials) defects and extrinsic dopants [56]. There could be a number of intrinsic defects with different ionization energies. The Kröger Vink notation uses: i = interstitial site, Zn = zinc, O = oxygen, and V = vacancy. The terms indicate the atomic sites, and superscripted terms indicate charges, where a dot indicates positive charge, a prime the negative charge, and a cross zero charge, with the

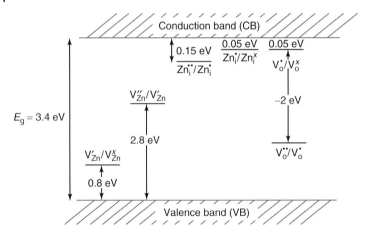

Figure 9.3.8 Energy levels of native defects in ZnO. The donor defects are $Zn_i{}^{\bullet\bullet}$, $Zn_i{}^{\bullet}$, Zn_i^x, $V_o{}^{\bullet\bullet}$, $V_o{}^{\bullet}$, and V_o, and the acceptor defects are $V_{Zn}{}''$ and $V_{Zn}{}'$. (Source: Figure from Book Ref. [59].)

charges in proportion to the number of symbols. A number of defect states within the bandgap of ZnO are clearly visible. The observed defects are $Zn_i{}^{\bullet\bullet}$, $Zn_i{}^{\bullet}$, Zn_i^x, $V_o{}^{\bullet\bullet}$, $V_o{}^{\bullet}$, and V_o, and the acceptor defects are $V_{Zn}{}''$, $V_{Zn}{}'$ (Figure 9.3.8). These defects are a function of dopant, concentration, shape, and structure of nanomaterials. The defect ionization energies vary from ~0.05–2.8 eV [57]. Zn interstitials and oxygen vacancies are known to be the predominant ionic defect types. However, which defect dominates in native, undoped ZnO is still a matter of great controversy [58].

Shalish *et al.* [60] demonstrated that the intensity of defect emission in an array of ZnO nanowires was directly proportional to the average surface-to-volume ratio of wire. Measurements of the polarization of band edge versus defect emission [61] and studies involving surfactant treatments in ZnO nanostructures [62, 63] also indicate that the visible emission originates from the surfaces of these materials.

The high exciton-binding energy at room temperature evoked the possibility of stimulated emission at room temperature, which has a lower threshold than electron–hole plasma recombination. However, as a matter of fact, still no report on electrically pumped lasing is reported, whereas optically pumped lasing and amplified spontaneous emission are available in several reports [28, 64, 65]. Amplified spontaneous emission was reported for a self-organized network of ZnO fibers [41], while lasing has been reported in nanowires [29, 66], tetrapods, [40, 67], and nanoribbons/combs [64].

Owing to these perplexing problems, the study of stimulated emission [38, 39, 42, 68] and temperature-dependent photoluminescence [69, 70] in ZnO is of great interest. Typically, the stimulated emission decay time is much faster than that of the spontaneous emission, so that it may be below the detection limit of few TRPL systems [71]. Emissions in the exciton–exciton (EE) and exciton–hole Plasma (EHP) regime exhibit different behaviors with time [45, 72]. The comparison

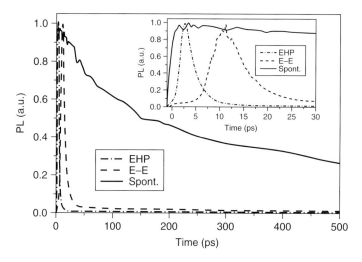

Figure 9.3.9 The radiative decay curves of the exciton–exciton (EE), exciton–hole plasma (EHP), and spontaneous (spon.). A magnified portion of −1 to 30 ps is given in the inset [73]. PL, photoluminescence.

between the decay curves of the spontaneous emission, EE, and EHP emissions from highly faceted rods is shown in Figure 9.3.9.

It is clear that although both types of stimulated emission have a shorter decay time compared to spontaneous emission, there are obvious differences in their temporal evolution. The EHP emission peak exhibits some shifting with time, which was established by direct measurements of the lasing spectra [45, 72–74] as well as by measuring transient profiles of the lasing dynamics as a function of wavelength [39]. EHP emission has a small rise time (1–2 ps) ascribed to the thermalization of the hot carriers [73, 75], and the decay time is of a few picoseconds [64, 72, 75]. The decay time of the lasing could be longer in case of long cavity length, lower losses at end facets, and lower defect concentrations [64]. In contrast to EHP emission, stimulated emission in the EE regime exhibits a long rise time as a longer time would be needed to achieve a high concentration of excitons in the excited state [72, 76]. However, the decay time of EHP emission and the EE emission is comparable [72, 77]. With respect to the evolution of the lasing spectra and any peak shifts in the EE regime, some spectral shifts of the peaks can be observed with time [78], but it is difficult to analyze the data because the measurements have been performed on an ensemble of the nanostructures.

The occurrence of stimulated emission in ZnO nanostructures at comparably lower threshold pump power density attracts attention [38, 39]. The high exciton binding energy of ZnO facilitates the lasing at a comparably lower threshold power density and due to that the excitation of the excitons is less probable, that is, low energy loss through nonradiative process [79]. The lasing in ZnO nanostructures could be achieved by two different methods: first, lasing from a cavity formed between two facets of individual ZnO nanostructures, and second, lasing due to coherent scattering in random media [80].

9.3.3.2 **Time-Resolved Spectroscopy of ZnO**

Similar to the confinement effect observed in absorption and emission spectra, confinement also affects the radiative recombination of excitons, as the exciton–photon coupling in nanosize (1–10 nm) is particularly strong and the exciton radiative recombination rate drastically varies with the size [12]. The size-dependent radiative lifetime of exciton in ZnO nanocrystals were calculated by Fonoberov and Balandin [81, 82]. The effect of the geometrical confinement of the ZnO quantum dots inside the SiO_x matrix on the exciton radiative lifetime is discussed by Zhang *et al.* [83]. They have observed that the confinement reduces the exciton radiative lifetime. The reduction of the exciton radiative lifetime is discussed in terms of exciton superradiance. Superradiance is the cooperative radiative decay of an initially inverted assembly of quantum oscillators. It generates from systems smaller then incident wavelength and when all oscillator strength is cooperated in one collective superradiant excited state. This leads to an enhancement of the optical transition oscillator strength and thus to a shortening of the excitation radiative lifetime. The effect of exciton superradiance has also been studied in semiconductor nanocrystals [84] and several other geometrically confined systems. Kayanuma [85] theoretically studied the quantum-size effects of Wannier excitons in semiconductor nanocrystals and found that the exciton radiative lifetime in the weak confinement region decreases rapidly with the size of the nanocrystal as a result of exciton coherence effect, in other words, the exciton superradiance. The decay time of luminescence is the subject of photon–matter interactions. These interactions include the radiative decay of the exciton polariton and various nonradiative processes, such as leak by deep-level traps, low-lying surface states, and multiphonon scattering. The exciton polariton luminescence is, however, quite sensitive to the concentration of defects and structural factors of the nanostructures, and it is not easy to separate the radiative recombination and nonradiative processes.

Size dependence of the TRPL has been investigated for the ZnO nanorods fabricated with catalyst-free metal–organic chemical vapor deposition [86]. The radiative recombination rate decreases as the length of the nanorods increases monotonically in the range of 150–600 nm. The variation in the decay time with the length of the rods is given in Table 9.3.1 and Figure 9.3.10. The coupling of the excitons with the electromagnetic wave; that is, the exciton–polariton effect is invoked to account for the results. This relation between the size of nanoparticles and radiative decay time appears when the size is comparable to the wavelength of the excitation light.

The ZnO tetrapods shows exceptional optical properties, that is, intense UV emission, no defect emission, and a photoluminescence lifetime in the range of tens of nanoseconds at room temperature, which is strongly dependent on the growth temperature. For individual tetrapods, the photoluminescence lifetimes were significantly longer than the ZnO, including nanorods ($\tau_1 = 190$ ps, $\tau_2 = 1.4$ ns) [87], epilayers ($\tau = 3.8$ ns) [88], and single crystals ($\tau_1 = 1$ ns, $\tau_2 = 14$ ns) [89]. ZnO single crystals exhibited biexponential decay, while the tetrapod structure exhibited single exponential or biexponential decay. In biexponential decay, the fast decay constant is typically attributed to nonradiative recombination, while the slow

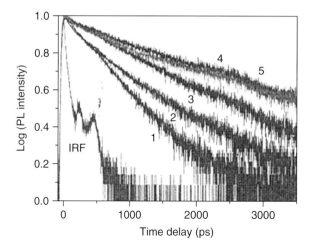

Figure 9.3.10 Decay curves of the samples (1–5 given in Table 9.3.1). Instrumental response function of the apparatus is also shown in Ref. [86]. PL, photoluminescence.

Table 9.3.1 Sizes of the ZnO nanorods and the time constants from the two exponential fits.

Sample	Diameter	Length (nm)	τ_1 (ps)	τ_2 (ps)
1	29	159	61 (44%)	291 (56%)
2	40	309	52 (61%)	447 (39%)
3	35	479	138 (42%)	786 (58%)
4	36	582	190 (37%)	1395 (63%)
5	40	1100	167 (51%)	1328 (49%)

Uncertainties in the diameter and length of the ZnO nanorods are 5 and 20 nm, respectively [86].

decay constant is attributed to the radiative lifetime of the free exciton [90]. If the recombination process was dominated by nonradiative channels, the lifetime would decrease with increasing temperature, since the nonradiative channels would play a more important role with temperature. These tetrapods exhibited an increase in the photoluminescence decay time with increasing temperature, indicating that the radiative recombination is the dominant recombination process.

Owing momentum conservation, usually all excitons cannot couple with the incident radiation field to yield photons. Theoretically, only those excitons at $k = 0$ can recombine radiatively; however, in practice, not only excitons at $k = 0$ contribute to the radiative recombination. It is observed that the homogeneous exciton linewidth has a certain spectral width ΔE at finite temperature. Actually, excitons with kinetic energy smaller than ΔE, the spectral linewidth of the transition, could recombine radiatively [91]. Exciton radiative lifetime is a function of temperature as the exciton redistribution occurs at elevated temperature. Effect of temperature on exciton radiative lifetime depends on dimensionality of the investigated structures and the ratio between homogeneous free exciton linewidth and thermal energy kT. If the exciton linewidth is less than the thermal energy, the radiative lifetime has been shown to change as $\tau_r = \tau_0/r(T)$ where τ_0 is the radiative lifetime at $T = 0$ and $r(T)$ is a fraction of free exitons with kinetic energy smaller than kT. On increasing temperature of the system, the average kinetic energy of the free excitons increased, hence the ratio $r(T)$ of excitons decreases. Therefore, the exciton radiative lifetime will increase with temperature.

For three-dimensional systems such as the studied ZnO tetrapods and assuming a Maxwell–Boltzmann distribution, $r(T)$ is roughly proportional to $T^{-3/2}$ and is given by

$$r(T) = \frac{2}{\sqrt{\pi}} \int_0^{\Gamma/kT} \sqrt{E} \exp\left(-E\right) dE \tag{9.3.15}$$

where E denotes the free exciton energy. This results in the characteristic $\tau_r \alpha\, T^{3/2}$ dependence, often reported in bulk semiconductors [92]. In the opposite case when thermal energy is less than exciton linewidth, however, τ_r is proportional to the homogenous linewidth and, therefore, changes with temperature. The exciton linewidth $\Gamma(T)$, dependence is mainly determined by interactions of excitons with acoustic and optical phonons as follows [91]

$$\Gamma(T) = \Gamma_0 + \gamma_a T + \frac{\Gamma_{\mathrm{LO}}}{\left[\exp\left(\hbar\omega_{\mathrm{LO}}/kT\right) - 1\right]} \tag{9.3.16}$$

Here, $\hbar\omega_{\mathrm{LO}}$ is the energy of a LO phonon, Γ_0 is the zero temperature broadening parameter, γ_a is the coupling strength of an exciton–acoustic phonon interaction, and Γ_{LO} is a parameter describing exciton–LO–phonon interaction.

In the case of ZnO nanorods, the exciton radiative lifetime increases with temperature (T^2). Furthermore, the spectral linewidth of the photoluminescence of the ZnO nanorods also increases with temperature as T^2, suggesting a linear dependence of exciton radiative lifetime on the spectral line width. A plot between the exciton radiative lifetime and the spectral line width is given in Figure 9.3.11, which shows a linear relationship between the spectral line width and radiative lifetime. Similar behavior is also reported by Feldmann *et al.* in GaAs quantum wells.

The physics behind is that the $k = 0$ oscillator strength is shared equally among all the states within the spectral line width ΔE [91]. That means the exciton radiative recombination rate is not solely determined by the exciton oscillator strength but depends on the coherence extension of an exciton, which decreases with temperature due to scattering by phonons, defects, or impurities.

Emission lifetime is also a function of excitation energies. The picosecond TRPL measurements of star-shaped ZnO nanostructures show a biexponential decay behavior, which is strongly dependent on the excitation intensity; the slow decay term decreased faster than the fast decay term as the excitation intensity increased and the emission decays were dominated by the fast one [94]. It was also reported that the emission decays decreased superlinearly before the appearance of the

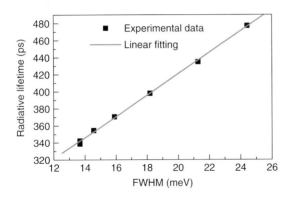

Figure 9.3.11 Spectral linewidth dependence of the exciton radiative lifetime. The solid squares stand for experimental data. The linear fit of the experimental data is given in Ref. [93]. FWHM, full width at half maximum.

stimulated emission (SE). This behavior may be used to deduce the threshold of SE or lasing. Wilkinson *et al.* [95] reported that a free exciton in ZnO single crystal has a lifetime of 403 ps at room temperature, and the lifetime of trapped carriers is approximated to 3.8 ns. Bauer *et al.* [96] reported that a direct radiative recombination of excitons can take place with a time constant of 12 ps in ZnO nanocrystalline thin films. Moreover, Guo *et al.* [97] reported that the time constant for the capture of free excitons at the band-tail states is on the order of 30 ps, and a slower decay term of 100–400 ps is associated with radiative recombination of free or localized excitons. In fact, different lifetime behaviors have been observed for ZnO with a fast time constant ranging from ∼10 to several tens of picoseconds and a slow term ranging from several hundreds of picoseconds to nearly 20 ns, and the reported time decay is usually smaller than 100 ps when a stimulated emission or a lasing action took place in ZnO nanocrystals [98, 99].

Study of ultrafast carrier dynamics in exciton–exciton scattering and electron–hole plasma regimes of ZnO nanoneedles shows very fast decay time (1 ps) of the photoluminescence. Even though no structure is detected in the time-integrated spectra of the electron hole plasma emission, the TRPL photoluminescence reveals the coexistence of the excitons and free carriers. The shortest decay time previously reported for ZnO lasing is 3 ps [100]. The similar rise time of 1 ps for the electron–hole plasma regime is reported in other cases too [101]. The decay curve of the stimulated emission in the exciton–exciton regime shows a similar rise time and no significant delay time, unlike ZnO tetrapod nanostructures [44] and nanoribbons and nanowires [40]. Longer delay time until the onset of exciton–exciton lasing was previously attributed to the longer time for the weak exciton–exciton interaction to produce sufficient scattering events [40].

Ozgur *et al.* [50] investigated the effect of annealing and stimulated emission on carrier dynamics in RF-sputtered ZnO thin films at room temperature and at 85 K using TRPL spectroscopy. The exciton densities were kept slightly below the threshold $I_{th}(\sim 30\ \mu J\ cm^{-2})$ to measure the spontaneous emission decay times, while high excitation densities ($\sim$200 µJ cm^{-2}) were used to observe the recombination dynamics under the influence of stimulated emission, single exponential decay fit revealed the spontaneous recombination times as 74, 59, and 30 ps for the samples annealed at 1000, 950, and 800 °C, respectively. Increment in the decay times with annealing temperature suggests a reduction in the density of nonradiative recombination centers. As expected, the stimulated-emission-induced recombination occurs very fast ($<$30 ps). The decay curve for above excitation threshold also show a much weaker and slower decaying component visible after the stimulated emission is over ($\sim$55 ps) with the characteristic decay times of the spontaneous recombination.

Reynolds *et al.* [102] measured the recombination lifetime of the allowed (Λ_5) and forbidden-free excitons (Λ_6 allowed by induced strain) at 2 K in a strained single crystal ZnO grown by the hydrothermal method as 259 and 245 ps, respectively. The lifetime of the Λ_5 exciton was slightly higher, 322 ps, for an unstrained sample. They noted that free exciton lifetimes are determined not only by the radiative decay but also by the nonradiative decay and capture processes leading to bound excitons. In general, pure ZnO shows n-type conductivity, while the Zn interstitials (Zn_i) and/or Oxygen vacancies (V_O) are the sources of free electrons. However, the role of neutral and negatively charged defects has not been clarified yet, although their presence has been pointed out [103]. In case of semiconductor positrons, the annihilation technique is mostly used for the study of neutral/charged defects [104]. In case of ZnO, Zn vacancy (V_{Zn}) is one of the most probable candidates. Because the momentum distribution of electrons in such defects differs from that in bulk materials, the annihilating positron–electron pairs show Doppler broadening for the 511 keV annihilation γ-rays. The resultant spectrum is characterized by the *S* parameter, which mainly reflects the fraction of annihilation of positron–electron pairs with a low momentum distribution. It was reported that the epitaxial ZnO layers exhibited shorter carrier lifetimes because of the higher defect densities compared to the bulk samples. Detailed study was done by Koida *et al.* [105]. They studied the influences of point defects on the nonradiative processes in ZnO using steady-state and time-resolved photoluminescence spectroscopy, making a connection with the results of positron annihilation measurement in bulk and epitaxial ZnO layers. The single crystal sample showed biexponential behavior with decay constants of 0.97 and 14 ns, which were suggested to represent the free exciton lifetime and the free carrier lifetime including trapping and emission processes. Pure ZnO epitaxial films were grown by combinatorial laser molecular-beam epitaxy [106]. The single exponential decay time, which is mainly governed by nonradiative process, increased from 46 to 110 ps with increasing growth temperature from 570 to 800 °C. Although homoepitaxial film exhibits the shortest decay time (36 ps) in spite of the smallest number of point defects among the epilayers, the relation between the photoluminescence decay time and the point defect density is ambiguous. It could be concluded that the nonradiative process was induced by certain defect species, which occurs because of the presence of Zn vacancies such as vacancy complexes.

Decay curve analysis of high-quality ZnO epitaxial layers grown on sapphire by low-pressure metal organic vapor-phase epitaxy was investigated by Jung *et al.* [107]. They reported biexponential decay behavior of ZnO and the decay times extracted was 180 ps and 1 ns, due to the nonradiative and radiative excitonic recombination times, respectively, consistent with the measurements of Teke *et al.* [108] and Koida *et al.* [105] in bulk ZnO samples.

9.3.4
Concluding Remarks

The advent of lasers 50 years ago has put the spectroscopic techniques on a very high pedestal. Technology has succeeded in gaining control over the pulse width of the laser and achieved fascinating autosecond pulse width, which may help to monitor even the shortest ever event at the atomic level. One of the applications of lasers is the understanding of the relaxation dynamics of the excited species, which is only possible when one has ultrashort laser pulses. In this chapter, a brief illustration of the importance of TRPL was presented especially in terms of nanomaterials. TRPL gives a detailed insight into the dynamics of excited species, and the many aspects of other processes that are highly in demand for the quality control of nanomaterial-based products. One such important nanomaterial, ZnO, which is gaining importance due to its commercial aspects, was also discussed.

References

1. Miller, R.C., Kleinman, D.A., Tsang, W.T., and Gossard, A.C. (1981) *Phys. Rev. B*, **24**, 1134.
2. Rossetti, R., Nakahara, S., and Brus, L.E. (1983) *J. Chem. Phys.*, **79**, 1086.
3. Brus, L.E. (1984) *J. Chem. Phys.*, **80**, 4403.
4. Misawa, K., Yao, H., Hayashi, T., and Kobayashi, T. (1991) *J. Chem. Phys.*, **94**, 4131.
5. Guo, L., Yang, S., Yang, C., Yu, P., Wang, J., Ge, W., and Wong, G.K.L. (2000) *Appl. Phys. Lett.*, **76**, 2901.
6. Morkoç, H. and Özgür, Ü. (2009) *Zinc Oxide: Fundamentals, Materials and Device Technology*, Wiley-VCH Verlag GmbH.
7. Gilliland, G.D. (1997) *Mater. Sci. Eng.*, **18**, 99–400.
8. Yacobi, B.G. and Holt, D.B. (1990) *Cathodoluminescence Microscopy of Inorganic Solids*, 1st edn, Plenum Press, New York.
9. Glinka, Y.D., Shahbazyan, T.V., Peraki, I.E., Talk, N.H., Liu, X.Y., Sasaki, Y., and Furdyna, J.K. (2002) *Appl. Phys. Lett.*, **81**, 3717.
10. Mikhailovsky, A.A., Malko, A.V., Hollingsworth, J.A., Bawendi, M.G., and Klimov, V.I. (2002) *Appl. Phys. Lett.*, **80**, 2380.
11. Klimov, V.I., Mikhailovsky, A.A., Xu, S., Malko, A., Hollingsworth, J.A., Leatherdale, C.A., Eisler, H.J., and Bawendi, M.G. (2000) *Science*, **290**, 314.
12. Gil, B. and Kavokin, A.V. (2002) *Appl. Phys. Lett*, **81**, 748.
13. Norris, D.J., Efros, A.L., Rosen, M., and Bawendi, M.G. (1996) *Phys. Rev. B*, **53**, 16347.
14. Colvin, V.L., Alivisatos, A.P., and Shank, C.V. (1994) *Phys. Rev. B*, **49**, 14435.
15. Chichibu, S., Azuhata, T., Sota, T., Amano, H., and Akasaki, I. (1997) *Appl. Phys. Lett.*, **70**, 2085.

16. Shikanai, A., Azuhata, T., Sota, T., Chichibu, S., Kuramata, A., Horina, K., and Nakamura, S. (1997) *J. Appl. Phys.*, **81**, 417.

17. Foreman, J.V., Li, J., Peng, H., Choi, S., Everitt, H.O., and Liu, J. (2006) *Nano Lett.*, **6**, 1126.

18. Reynolds, D.C., Look, D.C., Jogai, B., Hoelscher, J.E., Sherriff, R.E., and Molnar, R.J. (2000) *J. Appl. Phys.*, **88**, 1460.

19. Bergman, J.P., Holtz, P.O., Monemar, B., Sun Da Ram, M., Merz, J.L., and Gossard, A.C. (1991) *Phys. Rev. B*, **43**, 4765.

20. Lefebvre, P., Allegre, J., and Mathieu, H. (1999) *Mater. Sci. Eng. B*, **59**, 307.

21. O'Connor, D.V. and Phillips, D. (1984) *Time Correlated Single Photon Counting*, Academic Press, London.

22. Becker, W., Bergmann, A., Biskup, C., Kelbauskas, L., Zimmer, T., Klöcker, N., and Benndorf, K. (2003) *Proc. SPIE*, **4963**, 1–10.

23. Ballew, R.M. and Demas, J.N. (1989) *Anal. Chem.*, **61**, 30–33.

24. Mahr, H. and Hirsch, M.D. (1975) *Opt. Commun.*, **13**, 96.

25. Bagnall, D.M., Chen, Y.F., Zhu, Z., Yao, T., Shen, M.Y., and Goto, T. (1998) *Appl. Phys. Lett.*, **73**, 1038–1040.

26. Tang, Z.K., Wong, G.K.L., Yu, P., Kawasaki, M., Ohtomo, A., Koinuma, H., and Segawa, Y. (1998) *Appl. Phys. Lett.*, **72**, 3270–3272.

27. Zu, P., Tang, Z.K., Wong, G.K.L., Kawasaki, M., Ohtomo, A., Koinuma, H., and Segawa, Y. (1997) *Solid State Commun.*, **103**, 459.

28. Bagnall, D.M., Chen, Y.F., Zhu, Z., Yao, T., Koyama, S., Shen, M.Y., and Goto, T. (1997) *Appl. Phys. Lett.*, **70**, 2230–2232.

29. Reynolds, D.C., Look, D.C., and Jogai, B. (1996) *Solid State Commun.*, **99**, 873.

30. Hayashi, H., Ishizaka, A., Haemori, M., and Koinuma, H. (2003) *Appl. Phys. Lett.*, **82**, 1365.

31. Ng, H.T., Chen, B., Li, J., Han, J., Meyyappan, M., Wu, J., Li, S.X., and Haller, E.E. (2003) *Appl. Phys. Lett.*, **82**, 2023.

32. Sharma, P., Sreenivas, K., and Rao, K.V. (2003) *J. Appl. Phys.*, **93**, 3963.

33. Look, D.C. (2001) *Mater. Sci. Eng. B*, **80**, 383.

34. (a) Baranov, P.G., Romanov, N.G., Tolmachev, D.O., Orlinskii, S.B., Schmidt, J., de Mello Donega, C., and Meijerink, A. (2006) *JETP Lett.*, **84**, 400; (b) Full Text via CrossRef | View Record in Scopus | Cited By in Scopus (1) Baranov, P.G., Romanov, N.G., Tolmachev, D.O., Orlinskii, S.B., Schmidt, J., de Mello Donega, C., and Meijerink, A. (2006) *JETP Lett.*, **84**, 400.

35. Singh, S., Thiyagarajan, P., Mohan Kant, K., Anita, D., Thirupathiah, S., Rama, N., Tiwari, B., Kottaisamy M., and Ramachandra Rao, M.S. (2007) *J. Phys. D: Appl. Phys.*, **40**, 6312–6327.

36. Jung, K.Y., Kang, Y.C., and Park, S.B. (1997) *J. Mater. Sci. Lett.*, **22**, 1848.

37. Wang, Z.L. (2004) *Nanostruct. Zinc Oxide, Mater. Today*, **7**, 26–33.

38. Huang, M.H., Mao, S., Feick, H., Yan, H., Wu, Y., Kind, H., Weber, E., Russo, R., and Yang, P. (2001) *Science*, **292**, 1897.

39. Liu, C., Zapien, J.A., Yao, Y., Meng, X., Lee, C.S., Fan, S., Lifshitz, Y., and Lee, S.T. (2003) *Adv. Mater.*, **15**, 838.

40. Johnson, J.C., Yan, H., Yang, P., and Saykally, R.J. (2003) *J. Phys. Chem. B*, **107**, 8816.

41. Yan, H., Johnson, J., Law, M., He, R., Knutsen, K., McKinney, J.R., Pham, J., Saykally, R., and Yang, P. (2003) *Adv. Mater.*, **15**, 1907.

42. Yu, S.F., Yuen, C., Lau, S.P., Park, W.I., and Yi, G.-C. (2004) *Appl. Phys. Lett.*, **84**, 3241.

43. Choy, J.-H., Jang, E.-S., Won, J.-H., Chung, J.-H., Jang, D.-J., and Kim, Y.-W. (2003) *Adv. Mater.*, **15**, 1911.

44. Qiu, Z., Wong, K.S., Wu, M., Lin, W., and Xu, H. (2004) *Appl. Phys. Lett.*, **84**, 2739.

45. Leung, Y.H., Kwok, W.M., Djurisic, A.B., Phillips, D.L., and Chan, W.K. (2005) *Nanotechnology*, **16**, 579.

46. Sun, X.W., Yu, S.F., Xu, C.X., Yuen, C., Chen, B.J., and Li, S. (2003) *Jpn. J. Appl. Phys.*, **42**, L1229.

47. Kwok, W.M., Djurişiæ, A.B., Leung, Y.H., Chan, W.K., and Phillips, D.L. (2005) *Appl. Phys. Lett.*, **87**, 093108.

48. Kong, X.Y. and Wang, Z.L. (2003) *Nano Lett.*, **3**, 1625.

49. Yan, H., He, R., Johnson, J., Law, M., Saykally, R.J., and Yang, P. (2003) *J. Am. Chem. Soc.*, **125**, 4728.

50. Ozgur, U., Teke, A., Liu, C., Cho, S.J., Morkoc, H., and Everitt, H.O. (2004) *Appl. Phys. Lett.*, **84**, 3223–3225.

51. Wang, L. and Giles, N.C. (2003) *J. Appl. Phys.*, **94**, 973.

52. Leiter, F., Alves, H., Pfisterer, D., Romanov, N.G., Hofmann, D.M., and Meyer, B.K. (2003) *Phys. B*, **340–342**, 201.

53. Jung, S.W., Park, W.I., Cheong, H.D., Gyu-Chul, Y., Hyun, M.J., Hong, S., and Joo, T. (2002) *Appl. Phys. Lett.*, **80**, 1924–1926.

54. Vanheusden, K., Warren, W.L., Seager, C.H., Tallant, D.R., Voigt, J.A., and Gnade, B.E. (1996) *J. Appl. Phys.*, **79**, 7983.

55. van Dijken, A., Meulenkamp, E.A., Vanmaekelbergh, D., and Meijerink, A. (2000) *J. Phys. Chem. B*, **104**, 1715.

56. Mende, L.S. and Driscoll, J.L.M. (2007) *Mater. Today*, **10**, 40–48.

57. Han, J. *et al.* (2002) *J. Eur. Ceram. Soc.*, **22**, 49.

58. Hagemark, K.I. (1976) *J. Solid State Chem.*, **16**, 293.

59. Kröger, F.A. (1974) *The Chemistry of Imperfect Crystals*, 2nd edn, North Holland, Amsterdam, p. 73.

60. Shalish, I., Temkin, H., and Narayanamurti, V. (2004) *Phys. Rev. B*, **69**, 245401.

61. Hsu, N.E., Hung, W.K., and Chen, Y.F. (2004) *J. Appl. Phys.*, **96**, 4671.

62. Li, D., Leung, Y.H., Djurisic, A.B., Liu, Z.T., Xie, M.H., Shi, S.L., Xu, S.J., and Chan, W.K. (2004) *Appl. Phys. Lett.*, **85**, 1601–1603.

63. Djurisic, A.B., Leung, Y.H., Choy, W.C.H., Cheah, K.W., and Chan, W.K. (2004) *Appl. Phys. Lett.*, **84**, 2635–2637.

64. Song, J.K., Szarko, J.M., Leone, S.R., Li, S., and Zhao, Y. (2005) *J. Phys. Chem. B*, **109**, 15749.

65. Kwok, W.M., Leung, Y.H., Djurisic, A.B., Chan, W.K., and Phillips, D.L. (2005) *Appl. Phys. Lett.*, **87**, 093108.

66. Xu, C.X., Sun, X.W., Yuen, C., Yu, S.F., and Dong, Z.L. (2005) *Appl. Phys. Lett.*, **86**, 011118.

67. Sarko, J.M., Song, J.K., Blackledge, C.W., Swart, I., Leone, S.R., Li, S., and Zhao, Y. (2005) *Chem. Phys. Lett.*, **404**, 171.

68. Djurisic, A.B., Kwok, W.M., Leung, Y.H., Chan, W.K., and Phillips, D.L. (2005) *J. Phys. Chem. B*, **109**, 19228.

69. Kato, H., Sano, M., Miyamoto, K., Yao, T., Zhang, B.-P., Wakatsuki, K., and Segawa, Y. (2004) *Phys. Status Solidi B*, **241**, 2825.

70. Meyer, B.K., Alves, H., Hofmann, D.M., Kreigseis, W., Forster, D., Bertram, F., Christen, J., Hoffmann, A., Strasburg, M., Dworzak, M., Haboeck, U., and Rodina, A.V. (2004) *Phys. Status Solidi B*, **241**, 231.

71. Cao, L., Zou, B., Li, C., Zhang, Z., Xie, S., and Yang, G. (2004) *Europhys. Lett.*, **68**, 740.

72. Kwok, W.M., Leung, Y.H., Djurišić, A.B., Chan, W.K., and Phillips, D.L. (2005) *Appl. Phys. Lett.*, **87**, 093108.

73. Kwok, W.M., Djurisic, A.B., Leung, Y.H., Chan, W.K., Phillips, D.L., Chen, H.Y., Wu, C.L., Gwo, S., and Xie, M.H. (2005) *Chem. Phys. Lett.*, **412**, 141.

74. Takeda, J., Kurita, S., Chen, Y., and Yao, T. (2001) *Int. J. Mod. Phys. B*, **15**, 3669.

75. Yamamoto, A., Kido, T., Goto, T., Chen, Y., Yao, T., Kasuya, A., and Cryst, J. (2000) *Growth*, **214**, 308.

76. Johnson, J.C., Knutsen, K.P., Yan, H., Law, M., Zhang, Y., Yang, P., and Saykally, R.J. (2004) *Nano Lett.*, **4**, 197.

77. Kwok, W.M., Djurišića, A.B., Leung, Y.H., Chan, W.K., and Phillips, D.L. (2005) *Appl. Phys. Lett.*, **87**, 223111.

78. Djurišić, A.B., Kwok, W.M., Leung, Y.H., Chan, W.K., and Phillips, D.L. (2005) *J. Phys. Chem. B*, **109**, 19228.

79. Zhang, X.Q., Suemune, I., Kumano, H., Yao, Z.G., and Huang, S.H. (2008) *J. Lumin.*, **122–123**, 828.

80. Hsu, H.-C., Wu, C.-Y., and Hsieh, W.-F. (2005) *J. Appl. Phys.*, **97**, 064315.

81. Fonoberov, V.A. and Balandin, A.A. (2005) *Appl. Phys. Lett.*, **86**, 226101.

82. Fonoberov, V.A. and Balandin, A.A. (2004) *Appl. Phys. Lett.*, **85**, 5971.

83. Zhang, X.H., Chua, S.J., Yong, A.M., Chow, S.Y., Yang, H.Y., Lau, S.P., and Yu, S.F. (2006) *Appl. Phys. Lett.*, **88**, 221903.

84. Itoh, T., Furumiya, M., Ikehara, T., and Gourdon, C. (1990) *Solid State Commun.*, **73**, 271.

85. Kayanuma, Y. (1988) *Phys. Rev. B*, **38**, 9797.

86. Hong, S., Joo, T., Park, W., Ho Jun, Y., and Yi, G.C. II (2003) *Appl. Phys. Lett.*, **83**, 4157.

87. Hong, S., Joo, T., Park, W.I., Jun, Y.H., and Yi, G.-C. (2003) *Appl. Phys. Lett.*, **83**, 4157.

88. Chichibu, S.F., Onuma, T., Kubota, M., Uedono, A., Sota, T., Tsukazaki, A., Ohtomo, A., and Kawasaki, M. (2006) *J. Appl. Phys.*, **99**, 093505.

89. Koida, T., Uedono, A., Tsukazaki, A., Sota, T., Kawasaki, M., and Chichibu, S.F. (2004) *Phys. Status Solidi A*, **201**, 2841.

90. Ozgur, U., Alivov, Y.I., Liu, C., Teke, A., Reshchikov, M.A., Dogan, S., Avrutin, V., Cho, S.-J., and Morkoc, H. (2005) *J. Appl. Phys.*, **98**, 041301.

91. Feldmann, J., Peter, G., Göbel, E.O., Dawson, P., Moore, K., Foxon, C., and Elliott, R.J. (1987) *Phys. Rev. Lett.*, **59**, 2337.

92. p'tHooft, G.W., van der Poel, W.A.J.A., Molenkamp, L.W., and Foxon, C.T. (1987) *Phys. Rev. B*, **35**, 8281.

93. Zhang, X.H., Chua, S.J., Yong, A.M., Yang, H.Y., Lau, S.P., Yu, S.F., Sun, X.W., Miao, L., Tanemura, M., and Tanemura, S. (2007) *Appl. Phys. Lett.*, **90**, 013107.

94. Li, C.P., Guo, L., Wu, Z.Y., Ren, L.R., Ai, X.C., Zhang, J.P., Lv, Y.Z., Xu, H.B., and Yu, D.P. (2006) *Solid State Commun.*, **139**, 355–359.

95. Wilkinson, J., Ucer, K.B., and Williams, R.T. (2004) *Rad. Meas.*, **38**, 501.

96. Bauer, C., Boschloo, G., Mukhtar, E., and Hagfeldt, A. (2004) *Chem. Phys. Lett.*, **387**, 176.

97. Guo, B., Ye, Z.Z., and Wong, K.S. (2003) *J. Cryst. Growth*, **253**, 252.

98. Soukoulis, C.M., Jiang, X.Y., Xu, J.Y., and Cao, H. (2002) *Phys. Rev. B*, **65**, 041103.

99. Huang, M.H., Mao, S., Feick, H., Yan, H.Q., Wu, Y.Y., Kind, H., Weber, E., Russo, R., and Yang, P.D. (2001) *Science*, **292**, 1897.

100. Sarko, M., Song, J.K., Blackledge, C.W., Swart, I., Leone, S.R., Li, S., and Zhao, Y. (2005) *Chem. Phys. Lett.*, **404**, 171.

101. Yamamoto, A., Kido, T., Goto, T., Chen, Y., Yao, T., and Kasuya, A. (2000) *J. Cryst. Growth*, **214**, 308.

102. Reynolds, D.C., Look, D.C., Jogai, B., Litton, C.W., Cantwell, G., and Harsch, W.C. (1999) *Phys. Rev. B*, **60**, 2340.

103. Kohan, F., Ceder, G., Morgan, D., and Van de Walle, C.G. (2000) *Phys. Rev. B*, **61**, 15019.

104. Krause-Rehberg, R. and Leipner, H.S. (1999) *Positron Annihilation in Semiconductors*, Solid-State Sciences, Vol. **127**, Springer, Berlin.

105. Koida, T., Chichibu, S.F., Uedono, A., Tsukazaki, A., Kawasaki, M., Sota, T., Segawa, Y., and Koinuma, H. (2003) *Appl. Phys. Lett.*, **82**, 532.

106. Ohnishi, T., Komiyama, D., Koida, T., Ohashi, S., Stauter, C., Koinuma, H., Ohtomo, A., Lippmaa, M., Nakagawa, N., Kawasaki, M., Kikuchi, T., and Omote, K. (2001) *Appl. Phys. Lett.*, **79**, 536.

107. Jung, S.W., Park, W.I., Cheong, H.D., Yi, G.-C., Jang, H.M., Hong, S., and Joo, T. (2002) *Appl. Phys. Lett.*, **80**, 1924.

108. Teke, A., Özgür, Ü., Doğan, S., Gu, X., Morkoç, H., Nemeth, B., Nause, J., and Everitt, H.O. (2004) *Phys. Rev. B*, **70**, 195207.

10
Photoacoustic Spectroscopy and Its Applications in Characterization of Nanomaterials

Kaushal Kumar, Aditya Kumar Singh, and Avinash Chandra Pandey

10.1
Introduction

Photoacoustic (also referred to as *optoacoustic*) spectroscopy is one among the optical characterization tools and is by far not as popular as other photometry techniques (such as absor ption or fluorescence). However, in the literature, it has been described as a gradually progressing method for materials characterization [1]. This technique has great potential to give an insight into the energetic balance of the photothermal and photochemical processes and also into some other aspects such as phase changes and defects in the materials. Nonetheless, the popularity of this technique has been obscured by the extensive mathematical descriptions, high sensitivity to the environment, and lack of suitable high-power tunable excitation sources. Other useless complications have also resulted from the number of different names such as PAC (photoacoustic calorimetry), PAS (photoacoustic spectroscopy), LIOAC (laser-induced optoacoustic calorimetry), LIOAS (laser-induced optoacoustic spectroscopy), PTRPA (pulsed time-resolved photoacoustics), and LPAS (laser photoacoustic spectroscopy) assigned to the same technique, using the same physical principle: the photoacoustic (PA) effect. Inappropriate names or nonstandardized language confuse the readers.

In the term *photoacoustic*, the prefix *photo* refers to *photons as excitation source* and *acoustic* refers to *generation of sound*. What, sound? Yes!! There is really a sound. The periodic light excitation of the specimen generates sound waves, and the technique that senses these sound waves is known as *photoacoustic spectroscopy*. In short, this is a technique of *listening to the interaction of radiation with matter*. The first attempt in this path was done by Alexander Graham Bell in 1881 [2, 3]. He discovered that when periodically chopped sunlight is focused on a sample in an airtight cell to which a hearing tube was connected, a sound effect was produced. He named this device as photophone. In addition, he noted a similar effect when infrared or ultraviolet light was used.

The heat (and hence rise in temperature) in the samples originates because of the nonradiative processes that occur after photon excitation. Any kind of sample that absorbs energy will liberate at least a part of the excitation energy in this way,

Nanomaterials: Processing and Characterization with Lasers, First Edition.
Edited by Subhash Chandra Singh, Haibo Zeng, Chunlei Guo, and Weiping Cai.
© 2012 Wiley-VCH Verlag GmbH & Co. KGaA. Published 2012 by Wiley-VCH Verlag GmbH & Co. KGaA.

and hence the methods to measure this nonradiative part is applicable to almost all types of samples. The liberated heat energy not only carries the information regarding the absorbed energy but also contains details regarding the thermal properties of the sample. A group of such spectroscopic methods based on the measurement of photoinduced heating of the sample are called the *photothermal methods* [4–9]. Hence, on the basis of photothermal spectroscopy one can measure the liberated heat. The thermalization of a sample or the medium as a result of nonradiative relaxation not only results change in temperature of the sample but also brings about changes in many other parameters such as density, pressure, refractive index, and so on. Hence, there exist a number of photothermal techniques depending on the mode of detection. Only the absorbed light energy contributes to the energy liberated in the form of heat. Scattered or reflected light will not contribute to the photothermal signals. Consequently, photothermal spectroscopy more accurately measures optical absorption in highly scattering solutions, in solids and at interfaces. PAS is one such photothermal technique.

Earlier, Bell's demonstration was evidently regarded as a curiosity of no functional or scientific value and was quickly forgotten. The discovery of laser in the year 1960 is credited with popularizing this technique, and at present, this technique has become one of the important tools to characterize the materials, especially nanomaterials. This novel technique has no restriction on the sample's physical state and is equally suitable for the study of biological systems and surfaces of solids. The technique can be applied to investigate absorption spectra, fluorescence yields, depth profiles, phase transitions, thermal conductivities, and many others [9–12]. The characterization of nanomaterials through this technique is very important, and this spectroscopy can visualize the dark part of the nanomaterials. The nanomaterials possess higher electron–phonon coupling that gives rise to large nonradiative relaxations, and hence nanomaterials have become an important tool to identify maligned cells and their treatment in the living body. The PA microscopy of the living body is the hottest topic of research nowadays. The future is for nanomaterial-based devices, so it is very important to study light-to-heat conversion in nanomaterials using the PA technique. This technique is very simple, easy to fabricate in the laboratory with least instrumentation, applicable to materials in any phase (solid, liquid, and gas), and finally has a large impact.

This chapter briefly describes the PA technique and its application in material characterization. The author's main aim is to familiarize the reader with this technique, its fabrication in the laboratory, and characterization of solid samples.

10.1.1
Theory of the Signal Generation

Rosencwaig and Gersho (R–G) modified earlier theories and accounted the universally accepted one for all observed effects in PAS of solids. The detailed discussion of the theory can be found in the original work [13]. Here, the author has given only the main outcome of the theory and the basic formulae useful for the analysis of solid-phase samples.

R–G suggested a simple model based on previous arguments and derived a mathematical expression for the resulting pressure fluctuations. The model is sufficient to describe the PA signal generation in condensed matter. According to the R–G theory, with a gas microphone detection of the PA signal, the signal depends on the generation of an acoustic pressure disturbance at the sample–gas interface. The generation of the surface pressure disturbance, in turn, depends on the periodic temperature at the sample–gas interface. Exact expressions for this temperature are derived in the R–G theoretical model, but the transport of the acoustic disturbance in the gas is treated in an approximate heuristic manner, which is, however, valid in most experimental conditions. The formulation of the R–G model is based on the light absorption and thermal wave propagation in an experimental configuration as shown in Figure 10.1. Here, the sample is considered to be in the form of a disc of thickness l. It is assumed that the back surface of the sample is in contact with a backing material of thickness l_b and the front surface is in contact with a gas column of length l_g. It is further assumed that both gas and backing material are not light absorbing. The following are the parameters used in the R–G model

- k: the thermal conductivity (cal cm^{-1} s^{-1} °C^{-1})
- ρ: the density (g cm^{-3})
- C: the specific heat capacity (cal g^{-1} °C^{-1})
- $\alpha = k/\rho C$: thermal diffusivity (cm^2 s^{-1})
- $a = \sqrt{\frac{\omega}{2\alpha}}$: thermal diffusion coefficient (cm^{-1})
- $\mu = 1/a$: the thermal diffusion length (cm)

where $\omega = 2\pi f$, with f the modulation frequency of the incident light beam.

The intensity I of the incident radiation on the surface is assumed to have the form

$$I = \frac{1}{2} I_0 \left(1 + \cos \omega t\right) \tag{10.1}$$

When this sinusoidally modulated light beam of intensity I_0 is incident on a solid sample having an absorption coefficient β, the heat density generated at any point

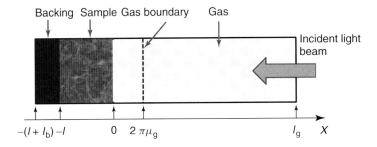

Figure 10.1 Schematic representation of photoacoustic experimental configuration.

due to the light absorbed can be represented by

$$H = \frac{1}{2}\beta I_0 e^{\beta x} \left(1 + \cos \omega t\right) \tag{10.2}$$

The thermal diffusion equation in the solid (taking into account the distributed heat source) can be written as

$$\frac{\partial^2 \theta}{\partial x^2} = \frac{1}{\alpha}\frac{\partial \theta}{\partial t} - \frac{\beta I_0 \eta}{2k} e^{\beta x} \left(1 + e^{i\omega t}\right) \qquad \text{for } -l \le x \le 0 \tag{10.3}$$

For the backing material and the gas, the heat diffusion equations may be written as

$$\frac{\partial^2 \theta}{\partial x^2} = \frac{1}{\alpha_b}\frac{\partial \theta}{\partial t} \qquad \text{for } -\left(l_b + l\right) \le x \le -l \tag{10.4}$$

$$\frac{\partial^2 \theta}{\partial x^2} = \frac{1}{\alpha_g}\frac{\partial \theta}{\partial t} \qquad \text{for } 0 \le x \le l_g \tag{10.5}$$

where θ is the temperature and η the light-to-heat conversion efficiency. Here, the subscripts b and g represent the backing and gas, respectively. The real part of the complex-valued solution $\theta(x, t)$ of Eqs. (10.3–10.5) is the solution of physical interest and represents the temperature in the cell relative to ambient temperature as a function of position and time.

After imposing appropriate boundary conditions for the temperature and heat flux continuity and neglecting heat flow through convection in the gas under steady-state conditions, the explicit solution to the complex amplitude of the periodic temperature at the solid–gas boundary can be obtained as

$$\theta_0 = \frac{\beta I_0}{2k \left(\beta^2 - \sigma^2\right)}$$
$$\times \left[\frac{(\gamma - 1)(b + 1) e^{\sigma l} - (\gamma + 1)(b - 1) e^{-\sigma l} + 2(b - \gamma) e^{-\beta l}}{(g + 1)(b + 1) e^{\sigma l} - (g - 1)(b - 1) e^{-\sigma l}}\right] \tag{10.6}$$

where

$$b = \frac{k_b a_b}{ka}, \qquad g = \frac{k_g a_g}{ka}, \qquad r = (1 - i)\frac{\beta}{2a}, \qquad \sigma = (1 + i) a$$

Periodic heat flow from the solid to the surrounding gas produces a periodic temperature variation in the gas. The time-dependent component of the temperature in the gas attenuates rapidly to zero with increasing distance from the surface of the solid. At a distance of $2\pi\mu_g$, where μ_g is the thermal diffusion length of the gas, the periodic temperature variation in the gas is effectively fully damped out. Thus, there is a boundary layer of gas, which is only capable of responding thermally to the periodic temperature on the surface of the sample. This layer of gas expands and contracts periodically and thus can be thought of as acting as an acoustic piston on the rest of the gas column, producing an acoustic pressure signal that travels through the entire gas column. Assuming that the rest of the gas responds adiabatically to the action of the acoustic piston, the adiabatic gas law can be used to derive an expression for the complex envelope of the sinusoidal pressure variation Q as

$$Q = \frac{\gamma P_0 \theta_0}{\sqrt{2} T_0 l_g a_g} \tag{10.7}$$

with θ_0 given by Eq. (10.6). Symbols γ, P_o, and T_o are the ratio of heat capacities of air, ambient pressure, and temperature, respectively.

Equation (10.7) can be used to evaluate the magnitude and phase of the acoustic pressure wave produced in the cell due to the PA effect. However, a useful interpretation of the above equation is rather difficult in the present form. Hence, some special cases, according to the experimental conditions, have to be considered to get a clear physical insight. In fact, three lengths related to the sample, namely, the physical length l, thermal diffusion length μ, and optical absorption length $l_\beta = (1/\beta)$ can be made use of in arriving at different special cases.

10.1.2
Optically Transparent Solids ($l_\beta > l$)

- **Case 1(a): Thermally thin solids ($\mu \gg l$; $\mu > l_\beta$)** This condition results in

$$Q = \frac{(1-i)\beta l}{2a_g} \left(\frac{\mu_b}{kb}\right) Y \tag{10.8}$$

with

$$Y = \frac{\gamma P_0 I_0}{2\sqrt{2} T_0 l_g} \tag{10.9}$$

Here, the acoustic signal is proportional to βl and varies as f^{-1}. Moreover, the signal is now determined by thermal properties of the backing material.

- **Case 1(b): Thermally thin solids ($\mu > l$; $\mu < l_\beta$)** This case is different from the earlier one in that while the thermal diffusion length of the sample is larger than the length of the sample it is smaller than the absorption length in the sample. Under approximations, the signal behaves in the same manner as in the previous case.

$$Q = \frac{(1-i)\beta l}{2a_g} \left(\frac{\mu_b}{kb}\right) Y \tag{10.10}$$

- **Case 1(c): Thermally thick solids ($\mu < l$; $\mu \ll l_\beta$)**

$$Q = -i\frac{\beta\mu}{2a_g} \left(\frac{\mu}{k}\right) Y \tag{10.11}$$

Only light absorbed within the first thermal diffusion length contributes to the signal in spite of the fact that light is being absorbed throughout the length of the sample. The signal varies with frequency as $f^{-3/2}$.

10.1.3
Optically Opaque Solids ($I_\beta \ll I$)

In this case, most of the light is absorbed within a very small thickness of the sample near the front surface.

- **Case 2(a): Thermally thin solids ($\mu \gg I$; $\mu \gg I_\beta$)**

$$Q = \frac{(1-i)}{2a_g}\left(\frac{\mu_b}{kb}\right)Y \tag{10.12}$$

This expression shows that the acoustic signal is independent of β of the absorption coefficient as the whole radiation energy is absorbed. The frequency dependence is of the type f^{-1}.

- **Case 2(b): Thermally thick solids ($\mu < I$; $\mu > I_\beta$)**

$$Q = \frac{(1-i)}{2a_g}\left(\frac{\mu}{k}\right)Y \tag{10.13}$$

This is analogous to Eq. (10.12), but the sample now depends on the thermal properties of the sample rather than on the backing material as in previous case. Again the signal varies as f^{-1}.

- **Case 2(c): Thermally thick solids ($\mu \ll I$; $\mu < I_\beta$)**

$$Q = -i\frac{\beta\mu}{2a_g}\left(\frac{\mu}{k}\right)Y \tag{10.14}$$

It shows that the acoustic signal is proportional to the absorption coefficient and the thermal properties of the sample. It also varies as $f^{-3/2}$. A schematic representation of the different cases is given in Figure 10.2.

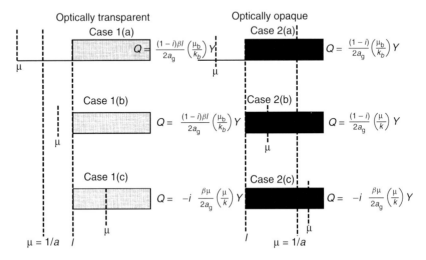

Figure 10.2 Schematic representation of various cases of optically transparent solids and optically opaque solids.

The cases discussed so far can be used in the PA study of any kind of sample. One of the important results of the R–G theory is that the PA signal is always linearly proportional to the incident light intensity, irrespective of the sample properties and cell geometry. In cases 2(a) and 2(b), we have seen that the PA signal is independent of the optical absorption coefficient of the sample. In these cases, therefore, the only term in Eqs. (10.12) and (10.13) that depends on the wavelength of the incident radiation is the light source intensity I_0. Thus, it is clear that the PA spectrum of an optically opaque sample ($\mu > l_\beta$) is simply the power spectrum of the light source.

10.1.4
Three-Dimensional Heat Flow Model

The R–G theory gives a simplified one-dimensional heat flow model. Quimby and Yen [14] recently studied the heat flow problem in three dimensions both theoretically and experimentally. They concluded that if the lateral dimensions of the cell are equal to or less than a few times the thermal diffusion length of the gas, the one-dimensional treatment is not satisfactory. A lower limit is obtained for the chopping frequency for a given cell diameter below which the one-dimensional theory is no longer adequate. This minimum frequency is

$$f_{min} = \frac{1}{2} \frac{\alpha_g}{D^2} \tag{10.15}$$

where D is the diameter of the sample chamber and α_g is the thermal diffusivity of the gas. Quimby and Yen found that the one-dimensional theory of the PA effect in solid samples is applicable only if the thermal diffusion length of the gas is lesser than the radius of the sample chamber.

10.1.5
Thermal Diffusivity

According to the one-dimensional heat flow model of R–G, the pressure fluctuation in the air inside the chamber is given by

$$P = \frac{\gamma P_0 I_0 \left(a_g a_s\right)^{1/2}}{2\pi l_g T_0 k_s f \sin h \left(l_s \sigma_s\right)} e^{i(\omega t - \pi/2)} \tag{10.16}$$

where γ is the ratio of specific heat capacities of air, P_0 and T_0 are the ambient pressure and temperature, I_0 is the radiation intensity, f is the modulation frequency, and l_j, k_j, and α_j are the length, thermal conductivity, and the thermal diffusivity of the medium. j = g refers to the gas, and j = s refers to the solid sample. $\sigma_j = (1 + i)a_j$, where $a_j = (\pi f / \alpha_j)^{1/2}$ is the thermal diffusion coefficient of the medium j.

For a thermally thin sample (i.e., $l_s a_s <<< 1$), under approximation, the above expression implies that the PA signal amplitude for the thermally thin sample varies as $f^{-3/2}$, and the phase is insensitive to the variation in the modulation frequency. For a thermally thick sample, the amplitude of the PA signal decreases exponentially with the modulation frequency, while the phase decreases linearly with $b\sqrt{f}$. Hence, thermal diffusivities of sample (α_s) can be evaluated either from the amplitude data or from the phase response with respect to the modulation frequency.

10.1.6
Saturation Effect in PAS

In PAS, the problem of saturation of the signal tends to obscure the relative intensities of the signals at different wavelengths. If the sample has a high value of the absorption coefficient, the PA spectrum becomes almost independent of β. According to the theoretical model suggested by R–G, only that part of the heat produced in the sample contributes to the signal, which is generated within a depth $\leq \mu$ (μ is the thermal diffusion length of the sample) below the surface. The incident light also varies as $1/\beta$ in going inside the sample, and they observed that if the whole incident light energy is absorbed in the first thermal diffusion length (μ) then the PA signal becomes independent of the optical absorption coefficient of β . This accounts for the saturation. The saturation effect can be removed in either of the following ways:

1) By increasing the effective absorption length in the sample, that is, by increasing $l_\beta = (1/\beta)$.
2) By decreasing μ, the thermal diffusion length in the sample.

Dilution of the sample with nonabsorbing materials such as Al_2O_3, silica gel, and MgO has also been used to decrease the saturation effect. It is found that the saturation is decreased to a larger extent if the sample is ground many times the dilution.

Fuchsman and Silversmith [15] have shown that in the process of grinding, the solid samples get coated on the surface of the diluent particles. In addition, the sample is divided into smaller particles through grinding as revealed by electron microscope studies of the ground sample. The effect of the coating is to reduce the number of absorbing particles in the path of the incoming light, which results in an increase in the effective absorption length $l_\beta = (1/\beta)$. With suitable manipulation of $l_\beta > \mu$, the saturation disappears, and the characteristic peaks of the sample are revealed. Fuchsman and Silversmith have used this technique to remove saturation in tetraphenylporphyrin employing Al_2O_3 as diluents.

10.1.7
Photoacoustic versus Absorption Spectroscopy

The PAS is equivalent to absorption spectroscopy if all the absorbed energy at each wavelength is lost completely by nonradiative processes. There are many differences, however, both in the two spectra as well as in their detection method.

The absorption spectrum is recorded using photomultiplier tubes whose response is proportional to the photon flux, that is, number of photons per unit area, whereas the PA spectrum is obtained indirectly from a microphone reading. The latter is thus proportional to the power of the incident radiation, that is, it depends not only on the number of photons per unit area but also on the energy of the individual photon. It can be easily verified that a photon at 3000 Å can generate twice as much heat as a photon of 6000 Å. The PAS spectrum of carbon black for a given continuous source has been compared with the spectrum of the same source recorded using a radiometer detector. The two spectra are found completely identical. This clearly demonstrates that the PA spectrum is a power spectrum. Since carbon black is an almost perfect absorber, it is normally used for the production of a reference signal for normalization; when the spectrum obtained by a PA spectrometer is divided by the power spectrum of the source then the actual signal is obtained. The PAS therefore provides much more information than the optical absorption spectrum.

10.2
Instrumentation

A common PA spectrometer consists of four parts: a periodic (modulated or pulsed) source of illumination, sample chamber, a means of detecting the acoustic signal, and the data acquisition system. A typical spectrometer is shown in Figure 10.3. Various modifications of this fundamental instrumentation have been used to perform a wide variety of PA experiments. There are two kinds of excitation sources available: continuous and pulsed. The measurement can be taken either in scanning mode (obtaining PAS intensity vs wavelength spectra) or in fixed wavelength mode (obtaining PAS intensity vs modulation frequency graph). The PA spectrometer can work either in modulated continuous wave source excitation mode or pulsed source excitation mode.

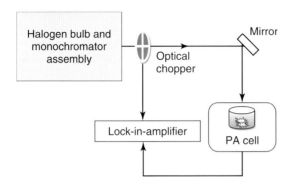

Figure 10.3 A typical spectrometer design.

10.2.1
Modulated Continuous Wave Source Spectrometer

This spectrometer uses a continuous wave radiation source modulated by an external chopping device that can measure the absorption spectra of the samples over the wavelength region of the source. This type of spectrometer is very easy to fabricate in the laboratory. With some modifications in the sample cell, the spectrometer can also measure the thermal conduction property of the samples. The main parts of the spectrometer are discussed below.

10.2.1.1 Radiation Sources

The conventional excitation source for obtaining the PAS intensity versus wavelength spectrum is a lamp and monochromator assembly and is the most popular. To obtain the spectra in the ultraviolet-visible-near-infrared region, one can use either a xenon arc lamp or a tungsten halogen bulb in conjunction with a low f number monochromator. The lamp with monochromator can provide continuous tunability from ultraviolet to infrared. The major limitations of this source are low bandwidth and low output power. But lamp–monochromator assembly is mostly used for obtaining spectra in broad range with low resolution. Since signal-to-noise ratio depends on the amount of light falling on the sample, high-power light sources are highly needed. Also, a low f number monochromator that gives high-power light throughout is desirable. In Table 10.1, some suitable lamps for a spectrometer are given.

Lasers have high spectral resolution and power, but they have limited tunability (and fixed wavelength). A laser requires no monochromator, and if it is operated in pulsed mode, it requires no chopper. The laser beam is highly collimated and is advantageous for many cell configurations. The lasers are available in the continuous wave and pulsed wave mode. In the visible region, dye lasers provide a fairly large wavelength range, and with the aid of frequency doubling crystals the range can be extended to the UV region. For the IR region, the recently developed semiconductor and quantum cascade lasers are suitable. There are a lot of commercially available lasers in the market. Figure 1.20 shows wavelengths of some commercially available lasers. Some suitable lasers (continuous wave and

Table 10.1 Incoherent sources of continuum radiation for PAS spectra.

Lamp name	Wavelength region
Tungsten halogen lamp (600 W)	400 to >1000
Xenon arc lamp (1 kW)	400–2500
Mercury arc lamp	230–650
Carbon arc (cerium core)	350–700
Xenon continuous lamp	150–225

Table 10.2 Some suitable lasers for PAS.

Laser type	Operation wavelength(s)	Mode and power
Helium–neon laser	632.8 nm, 543.5 nm, 593.9 nm, 611.8 nm, 1.1523 μm.	CW, low power
Argon laser	454.6 nm, 476.5 nm, 488.0 nm, 514.5 nm, also frequency doubled to provide 244 nm, 257 nm	CW with line tunable, high power up to 10 W
Nitrogen laser	337.1 nm	Pulsed
Carbon dioxide laser	10.6 μm(9.4 μm)	Line tunable
Carbon monoxide laser	2.6–4 μm, 4.8–8.3 μm	–
Dye lasers	390–435 nm (stilbene), 460–515 nm (coumarin 102), 570–640 nm (rhodamine 6G), many others including harmonic generation	CW, pulsed and high power tunable
Titanium sapphire (Ti:sapphire) laser	600–1100 nm	CW, pulsed and high power tunable
Nd:YAG laser	1064, 532, 355, and 266 nm	CW, pulsed and high power line tunable
Optical parametric oscillator (OPO)	300–3000 nm	CW, pulsed and high power tunable

pulsed) for PAS are given in Table 10.2. Detailed information on the lasers is given in Chapter 1 and in Refs. [16–18].

10.2.1.2 Sample Cell

In any PA spectrometer, the PA cell is one of the most important components of the system. The design of the sample cell is very important and depending on whether one is using a single beam system or a double beam system contains two cells. The PA cell design must meet some requirements [5] according to the theoretical studies made by R–G [19]. These requirements have been met with care during the fabrication of the cell. A PA cell should obey the following criteria

1) The materials of the cell should exhibit good thermal property and be of sufficient thickness to form a good acoustic barrier. Acoustic isolation from the environment can be done by placing the cell in a sound-proof box mounted on springs for isolation from outside vibrations. The table on which the spectrometer is kept could also be isolated.

2) The window above the sample should be transparent in the region of interest and also be of sufficient thickness. The fused quartz window is ideal for this purpose.

3) The light scattered from the sample should not be allowed to fall on the cell walls and microphone diaphragm. For this purpose, the geometry of the cell should be such that the microphone is kept away from the beam path and the sample is directly below the window, with the window being larger than the sample dimensions so that the scattered light or unabsorbed light is reflected out of the cell. Impurities in the sample holder or in the cell that absorb light should be thoroughly eliminated. The cell walls should, therefore, be highly polished, and the sample holder should be easy to clean or replace. It should also be easy to clean the window of the cell periodically.

4) The cell dimensions should be such as to minimize its volume since the PAS signal varies inversely with the cell volume.

5) Thermoviscous damping is an important parameter [20]. The thermoviscous damping coefficient varies as $f^{1/2}$ and becomes important at high frequencies. The cell should, therefore, have a minimum distance between the sample and the window and maximum passageway dimensions between the sample region and the microphone. The dimensions suggested by Rosencwaig are typically 1–3 mm.

6) In order to improve the acoustic signal, it is possible to work with Helmholtz resonance cells or with specially designed cylindrical microphones. Limitations in these cases are the frequency response of the microphone is not fiat and the cell cannot be used at frequencies other than the resonance frequency. Typical cell designs are shown in Figure 10.4. The International School of Photonics (ISP), Kochi, India [21] has also developed some PA cells keeping in mind the optimum signal-to-noise ratio.

In general, the microphone section is connected through a narrow passage with sample section. The resonance effects between volumes connected by a narrow passage should be avoided. In case of resonance condition, the signal versus frequency response may not meet the theoretical requirements as discussed earlier for the one-dimensional model. The usual cell geometry is such that the flat electret microphone is mounted perpendicular to the direction of incidence of light (Figure 10.4). In the one-dimensional model, the thermal waves in the gas phase propagate in the direction directly opposite to that of the incident light. The closest approximation to such a cell design is given by Aamodt *et al.* [22]. These authors have shown that the signal strength is maximum when $\mu = l$, the length of the gas phase. Ferrell and Haven [23] have used a configuration in which the microphone is placed opposite the incident light direction (Figure 10.4c). The sample is placed on a paper positioned between the window and the microphone. A pinhole in the paper is sufficient to connect the microphone and sample compartments to obtain signals without distortion or diminution.

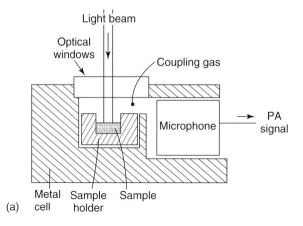

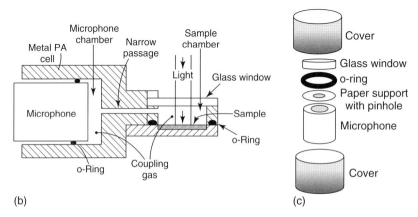

Figure 10.4 Examples of indirect PA cells: (a) indirect cell configuration (Rosencwaig [20]), (b) separate microphone and sample chamber (Helmholtz), and (c) cell design from Ferrell and Haven.

10.2.1.3 Modulation Techniques

Light beam modulation is required in order to observe a PA signal. Several modulation methods that fall under either amplitude modulation or frequency modulation are used. The amplitude modulation is the most commonly used technique.

The use of a mechanical chopper is an inexpensive and efficient way of amplitude modulation. The high-precision, variable-speed and low-vibration noise choppers are available commercially. The depth of modulation of mechanical choppers is 100%. It can also be easily fabricated in the laboratory.

The electro-optic method is also used and is superior to the mechanical method. It involves changing the plane of polarization of the incoming polarized laser beam in a nonlinear crystal ammonium dihydorgen phosphate (ADP), potassium dihydorgen phosphate (KDP) with the application of a modulated electrical field to the crystal. The beam can be modulated in square, triangular, sinusoidal, or any other form by this method. The modulation rage is 0–20 MHz, and depth is as high as 100%. The electro-optic modulators are expansive and wavelength specific. Acousto-optic modulators are also in use. When high spectral resolution is required with tunable laser sources, frequency modulation is preferred. Amplitude modulation has the disadvantage of window absorption that can be eliminated by using this technique. Other details can be found in Refs. [5, 6, 9].

10.2.1.4 Signal Detectors

There are several methods available for detecting the acoustic signal. These can be broadly classified into three groups: pressure sensors, piezoelectric sensors, and temperature sensors. The most utilized method in PAS is pressure sensors, and the most common device is the microphone.

Microphones are available in two types: condenser microphone and electret microphone. The condenser microphone needs a bias voltage to operate, but the electret does not require it, and hence simplify the apparatus. Many commercial bands are available with the choice of sensitivity, bandwidth, size, and noise. Sensitivity as high as $\sim$100 mV Pa^{-1} is available (e.g., Bruel and Kjaer, Knowles). These microphones are more sensitive than the condenser microphones. The details of the working of the microphone are available in the literature [5, 9].

For detecting the direct PA signal, microphones are typically not suitable because of large acoustic impedance mismatch at the sample–gas interface. Owing to this impedance mismatch, only a small part of the acoustic energy is transferred from sample to coupling gas. To overcome this deficiency, piezoelectric transducers, which are commercially available, are utilized. Examples of such devices are lead zirconate titanate (PZT), lithium niobate, lead metaniobate, quartz, and so on. The sensitivity of piezoelectric PZT transducers is $\sim$3 V Pa^{-1}, which is much smaller than that of the microphone. However, PZT transducers are preferred over microphones in pulsed excitation because of their faster rise times and better acoustic impedance matching. The procedure of mounting piezoelectric devices is described in Ref. [24].

10.2.1.5 Design of the Low-Cost Continuous Wave PA Spectrophotometer

In this section, a simple design of the low-cost PA spectrophotometer is described, which can easily be fabricated in the laboratory. This spectrometer is able to record wavelength-dependent spectra as well as frequency-dependent PA signal with only some modifications in the sample chamber. The following accessories are needed for it:

1) Halogen bulb (1000 W) and diode laser ($\sim$50 mW)
2) A combination of two short focal length plano-convex lenses
3) Optical chopper
4) Low *F*-number (focal ratio) monochromator
5) Plano-concave mirror
6) Lock-in amplifier

7) Aluminum block, quartz window, condenser microphone, resistances, capacitances, and so on, for sample cell and amplifier section.
8) Recording system.

10.2.1.5.1 Fabrication of Conventional Photoacoustic Cell for Wavelength versus PA Intensity Spectra

On the basis of theoretical explanation of signal generation in a PA cell, which houses the sample chamber and the microphone chamber, the cell must have certain characteristics for optimum use. The design of the cell involves a complex optimization procedure in order to achieve the necessary high signal-to-noise ratio in PA studies. Since the signal amplitude in the PA cell used for solid samples varies inversely with the gas volume, one should attempt to minimize the gas volume. Furthermore, the distance between the sample and the cell window should always be greater than the thermal diffusion length of the gas, since it is the boundary layer of the gas that acts as an acoustic piston generation of the PA signal in the cell. The requirements that give good signal-to-noise ratio are summarized below.

1) The material of the body of the cell should be chosen with good acoustic seal and with walls of sufficient thickness to form a good acoustic barrier. Hence, aluminum metal has been used as the building material.
2) The cell must also be acoustically isolated from any external signal source (e.g., from chopper, room vibrations, etc.).
3) Stray light absorbed at the cell walls results in temperature rise of the cell walls. This results in the generation of large spurious signals that increase noise in the cell.
4) According to Patel and Tam [25], the increases in the temperature of walls (ΔT_{WALL}), due to the absorption of stray light is given as

$$\Delta T_{\mathrm{WALL}} = \frac{\mathrm{Constt}}{\alpha^{1/2} \rho C}$$

where α is the thermal diffusivity of the material, ρ the density of the material, and C the specific heat of the material
5) The constant contains surface optical absorption and reflection factors. Hence, constriction materials for the PA cell should have small surface optical absorption, large thermal diffusion length, high density, and high specific heat.
6) The window should be optically transparent throughout the wavelength region of interest and should also be a good attenuator of sound; hence, quartz window has been used.
7) Absorption of scattered light by the cell walls and the microphone diaphragm should be minimum; hence, the microphone is positioned away from the beam path and the internal walls of the cell are polished to avoid absorption. The area of the window is large in comparison to the sample area so that any unabsorbed light or scattered light is reflected out of the cell.

8) Since the signal in the PA cell varies inversely with the gas volume in the cell, the gas volume should be minimized. Thus, the air volume contained in the cell is kept small so as to get a large PA signal.

Keeping these requirements in mind, the PA cell may be constructed as per the method described below. To fabricate a sample chamber, take an aluminum block of 10 cm length, 6 cm breadth, and 4 cm height. Drill holes at four corners of the block to tighten the screw and then cut the block into two parts 3 : 1 ratio in height. The larger part contains the sample chamber and microphone housing. Drill a cylindrical cavity in this block, which will hold a stainless steel sample holder with 1 cm height and 2.0 cm diameter. Construct a sample holder of depth 2.0 mm, which fits into the sample cavity. Make another hole up to the end and about 1.5 cm away from the center of the first hole having the diameter equal to the microphone. Fit the wire-connected microphone in the hole. The surfaces of the microphone and plane of block must be in the same level. Fix the microphone with glue and make sure it is airtight toward the back. The acoustic signal through the wires from the microphone is taken out from the end of the microphone housing to be fed into the input of the preamplifier section. The smaller aluminum plate contains a quartz window of diameter somewhat larger than 2.0 cm and must be just above the sample holder. The window will allow light to fall on the sample. A rectangular channel of 2 mm depth and 1 mm wide is cut into the rubber gasket extending across the sample chamber and reaching to the middle of the microphone chamber. The microphone chamber is about 1 mm deep and is formed in the region between the electret diaphragm and the bottom of the rubber gasket. The above rectangular channel permits air to pass from the sample chamber into the microphone chamber and has a 2 mm^2 cross-sectional area and is about 25 mm long. The minimum separation between the window and the sample surface is kept to 2 mm to remove any thermoviscous damping of the acoustic signal. Keep the volume of the air column connecting sample chamber to the microphone less than 1.0 cm^3 for maximum signal intensity (R–G theory). The whole assembly is tightened into one compact unit with nuts and bolts to form the complete PA cell. Figure 10.5 shows views of a PAS cell.

10.2.1.5.2 Open Photoacoustic Cell Configuration

Open photoacoustic cell (OPC) configuration is a modified and more convenient form of the conventional PA configuration discussed above. In OPC, usually, the solid sample will be mounted directly on top of the microphone, leaving a small volume of air in between the sample and the microphone. It is an open cell detection configuration in the sense that the sample is placed on top of the detection system itself, as in the case of piezoelectric and pyroelectric detection. Consequently, this configuration is a minimum volume PA detection scheme, and hence the signal strength will be much greater than the conventional PA configurations. The major advantage of this configuration is that samples having large area can be studied, whereas in conventional PA cells sample size should be small enough to be contained inside the PA cavity. A schematic representation of a typical OPC is shown in Figure 10.6.

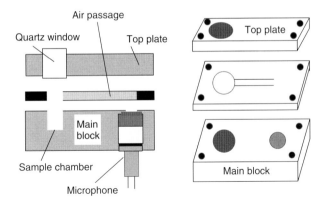

Figure 10.5 Design of a simple PA cell. (a) Cross-sectional view of the cell. (b) Vertical view of the cell.

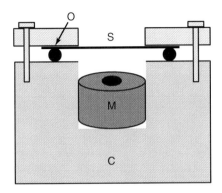

Figure 10.6 A general schematic representation of an open photoacoustic cell. Here, S is the sample, O the o-ring, M the microphone, and C the cell body. R–G theory can be used to derive an expression for the periodic pressure variation inside the air chamber.

10.2.1.5.3 For Frequency versus PA Intensity Spectra

The above-discussed OPC can be used to determine thermal diffusivity of solids. The heat transmission mechanism in OPC is depicted in Figure 10.7. For an optically opaque sample, the entire light is absorbed by the sample at $x = 0$, and the periodic heat is generated at the sample surface. The thermal waves generated

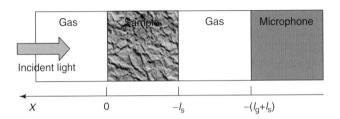

Figure 10.7 Schematic representation of an open photoacoustic cell configuration.

at $x = 0$ will penetrate through the sample to its rear surface. The heat thus reaching the sample air interface at $x = -l_s$ will get attenuated after traveling a small distance called the *first thermal diffusion length* (μ) in the air. The thermal diffusion length is given by $\mu = \sqrt{2\alpha/\omega}$, where α is thermal diffusivity of air and ω the modulation frequency of the incident light. Consequently, this periodic heating process arising as a result of the periodic absorption of light at the interface at $x = 0$ results in an acoustic piston effect in the air column in between the sample and the microphone. Pressure fluctuation in air inside the cell for a thermally thin sample can be written in terms of Eq. (10.16), which implies that the amplitude of PA signal from a thermally thin sample varies as $f^{-3/2}$. When the sample becomes thermally thick at high chopping frequencies, the signal varies as $(1/f) \exp(-bf)$, where b is constant. Hence, thermal diffusivity α can be evaluated from signal amplitude data with respect to modulation frequency, provided that the sample becomes thermally thick from thermally thin in the frequency region of interest.

For the measurement of frequency-dependent characteristics of materials, a different configuration is selected. The main aluminum block has size 5.5 cm × 5.0 cm × 2.5 cm, and the top plate has size 5.5 cm × 5.0 cm × 1.5 cm. Two quartz windows, one in the sample chamber and the other in the top plate, are fitted. In this configuration, the microphone is inserted inside and permits front as well as back illumination of the sample films. The view of this cell is shown in Figure 10.8.

The PA signal obtained by a condenser microphone mounted in the microphone chamber of the PA cell is very weak. Therefore, a large amplification is required before feeding it to the lock-in amplifier. A single-stage preamplifier based on IC 741 can be used for the measurements. The circuit diagram of the single stage preamplifier is shown in Figure 10.9. It is operated by a 9 V DC supply. The output impedance of the condenser microphone is very high (135 kΩ); therefore, a voltage follower may be added before the amplifier for impedance matching. Depending on the strength of the PA signal obtained from the microphone, the output of the amplifier can be varied by using a 47 kΩ potential divider (preset) to prevent the overloading of the lock-in amplifier. In order to eliminate the external noise,

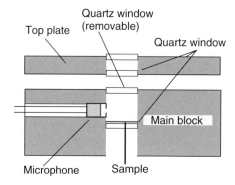

Figure 10.8 Cross-sectional view of cell for thermal diffusivity measurements.

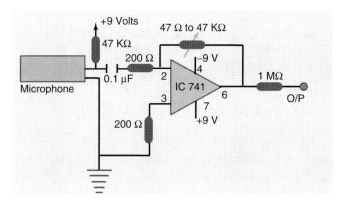

Figure 10.9 Microphone biasing and single-stage preamplifier circuit for signal amplification.

shielded cables British Naval Connector (BNCs) have been used for carrying the signals from the output of the microphone to the input of the preamplifier and the output of the preamplifier to the input of the lock-in amplifier. The use of extended two-stage amplification with a largely controlled (variable) gain enables one to record the PA spectra of any kind of samples. Usually, overloading of the lock-in amplifier occurs when a two-stage preamplifier is used; therefore great care is necessary to eliminate overloading of the lock-in amplifier.

A simple biasing power supply can be made to provide biasing voltage of ± 12 V for operating the preamplifiers.

10.2.1.5.4 Performance Studies

After designing the complete setup according to the diagram shown in Figure 10.3, the next step is parametric studies of the designed setup to ascertain the effects of source power, chopping frequency, gas coupling volume, and so on. Procedures of the parametric studies for optimization of signal-to-noise ratio can be found in Ref. [26]. In order to obtain the wavelength spectra of the sample, there is a need for a reference sample. The fine carbon black sample is used as a reference. It absorbs all the radiation falling on it and converts it into heat energy, which is sensed by the spectrometer. The spectrum obtained from a carbon black sample closely reproduces the power spectrum of the halogen lamp. In Figure 10.10, the power spectrum of the 650 W tungsten halogen bulb is shown in the 340–850 nm region using the carbon black sample.

10.2.2
Pulsed Photoacoustic Spectroscopy

The apparatus needed to perform pulsed PA studies consists again of a light source, sample cell, microphone, and signal processing unit. The difference between pulsed and continuous wave (CW) mode is in the excitation source. The pulsed

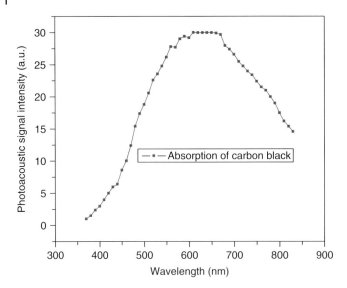

Figure 10.10 Power spectrum of the 650 W tungsten halogen bulb in the 340–850 nm region using the carbon black sample.

spectrophotometer uses light sources that deliver very high peak power radiation. The high peak power makes it possible to perform a variety of applications that are not possible using continuous sources. Accordingly, some modifications in the cell design and in the detectors and electronics have been made.

The pulsed sources are mostly lasers that deliver high peak powers in the range $10^{-3} - 1$ J per pulse. Such lasers include continuously tunable and fixed frequency or line tunable. The popular wavelength tunable lasers are Nd:YAG and N_2 lasers, pumped dye lasers, Ti:sapphire laser, and OPOs (optical parametric oscillators). The tuning range of OPOs is somewhat larger than that of the dye lasers. The commercial OPOs provide a tunable range over the large region, but these systems are very costly. The fixed frequency or line tunable lasers are Nd:YAG laser, N_2 laser, and CO_2 laser.

The sample cell for pulsed PAS must be of a design different from those used in the CW technique in order to minimize extraneous signals generated by the high energy pulses. In general, nonresonant cell configuration is used for low repetition rate lasers, whereas resonant cells are used for high repetition rate lasers. Dewey and Flint [27] have used resonance cell with an excitation frequency that was subharmonic of the cell's resonance frequency for high repetition rate lasers. This frequency excites only the cell's lowest radial mode. More details can be found in Refs. [27, 28].

The other factor that influences pulsed PAS cell design is the ballistic acoustic waves generated by sample adsorption. To avoid this, the sample window should be of the highest optical quality and properly cleaned. Poor windows can lead to two types of background signals: adsorption and scattering. The scattering from window imperfections can cause light to strike on/near microphone. This type of background signal can be suppressed by the use of transparent material for cell construction.

10.3
Applications of PA Spectroscopy to the Nanomaterials

The PA technique has proved to be a very useful nondestructive characterization technique in the fields of physics, chemistry, and recently in biological specimens. The studies can be divided into three main areas – bulk studies, surface studies, and de-excitation studies. The PA technique provides absorption spectra of solid samples. An unpolished sample surface also poses no problems, and the spectra of strongly scattering samples can be easily measured. It has spectral range from the ultraviolet to far IR based on the availability of excitation sources. For semiconductor materials, their band gaps can easily be calculated directly from the absorption edges in the PA spectra of semiconductors. In recent years, nanomaterials have attracted a lot of attention and several novel materials have been developed for various technological applications. One of the most challenging problems in nanomaterials research is their accurate characterization, which is very important for the efficient use of these technologically promising materials. For example, absolute absorption, quantum efficiency, thermal conductivity, and the elastic constants are important parameters for photonic applications. Although the conventional absorption or emission techniques can provide the absorption coefficient, determining the absolute absorption is not possible due to the presence of scattered light. Therefore, a PA technique that is sensitive and immune to scattered or reflected light is required to determine the absolute absorption in the wavelength region of interest. However, the studies of nanomaterials through this technique are very limited and need more research efforts. The studies can reveal the hidden character of the nanomaterials. Some interesting cases are discussed in the subsequent sections.

10.3.1
Determination of Optical Band Gap

The band gap of any material can be obtained using the PA technique. Xiong *et al.* [29] studied the PA spectra of nanoclusters of ZrO_2 of different sizes and found blueshift in the cut off wavelength of the PA absorption edge as the particle size decreases. The authors also found that the thermal constant of the nanoclusters increases significantly with the decrease in average grain size. Inoue *et al.* [30] applied the PA spectroscopy to evaluate the nonradiative transition of the Pr^{3+} in ZnO powders with various Pr concentrations and sintering temperatures. They found that PA spectra depend on the inner state of the Pr^{3+} in ZnO matrix, where the Pr can easily segregate to the ZnO grain boundaries and form intergranular-phase-based Pr_2O_3. Figure 10.11 shows the PA spectra of the sample with different ion concentrations. Peaks A_1, A_2, and B occur because

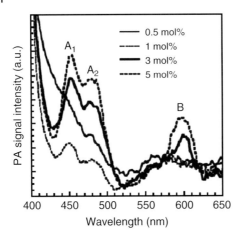

Figure 10.11 PA spectra Pr^{3+}-doped ZnO having 0.5, 1, 3, and 5 mol% Pr concentration. (Source: With kind permission from Elsevier Science.)

of nonradiative transitions from Pr^{3+} ions. Peak A_1 shows large concentration dependence, whereas peak B appears only for heavily Pr^{3+} doped system. The authors concluded that peak A_1 is related to the presence of Pr^{3+} ions in the ZnO grains, whereas peak B occurs because of the presence of Pr^{3+} at grain boundaries. This study is very informative and can be applied to other rare-earth-doped ZnO samples to identify doping sites of ions.

Zhang [31] studied the PA spectra of $BaFBr:Eu^{2+}$ after glow phosphor was prepared in different environments. The author observed that the PA intensity changes with the change in preparation environment. The study concluded that the sample that gives the least PA signal is the most suitable for fluorescence emission.

The semiconducting nanoparticles are very useful for several applications. Cadmium sulfide (CdS) is one of the direct band gap materials. Kuthirummal [10] studied the PA spectra of bulk and CdS nanowires. Figure 10.12 compares the PA spectra between bulk and nanophases. The band gap of bulk CdS powder occurs at 2.39 ± 0.04 eV, which agrees with the literature value of 2.42 eV. The absorption edge of CdS nanowires is much steeper and occurs at a slightly larger value of 2.49 ± 0.04 eV. These data show that there is no significant contribution from quantum-confinement effects because the average diameter of the nanowires was about 50 nm, which is much larger than the calculated Bohr radius of 2.8 nm. The increased steepness might be attributed to the relatively well-ordered structure and size. El-Brolossy *et al.* [32] have reported the optical absorption spectra of CdSe quantum dots. They found that by increasing the growing time, the redshift of the PA spectra was clearly observed.

Figure 10.13 shows the PA spectra of 8 and 350 nm $La_{0.8}Sr_{0.2}FeO_3$ nanoparticles, and 300 nm α-Fe_2O_3 commercial crystals, respectively [33]. In these three samples, the Fe^{3+} cations have a similar coordination environment in the crystal structures. Comparing the three curves, the same shape was observed; except the intensity

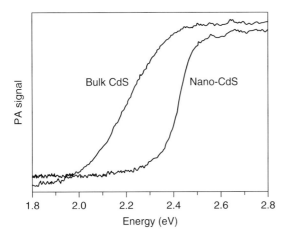

Figure 10.12 PA spectra of bulk and nano-CdS. Nano-CdS clearly shows increase in band gaps. (Source: With kind permission from the American Chemical Society.)

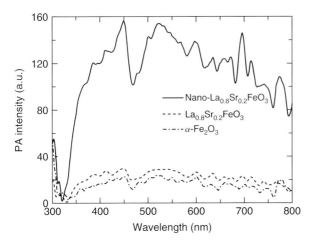

Figure 10.13 Comparison of photoacoustic spectra of the nanocrystalline $La_{0.8}Sr_{0.2}FeO_3$ (solid line), the conventional crystalline $La_{0.8}Sr_{0.2}FeO_3$ (dashed line), and the commercial crystalline α-Fe_2O_3 (dashed-dotted line) with the average particle size of 8 nm, 0.35 and 0.30 μm, respectively, with the inset of the surface photovoltaic spectroscopy (SPS) of the three samples (the meanings of the lines are the same as that above). (Source: *Phys. Rev. B*, **75** (2007) 045403, Copyright (2007) of The American Physical Society.)

of the PA signals, the intensity of 8 nm $La_{0.8}Sr_{0.2}FeO_3$ nanoparticle is higher than that of the other two by five to six times. Since the intensities of the PA signals of 350 nm $La_{0.8}Sr_{0.2}FeO_3$ nanoparticles and 300 nm α-Fe_2O_3 commercial crystals were nearly equal, the effect of both the La^{3+} and Sr^{2+} cations on electron–phonon interaction responsible for the PA signals in the region below the photoelectric threshold of the nanomaterial should be negligible. The most probable reason for the intensity difference is that there is a high density of surface localized states associating with the Fe^{3+} cation so far as the nano-$La_{0.8}Sr_{0.2}FeO_3$ is concerned.

Morais [34] has done PA characterization of various magnetic nanoparticles for magnetic drug delivery systems. He found enhancement in the PA intensity of Band-C due to charge confinement as particle size decrease from 10.7 nm to 3.8 nm (Figure 10.14).

10.3.2
Determination of Absolute Quantum Efficiency

PA spectroscopy can also be easily used to measure the absolute quantum efficiency (η) of the luminescent nanomaterials. The conventional luminescence tools have proved to be very difficult to measure the efficiency. In a luminescence measurement, the number of quanta absorbed from a beam of monochromatic light has to be compared with the number of quanta emitted in a polychromatic light whose distribution in the space is complicated, and therefore a number of correction factors are involved. Another way is the measurement of fluorescence lifetime; this method again suffers from several experimental difficulties such as separate measurement of the nonradiative contribution to the lifetime of the emitting state [35]. The PA method gives the quantum yield by determining the nonradiative part of the absorbed energy. The measurement of the absolute optical energy absorbed W_0 and the absolute heat energy generated W_{heat} provides the fluorescence quantum yield (η) for a two-level system as $\eta = (W_0 - W_{heat})/W_0$.

A key issue in measuring fluorescence quantum yield by the above equation is that absolute heat energy is involved. But the PA signal is proportional only to the modulated heat generated. The best way to avoid this is to perform the

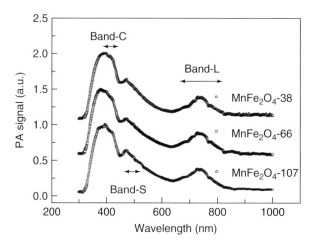

Figure 10.14 PA spectra of the manganese-ferrite-based ionic magnetic fluid samples containing nanoparticles with different average diameters (3.8, 6.6, and 10.7 nm). (Source: With kind permission from Elsevier Science.)

PA measurements twice, first with the desired luminescence quantum yield η_r and second with the quantum yield altered in a known way. This provides two equations with two unknowns and so η_r can be obtained. The details of mathematical formulation and procedure are given in [35–37]. Rosencwaig and Hildum [35] measured the quantum efficiency of the $^4F_{3/2}$ level of Nd^{3+} in silicate glass using this method. A comparison with the lifetime and luminescence method shows good agreement. Figure 10.15 shows PA and relative fluorescence quantum yield versus Nd_2O_3 concentration in silicate glasses. A more refined experimental procedure has been proposed by Rodriguez et al. [38]. This method does not require any internal standard.

The quantum yield of any material in any form can be obtained using the PA technique.

10.3.3
Determination of Thermal Diffusivity/Conductivity

PAS is a very powerful technique for measuring thermal diffusivity and hence the thermal conductivity of any material. The thermal conductivity decides the applicability of any material in various applications. The thermal conductivity of the nanomaterials changes very differently and that can be monitored with this technique. As thermal diffusion length is the function of chopping frequency, by varying the chopping frequency of optical radiation and consequently the amplitude of the thermal waves, the transition frequency at which a sample changes from thermally thin to thermally thick regime can be known from the amplitude spectrum of the PA signal. By knowing the transition frequency and the thickness

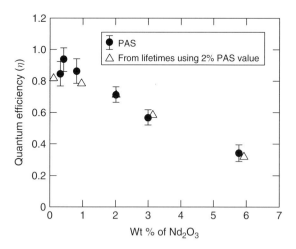

Figure 10.15 Photoacoustic and relative fluorescence quantum yield versus Nd_2O_3 concentration in silicate glasses. (Source: Reproduced from A. Rosencwaig, Edward A Hildum, *Phys. Rev. B*, **23** (1981), Copyright (1981) by The American Physical Society.)

of the specimen under investigation, the thermal diffusivity (α) can be evaluated using the expression

$\alpha = f_c l^2$, where f_c is the critical frequency and l is the sample thickness.

El-Brolossy *et al.* [32] obtained the thermal diffusivities of CdSe quantum dots and compared them with bulk CdS. The authors found one order enhancement in thermal conductivity of quantum dot CdSe than bulk CdSe. Figure 10.16 shows the plot of PA amplitude versus the inverse of chopping frequency variation for the CdSe quantum dot sample. A distinct change in slope (shown by arrow) determines the critical frequency where the sample goes thermally thin to thermally thick.

George *et al.* [39] obtained the thermal diffusivity of Al_2O_3-Ag nanocomposites. Chandra *et al.* [40] obtained thermal diffusivities of $AgI-Al_2O_3$ $AgI-Ba_{0.70}Sr_{0.30}TiO_3$ composite materials and the correlation between the thermal diffusivity and electrical conductivity of the composites.

10.3.4
Photoacoustic Spectroscopy in Biology

Use of nanomaterials in biology or medicine is a very new and vast area of research. In the past decade, a number of new works representing the interest of laser PA in biology were published. The applications of PA methods in biophysics and medicine, say, for diagnosis of carcinoma in body, photothermal treatment, and others seem prospective. Bioconjugated nanocontrast agents together with the PA imaging technique can reliably detect, diagnose, and characterize carcinoma cells. A list of some of the applications of nanomaterials in biology or medicine is given in Ref. [41].

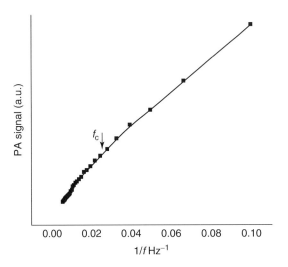

Figure 10.16 Variation in amplitude of PA signal with chopping frequency. (Source: With kind permission from Springer.)

The fact that nanoparticles exist in the same size domain as proteins makes them suitable for biological applications. However, size is just one of the many characteristics of nanoparticles that make them applicable in biology. Interaction of light with nanomaterials can induce several phenomena such as photon emission, heat generation, and photodissociation. The significance of PA imaging is that it overcomes the problems faced by conventional imaging techniques such as optical imaging and ultrasonic imaging and yields images of high contrast and high resolution in relatively large volumes of biological tissues [42]. In PA imaging, nonionizing laser pulses are delivered into biological tissues. Some of the delivered energy will be absorbed by the tissues and a part of that is converted into heat, leading to transient thermoelastic expansion, and thus wideband (e.g., MHz) ultrasonic emission may be termed as *laser-induced ultrasonics*. The generated ultrasonic waves are then detected by ultrasonic transducers to form images. It is known that optical absorption is closely associated with physiological properties, such as hemoglobin concentration and oxygen saturation. As a result, the magnitude of the ultrasonic emission (i.e., PA signal), which is proportional to the local energy deposition, reveals physiologically specific optical absorption contrast.

Gold nanoparticles and carbon nanotubes (CNTs) have gained popularity as nanosized contrast agents in PA imaging, and search for others is going on. The PA spectroscopy is a key technique to characterize nanoparticles that can generate enough heat on photon excitation. Figure 10.17 shows application of CNTs in the enhancement of PA contrast. El-Brolossy *et al.* [43] prepared gold nanoparticles in various shapes and characterized them through PA spectroscopy. PA spectra of gold nanoparticles having rodlike and spherelike structures are illustrated in Figure 10.18. A strong PA signal is observed in the case of the shepherd shape.

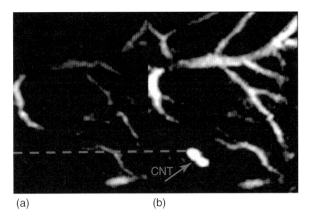

(a)   (b)

Figure 10.17 PA image of rat blood vessels; acquired (a) before carbon nanotube (CNT) injection and (b) after nanotube injection. Bright parts represent optical absorption, here, from blood vessels.

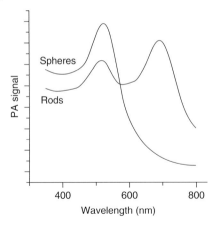

Figure 10.18 Photoacoustic spectra of gold nanoparticles in two different shapes. (Source: With kind permission from Springer.)

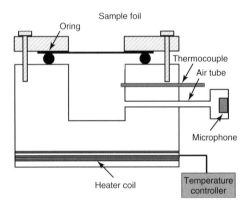

Figure 10.19 Cross-sectional view of sample cell with temperature variation facility.

10.3.5
Determination of Phase Transition with Temperature

PA spectroscopy can identify temperature-induced phase transitions in nanomaterials. To obtain the phase transitions, a variable temperature cell that can be fabricated according to Figure 10.19 is needed. This cell is of resonant type and contains a separate chamber for the microphone.

References

1. Fork, D.C. and Herbert, S.K. (1993) *Photochem. Photobiol.*, **57**, 207.
2. Bell, A.G. (1880) *Am. J. Sci.*, **20**, 305.
3. Bell, A.G. (1881) *Philos. Mag.*, **11**, 510.
4. Demtroder, W. (2003) *Laser Spectroscopy: Principles and Applications*, Springer.

5. Pao, Y.H. (ed.) (1977) *Optoacoustic Spectroscopy and Detection*, Academic Press, California.

6. Zharov, V.P. and Letokhov, V.S. (1986) *Laser Optoacoustic Spectroscopy*, Springer-Verlag, Berlin.

7. George, N.A. (1992) Photoacoustic and photothermal detection studies on certain photonic materials. PhD thesis. Cochin University of Science and Technology.

8. Hess, P. (ed.) (1989) *Photoacoustic, Photothermal and Photochemical Processes in Gases*, Springer-Verlag, Berlin.

9. Luscher, E., Korpiun, P., Coufal, H., and Tilgner, R. (1984) *Photoacoustic Effect: Principles and Applications*, Friedrich Vieweg & Sohn, Braunschweig.

10. Kuthirummal, N. (2009) *J. Chem. Educ.*, **86**, 1238.

11. Somoano, R.B. (1978) *Angew. Chem., Int. Ed. Engl.*, **17**, 238.

12. Petrov, V.V., Genina, E.A., and Lapin, S.A. (1999) *SPIE*, **3726**, 200.

13. Rosencwaig, A. (1978) *J. Appl. Phys.*, **45**, 2905.

14. Quimby, R.S. and Yen, W.M. (1979) *Appl. Phys. Lett.*, **35**, 43.

15. Fuchsmen, W.H. and Silversmith, A.J. (1979) *Anal. Chem.*, **51**, 589.

16. Nambiar, K.R. (2006) *Lasers: Principles, Types and Applications*, New Age International Publishers, New Delhi.

17. Csele, M. (2004) *Fundamentals of Light Sources and Lasers*, Wiley Interscience Publication, New Jersey.

18. Svelto, O. (2009) *Principles of Lasers*, 5th edn, Springer.

19. Rosencwaig, A. (1980) *Photoacoustics and Photoacoustic Spectroscopy*, John Wiley & Sons, Inc., New York.

20. Rosencwaig, A. (1977) *Rev. Sci. Instrum.*, **48**, 1133.

21. George, S.D., Kumar, B. A., Radhakrishnan, P., Nampoori, V.P.N., and Vallabhan, C.P.G. (2004) *Opt. Eng.*, **43**, 3114.

22. Aamodt, J.C., Murphy, L.C., and Parker, G.J. (1977) *J. Appl. Phys.*, **48**, 927.

23. Ferrell, N.C. and Haven, Y. (1977) *J. Appl. Phys.*, **48**, 3984.

24. Tam, A.C. and Patel, C.K.N. (1979) *Appl. Phys. Letts.*, **35**, 843.

25. Patel, C.K.N. and Tam, A.C. (1981) *Rev. Mod. Phys.*, **53**, 517.

26. Pandhija, S., Rai, N.K., Singh, A.K., Rai, A.K., and Gopal, R. (2006) *Prog. Cryst Growth Charact. Mater.*, **52**, 53.

27. Dewey, C.F. and Flint, J.H. (1979) *Technical Digest to the Topical Meeting on Photoacoustic Spectroscopy*, Optical Society of America, Ames, IA.

28. Mandelis, A. and Royce, B.S.H. (1979) *J. Appl. Phys.*, **50**, 4330.

29. Xiong, Y., Yu, K.N., and Xiong, C. (1994) *Phys. Rev. B*, **49**, 5607.

30. Inoue, Y., Okamoto, M., Kawahara1, T., and Morimoto, J. (2006) *J. Alloys Comp.*, **408–412**, 1234.

31. Zhang, Y. (1997) *Chem. Phys.*, **219**, 353.

32. El-Brolossy, T.A., Abdallah, S., Abdallah, T., Mohamed, M.B., Negm, S., and Talaat, H. (2008) *Eur. Phys. J. Spec. Top.*, **153**, 365.

33. Li, K.Y., Yang, F., Tian, Y.J., Liu, R.P., Wang, W.K., and Wang, H.T. (2007) *Phys. Rev. B*, **75**, 045403.

34. Morais, P.C. (2009) *J. Alloys Comp.*, **483**, 544.

35. Rosencwaig, A. and Hildum, E.A. (1981) *Phys. Rev.*, **B 23**, 3301.

36. Murphy, J.C. and Aamodt, L.C. (1977) *J. Appl. Phys.*, **48**, 3502.

37. Quimby, R.S. and Yen, W.M. (1978) *Opt. Lett.*, **3**, 181.

38. Rodriguez, E., Tocho, J.O., and Cusso, F. (1993) *Phys. Rev.*, **B 47**, 14049.

39. George, S.D., Anappara, A.A., Warrier, P.R.S., Warrier, K.G.K., Radhakrishnane, P., Nampoorie, V.P.N., and Vallabhan, C.P.G. (2008) *Mater. Chem. Phys.*, **111**, 38.

40. Chandra, S., Rai, S.B., Singh, P.K., Kumar, K., and Chandra, A. (2006) *J. Phys. D: Appl. Phys.*, **39**, 3680.

41. Salata, O.V. (2004) *J. Nanobiotechnol.*, **2**.

42. Xua, M. and Wang, L.V. (2006) *Rev. Sci. Instrum.*, **77**, 041101 (22).

43. El-Brolossy, T.A., Abdallah, T., Mohamed, M.B., Abdallah, S., Easawi, K., Negm, S., and Talaat, H. (2008) *Eur. Phys. J. Spec. Top.*, **153**, 361.

11
Ultrafast Laser Spectroscopy of Nanomaterials

Subhash Chandra Singh and Yashashchandra Dwivedi

11.1
Introduction

Optical spectroscopy has innumerable applications in the investigation of electronic, vibrational, optical, and other properties of atomic, molecular, nano and bulk scale systems. Applications of linear optical spectroscopic techniques such as UV−visible absorption/transmission, reflection, photoluminescence, scattering have proved their versatility in providing a large amount of invaluable information on many diverse aspects of nanoparticles (NPs) such as surface plasmon resonance (SPR) absorption, size, plasmonics, and other properties for metal NPs, while electron band structure, phonons, coupled phonon−plasma modes, single-particle excitation spectra of electrons and holes, surface properties, defects, and interfaces for semiconductor nanosystems. In addition to these static linear optical properties of nanodimensional systems, optical spectroscopy also has the potential for diagnosing fundamental but new and exciting information about the nonequilibrium, nonlinear, and transport properties of nanomaterials.

Linear and steady state laser spectroscopy of nanomaterials such as conventional steady state photolumniscence, scattering, reflection, and absorption can diagnose only steady state optical and electronic properties. For example, they can measure bandgap energy, defect level energy, and electron and hole carrier densities in valence and conduction bands, respectively, as well as in electron and hole acceptor defect levels but cannot measure the lifetime of different electronic state, times for radiative and nonradiative pathways, and so on. Lifetimes of electrons in the excited states and charge carriers in the bands and defect levels are of the order of 10^{-9} (ns) to 10^{-12} (ps) s, and it further decreases for higher energy states. The lifetimes of various dynamical processes in nanomaterials such as electron−electron, electron−phonon, and phonon−phonon interactions lie in the range of 10^{-15} (fs) to 10^{-9} s. Similarly, lifetime of electronic motion in the Bohr orbitals is of the order of 10^{-18} s (attoseconds; $\sim150 \times 10^{-18}$ s), while periods of protons and neutrons in the nuclei are of the order of 10^{-23} s (zeptoseconds). Diagnosis and monitoring of these ultrafast electronic and optical processes are termed as *ultrafast spectroscopy*. Measurement of any optical and electronic process

Nanomaterials: Processing and Characterization with Lasers, First Edition.
Edited by Subhash Chandra Singh, Haibo Zeng, Chunlei Guo, and Weiping Cai.
© 2012 Wiley-VCH Verlag GmbH & Co. KGaA. Published 2012 by Wiley-VCH Verlag GmbH & Co. KGaA.

with the lifetime (t) requires excitation of the NPs, with the laser having pulse width of $t_p (t_p < t)$ and detector having temporal resolution higher than the lifetime of the process. For example, if the lifetime of phonon–phonon interaction is of the order of 100 ps, a laser having pulse width shorter than 100 ps and a detector with temporal resolution of picoseconds are essential to diagnose the process. Similarly, an attosecond light pulse and detector with attosecond resolution are required to diagnose time period of electronic motion in different Bohr orbitals and transition times from one Bohr orbital to another.

Optical excitation of the nanoscale materials generates nonequilibrium distribution of charge carriers and excitons, and optical spectroscopy has the capability to diagnose these distribution functions. Combining the unique capability of optical spectroscopy with the ultrafast (picoseconds, femtoseconds, attoseconds) lasers and spatial and temporal imaging has made ultrafast spectroscopy a prevailing means for the diagnosis of a wide range of phenomenon associated with the carrier transport and relaxation dynamics in nanoscale materials. Ultrafast spectroscopy is an active field of research that has brought several new insights into the basic science of nanomaterials, atoms/molecules, electro-optics, and optoelectronics. It has the ability to precisely diagnose decaytimes of various carrier relaxation and transport processes of nanomaterials in the time scales of electrons and atoms. Ultrafast coherence phenomenon in nanomaterials and other molecular systems is of great importance, but still unexplored.

Ultrafast spectroscopy, which measures carrier relaxation and transport processes after excitation with ultrafast laser pulses, has made impressive progress since the introduction of ultrashort lasers (femtosecond, attoseconds) and detectors. Several ultrafast spectroscopic techniques such as pump-probe spectroscopy, time-resolved spectroscopy, streak camera detection, upconversion photolumniscence, four-wave mixing (FWM), photon echoes, and so on have been developed or improved for higher resolution and better precision measurements. In the following sections, we discuss various ultrafast laser spectroscopic techniques and their applications.

11.2
Ultrafast Time-Resolved Spectroscopy

Observation of the transient behavior of excited states is of great importance and interest in studying carrier dynamics and relaxation processes of materials. Ultrafast time-resolved spectroscopy (UTRS) is an efficient tool to analyze and explore various dynamical processes in photosynthesis and photochromics, which usually take place in the ultrashort time range (10^{-15}–10^{-14} s). Nowadays, due to easy availability of convenient and stable femtosecond lasers (<50 fs) with repetition rate 1 KHz, these lasers are now common and being used in UTRS, while attosecond lasers although not so common, are being used by a few groups to explore different exciting dynamical process. UTRS can be categorized into different groups according to the fundamental spectroscopic processes involved. For example, UV–Vis absorption principle is used in transient absorption measurement. Similarly, infrared (IR), fluorescence, Raman, and Faraday rotation techniques are used for the time-resolved ultrafast IR, time-resolved ultrafast fluorescence (TRUF), time-resolved ultrafast Raman, and time-resolved Faraday rotation (TRFR), respectively. In the following sections, we attempt to briefly present the basics and applications of these techniques.

11.2.1
Transient Absorption Spectroscopy

Transient absorption spectroscopy (TAS) is used to explore several ultrafast dynamical process such as excited state energy migration, electron and/or proton transfer processes, isomerization, intersystem crossing, generation, relaxation, recombination processes of electrons and holes, several photosynthetic, and photochromic reactions [1, 2]. The basic process of TAS technique is given in Figure 11.1. Measurement of transient absorption spectrum requires the use of pulsed lasers for generating electronic population in excited states, and another beam, continuous wave (CW) or pulsed, for probing their absorption. The TAS technique utilizes two pulses, namely, pump and probe pulses. The pump beam pulse should be of high intensity, while the intensity of probe beam pulse should be lower so that it does not produce any high-energy consequences such as multiphoton absorption. Intensity of probe beam is usually controlled by using neutral density filters, calcite polarizers, and half waveplates. A probe beam may be single wavelength or white light, which can be generated using sapphire plate or water cell. The probe beam can be pulsed or continuous depending on the temporal delay of the dynamical process. A time difference is initially present; absorbance of probe beam in the sample is scanned and then a pump beam is use to focus at the same probe beam spot, which excites the ground state electron and builds population in the excited electronic state. Absorbance of the probe beam is usually monitored in the presence of pump beam, and difference of absorbance can be plotted with respect to probe beam intensity, wavelength, and time. To monitor the spectrum, the time

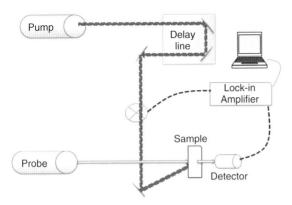

Figure 11.1 Experimental setup for transient absorption spectroscopy.

delay between the pump and probe beam is varied and the difference between absorbance is plotted against variation in delay. At present, femtosecond TAS is extended to shorter probe wavelengths by using laser-produced Bremsstrahlung X-rays [3] and extreme ultraviolet high-order harmonics [4, 5], as well as femtosecond synchrotron X-rays [6]. Recently, by applying theory with experimental TAS, the diagonal elements of the density matrix of strong-field-generated atomic ions were measured [7, 8]. The entire ion density matrix – including the coherences of strong-field-generated Kr^+ ions, is now measured by attosecond TAS [9]. For details of experimental arrangements, readers are advised to read Refs. [10, 11].

Although the basic process is absorption, the ultimate curve obtained contains information about several processes such as stimulated emission (due to probe beam, weak proportion), excited state absorption, and formation of transient or long-lived states other than ground state bleach [11], which affect the intensity of the resultant absorbance.

Application of TAS is well established to explore fast processes such as chemical reactions; transfer of excitation energy between molecules, parts of molecules, or molecules and their environment; charge separation dynamic processes that occur in the photosynthetic system, such as excited state energy migration, electron and/or proton transfer processes, isomerization, and intersystem crossing and transfer; and several other photophysical and photochemical processes. The major advantage of this technique over conventional time-resolved fluorescence (TRF) is the analysis of nonemissive states and dark states, which is invisible for TRF.

Niedzwiedzki *et al.* [12] have investigated the effect of photogenerated excited state on energy transfer and nonphotochemical quenching in xanthophyll molecules. Yeremenko *et al.* [13] compared the photocycles of photoactive yellow protein (PYP) in the crystalline state and solution and explained the observed differences in photocycle in both media due to the combined effect of dynamic steric hindrance imposed by intermolecular interactions of the protein molecules in the crystal lattice and the relatively low water content in a crystal. Tamai *et al.* [14] studied femtosecond dynamics of the photochromic reaction of a thiophene oligomer with a diarylethene structure in solution and reported formation of intermediate species. Alfano *et al.* [15] reported near-IR band of the equilibrated aqueous solvated electron and suggested involvement of rather complex processes such as solvation and solvent cooling on the observed spectral dynamics.

Luminescence quenching in semiconductors is one of the major hurdles in its applications. One of the crucial factors is surface trapping, which occurs due to imperfect surface, impurity, adsorbents, organic coated on the surface, and so on, which is among the fastest relaxation ($\sim 10^{-15} - 10^{-12}$) events in photoexcited semiconductor NPs. Surface trapping offers a very probable relaxation process for the electron and holes in semiconductor NPs. The problem in investigating these charge killer/trapping sites are that they are usually not observable with direct spectroscopic techniques and might only be assigned to the long-wavelength tail of the absorption spectra [16, 17]. Yoshihara *et al.* [18] have identified the reactive species involved in the very broad transient absorption spectrum of nanocrystalline TiO_2 films, using TAS.

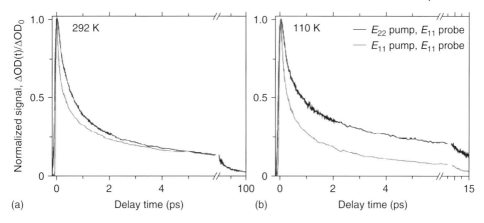

Figure 11.2 Transient absorption kinetics measured on resonant excitation of the E_{11} state at 988 nm and the E_{22} state at 570 nm, respectively, for the (6,5) tube species. (a) Data acquired at 292 K for an aqueous solution sample and (b) at 110 K for a PVP polymer composite film. (Source: Reprinted with the permission from Graham et al. [19] copyright @ American Chemical Society.)

Graham *et al.* [19] have investigated (6, 5) single-walled carbon nanotubes (CNTs) and the (7, 5) inner tubes of a dominant double-walled CNT species using femtosecond transient absorption technique. The authors reported slower decay in the first transition-allowed state (E_{11}) of a given tube type when the second transition-allowed state (E_{22}) is excited. A linear intensity dependence of the maximal amplitude of the transient absorption signal is found for the E_{22} excitation, whereas the corresponding amplitude scales linearly with the square root of the E_{11} excitation intensity (Figure 11.2a). A representative set of transient absorption profiles measured at 110 K is plotted in Figure 11.2b, where a remarkably faster decay is seen for the E_{11} pump case than for the E_{22} pump. The lifetimes (relative amplitudes) determined from deconvolution fitting are 140 fs (63%), 1.5 ps (29%), and 19 ps (8%) for the E_{11} excitation data, whereas the corresponding results are 310 fs (52%), 2.1 ps (28%), and 22.4 ps (20%) for E_{22} excitation.

Stepanov *et al.* [20] measured transient absorption spectra of C60 molecules trapped in cryogenic rare gas matrices and dissolved in toluene using two, 402 and 800 nm, wavelengths. They reported an increase in the transient absorption as time delay between the pump (402 nm) and probe (800 nm) beams was observed. For the cross polarization of the pump and probe pulses, they found a buildup time of 200 fs, similar for C60 in toluene solution or trapped in neon matrices.

Boucher *et al.* [21] reported ultrafast relaxation of fullerenes and fullerites by broadband femtosecond laser spectrometry. The intensity was varied over the range $10^{10}-10^{12}$ W cm^{-2}, the energy from 1.57 to 3.14 eV, and the pulse duration from 60 to 350 fs. The differential absorption was used to probe both the time- and energy-domain information. An increase in absorption was observed in solutions with a time constant of 1.2 ps.

11.2.2
Time-Resolved Ultrafast Fluorescence Spectroscopy

In conventional TRUF spectroscopy, a low-intensity radiation was used to resonantly excite the active species and monitor its decay profile using single channel detector photomultiplier (PMT) or photodiode and oscilloscope, even though the time resolution can be obtained in approximately nanosecond range only. Another technique named time-correlated single-photon counting is now commonly utilized [22] because of its better time resolution ($\sim$10 ps) and single-photon detection. The basic concept is the detection of fluorescence signal using detectors with fast response time such as PMT, microchannel plate, or avalanche photodiode. In this technique, electric signal generated by the detector is amplified and time delay between excitation pulse and signal is monitored by time–amplitude converter (TAC). The resulting output of TAC is fed to a multichannel analyzer, which traces the histogram of photon numbers against the time delay and provides a TRUF trace. This technique has shown its potential to detect single photon, and it can be achieved by keeping low excitation intensity and high repetition rate [23]. Higher time resolution can be achieved by using streak camera, although time resolution is not more than few picoseconds.

One other beneficial technique that was first introduced by Mahr and Hirsch [24], widely known as *"time-resolved fluorescence upconversion,"* to monitor fluorescence decay time with high temporal resolution. In this technique, an ultrashort pulse (gate pulse, g) is focused on an optical nonlinear crystal, which then combines with fluorescence (f) from the sample, which is separately excited with a required radiation laser (pump) (Figure 9.3.6). Usually, thin nonlinear crystals of potessium dihydrogen phosphate (KDP), barium borate (BBO), or lithium iodate (LiIO$_3$), are used. As the spectral bandwidth of the nonlinear crystal is inversely proportional to the crystal thickness, thin nonlinear crystals are used. The ultrashort gate pulse can be obtained from dye laser or Ti : Al$_2$O$_3$ lasers. The upconversion or sum-frequency ($u = g \pm f$) pulses are generated through second-order nonlinear process when phase matching condition is satisfied. The sum-frequency signal is generated only when the fluorescence and gate pulse temporally overlap and its intensity is proportional to the intensity of the two inputs. Therefore, the intensity of the u signal is proportional to the intensity of the portion of the fluorescence that is temporally overlapped with the gate pulse. Intensity of the upconverted pulse is detected, and variation in its intensity and the fluorescence band are monitored with respect to time delay between gate and pump pulses, which gives a replica of the temporal change of the fluorescence. The output from the crystal is a mixture of gate pulse, generated harmonic pulse, and sum-frequency, which can be spatially, separated using appropriate geometry to reduce the background. However, if the angle between fluorescence and gate beams is increased, for better spatial separation, time resolution is lost because of different arrival times of the two outer gate rays at the crystal. Upconversion method is good when small spectral changes are to be observed for fluorescent compounds. A good time resolution up to $\sim$100 fs can be achieved with this technique. Time resolution and accuracy can be tuned using modification in experimental setup [25]. Readers are advised to go through Refs. [26–30] for detailed experimental technique.

The nature and magnitude of the interaction strengths in branched structures are very important for applications in light emission and light harvesting. Varnavski *et al.* [31] have investigated coherent effects in energy transport in model dendritic structures using ultrafast fluorescence anisotropy spectroscopy and presented significance of coherence in energy transport by strong interchromophore interactions in branched fluorescent macromolecules of various symmetries. Fluorescence quenching due to electron transfer has been studied by Iwai *et al.* [32]. The fluorescence decay, at quencher concentration, can be fit by combining the diffusion equation with the Marcus equation of electron transfer. At higher concentrations, an additional component with a time constant of 250 fs appears, which is explained by the competition between the radiative transition $S_2 \rightarrow S_0$ and the nonradiative internal conversion $S_2 \rightarrow S_1$. The fast $S_2 \rightarrow S_1$ nonradiative transition can be regarded as a charge separation process.

Chosrowjan *et al.* [33, 34] measured the time profile of the fluorescence of photoactive yellow protein (PYP) and reported a fast decay component both on the red and blue sides of the fluorescence peak, and that the decay time is shortest for the wild-type PYP among the examined samples including mutants. It was proposed that this ultrafast component is related to the dynamic motion of the wave packet in the excited state [35].

Advancements in ultrafast laser technique make it possible to explore spectroscopic characteristics of single-walled nanotube (SWNTs). Excitation dynamics in the SWNT was analyzed using ultrafast transient absorption [36–39], and TRF techniques with sub-100-fs time resolution [40, 41] revealed a very fast kinetic decay component in the time scale of a few hundred femtoseconds. Through analysis of the nonexponential kinetic decay component and the dependence of the corresponding amplitude on excitation intensity, it was concluded that exciton–exciton annihilation in semiconducting SWNTs is the dominant relaxation process [42].

Prabhu *et al.* estimate a 7–10 ps exciton formation constant for bulk CdSe using an upconversion technique. The same technique has been utilized by Underwood *et al.* [43] to probe the dynamics of photogenerated excitons in high-quality CdSe nanocrystals (5.2 nm). These crystals show rise time in the decay curve of the lowest excited state on increasing the excitation energy from 575 to 525 nm (2.15–2.36 eV), yielding information on the timescale for relaxation from higher-lying electron or hole states to the emitting state.

Rubtsov *et al.* [44] have studied spectral sensitization and supersensitization of silver bromide nanocrystals by the femtosecond fluorescence upconversion technique. Very fast nonexponential fluorescence decays were observed with a fast component from 0.4 to 2.5 ps, and an average decay time from 1.1 to 5.5 ps, which depend on the type and size of AgBr grains. The average fluorescence decay time is several times longer in the cubic grains than in the octahedral grains. On addition of a supersensitizer (3,3′-disulfopropyl-9-ethyl-4,5,4′,5′-dibenzothiacarbocyanine), which is coadsorbed on the surface of silver bromide grains, the fluorescence decay became several times faster, with a mean decay time of 0.60–1.3 ps.

The major drawbacks of the upconversion technique are that there is light mixing of second-order nonlinear optical process, the transmission efficiency as a gate is not so high, and the phase matching condition should be satisfied one by one at each wavelength. These drawbacks restrict its applications. These drawbacks may be solved upto some extent using the instrument proposed by Gustavsson *et al.* [45], which does well-automated scanning over wavelengths at fixed time delays and in which the phase-matching angles of the upconversion crystal, the monochromator, and the delay stage are simultaneously computer controlled. Broadband fluorescence upconversion detection technique was proposed by Ernsting and coworkers [46], with near-IR gating and improved light collection geometry to record the entire fluorescence band without readjusting optical elements with a single pump-gate scan.

To resolve some of the problems in ultrafast upconversion technique, another technique called "optical Kerr gate" (OKG) has come up as a potentially viable alternative; it does not require phase matching and is therefore simpler than gated upconversion. However, the OKG and upconversion techniques are suited for different applications. The OKG technique is based on "optical Kerr shutter," which works on Kerr effect [47]. Optical Kerr shutter has been known to be the most effective method to obtain temporal sections of various ultrashort phenomena such as transient luminescence [48], light absorption, photoconductivity, photoimaging [49], and so on. The Kerr gate technique uses the transient birefringence induced in a transparent Kerr medium. Selection of Kerr medium is in the heart of the sensitivity and temporal resolution of OKG experiment. One of the materials with a strong optical Kerr effect is CS_2 liquid whose approximate response time is 0.8 ps [50]. Since the response time of CS_2 is of the order of approximately picoseconds, its application is limited in advanced ultrafast instruments. A series of oxide, Chelcogenide, and hybrid (lead phthalocyanine [51]) glassy materials are now reported with improved nonlinearity [52–57]. Until now, chalcogenide glasses [58] (nonlinear response time <200 fs), antimony polyphosphate glasses [59] (~155 fs), and heavy oxide glasses such as Bi_2O_3-B_2O_3-SiO_2 (<90 fs) [60] has been reported. Time resolution better then 100 fs was obtained from 1 mm thick fused silica, which has relatively small dispersion and UV–Vis transparency. In OKG, sample is excited with ultrashort laser pulses and its fluorescence is focused into a Kerr shutter. The Kerr shutter is consists of a pair of highly efficient crossed polarizers and Kerr medium (see Figure 11.3). The polarizers are kept orthogonal to each other so that fluorescence from the sample passing through the first polarizer is blocked by the second polarizer. A strong, linearly polarizes ultrashort pulse (gate) is focused into Kerr medium with certain delay time, which introduces transient birefringence in the medium and enables momentary passage of fluorescence. The polarization of the gate pulse is tilted ($\sim45°$) against the polarizations of the two polarizers, so that the polarization of the fluorescence is changed from linear to elliptical when it passes through the Kerr medium. This way, fluorescence can pass through the second polarizer and be detected by the spectrometer.

OKG has several advantages such as it rejects unwanted emission due to cross polarizer and allows all emission frequencies to pass simultaneously so

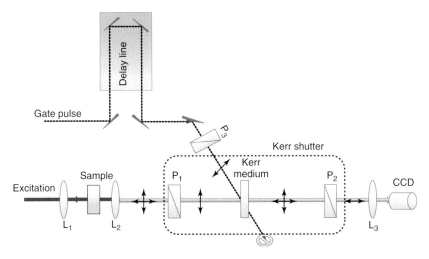

Figure 11.3 Typical experimental setup for time-resolved fluorescence based on Kerr effect.

that true broadband operation is achieved. The OKG technique is well suited for low-fluorescence samples. Arzhantsev and Maroncelli [61] have described the design and characterization of a general purpose spectrometer for recording time-resolved emission spectra of typical fluorescent species.

Analysis of solvation dynamics [62] was an important field of study that was strongly affected by ultrafast chemical reactions such as electron transfer reactions and proton transfer [63]. Kimura *et al.* [64] used OKG technique to study solvation dynamics of room temperature ionic liquids (4′-N,N-diethylamino-3-hydroxyflavone (DEAHF) and 4′-N,N-diethylamino-3-methoxyflavone (DEAMF)) and concluded that the excited state intramolecular proton transfer kinetics in these liquids are similar to those in conventional liquid solvents such as acetonitrile. Jimenez *et al.* [65] reported ultrafast solvation dynamics of a coumarin salt in water and demonstrated that a solvent response faster than 50 fs can dominate aqueous solvation dynamics.

Takeda *et al.* [66] used a quartz plate as a Kerr medium and found the typical Kerr efficiency and the instrumental response function to be ∼5% and ∼250 fs, respectively. The time resolution of our OKG technique is better than 100 fs, after the convolution procedure. The method has been applied to measure the ultrafast internal conversion from S_2 to S_1 state of β-carotene in n-hexane solution and the lifetime of the S_2 state has been determined to be 220 fs. Talapatra *et al.* [67] performed a degenerate four wave mixing (DFWM) experiment (400 fs pulse) and an OKG experiment (60 fs pulse) at a wavelength of 602 nm. The second-order complex hyperpolarizability was found to be $[-(5 \pm 2) + i(9 \pm 1)] \times 10^{-33}$ esu, but their theoretical results are 3 orders of magnitude smaller. Torres *et al.* [68] have proposed a two-stage nanostructured system for optical phase modulation and reported strong refractive nonlinearities with a dependence on polarization in the samples by a picosecond self-diffraction method and a femtosecond time-resolved

OKG technique. Control of the optical phase is a consequence of the separate excitation of the self-focusing and self-defocusing phenomena exhibited by Ag and Au nanocomposites, respectively. Hanada *et al.* [69] used optical Kerr shutter method to monitor room temperature TRF spectra of PYP and reported high time resolution of ~300 fs. For the first time, spectral narrowing in the picosecond time region was directly observed, which is consistent with the previously reported results of faster decay on both sides of the fluorescence peak.

11.2.3
Time-Resolved Ultrafast Infrared Spectroscopy

IR spectroscopy is a versatile tool as it provides a wealth of information about the presence of specific atom groups, nature of bonding, arrangement of atoms, and so on. The vibrational spectra with the associated modes of vibration and their magnitude have been studied for a large number of inorganic and organic compounds and are available as an extensive database. So the IR bands in the spectrum of the unknown molecule can be easily correlated with the presence of known groups or bonds. Any change of these values after some biological processes, for example, mutation, metabolic change, or any disease can be easily detected [70]. In the field of medicine, near-infrared reflectance spectroscopy (or NIR spectroscopy) holds a vital place. As far as the use of IR spectroscopy of NP is concerned, the basic hurdle is the resolutions of the incident beam to 5–10 μm. The FTIR technique has been suitably modified to enable it to study nanoscale objects. It has provided important information about chemical bonding in many newly developed futuristic materials, for example, CNT, graphene, and fullerene.

Although time-resolved IR spectroscopy was introduced in 1980s [71, 72], it gained popularity with the advent of ultrafast lasers, which provide rich molecular structural information, and transient chemical species in dynamic systems, which occur over a temporal range extending from minutes to femtoseconds. It provided valuable information on the processes of ligand binding, dissociation, protein folding, photocycles, reaction kinetics, and excited-state electronic structures, and a host of other dynamic chemical phenomena can be measured and characterized. Another important need of this technique is the information about hydrogen bonding because it has profound effects on molecular properties. Information about hydrogen bonding can be probed by analyzing −OH stretching vibration.

For ultrafast IR spectroscopy, materials are excited with two ultrafast laser pulses. The first is an IR pulse corresponding to a particular vibration (probe), and the second beam from an intense ultrafast laser of, usually, UV wavelength, acts as a pump beam. The probe beam is tuned to the UV-NIR regions for resonance Raman studies and to the mid-infrared (MIR) region to probe vibrational transitions. The depletion in vibrational population is monitored with time using either IR or Raman probe, depending on the vibrational mode of interest. Another technique uses a pump pulse to trigger a photochemical reaction and then a second laser pulse to spectrally probe the progress of the reaction after a known, controlled time delay. These techniques provide information about the kinetics of excited dark

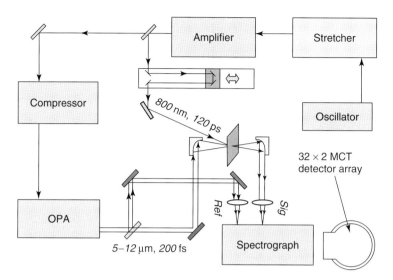

Figure 11.4 Experimental arrangement for ultrafast time-resolved IR spectroscopy. MCT, mercury cadmium telluride; OPA, optical parametric amplifier. (Source: Reprinted with the permission from [73] copyright @ American Chemical Society.)

and radiative states. Both the techniques (IR and Raman) facilitate probing of the individual mode of vibration and the distribution of vibrational energy among the internal modes as well. Experimental arrangement for ultrafast time-resolved IR spectroscopy is depicted in Figure 11.4 [73].

Broadband IR pump probe uses identical broadband IR pulses for pumping as well as probing and thus, exciting all modes within the measurement window. Optical parametric amplifier divides MIR pulses into a strong pump pulse and a weak probe pulse, which are spatially overlapped in the sample. The pump beam is chopped at half the repetition rate of the laser system, and for each pump-probe delay, a "transient spectrum" is measured as the difference in absorption between the two states: pump on and pump off. Measuring transient spectra at a sequence of delays between pump and probe allows one to construct "transient traces" as the signal at one IR probe frequency versus delay time between pump and probe. The details about the fundamental aspects of time-resolved IR spectroscopy can be obtained in Ref. [74]. IR pump probe technique was used to explore several dynamical processes. Zhang *et al.* [75] have reported observation of carbene and diazo formation from aryldiazirines within a few picoseconds of the laser pulse by ultrafast IR spectroscopy. Ultrafast IR spectroscopy is extensively used to explore the structure and dynamics of hydrogen-bonded systems in the condensed phase by several groups [76–80]. The vibrational population relaxation of the OH stretching mode of hydrogen-bonded phenol complexes (phenol–acetone, phenol–diethylether, and phenol–tetrahydrofuran) has been discussed by Ohta and Tominaga [81]. They reported the time scales of vibrational population relaxation for these complexes and correlated them with hydrogen bond strength, that is, stronger hydrogen bonding leads to faster vibrational population relaxation. Banno *et al.* [82]

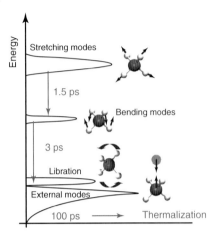

Figure 11.5 Different relaxation processes and time constant for LiBH$_4$. (Source: Reprinted with the permission from Ref. [85] copyright @ American Chemical Society.)

have shown that in phenol-base complexes the vibrational energy relaxation time strongly correlates with the strength of hydrogen bonding. The observed variation is proposed to be due to increased coupling between the OH stretching mode and the accepting mode of the vibrational energy relaxation. Heilweil *et al.* [83, 84] have measured the dependence of vibrational energy relaxation on hydrogen bond strength. The vibrational energy relaxation time of the free pyrrole in CCl$_4$ is 49 ps, whereas this time ranges from 13 ps for a complex with acetone to 0.7 ps for a complex with triethylamine in CCl$_4$. They concluded that the vibrational energy relaxation time decreases monotonically as the basicity of the proton acceptor increases, which indicates that the magnitude of the enhanced vibrational energy relaxation is mostly dictated by the strength of the hydrogen bond and is relatively independent of the vibration. Vibrational dynamics of LiBH$_4$ and several of its deuterium isotopomers were discussed in detail by Andresen *et al.* [85]. Vibrational lifetime of various BH and BD stretching modes were found to be ≈1.5 ps. They also reported different vibrational relaxation steps and time constants in LiBH$_4$ (Figure 11.5).

One of the most important concerns with nanomaterials is the investigation of sophisticated surface properties, which is important for both bare and coated NPs. Investigations are primarily centered on a search for intrinsic broadening mechanism in the light of the so-called T_1 (energy) and T_2 (phase) relaxation processes of the vibrational excited states of adsorbates. Those are the vibrational energy dissipation into the elementary excitation, such as phonons or electron–hole pairs in the metal substrate, and pure dephasing due to the energy exchange with the surroundings. Characterization of surface includes electron transfer dynamics of the molecule coated on the NP and in the semiconductor NP, which is well explored by femtosecond IR spectroscopy. The absorption of electron–holes in semiconductors falls in the IR region.

Ultrafast spectroscopy has an interesting application in semiconductor nanostructure materials as an electron may move ballistically at a velocity of $107 \, cm \, s^{-1}$ and hence can travel 10 nm in 100 fs. These length and time scales have become of great importance in real devices. In most recent studies, it was found that the forward electron injection from the excited state of strongly adsorbed dye to TiO_2 occurs in ultrafast time scale [86–90]. For example, in $Ru(dcbpy)_2(NCS)_2$-sensitized TiO_2 thin film [91, 92], an injection time of about 50 fs or faster has been reported. Charge carrier (electron and holes) dynamics in semiconductor materials, which play a critical role in determining the properties of the NPs, including transport, recombination, and trapping, has been discussed using IR technique by several groups [93–95]. The absorption coefficient of photogenerated electrons increases as the wavelength of the probe light becomes longer from visible to IR [96], and it is not altered by the emission of visible light during recombination. Yamakata and coworkers reported photogenerated charge carrier dynamics such as recombination kinetics and charge transfer to reactant molecules (water, O_2, CH_3OH) in TiO_2 [97], $NaTaO_3$ [98], and $BiWO_3$ [99] photocatalysts. Transient species created by charge transfer to reactant molecules [100] and electron transfer from a GaN photocatalyst to a Pt cocatalyst [101], an active center of hydrogen evolution, have also been examined.

Heimer *et al.* [95] reported excited state properties of $[Ru(4,4'-(COOCH_2CH_3)2-2, 2'-bipyridine)(2,2'-bipyridine)2]^{+2}$ and $[Ru(4,4'-(COOCH_2CH_3)^{2-}2,2'-bipyridine)-(4,4'(-CH_3)^{2-}2,2'-bipyridine)2]^{+2}$ in solution and anchored to nanostructured thin films of TiO_2 and ZrO_2 using time-resolved IR spectroscopy. For these molecules attached to TiO_2 semiconductor films, a transient absorption appears due to electrons injected into TiO_2 within $\sim$20 ps time constant. Photoinduced electron-transfer dynamics in $Fe(II)(CN)^{64-}$-sensitized TiO_2 NPs in D_2O solution were studied by Lian *et al.* [102, 103]. It was reported that the forward electron injection from $Fe(II)(CN)^{64-}$ to TiO_2 occurs in <50 fs, indicating a direct photoinduced charge-transfer process, while the back electron transfer decay is composed of three-exponential fit yielding time constants of 3 ps (35%), 40 ps (30%), and >1 ns (35%). The injection time for coumarin-343 sensitized TiO_2 NPs in D_2O is determined to be 125 ± 25 fs [104].

11.2.4
Time-Resolved Ultrafast Raman Spectroscopy

Although ultrafast IR spectroscopy provides a great deal of information about vibrational modes and their dynamics, all modes could not be explored due to restriction of selection rule. To explore the transient vibrational analysis, it is worth to use IR and Raman probes simultaneously. The Raman technique [105], first considered as an alternate for the IR spectroscopy, lost some of its charm after the development of good IR spectrophotometers. The discovery of the laser in 1960 and especially the development of tunable, high-power, and short pulse width lasers rejuvenated this technique to the extent that it is one of the most widely used spectroscopic techniques in chemistry, physics, and biology. Raman spectroscopy basically detects how objects scatter light and how each type of molecule produces a unique Raman signature. The basic Raman process involves interaction of the incident photon with the molecule, which is then raised to a nonstationary (virtual) state. This state decays quickly to a lower state, often a vibrationally excited state. The theory of Raman scattering, which now is well understood, shows that the spectrum results from the same type of quantized vibrational changes that are associated with IR absorption. Thus, the difference in wavelength between the incident and scattered radiation corresponds to wavelengths in the MIR region. Indeed, the Raman scattering spectrum and IR spectrum for a given species often resemble one another quite closely. There are, however, enough differences between the kinds of groups that are IR active and those that are Raman active to make the techniques complementary rather than competitive. For some problems, the IR method is the superior tool, while for others, the Raman procedure is more useful. An important advantage of Raman spectrum over IR spectrum lies in the fact that presence of water does not interfere with the spectrum of the sample; indeed, Raman spectra can be obtained from aqueous solutions. In addition, glass or quartz cells can be used, thus avoiding the inconvenience of working with sodium chloride or other atmospherically unstable confinements. This advantage is particularly important for biological and inorganic systems and in studies dealing with water pollution problems. Despite these advantages, Raman spectroscopy is subject to interference by fluorescence or impurities in the sample. Until now, there are several techniques based on the basic concept of Raman scattering such as surface-enhanced Raman spectroscopy (SERS) [106], coherent anti-Stoke's Raman spectroscopy (CARS) [107] hyper Raman spectroscopy [108], stimulated Raman spectroscopy [109], and resonance Raman spectroscopy [110]. Several reviews and books are available on ultrafast Raman spectroscopy [111].

Raman scattering is a weak process; however, the low sensitivity of conventional Raman scattering is overcome by resonance effect, which enhances a Raman signal by 10^6–10^8 times and facilitates the detection of signal from dilute solutions. Resonance Raman spectrum is obtained when the frequency of the incident laser light matches with or is close to that of an electronic transition of the scattering species. The intensity of resonance Raman scattering is a function of dynamics in the excited electronic state, and hence this technique can provide subpicosecond time-domain information about the excited chromophore and the solvent motion, which is coupled to the electronic transition on excitation with continuous pulse [112]. To monitor the dynamics of the transient species, pump-probe method is mainly used.

The experimental setup for pump-probe Raman spectroscopy is given in Figure 11.6. For pump-probe Raman spectroscopy experiment two laser pulses are used: one of them is the pulse having frequency suitable to pump the sample to the excited state. Pump beam creates population in the excited electronic state and vibrational coherences in both the ground and excited states. Another pulse, probe, with determined time delay may get attenuated, transmitted, or amplified by the sample before it reaches a detector. Spectra can be obtained at a series of time delays, enabling a kinetic "profile" of the excited state decay to be constructed.

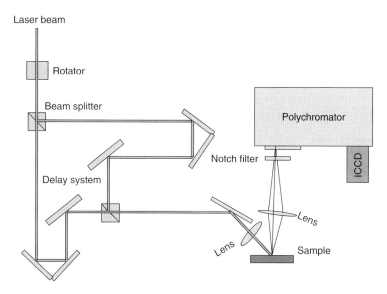

Figure 11.6 Experimental arrangement for ultrafast time-resolved Raman spectroscopy. ICCD, imaging charge coupled device.

The spectrum analyzing system is basically the same as that for conventional Raman spectroscopy. The Raman light is collected by a lens on the entrance slit of a single polychromator equipped with a liquid-nitrogen-cooled CCD camera. The pump pulse needs to be tuned to the appropriate absorption band to create the intermediate population, while the probe pulse must be coincident with an electronic absorption of the excited state. If the latter requirement is not fulfilled, that is, the excited state does not absorb at the probe wavelength, and then as the power of the excitation pulse is increased, the intensity of the Raman signals may actually decrease as the concentration of the analyte in the ground state becomes increasingly depleted. Species produced by a pump pulse are observed by Raman scattering; a pump-probe exploration of the initial electronically excited state can be done if the probe is prompt and relaxation processes can be performed if it is delayed. However, the isomerization and subsequent relaxation of molecules in an excited state is monitored by a variable-delay pulse. Figure 11.7 illustrates subpicosecond time-resolved Raman spectra of GaAs at different delay of probe pulse. If we have a small intensity in the second (probe) pulse and a strict polarization selection rule, which allows only the Raman signal from the second pulse, we can get Raman scattering of the carriers excited by the first (pump) pulse and detected by the second pulse. However, in actual situation, the Raman signal induced by the first pulse and the carriers produced by the second pulse cannot be ignored. Therefore, extracting correct time-resolved spectra (b–d) after the subtraction of (a) is shown in Figure 11.7 [113].

Time-resolved resonance Raman spectroscopy tends to suffer, however, from complications such as interference with spontaneous fluorescence that most

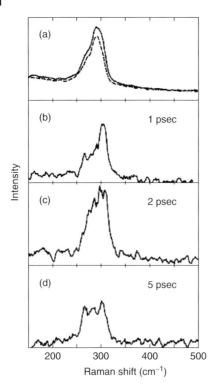

Figure 11.7 (a) Time-resolved anti-Stokes Raman spectrum for GaAs at room temperature. The instantaneous carrier concentration induced by pump pulse was almost 5×10^{16} cm^{-3}. The solid curve corresponds to the probe delayed by 1ps, while the dashed curve is attributed to the probe pulse that precedes the pump pulse by 20 ps. (b–d) Difference spectra at 1, 2, and 5 ps after the subtraction of Raman spectra like (a). (Reprinted with the permission from [113] copyright @ American Physical Society.)

molecules are capable of emitting in the excited state. The problem of strong background and fluorescence can be solved up to some extent by using "Kerr gate" method (similar to that described in section 11.2.2). It takes advantage of the instantaneous nature of Raman scattering in contrast to relatively slow emission of light in fluorescence. Hence, in this case when gate (in Kerr medium excited by gate pulse) is open for a short time of ~2 ps, only the signal can reach the detector and not the fluorescence as the most of the fluorescence that occurs in the time range of ~100 ps to ns does not reach the detector [114].

In case of the CARS technique, a strong directional coherent beam emerges out in the phase-matching direction at a blueshifted frequency of the sample when irradiated by two incident beams one of which corresponds to a Stoke's line. The intense signals are produced at resonance if the difference between pump and Stokes beams matches a Raman-active transition of the medium. CARS technique has been extensively used to map the temperature distribution in an internal

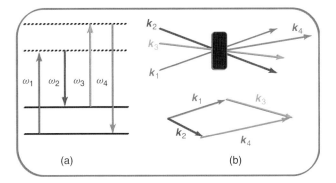

(a)   (b)

Figure 11.8 Schematic diagram of the molecular states showing the CARS process. Overlapping of wave vectors k_1 (pump), k_2 (Stokes), and k_3 (probe) to generate CARS signal along k_4 and the vectorial representation of the phase-matching condition.

combustion engine and *in situ* measurements of species concentrations. Because of its selectivity, the analytical potential is also high. CARS signals are often orders of magnitude stronger than those produced by spontaneous Raman scattering.

Until now, simple CARS technique has been extended to dual-pump [115], dual-broadband rotational [116], triple pump [117], and dual-pump dual-broadband CARS [118] techniques. There are several advantages of CARS technique including (i) reduction or elimination of the nonresonant contribution to the CARS signal when the probe beam is delayed with respect to the pump beam; (ii) reduction or elimination of the effects of collisions on the CARS signal, thereby reducing modeling uncertainty and increasing signal-to-noise ratio; and (iii) generation of signals at rates of 1 kHz or greater. The reduction or elimination of the nonresonant background and collisional effects will greatly simplify the modeling of CARS spectra and improve accuracy by eliminating the need for information concerning Raman linewidths. Owing to the inherently broadband spectral bandwidth of femtosecond laser, several modes including those falling within the femtosecond laser spectrum also get excited, which limits its use for CARS. For selective excitation, an additional treatment for pulse shaping [119] is required. Figure 11.8 is the schematic diagram of the molecular states for the CARS process and overlapping of wave vectors k_1 (Pump), k_2 (Stoke's), and k_3 (probe) to generate CARS signal along k_4 with the vectorial representation of the phase-matching condition.

Time-resolved CARS is a widely used technique for monitoring the changes in vibrational frequency in the femtosecond timescale [120]. Although CARS is linked with homodyne detection and shows high sensitivity, the phase-matching condition dictates that the direction of its signal beam varies constantly as the Raman frequency changes [121]. Therefore, it is difficult to obtain a dispersed Raman spectrum in a fixed optical configuration. A few other drawbacks of CARS are (i) the presence of the nonresonant background, which limits the detectivity in the range of parts per million to 1% depending on thermodynamic conditions and species studied; (ii) the sensitivity to laser instabilities like the other techniques; and (iii) being subject to saturation at the higher laser power density levels

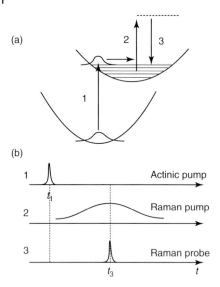

Figure 11.9 Energy level diagram (a) and pulse timing schematic (b) for femtosecond stimulated Raman spectroscopy. (Source: Reprinted with the permission [125] copyright @ American Chemical Society.)

(10^{10} W cm^{-2} or more). To overcome many of these shortcomings, several research groups attempted to improve the sensitivity of Raman spectroscopy in the 1970s by developing CW and nanosecond stimulated Raman spectroscopy in combination with optical heterodyne detection [122].

The concept of femtosecond stimulated Raman spectroscopy has been proposed by several groups [123]. Detailed experimental approach [124] of this technique enables the acquisition of resonance Raman spectra over a 1500 cm^{-1} window, free from fluorescence interference, with simultaneous high time ($<$100 fs) and frequency ($\sim$15 cm^{-1}) resolution. The stimulated Raman gain (SRG) is a third-order nonlinear optical process. Through the temporal and spatial interaction of the Raman pump pulse and the Stokes-shifted continuum pulse, photon creation can occur at frequencies resonant with vibrational energy gap. As can be seen from Figure 11.9, a photochemical reaction is triggered by a femtosecond ($\sim$50 fs) pump and then two laser pulses arrive simultaneously at the sample, which induces the Raman transition, resulting in a picoseconds "Raman pulse" and a femtosecond broadband continuum "probe pulse" that stimulate the scattering of any vibrational modes with frequencies between 600 and 2100 cm^{-1}. This method is very useful for acquisition of very high time-resolved vibrational structural information that is not accessible to the incoherent, spontaneous Raman technique.

The above-described techniques are not replacements for others and are widely in practice to evaluate several transient processes that can only be accessed by using these techniques. A brief review of the success of these techniques is given in the following paragraphs. The electronic Raman scattering time-resolved measurement has been done by using a single-pulse method by Kim *et al.* [126]

in bulk GaAs. They obtained the temperature of the photoexcited electrons from an analysis of Stokes to anti-Stokes intensity ratio of single particle excitation as a function of the excitation pulse width. Time-resolved Raman spectroscopy in subpicosecond region was applied to the optical phonons in GaAs by Kash *et al.* [113]. They used a pump-probe regime to observe Stokes to anti-Stokes ratio of the longitudinal optical (LO) phonon, which reflects an effect of photogenerated carriers. Butler *et al.* [127] have presented picosecond transient Raman spectra of trans-4,4′-diphenylstilbene and have analyzed the spectra on the basis of comparison with the Raman spectra of the ground and excited states and anion radical spectra of *trans*-stilbene and biphenyl. Taking the advantage of resonance Raman spectroscopy Woodruff, Dallinger, and coworkers have identified the lowest metal-to-ligand charge transfer (MLCT) state as involving electron localization on a single paradigm complex $[Ru(bpy)3]^{2+}$ ligand rather than delocalization over all three ligands [128, 129]. Schoonover *et al.* [130] have presented the first example of the application of transient Raman technique to intramolecular electron transfer in a series of chromophore quencher complexes based on MLCT excited states. The Raman data obtained supplement transient absorption providing vibrational spectra when transient absorption spectra are inconclusive and allow for the structures of short-lived intermediates to be inferred. Iwata and Hamaguchi have obtained the transient Raman spectra of S_1 *trans*-stilbene in *n*-heptane at one excitation wavelength using a transform-limited picosecond Raman apparatus [131]. They observe a distinct difference in the dynamics of the peak position and bandwidth at very early times. They attribute their results both to intramolecular effects and to structural relaxation involving the motion of the surrounding solvent. Petrich and Martin [132] have used utrafast Raman spectroscopy along with absorption spectroscopy to explore photophysics and the dynamics of hemoglobin and myoglobin. McMahon *et al.* [133] have also followed intramolecular electron transfer in a covalently linked porphyrin–viologen complex by transient Raman spectroscopy. A direct measure of quaternary structure change (∼20 μs) in photolyzed carbonmonoxyhemoglobin using transient UV resonance Raman spectroscopy has been reported by Su *et al.* [134].

Applications of coherent anti-Stokes Raman scattering in Raman lasers and Raman wavelength converters have been reviewed by Vermeulen *et al.* [135]. Kataoka *et al.* [136] have reported CARS spectra of *p*-xylene and N_2 gas, and the results were compared with the simulation curves based on two extreme cases of the statistical property of the laser source. It was shown that the effect of laser linewidth becomes remarkable when the Raman linewidth is comparable to or smaller than the laser linewidth and that the degree of coherence among the frequency components within the laser linewidth is an important factor in determining the spectral profile. Nikitin [137] has derived formulas for evaluating the diffusion coefficient and size of gas molecules from transient coherent anti-Stokes Raman scattering measurements. Schmitt *et al.* [138] have applied CARS to explore molecular dynamics in the gas phase (iodine vapor). It was reported that by changing the timing of the laser pulses, the wavepacket motion on both the electronically excited and ground states can be detected as oscillations in the coherent anti-Stokes signal.

Hayden and Chandler [139] reported the application of femtosecond time-resolved coherent Raman techniques to excite and monitor the evolution of vibrational coherence in gas-phase samples of benzene as well as of 1,3,5-hexatriene. Wang *et al.* [140] have investigated the vibrational dynamics in H_2O and D_2O using a pump frequency of 3400 cm^{-1}. The Stokes Raman method is used to study the delay between the excited-state decay and the ground state recovery, the vibrational Stokes shift, and the generation of weakened hydrogen bonding due to heat released by vibrational relaxation. Moger *et al.* [141] have shown the potential application of CARS microscopy to provide excellent label-free contrast of harmful, metal oxide NPs deep within a biological structure. They found that tuning $\omega_p - \omega_s$ to match the lipid CH stretch gave excellent contrast of the metal oxides against the surrounding biological structure. Wijekoon and Hetherington [142] reported hydrogenation of ethylene adsorbed on ZnO(0001) surface using a planar optical waveguide geometry. The CARS signal from the adsorbate layer was enhanced by choosing a combination of guided waves to minimize the contributions from the ZnO film. Spectra in the region of the ν_1 stretch provided evidence of the presence of several different types of adsorbed ethylene on the ZnO surface. Tong and coworkers [143] have visualized receptor-mediated endocytosis and intracellular trafficking with the aid of a CARS probe. By tuning the laser pulses to 3045 cm^{-1} corresponding to the aromatic C–H stretching vibration, the polystyrene NPs with a high density of aromatic C–H bonds were detected, while the epidetected CARS signal from cellular organelles was canceled by the destructive interference between the resonant contribution from the aliphatic C–H vibration and the non-resonant contribution. A detailed review on CARS characterization of CNT is given in Ref. [144] and has reported the coherent phonon dynamics in single-walled CNTs using broadband CARS in time–frequency domain. Zone-center G-phonons were coherently created by optical excitation with the relaxation time of 1.1 (0.1 ps), while zone-boundary D-phonons showed incoherent behavior, indicating electron–phonon, electron–defect, and phonon–phonon interactions in semiconducting nanotubes. Kang *et al.* [145] have estimated the lifetimes of optical phonons in single-walled CNTs by time-resolved incoherent anti-Stokes Raman scattering using a subpicosecond pump-probe method. Lifetimes in semiconducting and metallic nanotubes at room temperature are similar, 1.2 and 0.9 ps, respectively. The lifetimes of optical phonons decrease with increasing temperature, approximately scaling as $1/T$, consistent with anharmonic processes being the dominant decay mechanism for both semiconducting and metallic nanotubes. Kim *et al.* [146] have observed intrinsic coherent anti-Stokes emission in lithographically patterned gold nanowires. Polarization-dependent measurements reveal that the nanostructure's anti-Stokes response is polarized in the direction of the transverse SPR of the wire. Ichimura *et al.* [147] have shown that an electric field enhanced by a metallic nanoprobe has locally induced coherent anti-Stokes Raman scattering of adenine molecules in a nanometric DNA network structure. They visualized the DNA network structure at a specific vibrational frequency ($\sim$1337 cm^{-1}) corresponding to the ring-breathing mode of diazole of adenine molecules.

The laser, detection system, and methods that enable femtosecond broadband stimulated Raman spectroscopy are presented in detail by McCamant et al. [148]. Jin et al. [149] have developed a new femtosecond probe technique by using stimulated Raman spectroscopy. The SRG of neat cyclohexane was obtained to demonstrate the feasibility of the technique. Femtosecond stimulated Raman spectroscopy was proposed as a highly useful probe in time-resolved vibrational spectroscopy. Kukura et al. [150] have estimated that the actual time constant for high-spin formation was obtained from femtosecond stimulated Raman scattering experiments. They further obtained the vibrational spectrum of the short-lived ($\sim$160 fs) second excited singlet state, S_2, of β-carotene. Broad, resonantly enhanced vibrational features are also observed at $\sim$1100, 1300, and 1650 cm^{-1}, which decay with a time constant corresponding to the electronic lifetime of S_2 [151]. Smeigh et al. [152] analyzed the photoinduced spin crossover in [Fe(tren(6-R-py)3)]$^{2+}$ (R=H, CH$_3$)]. Dasgupta et al. [153] have investigated ultrafast excited-state isomerization in phytochrome and identified the timescale of formation for the primary photoproduct Lumi-R and assigned the key vibrational modes that are involved in the photochemistry. After excitation, the system evolves on the multidimensional excited-state surface, leading to the formation of the intermediate I* in 500 fs, which is significantly distorted along the C_{14}–C_{15}=C_{16} moiety. Wynne et al. [154] have found ground state coherence in the complex stretching mode at 165 cm^{-1} due to excitation by impulsive stimulated Raman scattering as well as decay of the ground state bleaching signal due to return electron transfer to the ground state. It is shown that a breakdown of the Born–Oppenheimer approximation can give rise to a coupling that leads to the observed electron-transfer reaction. Frontiera et al. [155] have probed interfacial electron transfer in coumarin-343-sensitized TiO$_2$ NPs using combination of ground and excited state stimulated Raman techniques, which clearly shows that the injected electron comes from the π backbone structure of coumarin 343, that the donation mechanism involves multiple nuclear coordinates including ring breathing and carbon bond stretching motions, and that the delocalized hole resides in the carbon backbone. Lee and coworkers [156] find a lifetime of 29.2 ps at room temperature in wurtzite ZnO, which is about an order of magnitude longer than typical lifetimes for Raman phonons in semiconductors [157]. Similarly, the corresponding E$_2$ mode in GaN was found to have a lifetime of $t = 70$ ps [158]. Aku-Leh et al. [159] have measured anharmonic properties of the low-frequency E$_2$ phonon in ZnO using impulsive stimulated Raman scattering. At 5 K, the frequency and lifetime are 2.9789 ± 0.0002 THz and 211 ± 7 ps, respectively. Bragas et al. [160] reported acousticlike vibrations in spontaneous Raman measurements on CdTe$_{0.68}$ Se$_{0.32}$ quantum dots (QDs) embedded in a borosilicate matrix and ascribed them to the $l = 0, l = 2$, and $l = 1$ modes of a spherical particle, whereas pump-probe stimulated Raman measurements on the same sample reveal only the symmetric $l = 0$ mode, the frequency of which increases as the laser central energy moves to higher energies. Consistent with the strong dependence of the exciton energy on the particle size, such a behavior is attributed to resonant size-selective excitation of nanocrystallites, whose absorption edge coincides with the central energy of the optical pulses.

11.2.5
Time-Resolved Ultrafast Faraday Rotation (TRFR) Spectroscopy

Optically active materials have the capability of rotating polarization vector of linearly polarized light in the presence of longitudinal (Faraday geometry) magnetic field. The refractive indices of material for left and right circularly polarized light are unequal, and the magnitude of this difference depends on the degree of magnetization of the materials and wavelength of the light being used. Circular birefringence induced by external magnetic field results in rotation of the polarization vector of linearly polarized light and is the basis of Faraday rotation. A sample placed in a longitudinal magnetic field induces Faraday rotation per unit length, which is proportional to the magnetization (M) of the sample. Therefore, materials having ordered spins (ferromagnetic or antiferromagnetic) or frustrated magnetic behavior exhibit Faraday rotation according to their degree of magnetization. Faraday rotation is a nondestructive, noninvasive spectroscopic technique that can measure small changes in sample magnetizations [161].

TRFR is a pump-probe spectroscopic technique that provides a unique way to determinate dopant–electron and exciton–exciton interaction strengths in spintronic materials by probing the transient magnetism of the sample. Here, a circularly polarized (right or left) pump beam of femtosecond pulse duration photoexcites spin polarized electrons from the sample placed in an external longitudinal magnetic field. In most of the reported experiments, the direction of magnetic field was perpendicular to the direction of laser beams and both pump and probe beams propagated collinearly. The magnetic response of the excited electrons are measured on the basis of how they rotate a linearly polarized probe beam of ultrafast duration. In this pump-probe technique, carriers are generated as well as probed by ultrafast laser pulses in resonance with the fundamental absorption band of bulk semiconductor or with a confined energy level of semiconductor nanostructures. Experimental arrangement for TRFR is depicted in Figure 11.10.

Whitaker *et al.* [162] pumped ZnO and $Zn_{1-x}Co_xO$ sol–gel thin films placed in external magnetic field of <1.4 T using the femtosecond laser pulse operating at the wavelength close to the bandgap of ZnO (368 nm at 10 K and 375 at 298 K). The rotation of the linear polarization of the probe beam (Faraday effect) at different time delays was measured after it passed through the sample. The pump beam was switched between left and right circular polarization. TRFR has previously been used to study epitaxial n-type ZnO films, where ensemble electron spin dephasing times up to $T_2^* = 2$ ns at 10 K were measured [163]. In another reports, TRFR was applied to study coherent spin dynamics in colloidal CdSe and related chalcogenide nanocrystals [164]. Janßen *et al.* [165] described application of TRFR spectroscopy in the ultraviolet region to diagnose the ultrafast spin dynamics of electrons in the colloidal ZnO QDs. Faraday rotation signals were observed at frequencies corresponding to an effective g factor of $g^* = 1.96$. Biexponential oscillation decay was reported due to (i) rapid depopulation of the fundamental exciton ($\tau = 250$ ps) and (ii) slow electron spin dephasing ($T_2^* = 1.2$ ns) within a metastable state formed by hole trapping at the QD surface. Figure 11.11 describes TRFR signal, magnetic

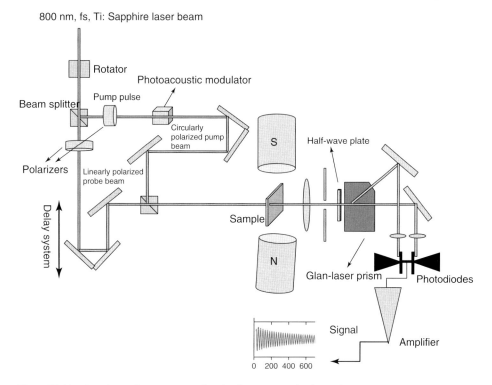

Figure 11.10 Experimental arrangement for ultrafast time-resolved Faraday rotation spectroscopy.

field dependence of the TRFR oscillatory frequency, and wavelength-dependent amplitude of TRFR signal for 7 nm diameter ZnO QDs at 10 K.

11.3
Other Multiple Wave Ultrafast Spectroscopic Techniques

11.3.1
Photon Echoes

The techniques that we have discussed earlier basically give frequency-domain information on the ensemble-averaged interactions that couple to the states involved in the transition. Usually, a spectroscopic peak contains four main characteristics, viz., peak position, intensity, width, and shape. The position of a peak and intensity describe the energy and the strength of an electronic transition, whereas the peak shape and width explore the local environment of the molecules and the intramolecular dynamics. Interaction between molecule and the solute includes microscopic dynamics, intermolecular couplings, and static structural perturbations associated with the distribution of local solvent configurations and involves

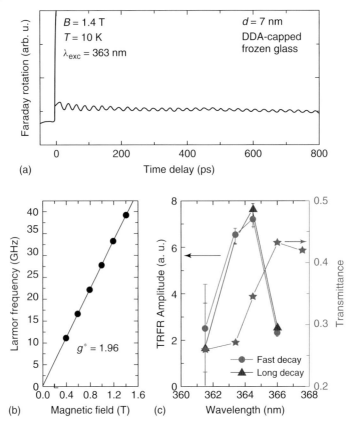

Figure 11.11 (a) TRFR signal, (b) magnetic field dependence of the TRFR oscillatory frequency, and (c) wavelength-dependent amplitude of TRFR signal for 7 nm diameter ZnO QDs at 10 K. (Source: Reprinted with the permission from Ref. [165] copyright @ American Chemical Society.)

the timescales of solvent evolution that modulate the energy of a transition. In general, when a molecule gets coupled to a weakly interacting local environment, the local environment alters the system to depart from a perfect sinusoidal behavior as each two-level system interacts differently with the surrounding environment. This process is called *optical dephasing*. As a consequence of optical dephasing, homogeneous or inhomogeneous broadening in absorption and fluorescence peaks occur.

Although the explanation of broadening depends on sophisticated processes, generally, two proximate situations may be considered. First condition is if the surroundings of each oscillator change rapidly to produce a similar environment for each oscillator (homogeneous dephasing) and second condition if each oscillator has a slightly different local environment, which does not change significantly over the course of time (inhomogeneous dephasing). Owing to inhomogeneous dephasing, a peak shift (shift in energy from unperturbed peak) occurs. Photon echo is a well-known nonlinear optical phenomenon, and it usually takes place as a transient coherence process in an inhomogeneous broadening atomic system using multiple pulse technique. The theoretical concept of photon echo was described by Zinth and Kaiser [166].

Ideally, in the limit of inhomogeneous broadening, photon echo experiments have shown to effectively eliminate the inhomogeneous contribution of the optical spectrum [167]. It includes two-pulse photon echo [168], three-pulse photon echo [169], and accumulated photon echo [170]. The basic concept of photon echo is analogous to that of spin echoes used in NMR or MRI technique. Initially, all the molecules have slightly different transition frequencies, and when these are excited with the ultrashort pulse (t_p), under the effect of electric field component, molecules are transferred to electronic coherence state. The time of the pulse (t_p) is shorter than the coherence time T_2. Immediately after the pulse, the molecules are oscillating in phase and strong coherent emission in the direction of the applied pulse, which is due to inhomogeneous broadening, that is, all molecules have slightly different transition frequency, the corresponding coherences will evolve at a different rate, the coherence quickly dephases, causing the coherent emission to decay. The time of decay depends approximately on the exciting pulse duration. When another pulse (π) is incident, it reverses the time evolution of the individual coherences, and all coherences regain their perfect phase relation such that the ions once again oscillate in phase and strong coherent emission results in the direction of phase matching. This coherent emission is called as "*echo*." Monitoring the variation in echo intensity versus delay time gives a direct probe of the homogeneous fluctuation of the transition dipole moment, which is in turn determined by the fluctuations of the environment. A typical two photon echo experimental setup is given in Figure 11.12. The two-photon echo technique can be used to store spatial and temporal information of a laser pulse [171]. Further, the

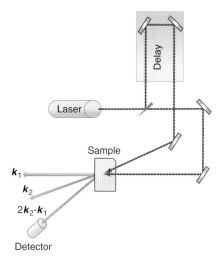

Figure 11.12 Experimental setup for two-beam photon echo.

bandwidth of the two-photon echo can be broad. The bandwidth of operation of the two-photon echo is dependent on the Fourier width of the pulse and the frequency distribution of the storage medium.

There are several interaction processes in liquid of the order of a few tens of femtoseconds and cannot be analyzed solely by two-pulse photon echo technique. To explore such low timescale processes, another pulse that extends the dynamical range on which transition frequency modulation can be studied is used. In this technique, similar to two-pulse technique, one pulse (k_1) creates coherence between the ground and excited state and another pulse (k_2) ends the coherence period (t) and initiates a population period (ground or excited state) T where spectral diffusion can occur. The role of third pulse (k_3) is to rephase the population period back into another coherence period. The resulting echo signal (k_e), which occurs in the direction of phase matching, that is, $k_1 - k_2 + k_3$ and $-k_1 + k_2 + k_3$, corresponds to specific spatial locations because of momentum conservation. By introduction of an additional controllable time interval, in comparison with the two-pulse photon echo technique, the timescales of the fluctuations of the transition frequencies can be characterized over a wide dynamic range. The peak shift for a particular T can be obtained by determining half the difference between the maximum of the peak positions obtained by scanning "t." A nonzero peak shift indicates rephasing ability, that is, retention of optical transition frequency information. As we know, three-pulse photon echo is a significant technique for the quantum communication in the future, and the intense light pulses can be effectively stored with this technique in an inhomogeneously broadened atomic ensemble.

Optical dephasing measurements such as photon echo and hole burning were originally used to study line-broadening mechanisms in low-temperature solids [172]. The storage of light at the single-photon level using three-pulse photon echo technique was reported by Sangouard *et al.* [173]. The quantum communication and computation based on three-pulse photon echo technique was experimentally investigated by Abazari *et al.* [174], and the dephasing time of a pure exciton at room temperature in a single-walled CNT embedded in polymer matrix was experimentally measured ($\sim$78 fs) with the three-pulse photon echo technique by Graham *et al.* [175].

The solvation process in liquids and the line-broadening dynamics in low-temperature glass can be divided into two components, that is, fast (optical dephasing approximately picoseconds) and slow (spectral diffusion, approximately microseconds or milliseconds) [172]. Nagasawa *et al.* [176] have reported the integrated three-pulse photon echo signal of Nile blue dye doped in poly(methyl methacrylate) two dimensionally by varying the first coherence period and the subsequent population period. The electronic dephasing time was in the femtosecond time range even at low temperatures, whereas rephrasing and echo formation were still possible even when the population period was extended to >100 ps.

Becker *et al.* [177] were the first to observe femtosecond photon echoes from the band-to-band transitions in a bulk semiconductor. The time decay of the echo, found to vary from 3.5 to 11 fs, has allowed us to determine the polarization dephasing rate in GaAs. This rate was found to depend on the carrier density in the experimental range covered, 1.5×10^{17} to 7×10^{18} cm^{-3}, indicating the dominance of carrier–carrier scattering as the principal dephasing mechanism. Noll and his colleagues [178] have reported picosecond photon echoes due to intrinsic optical excitations of a CdSSe semiconductor. Phase relaxation times up to several hundred picoseconds are found for excitons localized by chemical disorder in CdS$_x$Se$_{1-x}$ mixed crystals. These long dephasing times are attributed to reduced scattering of the localized excitons as compared to free excitons. Dephasing is identified to be due to energy relaxation of the excitons. Nanocrystals of cadmium selenide, having sizes in the QD region, have been studied by two-pulse photon echo and three-pulse photon echo peak shift experiments as a function of size [179] and temperature [180]. McKimmie *et al.* [181] have reported three-pulse photon echo in a series of cadmium selenide core-shell nanocrystals. The peak shift data were found to be highly dependent on the shell compositions and particle morphologies. Gong *et al.* [182] theoretically investigated the optical transition of $1s_e 1s_h$ exciton in a core-shell CdSe/ZnS nanocrystal QD based on the optical Bloch equations. The numerical calculation results reveal that three-pulse photon echo signals are sensitive to the variation of the size and structure of QD.

Bai and Fayer [183] have theoretically investigated the three nonlinear optical dephasing experiments, the incoherent photon echo, the accumulated grating echo, and the two-pulse photon echo, and have concluded that these three techniques are not equivalent. While the two-pulse photon echo measures homogeneous dephasing, the other two techniques are influenced by spectral diffusion. In general, the incoherent echo and the accumulated grating echo will measure dephasing rates that are faster than the two-pulse photon echo. The differences among the methods are calculated using a standard two-level system model of glasses. It is found that the differences depend on factors such as the pulse duration in the incoherent echo and the triplet lifetime in the accumulated grating echo. Tokmakoff *et al.* [184] have carried out the picosecond IR photon echo experiments and pump-probe experiments upto 10 K temperature and examined the vibrational dynamics of polyatomic solutes in polyatomic liquid and glassy solvents in detail. It is possible to selectively address the atoms within different frequency intervals within the inhomogeneously broadened absorption line and then store or read a single bit of information in this frequency interval. Rare earths are known to be of inhomogeneous broadened peaks ($>10^6$ [185] than homogeneous) and reported to be used for data storage. Storage densities in the range of terabits per square centimeter have been predicted for such materials [186]. In addition to storage density, photon echo techniques have other unique features, especially with respect to optical processing and manipulation of temporal light sequences [187]. Macfarlane [188] has presented a review of the measurement of photon echoes in rare-earth-doped solids and described how the choice of rare earth ion and host crystal affects the observation of the echo and the dephasing time measured. Luo *et al.* [189] have reported that a Pr-doped ZBLAN fiber is able to produce significant amplification of photon echo signals generated in Pr-doped Y$_2$SiO$_5$ at 606 nm. Asaka and colleagues [190], have observed accumulated photon echoes in Nd^{3+}-doped silicate glass by using coherent and incoherent (CW) light, and it was theoretically and experimentally verified that the time resolution is determined not by the pulse width but by the correlation time of the excitation field. Zhang *et al.* [191] have reported the dephasing time $\sim$200 fs at 7.5 and 300 K, which confirmed that the dephasing time for this transition was dominated by the vibrational state lifetime T_{11} for no coupling of the $^1S_0 \rightarrow {}^1F_3$ transition of Ce^{3+} ion to the ballistic motion of the glass. Meltzer *et al.* [192] monitored the optical dephasing of Pr^{3+} ions doped in LaF_3 nanocrystals (30–40 nm) embedded in the glass ceramics using two-pulse photon echo and reported that the dephasing time is 30–50 ns for $^3H_4 \rightarrow {}^3P_0$ transition.

11.3.2
Four-Wave Mixing

Interaction of three electromagnetic (EM) fields in a nonlinear optical medium for the generation of fourth EM field is the basis of FWM technique. One can understand the FWM process by considering independent interaction of individual beams with a dielectric medium. According to conventional Rayleigh scattering of linear optics, the first input EM field of wave vector $\boldsymbol{k}_1$ creates an oscillating polarization in the dielectric medium, which itself radiates a wave vector $\boldsymbol{k}_1'$ with some phase shift determined by the damping of the individual dipoles. The second input EM field $\boldsymbol{k}_2$ also drives the polarization of the dielectric and consequently reradiation of new EM wave $\boldsymbol{k}_2'$. The interference of the two waves ($\boldsymbol{k}_1'$ and $\boldsymbol{k}_2'$) produced through the interaction of first ($\boldsymbol{k}_1$) and second ($\boldsymbol{k}_2$) input EM fields with dielectric medium cause harmonics in the polarization at the sum and difference frequencies ($\boldsymbol{k}_1' \pm \boldsymbol{k}_2'$). Now, application of a third field of wave vector $\boldsymbol{k}_3$ drives the polarization, and this will beat with the input fields, $\boldsymbol{k}_1$ and $\boldsymbol{k}_2$, as well as the sum and difference frequencies $\boldsymbol{k}_1' \pm \boldsymbol{k}_2'$. This beating with the sum and difference frequencies is what gives rise to the fourth field in FWM. Figure 11.13 represents experimental arrangement for FWM with vectorial representation for phase mismatch and phase matching conditions. Two-wave photon echo, coherent anti-Stokes Raman spectroscopy (CARS), coherent Stokes Raman spectroscopy (CSRS), stimulated Raman gain spectroscopy (SRS), inverse Raman effect spectroscopy (IRES), and Raman-induced Kerr effect spectroscopy (RIKES) are special cases of FWM, and are cases in which two or more of the frequencies are degenerate.

FWM is usually reserved for the interaction of four spatially and spectrally distinct EM fields. FWM may be used for the probing either one-photon resonance or two-photon resonance in a material by measuring resonant enhancement as one or more of the frequencies are tuned. By tuning the frequencies to multiple resonances in the material, excited state cross sections, lifetimes, and linewidths may be measured [193]. FWM possesses the unique advantage of the direct comparison of Raman cross sections by tuning the two difference frequencies to the two Raman transitions of interest ($\omega_1 - \omega_2$ and $\omega_3 - \omega_2$). FWM also has the advantage of eliminating the nonresonant background signals present in

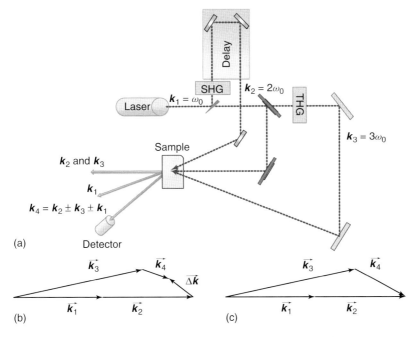

Figure 11.13 (a) Experimental setup for four-wave mixing and vectorial representation for (b) phase mismatch by Δk, and (c) perfectly phase-matched condition of FWM. SHG (second harmonic generation); THG (third harmonic generation) crystals.

the other methods. Complications involved in simultaneously overlapping three coherent beams while maintaining the phase matching condition is the main disadvantage of FWM. Sum frequency mixing $\omega_4 = 2\omega_1 + \omega_2$, third harmonic generation $\omega_4 = 3\omega_1$, difference frequency mixing (CARS) $\omega_4 = 2\,\omega_1 - \omega_2$, FWM $\omega_4 = \omega_1 + \omega_2 - \omega_3$, RIKE, TIRES, and SRS $\omega_2 = \omega_1 + \omega_2 - \omega_1$ are some of the special cases of FWM.

11.4
Measurement of Charge Carrier Dynamics

A fascinating feature of NPs is that their small size results in spatial confinement of the wave function of charge carriers, which is termed as *quantum size confinement* [10–13]. This effect becomes significant especially when the size is smaller than the Bohr exciton radius. Electrons in the particles interact with phonons, other charge carriers, defects, and the surfaces and interfaces. NPs have large percentage of surface-residing atoms with dangling bonds, surface defects, and adsorbants, which introduce large number of surface states. These states often fall within the bandgap of NPs and act as trap states for the electron and hole charge carriers. For example, in zinc oxide nanostructures, zinc atoms/ions at interstitial positions,

Z_i (neutral zinc), Z_i' (single charge zinc ions), and Z_i''(double charged zinc ions) and oxygen atoms/ions vacancies, V_O (neutral oxygen), V_O' (single charge oxygen ions), and $V''O$ (double charged oxygen ions) are donor defects or electron trap levels, while zinc neutral (V_{Zn}) and ions (V_{Zn}'/V_{Zn}'') vacancies are acceptor defects or hole trap levels [194]. Time-resolved spectroscopic techniques mentioned above provide us basic information on how charge carriers of one type interact with other charge carriers, phonons, surface, and defects. Surface defects of the NP along with its size affects charge carrier dynamics. Electron lifetime is calculated from the radiative and nonradiative decay pathways, which strongly depend on the size and surface defects. Ultrafast laser spectroscopy has the capability of the direct measurement of charge carrier dynamics. TAS is one of the popular ultrafast laser spectroscopic techniques for the measurement of charge carrier dynamics, where a femtosecond laser pulse is used as a pump to generate excited state electrons and a probe femtosecond laser beam excites populated electrons. Electron dynamics is measured by the change in the absorbance of probe beam as a consequence of decay in excited state electronic population density and can be measured as a function of delay between two pulses. Transient-bleaching spectroscopy is a similar approach for the measurement of charge carrier dynamics [195, 196]. Here a pump femtosecond beam excites electrons from ground state to the excited state, and electronic density of ground state recovers with time, which depends on the history and lifetime of various radiative and nonradiative decay processes. Carrier dynamics is measured by the change in the absorbance, as a consequence of ground state recovery of probe beam with the variation in delay between pump and probe pulses. Decrease in the oscillator strength of the excitonic transitions as a result of trapped electrons and holes [197] or filling of bands due to excess electrons in conduction band may cause large bleach at the first exciton energy [196].

Time-resolved photoluminescence (TRPL) is another ultrafast spectroscopic technique that assists in the determination of charge carrier dynamics. Radiative recombination of charge carriers emits photons of near-band-edge and subbandgap energies. Subbandgap emission corresponds for surface-trapped charge carriers, while near-band-edge emission attributes to bandgap excitons and shallow charge carriers. As mentioned above, time-resolved IR and time-resolved Raman are ultrafast spectroscopic techniques for the determination of electron–phonon and phonon–phonon relaxation dynamics.

11.4.1
Effect of Size and Surface on Charge Carrier Dynamics in Semiconductor NPs

As a result of quantum confinement in semiconductor NPs, their effective bandgap and redox potential increases with the decrease in size, which causes shorter wavelength shift in excitonic absorption and emission maximum. Increase in the size of semiconductor NPs results in decrease in density of states (DOSs) for electrons as well for the phonons, and hence decrease in the electron-phonon interaction and nonradiative decay processes. Weaker electron–phonon interaction in smaller sized NPs may result in longer electron lifetimes. On the other hand, spatial confinement in smaller sized NPs result in stronger electron–hole interaction and hence shorter lifetimes for electrons. These two competitive processes weaken the effect of quantum confinement on the charge carrier dynamics in semiconductor NPs.

Mittleman et al. [198] reported slightly shorter lifetime for smaller CdSe NPs and concluded that stronger electron–hole interaction process dominates in smaller NPs. Nosaka et al. [199] observed that the rate of electron transfer across the interface increases with the decrease in size and assumed that the electron–hole recombination rate is independent of the size of CdS NPs. In a similar work on CdS NPs, Zhang et al. [200] did not observe any size dependence of electron dynamics for sizes between 2 and 4 nm. Higher surface to volume ratio in smaller sized NPs and high density of trap states that arise from the extrinsic defects and dangling bonds often significantly affect carrier dynamics. Klimov et al. [201] observed fast trapping of holes by hole trapping state using ultrafast TRPL spectroscopy. For Cds NPs, trapping of electrons dominates in early (<100 fs) dynamics [202]. However, photon echo experiment resulted in longer trapping times of 500 fs to 8 ps for CdSe NPs [198]. NPs produced by different physical and chemical methods have different types and density of defect states; therefore, NPs of same chemical composition prepared by different approaches may possess different carrier dynamics. However, effects of chemical nature of trap states on the time of carrier trapping are unexplored and needs more research and development. The electron dynamics in NPs is highly sensitive to the adsorbents, surfactants, solvent, pH, hole and electron scavengers, and interaction of particle of interest with other particles or colloids [203]. Charge carrier dynamics of the core/shell nanostructures depends on the physical size and chemical nature of the shell layer. Change in surface defects density by chemical modification of surface of CdS NPs causes almost 200-fold enhancement in fluorescence, but almost no any change in early electron dynamics [204]. The lifetime of deep trap (DT) state having at least two trap states (Figure 11.14a) is given by the following equation:

$$\tau_{DT} = \phi_{DT}^{R}\tau_{DT}^{R}\left(1 + \frac{\tau_C}{\tau_{ST}^{R}} + \frac{\tau_C}{\tau_{ST}^{NR}}\right) \tag{11.1}$$

where τ_{DT}, τ_{DT}^{R}, and ϕ_{DT}^{R} are the observed lifetime, radiative lifetime, and fluorescence quantum yield, respectively, while τ_C, τ_{ST}^{R}, and τ_{ST}^{NR} are crossing time constant from shallow trap (ST) to DT, radiative, and nonradiative time constants, respectively. Figure 11.14b illustrates a sketch of electron dynamics in particles. If surface treatment occurs for longer τ_{ST}^{NR} and simultaneously, higher ϕ_{DT}^{R}, then it may be possible that there will be no any significant change in τ_{DT}. Another possible reason for no any change in electron dynamics as a consequence of surface modification may be the fact that surface treatment mainly affects the lifetime of DTs, which is comparatively longer; therefore it does not show significant change in early electron dynamics [205]. The early electron dynamics has a major contribution from lifetime of ST states, which is shorter and usually unaffected by surface treatment [205]. In transient absorption measurement, the sample is excited with the pump laser having wavelength corresponding to excitonic absorption, and the

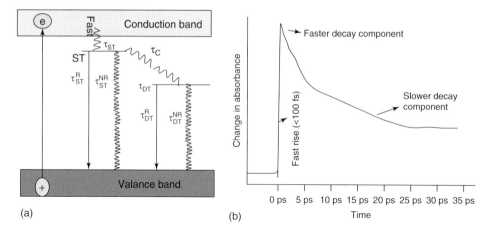

(a) (b)

Figure 11.14 (a) Energy level diagram and electronic transitions for two trapping states in semiconductor particles. ST stands for shallow states, while DT denotes deep trap. (b) Sketch of transient absorption process.

probe beam excites electron from this state. Population density of electrons in the excitonic level rises rapidly within <100 fs after the excitation of the sample with the pump laser, which results in a fast rise (100 fs) in the transient absorption. After that, electronic density in the excitonic level decreases exponentially, with double exponential decay function having a faster (2–10 ps) and comparatively slower (20–100 ps) time constants. Faster decay corresponds to the decrease in the population density of excitonic level because of the faster electronic transition from excitonic level to the ST state.

11.4.2
Effect of Excitation Power on Charge Carrier Dynamics: Picosecond Dynamics

Faster as well as slower decay processes and hence electron dynamics have significant dependence on the power of excitation laser. Zhang *et al.* [206] observed strong dependence of electron dynamics on excitation power for CdS as well as TiO$_2$ colloidal NPs. Higher excitation power corresponds to faster decay for both the components. Figure 11.15 illustrates transient absorption data for CdS NPs excited with different power densities. At lower excitation intensity, electron relaxation is dominated by 50 ps decay, followed by a slower decay of almost 1 ns. At higher excitation intensities, faster decay components having decay constant in the range of 1–10 ps start to appear. It is observed that the faster decay component is more sensitive to the power of excitation laser. Zhang *et al.* [207] reported that when the pump power is varied from 3 µJ/pulse (10 photons/CdS particle) to 0.3 µJ/pulse (1 photon/CdS particle) the amplitude of faster decay component decreases almost linearly with the pump power and becomes negligible for 0.3 µJ/pulse (Figure 11.15). The Slower decay component also decreases with the decrease of laser power, but it is also weakly visible at 0.3 µJ/pulse. Observation

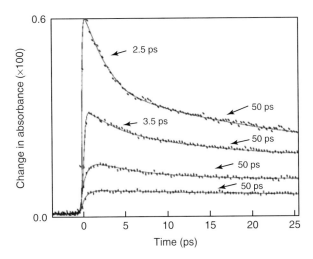

Figure 11.15 Power dependence of electron dynamics in colloidal CdS NPs using excitation (pump) wavelength of 390 nm and probe wavelength of 780 nm at four excitation intensities 0.12, 0.20, 0.59, and 1.18 photons $Å^{-2}$. (Source: Reprinted with the permission from [207] copyright @ American Chemical Society.)

of fast decay component at higher excitation intensities suggests that it is the consequence of exciton–exciton interactions as a result of multiple excitons created per particle. A number of predictions have been proposed for the observation of faster decay component including nongeminate recombination, second-order recombination, enhanced tunneling for secondary recombination, and Auger processes. Figure 11.15 shows power dependent transient absorption spectra of colloidal solution of CdS NPs using ultrafast pump-probe spectroscopy.

11.4.3
Effects of Size and Surface on the Electron Relaxation Dynamics in Metal NPs

In metals, quantum confinement effect is expected to occur at much smaller sizes as compared to those of the semiconductors. This difference has made fundamental differences in electronic relaxation processes in metals and semiconductors. Electron dynamics for metal NPs is quietly unexplored. Similar to the semiconductors, early studies on metal NPs were also focused on the equilibrium properties such as SPR absorption and structural characteristics, while nonequilibrium properties such as electron dynamics are almost unexplored. An early study by Heiweil and Hochstrasser [208] reported electron relaxation dynamics of >2 ps for Au NPs. Roberti et al. [209] carried out the first femtosecond study of electron dynamics on metal particles for Ag NPs in water. Following this report, electron relaxation dynamics for gold [210] and platinum [211] NPs were reported. The early relaxation dynamics for Ag and Au NPs was dominated by electron–phonon interaction with relaxation times of 2.5 and 7 ps, respectively. Relaxation times for both Ag and Au NPs were longer as compared to their corresponding bulk (670 fs for Ag and 1 ps for Au) metal films [212, 213] because of the weaker electron–phonon interaction in NPs. Stella *et al.* [214] reported that the relaxation time for Sn NPs decreases with size because of the enhanced surface collision. Relation between size and relaxation time for metal NPs is expressed by following equation:

$$\frac{1}{\tau_{Obs}} = \frac{aR^{\alpha}}{\tau_{bulk}} + \frac{bv_F}{R} \tag{11.2}$$

where τ_{obs} and τ_{bulk} are relaxation times observed for a particle of radius R and for bulk metal, respectively, v_F is the Fermi velocity of electrons in the metal, a and b are constants that depend on the characteristic of metals, α is a constant that can be derived from theoretical models of Belotskii and Tomchuk [215], and $1/b$ is a measure of the average number of surface collisions [214]. The first term on the right-hand side of Eq. (11.2) represents electron–phonon interaction, while the second term stands for surface collision [207]. According to Eq. (11.2), decrease in the particle size first causes increase in the relaxation time as compared to the bulk value due to the dominance of first term in the right-hand side of equation. With the further decrease in size, the second term of the expression dominates and causes decrease in relaxation time with size.

11.5
Conclusion and Future Prospects

In this chapter, we have described various ultrafast pump-probe/time-resolved spectroscopic techniques such as transient absorption or time-resolved pump-probe absorption, ultrafast TRF, ultrafast time-resolved IR, ultrafast time-resolved Raman, and ultrafast TRFR spectroscopy, and their applications in the diagnosis of various dynamical processes in atomic, molecular, and nanoscale systems. In most of the discussed ultrafast spectroscopic techniques, femtosecond or picoseconds laser systems were used for the diagnosis of dynamical processes of picosecond time resolution. However, dynamical processes of shorter time scales such as electronic transition times from one atomic/molecular orbital to the other or from valence band to the conduction band in nanoscale particles are still unexplored. How much time is spent in the emission of photons or photoelectrons after the excitation of atomic/molecular or nanoscale systems is still a question. Diagnosis of dynamical processes of atomic/molecular systems requires an attosecond laser beam and a detector of attosecond time resolution. Very recently, Schultze *et al.* [216] reported a delay of 21 ± 5 as in the emission of electrons from the 2p orbitals of neon atoms with respect to those released from the 2s orbital by excitation using the same 100 eV light pulse of <100 as duration. This observation breaks the assumption that the photoemission from atoms occurs instantly in response to incident radiation, which is used in setting the zero time in clocking atomic-scale electronic motion. Attosecond laser beam is the future of ultrafast spectroscopy of atomic, molecular, optical, and nanoscale physics, which might add several new insights into basic sciences and improve/modify or discard several available basic principles of electronic motions in atoms and molecules.

Acknowledgments

Dr. Singh is thankful to Irish Research Centre for Science, Engineering and Technology (IRCSET) for providing EMPOWER postdoctoral fellowship for financial assistance during compilation of this chapter.

References

1. Yoshihara, T., Katoh, R., Furube, A., Tamaki, Y., Murai, M., Hara, K., Murata, S., Arakawa, H., and Tachiya, M. (2004) *J. Phys. Chem. B*, **108**, 3817–3823.

2. Bahnemann, D., Henglein, A., and Spanhel, L. (1984) *Faraday Discuss. Chem. Soc.*, **78**, 151–163.

3. Raksi, F., Wilson, K.R., Jiang, Z., Ikhlef, A., Cote, C.Y., and Kieffer, J.-C. (1996) *J. Chem. Phys.*, **104**, 6066.

4. Loh, Z.-H., Khalil, M., Correa, R.E., Santra, R., Buth, C., and Leone, S.R. (2007) *Phys. Rev. Lett.*, **98**, 143601.

5. Seres, E. and Spielmann, C. (2007) *Appl. Phys. Lett.*, **91**, 121919.

6. Cavalleri, A., Rini, M., Chong, H.H.W., Fourmaux, S., Glover, T.E., Heimann, P.A., Kieffer, J.C., and Schoenlein, R.W. (2005) *Phys. Rev. Lett.*, **95**, 067405.

7. Santra, R., Yakovlev, V.S., Pfeifer, T., and Loh, Z.H. (2011) *Phys. Rev. A*, **83**, 033405.

8. Southworth, S.H., Arms, D.A., Dufresne, E.M., Dunford, R.W., Ederer, D.L., Höhr, C., Kanter, E.P., Krässig, B., Landahl, E.C., Peterson, E.R., Rudati, J., Santra, R., Walko, D.A., and Young, L. (2007) *Phys. Rev. A*, **76**, 043421.

9. Goulielmakis, E., Loh, Z.-H., Wirth, A., Santra, R., Rohringer, N., Yakovlev, V.S., Zherebtsov, S., Pfeifer, T., Azzeer, A.M., Kling, M.F., Leone, S.R., and Krausz, F. (2010) *Nature*, **466**, 739.

10. Klimov, V.I. and McBranch, D.W. (1998) *Opt. Lett.*, **23**, 277–279.

11. Berera, R., van Grondelle, R., and Kennis, J.T.M. (2009) *Photosynth. Res.*, **101**, 105–118.

12. Niedzwiedzki, D.M., Sullivan, J.O., Polivka, T., Birge, R.R., and Frank, H.A. (2006) *J. Phys. Chem. B*, **110**, 22872–22885.

13. Yeremenko, S., van Stokkum, I.H.M., Moffat, K., and Hellingwerf, K.J. (2006) *Biophys. J.*, **90**, 4224–4235.

14. Tamai, N., Saika, T., Shimidzu, T., and Irie, M. (1996) *J. Phys. Chem.*, **100**, 4689–4692.

15. Alfano, J.C., Walhout, P.K., Kimura, Y., and Barbara, P.F. (1993) *J. Chem. Phys.*, **98**, 5996–5998.

16. Burda, C. and El-Sayed, M.A. (2000) *Pure Appl. Chem.*, **72**, 165–177.

17. Klimov, V.I., Schwarz, Ch.J., McBranch, D.W., Leatherdale, C.A., and Bawendi, M.G. (1999) *Phys. Rev. B*, **60**, 2177.

18. Yoshihara, T., Katoh, R., Furube, A., Tamaki, Y., Murai, M., Hara, K., Murata, S., Arakawa, H., and Tachiya, M. (2004) *J. Phys. Chem. B*, **108**, 3817–3823.

19. Graham, M.W., Chmeliov, J., Ma, Y.-Z., Shinohara, H., Green, A.A., Hersam, M.C., Valkunas, L., and Fleming, G.R. (2011) *J. Phys. Chem. B*, **115** (18), 5201–5211.

20. Stepanov, A.G., Portella-Oberli, M.T., Sassara, A., and Chergui, M. (2002) *Chem. Phys. Lett.*, **358**, 516.

21. Boucher, D., Chekalin, S.V., Kovalenko, S.A., Matveets Yu, A., Masselin, P., Novikov, M.G., Ragulsky, V.V., and Stepanov, A.G. (1997) *SPIE Proc.*, **3239**, 302.

22. (a) McLoskey, D., Campbell, D., Allison A., and Hungerfor, G. (2011) *Meas. Sci. Technol.*, **22**, 067001; (b) O'Connor, D.V. and Phillips, D. (1984) *Time Correlated Single Photon Counting*, Academic Press, London; (c) Becker, W., Bergmann, A., Biscotti, G., and Rüc, A. (2004) *Proc. SPIE*, **5340**, 1–9.

23. (a) Carlsson, K. and Philip, J.P. (2002) *Proc. SPIE*, **4622**, 70–78; (b) Philip, J.P. and Carlsson, K. (2003) *J. Opt. Soc. Am. A*, **20**, 368–379.

24. Mahr, H. and Hirsch, M.D. (1975) *Opt. Commun.*, **13**, 96.

25. Zhang, X.X., Würth, C., Zhao, L., Genger, U.R., Ernsting, N.P., and Sajadi, M. (2011) *Rev. Sci. Instrum.*, **82**, 063108.

26. Kahlow, M.A., Jarzeba, W., DuBruil, T.P., and Barbara, P.F. (1988) *Rev. Sci. Instrum.*, **59**, 1098.

27. (a) Zhao, L., Lustres, J.L.P., Farztdinov, V., and Ernsting, N.P. (2005) *Phys. Chem. Chem. Phys.*, **7**, 1716–1725; (b) Damen, T.C. and Shah, J. (1988) *Appl. Phys. Lett.*, **52**, 1291.

28. Mokhtari, A., Chebira, A., and Chesnoy, J. (1990) *J. Opt. Soc. Am. B*, **7**, 1551.

29. (a) Shah, J. (1988) *IEEE J. Quant. Electr.*, **24**, 276–288; (b) Kahlow, M.A., Jarzeba, W., DuBruil, T.P., and Barbara, P.F. (1988) *Rev. Sci. Instrum.*, **59**, 1098–1109.

30. (a) Mialocq, J.-C. and Gustavsson, T. (2001) in *New Trends in Fluorescence Spectroscopy* (eds B. Valeur and J.-C. Brochon), Springer, Berlin; (b) Glasbeck M. and Zhang, H. (2004) *Chem. Rev.*, **104**, 1929–1954.

31. Varnavski, O.P., Ostrowski, J.C., Sukhomlinova, L., Twieg, R.J., Bazan, G.C., and Goodson, T. III (2002) *J. Am. Chem. Soc.*, **124**, 1736–1743.

32. Iwai, S., Murata, S., and Tachiya, M. (1998) *J. Chem. Phys.*, **109**, 5963.

33. Chosrowjan, H., Mataga, N., Nakashima, N., Imamoto, Y., and Tokunaga, F. (1997) *Chem. Phys. Lett.*, **270**, 267.

34. Mataga, N., Chosrowjan, H., Shibata, Y., Imamoto, Y., and Tokunaga, F. (2000) *J. Phys. Chem. B*, **104**, 5191.

35. Mihara, K., Hisatomi, O., Imamoto, Y., Kataoka, M., and Tokunaga, F. (1997) *J. Biochem.*, **121**, 876.

36. Ma, Y.-Z., Valkunas, L., Dexheimer, S.L., Bachilo, S.M., and Fleming, G.R. (2005) *Phys. Rev. Lett.*, **94**, 157402.

37. Korovyanko, O.J., Sheng, C.-X., Vardeny, Z.V., Dalton, A.B., and Baughman, R.H. (2004) *Phys. Rev. Lett.*, **92** 017403.

38. Ostojic, G.N., Zaric, S., Kono, J., Strano, M.S., Moore, V.C., Hauge, R.H., and Smalley, R.E. (2004) *Phys. Rev. Lett.*, **92**, 117402.

39. Huang, L., Pedrosa, H.N., and Krauss, T.D. (2004) *Phys. Rev. Lett.*, **93**, 017403.

40. Ma, Y.-Z., Stenger, J., Zimmermann, J., Bachilo, S.M., Smalley, R.E., Weisman, R.B., and Fleming, G.R. (2004) *J. Chem. Phys.*, **120**, 3368.

41. Wang, F., Dukovic, G., Knoesel, E., Brus, L.E., and Heinz, T.F. (2004) *Phys. Rev. B*, **70**, 24140.

42. (a) Valkunas, L., Ma, Y.-Z., and Fleming, G.R. (2006) *Phys. Rev. B*, **73**, 115432; (b) Huang, L. and Krauss, T.D. (2006) *Phys. Rev. Lett.*, **96**, 057407.

43. Underwood, D.F., Kippeny, T., and Rosenthal, S.J. (2001) *J. Phys. Chem. B*, **105**, 436–443.

44. Rubtsov, I.V., Ebina, K., Satou, F., Oh, J.W., Kumazaki, S., Suzumoto, T., Tani, T., and Yoshihara, K. (2002) *J. Phys. Chem. A*, **106**, 2795–2802.

45. Gustavsson, T., Cassara, L., Gulbinas, V., Gurzadyan, G., Mialocq, J.-C., Pommeret, S., Sorgius, M., and van der Muelen, P. (1998) *J. Phys. Chem. A*, **102**, 4229.

46. Schanz, R., Kovalenko, S.A., Kharlanov, V., and Ernsting, N.P. (2001) *Appl. Phys. Lett.*, **79**, 566.

47. (a) Schmidt, B., Laimgruber, S., Zinth, W., and Gilch, P. (2003) *Appl. Phys. B*, **76**, 809–814; (b) Gires, F. and Mayer, G. (1964) *C. R. Acad. Sci. URSS*, **258**, 2039.

48. Rentzepis, P.M., Topp, M.R., Jones, R.P., and Jortner, J. (1970) *Phys. Rev. Lett.*, **25**, 1742.

49. Wang, L., Ho, P.P., and Alfano, R.R. (1993) *Appl. Opt.*, **32**, 5043.

50. Tkachenko, N.V. (2006) *Optical Spectroscopy Methods and Instrumentations*, Elsevier.

51. Liu, H., Tan, W., Si, J., and Hou, X. (2008) *Opt. Express*, **16**, 13486–13491.

52. Jha, A., liu, X., Kar, A.K., and Bookey, H.T. (2001) *Science*, **5**, 475–479.

53. Lenz, G., Zimmermann, J., Katsufuji, T., Lines, M.E., Hwang, H.Y., Spälter,

S., Slusher, R.E., Cheong, S.W., Sanghera, J.S., and Aggarwal, I.D. (2000) *Opt. Lett.*, **25**, 254–256.

54. Kinoshita, S., Ozawa, H., Kanematsu, Y., Tanaka, I., Sugimoto, N., and Fujiwara, S. (2000) *Rev. Sci. Instrum.*, **71**, 3317–3322.

55. Tan, W., Yang, Y., Si, J., Tong, J., Yi, W., Chen, F., and Hou, X. (2010) *J. Appl. Phys.*, **107**, 043104.

56. Albrecht, H.S., Heist, P., Kleinschmidt, J., Lap, D.V., and Schröder, T. (1993) *Appl. Opt.*, **32**, 6659.

57. Brun, A., Georges, P., Saux, G.L., and Salin, F. (1991) *J. Phys. D*, **24**, 1225.

58. Kanbara, H., Fujiwara, S., Tanaka, K., Nasu, H., and Hirao, K. (1997) *Appl. Phys. Lett.*, **70**, 925.

59. Falcao Filho, E.L., Bosco, C.A.C., Maciel, G.S., de Araujo, Cid.B., Acioli, L.H., Nalin, M., and Messaddeq, Y. (2003) *Appl. Phys. Lett.*, **83**, 1292.

60. Lin, T., Yang, Q., Si, J., Chen, T., Chen, F., Wang, X., Hou, X., and Hirao, K. (2008) *Opt. Commun.*, **275**, 230–233.

61. Arzhantsev, S. and Maroncelli, M. (2005) *Appl. Spectrosc.* **59**, 206–220.

62. Maroncelli, M. (1993) *J. Mol. Liq.*, **57**, 1.

63. Nitzan, A. (2006) *Chemical Dynamics in Condensed Phases Relaxation, Transfer, and Reactions in Condensed Molecular Systems*, Oxford University Press, Oxford.

64. Kimura, Y., Fukuda, M., Suda, K., and Terazima, M. (2010) *J. Phys. Chem. B*, **114**, 11847–11858.

65. Jimenez, R., Fleming, G.R., Kumar, P.V., and Maroncelli, M. (1994) *Nature*, 471.

66. Takeda, J., Nakajima, K., Kurita, S., Tomimoto, S., Saito, S., and Suemoto, T. (2000) *J. Lumin.*, **87–89**, 927–929.

67. Talapatra, G.B., Manickam, N., Samoc, M., Orczyk, M.E., Karna, S.P., and Prasad, P.N. (1992) *J. Phys. Chem.*, **96**, 5206.

68. Torres, C., Rivera, L.T., Rojo, R.R., Martínez, R.T., Silva-Pereyra, H.G., Esqueda, J.A.R., Fernández, L.R., Sosa, A.C., Wong, J.C.C., and Oliver, A. (2011) *Nanotechnology*, **22**, 355710.

69. Hanada, H., Kanematsu, Y., Kinoshita, S., Kumauchi, M., Sasaki, J., and Tokunaga, F. (2001) *J. Lumin.*, **94–95**, 593–596.

70. Wolf, M., Ferrari, M., and Quaresima, V. (2007) *J. Biomed. Opt.*, **12**, 062104.

71. (a) Chesnoy, J. and Ricard, D. (1980) *Chem. Phys. Lett.*, **73**, 433–437; (b) Heilweil, E.J., Casassa, M.P., Cavanagh, R.R., and Stephenson, J.C. (1985) *Chem. Phys. Lett.*, **117**, 185.

72. Graener, H., Dohlus, R., and Laubereau, A. (1986) in *Ultrafast Phenomena V* (eds G.R. Fleming and A.E. Siegman), Springer, Berlin.

73. Zamkov, M.A., Conner, R.W., and Dlott, D.D. (2007) *J. Phys. Chem. C*, **111**, 10278–10284.

74. (a) Chabal, Y.J. (1988) *Surf. Sci. Rep.*, **8**, 211; (b) Ueba, H. (1997) *Prog. Surf. Sci.*, **55**, 115–179; (c) Fayer, M.D. (2001) *Ultrafast Infrared and Raman Spectroscopy*, Marcel Dekker, Inc., NewYork.

75. Zhang, Y., Burdzinski, G., Kubicki, J., and Platz, M.S. (2008) *J. Am. Chem. Soc.*, **130**, 16134–16135.

76. Nibbering, E.T.J. and Elsaesser, T. (2004) *Chem. Rev.*, **104**, 1887–1914.

77. Woutersen, S., Emmerichs, U., and Bakker, H.J. (1997) *Science*, **278**, 658–660.

78. Stenger, J., Madsen, D., Dreyer, J., Nibbering, E.T.J., Hamm, P., and Elsaesser, T. (2001) *J. Phys. Chem. A*, **105**, 2929–2932.

79. Levinger, N.E., Davis, P.H., and Fayer, M.D. (2001) *J. Chem. Phys.*, **115**, 9352–9360.

80. Ashihara, S., Huse, N., Espagne, A., Nibbering, E.T.J., and Elsaesser, T. (2006) *Phys. Lett.*, **424**, 66–70.

81. Ohta, K. and Tominaga, K. (2007) *Chem. Phys.*, **341**, 310–319.

82. Banno, M., Ohta, K., Yamaguchi, S., Hirai, S., and Tominaga, K. (2009) *Acc. Chem. Res.*, **42**, 1259–1269.

83. Heilweil, E.J., Casassa, M.P., Cavanagh, R.R., and Stephenson, J.C. (1986) *J. Chem. Phys.*, **85**, 5004–5018.

84. Grubbs, W.T., Dougherty, T.P., and Heilweil, E.J. (1995) *J. Phys. Chem.*, **99**, 10716–10722.

85. Andresen, E.R., Gremaud, R., Borgschulte, A., Cuesta, A.J.R., Zuttel, A., and Hamm, P. (2009) *J. Phys. Chem. A*, **113**, 12838–12846.

86. Asbury, J.B., Ellingson, R.J., Ghosh, H.N., Ferrere, S., Nozik, A.J., and Lian, T. (1999) *J. Phys. Chem. B*, **103**, 3110–3119.

87. Cherepy, N.J., Smestad, G.P., Gratzel, M., and Zhang, J.Z. (1997) *J. Phys. Chem.*, **101**, 9342.

88. Martini, I., Hodak, J.H., and Hartland, G.V. (1998) *J. Phys. Chem. B*, **102**, 9508.

89. Tachibana, Y., Moser, J.E., Gratzel, M., Klug, D.R., and Durrant, J.R. (1996) *J. Phys. Chem.*, **100**, 20056–20062.

90. Murakoshi, K., Yanagida, S., Capel, M., and Castner, J.E.W. (1997) *Interfacial Electron-Transfer Dynamics of Photosensitized Zinc Oxide Nanoclusters*, ACS Symposium Series, Vol. **679**, American Chemical Society, Washington, DC.

91. Ellingson, R.J., Asbury, J.B., Ferrere, S., Ghosh, H.N., Lian, T., and Nozik, A.J. (1998) *J. Phys. Chem. B*, **102**, 6455.

92. Hannappel, T., Burfeindt, B., Storck, W., and Willig, F. (1997) *J. Phys. Chem. B*, **101**, 6799.

93. Skinner, D.E., Colombo, D.P., Cavaleri, J.J., and Bowman, R.M. (1995) *J. Phys. Chem.*, **99**, 7853.

94. Yamakata, A., Yoshida, M., Kubota, J., Osawa, M., and Domen, K. (2011) *J. Am. Chem. Soc.*, **133**, 11351–11357.

95. (a) Heimer, T. and Heilweil, E.J. (1998) Proceedings for Ultrafast Phenomena XI. Berlin: Springer-Verlag, pp. 505–507. (b) Heimer, T.A. and Heilweil, E.J. (1997) *J. Phys. Chem. B*, **101**, 10990.

96. (a) Pankove, J.I. (1975) *Optical Processes in Semiconductors*, Dover, New York; (b) Basu, P.K. (1997) *Theory of Optical Processes in Semiconductors*, Oxford University Press, New York.

97. (a) Yamakata, A., Ishibashi, T., and Onishi, H. (2001) *J. Phys. Chem. B*, **105**, 7258; (b) Yamakata, A., Ishibashi, T.A., and Onishi, J. (2003) *J. Phys. Chem. B*, **107**, 9820.

98. Yamakata, A., Ishibashi, T., Kato, H., Kudo, A., and Onishi, H. (2003) *J. Phys. Chem. B*, **107**, 14383.

99. Amano, F., Yamakata, A., Nogami, K., Osawa, M., and Ohtani, B. (2008) *J. Am. Chem. Soc.*, **130**, 17650.

100. Yamakata, A., Ishibashi, T., and Onishi, H. (2003) *Chem. Phys. Lett.*, **376**, 576.

101. Yoshida, M., Yamakata, A., Takanabe, K., Kubota, J., Osawa, M., and Domen, K. (2009) *J. Am. Chem. Soc.*, **131**, 13218.

102. Ghosh, H.N., Asbury, J.B., Weng, Y., and Lian, T. (1998) *J. Phys. Chem. B*, **102**, 10208–10215.

103. Weng, Y.X., Wang, Y.Q., Asbury, J.B., Ghosh, H.N., and Lian, T. (2000) *J. Phys. Chem. B*, **104**, 93–104.

104. Ghosh, H.N., Asbury, J.B., and Lian, T. (1998) *J. Phys. Chem. B*, **102**, 6482–6486.

105. Raman, V. and Krishnan, K.S. (1928) *Nature*, **121**, 501.

106. Kosuda, K.M., Bingham, J.M., Wustholz, K.L., and van Duyne, R.P. (2011) in *Comprehensive Nanoscience and Technology*, vol. **3** (eds D.L. Andrews, G.D. Scholes, and G.P. Wiederrecht), Academic Press, Oxford, pp. 263–301.

107. Roy, S., Gord, J.R., and Patnaik, A.K. (2010) *Prog. Energy Combust. Sci.*, **36**, 280–306.

108. (a) Cyvin, S.J., Rauch, J.E., and Decius, J.C. (1965) *J. Chem. Phys.*, **43**, 4083; (b) Yang, W.H., Hulteen, J., Schatz, G.C., and Duyne, R.P.V. (1996) *J. Chem. Phys.*, **104**, 4313.

109. (a) Kasevich, M. and Chu, S. (1991) *Phys. Rev. Lett.*, **67**, 181; (b) Gaubatz, U., Rudecki, P., Schiemann, S., and Bergmann, K. (1990) *J. Chem. Phys.*, **92**, 5363; (c) Carman, R.L., Shimizu, F., Wang, C.S., and Bloembergen, N. (1970) *Phys. Rev. A*, **2**, 60.

110. (a) Parker, F.S. (ed.) (1983) *Applications of Infrared, Raman, and Resonance Raman Spectroscopy in Biochemistry*, Plenum Press, New York; (b) Alivisatos, P., Harris, T.D., Carroll, P.J., Steigerwald, M.L., and Brus, L.E. (1989) *J. Chem. Phys.*, **90**, 3463.

111. (a) Suemoto, T., Tanaka, K., and Ohtake, H. (1996) *Prog. Cryst. Growth Charact. Mater.*, **33**, 57–63; (b) Roy, S., Gord, J.R., and Patnaik, A.K.

(2010) *Prog. Energy Combust. Sci.*, **36**, 280–306; (c) Dhar, L., Rogers, J.A., and Nelson,K.A. (1994) *Chem. Rev.*, **94**, 157; (d) Browne, W.R. and McGarvey, J.J. (2007) *Coord. Chem. Rev.*, **251**, 454–473.

112. McHale, J.L. (2006) Resonance Raman spectroscopy, in *Handbook of Vibrational Spectroscopy* (eds J. Chalmers and P. Griffiths), John Wiley & Sons, Ltd, pp. 534–556. ISBN: 978-3-527-41048-4.

113. Kash, J.A., Tsang, J.C., and Hvam, J.M. (1985) *Phys. Rev. Lett.*, **54**, 2151.

114. (a) Coates, C.G., Olofsson, J., Coletti, M., McGarvey, J.J., Onfelt, B., Lincoln, P., Norden, B., Tuite, E., Matousek, P., and Parker, A.W. (2001) *J. Phys. Chem. B*, **105**, 12653; (b) Olofsson, J., Onfelt, B., Lincoln, P., Norden, B., Matousek, P., Parker, A.W., Tuite, E., and Inorg, J. (2002) *Biochemistry*, **91**, 286.

115. (a) Roy, S., Meyer, T.R., Lucht, R.P., Belovich, V.M., Corporan, E., and Gord, J.R. (2004) *Combust. Flame*, **138**, 273–284; (b) Cheng, J., Volkmer, A., and Xie, X.S. (2002) *J. Opt. Soc. Am. B*, **19**, 1363; (c) Cheng, J. and Xie, X.S. (2004) *J. Phys. Chem. B*, **108**, 827–840.

116. (a) Brackmann, C., Bood, J., Afzelius, M., and Bengtsson, P.-E. (2004) *Meas. Sci. Technol.*, **15** (R13), 25; (b) Schenk, M., Seeger, T., and Leipertz, A. (2005) *Appl. Opt.*, **44**, 5582–5593.

117. Roy, S., Meyer, T.R., Brown, M.S., Velur, V.N., Lucht, R.P., and Gord, J.R. (2003) *Opt. Commun.*, **224**, 131–137.

118. (a) Meyer, T.R., Roy, S., Lucht, R.P., and Gord, J.R. (2005) *Combust. Flame*, **142**, 52–61; (b) Weikl, M.C., Cong, Y., Seeger, T., and Leipertz, A. (2009) *Appl. Opt.*, **48**, B43–B50.

119. Konradi, J., Singh, A.K., and Materny, A. (2006) *J. Photochem. Photobiol. A: Chem.*, **180**, 289.

120. Joo, T. and Albrecht, A.C. (1993) *J. Chem. Phys.*, **99**, 3244.

121. Eesley, G.L. (1981) *Coherent Raman Spectroscopy*, Chapter 3, Pergamon Press, New York.

122. (a) Owyoung, A. (1978) *IEEE J. Quantum Electron.*, **QE-14**, 192; (b) Owyoung, A. (1981) in *Chemical Applications of Nonlinear Raman Spectroscopy*, Chapter 7, (ed. A.B.

Harvey), Academic Press, London, pp. 281–320. (c) Shen, Y.R. (1984) *The Principles of Nonlinear Optics*, Chapter 10, John Wiley & Sons, Ltd, New York, pp. 141–186.

123. (a) McCamont, D.W., Kukura, P., and Mathies, R.A. (2003) *J. Phys. Chem. A*, **107**, 8208; (b) Yoshizawa, M. and Kurosawa, M. (1999) *Phys. Rev. A*, **61**, 013808; (c) Rondonuwu, F.S., Watanabe, Y., Zhang, J.-P., Furuichi, K., and Koyama, Y. (2002) *Chem. Phys. Lett.*, **357**, 376.

124. McCamant, D.W., Kukura, P., and Mathies, R.A. (2003) *Appl. Spectrosc.*, **57**, 1317.

125. McCamont, D.W., Kukura, P., and Mathies, R.A. (2003) *J. Phys. Chem. A*, **107**, 8208.

126. Kim, D. and Yu, P.M. (1991) *Phys. Rev.*, **B43**, 4158.

127. Butler, R.M., Lynn, M.A., and Gustafson, T.L. (1993) *J. Phys. Chem.*, **97**, 2609–2617.

128. Dallinger, R.F. and Woodruff, W.H. (1979) *J. Am. Chem. Soc.*, **101**, 4391.

129. Bradley, P.G., Kress, N., Hornberger, B.A., Dallinger, R.F., and Woodruff, W.H. (1981) *J. Am. Chem. Soc.*, **103**, 7441.

130. Schoonover, J.R., Strouse, G.F., Chen, P., Bates, W.D., and Meyer, T.J. (1993) *Inorg. Chem.*, **32**, 2618–2619.

131. Iwata, K. and Hamaguchi, H. (1992) *Chem. Phys. Lett.*, **196**, 462.

132. Petrich, J.W. and Martin, J.L. (1989) *Chem. Phys.*, **131**, 31–47.

133. (a) McMahon, R.J., Ford, R.K., Patterson, H.H., and Wrighton, M.S. (1988) *J. Am. Chem. Soc.*, **110**, 2670; (b) Ford, R.K., McMahon, R.J., Yu, J., and Wrighton, M.S. (1989) *Spectrochim. Acta*, **4SA**, 23.

134. Su, C., Park, Y.D., Liu, G.Y., and Spiro, T.G. (1989) *J. Am. Chem. Soc.*, **111**, 3457.

135. Vermeulen, N., Debaes, C., and Thienpont, H. (2010) *Laser & Photon. Rev.*, **4**, 656–670.

136. Kataoka, H., Maeda, S., and Hirose, C. (1982) *Appl. Spectrosc.*, **36**, 565.

137. Nikitin, S.Yu. (2009) *Quantum Electron.*, **39**, 649–652.

138. Schmitt, M., Knopp, G., Materny, A., and Kiefer, W. (1997) *Chem. Phys. Lett.*, **270**, 9–15.

139. Hayden, C.C. and Chandler, D.W. (1995) *J. Chem. Phys.*, **103**, 10465.

140. Wang, Z., Pang, Y., and Dana, D.D. (2004) *Chem. Phys. Lett.*, **397**, 40–45.

141. Moger, J., Johnston, B.D., and Tyler, C.R. (2008) *Opt. Express*, **16**, 3408.

142. Wijekoon, W.M.K.P. and Hetherington, W.M. III (1993) *J. Am. Chem. Soc.*, **115**, 2882–2886.

143. Tong, L., Lu, Y., Lee, R.J., and Cheng, J.X. (2007) *J. Phys. Chem. B*, **111**, 9980–9985.

144. (a) Lefrant, S., Baltog, I., and Baibarac, M. (2009) *Synth. Met.*, **159**, 2173–2176; (b) Ikeda, K. and Uosaki, K. (2009) *Nanoletters*, **9**, 1378.

145. Kang, K., Ozel, T., Cahill, D.G., and Shim, M. (2008) *Nano Lett.*, **8** 4642.

146. Kim, H., Taggart, D.K., Xiang, C., Penner, R.M., and Potma, E.O. (2008) *Nano Lett.*, **8**, 2373.

147. Ichimura, T., Hayazawa, N., Hashimoto, M., Inouye, Y., and Kawata, S. (2004) *Phys. Rev. Lett.*, **92**, 220801.

148. McCamant, D.W., Kukura, P., Yoon, S., and Mathies, R.A. (2004) *Rev. Sci. Instrum.*, **75**, 4971.

149. Jin, S.M., Lee, Y.J., Yu, J., and Kim, S.K. (2004) *Bull. Korean Chem. Soc.*, **25**, 1829.

150. Kukura, P., McCamant, D.W., and Mathies, R.A. (2007) *Annu. Rev. Phys. Chem.*, **58**, 461.

151. Kukura, P., McCamant, D.W., and Mathies, R.A. (2004) *J. Phys. Chem. A*, **108**, 5921.

152. Smeigh, A.L., Creelman, M., Mathies, R.A., and McCusker, J.K. (2008) *J. Am. Chem. Soc.*, **130**, 14105–14107.

153. Dasgupta, J., Frontiera, R.R., Taylor, K.C., Lagarias, J.C., and Mathies, R.A. (2009) *Proc. Nat. Acad. Sci.*, **106**, 1784–1789.

154. Wynne, K., Galli, C., and Hochstrasser, R.M. (1994) *J. Chem. Phys.*, **100**, 4797.

155. Frontiera, R.R., Dasgupta, J., and Mathies, R.A. (2009) *J. Am. Chem. Soc.*, **131**, 15630–15632.

156. Lee, I.H., Yee, K.J., Lee, K.G., Oh, E., Kim, D.S., and Lim, Y.S. (2003) *J. Appl. Phys.*, **93**, 4939.

157. Vallée, F. (1994) *Phys. Rev. B*, **49**, 2460.

158. Yee, K.J., Lee, K.G., Oh, E., Kim, D.S., and Lim, Y.S. (2002) *Phys. Rev. Lett.*, **88**, 105501.

159. Aku-Leh, C., Zhao, J., Merlin, R., Menéndez, J., and Cardona, M. (2005) *Phys. Rev. B*, **71**, 205211.

160. Bragas, A.V., Aku-Leh, C., and Merlin, R. (2006) *Phys. Rev. B*, **73**, 125305.

161. Crooker, S.A., Awschalom, D.D., and Samarth, N. (1995) *IEEE J. Sel. Top. Quantum Electron.*, **1**, 1082.

162. Whitaker, K.M., Raskin, M., Kiliani, G., Beha, K., and Ochsenbein, S.T. (2011) *Nano Lett.*, **11**, 3355–3360.

163. Ghosh, S., Sih, V., Lau, W.H., Awschalom, D.D., Bae, S.-Y., Wang, S., Vaidya, S., and Chapline, G. (2005) *Appl. Phys. Lett.*, **86**, 232507.

164. (a) Gupta, J.A., Awschalom, D.D., Peng, X., and Alivisatos, A.P. (1999) *Phys. Rev. B*, **59**, R10421–R10424; (b) Gupta, J.A, Awschalom, D.D, Efros, A.L., and Rodina, A.V. (2002) *Phys. Rev. B*, **66**, 125307; (c) Berezovsky, J., Gywat, O., Meier, F., Battaglia, D., Peng, X., and Awschalom, D.D. (2006) *Nat. Phys.*, **2**, 831.

165. Janßen, N., Whitaker, K.M., Gamelin, D.R., and Bratschitsch, R. (2008) *Nano Lett.*, **8**, 1991–1994.

166. Zinth, W. and Kaiser, W. (1988) *Ultrafast Coherence Spectroscopy*, Chapter 6, vol. **60**, Springer-Verlag, Heidelberg, pp. 235–277.

167. (a) Mossberg, T., Flusberg, A., Kachru, R., and Hartmann, S.R. (1977) *Phys. Rev. Lett.*, **39**, 1523; (b) Lee, H.W.H., Patterson, F.G., Olson, R.W., Wiersma, D.A., and Fayer, M.D. (1982) *Chem. Phys. Lett.*, **90**, 172; (c) Morsink, J.B.W., Hesselink, W.H., and Wiersma, D.A (1982) *Chem. Phys.*, **71**, 289; (d) Becker, P.C., Fragnito, H.L., Bigot, J.Y., Brito-Cruz, C.H., Fork, R.L., and Shank, C.V. (1989) *Phys. Rev. Lett.*, **63**, 505; (e) Nibbering, E.T.J., Wiersma, D.A., and Duppen, K. (1991) *Phys. Rev. Lett.*, **66**, 2464; (f) Joo, T. and Albrecht, A.C. (1993) *Chem. Phys.*, **176**, 233.

168. (a) Wong, C.Y. and Scholes, G.D. (2011) *J. Phys. Chem. A*, **115**, 3797–3806; (b) Yeremenko, S., Pshenichnikov, M.S., and Wiersma, D.A. (2003) *Chem. Phys. Lett.*, **369**, 107; (c) Brewer, R.G. and Hahn, E.L. (1975) *Phys. Rev. A*, **11**, 1641–1649; (d) Bai, Y.S. and Payer, M.D. (1988) *Chem. Phys.*, **128**, 135–155; (e) de Boeij, W.P., Pshenichnikov, M.S., and Wiersm, D.A. (1998) *Annu. Rev. Phys. Chem.*, **49**, 99–123; (f) Fleming, G.R. and Cho, M.-H. (1996) *Annu. Rev. Phys. Chem.*, **47**, 109–134.

169. (a) Agarwal, R., Prall, B.S., Rizvi, A.H., Yang, M., and Fleming, G.R. (2002) *J. Chem. Phys.*, **116**, 6243; (b) Pisliakov, A.V., Mancal, T., and Fleming, G.R. (2006) *J. Chem. Phys.*, **124**, 234505; (c) Oskouei, A.A., Tortschanoff, A., Bräm, O., van Mourik, F., Cannizzo, A., and Chergui, M. (2010) *J. Chem. Phys.*, **133**, 064506; (d) van Dao, L., Lincoln, C., Lowe, M., and Hannaford, P. (2004) *J. Chem. Phys.*, **120**, 8434.

170. (a) Schenzle, A., DeVoe, R.G., and Brewe, R.G. (1984) *Phy. Rev. A*, **30**, 1866–1872; (b) Fidder, H., de Boer, S., and Wiersma, D.A. (1989) *Chem. Phys.*, **139**, 317–326.

171. Mossberg, T.W. (1982) *Opt. Lett.*, **7**, 77–79.

172. (a) Jankowiak, R., Hayes, J.M., and Small, G.J. (1993) *Chem. Rev.*, **93**, 1471; (b) Narasimhan, L.R., Littau, K.A., Pack, D.W., Bai, Y.S., Elschner, A., and Fayer, M.D. (1990) *Chem. Rev.*, **90**, 439

173. Sangouard, N., Simon, C., Minar, J., Afzelius, M., Chaneliére, T., Gisin, N., Le Gouët, J.L., Riedmatten, H. de, and Tittel, W. (2010) *Phys. Rev. A*, **81**, 062333.

174. Abazari, A.D., Saglamyurek, E., Ricken, W.S.R., Mela, C.L., and Titte, W. (2009) *adsabsharvardedu*, 4. Retrieved from http://arxiv.org/abs/0910.2457.

175. Graham, M.W., Ma, Y., and Fleming, G.R. (2008) *Nano Lett.*, **8**, 3936.

176. Nagasawa, Y., Seike, K., Muromoto, T., and Okada, T. (2003) *J. Phys. Chem. A*, **107**, 2431–2441.

177. Becker, P.C., Fragnito, H.L., Brito Cruz, C.H., Fork, R.L., Cunningham, J.E., Henry, J.E., and Shank, C.V. (1988) *Phys. Rev. Lett.*, **61**, 1647.

178. Noll, G., Siegner, U., Shevel, S., and Gobel, E.O. (1990) *Phys. Rev. Lett.*, **64**, 792.

179. Mittleman, D.M., Schoenlein, R.W., Shiang, J.J., Colvin, V.L., Alivisatos, A.P., and Shank, C.V. (1994) *Phys. Rev. B*, **49**, 14435–14447.

180. (a) Salvador, M.R., Wilson, M.W., and Scholes, G.D.J. (2006) *Chem. Phys.*, **125**, 184709-1–184709-16; (b) Takemoto, K., Ikezawa, M., and Masumoto, Y. (2003) *Phys. Status Solidi C*, **0**, 1279–1282. (c) Masumoto, Y., Ikezawa, M., Hyun, B.R., Takemoto, K., and Furuya, M. (2001) *Phys. Status Solidi B*, **224**, 613–619; (d) Scholes, G.D. (2004) *J. Chem. Phys.*, **121**, 10104–10110.

181. McKimmie, L.J., Lincoln, C.N., Jasieniak, J., and Smith, T.A. (2010) *J. Phys. Chem. C*, **114**, 82–88.

182. Gong, S., Yang, G., Ban, D., Fu, J., and Fu, Y. (2011) *Opt. Mater.*, **34**, 36–41.

183. Bai, Y.S. and Fayer, M.D. (1988) *Chem. Phys.*, **128**, 135–155.

184. (a) Tokmakoff, A., Zimdars, D., Urdahl, R.S., Francis, R.S., Kwok, A.S., and M.D. Fayer (1995) *J. Phys. Chem.*, **99** (13), 310; (b) Tokmokoff, A. and Fayer, M.D. (1995) *Acc. Chem. Res.*, **28**, 437.

185. Equall, R.W., Cone, R.L., and Macfarlane, R.M. (1995) *Phys. Rev. B*, **52**, 3963.

186. Babbitt, W.R. and Mossberg, T.W. (1994) *J. Opt. Soc. Am. B*, **11**, 1948.

187. (a) Bai, Y.S., Babbitt, W.R., Carlsson, N.W., and Mossberg, T.W. (1984) *Appl. Phys. Lett.*, **45**, 714; (b) Kroll, S. and Elman, U. (1993) *Opt. Lett.*, **18**, 1834; (c) Babbitt, W.R. and Mossberg, T.W. (1995) *Opt. Lett.*, **20**, 910; (d) Sonajalg, H., Debarre, A., Le Gouet, J.-L., Lorgere, I., and Tchenio, P. (1995) *J. Opt. Soc. Am. B*, **12**, 1448; (e) Graf, F.R., Bernd, H., Maniloff, E.S., Altner, S.B., Renn, A., and Wild, U.P. (1996) *Opt. Lett.*, **21**, 284.

188. Macfarlane, R.M. (1995) *Laser Phys.*, **5**, 567–572.

189. Luo, B., Elman, U., Kroll, S., Paschotta, R., and Tropper, A. (1998) *Opt. Lett.*, **23**, 442.

190. Nakatsuka, S.H., Fujiwara, M., and Matsuoka, M. (1984) *Phys. Rev. A*, **29**, 2286.

191. Zhang, S., Sun, Z., Yang, X., Wang, Z., Lin, J., Huang, W., Xu, Z., and Li, R. (2004) *Opt. Commun.*, **241**, 481–486.

192. Meltzer, R.S., Zheng, H., and Dejneka, M.J. (2004) *J. Lumin.*, **107**, 166–175.

193. (a) Oudar, J.-L. and Shen, Y.R. (1980) *Phys. Rev. A*, **22**, 1141–1158; (b) Ender, D.A. (1982) Doubly-resonant two-photon-absorption-induced four-wave mixing in Tb(OH)3 and LiTbF4. PhD Thesis at Montana State University.

194. (a) Özgür, Ü., Alivov, Y., Liu, C., Teke, A., Reshchikov, M., Dogan, S., Avrutin, V., Cho, S., and Morkoc, H. (2005) *J. Appl. Phys.*, **98**, 041301–041404; (b) Schmide-Mende, L. and MacManus-Driscoll, J.L. (2007) *Nano Today*, **10**, 40–48.

195. (a) Dimitrijevic, N.M. and Kamat, P.V. (1987) *J. Phys. Chem.*, **91**, 2096; (b) Haase, M., Weller, H., and Henglein, A.J. (1988) *Phys. Chem.*, **92**, 4706; (c) Wang, Y., Herron, N., Mahler, W., and Suna, A.J. (1989) *Opt. Soc. Am. B*, **6**, 808; (d) Eychmueller, A., Vobmeyer, T., Mews, A., and Weller, H. (1994) *J. Lumin.*, **58**, 223.

196. Kamat, P.V., Dimitrijevic, N.M., and Nozik, A.J. (1989) *J. Phys. Chem.*, **98**, 2873.

197. Hilinski, E.F., Lucas, P.A., and Wang, Y. (1988) *J. Chem. Phys.*, **89**, 3435.

198. Mittleman, D.M., Schoenlein, R.W., Shiang, J.J., Colvin, V.L. *et al.* (1994) *Phys. Rev. B: Condens. Matter*, **49**, 14435.

199. Nosaka, Y., Ohta, N., and Miyama, H. (1990) *J. Phys. Chem.*, **94**, 3752.

200. Zhang, J.Z., O'Neil, R.H., and Roberti, T.W. (1994) *J. Phys. Chem.*, **98**, 3859.

201. Klimov, V., Bolivar, P.H., and Kurz, H. (1996) *Phys. Rev. B*, **53**, 1463.

202. Skinner, D.E., Colombo, D.P., Cavaleri, J.J., and Bowman, R.M. (1995) *J. Phys. Chem.*, **99**, 7853.

203. Smith, B.A., Waters, D.M., Faulhaber, A.E., Kreger, M.A., Roberti, T.W., and Zhang, J.Z. (1997) *J. Sol-Gel Sci. Technol.*, **9**, 125.

204. O'Neil, M., Marohn, J., and McLendon, G. (1990) *Chem. Phys. Lett.*, **168**, 208.

205. Zheng, J.P., Shi, L., Choa, F.S., Liu, P.L., and Kwok, H.S. (1988) *Appl. Phys. Lett.*, **53**, 643.

206. Zhang, J.Z., O'Neil, R.H., and Roberti, T.W. (1994) *J. Phys. Chem.*, **98**, 3859–3864.

207. Zhang, J.Z. (1997) *Acc. Chem. Res.*, **30**, 423–429.

208. Heilweil, E.J. and Hochstrasser, R.M. (1985) *J. Chem. Phys.*, **82**, 4762.

209. Roberti, T.W., Smith, B.A., and Zhang, J.Z. (1995) *J. Chem. Phys.*, **102**, 3860.

210. Faulhaber, A.E., Smith, B.A., Andersen, J.K., and Zhang, J.Z. (1996) *Mol. Cryst. Liq. Cryst.*, **283**, 25.

211. Zhang, J.Z., Smith, B.A., Faulhaber, A.E., Andersen, J.K., and Rosales, T.J. (1996) *Ultrafast Processes Spectrosc.*, **9**, 561.

212. Groeneveld, R.H.M., Sprik, R., and Lagendijk, A. (1990) *Phys. Rev. Lett.*, **64**, 784.

213. Brorson, S.D., Fujimoto, J.G., and Ippen, E.P. (1987) *Phys. Rev. Lett.*, **59**, 1962.

214. Stella, A., Nisoli, M., de Silvestri, S., Svelto, O., Lanzani, G., Cheyssac, P., and Kofman, R. (1996) *Springer Ser. Chem. Phys.*, **62**, 439.

215. Belotskii, E.D. and Tomchuk, P.M. (1992) *Int. J. Electron.*, **73**, 955.

216. Schultze, M., Fieß, M., Karpowicz, N., Gagnon, J., Korbman, M., Hofstetter, M., Neppl, S., Cavalieri, A.L., Komninos, Y., Mercouris, Th., Nicolaides, C.A., Pazourek, R., Nagele, S., Feist, J., Burgdörfer, J., Azzeer, A.M., Ernstorfer, R., Kienberger, R., Kleineberg, U., Goulielmakis, E., Krausz, F., and Yakovlev, V.S. (2010) *Science*, **328**, 1658.

12
Nonlinear Optical Characterization of Nanomaterials

Rashid Ashirovich Ganeev

12.1
Influence of Laser Ablation Parameters on the Optical and Nonlinear Optical Characteristics of Colloidal Solution of Semiconductor Nanoparticles

12.1.1
Introduction

Colloidal solution of nanoparticles, that is, nanoparticles suspended in different insulator matrices, found application because of their fast response and large nonlinearities. In our previous studies, we measured nonlinear optical characteristics of metal (Ag, Au, Pt, Cu) colloidal solutions at different aggregated states and investigated their optical limiting properties [1]. Meanwhile, the semiconductor quantum dots and colloidal microcrystallites have attracted extensive attention over the past decade [2]. Different methods were used to measure their nonlinearities, including degenerate four-wave mixing (DFWM), Z-scan technique, optical interferometry, and nonlinear absorption [3].

Among a variety of preparation methods of nanoparticle solutions the most popular is the chemical one [4, 5]. The laser ablation method is another effective technique for preparation of nanofilms and nanoparticle solutions. The aqueous metal and carbon colloids prepared by laser ablation possess high stability and do not require the addition of stabilizers [6]. Local field enhancement of nanoparticles in colloidal suspensions [7] and considerable nonlinear optical susceptibilities of chalcogenide structures [8, 9] allow the expectation of new interesting features of such semiconductor nanoparticles prepared by laser ablation.

An interest in the synthesis, characterization, and application of colloidal semiconductor "quantum dot" materials has grown markedly [10, 11] due to strong size-related dependence of their optical and electronic properties. However, the dependence of the optical nonlinearities on the sizes of the semiconductor nanoparticles has yet to be determined.

In this section, the laser ablation method used for the preparation and investigation of aqueous colloidal solutions of As_2S_3 and CdS nanoparticles is discussed.

Nanomaterials: Processing and Characterization with Lasers, First Edition.
Edited by Subhash Chandra Singh, Haibo Zeng, Chunlei Guo, and Weiping Cai.
© 2012 Wiley-VCH Verlag GmbH & Co. KGaA. Published 2012 by Wiley-VCH Verlag GmbH & Co. KGaA.

The measurements of nonlinear refractive indices (n_2), nonlinear absorption coefficients (β), and third-order nonlinear susceptibilities ($\chi^{(3)}$) of these solutions by the Z-scan technique are presented using picosecond radiation at the wavelength of $\lambda = 532$ nm. We compare these data with the one obtained for As_2S_3 and CdS thin films.

12.1.2
Experimental Setup

The method described in [6] was used for the preparation of aqueous semiconductor colloidal solutions. Initially this technique was used mostly for preparation of thin films and nanoparticles and was performed under vacuum conditions to prevent contamination and oxidation of produced materials. However, 10 years ago it was shown that such a technique is applicable to targets in solutions (water and organic solvents; [12, 13]). This method allowed the preparation of stable colloids without the need for stabilizers.

The Q-switched Nd:YAG laser ($\lambda = 1064$ nm, $t = 20$ ns, $E = 15$ mJ) at 10 Hz pulse repetition rate was used for the laser ablation of bulk As_2S_3 and CdS semiconductors. The samples (bulk As_2S_3 or CdS) were immersed in a quartz cell containing distilled water (cell thickness of 5 cm). Laser radiation was focused by an 8 cm focal length lens onto the surface of the sample that was kept close to the cell's back window to prevent the optical breakdown of the front cell window. Samples were irradiated for 15 min durations. This method allowed synthesizing semiconductor nanoparticles, such as cadmium sulfide, in colors ranging from water white to orange depending on the particle's diameter in aqueous solutions (from 2 to 8 nm) with no change in chemical composition. Other optical properties of the particles, such as refractive index and nonlinear optical response, exhibited similar changes with particle size. Weight volume of nanoparticles in solution was estimated to be $(3–5) \times 10^{-5}$.

The linear absorption coefficients at $\lambda = 532$ nm were measured to be 0.37 and 0.52 cm^{-1} for As_2S_3 and CdS solutions, respectively. These solutions had not aged during one week. The effectiveness of laser ablation decreased during irradiation due to the cloud appearance of the nanoparticles near the ablated area. The solution was mixed in order to keep constant homogeneity. We tried to increase the concentration of solutions using longer ablation time. However, at high concentrations of nanoparticles the spontaneous clusterization led to the sedimentation of semiconductor nanoparticles.

In our studies, we focused 1064 nm radiation on the surface of targets. Such laser ablation increases the efficiency of colloid formation. Previous studies [6] have shown the sharp growth of ablation efficiency of 1064 nm radiation under tight focusing conditions with respect to unfocused radiation. The fluence at the target was 20 J cm^{-2}. The absorption of water solvent in the visible range increased with the growth of nanoparticle concentration, but no noticeable absorption of 1064 nm radiation was observed in the case of constant artificial mixing near the ablated area. We did not use the organic solvents in these experiments for stabilization of

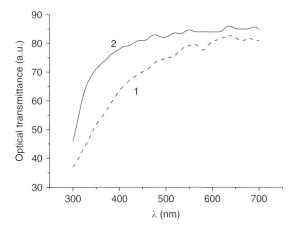

Figure 12.1 Transmission spectra of semiconductor colloidal solutions. (1) As$_2$S$_3$ and (2) CdS.

nanoparticles. The ablated material was moved each 100 shots for preventing the influence of target craterization on ablation and nanoparticle formation efficiency.

The influence of absorption of water solvent and nanoparticle solutions during ablation takes place in the case of shorter wavelength radiation (e.g., second and third harmonics of Nd:YAG laser radiation) due to closeness to plasmon bands of nanoparticles. In that case, melting, aggregation, and dissociation of nanoparticles were reported. These processes change the shape of the absorption spectra of colloids. So, longer wavelength laser radiation seems to be more appropriate for effective ablation.

Transmission spectra of semiconductor solutions are presented in Figure 12.1. The absorption edge of semiconductor nanoparticle solutions was located in UV range, whereas for bulk semiconductors it was located in the visible range. The observed UV absorption edge of synthesized CdS nanoparticles ($\lambda = 320$ nm, $E_g = 3.86$ eV) was strongly blueshifted as compared with the one of bulk CdS ($\lambda = 490$ nm, $E_g = 2.53$ eV), thus suggesting the formation of nanoparticles in our solutions. The same properties were observed in the case of As$_2$S$_3$ nanoparticle solutions.

The nanoparticle sizes were inspected using a transmission electron microscope (TEM) JEM-201OF. The aqueous composite was dissolved in ethanol. A drop of the solution was placed on a carbon-coated copper grid that was left to dry before transferring into the TEM sample chamber. Micrographs were taken at an acceleration voltage of 200 kV. The energy dispersive X-ray spectroscopy (EDX) analyzer was used for elemental analysis of the colloidal nanoparticle solutions.

The Z-scan technique was applied for the investigation of nonlinear optical characteristics of the samples. The detailed description of the experimental setup was published elsewhere [14]. The output laser characteristics were as follows: pulse duration 55 ps, $\lambda = 532$ nm, $E = 0.2$ mJ, 2 Hz pulse repetition rate. The solutions were contained in 5 mm thick quartz cells. The investigated samples were mounted on a translation stage controlled by a computer. The samples were moved along the Z-axis through the focal point of 25 cm focal length lens. Beam waist in focal plane was 102 µm. Laser pulse energy was measured by a calibrated photodiode. The one millimeter aperture was fixed at a distance of 120 cm from the focal plane (closed-aperture scheme). Output radiation propagated through this aperture was measured by a second photodiode. The closed-aperture scheme allowed to determine the sign and magnitude of n_2 and $Re\chi^{(3)}$.

The open-aperture Z-scan scheme was used for the measurements of β and $Im\chi^{(3)}$. The aperture in that case was removed. The second photodiode was kept at such a distance from the sample that it allowed measurement of all radiation transmitted through the cell. The experiments were carried out at intensities up to 5×10^{10} W cm^{-2}.

Chalcogenide CdS and As_2S_3 thin films were used for comparison with CdS and As_2S_3 nanoparticle solutions. The films were prepared by evaporation in vacuum of chalcogenide components (CdS, As_2S_3) onto the surface of BK7 glass substrates. The thickness of films was 10 µm.

12.1.3
Results and Discussion

12.1.3.1 Measurements of n_2 of Semiconductor Solutions

One of the advantages of the Z-scan technique is the possibility of distinguishing two nonlinear optical mechanisms when they are presented simultaneously. In general, when we have the contribution from both nonlinear refraction and nonlinear absorption, the $T(z)$ dependence can be presented as follows:

$$T(z) = 1 + \frac{4x}{(x^2+9)(x^2+1)}\Delta\Phi_0 - \frac{2(x^2+3)}{(x^2+9)(x^2+1)}\Delta\Psi_0 \tag{12.1}$$

where $T(z)$ is the normalized transmittance of the sample, $x = z/z_0$, $z_0 = k(\omega_0)^2/2$ is the diffraction length, $k = 2\pi/\lambda$ is the wave number, ω_0 is the beam waist radius, $\Delta\Phi_0$ and $\Delta\Psi_0$ are the phase changes due to nonlinear refraction and nonlinear absorption, respectively, $\Delta\Phi_0 = k\gamma I_0 L_{eff}$, $\Delta\Psi_0 = \beta I_0 L_{eff}/2$. Here, γ is the nonlinear refractive index (measured in SI units), I_0 is the laser radiation intensity in the focal plane, $L_{eff} = [1 - \exp(-\alpha_0 L)]/\alpha_0$ is the effective length of the sample, α_0 is the linear absorption coefficient, L is the sample length. n_2 (nonlinear refractive index measured in electrostatic units) and γ are connected through the relation $\gamma (m^2\ W^{-1}) = 40\pi n_2(esu)/c(m\ s^{-1})n_0$, where c is the light velocity and n_0 is the linear refractive index. By making the substitution $\rho = \beta/2\ k\gamma$, one can get the relation between $\Delta\Phi_0$ and $\Delta\Psi_0(\Delta\Psi_0 = \rho\Delta\Phi_0)$. In that case, Eq. (12.1) can be rewritten as follows:

$$T = 1 + \frac{2(-\rho x^2 + 2x - 3\rho)}{(x^2+9)(x^2+1)}\Delta\Phi_0 \tag{12.2}$$

This relation was used for theoretical dependence calculations, which is taken into account in our experimental conditions. After fitting of ρ and $\Delta\Phi_0$, we

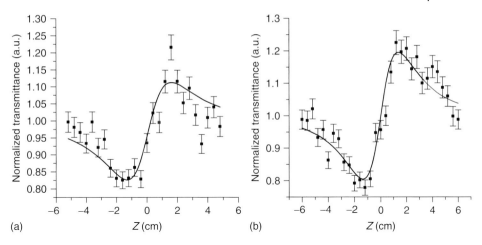

Figure 12.2 Normalized transmission dependencies of (a) As_2S_3 and (b) CdS nanoparticle solutions in the case of a closed-aperture scheme. Solid lines are the theoretical fits.

consequently found γ and β of nanoparticle solutions. We also made independent β measurements of these solutions using the open-aperture Z-scan technique. One can obtain the data on nonlinear refraction and nonlinear absorption coefficients of material in some particular cases from the closed-aperture Z-scan. Open-aperture Z-scan can be performed to test the results obtained from the nonlinear curve fitting using Eq. (12.2).

In Figure 12.2, the normalized transmission dependencies of As_2S_3 and CdS colloidal solutions are presented for a closed-aperture scheme. One can conclude from the obtained dependencies about the positive value of nonlinearity, that is, self-focusing of laser radiation. Relatively symmetric curves (respectively to the focal point, $z = 0$) indicate an insignificant influence of nonlinear absorption at this intensity level ($I = 4 \times 10^9$ W cm^{-2}), thus causing us to make the nonlinear absorption measurements using an open-aperture scheme at higher intensities when this nonlinear optical process appeared to be more indicatable.

The values of n_2 and $Re\chi^{(3)}$ of the CdS solution were calculated to be 1.72×10^{-12} and 2.74×10^{-13} esu, respectively, at the intensity of 4×10^9 W cm^{-2}. The solid lines in the figure are the theoretical fits calculated for the medium with third-order nonlinearity. Intensity-dependent measurements of n_2 and $Re\chi^{(3)}$ have shown the decrease in their values with intensity growth. Thus, at intensity increasing up to 1.17×10^{10} W cm^{-2}, the values of n_2 and $Re\chi^{(3)}$ were measured to be 6.58×10^{-13} and 1.05×10^{-13} esu that are 2.6 times smaller as compared with the previous case.

The unfixed value of the nonlinear refractive index in our studies indicates the influence of high orders of nonlinearity induced by the appearance of free carriers. The same feature was reported by Du *et al.* [5]. The importance of free carrier influence on the variations of nonlinear refraction was also discussed by Li *et al.* [15] in their studies of bulk CdS.

The intensity-dependent variation of the nonlinear refractive index is the proof of the existence of the fifth-order nonlinear optical properties. The refractive index in that case can be presented by Du *et al.* [5].

$$n = n_0 + n_{ef}I = n_0 + (n_2 + n_4 I)I \tag{12.3}$$

where n is the total refractive index, n_{ef} is the effective nonlinear refractive index, and n_4 denotes the fifth-order nonlinearity.

Our results on the variations of the nonlinear refractive index with the intensity growth have shown that the change of n_{ef} due to free carrier generation becomes negative ($n_4 < 0$). The same result was previously reported in [15]. Our results on intensity-dependent variations of nonlinear refractive index are also consistent with calculations [16] predicting the decrease of n_2 due to free carrier appearance.

The variation of refractive index can be determined as [16]

$$\Delta n = \gamma I + \sigma_r N(I) \tag{12.4}$$

The second term in Eq. (12.4) is the additional refractive index change due to the influence of the free carriers. $N(I)$ is the intensity-dependent density of free carriers. Parameter σ_r characterizes refractive index change due to the single free carrier appearance. σ_r in most semiconductors has a negative value. Free-carrier-induced process can compensate and even be larger than positive nonlinear refraction [16], thus changing the sign of nonlinear refraction (from self-focusing to self-defocusing).

The same measurements were carried using As_2S_3 solution. In that case, the dependence of n_2 on laser intensity was also observed. The values of n_2 and $Re\chi^{(3)}$ at $I = 2.94 \times 10^9$ W cm^{-2} were calculated to be 2.21×10^{-12} and 3.54×10^{-13} esu, while at the intensity of 3.32×10^{10} W cm^{-2} the 1.5 times decrease of n_2 and $Re\chi^{(3)}$ ($n_2 = 1.45 \times 10^{-12}$ esu, $Re\chi^{(3)} = 2.3 \times 10^{-13}$ esu) was observed.

Small values of nonlinear optical parameters of these solutions were predominantly due to low concentration of nanoparticles. Taking into account the volume part of semiconductor nanoparticles (4×10^{-5}), we can estimate that their nonlinearities exceed the ones of bulk materials. In particularly, n_2 of bulk CdS was previously measured to be -2×10^{-10} esu ($\gamma = -5.3 \times 10^{-17}$ m^2 W^{-1} [15]). Our data show that this parameter for semiconductor nanoparticles is of the order of 10^{-7} esu, taking into account the volume part of the active nonlinear medium.

One of the possible explanations of observed high nonlinearities of semiconductor nanoparticles is the influence of field localization in such structures. Surface-enhanced nonlinear spectroscopy of various nanosized structures gives valuable information about the nature of considerable growth of both optical and nonlinear optical characteristics of such media. Surface-enhanced hyper-Raman scattering and anti-Stokes Raman scattering, second- and third-harmonic generation, as well as other nonlinear optical responses of fractals and nanofilms were studied in [17–19]. The fractal geometry results in the localization of plasmon excitation in the "hot" spots, where the local field can exceed the applied field by several orders of magnitude. The high local fields of the localized fractal modes result in dramatic enhancement of nonlinear optical responses. Our estimations of nonlinear indices of CdS nanoparticles (taking into account their volume part in solutions) have shown that their n_2 exceeds the one of bulk CdS by more than 3 orders ($1.72 \times 10^{-12} \times 4 \times 10^{-5} \approx 7 \times 10^{-7}$, and 2×10^{-10} esu, respectively). This value (3.5×10^{-3}) is comparable with calculations by various authors predicting the 10^3 enhancement of nanoparticle nonlinearities with respect to the ones of bulk material. Our estimations of n_2 of dissolved CdS nanoparticles can be considered as a qualitative confirmation of an assumption on local-field-induced enhancement of observed nonlinearities.

Large third-order nonlinear susceptibilities of semiconductor nanoparticles were reported previously in various studies. Wang and Mahler [20] reported on large third-order nonlinearities of CdS nanocrystallites embedded in Nafion films. The self-focusing ($\lambda = 514.5$ nm) in colloidal CdS semiconductor quantum dots was observed by Shen *et al.* [21]. n_2 was found to be 10^5 times larger than the nonlinear refractive index of CS_2 (that has the nonlinearity of 2 orders smaller than bulk CdS). Analogous results indicating extremely high nonlinearities of such nanoparticles were reported recently in [5, 22, 23].

12.1.3.2 The Analysis of Self-Interaction Processes in Semiconductor Solutions

We attributed the observed nonlinearity to the CdS and As_2S_3 nanoparticles, since the quartz cells filled with pure water did not indicate the nonlinear optical properties at intensities used. Our separate measurements of quartz nonlinearity have shown that its n_2 was 7.2×10^{-14} esu, that is, close to the data reported early (7.8×10^{-14} esu, [24]), thus confirming the appropriate calibration of our Z-scan setup.

We have analyzed various possible mechanisms, to consider the origin of nonlinear processes observed in our studies. The first is the thermal effect, which is dominant for micro- or nanosecond pulses and takes place because of heat transfers from nanoparticles to the environment. However, it might be considerable for pico- or even femtosecond pulses if a high repetition rate of pulses is used. We performed similar scans with single shots. No difference was found in 2-Hz pulse repetition rate studies, indicating that the long-term thermal phenomena were negligible under our experimental conditions.

Previously, we reported measurements of nonlinear optical characteristics of the semiconductor As_2S_3 colloidal solution at the wavelength of Nd:YAG laser radiation ($\lambda = 1064$ nm, $t = 25$ ns) [25]. In the case of nanosecond radiation, the nonlinear refractive index was measured to be -3×10^{-11} esu. We attributed the appearance of nonlinear refraction in these studies to the thermal effect. It is known that the thermal effect appearance connected with the acoustic wave propagation. The time required for density reduction is determined by a ratio of beam waist radius (ω_0) to the velocity of sound (V_s) in the dielectric ($\tau = \omega_0/V_s$). Taking into account our experimental conditions ($\omega_0 = 50$ μm at the wavelength of 1064 nm, $V_s \sim 1500$ m s^{-1}) we found $\tau \sim 33$ ns, that is, the influence of thermal effect should not be neglected.

Note that the critical parameter of those experiments using nanosecond pulses was the absorption of the medium. A parameter describing thermal self-action of laser radiation (self-focusing or self-defocusing) is $M = (1/C\rho)(dn/dT)$, which is equal to -1.04×10^{-4} cm^3 cal^{-1} for water. Here C and ρ are the specific heat and the density of water, respectively. M is a crucial parameter for the determination of refractive index variations (Δn). The latter parameter connected with the absorbed energy (ΔE) in a unit volume by the relation $\Delta n = M \times \Delta E$, with M being independent of absorbed energy. The numerical analysis of this relation for water solution of gold nanoparticles in the field of nanosecond pulses was presented by Mehendale *et al.* [26]. It was shown that the thermal effect leads to the radiation self-defocusing in related materials where water is used as a solvent. So, the prevailing mechanism contributing to the refractive index variation was the thermal lens due to nanosecond laser radiation absorption in the investigated solution. The output signal in the closed-aperture Z-scan scheme depends on the variations of radiation transmitted through the aperture with the transmission factor of 3%. So, the 1/6 part of the focal spot is important for a self-defocusing appearance. It means that the self-defocusing becomes noticeable 6 ns after the beginning of laser pulse propagation through the medium. Such estimations of the self-defocusing process were confirmed in the experiments with colloidal metals, when the temporal shape of nanosecond pulses propagated through the aperture was analyzed [1].

In the case of picosecond or shorter pulses, the main origin of the nonlinearities in such semiconductor compounds are known to be two-photon and Raman transitions, AC-Stark shift, optical Kerr effect, and photoexcited free carriers. Picosecond Z-scan and DFWM measurements showed that the first four effects are responsible for the nonlinearities at the intensities up to 5×10^8 W cm^{-2}, while at higher intensities, the nonlinear effects originating from the two-photon excited free carriers proved not to be neglected [15]. Those observations also confirm our assumptions on the influence of free carrier generation on the decrease of n_2 at intensities used (well above the mentioned intensity level).

12.1.3.3 The Sign of Nonlinear Refraction of Semiconductor Nanoparticles

Let us now discuss the sign of nonlinear refraction of investigated semiconductor nanoparticle solutions. Such nanostructures, for instance, CdS, can exhibit another sign of n_2 in comparison with that of bulk CdS at $\lambda = 532$ nm. A scaling rule between the nonlinear refractive index, n_2, and the ratio of the photon energy to the band gap energy ($h\omega/E_g$) was analyzed previously in different materials. The nonlinear refractive index, n_2, was found to change its sign at 0.69 value of this ratio under different E_g or laser frequency ω [27]. The previous studies of bulk CdS showed the spectral dispersion of the Kerr nonlinearity sign [27]. Thus, n_2 had a negative sign at $\lambda = 610$ nm, and changed to positive at $\lambda = 780$ nm, that was in agreement with the calculations using the Kramers–Kronig relation. We believe that in our case the ratio between laser frequency and surface plasmon resonance of CdS nanoparticles is a crucial factor in the definition of the sign of nonlinear refraction.

Some previous studies of such nanoparticles show that they possess the analogous sign of n_2 as bulk media. Du *et al.* [5] demonstrated the negative sign of nonlinear refractive index for CdS nanoparticles at $\lambda = 532$ nm using 7 ns pulses and similar results were also reported by Rakovich *et al.* [23].

Large crystallites (>10 nm) show an optical absorption close to that of bulk crystalline material. However, smaller crystallites show a large blueshift in absorption edge that lead to the variations of effective band gap. These observations can be understood as quantum size effects resulting from confinement of the electron and hole in a small volume. Associated with the change in the absorption spectrum, and hence effective band gap, there is a change in the refractive index. As described above, the change in the refractive index can be determined via the Kramers–Kronig relationship. The relationship between particle sizes and the peak energy of excitons was demonstrated previously in [28].

As the radii of semiconductor particles are decreased below or close to the Bohr exciton radius, their optical and electronic properties can be tuned by changing the size of the particles. The linear properties, such as blueshift in absorption and fluorescence, have been well characterized for II–VI semiconductor nanoparticles for several years. However, the nonlinear optical responses in relation to particle size variations and charge carrier dynamics are less understood. The intrinsic measurement of size, at which variations in nonlinear optical processes begin to become important, is given by the diameter of the 1S exciton in bulk crystalline material. For CdS, this diameter is ≈ 6 nm. Rossetti *et al.* [29] calculated the shift in band gap energy at different sizes of CdS nanoparticles. Their calculations predict a wide range of absorption thresholds (effective band gaps), ranging from 490 nm ($E_g = 2.53$ eV for bulk CdS), then to 471 nm ($E_g = 2.63$ eV, at 6.6 nm diameter of CdS nanoparticles), 367 nm ($E_g = 3.36$ eV, at 3 nm diameter), and further to 286 nm ($E_g = 4.33$ eV at 2.1 nm diameter). Thus, our spectrum ($E_g = 3.86$ eV) is a sum of various spectra corresponding to the different sizes of CdS nanoparticles. The influence of small nanoparticles (with size diameters between 3.5 and 4.5 nm) can change the sign of the self-interacting process (from self-defocusing to self-focusing). In our case the $\hbar\omega/E_g$ value becomes equal to 0.65. If applicable, these assumptions and observations show, for the first time to our knowledge, the change of the sign of nonlinear optical process (in our case from self-defocusing for bulk material to self-focusing for semiconductor nanoparticle solutions).

Analogous consideration was carried out for As_2S_3 solution. Its energy band gap shifted in accordance with the absorption spectrum from 2.37 (bulk As_2S_3) to 3.19 eV (semiconductor solution). The absorption cutoff in that case was not so expressed as in the case of the CdS solution, indicating the broader size distribution of As_2S_3 nanoparticles.

The CdS nanoparticle sizes were estimated to be between 4 and 6 nm taking into account the shift of the absorption edge. The sizes of As_2S_3 nanoparticles were found to be between 4 and 9 nm. We carried out the TEM measurements of investigated solutions for comparison with the estimations of nanoparticle sizes depicted from absorption spectra. Figure 12.3 shows the TEM image of a single CdS nanoparticle. The size distribution of nanoparticles measured from the TEM images was found to be between 2.5 and 6 nm and mostly centered at 4 nm. The

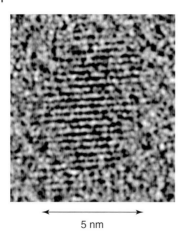

Figure 12.3 TEM image of a single CdS nanoparticle on a thin carbon support film. The nanoparticle are supported by a carbon film.

5 nm

same measurements were carried out for As_2S_3 nanoparticles. It was found to be of a broader size distribution (4–12 nm) in comparison with CdS nanoparticles and with the one depicted from spectral measurements.

Small nanoparticles (with sizes of 4 nm and smaller) can play a predominant role in the overall nonlinear refractive index of such solutions because of quantum confinement effects. Their positive sign of n_2 and the contribution in overall nonlinear refraction can prevail with respect to the negative sign of n_2 of higher sized nanoparticles. This peculiarity of small-sized nanoparticles can be considered as an explanation of the experimentally observed positive sign of chalcogenide solutions.

Local elemental characterization of the nanoparticles was carried out using EDX spectroscopy. In particular, the quantitative result of EDX had shown that the ratio of Cd and S was 55 : 45 (at.%), whereas the same for As and S was 35 : 65. The expected result for CdS was 50 : 50 because the EDX technique allows measuring the atomic ratio of the investigated compound. The same parameter for As_2S_3 was expected to be 2 : 3 or 40 : 60. The deviation from the atomic ratio of such semiconductor nanoparticles was reported previously in [30].

12.1.3.4 Nonlinear Absorption Measurements

The open-aperture Z-scan scheme was used for the investigation of nonlinear absorption. The reason for these studies was the asymmetric dependence observed at high intensities in the closed-aperture Z-scan scheme, indicating the presence of nonlinear absorption. Previously, the normalized transmittance dependence in the case of the open-aperture Z-scan scheme has shown the characteristic appearance of nonlinear absorption in the case of nanosecond 1064 nm pulses [25]. The same measurements were performed in the case of picosecond 532 nm pulses. The results of these investigations are presented in Figure 12.4. The normalized transmittance for the open-aperture conditions is given by Capple *et al.* [31].

$$T = q^{-1} \times \ln(1 + q) \tag{12.5}$$

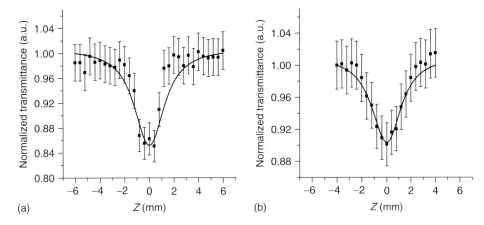

Figure 12.4 Normalized transmission dependencies for (a) As$_2$S$_3$ and (b) CdS in the case of open-aperture scheme. Solid lines are the theoretical fits.

where $q = \beta I_0 L_{eff}/(1 + z^2/z_0^2)$. Here I_0 is the laser intensity in the focal plane. The fits of Eq. (12.5) to the experimental data are depicted in Figure 12.4 by the solid lines. From these fits the β values were calculated to be 6×10^{-12} and 2.9×10^{-12} m W^{-1} for As$_2$S$_3$ and CdS solutions, respectively. The corresponding values of Im$\chi^{(3)}$ for As$_2$S$_3$ and CdS solutions were calculated to be 2.2×10^{-13} and 1×10^{-13} esu, respectively. Analyzing the obtained results for real and imaginary parts of $\chi^{(3)}$, we can conclude about their equal contribution in absolute value of $\chi^{(3)}$.

The closed-aperture data in some cases appeared to be sufficient for analysis of both nonlinear refraction and nonlinear absorption. We did not present an appropriate asymmetric $T(z)$ dependence at high intensities because of some uncertainties of β and n_2 measurements from these data. Our open-aperture $T(z)$ curves (Figure 12.4) allowed calculating the β within an accuracy of 30%, which was better than the data depicted from the closed-aperture scheme (50%).

The previously measured value of β for bulk cadmium sulfide ($\lambda = 532$ nm) was reported to be 5.4×10^{-11} m W^{-1} [17]. Taking into account a small volume of active particles we can again conclude about strong confinement effects in such semiconductor solutions, leading to the considerable enhancement of nonlinear absorption characteristics in comparison with bulk media. In studies at present, the CdS and As$_2$S$_3$ concentrations in solutions were approximately 0.06 g l^{-1}. Note that the nonlinear parameters of solutions can be increased by increasing the volume fraction of nanoparticles. The influence of the volume fraction of nanoparticles on their nonlinear optical properties was analyzed theoretically in [32] for semiconductor nanoparticles in various films, but in the case of semiconductor solutions an additional investigation should be done.

Our measurements of β for thin CdS and As$_2$S$_3$ films have also shown the considerable increase in nonlinear absorption in comparison with bulk material due to surface-enhanced effect ($\beta = 1.6 \times 10^{-7}$ m W^{-1} for As$_2$S$_3$ film ($\lambda = 1064$ nm); $\beta = 2 \times 10^{-8}$ m W^{-1} for CdS film ($\lambda = 532$ nm)).

It is worth noting the opportunity of using of organic solvents instead of water due to various reasons. One of the advantages of preparing nanoparticles in organic solvents is that the distribution of particle sizes is narrower than that in water [29], as it was shown using polyvinylpyrrolidone (PVP) solutions of CdS nanoparticles [22]. Stabilization of metal nanoparticle nonlinearities was shown previously also using PVP stabilizers [33]. The nonlinear parameters of semiconductor nanoparticle solutions in the case of organic solvents can be considerably increased because of the formation of clusters with controlled sizes.

12.2
High-Order Harmonic Generation in Silver-Nanoparticle-Contained Plasma

12.2.1
Introduction

High-order harmonic generation (HHG) from molecules has shown a higher efficiency compared with atoms, especially for pump lasers with elliptical polarization [34]. One can explain this phenomenon as due to the effective cross section for recombination being larger in the molecule than in the atom. One can therefore expect to increase the harmonic efficiency for targets with larger dimensions. Since clusters are much smaller than the laser wavelength, they contain many equivalent, optically active electrons at effectively the same point in the laser field. This leads to the possibility that each of these electron oscillators may contribute coherently to a global cluster dipole. In addition, these cluster electrons see a binding potential, which is modified from the single-atom case by the proximity of neighboring ions. Also, bound and free electrons in the cluster can result in shielding of the laser field inside the cluster. Theoretical simulations reveal a 2-order enhancement of low-order harmonics for argon clusters compared with monomer argon [35], which has not yet been demonstrated experimentally.

Only a few studies using 10^3-10^4 atoms/cluster media (with the cluster sizes of 2–8 nm) have been reported, with some indications on the optimal Hagena parameter (6×10^3 [36]) related to the sizes of nanoclusters when one applies the gas jet technique. Note that previous experimental evidence of frequency conversion in gas clusters were restricted by the generation of *low-order* harmonics [37]. No studies have been reported using larger sized particles ($\sim$100 nm, 10^7 atoms/cluster). One idea is to use these commercially available nanoparticle in pump-probe HHG experiments, by using laser ablation of the targets containing nanoparticles as the nonlinear medium. Recently, such studies were reported and showed some advantages and disadvantages of the application of clusters for the generation of harmonics [38–40].

In this section, we report the systematic studies of HHG from the laser ablation that contains nanostructured material. We study the influence of particle sizes in laser ablation plasma on the HHG efficiency. For these purposes, we compare harmonic yields from the laser plumes containing 110 nm spherical Ag particles, 100–1000 nm blocks of colloidal silver, 2–14 μm aluminum particles, and 50 nm thick, few micrometers long CuO nanorods. We compare these results with those from laser ablation of solid targets of silver, aluminum, and copper. These studies showed that smaller sized structures are favorable for high-order harmonic conversion efficiency, while its effect on harmonic cutoff was insignificant.

12.2.2
Experimental Arrangements

We show a schematic diagram of the experimental setup in Figure 12.5. Experiments were performed using the 10 Hz, 10 TW beam line of the Canadian Advanced Laser Light Source (ALLS). To create the ablation, a prepulse that was split from the uncompressed Ti:sapphire laser (210 ps, 800 nm, 10 Hz) was focused on a target placed in the vacuum chamber by using a plano-convex lens (focal length $f = 150$ mm). We adjusted the focal spot diameter of the prepulse beam to be 600 μm on the target surface. The intensity of the subnanosecond prepulse, I_{pp}, on the target surface was varied between 7×10^9 and 4×10^{10} W cm^{-2}. We chose this range of prepulse intensity variations from the previous studies with different ablated targets. After some delay (10–80 ns), the femtosecond main pulse ($E = 8$–25 mJ, $t = 35$ fs, $\lambda = 800$ nm central wavelength, 40 nm full width at half maximum (FWHM)) was irradiated onto the plasma from the orthogonal direction by using a MgF$_2$ plano-convex lens ($f = 680$ mm). Our experiments were performed with femtosecond main pulse intensities of up to $I_{fp} = 3 \times 10^{15}$ W cm^{-2}, above which HHG efficiency decreased.

The harmonics were spectrally dispersed by a homemade spectrometer with a flat-field grating (1200 lines per millimeter, Hitachi). The extreme ultraviolet (XUV) spectrum was then detected by a microchannel plate and finally recorded using a charge-coupled device (CCD).

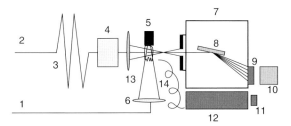

Figure 12.5 Experimental setup for the measurements of high-order harmonic generation from laser ablation of various targets. (1) Subnanosecond prepulse; (2) main pulse; (3) delay line; (4) compressor; (5) target; (6, 13) focusing lenses; (7) XUV spectrometer; (8) grating; (9) microchannel plate (MCP); (10, 11) CCD; (12) UV spectrometer; and (14) fiber waveguide.

In these studies, we first used the silver powder of spherical nanoparticles (Alfa Aesar). This powder was glued on a glass substrate, a paper tape, and a drop of superglue. We also used colloidal silver with larger sized particles to compare the harmonic yield from the particles of different sizes. The other sample was an aluminum powder with larger particle size (a few micrometers). We also used the targets containing CuO nanorods. These nanorods were synthesized by thermal annealing of thin copper foil [41]. We glued all these samples on different substrates in a way similar to that used for silver nanoparticles.

We compared the results of these studies with the HHG using laser ablation from the bulk targets of the above sample materials. For these purposes, the 4 mm long strips of bulk Ag, Al, and Cu were used as the targets.

12.2.3
Results and Discussion

To define which process governs the enhancement of high-order harmonic yield for small-sized nanoparticles, we first performed structural measurements of these samples. We imaged the size and shape of the particles under investigation, using the scanning electron microscope (SEM). Figure 12.6a shows the SEM microphotography of Ag nanoparticles. The mean size of nanoparticles was measured to

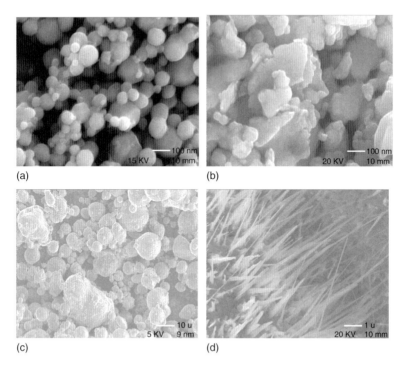

Figure 12.6 SEM images of (a) Ag nanoparticle, (b) Ag colloids, (c) Al powder, and (d) CuO nanorods.

be 110 nm. The particle sizes varied between 80 and 300 nm. The shape of the nanoparticles was almost spherical.

The second material we studied was the nanoblocks of colloidal silver. The sizes and shape of nanoblocks were different for colloidal silver. Colloidal silver prepared by chemical methods can be aggregated in various shapes and cause a strong nonlinear optical response of the medium containing such structures [1]. The shape of the colloidal nanoblocks we used was irregular (Figure 12.6b), and the size distribution was over a much broader range (from 100 to 1000 nm) compared to spherical Ag nanoparticles.

The third medium studied by us was Al powder. Our sample of Al powder had larger particle size and broad size distribution (Figure 12.6c), with a spherical shape. The size of the particles in the Al powder (which varied between 2 and 14 µm) was much larger than those of silver nanoparticles.

The fourth material was a nanorod-shaped CuO. CuO nanorods showed fragile features and their observed structural properties depended on the analysis method. The SEM image of thermally annealed thin Cu foil shows clear indication of an abundance of nanorods grown as thin films during the thermal heating (Figure 12.6d). The diameter of the rods was between 40 and 100 nm, while their length varied from 3 to 8 µm. Further, the process of fixing these thin structures on various substrates for laser ablation led to the destruction (probably mechanical) of these nanorods. Therefore, further SEM analysis of these films revealed a notable decrease in the concentration of the nanorods. Laser ablation of the nanorods also led to destruction of these structures because of melting.

To study the effects of the size and shape of nanostructured particles on the high-order nonlinearities, we performed experiments to generate high-order harmonics using the above-described structured targets. The results show an increase in the harmonic yield only for 110 nm spherical Ag nanoparticles, while no measurable enhancement was observed for the other structured targets.

We performed the first experiments using the 110 nm silver nanoparticles glued on different substrates. Ablation plume was generated using a low-intensity, 210 ps prepulse laser, which was followed by a femtosecond main pump laser to generate the harmonics. As we normally do in harmonic experiments using bulk targets, we irradiated a single position on the surface without moving the target. In experiments with bulk targets, one can generate harmonics with good shot-to-shot reproducibility with this method. However, the results using Ag nanoparticle targets showed a different tendency. The first shot on a fresh target surface produced strong harmonic generation, which was much stronger than those observed using bulk silver targets. This was followed by a sharp decrease in the harmonic yield after a few shots at the same target position. A similar tendency of the harmonic intensity to decrease after several consecutive irradiations was also observed for different structured targets, and was attributed to the fast evaporation of particles from the surface after a few shots.

We found that HHG from nanoparticles started to appear at smaller prepulse intensities compared with the case of bulk targets. We attribute this tendency to the lower ablation threshold and lower cohesive energy of the host substrate that

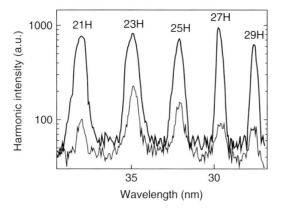

Figure 12.7 Harmonic distribution in the mid-plateau region in the case of the plasma produced on the surface of bulk Ag target (thin lineout) and the Ag-nanoparticle-contained plasma (thick lineout).

contains the nanoparticles, compared with those of bulk material. Under equal pump laser conditions, the harmonic yields from silver nanoparticles was at least six times higher compared with those generated from the plume of bulk targets under equal experimental conditions (Figure 12.7). For HHG with bulk silver targets, we can make a conservative estimate of the harmonic conversion efficiency to be 8×10^{-6}, which was measured using longer laser pulses (130 fs) [42]. We can therefore make a modest estimate of the harmonic conversion efficiency for silver nanoparticles to be about 4×10^{-5} in the plateau region. We should note that there is still space to further enhance this high conversion efficiency by choosing the appropriate sizes of nanoparticles, as well as by optimizing the plasma and experimental conditions.

The enhancement of harmonic generation efficiency from cluster media compared to that from monoparticle plasma was more pronounced at the small delay between the prepulse and main pulse. The optimal distance between the main beam and target was found to be in the range of 100–150 μm. The effect of the absorption of radiation by the medium was expected to be insignificant because of the small concentration of plasma particles estimated to be in the range of 10^{17} cm^{-3}.

The role of the molecules from the glue was analyzed in detail during these and the recently published [39, 40]. We carried out separate studies of the availability of the HHG during the propagation of the femtosecond pulse through the glue-molecule-contained plasmas, when no nanoparticles were inserted in this organic solvent. No harmonics were observed in these experiments. From these studies we concluded that the role of glue molecules was insignificant in our experiments with the mixtures of glue and nanoparticles.

To analyze the influence of sizes of the particles taking part in HHG on harmonic yield we studied this process in silver-colloid-containing jelly. This jelly was irradiated by prepulse, and the plasma created during this irradiation was

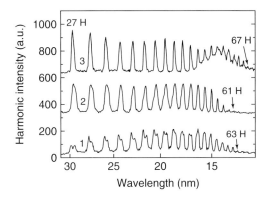

Figure 12.8 High-order harmonic spectra generated from plasma of (1) colloidal silver, (2) on the bulk silver target, and (3) glue containing 110 nm silver nanoparticles.

used as a nonlinear medium for harmonic generation. SEM analysis confirmed the presence of higher sized nanoparticles in this medium. In this case, the harmonic yield was decreased compared with the case of smaller silver nanoparticles, although we were able to observe approximately the same cutoffs in the case of these two media. For silver colloids we were able to generate harmonics as high as up to the 63rd order (Figure 12.8, curve 1).

At the same time, harmonic cutoff for 110 nm Ag nanoparticles was slightly extended compared with that obtained from the plasma created on the bulk Ag surface (Figure 12.8, curves 3 and 2). Harmonics as high as the 67th order were observed in these studies of silver nanoparticles, while, for bulk silver target, under the same conditions we observed the harmonics up to the 61st order. This small enhancement of harmonic cutoff corroborates with the observations reported in Ref. [37].

To investigate the influence of particle sizes on harmonic properties, we studied HHG from aluminum for bulk solid-state target and micrometer-sized powder. Figure 12.9 shows the results of these experiments. The results show that aluminum is an inefficient medium for HHG, both in terms of maximum cutoff energy and conversion efficiency. Plasmas from powder as well as bulk targets provide roughly the same cutoffs (the 33rd and 37th orders for powder and bulk aluminum, respectively, Figure 12.9a). Further, we did not observe any enhancement of conversion efficiency for powder targets compared to bulk. We observed the same result when a 400 nm pump laser (second harmonic of the Ti:sapphire laser) was used as the main pulse. In this case, the cutoff was at the 19th and 15th order for powder and bulk targets, respectively (Figure 12.9b). Interestingly, the cutoff energies from Al powder and bulk Al plumes for the two different pump laser wavelengths were similar ($\sim$56 eV).

Finally, HHG experiments using plasma produced on the surface of CuO nanorods did not show any enhancements compared to those of bulk copper targets. The HHG conversion efficiencies in these cases were approximately equal

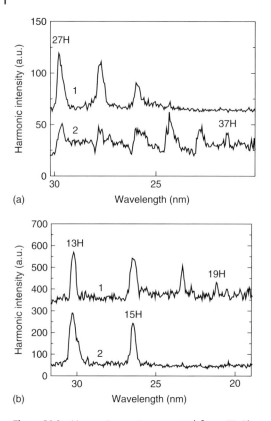

Figure 12.9 Harmonic spectra generated from (1) Al-powder-containing plume and (2) plume from the Al bulk target in the cases of (a) 800 nm and (b) 400 nm driving pulses.

to each other, and the plateau-like shape of harmonics had harmonic cutoffs at the 47th and 43rd order, respectively (Figure 12.10).

These studies showed that for large-sized particles no improvements in both the conversion efficiency and extension of cutoff occur, while for small-sized particles one can achieve the HHG with higher conversion efficiency.

To obtain maximum HHG conversion efficiency, it is essential to define the maximum tolerable particle size at which enhancement can be achieved, since increasing the dimension of particles increases its polarizability. A large polarizability of a medium is critical to efficient harmonic generation [43]. At the same time, the increase in cluster size leads to some mechanisms that can restrict harmonic generation (such as the use of only surface atoms for HHG, reabsorption of harmonics, recombination of electrons from inner parts of clusters, etc.).

The structure of ablated Ag plume produced from bulk silver target was investigated [44] by studying the Ag films deposited on silicon substrates. The aim was to study the presence of nanoclusters in these plumes, which could be responsible for the enhancement of harmonic conversion efficiency in our work, due to the quantum confinement effect. These structural studies showed that the "optimal

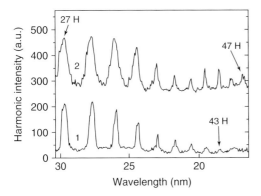

Figure 12.10 Harmonic spectra generated from (1) plume produced on bulk Cu target and (2) plume produced on CuO-nanorod-containing surface.

plumes" prepared at weak focusing conditions consisted of monoatomic particles, without the presence of clusters. The preparation of nanoparticles in the laser plumes produced on the surfaces of bulk targets requires a strong interaction of the laser prepulse with the target surface. However, such an interaction creates plasma, which is far from the optimal condition for efficient HHG. Note that in the reported studies we used the optimal plumes.

A comparison of low-order harmonic generation using single atoms and multiparticle aggregates has been reported for Ar atoms and clusters [37]. It was demonstrated that a medium of intermediate-sized clusters with a few thousand atoms of inert gas has a higher efficiency for generating the higher order harmonics, compared with a medium of isolated gas atoms of the same density. The reported enhancement factor for the third to ninth harmonics was about five. It was also shown that the dependence of harmonic efficiency on the pump laser intensity is much more prominent for clusters than for isolated atoms. We also observed these tendencies in the HHG from nanoparticles.

These studies have shown that, for any nanoparticle-contained plumes, no considerable improvements in the extension of the harmonic cutoff were observed. At the same time, an enhancement of the harmonic yield in the low-energy plateau range in the case of small-sized nanoparticle-contained plumes was achieved. The value of the enhancement factor could be attributed to the rather different concentration of clusters and monoparticles in the plume at different excitation conditions. Further studies are required to define the nature of these peculiarities.

The above observations give us a rough picture of the ablation process for nanoparticle targets. The material directly surrounding the nanoparticles is polymer (epoxy glue), which has a lower ablation threshold than metallic materials. Therefore, the polymer starts to ablate at relatively low intensities, carrying the nanoparticle with it, resulting in the lower prepulse intensity. Polymer also has a lower melting temperature than metals. Therefore, repetitive irradiation of the target leads to melting and change in the properties of the target. This results in the change in conditions of the plasma plume, resulting in a rapid reduction in the harmonic intensity with increased shots. The different shot-to-shot harmonic intensities for different substrates can be explained by the different adhesion properties of nanoparticles to the substrate.

Owing to such a rapid change in the conditions for harmonic generation with nanoparticle targets, it was difficult to precisely define the dependence of the harmonic yield on the prepulse and main pulse intensities. Nevertheless, rough measurements of the dependence of harmonic yield as the function of main pulse intensity for Ag nanoparticles have shown a saturation of this process at relatively moderate intensities ($I_{fp} \approx 8 \times 10^{14}$ W cm^{-2}).

Harmonics from plasma nanoparticles also exhibited several characteristics similar to gas harmonics. First, the harmonic intensity decreased exponentially for the lower orders, followed by a plateau, and finally a cutoff. Next, the harmonic intensity was strongly influenced by the focus position of the main pump laser, along the direction parallel to the harmonic emission. The strongest harmonic yield was obtained when the main pump laser was focused 4–5 mm behind the nonlinear medium. We observed the same tendency for harmonics using a bulk silver target. The typical intensity of the pump laser for maximum harmonic yield was between 5×10^{14} and 2×10^{15} W cm^{-2}. In our case, we needed to focus the pump laser considerably away from the medium, since the total intensity that would be produced at focus would considerably exceed the barrier suppression intensity of ions. This would result in overionization of the plasma, leading to the reduction in the harmonic yield that we have observed in our experiments.

For the intense laser pulses (of the order of 10^{16} W cm^{-2}) focused in the center of the plume, the harmonic radiation was considerably decreased. However, the harmonic intensity showed a maximum when the main laser beam was focused either before or after the plume, depending on the experimental conditions. For the 24 ns delay, the maximum harmonic output was obtained in the case of focusing behind the plasma area. In the case of nanoparticle-contained plumes, we made only the qualitative measurements, since the depletion of particles during multiple shots did not allow us to obtain the exact dependence of the absolute value of harmonic intensity on the position of focal plane in front of and behind the plasma plume.

This tendency remained unchanged irrespective of the harmonic order. However, the opposite feature was observed for the 80 ns delay, when the maximum harmonic generation efficiency was obtained for focusing main pulse before the plasma. Such a behavior is similar to those reported by several authors in laser-gas jet HHG experiments, which was explained by the depletion, phase-mismatching, and self-defocusing effects at high laser intensity. The presence of the abundance of free electrons leads to the dephasing of the HHG. This is true for the plasmas created on the surfaces of bulk targets. As for nanoparticle-contained plumes, their consistence is so fragile that the conditions of phase matching can change even after a few tens of shots on the cluster-contained surface because of evaporation and the change in the concentration of particles.

Another important issue is a relation between the coherence length and the length of the medium. It is preferable to match the coherence length ($L_c = \pi / \Delta k$)

and the plasma sizes for achieving the phase matching between the harmonic and fundamental waves. For the plasma densities used (a few units of 10^{17} cm^{-3}), the estimations result in $L_c \sim 0.7$ mm for harmonics in the range of the 21st order, with a further decrease in L_c for higher harmonic orders. We optimized the plasma size experimentally by defocusing the prepulse beam to expand the heated region of plasma up to 0.6 mm. Our studies with longer plumes (when we used a considerably defocused prepulse beam or a cylindrical focusing for the prepulse) failed to achieve better conversion efficiency. Furthermore, the harmonic yield in some cases considerably decreased, pointing out the less favorable conditions for the HHG using prolonged plasma. This can be attributed to both the mismatch with the coherence length and the increase in reabsorption. So, the phase-matching conditions should surely play a role in our experiments, which was proved by the observations of the dependence of the HHG yield on the position of focal plane of fundamental radiation relative to the nanoparticle-contained plasma.

One of the possible reasons for the relatively small harmonic enhancement in the case of nanoparticles, compared with the bulk material, could be the change in nanoparticle structure during the laser ablation. The laser intensity used to create the plume is a very sensitive factor that has to be carefully investigated before one can make statements about the integrity of nanoparticles in the plasma area after the ablation. One could expect the melting and aggregation of nanoparticles during the interaction of the laser prepulse with the target surface. The presence of nanoparticles in the plumes in our studies was confirmed by analyzing the spatial characteristics of the ablated material deposited on the nearby glass substrate. However, further studies of plasma components during the ablation of the nanoparticle-contained targets will benefit the statement about the presence and integrity of nanoclusters in the plumes at the moment when the femtosecond pulse arrives in the area of interaction.

The harmonic efficiency is determined by at least three factors: (i) the strength of the dipole response from a single particle, (ii) the number of emitting particles, and (iii) the phase-matching conditions. The first factor seems to not influence the efficiency of highest orders, since it plays the role for the low-order harmonics described by the perturbative theory. The second and third factors are important, as well as some other factors (laser intensity, closeness to each other of the harmonic wavelength and the resonances with strong oscillator strengths, role of dephasing at the focusing outside the plasma area, self-phase modulation, optical Kerr effect leading to the self-defocusing due to the abundance of free electrons, integrity and aggregation of nanoclusters, etc.). We did not carry out a thorough analysis of these factors because of the obvious difficulties in the definition of the exact quantitative parameters of the nanoparticle-contained plumes since the variation of nanoparticle density from shot to shot prevents exact measurements of this parameter.

The number of emitting particles remains an important issue of these studies and can be considered from the point of view of relative concentration of the 100 nm Ag particles and the Ag blocks from colloidal silver, as well as the bulk silver targets. To exactly define the concentration in the first two cases is a really difficult task. To resolve this problem, we analyzed the deposition of debris from the above three samples on the silicon wafer and Al foil placed close to the ablation area. We studied the structural properties as well as the thickness of the deposited debris by irradiating the targets under the same conditions (duration of ablation, laser intensity, pulse repetition rate, etc.). These studies have shown the approximately equal thickness of the deposited material, which can be interpreted as the roughly equal concentration of the particles in the area of interaction with a strong femtosecond pulse during the HHG experiments with the 100 nm particles and blocks. Our previous analysis and calculations of Ag particle concentration in the case of the ablation of the bulk silver targets by using the HYADES code gave us the concentration of Ag ions in the range of 2×10^{17} cm^{-3} [45]. Unfortunately, this simulation technique could not be used in the case of nanoparticle-contained ablated targets.

We can assume that in the case of Ag nanoparticles, the ejected electron, after returning to the parent particle, can recombine with any of the Ag atoms in the cluster, which enhances the probability of the emission of harmonics compared with the single ions. This can be a reason for the observed enhancement of the HHG yield from the nanoparticle-contained plume compared with the single-particle-contained plume. As for the Ag blocks, we assumed that the enhancement of the clusters above the appropriate sizes leads to the decrease in the high-order nonlinear optical response of the medium because of the losses caused by the recombination of electrons before the ejection from the blocks, reabsorption of the harmonics inside the high-sized structures, and so on. In our paper, we present the experimental evidence of the enhancement of the harmonics in the case of small-sized nanoparticles compared with the single particles and blocks. Further studies are helpful for the definition of the main factors influencing the HHG yield from the plumes containing particles of different sizes.

12.3
Studies of Low- and High-Order Nonlinear Optical Properties of BaTiO$_3$ and SrTiO$_3$ Nanoparticles

12.3.1
Introduction

Clusters produce strong low-order nonlinear optical response (e.g., nonlinear refraction and nonlinear absorption); they can also emit coherent XUV radiation through the HHG. These studies have shown that one can expect an improvement in HHG efficiency by switching to cluster media. However, previous studies of the HHG from such objects were limited to nanoclusters (Ar, Xe), which are formed in high-pressure gas jets because of rapid cooling by adiabatic expansion.

Meanwhile, there is a new emerging class of nanoparticles – nanocrystallites – that has attracted much interest. Among them, the nanoclusters of photorefractive crystals possess the advanced properties both from the point of view of low- and high-order nonlinearities. The photorefractive effect is a phenomenon in which the local index of refraction of a medium is changed by illumination, using a laser beam with spatial variation of the intensity. The crystals possessing photorefractive properties have attracted interest because of their various potential applications, such as optical limiting, spatial light modulation, online holographic recording, and so on. However, nonlinear optical properties of such nanoparticles are less explored.

Barium titanate is one of the most prominent ferroelectric photorefractive materials. The $BaTiO_3$ and $SrTiO_3$ crystals have often been used as the matrices for the nanoparticles [46, 47]. These materials possess strong dielectric constants, a large nonlinear optical effect, and ferroelectricity. At the same time, these materials could be prepared as the nanoparticles for various applications. In the past, the third-order nonlinear optical properties of bulk $BaTiO_3$ and $SrTiO_3$ crystals were analyzed [48–50]. In particular, the two-photon absorption coefficient of strontium titanate at the wavelength $\lambda = 0.69\ \mu m$ was found to be $3 \times 10^{-11}\ m\ W^{-1}$ [48], while the nonlinear refractive index at $1.064\ \mu m$ was measured to be $48.5 \times 10^{-20}\ m^2\ W^{-1}$ [49]. In the case of barium titanate, the two-photon absorption coefficient at $\lambda = 0.596\ \mu m$ was reported to be $0.1 \times 10^{-11}\ m\ W^{-1}$ [50]. However, no studies of the low-order nonlinear optical characteristics of $BaTiO_3$ and $SrTiO_3$ nanoparticles were reported so far. The same can be said about the investigation of high-order nonlinearities of these nanostructures, particularly with regard to their application for the HHG.

In this section, we present the results of systematic studies of the nonlinear optical properties of $BaTiO_3$ and $SrTiO_3$ nanoparticles, as well as their structural characteristics. We analyze the nonlinear refraction and nonlinear absorption of these nanoparticles using the 790 nm radiation of different pulse durations. Further, we discuss the results of studies of the HHG from the laser-ablated plumes containing barium titanate and strontium titanate nanoparticles. The comparison with the low- and high-order nonlinearities of $BaTiO_3$ bulk photorefractive crystal is discussed.

12.3.2
Experimental Arrangements

The $BaTiO_3$ and $SrTiO_3$ nanoparticles were purchased from the Wako Pure Chemical Industries, Ltd. (Japan) and used in experiments without further purification. In the case of the studies of third-order nonlinearities, these nanoparticles were dissolved in ethylene glycol at different concentrations varying from 5×10^{-4} to 5×10^{-3} weight parts (i.e., at the molar concentrations between 2 and 20 mM). The low-order nonlinear optical characteristics of $BaTiO_3$ and $SrTiO_3$ nanoparticles suspensions were studied using the conventional Z-scan technique [51]. The 1 mm thick silica glass cells filled with these suspensions were used in these studies. The 790 nm laser radiation of different pulse durations (120 fs or 210 ps) was focused by a 300 mm focal length lens (Figure 12.11). The samples were moved along the z-axis through the focal area to observe the variations of the

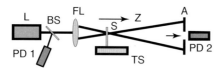

Figure 12.11 Experimental setup for the measurements of the low-order nonlinear optical characteristics of nanoparticle suspensions. L, Ti:sapphire laser; BS, beam splitter; PD1 and PD2, photodiodes; FL, focusing lens; S, sample; TS, translating stage; A, aperture.

phase and the amplitude of propagated radiation in the far field. The energy of laser pulses was measured by a calibrated photodiode. A 1 mm aperture was placed at a distance of 500 mm from the focal plane (closed-aperture scheme). The radiation propagating through this aperture was registered by a second photodiode. The closed-aperture scheme allowed for determining the sign and value of the nonlinear refractive indices of samples, as well as the value of their nonlinear absorption coefficients. The latter parameter could also be verified using the open-aperture scheme. The 2 mm thick $BaTiO_3$ crystal was used as a sample under investigation to compare the nonlinear optical response in the case of nanoparticles and bulk material. Our nonlinear optical measurements were calibrated using the CS_2. The application of 210 ps and 120 fs pulses results in drastically different nonlinear indices for the carbon disulfide. Previously, we measured the nonlinear refractive index of this standard medium for different pulse durations [52]. In the studies at present, we took into account the difference in the nonlinear optical response of CS_2 at the different timescales, when compared with the nonlinearities of nanoparticles.

To study the high-order nonlinearities through the HHG, the nanoparticles were glued on various substrates (drop of glue, tape, and glass) and placed inside the vacuum chamber to ignite the plume by laser ablation of these structures. The experimental setup of the HHG from laser plasma was described in [53] and in the previous section (Figure 12.5). Here, we present a few details of these experiments. The pump source was a commercial, chirped pulse amplification Ti:sapphire laser (Spectra Physics: TAS-10F), whose output was further amplified using a homemade three-pass amplifier operating at a 10 Hz pulse repetition rate. A prepulse of 210 ps pulse width was split from the amplified laser beam by a beam splitter before a main pulse compressor. A spherical or cylindrical lens focused the prepulse on a target comprising the $BaTiO_3$ or $SrTiO_3$ nanoparticles glued on different matrices (Figure 12.12). We also used the $BaTiO_3$ bulk crystal as a target to compare the HHG in the case of the plasma-plume-containing clusters and monoatomic particles of the same origin. The width of the laser line focused at the target surface in the case of the focusing by the cylindrical lens was adjusted between 50 and 200 μm at 2 mm plasma length, and the intensity of the 210 ps prepulse at the target surface was varied in the range of $I_{pp} = (0.7–7) \times 10^{10}$ W cm^{-2}. In the case of spherical focusing, the area of ablation was adjusted in the range of 0.6 mm and the 210 ps prepulse intensity at the target surface was varied in the range of $I_{pp} = (0.5–5) \times 10^{10}$ W cm^{-2} by changing the energy of the prepulse.

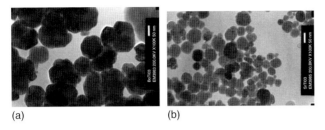

(a)   (b)

Figure 12.12 TEM micrographs of the (a) BaTiO$_3$ and (b) SrTiO$_3$ nanoparticles.

The parameters of the main pulse after propagating the compressor stage were as follows: center wavelength of 790 nm, pulse energy of 12 mJ, and pulse duration of 120 fs. After the proper delay compared with the prepulse irradiation this main pulse was focused by a spherical lens ($f/10$) on the ablation plume from the orthogonal direction. The maximum intensity of the femtosecond laser beam at the focal spot was 6×10^{16} W cm^{-2}. Since this intensity considerably exceeded the barrier suppression intensity of doubly charged particles, the position of laser focus was adjusted to be either before or after the laser plume to exclude the influence of the considerable ionization of laser plume and to maximize the harmonic yield. The intensity of the main pulse at the plasma area was adjusted between 5×10^{14} and 3×10^{15} W cm^{-2}. Our experiments were carried out under weak focusing conditions ($b > L_p$, b is the confocal parameter of the focused radiation and L_p is the plasma length; $b = 3$ mm, $L_p = 0.6$ mm (in the case of spherical focusing), and $L_p = 2$ mm (in the case of cylindrical focusing)). The optimal distance between the HHG interaction area and target surface was varied in the range of 100–200 μm. The average density of the plume in the area of interaction was calculated to be $(1–3) \times 10^{17}$ cm^{-3}.

The spectrum of generated high-order harmonics was analyzed by a grazing incidence XUV spectrometer with a gold-coated flat-field grating (Hitachi, 1200 grooves per millimeter). The XUV spectrum was detected using a microchannel plate with a phosphor screen read-out (Hamamatsu, model F2813-22P), and the optical output from the phosphor screen was recorded using a CCD camera (Hamamatsu, model C4880). The details of the absolute calibration of the harmonics measured by this XUV spectrometer were described elsewhere [53].

12.4
Results and Discussion

12.4.1
Structural Characterization of the Samples

The size and composition of nanoparticles were analyzed by the TEM and EDX, respectively. The size of the nanoparticle is of crucial importance when one considers the nonlinear optical processes, since the influence of quantum confinement

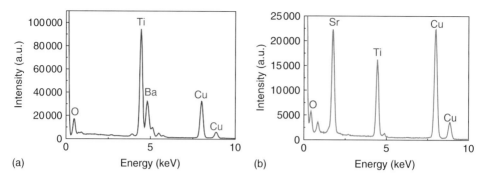

(a)   (b)

Figure 12.13 EDX spectra obtained from the (a) $BaTiO_3$ and (b) $SrTiO_3$ nanoparticles.

on the nonlinear optical properties of a medium strongly depends on the spatial extension of such a structure. Figure 12.12 shows the TEM images of $BaTiO_3$ and $SrTiO_3$ nanoparticles. The mean size of $BaTiO_3$ nanoparticles was defined to be 92 nm, with the particle size distribution varying between 50 and 160 nm. The TEM images of $SrTiO_3$ nanoparticles showed smaller sizes, with the mean size of 38 nm, and narrower size distribution compared to the $BaTiO_3$ nanoparticles. The shape of the nanoparticles was close to spherical. An elemental analysis of nanoparticles under investigation using EDX spectroscopy confirmed the presence of $BaTiO_3$ and $SrTiO_3$ particles in these suspensions (Figure 12.13; the copper lines originated from the grid material). These measurements suggest that no quantum-confinement-induced enhancement of the nonlinear optical response can be expected for such large (38 and 92 nm) particles.

Figure 12.14a illustrates a high-resolution transmission electron microscopic (HRTEM) image of the edge of the 105 nm $BaTiO_3$ cluster and the corresponding electron diffraction pattern, which clearly reveal a single crystal-like structure of the nanoparticle. Owing to the random position of nanoparticles in the grid, different electron diffraction patterns could be obtained from the same material. In particular, when we analyzed the $SrTiO_3$ nanoparticles, various diffraction patterns

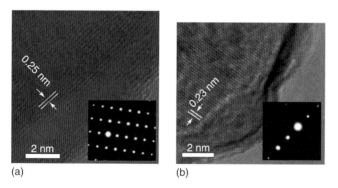

(a)   (b)

Figure 12.14 HRTEM images of the (a) $BaTiO_3$ and (b) $SrTiO_3$ nanoparticles. Insets: electron diffraction patterns of corresponding nanoparticles.

appeared in the electron diffraction images (see one of them in the inset in Figure 12.14b), while the HRTEM still showed the spacing at the edge of the 33 nm particle. The regular spacing of the lattice planes was about 0.25 and 0.23 nm in the case of $BaTiO_3$ and $SrTiO_3$ nanoparticles, which is consistent with the (111) lattice spacing of cubic bulk samples.

The aim of the structural studies at present was to define which kind of nanoparticle structure was presented in commercially available powders that could be used in the nonlinear optical studies. These studies revealed that the higher sized $BaTiO_3$ nanoparticles possess higher ordering of the structure compared to that of the $SrTiO_3$ nanoparticles. One can expect the influence of crystalline properties on the nonlinear optical response from the $BaTiO_3$- and $SrTiO_3$-nanoparticle-contained media. On the other hand, the role of nanoparticle structures in the definition of their nonlinear optical properties is not so obvious, since the variation of these properties mostly depends on the cluster sizes rather than the ordering of atoms in clusters, due to random distribution of nanocrystallites in the suspensions and plasma plumes. It is hard to expect that the photorefractive effect and polarization-dependent properties of nonlinear susceptibilities will influence the values of third-order nonlinear refraction and absorption, as well as the high-order nonlinearities responsible for the HHG.

12.4.2
Nonlinear Refraction and Nonlinear Absorption of $BaTiO_3$- and $SrTiO_3$-Nanoparticle-Contained Suspensions

The pulse duration of laser radiation has proved to be an important parameter for the determination of the nonlinear optical properties of different materials [52]. This peculiarity seems to be of special importance in the case of the photorefractive media where both the electronic Kerr effect and two- and three-photon absorption can be concurrent with the photorefractive effect, dynamic grating, and free carrier generation. The diffusion time, recombination time, diffusion length of photoexcited charge carriers, and pulse duration are the main parameters that can influence the nonlinear optical characteristics of photorefractive bulk materials. Although the influence of the above processes remains unclear in the case of the nanocrystallites of photorefractive materials, the temporal characteristics of laser pulse should play a role in the definition of the nonlinear absorption and refraction. We measured the parameters of these two nonlinear optical processes in the case of $BaTiO_3$- and $SrTiO_3$-nanoparticle-contained suspensions using the 120 fs and 210 ps pulses.

The fresh suspensions possessed the higher values of nonlinear optical parameters; however, these measurements have shown poor reproducibility. The sedimentation of higher sized nanoparticles caused variations of the nonlinear optical properties of suspensions. We analyzed the nonlinear optical characteristics of $BaTiO_3$- and $SrTiO_3$-nanoparticle-contained suspensions after the stabilization of their structural characteristics. Our studies revealed that a liquid with higher viscosity (ethylene glycol) is preferable from the point of view of the structural

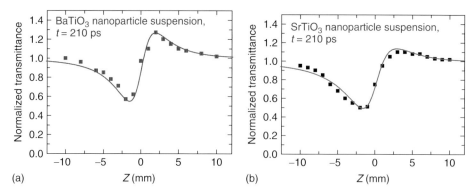

(a) (b)

Figure 12.15 Normalized transmittance dependencies in the case of (a) BaTiO$_3$- and (b) SrTiO$_3$-nanoparticle-contained suspensions measured using the 210 ps pulses. Solid curves are the theoretical fits.

stabilization of nanoparticle suspensions. The Z-scan technique suffers from being highly sensitive to sample scatter. The background scans were done to correct for this possibility. They did not show the influence of scattering on the definition of the nonlinearities of the particles under investigation, when one used the moderate concentrations of the nanoparticles in the suspensions (less than 10 mM).

Figure 12.15a shows the closed-aperture Z-scan of BaTiO$_3$-nanoparticle-contained suspension measured using the 210 ps pulses. It demonstrates the positive non-linear refraction, with some indication of the appearance of nonlinear absorption. The same pattern was observed in the case of the SrTiO$_3$-nanoparticle-contained suspension (Figure 12.15b). However, the asymmetry of the normalized transmittance curve in the latter case was more pronounced compared with the BaTiO$_3$-nanoparticle-contained suspension, pointing out the significant influence of the nonlinear absorption.

Equation (12.1) was used for fitting the theoretical dependencies and experimental data. After fitting the ρ and $\Delta\Phi_0$ we consequently found the γ and β. The γ and β of our samples were measured to be 3.4×10^{-19} m^2 W^{-1} and 7×10^{-15} m W^{-1}, respectively, in the case of the BaTiO$_3$ nanoparticles suspended in ethylene glycol (at 5 mM concentration of nanoparticles and $L = 1$ mm), and 9×10^{-19} m^2 W^{-1} and 2×10^{-14} m W^{-1}, respectively, for SrTiO$_3$-nanoparticle-contained suspension (at 5 mM concentration of nanoparticles and $L = 1$ mm), when the 210 ps pulses were used as a probe radiation. We verified that the nonlinear optical response of ethylene glycol ($\gamma < 5 \times 10^{-20}$ m^2 W^{-1}) was insignificant compared with the nonlinearities caused by the nanoparticles.

No difference in the nonlinear optical properties of suspensions was found in the case of the 10 Hz pulse repetition rate and single pulses, thus indicating the insignificant role of accumulative processes. The application of short laser pulses (in picosecond and femtosecond ranges) at low pulse repetition rate allows excluding the influence of slow thermal-related nonlinear optical processes and analyzing the self-interaction processes caused by the response from the nanoparticles.

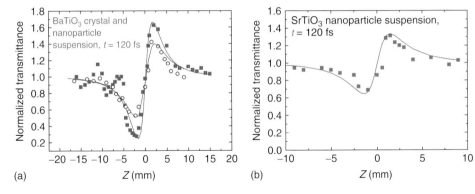

Figure 12.16 Normalized transmittance dependencies of the (a) BaTiO₃-nanoparticle-contained suspension (open circles) and BaTiO₃ crystal (filled squares) and

(b) SrTiO₃-nanoparticle-contained suspension measured using the 120 fs pulses. Solid curves are the theoretical fits.

Analogous measurements of the nonlinear optical characteristics of nanoparticle suspensions were performed using the shorter pulses (Figure 12.16). The calculations of the γ of these suspensions in the case of 120 fs pulses gave approximately the same results as in the case of longer pulses, while the nonlinear absorption became almost insignificant. These studies revealed that in the case of nanoparticle medium, the difference between the nonlinear refractive indices at different time scales (210 ps and 120 fs) is insignificant, thus pointing out the involvement of only the fast processes in this nonlinear optical process, contrary to the case of bulk materials (see [52] and references therein). The reason for the difference in the nonlinear absorption for the two timescales is not clear and requires additional studies.

To compare the nonlinear optical response from the nanoparticles and that from bulk sample of the same material, we measured the nonlinear refractive characteristics of a 2 mm thick BaTiO₃ crystal using the 120 ps pulses (Figure 12.16a, filled squares). This crystal showed the positive nonlinear refraction, as in the case of nanoparticle suspensions. These measurements revealed some asymmetry in the shape of the Z-scan, in addition to the observation of the nonlinear absorption, which causes the difference between the values of valley and peak in the $T(z)$ dependence. The asymmetry in the case of the left part of Z-scan probably points out the influence of crystal properties of the material.

The photorefractive effect takes approximately 1 μs to manifest itself and takes 10 ms to relax. Our femtosecond and picosecond pulses were too short for the influence of the photorefractive process to be observed, and the interval between pulses (100 ms) is far greater than the photorefractive relaxation time. Therefore, the contribution of this effect to the observed nonlinear optical refraction and absorption can be neglected. One can thus assume that in the case of photorefractive media, purely local phenomena are predominant in the femtosecond and picosecond timescales of the main pulse. Another mechanism that could contribute to the decrease in on-axis propagation through the BaTiO₃ crystal can be attributed to the time-resolved buildup of grating. However, a buildup time of grating in photorefractive crystals was previously measured to be 4 ns [54], which also shows that this process is too slow to influence the nonlinear optical refraction and absorption in the case of shorter pulses. For comparison of the nonlinearities of the BaTiO$_3$ crystal with those of BaTiO$_3$ nanoparticles, it is preferable to carry out experiments using shortest pulses in order to exclude the influence of slower processes, which could be manifested at subnanosecond (210 ps) timescale.

The calculated values of the nonlinear optical characteristics of the BaTiO$_3$ crystal at $\lambda = 790$ nm after fitting the theoretical curve and experimental data were as follows: $\gamma = 6 \times 10^{-20}$ m^2 W^{-1} and $\beta = 1.1 \times 10^{-13}$ m W^{-1}. We did not find in the literature the data on the nonlinear refractive index of the BaTiO$_3$ crystal, while the nonlinear absorption coefficient was almost 1 order of magnitude less than the reported value of this parameter measured at $\lambda = 596$ nm [50]. The latter difference can be attributed to both the measurements at different wavelengths and pulse durations (120 fs and 1 ps). BaTiO$_3$ has no electronic resonances between 596 and 790 nm that would cause such a large difference. Perhaps the closeness of the former wavelength to the transmission edge of the barium titanate (380 nm) caused the increase in the two-photon absorption coefficient with regard to the longer wavelength pulse.

From these studies one can compare the γ value of nanoparticles and the bulk structure of the same material. The difference between them is considerable and could not be attributed to the inaccuracy in the measurements of the molar concentration of the nanoparticles in suspension due to their sedimentation. Since it was mentioned that for the 92 nm particles the role of quantum size effect could be considered insignificant, one can expect the same values of the nonlinear optical properties of these two samples. However, our studies show that the nonlinear refractive indices of the BaTiO$_3$ crystal and those of the nanoparticles differ by 4 orders of magnitude (taking into account the 10^{-3} volume part of nanoparticles in the suspensions), which is a manifestation of the influence of size effect in the case of the nanoparticle-contained medium. The reason for such a difference between the nonlinearities of relatively high-sized nanoparticles and bulk material is not clear, and further studies have to be carried out using nanoparticles of different sizes. One explanation might be related to the presence of some amount of very small nanoparticles, which can cause a strong enhancement of the nonlinear optical response. However, their concentration seems insufficient, since the sizes of most particles were centered between 70 and 130 nm.

The sizes of SrTiO$_3$ nanoparticles were closer to the range where the quantum confinement starts to be an important factor. Although we did not compare the low-order nonlinear optical properties of these nanoparticles with those of the bulk crystal, one could expect the same difference in the nonlinear optical characteristics between the 32 nm SrTiO$_3$ nanoclusters and the bulk crystal, since the parameters of γ and β for the two nanoparticle samples (SrTiO$_3$ and BaTiO$_3$) were close to each other, while the reported values of γ and β for SrTiO$_3$ crystals measured under different experimental conditions [48, 49] were 2–3 orders of magnitude less than those observed in the experiments at present, which is a characteristic feature of the size-related enhancement of the nonlinearities of nanoparticles. One might expect that the nonlinearities of bulk materials were smaller compared to those of nanoparticles because of the disappearance of some crystal-related processes, which masked, or diminished, the nonlinear optical response of the bulk structure. At this condition when the role of size effect remains unclear, the pure electronic response from the nanoparticles could exceed the combined effect of different nonlinear optical processes in a well-ordered structure.

12.5
High-Order Harmonic Generation from the BaTiO$_3$- and SrTiO$_3$-Nanoparticle-Contained Laser Plumes

We performed the first experiments using the nanoparticles glued on different substrates. Ablation plume was generated using a low intensity 210 ps prepulse, which was followed by the propagation of the femtosecond main pulse through the plasma to generate the harmonics. As we normally do in harmonic experiments using bulk targets, we irradiated a single position on the surface of the nanoparticle-contained substrate without moving the target. In experiments with bulk targets, one can generate harmonics with good shot-to-shot reproducibility with this method. However, our previous results using Ag nanoparticle targets showed an opposite tendency. The first shot on a fresh target surface produced strong harmonic generation, which was much stronger than the one observed using bulk silver targets under the same conditions. This was followed by a steep decrease in the harmonic yield after a few shots at the same target position. A tendency for harmonic intensity to decrease after several consecutive irradiations was attributed to the fast evaporation of particles from the surface after a few shots.

The presence of rather large nanoparticles causes the diminishing of the role of the size-related effect on the nonlinear optical response of the plume containing the BaTiO$_3$ and SrTiO$_3$ multiatomic particles. For BaTiO$_3$-nanoparticle-contained laser plumes, we were able to generate the harmonics up to the 39th order (Figure 12.17a). The harmonic cutoff for the plume created on the surface of the BaTiO$_3$ bulk crystal (41st harmonic, Figure 12.17b, curve 1) was approximately the same compared with the one obtained from the plasma created on BaTiO$_3$-nanoparticle-contained substrates. Note the increase in plasma emission with further increase in prepulse intensity on the crystal surface (Figure 12.17b, curve 2). The harmonic conversion efficiencies at the plateau range were estimated to be 1×10^{-6} for both plasmas assuming the identity of the experimental conditions of the studies at present studies and the investigations of the HHG from silver plume, when the absolute calibration of harmonic yield was established [53]. This equality of harmonic cutoffs and efficiencies points out the absence of the influence of quantum-confinement-induced processes during the HHG from 92 nm nanoparticle-contained plumes and corroborates the observations reported in Ref. [37]. In the case of barium titanate, the influence of quantum confinement can be expected for the particles with the sizes less than 16 nm [46].

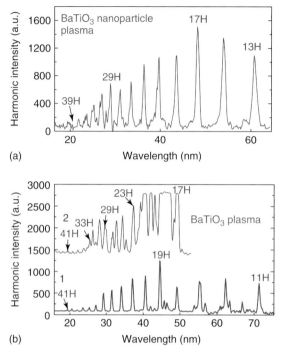

(a)

(b)

Figure 12.17 High-order harmonic spectra obtained from the plasmas produced on the surfaces of (a) BaTiO$_3$-nanoparticle-contained substrates ($I_{pp} = 8 \times 10^9$ W cm^{-2}) and (b) BaTiO$_3$ bulk crystal. Curves 1 and 2 on the Figure 12.8b correspond to the prepulse intensities of 3×10^{10} and 5×10^{10} W cm^{-2}, respectively.

These studies showed that for large-sized particles no improvements in both conversion efficiency and extension of the harmonic cutoff occur. To obtain maximum HHG conversion efficiency, it is essential to define the maximum tolerable particle size at which the enhancement can be achieved, since the increase in the particle sizes leads to the increase in their polarizability. A large polarizability of medium is critical to efficient harmonic generation. At the same time, the increase in cluster size leads to the appearance of some mechanisms that can restrict harmonic generation as stated in previous sections.

We observed the increase in conversion efficiency in the case of 38 nm SrTiO$_3$-nanoparticle-contained laser-produced plasma plumes (Figure 12.18), while the harmonic cutoff (39th harmonic) was not compared with that from the bulk strontium-titanate-laser-produced plasma plume. The higher conversion efficiency was observed in the case of SrTiO$_3$-nanoparticle-contained laser plumes compared with the case of BaTiO$_3$-nanoparticle-contained plasma. The ratio of harmonic yields was in the range of four and depended on the prepulse intensity at the nanoparticle-contained surface. Note that equal concentrations of SrTiO$_3$ and BaTiO$_3$ nanoparticles were glued on the substrates to prepare the targets. We initially verified that the harmonics generated from these substrates and glues itself

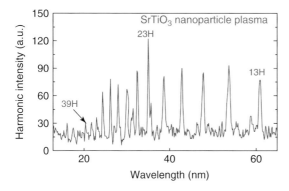

Figure 12.18 High-order harmonic spectra obtained from the plasma produced on the surface of SrTiO$_3$-nanoparticle-contained substrate. The measurements were carried out at the prepulse intensity of 8×10^9 W cm^{-2}.

(drop of glue, tape, and glass), without nanoparticles, were negligible compared with those from BaTiO$_3$ and SrTiO$_3$ nanoparticle-contained plasmas.

The harmonic pattern was stable from shot to shot until a few hundred shots in the case of BaTiO$_3$- and SrTiO$_3$-nanoparticle-contained plumes, contrary to the case of the target-contained Ag nanoparticles, when the harmonics disappeared after a few shots of the prepulse on the target surface. This attributed a much higher concentration of the former species at the target surface compared with that of the silver nanoparticles.

We found that the HHG from BaTiO$_3$ and SrTiO$_3$ multiatomic particles started to be efficient at considerably smaller prepulse intensities (8×10^9 W cm^{-2}) compared with the case of BaTiO$_3$ crystal (3×10^{10} W cm^{-2}). We attribute this tendency to the lower ablation threshold and lower cohesive energy of the host substrate that contains nanoparticles, compared with those of bulk material.

12.6
Conclusions

In this chapter, low- and high-order nonlinear optical characteristics of various nanostructures were discussed.

We presented the results of investigations of the real and imaginary parts of the third-order nonlinear susceptibilities of semiconductor colloidal solutions prepared by laser ablation. The dependencies for n_2 on laser intensity showed the influence of free carriers on the nonlinearities of As$_2$S$_3$ and CdS nanoparticle solutions. The observed positive sign of nonlinear refraction of such structures was discussed, and the explanation based on size-related effective band gap was offered. The analysis of observed nonlinear optical features of As$_2$S$_3$ and CdS solutions and thin films was carried out and discussed in comparison with bulk materials. High nonlinear susceptibilities of such structures were attributed to size-related confinement effects.

Influence of size of nanoparticles in laser ablation plasma on the HHG efficiency is discussed. Comparison of harmonic efficiencies from the laser plumes containing 110 nm spherical Ag particles, 100–1000 nm blocks of colloidal silver, 2–14 μm aluminum particles, and 50 nm thick, few micrometer long CuO nanorods is done. We also compared these results with those from the monoatomic plumes produced on the surfaces of bulk materials. We showed that small-sized structures are favorable for high-order harmonic conversion efficiency. The maximum conversion efficiency observed (for 110 nm Ag nanoparticles) was estimated to be 4×10^{-5}.

There was little difference in the harmonic cutoff order for the plumes that contain small multiatomic particles and those produced from bulk targets. In particular, the plasma with 110 nm Ag nanoparticles demonstrated 67th harmonic cutoff, compared to the 63rd and 61st harmonic cutoffs for colloidal silver and bulk silver, respectively. For longer particles (Al powder and CuO nanorods), the harmonic cutoffs were less then those obtained from ablation produced on the surfaces of bulk Al and Cu targets.

Our studies have shown that to obtain maximum HHG conversion efficiency, it is essential to define the maximum tolerable particle size that can be used, since particles with larger dimensions have higher polarizability. Some additional studies such as the influence of the dipole moment of the nanoparticles and cluster concentration on the phase matching, Z-scans of the cluster-contained plasma, intensity scaling, and so on, will benefit the understanding of the peculiarities of the application of the nanostructured media for the HHG.

The results of systematic studies of the nonlinear optical properties of $BaTiO_3$ and $SrTiO_3$ nanoparticles, as well as their structural characteristics, were reported. We measured the nonlinear refractive indices and nonlinear absorption coefficients of $BaTiO_3$- and $SrTiO_3$-nanoparticle-contained suspensions using the 790 nm femtosecond and picosecond pulses. The comparison of these measurements with those obtained using $BaTiO_3$ photorefractive crystal has demonstrated a considerable enhancement of third-order nonlinear optical characteristics in the case of nanoparticle-contained suspensions. At the same time, the structural measurements have shown that the sizes of nanoparticles used in these studies were greater than those at which one could expect the influence of quantum confinement processes. Further detailed studies of the variations of low-order nonlinearities should be carried out to distinguish the particles sizes at which the increase in nonlinear optical properties becomes significant.

We also reported the studies of high-order nonlinear optical processes in the plasma containing $BaTiO_3$ and $SrTiO_3$ multiatomic particles. We analyzed the influence of various experimental conditions on the HHG efficiency from the laser-ablated plasma plumes produced on the surface of targets containing these nanoparticles. We compared the HHG from nanoparticles with the analogous process obtained from the laser ablation of the $BaTiO_3$ solid crystal. The high-order harmonic efficiencies at the plateau region were 1×10^{-6} and 4×10^{-6} in the case of barium titanate- and strontium-titanate-nanoparticle-contained plasmas. While the harmonic cutoffs (39th harmonic) were equal for the two nanoparticle-contained species, and the fourfold enhancement in the conversion efficiency was observed in the case of smaller sized $SrTiO_3$ nanoparticles compared to the $BaTiO_3$-nanoparticle-contained plasma. These studies showed that for larger sized particles, no improvements in the extension of cutoff compared to the monoatomic plasma occur.

The nanoparticle size distribution for both the $BaTiO_3$ and $SrTiO_3$ particles was extremely broad. Thus, it is difficult to conclude with any certainty the role of different nanoparticles in the processes analyzed. Narrower nanoparticle sizes will benefit the conclusion of the role of particle sizes in the nonlinear optical response of the medium.

Acknowledgments

The author gratefully acknowledges the help of A. I. Ryasnyansky, R. I. Tugushev, T. Usmanov, M. Suzuki, M. Baba, H. Kuroda, L. B. Elouga Bom, and T. Ozaki during these studies. The author also acknowledges the support from the Fonds Québécois de la Recherche sur la Nature at les Technologies, Institut National de la Recherche Scientifique, Canada, and Japan Society for the Promotion of Science to carry out this work.

References

1. Ganeev, R.A., Ryasnyansky, A.I., Kamalov, S.R., Kodirov, M.K., and Usmanov, T. (2001) *J. Phys. D*, **43**, 1602.
2. Ma, H., Gomes, A.S.L., and De Araujo, C.B. (1993) *Opt. Lett.*, **18**, 414.
3. Kull, M. and Coutaz, J.-L. (1990) *J. Opt. Soc. Am. B*, **7**, 1463.
4. Kurth, D.G., Lehmann, P., and Lesser, C. (2000) *Chem. Commun.*, **11**, 949.
5. Du, H., Xu, G.Q., Chin, W.C., Huang, L., and Li, W. (2002) *Chem. Mater.*, **14**, 4473.
6. Tsuti, T., Iryo, K., Ohta, H., and Nashimura, Y. (2000) *Jpn. J. Appl. Phys.*, **39**, L981.
7. Uchida, K., Kaneko, S., Omi, S., Hata, C., Tanji, H., Asahara, Y., and Ikushima, A. (1994) *J. Opt. Soc. Am. B*, **11**, 1236.
8. Smektala, F., Quemard, C., Leneindre, L., Lucas, J., Barthelemy, A., De Angelis, C., and Non-Cryst, J. (1998) *Solids*, **239**, 139.
9. Ganeev, R.A., Ryasnyansky, A.I., Kodirov, M.K., and Usmanov, T. (2002) *J. Opt. A*, **4**, 446.
10. Alivisatos, A.P. (1996) *Science*, **271**, 933.
11. Schwerzel, R.E., Spahr, K.B., Kurmer, J.P., Wood, V.E., and Jenkins, J.A. (1998) *J. Phys. Chem. A*, **102**, 5622.
12. Henglein, A. (1993) *J. Phys. Chem.*, **97**, 5457.
13. Heddersen, J., Chumanov, G., and Cotton, T.M. (1993) *Appl. Spectrosc.*, **47**, 1959.
14. Ganeev, R.A., Kamalov, S.R., Kulagin, I.A., Usmanov, T., Ryasnyansky, A.I., Kodirov, M.K., and Kamanina, N.V. (2002) *Nonlinear Opt.*, **28**, 263.
15. Li, H.P., Kam, C.H., Lam, Y.L., and Ji, W. (2001) *Opt. Commun.*, **190**, 351.
16. Said, A.A., Sheik-Bahae, M., Hagan, D.J., Wei, T.H., Wang, J., Young, J., and Van Stryland, E.W. (1992) *J. Opt. Soc. Am. B*, **9**, 405.
17. Shalaev, V.M., Polyakov, E.Y., and Markel, V.A. (1996) *Phys. Rev. B*, **53**, 2437.
18. Tagakahara, T. (1987) *Phys. Rev. B*, **36**, 9193.
19. Neeves, A.E. and Birnboim, M.N. (1989) *J. Opt. Soc. Am. B*, **6**, 789.
20. Wang, Y. and Mahler, W. (1987) *Opt. Commun.*, **61**, 233.

21. Shen, Q., Liang, P., and Zhang, W. (1995) *Opt. Commun.*, **115**, 133.

22. Yao, H., Takahara, S., Mizuma, H., Kozeki, T., and Hayashi, T. (1996) *Jpn. J. Appl. Phys.*, **35**, 4633.

23. Rakovich, Y.P., Artemyev, M.V., Rolo, A.G., Vasilevskiy, M.I., and Gomes, M.J.M. (2001) *Phys. Stat. Sol. B*, **224**, 319.

24. DeSalvo, R., Said, A.A., Hagan, D.J., Van Stryland, E.W., and Sheik-Bahae, M. (1996) *IEEE J. Quantum Electron*, **32**, 1324.

25. Ganeev, R.A., Ryasnyansky, A.I., and Usmanov, T. (2003) *Opt. Quantum Electron*, **35**, 211.

26. Mehendale, S.C., Mishra, S.R., Bindra, K.S., Laghate, M., Dhami, T.S., and Rustagi, K.S. (1997) *Opt. Commun.*, **133**, 273.

27. Krauss, T.D. and Wise, F.W. (1994) *Appl. Phys. Lett.*, **65**, 1739.

28. Kayanuma, Y. and Momiji, H. (1990) *Phys. Rev. B*, **41**, 10261.

29. Rossetti, R., Ellison, J.L., Gibson, J.M., and Brus, L.E. (1984) *J. Chem. Phys.*, **80**, 4464.

30. Baykul, M.C. and Balcioglu, A. (2000) *Microelectron. Eng.*, **51**, 703.

31. Capple, P.B., Staromlynska, J., Hermann, L.A., and Mckay, T.J. (1997) *J. Nonlinear Opt. Phys. Mater.*, **6**, 251.

32. Hanamura, E. (1988) *Phys. Rev. B*, **37**, 1273.

33. Ganeev, R.A., Ryasnyansky, A.I., Kodirov, M.K., Kamalov, S.R., Li, V.A., Tugushev, R.I., and Usmanov, T. (2002) *Appl. Phys. B*, **74**, 47.

34. Flettner, A., König, J., Mason, N.B., Pfeifer, T., Weichmann, U., Düren, R., and Gerber, G. (2002) *Eur. Phys. J. D*, **21**, 115.

35. Hu, S.X. and Xu, Z.Z. (1997) *Appl. Phys. Lett.*, **71**, 2605.

36. Vozzi, C., Nisoli, M., Caumes, J.-P., Sansone, G., Stagira, S., De Silvestri, S., Vecchiocattivi, M., Bassi, D., Pascolini, M., Poletto, L., Villoresi, P., and Tondello, G. (2005) *Appl. Phys. Lett.*, **86**, 111121.

37. Donnelly, T.D., Ditmire, T., Neuman, K., Pery, M.D., and Falcone, R.W. (1996) *Phys. Rev. Lett.*, **76**, 2472.

38. Ganeev, R.A., Elouga Bom, L.B., Kieffer, J.-C., and Ozaki, T. (2007) *Phys. Rev. A*, **75**, 063806.

39. Ganeev, R.A., Suzuki, M., Baba, M., Ichihara, M., and Kuroda, H. (2008) *J. Phys. B: At. Mol. Opt. Phys.*, **41**, 045603.

40. Ganeev, R.A., Suzuki, M., Baba, M., Ichihara, M., and Kuroda, H. (2008) *J. Appl. Phys.*, **103**, 063102.

41. Kumar, A., Srivastava, A.K., Tiwari, P., and Nandedkar, R.V. (2004) *J. Phys.: Condens. Matter*, **16**, 8531.

42. Ganeev, R.A., Baba, M., Suzuki, M., and Kuroda, H. (2005) *Phys. Lett. A*, **339**, 103.

43. Liang, Y., Augst, S., Chin, S.L., Beaudoin, Y., and Chaker, M. (1994) *J. Phys. B*, **27**, 5119.

44. Ganeev, R.A., Chakravarty, U., Naik, P.A., Srivastava, H., Mukherjee, C., Tiwari, M.K., Nandedkar, R.V., and Gupta, P.D. (2007) *Appl. Opt.*, **46**, 1205.

45. Elouga Bom, L.B., Kieffer, J.-C., Ganeev, R.A., Suzuki, M., Kuroda, H., and Ozaki, T. (2007) *Phys. Rev. A*, **75**, 033804.

46. Wang, W., Qu, L., Yang, G., and Chen, Z. (2003) *Appl. Surf. Sci.*, **218**, 24.

47. Guan, D.Y., Chen, Z.H., Wang, W.T., Lu, H.B., Zhou, Y.L., Jin, K.J., and Yang, G.Z. (2005) *J. Opt. Soc. Am. B*, **22**, 1949.

48. Lotem, H. and de Araujo, C.B. (1977) *Phys. Rev. B*, **16**, 1711.

49. Adair, R., Chase, L.L., and Payne, S.A. (1989) *Phys. Rev. B*, **39**, 3337.

50. Boggess, T., White, J.O., and Valley, G.C. (1990) *J. Opt. Soc. Am. B*, **7**, 2255.

51. Sheik-Bahae, M., Said, A.A., Wei, T.-H., Hagan, D.J., and Van Stryland, E.W. (1990) *IEEE J. Quantum Electron*, **26**, 760.

52. Ganeev, R.A., Baba, M., Ryasnyansky, A.I., Suzuki, M., Turu, M., and Kuroda, H. (2004) *Appl. Phys. B*, **78**, 433.

53. Ganeev, R.A. (2007) *J. Phys. B: At. Mol. Opt. Phys.*, **40**, R213.

54. Jonathan, J.M.C., Roosen, G., and Roussignol, P. (1988) *Opt. Lett.*, **13**, 224.

13
Polarization and Space-Charge Profiling with Laser-Based Thermal Techniques

Axel Mellinger and Rajeev Singh

13.1
Introduction

13.1.1
Overview

For many years, space charges and electric dipoles in solids have garnered the attention of researchers. A dielectric material that can store space charges or contains oriented electrical dipoles to produce a permanent polarization is called "electret," a term coined by O. Heaviside [1, 2] in analogy to the well-known magnets. The invention of the electret condenser microphone [3] and the discovery of piezo- and pyroelectricity in polyvinylidene fluoride (PVDF) [4] opened up a wide market for electromechanical transducer applications. In addition, electrets are used in radiation dosimeters and gas filters [5]. In recent years, nonpolar *ferroelectrets* [6] with high piezoelectric coefficients have received considerable attention. On the other hand, injected space charge is known to have detrimental effects on the host polymer [7, 8]. This is an important issue in the failure of high-voltage cable insulations [9].

In view of the wide range of applications of electrets, nondestructive techniques to probe the distribution of space charges and electrical polarization are an obvious necessity. During the past four decades, numerous techniques for obtaining space charge and polarization depth profiles in insulating materials have been developed [10] and applied to a wide range of topics, such as the accumulation of space charge in high-voltage cable insulations [11], the development and optimization of pyroelectric and piezoelectric sensors [12, 13], and basic research into the mechanisms of charge storage [14] in electret polymers. All depth-profiling methods use an external stimulus (usually thermal, mechanical, or electrical) to generate a response (electrical or mechanical), which carries information on the spatial distribution of embedded electric dipoles or space charges. Each of these techniques has its particular strengths and weaknesses. This article focuses on thermal techniques with laser-generated thermal waves or pulses. Compared to their acoustic counterparts, these methods offer submicrometer depth resolution as

Nanomaterials: Processing and Characterization with Lasers, First Edition.
Edited by Subhash Chandra Singh, Haibo Zeng, Chunlei Guo, and Weiping Cai.
© 2012 Wiley-VCH Verlag GmbH & Co. KGaA. Published 2012 by Wiley-VCH Verlag GmbH & Co. KGaA.

well as the possibility of three-dimensional imaging. This work intends to provide students and researchers in the field of electrets with the tools to carry out their own investigations of polarization phenomena in electrets.

13.1.2
History of Thermal Techniques for Polarization and Space-Charge Depth-Profiling

In the study of electrets, an important problem is the determination of the spatial variation of polarization and space-charge distributions. The first general technique for either polarization or space charge utilized sectioning techniques in which successive layers of poled electrets were removed and analyzed for their charge account. Some early examples of this approach are used by Thiessen *et al.* [15] and by Walker and Jefimenko [16], both of whom sectioned carnauba wax electrets. An extension of this technique in conjunction with another method has been proposed by Collins [17]. A second general approach has been the use of multilayer or sandwich structures. Phelan *et al.* [18] poled sandwich structures consisting of four layers of polyvinyl fluoride (PVF) and polyvinylidene fluoride (PVDF or PVF_2) and then measured the pyroelectric coefficient in each layer. Marcus [19] constructed eight-layer sandwiches and measured the piezoelectric coefficient, achieving moderately high resolution. Both of these approaches have the serious drawback that the analytical technique is a destructive one.

Nondestructive thermal techniques use the absorption of a short light pulse or a periodically modulated laser beam by an opaque surface layer to create a time-dependent, spatially varying temperature distribution, as shown in Figure 13.1. In samples that are either pyroelectric or contain an electric space charge, this gives rise to a short-circuit current, which again carries information on the polarization or space-charge depth profile. Although initial insights into the potential of these techniques were obtained by Phelan and Peterson in their investigation of the frequency response of pyroelectric detectors [18, 20], the first dedicated study of space-charge profiles by means of thermal pulses was reported by Collins [21, 22]. A key difference from acoustic methods is the fact that the propagation of the thermal pulse or wave is a diffusion phenomenon rather than a linear propagation, which greatly complicates signal analysis. Initially, it was assumed that thermal pulses are only suitable for obtaining charge centroids [23], since extracting a charge or polarization profile from the measured current involves solving a Fredholm integral equation of the first kind, which is an ill-conditioned problem [24]. Later, however, several deconvolution methods were shown to yield polarization depth profiles [25–28], albeit with decreasing resolution at larger depths. While the first thermal pulse experiments were carried out with xenon flash lamps delivering pulse durations in the microsecond to millisecond range (which is not short compared to the diffusion time for distances in the micrometer range) [22], later implementations used mode-locked pulsed laser sources with pulse durations in the nanosecond range (and corresponding diffusion lengths well below 0.1 μm), which simplifies data analysis.

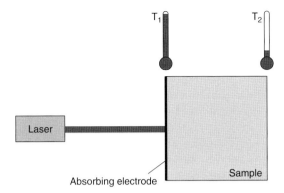

Figure 13.1 Principle of laser-based thermal depth-profiling techniques. The laser beam is either pulsed or intensity-modulated.

A frequency-domain counterpart to the thermal-pulse technique was introduced by Lang and Das-Gupta [29], now commonly referred to as *thermal wave technique* or *laser intensity modulation method* (LIMM). Here, the top electrode is irradiated with an intensity-modulated continuous-wave (CW) laser beam while the short-circuit current is recorded with a phase-sensitive lock-in amplifier. With relatively modest laboratory equipment, this technique is capable of achieving a near-surface depth resolution of $< 0.1\ \mu m$ [30]. The deconvolution process of the signal is similar to that used for thermal pulses. The LIMM equation [31], which relates the polarization or space-charge depth profile to the measured current, was solved using, for example, Tikhonov regularization [32, 33], polynomial expansion [34], and neural networks [35]. A particularly simple and straightforward scale transformation technique was introduced by Ploss *et al.* who demonstrated that the difference between the real and imaginary part of the pyroelectric current is proportional to the pyroelectric coefficient [36]. However, this method is limited to regions near the heated surface. For depths larger than one-fourth of the sample thickness, the accuracy decreases significantly.

Modern computational power permitted a new approach using an iterative method [37], previously known from digital image processing and signal reconstruction. Simulations showed slightly smaller residuals than those obtained using Tikhonov regularization. In addition, this technique also appeals from an educational point of view, since it visualizes how the "true" depth profile gradually emerges from a starting solution as the iteration progresses. One suitable initial distribution is the scale transformation solution. With increasing number of iterations, its deficiencies at larger depths are gradually corrected.

In recent years, a *Monte Carlo*-based technique [38] has been shown to reproduce a given pyroelectric profile with better accuracy than regularization [39]. Although the computational effort can be substantial, this method lends itself to parallel computing in multicore or multiprocessor computers [40].

A direct comparison between the time- and frequency-domain approaches was published by Mellinger *et al.* in 2005 [41] and showed excellent agreement for a space-charge profile in corona-charged polytetrafluoroethylene (PTFE) (cf. p. 10). However, the thermal-pulse data was acquired up to 50 times faster than the LIMM measurement. This allows the real-time monitoring of thermally or optically induced charge decay on a timescale of minutes, rather than hours. The fast data acquisition of the thermal-pulse technique was used for obtaining a three-dimensional polarization profile of a corona-poled PVDF film [42] by means of a tightly focused laser beam. The lateral resolution of 40 μm (combined with a depth resolution of better than 0.5 μm) represents a significant advance over previous attempts of multidimensional polarization or space-charge mapping. Acoustic techniques have not yet resolved features less than 0.5 mm wide [43]. With focused LIMM (FLIMM) [44], a resolution in the 10 μm range has been reported, but the long measurement times in the frequency domain either severely limited the number of beam pointings [44] or constrained the number of frequencies to a few selected values [45]. However, modeling of the focused thermal-pulse propagation through the two-layer system consisting of the metal electrode and the polymer film (and its influence on the pyroelectric current) is a nontrivial problem [46]. There is evidence, however, that metal electrodes with their high thermal diffusivity[1] may limit the lateral resolution to several times the sample thickness [42], unless lateral thermal diffusion is taken into account when analyzing the data.

A review of thermal depth-profiling techniques was published in Ref. [47]. Not only their strengths but also their limitations are highlighted by a discussion of the spatial resolution [48]. Comparisons between thermal and acoustic techniques [49, 50] have shown good agreement. All-optical techniques, such as thermal-pulse technique and LIMM, are the methods of choice for investigating soft, piezoelectric polymer foams [51], which are easily deformed on applying a mechanical stress. They can be applied *in situ* [52], even when the sample is stored in vacuum. In addition, they have been successfully applied to study heat diffusion processes in solids. Even in the absence of polarization or space charge, information on the thermal diffusivity of polymers has been obtained by applying a bias voltage and detecting the electric-field-induced pyroelectricity [53]. Alternatively, diffusivity values have been calculated by recording the phase shift of the transient temperature at the front and back electrodes as a function of the modulation frequency [52, 54].

13.2
Theoretical Foundations and Data Analysis

This section outlines the theoretical framework for understanding and analyzing laser-based polarization measurements. There is a substantial difference between analyzing data from acoustic and thermal experiments. While acoustic pulses propagate through the sample at a constant speed of sound, heat flow is a thermal *diffusion* process. As a consequence, thermal techniques require a more sophisticated analysis than the relatively straightforward acoustic methods.

---

1) Typical thermal diffusivity values are $\approx 10^{-4}$ m^2 s^{-1} for metals and $\approx 10^{-7}$ m^2 s^{-1} for polymers.

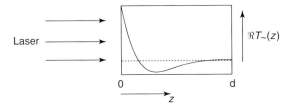

Figure 13.2 Thermal wave inside the sample at a fixed modulation frequency. The real part of the temperature is shown as a function of depth z.

Thermal polarization probing can be carried out in the following two ways:

1) In the time-domain approach, the opaque front electrode is illuminated by a short light pulse, usually from an Nd:YAG laser, and the transient short-circuit current between the front and back electrodes is recorded as a function of time.
2) In the frequency-domain approach, the front electrode is periodically heated by an intensity-modulated light source (usually a diode laser). The short-circuit current is fed into a lock-in amplifier that extracts the signal component having the same frequency as the laser modulation.

The two approaches are related to each other via a Fourier transform, so it suffices to discuss one of them. As we shall see later, the frequency-domain approach provides an easy way to compensate phase shifts created by the data acquisition instrumentation. Therefore, the following sections discuss this framework. For a discussion of thermal-pulse analysis, see Refs [25, 26].

13.2.1
One-Dimensional Heat Conduction

Figure 13.2 shows a sample of thickness d, which is illuminated from the $z = 0$ side by a sinusoidally modulated laser beam with a time-dependent intensity

$$j(t) = j_0 + j_{\sim}e^{i\omega t} = j_0 + j_{\sim}e^{2\pi i f t} \tag{13.1}$$

where f is the modulation frequency [2]. A fraction η of the light is absorbed and heats up the sample. The temperature distribution inside the polymer film is given by the one-dimensional heat conduction equation (also known as *Fick's law*):

$$\kappa \nabla^2 T(t, x, y, z) = c\rho \frac{\partial T(t, x, y, z)}{\partial t} \tag{13.2}$$

Here, $T(t, x, y, z)$ is the transient temperature distribution, κ the thermal conductivity [W m^{-1} K^{-1}], c the heat capacity per unit mass [J kg^{-1} K^{-1}], and ρ the mass density. The three material constants are often combined into the thermal diffusivity

$$D = \kappa/(c\rho) \; (\text{m}^2 \, \text{s}^{-1}) \tag{13.3}$$

---

2) For reasons of simplicity, ω is sometimes called "modulation frequency," even though the laser is modulated at $f = \omega f/(2\pi)$.

If the diameter of the laser beam is large compared to the sample thickness, Eq. (13.2) reduces to its one-dimensional form, as shown by Emmerich *et al.* [55] and Lang [56]:

$$D\frac{\partial^2 T(t, z)}{\partial z^2} = \frac{\partial T(t, z)}{\partial t} \tag{13.4}$$

where z is the coordinate perpendicular to the film surface. Note that Eq. (13.4) only describes the *time-dependent* part of the solution. The complete solution contains a time-independent term, which is a function of the average laser intensity j_0. Since we heat the front surface of the sample and have thermal losses (by wave thermal conductivity, radiation, and convection) at the rear side and edges of the heated area, a temperature gradient will form inside the polymer. However, this temperature distribution is stationary and does not contribute to the pyroelectric current. Equation (13.4) is a good approximation even for some multilayer systems. For example, in polymer films with evaporated metal electrodes, only heat conduction in polymer itself needs to be considered, as the electrodes are usually two–three orders of magnitude thinner than the polymer film.

If the sinusoidally modulated light is absorbed at $z = 0$, the boundary conditions are [57]

$$\eta j_\sim e^{i\omega t} - G_0 T = -\kappa \left. \frac{\partial T}{\partial z} \right|_{z=0} \tag{13.5}$$

$$\text{and} \quad G_d = \kappa \left. \frac{\partial T}{\partial z} \right|_{z=d} \tag{13.6}$$

where G_0 and G_d are the heat loss coefficients at the front and rear surfaces, κ is the thermal conductivity of the sample, and $\eta j_\sim e^{i\omega t}$ is the absorbed power per unit area. For a free-standing film, both G_0 and G_d can usually be neglected, whereas samples thermally connected to a substrate may exhibit a substantial heat loss G_d. Under these conditions, Eq. (13.4) is solved by [57–59]

$$T(t, z) = T_\sim(\omega, z)e^{i\omega t} = \frac{\eta j_\sim}{\kappa k} \frac{\cosh[k(d - z)] + \frac{G_d}{\kappa k}\sinh[k(d - z)]}{\left(1 + \frac{G_0 G_d}{\kappa^2 k^2}\right) \sinh(kd) + \frac{G_0 + G_d}{\kappa k}\cosh(kd)} e^{i\omega t} \tag{13.7}$$

where the frequency-dependent complex thermal wave vector is given by

$$k = (1 + i)k_r \tag{13.8}$$

with

$$k_r = \sqrt{\omega/2D} \tag{13.9}$$

The penetration depth z_r of the thermal wave into the sample is given by the inverse of the thermal wave vector k_r, that is

$$z_r = \sqrt{2D/\omega} \tag{13.10}$$

$T_\sim(z)$ is a complex quantity, indicating that the temperature oscillation inside the sample may be phase shifted with respect to the laser beam.

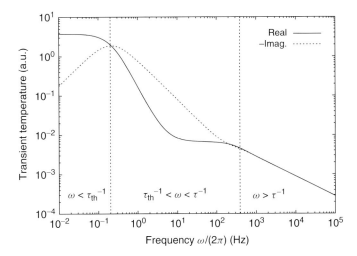

Figure 13.3 Transient temperature at the front surface ($z = 0$), calculated from Eq. (13.7) with $d = 20\ \mu m$, $\kappa = 0.15\ Wm^{-1}K^{-1}$, $D = 10^{-7}m^2s^{-1}$, and $G_0 = G_d = 20\ Wm^{-2}K^{-1}$. See the text for the explanation of the three frequency regimes. (Source: After Bauer and Ploss [54].)

To understand the significance of Eq. (13.7), it is helpful to study the magnitude of the surface temperature ($z = 0$) as a function of frequency, as was done by Bauer and Ploss [54]. Figure 13.3 shows three distinct regimes.

1) Low frequencies ($\omega < 1/\tau_{th}$): The sample is in internal thermal equilibrium, as well as in thermal equilibrium, with its environment. The sample temperature $T(t, z = 0)$ follows the laser intensity $j(t)$ without phase lag so that the real part $\Re T(0, t)$ approaches a constant value and the imaginary part $\Im T(t, 0) \to 0$.

2) Intermediate range ($1/\tau_{th} < \omega < 1/\tau = (\pi^2 D)/d^2$): The sample is no longer in thermal equilibrium with its environment but still in internal equilibrium (the thermal wave penetrates the entire sample). When the laser intensity $j(t)$ is higher than its average value j_0, excess heat is deposited in the sample, causing the temperature to increase. When $j(t)$ is less than j_0, the sample temperature decreases. Thus, the temperature $T(t, 0)$ lags behind the laser intensity by a phase angle of up to $-\pi/2$.

3) High frequencies ($\omega \gg 1/\tau$): The sample is not in thermal equilibrium; the thermal wave does not penetrate the film. The phase angle between $T(t, 0)$ and $j(t)$ approaches $-\pi/4$. This frequency range will later be used for an approximate solution of the LIMM equation.

As thermal wave methods require an inhomogeneous temperature distribution in order to probe the polarization depth profile, measurements are carried out mostly in the high-frequency regime. At high frequencies, the thermal wave vector is significantly larger than the inverse sample thickness (i.e., $k_r \gg d^{-1}$), and

Eq. (13.7) can be approximated by

$$T_\sim(z) = \frac{\eta j_\sim}{\kappa k} \, \mathrm{e}^{-kz} \tag{13.11}$$

This approximation will later be needed for the analysis of experimental LIMM data.

13.2.2
The One-Dimensional LIMM Equation and its Solutions

With the known temperature distribution inside the sample, we can now calculate the pyroelectric signal generated by the oscillating temperatures. The experimental pyroelectric coefficient

$$p = \frac{1}{A} \frac{\partial Q}{\partial T} \tag{13.12}$$

describes the change ΔQ in surface charge resulting from a temperature change ΔT. The exact description of all first-order terms contributing to the pyroelectric coefficient is given by

$$g(z) = \alpha_p P(z) - (\alpha_z - \alpha_\epsilon)\epsilon_0 \epsilon \, E(z) \tag{13.13}$$

where ϵ_0 is the permittivity of vacuum, ϵ the relative permittivity of the sample, and $E(z)$ the electric field. The three coefficients α_p, α_z, and α_ϵ denote the relative temperature dependencies of the polarization, the thermal expansion, and the permittivity, respectively. If the polarization is locally compensated by a charge density $\rho(z)$, Eq. (13.13) reduces to [36]

$$g(z) = \alpha_p P(z) \equiv p(z) \tag{13.14}$$

The contribution to the pyrocurrent from a thin slice dz of the sample depends on the local pyroelectric coefficient $p(z)$ and the rate of change of the transient temperature. For calculating the pyrocurrent measured between the electrodes on the end surfaces we need to integrate over the sample thickness d.

$$I(t) = \frac{A}{d} \int_0^d g(z) \frac{\partial T(t, z)}{\partial t} \, dz$$

Changing from the time to the frequency domain, we obtain the one-dimensional *LIMM equation* [60]

$$I_\sim(\omega) = \frac{i\omega A}{d} \int_0^d g(z) T_\sim(k, z) \, dz \tag{13.15}$$

with the distribution function

$$g(z) = (\alpha_\epsilon - \alpha_z) \int_0^z \rho(\zeta) \, d\zeta + p(z) \tag{13.16}$$

where α_ϵ is the temperature coefficient of the permittivity, α_z is the thermal expansion coefficient, and $p(z)$ is the pyroelectric coefficient. For free-standing samples, Eq. (13.7) (with $G_0 = G_d = 0$) can be substituted into Eq. (13.15), yielding

$$I_\sim = \frac{A}{d} \frac{\eta j_\sim}{c\rho} \frac{k}{\sinh(kd)} \int_0^d g(z)\cosh[k(d-z)]\,dz \tag{13.17}$$

In a LIMM experiment, one measures the pyroelectric current $I_\sim$ as a function of the modulation frequency ω. The key problem in the analysis of LIMM data is the reconstruction of the distribution function $g(z)$ from the experimentally observed frequency spectrum $I_\sim(\omega)$. The LIMM equation (Eq. 13.15) is a Fredholm integral equation of the first kind, for which finding a solution is an ill-conditioned problem [61]. Within the experimental errors, the observed current $I_\sim(\omega)$ can result from an infinite number of distributions $g(z)$, most of which are strongly oscillating functions. Over the past two decades, a number of techniques have been applied to obtain a unique solution of the LIMM equation [47]. In this article, we shall review three of these techniques in some detail: the scale transformation method (being the simplest approach), Tikhonov regularization (as it illustrates the matrix operations that are fundamental to many other techniques as well), and the recently invented Monte Carlo technique, which has been shown to yield slightly more accurate results than Tikhonov regularization.

13.2.2.1 Scale Transformation
The most simple solution is a scale transformation technique, developed by Ploss *et al.* [36]. Let us revisit the high-frequency approximate solution (Eq. 13.11) of the heat conduction equation. Inserting this approximation into Eq. (13.17), we obtain

$$I_\sim = \frac{\eta j_\sim A}{c\rho d} \int_0^\infty g(z)ke^{-kz}\,dz \tag{13.18}$$

Since $g(z) = 0$ outside the sample, the upper integration limit was changed to ∞ without affecting the result. Splitting this result into its real and imaginary parts yields

$$\Re I_\sim = \frac{\eta j_\sim A}{c\rho d} \int_0^\infty g(z)k_r[\cos(k_r z) + \sin(k_r z)]e^{-k_r z}\,dz \tag{13.19}$$

$$\Im I_\sim = \frac{\eta j_\sim A}{c\rho d} \int_0^\infty g(z)k_r[\cos(k_r z) - \sin(k_r z)]e^{-k_r x}\,dz \tag{13.20}$$

We now define the *penetration depth*

$$z_r = 1/k_r = \sqrt{2D/\omega} \tag{13.21}$$

and the function

$$g_a(z_r) = \frac{c\rho d}{\eta j_\sim A}\left[(\Re - \Im)I_\sim(\omega = 2D/z_r^2)\right] \tag{13.22}$$

Inserting Eqs. (13.19) and (13.20), we obtain

$$g_a(z_r) = \int_0^\infty g(z)\frac{2}{z_r}\sin\left(\frac{z}{z_r}\right)e^{-z/z_r}\,dz \tag{13.23}$$

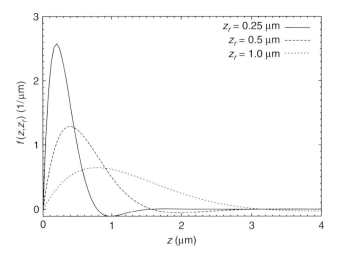

Figure 13.4 Plot of the scanning function $f(z, z_r)$ for different values of z_r.

This is an integral equation with the kernel

$$f_a(z, z_r) = \frac{2}{z_r} \sin\left(\frac{z}{z_r}\right) e^{-z/z_r} \tag{13.24}$$

(also called the *thermal window function* [62]), which vanishes for both $x = 0$ and $x \to \infty$ and has the property

$$\int_0^\infty z^n f_a(z, z_r) \, dz = z_r^n \qquad \text{for } n \in \{0, 1, 2\} \tag{13.25}$$

Therefore, $f_a(z, z_r)$ is a normalized scanning function that retrieves an approximate value of $g(z)$ at the location $z = z_r$. In particular, it *exactly* retrieves the coefficients of a Taylor expansion up to the second order. We can thus regard $g_a(z_r)$ as an approximation to the true distribution $g(z)$. [3] The penetration depth z_r is directly related to the modulation frequency ω through Eq. (13.21).

The spatial resolution at a given depth z_r is determined by the width of the kernel $f_a(z, z_r)$. Figure 13.4 shows that the resolution increases as we move closer to the heated surface of the sample (which is equivalent to increasing the frequency ω). Beyond the center of the sample, the spatial resolution is insufficient to obtain a meaningful result. Moreover, the approximate solution of the heat conduction equation no longer holds. Therefore, to obtain a complete depth profile, we have to do a second measurement from the rear side of the sample.

The spatial resolution can be improved by applying differential operators to Eq. (13.23). For further details, see Ref. [36].

---

3) If Eq. (13.25) was true for all $n \geq 0$, we would have $g_a(z) = g(z)$ and the kernel $f_a(z, z_r)$ would be the delta function $\delta(z - z_r)$.

13.2.2.2 Tikhonov Regularization

Tikhonov regularization [32, 33] is another common technique for solving the LIMM equation (13.15). Here the ambiguity in the solution of the LIMM equation is removed by introducing additional constraints. For example, if we seek to obtain a smooth function, we would require that the integral over the second derivative of the pyroelectric distribution function be as small as possible, that is,

$$\mathcal{B}[g] = \int_0^d |g''(x)|^2 \, dx \stackrel{!}{=} \text{Min} \tag{13.26}$$

To implement Tikhonov regularization, we rewrite the integral in the one-dimensional LIMM equation (Eq. 13.15) as a set of linear equations

$$I_m = \sum_{n=0}^M T_{mn} g_n \tag{13.27}$$

where g_n is the value of distribution function at a suitably chosen set of depths z_n and $T_{mn} = (A/d) i \omega_m T(\omega_m, z_n)$ describes the temperature distribution at frequency ω_m and depth z_n. Often, z_n are chosen to be evenly spaced on a logarithmic scale, in order to account for the higher resolution of LIMM near the surface.

To solve for the unknown g_n, we calculate the differences between the observed and calculated I_m,

$$\chi^2[g] = \sum_n \left| \frac{I_m - \sum_{n=0}^N T_{mn} g_n}{\sigma_m} \right|^2 \tag{13.28}$$

which can be written in matrix notation as

$$\chi^2[g] = (b - Ag)^\dagger (b - Ag) \tag{13.29}$$

(the operator † denotes the transpose, complex conjugate matrix, i.e., $A^\dagger = \tilde{A}^*$). The $M \times N$ matrix A and the column vector b are defined as $A_{mn} = T_{mn}/\sigma_m$ and $b_m = I_m/\sigma_m$, respectively, with σ_m being weight factors (often set equal to 1). In order to obtain a unique solution, we minimize $\chi^2[g]$ together with the constraint from Eq. (13.26)

$$\chi^2[g] + \lambda \mathcal{B}[g] = \text{min} \tag{13.30}$$

where the *regularization parameter* λ is a weight factor that determines whether the minimization of the first term or the regularization condition $\mathcal{B}$ has the greater priority. The regularization constraint $\mathcal{B}[g]$ can be expressed in matrix notation as

$$\mathcal{B}[g] = |B \cdot g|^2 = \tilde{g}(B^\dagger B)g = \tilde{g}Hg \tag{13.31}$$

with the second-derivative operators

$$B = \begin{pmatrix} 1 & -2 & 1 & 0 & 0 & 0 & 0 & \cdots & 0 \\ 0 & 1 & -2 & 1 & 0 & 0 & 0 & \cdots & 0 \\ 0 & 0 & 1 & -2 & 1 & 0 & 0 & \cdots & 0 \\ \vdots & & & & \ddots & & & & \vdots \\ 0 & \cdots & 0 & 0 & 0 & 1 & -2 & 1 & 0 \\ 0 & \cdots & 0 & 0 & 0 & 0 & 1 & -2 & 1 \end{pmatrix} \tag{13.32}$$

and

$$H = \begin{pmatrix} 1 & -2 & 1 & 0 & 0 & 0 & 0 & \cdots & 0 \\ -2 & 5 & -4 & 1 & 0 & 0 & 0 & \cdots & 0 \\ 1 & -4 & 6 & -4 & 1 & 0 & 0 & \cdots & 0 \\ \vdots & & & & \ddots & & & & \vdots \\ 0 & \cdots & 0 & 0 & 1 & -4 & 6 & -4 & 1 \\ 0 & \cdots & 0 & 0 & 0 & 1 & -4 & 5 & -2 \\ 0 & \cdots & 0 & 0 & 0 & 0 & 1 & -2 & 1 \end{pmatrix}$$ (13.33)

The expression to be minimized is then

$$(\mathbf{b} - \mathbf{Ag})^{\dagger}(\mathbf{b} - \mathbf{Ag}) + \lambda \tilde{\mathbf{g}} \mathbf{Hg} \overset{!}{=} \min$$ (13.34)

We know that we are at a local extremal point when the partial derivatives with respect to the components g_m of $\mathbf{g}$ vanish. Thus, with some algebraic rearrangements and the definition

$$\mathbf{X} = \mathbf{A}^{\dagger}\mathbf{A} + \lambda \mathbf{H}$$ (13.35)

we obtain the linear set of equations

$$(\tilde{X} + \mathbf{X})\mathbf{g} = \mathbf{A}^{\dagger}\mathbf{b} + \tilde{A}\mathbf{b}^{*}$$ (13.36)

which is then solved for g.

Selecting the proper value of the regularization parameter λ is critical for reconstructing g. When λ is chosen too small, the solution will be unstable and shows strong oscillations. If λ is too large, the solution will be oversimplified by the constraint B. Several methods for determining λ have been described in the literature [60], most notably the L-curve method [47, 63]. A reasonable "first guess" is the value that makes both terms in Eq. (13.35) of equal magnitude.

$$\lambda = \mathrm{Tr}(\mathbf{A}^{\dagger}\mathbf{A})/\mathrm{Tr}(\mathbf{H})$$ (13.37)

13.2.2.3 Monte Carlo Technique

In 2005, Tuncer and Lang developed one of the newest techniques to solve the LIMM equation [38]. The steps were described in ref. [39] as follows:

1) Using a random number algorithm, generate a large number of depth values z_n between 0 and the sample thickness d. Their number N should be ~90% of the number of frequencies M.
2) Calculate $i\omega_m \mathbf{T}(\omega_m, z_n)$ for each combination of z_n and ω_m values, according to Eqs. (13.7)–(13.9).
3) Insert these values into a $2M \times M$ matrix $\mathbf{T}$. Rows 1 through M are the real parts, and rows $M + 1$ through $2M$ are the imaginary parts.
4) Create a $2M$ vector $\mathbf{j}$ holding the values of the short-circuit current as a function of frequency. Again, rows 1 through M are the real parts, and rows $M + 1$ through $2M$ are the imaginary parts.

5) Solve the matrix equation $\mathbf{j} = \mathbf{T} \cdot \mathbf{g}$ for $\mathbf{g}$ using singular value decomposition, discarding the smaller singular values. If too many values are kept, the solution will start to oscillate.

6) Repeat steps 1–5 at least several hundred times.

7) Sort the z-axis values from the Monte Carlo iterations into a number of bins, for example, 100, and average the g values in each bin. The result is a histogram of the polarization distribution.

Even though this technique is relatively slow because of the large number of iterations, it was shown to yield the best agreement between a simulated polarization distribution and the reconstructed curves [39]. In addition, the necessary choice in the number of singular values is more robust than the estimate of the regularization parameter λ. Since the Monte Carlo iterations can be carried out in parallel, modern multicore CPUs or multiprocessor computer clusters can greatly speed up the calculations [40].

Figure 13.5 shows a comparison between the three techniques. Using a "soft bimorph" polarization profile, the pyroelectric current was calculated for a PVDF film with a thickness of $5\,\mu m$ using Eq. 13.15. Next, the pyroelectric profile was reconstructed using scale transformation, Tikhonov regularization, and the Monte Carlo approach.

13.2.2.4 Other Techniques

Several other techniques have been developed that address some of the shortcomings of scale transformation and Tikhonov regularization, such as Fourier series, an iterative approach [37], and polynomial regularization [47, 64]. See Refs [47] and [39] for further references.

13.2.3
Two- and Three-Dimensional Analysis

All previously discussed techniques for reconstructing polarization profiles from LIMM or thermal-pulse measurements are based on the one-dimensional analytical solution of Fick's second law. For tightly focused laser beams, however, the fast lateral thermal diffusion in the (usually metallic) top electrode needs to be taken into account, or else, the lateral resolution will deteriorate, as was shown experimentally [42, 65]. Since typical thermal diffusivities of metals are of the order of $[10^{-4}]m^2s^{-1}$ (~ 3 orders of magnitude higher than in polymers), the lateral spread of the heat pulse can significantly exceed the diffusion length in the depth direction. As an example, consider a material with a laterally structured polarization, as shown in Figure 13.6. Even though the laser beam is focused on an area with zero polarization, the thermal pulse probes part of the poled region, leading to a nonzero pyroelectric current. When the polarization profile is then reconstructed using the traditional one-dimensional model, one would mistakenly obtain a nonzero polarization at the location of the laser beam spot. Consequently, the

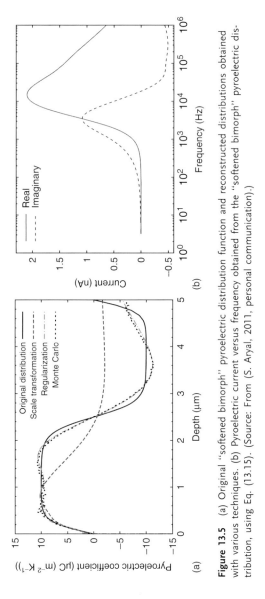

Figure 13.5 (a) Original "softened bimorph" pyroelectric distribution function and reconstructed distributions obtained with various techniques. (b) Pyroelectric current versus frequency obtained from the "softened bimorph" pyroelectric distribution, using Eq. (13.15). (Source: From (S. Aryal, 2011, personal communication).)

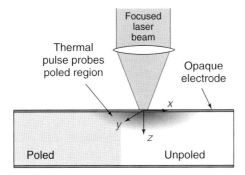

Figure 13.6 Diffusion of a thermal pulse in a sample with a laterally structured polarization. Owing to its fast lateral spreading, the thermal pulse probes part of the poled region, even though the laser beam is focused on an unpoled part of the sample. (Source: From ref. [40].)

lateral resolution can be significantly worse than the limit imposed by the laser spot size.

Very recently, some progress was made toward coupling the data obtained from neighboring beam pointings [40], using numerically calculated temperature distributions for a bilayer system (metal electrode + polymer sample) to reconstruct the polarization profile of a two-dimensional model system consisting of a pyroelectric distribution $g(X, z)$ with a "soft bimorph" shape in the z direction and a sudden transition from $g \neq 0$ for $X < 0$ to $g = 0$ for $X > 0$ (Figure 13.7a). The preliminary results show good agreement of the reconstructed profile (Figure 13.7b) with the original distribution. Most importantly, the new approach reduces the smoothing in the lateral direction observed for the traditional one-dimensional analysis (Figure 13.7c). It is expected that the results can easily be adapted to a real-world three-dimensional case.

13.3
Experimental Techniques

The key to nondestructive measurement of space charge and polarization profiles is the use of an external stimulus (usually thermal, mechanical, or electrical) to generate a response (electrical or mechanical) that carries information on the spatial distribution of embedded electric dipoles or space charges. The absorption of a laser pulse by an opaque target (Figure 13.1) at the surface of the material under investigation results in fast heating and hence thermal expansion of the target. Frequently, the target is identical to the conductive electrode, as thin-film metal layers (thickness ≥ 50 nm) provide sufficient absorption. The hot target has two effects, which is discussed in the following subsection.

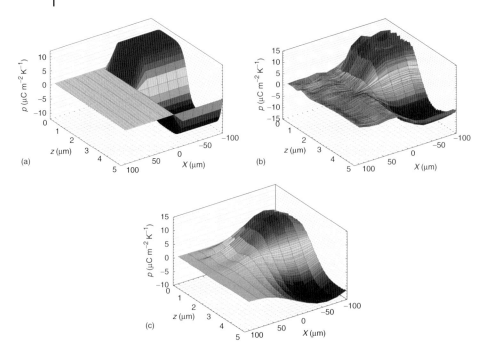

Figure 13.7 Two-dimensional pyroelectric distribution as a function of depth z and the in-plane coordinate X. (a) Original distribution, (b) reconstructed distribution, calculated by numerically coupling the data from up to 5 beam pointings on each side, and (c) reconstructed distribution using the one-dimensional model (no coupling of data from neighboring beam pointings). (Source: From ref. [40].)

13.3.1
Basic Principle

1) The rapid thermal expansion sends a pressure pulse into the sample, propagating at the speed of sound and locally compressing the material. Any electric polarization will change accordingly, as molecular dipole density changes. Likewise, embedded space charges will experience a local density change. This effect is the basis for a range of acoustic depth-profiling techniques, which are beyond the scope of this article. For an overview, see Refs [5, 66, 67]. As typical speeds of sound in solids are of the order of several 1000 m s^{-1}, these changes take place on a timescale of <1 ns for each micrometer of sample thickness. Owing to the band width limits of the detection circuits, and the RC time constant (where R is the resistance and C the capacitance) of the sample and preamplifier, achieving a spatial resolution of <2 μm is difficult.

2) The heat deposited in the target will diffuse into the sample. This process occurs on a timescale of micro- and milliseconds and is hence a much slower process than the propagation of pressure pulses. The thermal pulse

or wave leads to nonuniform heating of the material under investigation, and hence a nonuniform thermal expansion. The heating will lead to a change in polarization via the primary or secondary pyroelectric effect, or may change the local space charge density via thermal expansion.

Lasers are not the only possible heat source for thermal polarization profiling; samples with thicknesses in the millimeter range have successfully been studied using the thermal step method [68, 69], where a cold liquid is suddenly brought into contact with the sample via a heat exchanger.

13.3.2
Laser Intensity Modulation Method (LIMM)

In a typical LIMM experiment, the sample is coated with conductive surface electrodes, such as aluminum or copper, and illuminated with an intensity-modulated diode laser (Figure 13.8) with a power in the range of 1–50 mW.

The short-circuit current is amplified and converted to a voltage signal with a low-noise current preamplifier. Variable-gain preamplifiers, such as the FEMTO DLPCA-200, offer the possibility to adjust the band width to the desired frequency range. Since the pyroelectric current is usually small compared to radio-frequency noise from a variety of sources, lock-in signal detection is required. Both the real (in-phase) and imaginary ($90°$ out of phase) parts must be recorded. Modulation frequencies typically range between 10 and 100 kHz or more, where the lower limit is chosen such that the thermal penetration depth given by Eq. (13.10) is equal to or larger than the sample thickness, while the upper limit is usually determined by the band width of the lock-in amplifier. In polymers with a typical diffusivity of $[10^{-7}]$ m^2s^{-1}, the penetration depth at $f = 10$ Hz and 100 kHz is ~50 and 0.5 μm, respectively.

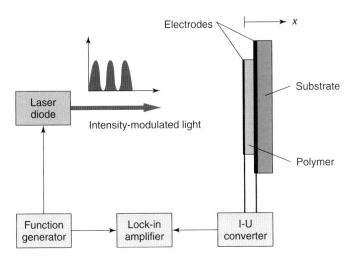

Figure 13.8 Schematic setup of a LIMM experiment.

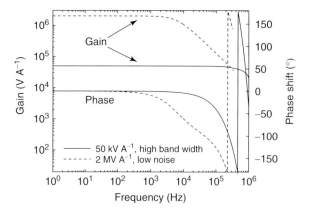

Figure 13.9 Frequency dependence for the gain, and phase shift of a Stanford Research SR 570 current preamplifier. (Source: From ref. [41].)

The preamplifier may introduce a nonnegligible phase shift if the modulation frequency is larger than one-tenth of its band width. In this case, it may be necessary to correct the measured current $I(f)$ with the experimentally determined complex gain spectrum $\tilde{\alpha}(f)$, such as the one shown in Figure 13.9

$$I_{\text{corr}}(f) = \frac{I(f)}{\tilde{\alpha}(f)} \tag{13.38}$$

13.3.3
Thermal Pulses

The setup for a thermal-pulse experiment is similar to that for a LIMM setup, with the intensity-modulated CW laser replaced by a pulsed one and the lock-in amplifier replaced by a fast analog-to-digital converter, preferably with a large bit-depth (14 bits or higher). According to the Nyquist sampling theorem, the sampling rate must be at least twice the highest frequency to be recorded and the pulse duration should be less than the inverse sampling rate. Compact Nd:YAG lasers with pulse energies of a few milliJoules are well suited to this task; their pulse width of 4–5 ns would theoretically allow a band width of 100 MHz. Frequency-doubled lasers are preferred, since their visible light output greatly facilitates beam alignment. A potential problem (especially in experiments with focused laser light) is the thermal stress on the electrode. Since the thermal diffusion length $\ell = \sqrt{Dt}$ in a metal is only $\sim$20 nm after 5 ns, the entire pulse energy gets deposited into the thin electrode layer. At sufficiently high pulse energies, the electrode temperature may rise by several hundreds of Kelvin, causing ablation and damage to the sample. As the detection circuit often imposes band width limits of no more than 100–200 kHz, a laser with a pulse duration in the microsecond range would fulfill the Nyquist theorem while spreading the heat deposition process out over a longer time, thus reducing thermal stress. Unfortunately, relatively few laser systems with pulse lengths in the micrometer range are available in the market.

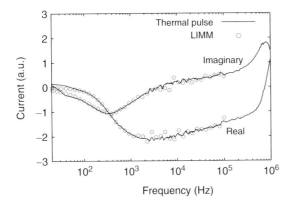

Figure 13.10 Fourier transformed displacement current of a charged PTFE film, measured with two different gain/band width settings of the preamplifier. (Source: From ref. [41].)

The recorded transient current $I(t)$ is Fourier transformed into the frequency domain according to

$$J_{corr}(f_m) = \frac{\Delta t}{\tilde{\alpha}(f_m)} \sum_{k=0}^{M-1} I(t_k) e^{-2\pi i km/M} \tag{13.39}$$

where M is the number of data points. Frequencies are given by

$$f_m = \frac{m}{M\Delta t}, \qquad m = 1 \dots \frac{M}{2} \tag{13.40}$$

and $\tilde{\alpha}(f_m)$ is the preamplifier gain correction discussed in the previous section.

In a direct comparison between the time- and frequency-domain approaches [41], there was excellent agreement for the current spectrum resulting in a corona-charged PTFE sample with embedded space charge (Figure 13.10). However, the acquisition time for the thermal-pulse data was 40–50 times less than that for the LIMM measurement.

13.3.4
Three-Dimensional Mapping

While both LIMM and the thermal-pulse method were originally designed to provide one-dimensional polarization depth profiles, significant progress toward a full three-dimensional mapping method has been reported in recent years. By focusing the laser to a tight spot size (typically between 5 and 200 μm) and scanning the beam across the sample surface (or mounting the sample on an X-Y translation stage), 3D polarization maps were obtained both in the frequency domain (FLIMM) [44, 70] and in the time domain (thermal-pulse tomography (TPT)) [42, 46, 65]. With a lateral resolution ranging from 10 to 400 μm, these techniques bridge the gap between bulk-only methods and piezoelectric force microscopy [71]. A typical TPT setup is shown in Figure 13.11.

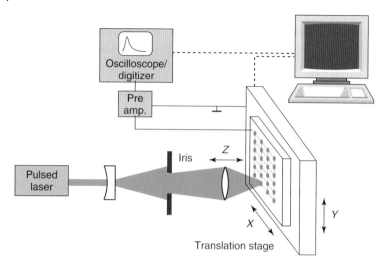

Figure 13.11 Thermal-pulse tomography setup.

Scanning a sample with a large number of beam pointings can be time consuming, especially when using the slower FLIMM technique. To minimize the data acquisition time, scans in the *X-Y* direction are sometimes carried out at a few selected number of frequencies [45, 72].

13.4
Applications

13.4.1
Films of Polyvinylidene Fluoride and its Copolymers

13.4.1.1 Comparison of Focused LIMM and TPT

The recent advances in focused LIMM and TPT have been compared in several papers. The lateral resolution of TPT was studied by performing a series of one-dimensional scans across the edge of the 'T'-shaped electrode [42]. The width of the transition region between the poled and unpoled material is determined by electrical stray fields and is expected to be less or equal to the sample thickness of 11 μm, significantly smaller than the 40 μm laser beam waist. The effective lateral resolution, however, shows a strong dependence on depth (Figure 13.12a) because of the fast lateral thermal diffusion in the copper electrode, which has a thermal diffusivity $D = \kappa/(c\rho)$ ~3 orders of magnitude larger than that of the polymer. As the diffusion length at a given time t is $\ell = \sqrt{Dt}$, the lateral diffusion speed is ~$\sqrt{1000} \approx 30\times$ faster than the vertical diffusion speed into the polymer. As a consequence, the heated front electrode cools more rapidly than the polymer, thus turning from a heat source to a heat sink, so that the point of maximum temperature starts moving into the polymer (Figure 13.12b). Enhancing the lateral

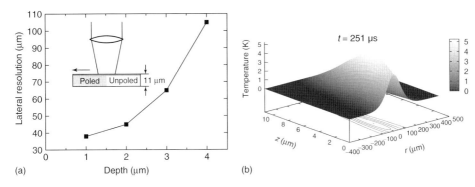

Figure 13.12 (a) Depth dependence of the lateral resolution in PVDF at a beam spot size of ~40 μm. The data were obtained by scanning the poled/unpoled transition region, as shown in the inset. (b) Temperature distribution in the electrode and polymer 251 μs after the laser pulse, calculated by means of the finite element method [46]. Note the different diffusion lengths in the radial (r) and depth (z) directions. (Source: From ref. [65].)

resolution thus depends on the ability to find a surface electrode with good electrical conductivity to maintain the short-circuit condition, but low thermal diffusivity.

13.4.1.2 Poling Dynamics

TPT was used to study the poling dynamics of PVDF. The samples were subjected to slowly varying bipolar electric fields and the poling cycle was interrupted at various levels of the electric field. Figure 13.13 shows an incomplete switching in the bulk, and a polarization pinning at depths up to 1 μm. Polarization pinning has been observed in PVDF by other authors but has been attributed to material fatigue after more than 10^4 poling cycles [73]. Further studies are needed to determine the cause of the incomplete switching.

13.4.2
PVDF–TrFE Coaxial Sensor Cables

Coaxial sensor cables with an active layer made of piezoelectric PVDF–TrFE copolymer (cf. Figure 13.14a) have found important applications in traffic monitoring and intrusion detection. Typically, they are poled (and thus given their piezoelectric properties) by subjecting the extruded cables to a corona discharge formed by a series of high-voltage points around the cable [13]. Recently, polarization maps were obtained with the new tomographic technique from cables poled either in the above-described continuous process or in the laboratory with a single stationary needle [74]. The cables were prepared for TPT measurements by replacing their soft wire core with a stainless steel pin, removing their protective cladding and shielding, and coating the active PVDF–TrFE layer with a Cu electrode (Figure 13.14b). Samples poled in the continuous process show a rather uneven distribution that can be attributed to the fact that any point on the cable was exposed to the corona discharge for no more than 300 ms at the given drawing speed. The laboratory-poled

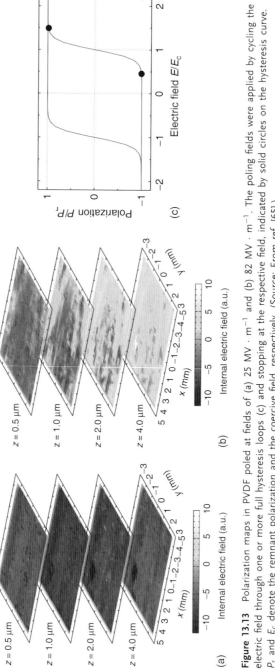

Figure 13.13 Polarization maps in PVDF poled at fields of (a) 25 MV · m^{-1} and (b) 82 MV · m^{-1}. The poling fields were applied by cycling the electric field through one or more full hysteresis loops (c) and stopping at the respective field, indicated by solid circles on the hysteresis curve. P_r and E_c denote the remnant polarization and the coercive field, respectively. (Source: From ref. [65].)

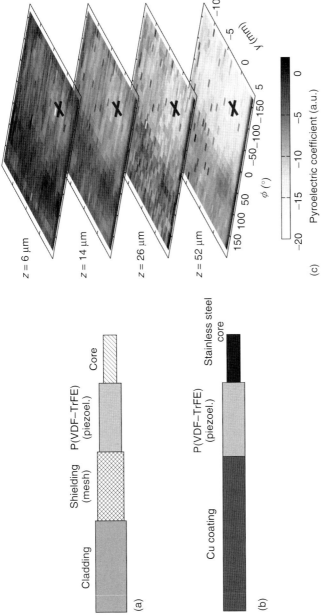

Figure 13.14 Schematic view of a piezoelectric sensor cable (a) as received and (b) after preparation for thermal-pulse tomography. The stainless steel pin serves both as inner electrode and as mounting point. (c) Polarization map of a sensor cable poled with a single stationary needle at −25 kV. The crosses mark the needle position. (Source: From ref. [74].)

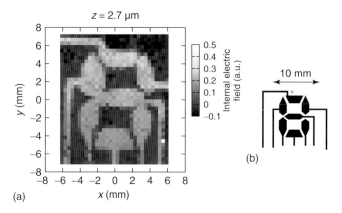

Figure 13.15 Internal electric field in electron-beam-irradiated PTFE film. (a) Field distribution at a depth of $z = 2.7\,\mu m$ and (b) shadow mask used for charging. (Source: From ref. [46].)

cables, on the other hand, show a smooth polarization centered on the needle position in a region ~50 ... 100° wide (cf. Figure 13.14c), depending on the poling voltage [74]. It was thus shown that optimum poling can be achieved with a set of four to six corona needles at slow drawing speeds.

13.4.3
Space-Charge Electrets

TPT is also suitable for mapping the charge distribution in space-charge electrets. For example, films of PTFE (thickness 17 μm) were irradiated with a monoenergetic electron beam (15 keV) through a shadow mask (cf. Figure 13.15b). The resulting patterned space-charge distribution is shown in Figure 13.15a. The space-charge pattern is an exact copy of the mask structure. For maximum signal-to-noise ratio, the samples had to be glued to a substrate in order to avoid thermoelastic resonances [59].

Very recently, TPT was successfully used to image space charge in voided materials, the so-called ferroelectrets [6, 75]. In one study, a double-sided adhesive mesh was sandwiched between two polycarbonate films and the voids were internally charged through dielectric barrier discharges. The TPT maps clearly reproduced the charge pattern in the artificial voids [76].

13.4.4
Polymer-Dispersed Liquid Crystals

Polymer-dispersed liquid crystals (PDLCs) [77] consist of micrometer-size liquid crystal (LC) droplets homogeneously dispersed in a host polymer and have electro-optical properties that allows their use as "light valves." An applied electric bias field orients the optical axes of the droplets, thus changing the LC refractive index. If the LC refractive index–either in the "on" or the "off" state–matches that of the polymer matrix, light will be transmitted through the PDLC, while a mismatch in the other state leads to strong scattering.

When the host polymer is ferroelectric, the local electric field across the LC-filled cavities acts as a bias field to the LC, opening up a range of potential applications, from bistable electro-optical switches to pressure- and temperature-sensitive light valves [78]. In a recent experiment, nematic LC droplets were embedded into a copolymer of vinylidene fluoride with trifluoroethylene (PVDF–TrFE) [79]. After poling the polymer matrix in direct contact with an electric field well above the coercive field of the material, a cluster of LC droplets could be imaged using an upgraded TPT setup with a spot size of 10 µm (Figure 13.16). The droplets had major axes (parallel to the film surface) between 2 and 5 µm and minor axes (perpendicular to the film surface) between 0.5 and 2 µm. Thus, the high near-surface depth resolution allowed the identification of the droplets in the z direction.

13.4.5
Nanomaterials

LIMM and the thermal-pulse technique are well suited to study polarization and space-charge effects in nanomaterials. In recent years, nanosized MgO filler particles were reported to improve the breakdown strength in polyethylene [80], which is widely used as insulator for electrical power transmission cables, presumably because electrical tree growth is slowed down or hindered by the nanofillers. Thus, there is a demand for the study of space charge in polymer nanocomposites. In a recent study [81] of micrometer- and nanometer-sized ZnO particles in low-density polyethylene (LDPE), near-surface space charge densities of ~150–200 C/m^3 were observed in nanofilled LDPE (10% w/w) using the LIMM technique, after the sample had been subjected to a poling field of 30 kV mm^{-1}. This represents a slight decrease from the 350 Cm^{-3} charge density in pure LDPE.

13.5
Summary and Outlook

Laser-based thermal polarization and space-charge profiling techniques have become a standard tool in electret research. In recent years, the trend has been toward higher lateral resolution, allowing true three-dimensional mapping. Thus these techniques are bridging the gap between bulk (or one-dimensional) measurements on the one hand and piezoelectric force microscopy [82] on the other hand. With improving signal processing and data acquisition hardware, as well as the ongoing theoretical efforts toward a three-dimensional solution of the LIMM equation, further gains in lateral resolution can be expected. In addition, work is underway to improve the depth resolution of laser-based acoustic methods to the 10 nm range, using femtosecond laser pulses and electro-optic detection [67].

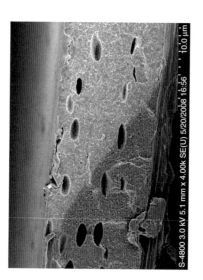

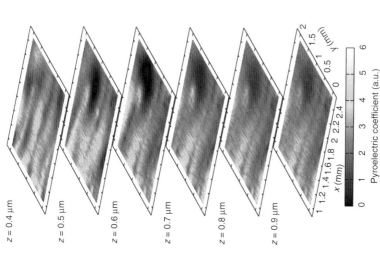

Figure 13.16 (a) Three-dimensional polarization in a PDLC film with a ferroelectric PVDF–TrFE polymer matrix. The dark area represents the liquid crystal (LC) droplets, while the stripes in the x direction are due to polarization inhomogeneities in the polymer. (a) SEM image of the PDLC film with 10% w LC content. (From ref. [79].)

References

1. Heaviside, O. (1885) Electromagnetic induction and its propagation. Electrization and Electrification. Natural Electrets. *The Electrician*, August 7, 1885, 230–231.

2. Heaviside, O. (1892) Electromagnetic induction and its propagation. Electrization and Electrification. Natural Electrets, in *Electrical Papers*, vol. 1, (ed. O. Heaviside), Chelsea, New York, pp. 488–493.

3. Sessler, G.M. and West, J.E. (1962) Self-biased condenser microphone with high capacitance. *J. Acoust. Soc. Am.*, **34**, 1787.

4. Kawai, H. (1969) The piezoelectricity of poly(vinylidene fluoride). *Jpn. J. Appl. Phys.*, **8**, 975–977.

5. Gerhard-Multhaupt, R. (ed) (1999) *Electrets*, vol. 2, 3rd edn, Laplacian Press, Morgan Hill, CA.

6. Bauer, S., Gerhard-Multhaupt, R., and Sessler, G.M. (2004) Ferroelectrets: soft electroactive foams for transducers. *Phys. Today*, **57**, 37–43.

7. Dissado, L.A. and Fothergill, J.C. (1992) *Electrical Degradation and Breakdown in Polymers*, Peter Peregrinus Ltd., London.

8. Zhang, Y., Lewiner, J., and Alquié, C. (1996) Evidence of strong correlation between space-charge buildup and breakdown in cable insulation. *IEEE Trans. Diel. Electr. Insul.*, **3**, 778–783.

9. Dissado, L.A., Mazzanti, G., and Montanari, G.C. (1997) The role of trapped space charges in the electrical aging of insulating materials. *IEEE Trans. Diel. Electr. Insul.*, **4**, 496–506.

10. Bauer, S. and Bauer-Gogonea, S. (2003) Current practice in space charge and polarization profile measurements using thermal techniques. *IEEE Trans. Diel. Electr. Insul.*, **10**, 883–902.

11. Bambery, K.R. and Fleming, R.J. (1998) Space charge accumulation in two power cable grades of XLPE. *IEEE Trans. Diel. Electr. Insul.*, **5**, 103–109.

12. Bauer, S. (1996) Poled polymers for sensors and photonic applications. *J. Appl. Phys.*, **80**, 5531–5558.

13. Wegener, M. and Gerhard-Multhaupt, R. (2004) Electric poling and electromechanical characterization of 0.1-mm-thick sensor films and 0.2-mm-thick cable layers from piezoelectric poly(vinylidene fluoride-trifluoroethylene). *IEEE Trans. Ultrason. Ferroelectr. Freq. Contr.*, **50**, 921–931.

14. Mellinger, A., Camacho Gonzá, F., and Gerhard-Multhaupt, R. (2004) Photostimulated discharge in electret polymers: an alternative approach for investigating deep traps. *IEEE Trans. Diel. Electr. Insul.*, **11**, 218–226.

15. Thiessen, V., Winkel, A. and Hermann, K. (1936) Elektrische Nachwirkungen im erstarrten Dielektrikum. *Phys. Z.*, **37**, 511–520.

16. Walker, D.K. and Jefimenko, O. (1973) Volume charge distribution in carnauba wax electrets. *J. Appl. Phys.*, **44**, 3459–3464.

17. Collins, R.E. (1981) Thermal pulsing technique applied to polymer electrets. *Ferroelectrics*, **33**, 65–74.

18. Phelan, R.J. Jr., Peterson, R.L., Hamilton, C.A., and Day, G.W. (1974) The polarization of PVF and PVF$_2$ pyroelectrics. *Ferroelectrics*, **7**, 375–377.

19. Marcus, M.A. (1981) Controlling the piezoelectric activity distribution in poly(vinylidene fluoride). *Ferroelectrics*, **32**, 149–155.

20. Peterson, R.L., Day, G.W., Gruzensky, P.M., and Phelan, R.J. Jr. (1974) Analysis of response of pyroelectrical optical detectors. *J. Appl. Phys.*, **45**, 3296–3303.

21. Collins, R.E. (1975) Distribution of charge in electrets. *Appl. Phys. Lett.*, **26**, 675–677.

22. Collins, R.E. (1976) Analysis of spatial distribution of charges and dipoles in electrets by a transient heating technique. *J. Appl. Phys.*, **47**, 4804–4808.

23. von Seggern, H. (1978) Thermal-pulse technique for determining charge distributions: effect of measurement accuracy. *Appl. Phys. Lett.*, **33**, 134–137.

24. Weisstein, E.W. Fredholm integral equation of the first kind. From Math-World – A Wolfram Web Resource,

http://mathworld.wolfram.com/ Fred-holmIntegralEquationoftheFirstKind.html. Retrieved July 2011.

25. Mopsik, F.I. and DeReggi, A.S. (1982) Numerical evaluation of the dielectric polarization distribution from thermal-pulse data. *J. Appl. Phys.*, **53**, 4333–4339.

26. Bauer, S. (1993) Method for the analysis of thermal-pulse data. *Phys. Rev. B*, **47**, 11049–11055.

27. Ploss, B. (1994) Probing of pyroelectric distributions from thermal wave and thermal pulse measurements. *Ferroelectrics*, **156**, 345–350.

28. Bloß, P., DeReggi, A.S., and Schäfer, H. (2000) Electric-field profile and thermal properties in substrate-supported dielectric films. *Phys. Rev. B*, **62**, 8517–8530.

29. Lang, S.B. and Das-Gupta, D.K. (1981) A technique for determining the polarization distribution in thin polymer electrets using periodic heating. *Ferroelectrics*, **39**, 1249–1252.

30. Sandner, T., Suchaneck, G., Koehler, R., Suchaneck, A., and Gerlach, G. (2002) High frequency LIMM – a powerful tool for ferroelectric thin film characterization. *Integr. Ferroelectr.*, **46**, 243–257.

31. Lang, S.B. and Das-Gupta, D.K. (1986) Laser-intensity-modulation method: a technique for determination of spatial distributions of polarization and space charge in polymer electrets. *J. Appl. Phys.*, **59**, 2151–2160.

32. Tikhonov, A.N., Goncharskii, A.V., Stepanov, V.V., and Kochikov, I.V. (1987) Ill-posed image processing problems. *Sov. Phys. Dokl.*, **32**, 456–458.

33. Weese, J. (1992) A reliable and fast method for the solution of Fredholm integral equations of the first kind based on Tikhonov regularization. *Comput. Phys. Commun.*, **69**, 99–111.

34. Lang, S.B. (2001) Polynomial solution of the Fredholm integral equation in the laser intensity modulation method (LIMM). *Integr. Ferroelectr.*, **38**, 111–118.

35. Lang, S.B. (2000) Use of neural networks to solve the integral equation of the laser intensity modulation method (LIMM). *Ferroelectrics*, **238**, 281–289.

36. Ploss, B., Emmerich, R., and Bauer, S. (1992) Thermal wave probing of pyroelectric distributions in the surface region of ferroelectric materials: a new method for the analysis. *J. Appl. Phys.*, **72**, 5363–5370.

37. Mellinger, A. (2004) Unbiased iterative reconstruction of polarization and space-charge profiles from thermal-wave experiments. *Meas. Sci. Technol.*, **15**, 1347–1353.

38. Tuncer, E. and Lang, S.B. (2005) Numerical extraction of distributions of space-charge and polarization from laser intensity modulation method. *Appl. Phys. Lett.*, **86**, 071107.

39. Lang, S.B. and Fleming, R. (2009) A comparison of three techniques for solving the fredholm integral equation of the laser intensity modulation method (LIMM). *IEEE Trans. Diel. Electr. Insul.*, **16**, 809–814.

40. Aryal, S. and Mellinger, A. (2011) Two-and three-dimensional analysis of polarization profiles in electret materials with thermal pulses, in *Ann. Rep. Conf. Electr. Insul. Dielectr. Phenom.*, IEEE Service Center, Piscataway, NJ, pp. 243–246.

41. Mellinger, A., Singh, R., and Gerhard-Multhaupt, R. (2005) Fast thermal-pulse measurements of space-charge distributions in electret polymers. *Rev. Sci. Instrum.*, **76**, 013903.

42. Mellinger, A., Singh, R., Wegener, M., Wirges, W., Gerhard-Multhaupt, R., and Lang, S.B. (2005) Three-dimensional mapping of polarization profiles with thermal pulses. *Appl. Phys. Lett.*, **86**, 082903.

43. Maeno, T. (2001) Three-dimensional PEA charge measurement system. *IEEE Trans. Diel. Electr. Insul.*, **8**, 845–848.

44. Marty-Dessus, D., Berquez, L., Petre, A., and Franceschi, J.L. (2002) Space charge cartography by FLIMM: a three-dimensional approach. *J. Phys. D: Appl. Phys.*, **35**, 3249–3256.

45. Quintel, A., Hulliger, J., and Wübbenhorst, M. (1998) Analysis of the polarization distribution in a polar perhydrotriphenylene inclusion compound by scanning pyroelectric microscopy. *J. Phys. Chem. B*, **102**, 4277–4283.

46. Mellinger, A., Singh, R., Wegener, M., Wirges, W., Suárez, R.F., Lang, S.B., Santos, L.F., and Gerhard-Multhaupt, R. (2005) High-resolution three-dimensional space-charge and polarization mapping with thermal pulses. Proceedings, 12th International Symposium on Electrets, Salvator, Brazil, pp. 212–215.

47. Lang, S.B. (2004) Laser intensity modulation method (LIMM): review of the fundamentals and a new method for data analysis. *IEEE Trans. Diel. Electr. Insul.*, **11**, 3–12.

48. Ploss, B. (2002) The resolution of thermal profiling techniques, in *Proceedings, 11th International Symposium on Electrets*. IEEE Service Center, Piscataway, NJ, pp. 177–180.

49. Das-Gupta, D.K., Hornsby, J.S., Yang, G.M., and Sessler, G.M. (1996) Comparison of charge distributions in FEP measured with thermal wave and pressure pulse techniques. *J. Phys. D: Appl. Phys.*, **29**, 3113–3116.

50. Bloß, P., Steffen, M., Schäfer, H., Yang, G.-M., and Sessler, G.M. (1996) Determination of the polarization distribution in electron-beam-poled PVDF using heat wave and pressure pulse techniques. *IEEE Trans. Diel. Electr. Insul.*, **3**, 182–190.

51. van Turnhout, J., Staal, R.E., Wübbenhorst, M., and de Haan, P.H. (1999) Distribution and stability of charges in porous polypropylene films, in *Proceedings, 10th International Symposium on Electrets*, IEEE Service Center, Piscataway, NJ, pp. 785–788.

52. Mellinger, A., Singh, R., Camacho González, F., Szamel, Z., and Głowacki, I. (2004) In situ observation of optically and thermally induced charge depletion in chromophore-doped cyclic olefin copolymers, in *Annual Report, Conference on Electrical Insulation and Dielectric Phenomena*, IEEE Service Center, Piscataway, NJ, pp. 498–501.

53. Bauer, S. and DeReggi, A.S. (1996) Pulsed electrothermal technique for measuring the thermal diffusivity of dielectric films on conducting substrates. *J. Appl. Phys.*, **80**, 6124–6128.

54. Bauer, S. and Ploss, B. (1990) A method for the measurement of the thermal, dielectric, and pyroelectric properties of thin pyroelectric films and their applications for integrated heat sensors. *J. Appl. Phys.*, **68**, 6361–6367.

55. Emmerich, R., Bauer, S., and Ploss, B. (1992) Temperature distribution in a film heated with a laser spot: theory and measurement. *Appl. Phys. A*, **54**, 334–339.

56. Lang, S.B. (2001) Two-dimensional thermal analysis of thin-film pyroelectric infrared detectors. *Ferroelectrics*, **258**, 297–302.

57. Ploss, B., Bauer, S., and Bon, C. (1991) Measurement of the thermal diffusivity of thin films with bolometers and with pyroelectric temperature sensors. *Ferroelectrics*, **118**, 435–450.

58. van der Ziel, A. (1973) Pyroelectric response and d* of thin pyroelectric films on a substrate. *J. Appl. Phys.*, **44**, 546–549.

59. Bloß, P. and Schäfer, H. (1994) Investigations of polarization profiles in multilayer systems by using the laser intensity modulation method. *Rev. Sci. Instrum.*, **65**, 1541–1550.

60. Lang, S.B. (1991) Laser intensity modulation method (LIMM): experimental techniques, theory and solution of the integral equation. *Ferroelectrics*, **118**, 343–361.

61. Press, W.H., Flannery, B.P., Teukolsky, S.A., and Vetterling, W.T. (2007) *Numerical Recipes: the Art of Scientific Computing*, Cambridge University Press, Cambridge.

62. Lang, S.B. (1998) An analysis of the integral equation of the surface laser intensity modulation method using the constrained regularization method. *IEEE Trans. Diel. Electr. Insul.*, **5**, 70–76.

63. Hansen, P.C. and O'Leary, D.P. (1993) The use of the L-curve in the regularization of discrete ill-posed problems. *SIAM J. Sci. Comput.*, **14**, 1487–1503.

64. Lang, S.B. (2006) Fredholm integral equation of the laser intensity modulation method (LIMM): solution with the polynomial regularization and l-curve methods. *J. Mater. Sci.*, **41**, 147–153.

65. Mellinger, A., Suárez, R.F., Singh, R., Wegener, M., Gerhard-Multhaupt, R., Lang, S.B., and Wirges, W. (2008) Thermal-pulse tomography of space-charge and polarization distributions in electret polymers. *Int. J. Thermophys.*, **29**, 2046–2054.

66. Fleming, R.J. (2005) Space charge profile measurement techniques: Recent advances and future directions. *IEEE Trans. Diel. Electr. Insul.*, **12**, 967–978.

67. Holé, S. (2009) Recent developments in the pressure wave propagation method. *IEEE Trans. Electr. Insul. Mag.*, **25**, 7–20.

68. Toureille, A., Notingher, P. Jr., Vella, N., Malrieu, S., Castellon, J., and Agnel, S. (1998) The thermal step technique: an advanced method for studying the properties and testing the quality of polymers. *Polym. Int.*, **46**, 81–92.

69. Notingher, P. Jr., Agnel, S., and Toureille, A. (2001) Thermal step method for space charge measurements under applied dc field. *IEEE Trans. Diel. Electr. Insul.*, **8**, 985–994.

70. Petre, A., Marty-Dessus, D., Berquez, L., and Franceschi, J.L. (2006) Space charge cartography by FLIMM on SEM-irradiated PTFE thin films. *J. Electrostat.*, **64**, 492–497.

71. Gruverman, A. and Kalinin, S. (2006) Piezoresponse force microscopy and recent advances in nanoscale studies of ferroelectrics. *J. Mater. Sci.*, **41**, 107–116.

72. Stewart, M. and Cain, M. (2009) Use of scanning LIMM (laser intensity modulation method) to characterise polarisation variability in dielectric materials. *J. Phys. Conf. Ser.*, **183**, 012001.

73. Sakai, S., Date, M., and Furukawa, T. (2002) Development of polarization distribution in fatigued films of ferroelectric vinylidene fluoride/trifluoroethylene copolymer. *Jpn. J. Appl. Phys.*, **41**, 3822–3828.

74. Suárez, R.F., Mellinger, A., Wegener, M., Wirges, W., Gerhard-Multhaupt, R., and Singh, R. (2006) Thermal-pulse tomography of polarization distributions in a cylindrical geometry. *IEEE Trans. Diel. Electr. Insul.*, **13**, 1030–1035.

75. Qiu, X. (2010) Patterned piezo-, pyro-, and ferroelectricity of poled polymer electrets. *J. Appl. Phys.*, **108**, 011101.

76. Qiu, X., Holländer, L., Suárez, R.F., Wirges, W., and Gerhard, R. (2010) Polarization from dielectric-barrier discharges in ferroelectrets: mapping of the electric-field profiles by means of thermal-pulse tomography. *Appl. Phys. Lett.*, **97**, 072905.

77. Drzaic, P.S. (1986) Polymer dispersed nematic liquid-crystal for large area displays and light valves. *J. Appl. Phys.*, **60**, 2142–2148.

78. Ganesan, L.M., Wirges, W., Mellinger, A., and Gerhard, R. (2010) Piezo-optical and electro-optical behaviour of nematic liquid crystals dispersed in a ferroelectric copolymer matrix. *J. Appl. Phys. D*, **43**, 015401.

79. Suárez, R.F., Ganesan, L.M., Wirges, W., Gerhard, R., and Mellinger, A. (2010) Imaging liquid crystals dispersed in a ferroelectric polymer matrix by means of thermal-pulse tomography. *IEEE Trans. Diel. Electr. Insul.*, **17**, 1123–1127.

80. Danikas, M.G. and Tanaka, T. (2009) Nanocomposites-a review of electrical treeing and breakdown. *IEEE Trans. Electr. Insul. Mag.*, **25**, 19–25.

81. Fleming, R.J., Ammala, A., Casey, P.S., and Lang, S.B. (2008) Conductivity and space charge in LDPE containing nano- and micro-sized ZnO particles. *IEEE Trans. Diel. Electr. Insul.*, **15**, 118–126.

82. Batagiannis, A., Wübbenhorst, M., and Hulliger, J. (2010) Piezo- and pyroelectric microscopy. *Curr. Opin. Solid State Mater. Sci.*, **14**, 107–115.

Index

Nanomaterials: Processing and Characterization with Lasers, First Edition.
Edited by Subhash Chandra Singh, Haibo Zeng, Chunlei Guo, and Weiping Cai.
© 2012 Wiley-VCH Verlag GmbH & Co. KGaA. Published 2012 by Wiley-VCH Verlag GmbH & Co. KGaA.